PERIODIC TABLE OF THE ELEMENTS

Atomic masses are based on ^{12}C. Atomic masses in parentheses are for the most stable isotope.

Legend:
- 6 — Atomic number
- H — Symbol
- 12.011 — Atomic mass

Groups

Period	IA(1)	IIA(2)	IIIB(3)	IVB(4)	VB(5)	VIB(6)	VIIB(7)	VIIIB(8)	VIIIB(9)	VIIIB(10)	IB(11)	IIB(12)	IIIA(13)	IVA(14)	VA(15)	VIA(16)	VIIA(17)	VIIIA(18)
1	1 H 1.0079																	2 He 4.00260
2	3 Li 6.941	4 Be 9.01218											5 B 10.81	6 C 12.011	7 N 14.0067	8 O 15.9994	9 F 18.998403	10 Ne 20.179
3	11 Na 22.98977	12 Mg 24.305											13 Al 26.98154	14 Si 28.0855	15 P 30.97376	16 S 32.06	17 Cl 35.453	18 Ar 39.948
4	19 K 39.0983	20 Ca 40.08	21 Sc 44.9559	22 Ti 47.90	23 V 50.9415	24 Cr 51.996	25 Mn 54.9380	26 Fe 55.847	27 Co 58.9332	28 Ni 58.70	29 Cu 63.546	30 Zn 65.38	31 Ga 69.72	32 Ge 72.59	33 As 74.9216	34 Se 78.96	35 Br 79.904	36 Kr 83.80
5	37 Rb 85.4678	38 Sr 87.62	39 Y 88.9059	40 Zr 91.22	41 Nb 92.9064	42 Mo 95.94	43 Tc (98)	44 Ru 101.07	45 Rh 102.9055	46 Pd 106.4	47 Ag 107.868	48 Cd 112.41	49 In 114.82	50 Sn 118.69	51 Sb 121.75	52 Te 127.60	53 I 126.9045	54 Xe 131.30
6	55 Cs 132.9054	56 Ba 137.33	57 La 138.9055	72 Hf 178.49	73 Ta 180.9479	74 W 183.85	75 Re 186.207	76 Os 192.2	77 Ir 192.22	78 Pt 195.09	79 Au 196.9665	80 Hg 200.59	81 Tl 204.37	82 Pb 207.2	83 Bi 208.9804	84 Po (209)	85 At (210)	86 Rn (222)
7	87 Fr (223)	88 Ra 226.0254	89 Ac 227.0278	104 Unq (261)	105 Unp (262)	106 Unh (263)	107 Uns (262)	108	109 Une (266)									

* Lanthanide series

58 Ce 140.12	59 Pr 140.9077	60 Nd 144.24	61 Pm (145)	62 Sm 150.4	63 Eu 151.96	64 Gd 157.25	65 Tb 158.9254	66 Dy 162.50	67 Ho 164.9304	68 Er 167.26	69 Tm 168.9342	70 Yb 173.04	71 Lu 174.967

† Actinide series

90 Th 232.0381	91 Pa 231.0359	92 U 238.029	93 Np 237.0482	94 Pu (244)	95 Am (243)	96 Cm (247)	97 Bk (247)	98 Cf (251)	99 Es (252)	100 Fm (257)	101 Md (258)	102 No (259)	103 Lr (260)

Introduction To College Chemistry

SECOND EDITION

Introduction To College Chemistry

DREW H. WOLFE
Hillsborough Community College

McGraw-Hill Publishing Company

New York St. Louis San Francisco Auckland Bogotá
Caracas Hamburg Lisbon London Madrid Mexico Milan
Montreal New Delhi Oklahoma City Paris San Juan
São Paulo Singapore Sydney Tokyo Toronto

INTRODUCTION TO COLLEGE CHEMISTRY

4567890   DOCDOC   93210

ISBN 0-07-071427-4

This book was set in Baskerville by York Graphic Services, Inc.
The editors were Karen S. Misler and Jack Maisel;
the designer was Merrill Haber;
the production supervisor was Joe Campanella.
The cover photograph is courtesy of Henry Ries ©1986
The Stock Market.
The photo editor was Rosemarie Rossi.
R. R. Donnelley & Sons Company was printer and binder.

Wolfe, Drew H.
  Introduction to college chemistry.

  Includes index.
  1. Chemistry.   I. Title.
QD33.W84 1988   540 87-17131
ISBN 0-07-071427-4

To Cynthia and Natasha
and
to the memory of my father

Contents

To the Student

Many students think that chemistry is studied simply to acquire chemistry facts; this is not correct. Learning the facts of chemistry is a relatively minor outcome of completing a chemistry course, considering that a rather large percent of the facts learned in an introductory course are soon forgotten. Instead, studying chemistry should help you to develop long-lasting skills and abilities that will be useful to you both in your future studies, and in many aspects of your day-to-day life.

Some of the important outcomes of a careful study of chemistry are the ability to solve scientific problems systematically, the development of greater abstract reasoning skills, the ability to organize and grasp a large quantity of factual information, and a better understanding of the language spoken by chemists. These outcomes are in addition to those objectives that are customarily thought of as the primary chemistry course goals: the development of an appreciation of chemistry and what chemists do; an understanding of the most important laws, principles, concepts, and facts of chemistry; and the attainment of some special skills unique to the field of chemistry.

One of the principal goals of education is the development of problem-solving ability. A person who can solve problems is much more effective, and is usually a more valuable employee, than one who only can recite facts. Students who develop good problem solving skills can apply these skills to their future studies, and eventually to their professions. Studying chemistry is an excellent way to expand your problem solving skills. This text presents a general problem-solving method that can be used to solve many different types of chemistry problems.

Chemistry involves many abstract concepts because it deals, for the most part, with things we cannot see or experience directly. For this reason, many students think of chemistry as "hard." However, throughout our lives we are faced with many abstract ideas such as the concepts of good, evil, existence, right, wrong, and many others. Studying chemistry can help you to expand your abstract reasoning skills, and these skills will allow you to make more intelligent decisions in all areas of your life.

Often, students confronted with a large body of facts or ideas, such as those encountered in a chemistry course, are overwhelmed by the magnitude of what is to be learned. This task can become more manageable if related facts and ideas are grouped together. The material to be learned is thus organized into manageable portions. Chemists organize information about substances in this way, identifying regularities which are used to link a large number of substances together. Acquiring methods of systematically organizing facts and ideas is a desirable outcome of studying chemistry.

Your study of chemistry will also give you an understanding of the specific language of the chemist. Chemists and other physical scientists use words carefully, and often use the fewest words possible to express a concept. Chemists generally use words that have exact meanings, and the interchanging of terms is rare. When possible, words are not used at all. Instead, chemists write chemical and mathematical expressions. These are the most concise means for expressing chemical relationships. Knowledge of the language of chemistry can help you to make better decisions about the medicines, chemicals, and foods you put into your body, and will help greatly in any future study of chemistry or the other sciences.

Drew H. Wolfe

Preface

Introduction to College Chemistry, second edition, is a comprehensive preparatory college chemistry textbook that can most effectively be used in beginning chemistry courses to prepare students for general college chemistry and in chemistry courses designed for nonscience students. Additionally, this book may be used in the first semester of some allied health chemistry courses.

For this revised textbook, the author has added missing material, deleted superfluous material, rewritten passages to make them easier to understand, reorganized material within chapters to strengthen them, and added features that make the book more interesting to the reader.

As in the first edition the major emphasis is on problem solving. The introductory chapter discusses the scope of chemistry, Chapter 2, devoted totally to chemical problem solving, discusses scientific problems and introduces the factor-label method, wherein a systematic stepwise procedure is proposed for students to follow when they solve a chemistry problem. The principles and procedures regarding problem solving in Chapter 2 are applied and further developed throughout the book.

Chapter 3, Chemical Measurement, has been reorganized so that the coverage of uncertainty in measurements and significant figures comes before the discussion of the SI system of units so that students can more easily apply their knowledge of significant figures when they learn how to solve unit-conversion problems. Because of the importance of significant figures and the difficulty students have in understanding them, the discussion of them has been expanded. The author has even taken greater care in the second edition to totally eliminate significant figure errors in the book.

Chapter 4, Matter and Energy, has been slightly expanded to include a discussion of natural abundances of elements in the earth's crust and the whole earth. But the most significant change in this chapter is an expanded discussion of energy; the sections that cover heat versus temperature and specific heats have been markedly improved.

Chapter 5, Atoms, has one major addition as a result of reviewers comments—a discussion of the history of the development of the atomic theory. In this section, the most salient experiments leading to the modern atomic theory are briefly discussed—e.g., discharge tube experiments, gold-leaf experiment, and atomic spectroscopy. Also included is a very brief discussion of the Bohr model of the atom and how it led to the modern quantum mechanical model.

Chapter 6, Periodic Properties, and Chapter 7, Mole Concept and Calculations, were changed only slightly. In both chapters, additional Example Problems have been included to provide more help for students. Chapter 8, Molecules, Compounds, and Chemical Bonding, has a number of significant changes. The discussion of electronegativity has been moved so that electronegativity arguments can be applied to ionic bonding, and a discussion of the valence shell electron pair repulsion (VSEPR) method is included. Also, more relevant molecules are given as examples.

Chapter 9, Chemical Nomenclature of Inorganic Compounds, remains essentially the same; however, less emphasis is placed on the old names of compounds (e.g., cuprous nitrate and plumbous sulfide). The old names have not been totally deleted, because they are still widely used outside of chemistry and in industry. A brief subsection on the names of hydrates was included in response to some reviewers' requests. Chapter 10, Chemical Equations, now introduces redox reactions along with the four major classes of inorganic reactions, and the discussion on how to predict the products of reactions has been rewritten with more examples and greater emphasis on the logic used to solve such problems.

Chapter 11, Stoichiometry, Chapter 12, Gases, Chapter 13, Liquids, Solids, and Changes of States, and Chapter 14, Descriptive Inorganic Chemistry, have only minor substantive changes, but Chapter 15, Solutions, has been rewritten for greater clarity and a section on solution stoichiometry has been included. The final five chapters have been updated with more relevant examples added to each; for example, the discussion of uses of radioactive materials in Chapter 19 has been expanded to include military applications and additional medical examples. In Chapter 20, Overview of Organic and Biologically Important Compounds, each different class of organic compounds now contains a brief discussion of their most important reactions.

Note that Chapter 14 is devoted entirely to descriptive inorganic chemistry of representative metals and nonmetals. The chapter can be used effectively to conclude a one-semester course. In Chapter 14, most of the important principles presented in earlier chapters are applied. Exercises at the end of the chapter test students' knowledge of periodic properties, stoichiometry, molecular structure, physical states, and reactions.

Two new features appear in this edition of the book. A list of Key Terms now follows the Summary. The Key Terms section contains an alphabetical list of the most important new terms that are introduced in a chapter. Additionally, throughout the book the students will find special topics of interest called Chem Topics. These brief discussions were selected to make the task of learning chemistry more interesting, and at the same time to present relevant applications of the concepts in the chapter.

All the Example Problems have been carefully scrutinized: some have been rewritten entirely, some have been modified, and some have been replaced with better problems. A significant number of new Example Problems have been included. The Exercises in the chapter and those that come after the chapter have been extensively changed. Almost all numerical problems have been modified in one way or another, and many of the discussion questions have been changed. The final subsection of the Exercises, Additional Exercises, has been greatly expanded with problems that are "unclassified" and those that are more difficult. Unlike the first edition, only the answers to selected Exercises are included at the end of the book. These questions are color-coded throughout the text. An attempt was made to provide the answers to a wide variety of Exercises so that students could check to see that they were solving them correctly, and at the same time leaving enough unanswered Exercises for professors to assign.

The second edition has been totally redesigned for greater organization and clarity. Many figures have been redrawn and new figures have been added. Moreover, a four-color insert that shows interesting pictures that relate to topics discussed throughout the book has been added to this edition.

The general organization of the chapters remains the same in the second edition. Each chapter contains the following pedagogical aids:

1. *Study Guidelines* that list what the students should be able to do after successfully completing the chapter.
2. *Margin notes* that amplify and expand topics in the body of the text.
3. Many *diagrams and photographs* that illustrate abstract chemical principles, procedures for solving problems, chemical structures, and reactions.
4. *Example Problems* that contain carefully worded explanations of each step. Significant figures are emphasized in each Example Problem.
5. Chapter *Summaries* that contain the most important ideas and concepts in the chapter. All important terms are boldfaced.
6. In excess of 1500 *Exercises,* ranging from trivial to very difficult. Many are composed of four or more parts. Some are located after each section for review purposes and others are located at the end of the chapters.

The appendixes include

1. *Review of Math Skills,* which discusses basic algebra, exponential and scientific notation, and graphing skills. Each section of this appendix contains Exercises.

2. *Chemistry Calculations Using Calculators,* which discusses both algebraic notation and reverse polish notation.
3. *Physical Constants and Conversion Factors,* which lists common physical constants and frequently used conversion factors.
4. *Table of Ions and Their Formulas,* which is a comprehensive list of all polyatomic ions that the student may encounter.
5. *Logarithms to Base 10,* which has an explanation of how to use a common log table along with the table itself.
6. *Answers to Selected Exercises,* which has the answers to all problems with numbers that are printed in color.

The volume concludes with a Glossary, which gives succinct definitions and descriptions of the most important words and terms in the book, and a comprehensive, carefully prepared index.

Besides the textbook, a *Study Guide/Solution Manual* prepared by Dr. Robert Buckley and Ms. Elsie Gross of Hillsborough Community College is available. In the *Study Guide* additional explanations are given for difficult topics, as well as added Example Problems. This work also has practice tests and shows setups of how to calculate all Exercises that have answers in the textbook.

The *Instructor's Guide* contains the answers and setups to the Exercises that are not found in the text. A new and expanded test bank of multiple-choice and fill in the blanks questions is also included in this guide. In addition a computerized test bank of the problems is available. Finally, a set of overhead transparencies of selected illustrations in the text is available.

Acknowledgements
I would like to express my appreciation and thanks for the following reviewers whose comments, suggestions, and criticisms helped fine-tune this revision. Wayne L. Felty, Pennsylvania State University; Larry W. Houk, Memphis State University; William B. Huggins, Pensacola Junior College; John T. Ohlsson, University of Colorado–Boulder; Gordon Parker, University of Toledo; Robert M. Perrone, Community College of Philadelphia; Quentin R. Peterson, Central Michigan University; Nancy Reitz, American River College; Leo Spinar, South Dakota State University; Eugene D. Thomas, California State University–Chico; and Patricia Milliken Wilde, Triton College.

I would like to especially thank those in the college division of McGraw-Hill who have worked hard to produce the best possible textbook. Karen Misler, the chemistry editor, was involved in all aspects of bringing the second edition to life. Jack Maisel, the supervising editor, took the raw manuscript and converted it into a textbook. Merrill Haber produced an excellent design, Joseph Campanella was the production supervisor, Rosemarie Rossi found all the new photographs for this edition, and Ruth Mendelsohn was the copy editor.

This project would not have been completed without the love and understanding that I received from my wife, Cynthia, and daughter, Natasha. I cannot thank them enough for their strength and support during my recent illness.

Drew H. Wolfe

Introduction To College Chemistry

CHAPTER
One

Introduction to Chemistry

STUDY GUIDELINES

After completing Chapter 1, you should be able to

1. Write a simple definition of chemistry and explain the relationship of chemistry to the other sciences

2. Define and give examples of different types of matter

3. Identify and describe the main divisions of chemistry

4. Explain the role that the discovery of new materials like bronze and iron played in the development of civilization

5. Discuss the contributions of the ancient Greek philosophers to the development of a better understanding of the nature of matter

6. Describe the contributions made by the alchemists to the development of chemistry

7. Identify the scientific contributions of Boyle, Stahl, Priestley, and Lavoisier

8. List contributions by chemists to modern society

9. Begin to develop your own successful strategy for learning chemistry

Chemistry is one of the most exciting and relevant areas of study because all aspects of life are to some degree related to chemistry. From this textbook you will learn the most fundamental chemical concepts in a form that will allow you to apply them to your specific areas of interest. To begin our study, let's state a simplified working definition of chemistry. The study of **chemistry** is concerned with the composition of matter, the changes that matter undergoes, and the relation between these changes in matter and changes in energy. A **chemist** is a person who studies the composition, structure, and properties of matter and seeks to explain the changes that matter undergoes.

1.1 THE SCOPE OF CHEMISTRY

Chemistry is the science concerned principally with matter and its interactions.

1

But what is matter? Put simply, **matter** is anything that has mass and occupies space. The earth, and everything on it, is composed of matter. We use terms such as substances, materials, objects, and bodies when referring to matter. Examples of matter are as far ranging as the air you breathe, the food you eat, the objects that you own, and the ground upon which you walk. As we will discover, and as we will stress throughout the textbook, matter is closely associated with energy; in fact, in some rare instances it cannot be easily distinguished from energy. At this time we will defer our discussion of matter and energy to complete our brief look at the scope of chemistry.

Chemistry overlaps with and is an integral part of the other sciences (Fig. 1.1). Biology, the study of living systems, applies chemical principles to help understand the functioning of cells, the basic units of life. Geology, the study of the earth, incorporates chemical observations in order to elucidate the processes that occur on earth.

Chemistry is sometimes called the central science because it is closely tied to all of the other sciences.

Physics and chemistry, both physical sciences, overlap to a large degree since physics also deals with matter, energy, and the interaction of the two. The principal difference between physics and chemistry is that physicists are more interested in the most fundamental components and regularities of nature and how they fit together to yield our universe. Chemists and physicists make use of the same laws of nature to gain a better understanding of the properties and behavior of matter.

Physics is the natural science that deals with subjects such as light, heat, motion, electricity, optics, and the most basic structure of matter.

The study of chemistry may be separated into artificial divisions or branches that categorize the most significant areas of study. These overlapping divisions include (1) analytical, (2) inorganic, (3) organic, (4) biological, (5) physical, and (6) geological chemistry. An insight into each division is gained by considering what the various types of chemists study.

Analytical chemists examine matter to find the identity and amount of its components. Most chemists use analytical procedures and techniques to some degree in their studies. Inorganic chemists study the properties, structures, and reactions of all elementary substances. As we will discover, one of these substances, carbon, has a special set of properties. Therefore, a division of chemistry is entirely devoted to this vast topic. It is called organic chemistry. Organic chemists investigate the properties of carbon-containing substances, and produce new ones, in the laboratory. Biological chemists, also called biochemists, study the compounds that compose living things and the ways they interact to produce living systems. Physical chemists apply the concepts of physics to the behavior of matter in an effort to better understand that behavior. Finally, geochemists investigate the structures, properties, and reactions of substances found in the earth's crust, atmosphere, and oceans.

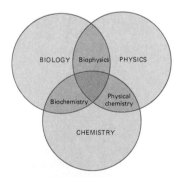

Figure 1.1
Science is divided into three major overlapping disciplines: chemistry, physics, and biology.

REVIEW EXERCISES

1.1 Use a dictionary to find the definitions of each of the following: *(a)* chemistry, *(b)* physics, *(c)* biology, *(d)* geology.

1.2 List and describe the major divisions of chemistry.

The birth of chemistry coincides with the first time people became aware that they could improve upon what nature offered. By observing lightning, fire, decay, and other natural phenomena, primitive people eventually discovered that the properties of objects could be changed.

After harnessing fire, ancient people solved day to day problems more efficiently. Pottery or bricks were formed from baked clay. Through trial and error, ceramics, glazes, and glasses were discovered. Advances in the "chemistry" of these times helped people develop the foundation for civilization.

During this early period of history, now called the stone age, people found that they could improve their lives by developing new materials. The search for new substances to replace their stone tools led to the discovery of metals. Metals added a new dimension to life. Unlike stone, metals could be hammered and shaped into a multitude of forms. Weapons created from metals remained sharper longer than stone counterparts, and metal weapons could be resharpened. Copper, gold, and tin were the first metals used—a copper cooking pan was found in an Egyptian tomb dating back to 3200 B.C.

Sometime around 3000 B.C. a startling discovery was made. If copper and tin ores were heated and mixed, a new metal (an alloy) was formed which was much harder than either copper or tin. The metal, bronze, ushered in a new era—the bronze age.

A thousand years elapsed before the bronze age ended. It was common knowledge throughout the world that a superior metal, iron, existed. But it was rare, and no method for extracting the iron from its ore was readily available. In approximately 1500 B.C. the Hittites, a group of people who lived in Asia Minor, found a means of liberating iron from its ore, using a method that is a forerunner of our present day smelting process. By chance, the Hittites heated iron ore in a charcoal (a form of carbon) smelter, producing an iron-carbon mixture that resembles steel. Thus the world was thrust into the iron age about 1000 B.C.

During the iron age, other practical chemical advances were made. In Egypt, various chemicals were incorporated into all aspects of life and death. The ancient Egyptians made alcoholic beverages by the fermentation of fruits, concocted embalming fluids to preserve the dead, and developed pigments, dyes, and paints that have lasted to modern times.

The ancient Greeks, in about 600 B.C., were one of the first peoples to pose important questions about why matter behaved as it did. They wanted to know the effect of heat on metals; what were the most basic forms of matter; and if one metal could be changed into another. Greek scholars sought to determine the composition of the universe.

Thales of Miletus (640–546 B.C.) was one of the first Greek philosophers to conceive the idea of "elements." Thales suggested that elements were the most fundamental forms of matter. Other Greek thinkers looked to nature and speculated that the entire universe was composed of four elements: earth, air, fire, and water (Fig. 1.2).

1.2 THE EARLY HISTORY OF CHEMISTRY

Ancient Peoples

Our word "metal" is thought to be derived from a Greek word that means "to search for."

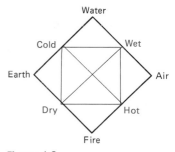

Figure 1.2
The ancient Greeks believed that all matter was composed of four "elements:" earth, air, fire, and water.

Steel is an alloy of iron that contains other metals and less than 0.5% carbon. Alloys are most frequently produced when two or more metals are mixed.

The Ancient Greeks

Figure 1.3
Aristotle. *(Burndy Library.)*

Figure 1.4
An alchemist and his assistant are shown working in their "laboratory." Alchemists developed many different types of glassware, equipment, and procedures in their quest for gold. *(New York Public Library, Picture Collection.)*

Aristotle (384–322 B.C.) (Fig. 1.3), one of the greatest ancient Greek philosophers, accepted and advanced the four-element theory. He suggested that in addition to the four basic elements there could be two pairs of opposite qualities: hot-cold and wet-dry. Aristotle also believed that each element had its own set of properties; e.g., earth should fall and fire should rise. With each modification of the four-element theory, a greater number of plausible explanations were proposed to questions and problems that had puzzled people for centuries.

Today, some people might think of these ideas as humorous or simple-minded solutions to complex problems. Nevertheless, the four-element theory, in a variety of forms, lasted for approximately 2000 years. In part, the longevity of this theory can be attributed to its simplicity. Many problems of nature are explained effortlessly in terms of earth (solid), air (gas), fire (energy), and water (liquid).

Aristotle was the most famous student of Plato. Plato thought that Aristotle was "the intelligence of the school."

Ancient Greek philosophers also attempted to understand the problem of the decomposition of matter into smaller parts. If a stone was broken or crushed, could the fragments of stone be further subdivided? If so, was there a limit beyond which the fragment could no longer be divided? About 450 B.C. **Leucippus** and his disciple **Democritus** proposed the idea of the *atom*. The word "atom" is derived from the Greek term for "indivisible." Democritus suggested that the smallest particle of matter was an atom—a unit of matter that was indivisible.

Democritus also speculated that atoms in different substances varied with respect to size and shape. This was an incredible proposition, considering that Democritus was a philosopher—one who proposes ideas about nature through logic—rather than a true scientist—one who conducts controlled systematic experiments based on observable facts.

To understand the very large, we must understand the very small.
Democritus

From approximately A.D. 300 to A.D. 1100 the Dark Ages prevailed in Europe and chemical advancements almost came to a standstill. However, in Africa and the Middle East, Arab cultures continued to make significant chemical contributions during this period. A small group of Arabs tried to find a way to convert (they said "transmute") cheap, abundant metals to gold. This period in the history of chemistry is now known for these dedicated men who searched for gold—the **alchemists** (Fig. 1.4).

As part of their quest for a way to change base metals to gold, the alchemists sought to find the magic elixir of life, or as it is sometimes called, the philosopher's stone. They thought that the magic elixir could rid one's body of disease and was the key to eternal life. Thousands of alchemists searched in vain for gold and the magic elixir. Even though they never achieved their principal goals, they did uncover a vast amount of chemical knowledge. Various contemporary laboratory techniques and glassware are traced to the alchemists. Some historians believe that the term "chemistry" is derived from the alchemists' term for the mixing of chemicals.

The practice of alchemy continued for more than 2000 years, from the period before the birth of Christ until the eighteenth century. Alchemy died and chemistry emerged because curious people started to ask more probing questions about matter. What explains the behavior of matter? Is matter composed only of earth, air, fire, and water? Do all substances act in a predictable, regular manner?

Robert Boyle, an Irishman who lived from 1627 to 1691 (Fig. 1.5), saw the shortcomings of alchemy and decided to apply what is now known as scientific reasoning to the study of chemistry. Boyle followed the lead of other great scientific investigators of his time: Galileo Galilei (1564–1642), Jan Baptista Van Helmont (1577–1644), Evangelista Torricelli (1608–1647), and Otto von Guericke (1602–1686).

Boyle's exacting studies of gases and their properties supported the idea proposed 1000 years before by certain Greek philosophers—that matter is composed of atoms. In his famous book *The Sceptical Chymist*, published in 1661, Boyle attacked the idea that matter is composed of only four elements. Instead, Boyle proposed that if a substance thought to be an element is capable of being broken down into simpler forms, then it is not an element. One of the most significant outcomes of Boyle's work is the idea of careful experimentation as a vital component of science—the idea that any propositions regarding matter must be supported by reproducible observations.

Other scientists of the seventeenth century were concerned with the nature of energy, which they called "fire." Their interest was spurred by the invention of the steam engine and the possibility of developing more efficient engines capable of performing heavy work. Scientists wanted the answers to questions like these: Why do certain substances burn while others do not? How is heat transferred from one object to another? What is the nature of the combustion process?

The Alchemists

Figure 1.5
Robert Boyle. *(National Portrait Gallery, Smithsonian Institution, Washington, D.C.)*

Early Scientists

Figure 1.6
Joseph Priestley. *(National Portrait Gallery, Smithsonian Institution, Washington, D.C.)*

A German physician and chemist, **Georg Ernest Stahl** (1660–1734), proposed the *phlogiston theory* to help answer some of these questions about "fire." The term "phlogiston" was derived from the Greek term that means "to set on fire." Stahl's phlogiston theory described combustible objects as those that contain a large quantity of phlogiston. As an object burned, Stahl suggested, phlogiston flowed from the object; the object stopped burning when all of the phlogiston was released. A log burned because it contained phlogiston. The resulting ashes lacked phlogiston; consequently, ashes were noncombustible.

Joseph Priestley's (1733–1804) (Fig. 1.6) discovery of oxygen as a component of air and of its ability to support combustion (burning), brought about the end of the phlogiston theory. Priestley informed the French scientist **Antoine Laurent Lavoisier** (1743–1794) (Fig. 1.7) of his discovery of oxygen. Lavoisier immediately repeated Priestley's experiments and found that oxygen was truly formed when he performed them. But Lavoisier saw something much more important—that mathematics could explain the decomposition of matter. When dealing with matter, Lavoisier found that the whole equaled the sum of its parts. He then conducted a classic experiment, heating mercury and oxygen to conclusively show that oxygen and not phlogiston supported combustion.

Stahl was physician to King Frederick William I of Prussia. Besides advancing the phlogiston theory, Stahl was a leading proponent of vitalism—the idea that a different set of natural laws governs living systems.

Priestley, a Unitarian minister, befriended Ben Franklin and Thomas Jefferson in the United States after leaving England because of religious persecution.

Figure 1.7
Antoine Laurent Lavoisier.
(New York Public Library, Picture Collection.)

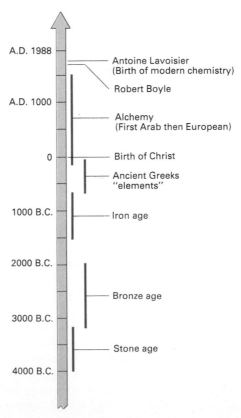

Figure 1.8
A time line of the history of chemistry.

- A.D. 1988 — Antoine Lavoisier (Birth of modern chemistry)
- Robert Boyle
- A.D. 1000 — Alchemy (First Arab then European)
- 0 — Birth of Christ
- Ancient Greeks "elements"
- 1000 B.C. — Iron age
- 2000 B.C. — Bronze age
- 3000 B.C. — Stone age
- 4000 B.C.

Lavoisier is considered the father of modern chemistry. His textbook, *Elementary Treatise on Chemistry,* published in 1789, indicated to the world that chemistry was a science based on theories supported by reproducible experiments. In the *Treatise* he discussed 33 elements known at that time. To his credit, all but two, caloric and light, are considered elements today. Lavoisier's contributions to chemistry are comparable to those of Isaac Newton to physics. Lavoisier possessed a rare talent found in few people who have ever lived: he correctly organized and interpreted a large body of facts, yielding a completely new area of human concern, that of modern chemistry. Figure 1.8 shows a time line of the history of chemistry.

Lavoisier's text was published in the year that the French Revolution began. He was arrested and tried for crimes against the people. Lavoisier pleaded that he was a scientist and not an aristocrat. The reply from the revolutionaries was, "The Republic has no need for scientists." He was guillotined on May 8, 1794.

REVIEW EXERCISES

1.3 Write a brief explanation of how the following materials changed the lives of people: *(a)* stone during the stone age, *(b)* bronze during the bronze age, *(c)* iron during the iron age.

1.4 What contributions did the ancient Greeks make to the advancement of people's understanding of matter?

1.5 Write the name of the scientist responsible for each of the following: *(a)* phlogiston theory, *(b)* discovery of oxygen, *(c)* proposition that matter was not solely composed of the four elements.

1.6 What were the main goals of the alchemists?

Advances in chemistry make a valuable contribution to our modern lifestyle. For example, deadly diseases are controlled by potent antimicrobial agents (commonly called antibiotics). Tetracyclines, sulfa drugs, and penicillins control diseases that were incurable 50 years ago. Many people take for granted that no matter what illness we contract, it can be treated with a drug (Fig. 1.9).

1.3 CHEMISTRY IN THE MODERN WORLD

In 1910, Paul Ehrlich was the first to use antimicrobial agents. Penicillin was discovered by Sir Alexander Fleming in 1929. Sulfa drugs were first identified by Domagk in 1935.

We see chemistry in action every day in our homes. The "art" of cooking involves chemical techniques for frying, baking, roasting, and boiling. Cleaning involves selecting the appropriate soap, detergent, or cleaning fluid to best remove dirt and stains. Waxes are rubbed or sprayed on surfaces to protect and beautify. Pesticides are used to rid our houses and gardens of insect pests, and herbicides remove unwanted plants from our lawns.

Home recreation incorporates many advances in chemical technology. Picture tubes in color television sets contain substances called phosphors that glow various colors when struck by an electron beam. Magnetic recording tape is manufactured by affixing specific metallic substances to a strong plastic tape that does not stretch or shrink. Advances in the chemical processes used to manufacture transistors and integrated circuits (ICs) have resulted in personal computers, videotape recorders, giant television screens, electronic games, and excellent digital high-fidelity stereophonic sound systems.

Modern means of transportation also rely on the chemical industry. For example, passenger jets can fly faster and carry more people eco-

Figure 1.9
Today many people rely on medicines and drugs from birth to death. *(Chemical Manufacturers Association.)*

nomically as a result of strong new metals and plastics used in the bodies of aircraft. Plastic parts have been developed that are superior in many ways to the heavier, more costly metal parts they replace. In part, the increased gas mileage of new cars is due to the decreased weight of the cars, a result of the replacement of heavier metal parts with lighter plastic parts.

Modern society has become very dependent upon the continuing advances in chemistry. To a degree, many people expect chemists to rescue them from the clutches of life's problems and annoyances. But, unfortunately, for each new advance we must pay some price. Environmental problems result from by-products of the chemical industry. New health problems are caused by lifesaving drugs. Pesticides used to protect our crops accumulate in our bodies, causing serious health problems.

There are no "free lunches" in this world.

1.7 List three ways in which your life would change if all plastic goods were no longer available.

1.8 Write a short paragraph describing the role of chemists in the modern world.

REVIEW EXERCISES

I n order to be most successful in your study of chemistry, you should have some sort of learning strategy or plan. Consider the following steps as a model for developing your own strategy.

1. Approach the study of chemistry in a positive way. If you think chemistry is impossible and that you are going to fail—you will! Instead, try to regard chemistry as an important, central subject that will not only assist you in obtaining your educational goals but will give you a deeper understanding of yourself and your environment.

2. Begin your study of each new chemistry topic by quickly reading the material in the textbook before attending the lecture. It is not important to memorize new terms or solve problems at this time. Use the Study Guidelines found at the beginning of each chapter to identify the most significant topics in the chapter.

3. After attending the lecture and obtaining a good set of lecture notes—the highlights of what your professor thinks is most important— reread the chapter, paying careful attention to those topics discussed in the lecture. While doing this second reading, answer all Review Exercise questions; these are located at the end of each chapter section. Strive to understand as much as possible of one section of the textbook before proceeding to the next one. On some occasions you may want to answer appropriate Exercises at the end of the chapter before moving to the next section.

4. After you have reread the chapter and answered the Review Exercises, complete all assigned Exercises at the end of each chapter. **The only way to successfully learn chemistry is to do chemistry by solving problems and answering questions.** If you can solve the problems and answer the questions at the end of the chapter, then, and only then, will you have an adequate knowledge of the material in the chapter. Some helpful hints to consider when solving problems are:

(a) Read each problem carefully and determine if you understand what the problem asks. If not, reread the appropriate section in the chapter or refer to your lecture notes.

(b) Determine what is to be found (what is unknown), and write it on paper.

(c) Extract from the problem all relevant information, and write it below your statement of what is to be found.

(d) Use the factor-label method (discussed in Chap. 2), when appropriate, to solve the problem. When the factor-label method cannot be used, you must rely on logic or knowing the correct relationship or equation. Also, **never forget units.** Knowing what unit belongs with a number is usually more important than knowing the number itself.

(e) If possible, check to see if the answer corresponds to the correct answer listed in Appendix 6. If the answer is incorrect, check for arithmetic errors, and if none are found, compare the numbers used to those given and carefully examine the units for possible errors.

(f) Instead of wasting a lot of time on a problem that you are unable to solve at the moment, leave the problem and try others. Come

back to the unsolved problem later, perhaps in a few days. After returning to the problem, you may find that you can solve it with ease (brains work in mysterious ways).

(g) Seek help any time you are unable to solve a problem after a reasonable attempt to obtain the correct answer. Do not be afraid to ask questions; this is one of the most effective ways to learn chemistry.

The important thing is to not stop questioning.
Albert Einstein

5. To ascertain how well you understand the material presented in the chapter, carefully go through the Study Guidelines listed at the beginning of the chapter to see if you have learned the material completely. Alternatively, find the practice test located in the Study Guide that accompanies the textbook, and take the test without consulting your notes or textbook. After grading the test, try to re-solve any problems that you were unable to complete under the simulated test conditions. Return to the chapter section and review those topics that you have not mastered.

The list above outlines only a suggested learning strategy. Although it is not necessary to follow each step given here exactly, it is necessary to develop a successful strategy for learning chemistry, whatever that strategy may be. If you do not have a better method, use the one presented until you have developed your own plan.

A few words concerning your lecture notes are in order. Lecture notes are very important because they give an added perspective (your professor's) on the material in the textbook. But, in your zeal to obtain a complete set of notes, do not forget to listen to what the professor is saying. Listening is a critical skill to learn. Shortly after each lecture, rewrite your lecture notes. Rewriting your notes will allow you to rethink what was said, correct errors, clear up undecipherable passages, and add any extra thoughts.

Repetition is another key to learning chemistry efficiently. The amount of study time necessary for learning chemistry is normally greater than that needed for many other subjects. A large proportion of your study time should be spent working and reworking problems and transferring thoughts from your brain to paper. Correctly solving a problem once does not guarantee that you have totally learned how to solve problems of this type. Work many similar problems before proceeding to new problems. When learning new ideas, definitions, concepts, or rules, write them down on paper, over and over again. Chemistry is best learned through repetition!

To maximize your achievement, plan to spend time each day studying chemistry. Cramming at the last minute rarely works in chemistry. Shorter, less intense periods of study spread out over a longer time are most effective. Using this technique, you will not get bored as readily, and you will not be frustrated by trying to accomplish too much at the last moment.

Finally, you must learn that when studying chemistry, you must know and understand the previously introduced concepts before you can tackle new material. Failure to learn more basic material will prevent you from understanding new topics. Thus, if you find yourself in this situation, go back and learn the more fundamental concepts before trying to comprehend the new concepts.

SUMMARY

The study of **chemistry** is concerned with the composition of matter, the changes that matter undergoes, and the relation between the changes in matter and changes in energy. Six major divisions of chemistry are analytical chemistry, inorganic chemistry, organic chemistry, physical chemistry, biochemistry, and geochemistry. Chemistry is sometimes considered the central science because all other sciences, to a degree, deal with matter. Chemistry's domain extends into every aspect of life, from birth to death.

Chemistry began in ancient times, when people saw that matter could be changed and used to improve the quality of life. Discoveries at this time were made through trial-and-error methods, leading people through the stone, bronze, and iron ages. The ancient

Egyptians and Greeks were among the first civilizations to question why matter behaved as it did. Early Greek philosophers proposed explanations for the composition and structure of matter. About 450 B.C. Democritus suggested that matter was composed of tiny particles called atoms.

Modern chemistry grew out of the pseudoscience called alchemy. Alchemists searched for methods to convert base metals to gold. Robert Boyle was one of the first scientists to suggest that ideas and thoughts about matter must be supported by reproducible experiments. Antoine Lavoisier is credited with being the father of modern chemistry as a result of his pioneering experiments on the properties of matter.

EXERCISES*

Scope of Chemistry

1.9 (a) What is the dictionary definition of chemistry? (b) Compare the dictionary definition of chemistry to the one proposed in the chapter.

1.10 (a) What is the name of the discipline in which chemistry overlaps with biology? (b) What are some topics that might be studied in this discipline?

1.11 What are 10 different examples of matter found in your house?

1.12 List two different areas that each of the following might study: (a) chemists, (b) biologists, (c) physicists, (d) geologists.

1.13 Explain how the sciences of physics and chemistry are similar and different.

1.14 Explain what each of the following chemists investigates: (a) analytical, (b) inorganic, (c) organic, (d) biological, (e) physical, (f) geological.

1.15 What type of chemist would study the following: (a) rocks and minerals, (b) synthesis of a new carbon compound, (c) antibiotics, (d) struc-

ture of metals, (e) amount of pollution in the air?

1.16 List five household consumer products that help simplify the task of living.

History of Chemistry

1.17 What effect did the harnessing of fire by early peoples have on the development of civilization?

1.18 For prehistoric peoples, what advantages were afforded by the discovery of metals to replace stone objects?

1.19 What chemical advances were made during the: (a) bronze age, (b) iron age?

1.20 (a) What metals were used by ancient peoples? (b) Why did the discovery of iron (actually steel) significantly change the course of history?

1.21 What are the four elements of matter suggested by the Greeks?

1.22 Explain how Aristotle modified the four-element theory.

*For exercise numbers printed in color, answers can be found at the end of the book.

1.23 The four-element theory was used for thousands of years. List two plausible reasons for its longevity.

1.24 How do philosophers differ from scientists when approaching and solving problems?

1.25 *(a)* Who is credited with the idea that matter is composed of small particles called atoms? *(b)* What term did he use to describe atoms?

1.26 What contributions did the alchemists make to modern chemistry?

1.27 What was the importance of the magic elixir to the alchemists?

1.28 Propose a reason why Boyle entitled his book *The Sceptical Chymist.*

1.29 Who investigated the following: *(a)* application of mathematics to chemical changes, *(b)* release of phlogiston when objects burned, *(c)* divisibility of particles, *(d)* properties of gases?

1.30 How did the early scientists explain the loss of weight by an object when it burns?

1.31 Why is Antoine Lavoisier considered the father of modern chemistry?

1.32 If human beings still exist in A.D. 3000, what label (stone age, bronze age, iron age, . . .) might they attach to the middle to late twentieth century?

Additional Exercises

1.33 *(a)* Go through a current issue of the newspaper and list all topics that directly concern chemistry or closely related topics. *(b)* Explain how the topics found in part *(a)* affect your life.

1.34 List and discuss three ways in which chemical advances have produced environmental or societal problems.

1.35 Use newspaper and magazine articles to obtain information on how chemistry affects you within the community in which you live. Write a summary of the information that you find.

1.36 Compare the learning strategy for studying chemistry presented in Sec. 1.4 to that which you presently use for other subjects. *(a)* How are they similar and different? *(b)* What changes must you make to most effectively learn chemistry?

1.37 Go to the library and find out what the most important scientific accomplishments of each of the following scientists were: *(a)* Galileo Galilei (1564–1642), *(b)* Jan Baptista Van Helmont (1577–1644), *(c)* Evangelista Torricelli (1608–1647), *(d)* Otto von Guericke (1602–1686).

CHAPTER Two

Problem Solving in Chemistry

After completing Chapter 2, you should be able to

1. List and explain the six principal steps of the scientific method

2. Define fact, data, hypothesis, theory, and law

3. Distinguish among facts, theories, and laws

4. Discuss limitations of the scientific method

5. Develop a personal strategy for approaching and solving chemistry problems

6. List and apply all steps required to solve a given chemistry problem

7. Write conversion factors, given equalities

8. Apply the factor-label method in solving unit-conversion problems

A systematic, logical procedure called the **scientific method** is used by scientists to solve problems. In reality, no one method applies in all cases. You should think of the scientific method as a general set of rules that guide scientists when they pursue a problem. Not all of the rules are followed all of the time, but in general the steps are as follows:

1. State the problem precisely in terms of the most relevant variable factors.

2. Obtain facts pertinent to the problem through carefully controlled experiments.

3. Organize, analyze, and evaluate the collected facts, keeping in mind the problem being solved, and try to find a pattern in the facts. If no pattern exists, reevaluate the problem; possibly the wrong problem is being investigated, or the problem is not stated clearly.

2.1 TECHNIQUES FOR SOLVING SCIENTIFIC PROBLEMS

4. Propose an explanation (hypothesis) to account for the pattern found in the data.

5. Conduct experiments to determine if the proposed hypothesis applies in similar situations.

6. If the experiments support the hypothesis, publish the findings to inform the rest of the world; however, if the experiments do not support the hypothesis, then modify the hypothesis, experimental procedures, or problem and start again.

Let's take a closer look at each step of the scientific method (Fig. 2.1). The first step is to state the problem precisely, in terms of what it is you are trying to solve. An exact statement of the problem should indicate the direction to follow when solving the problem. A large quantity of information (facts, laws, and theories) is collected. This information is organized, evaluated, and analyzed prior to stating the problem.

After the problem is defined, one or more experiments are performed to collect additional facts (data) that will be used to help solve the problem. A **fact** is an accepted truth—something that everyone accepts as correct. **Data** are facts collected during an experiment; from the data, the problem is solved.

When collecting data, experimenters are careful to manipulate, or change, one variable quantity at a time. In other words, they do a **controlled experiment.** The variable of interest, called the *independent variable,* is changed, and the effect on the outcome, or *dependent variable* is observed. All other potentially variable quantities are held constant; i.e., they are not allowed to change. By performing controlled experiments the investigator can discover relationships that exist between two variables.

After all of the data are collected, they are analyzed to find regularities and patterns. Experimenters then ask themselves what accounts for the regularities in the data. After answering this question, they formulate a **hypothesis,** which is a tentative guess that explains the patterns in the data. Outcomes are then predicted for new experiments that will test the hypothesis. New experiments are conducted and their results analyzed to determine if the hypothesis is supported or not. If it is supported, the hypothesis is labeled valid; if not, the hypothesis is modified.

A hypothesis or a group of closely related hypotheses that are supported through tight, controlled experiments may be elevated to the level of a **theory.** It is important to note that a theory mainly differs from a hypothesis in that a theory to some degree has been substantiated through good experiments and is generally more broad and encompassing than a hypothesis.

Theories are sometimes called *models* because they are used to create "pictures" of phenomena that cannot be observed directly. For example, a theory is used by chemists to explain the behavior and structure of atoms, which are infinitesimal bodies that cannot be seen clearly. From this theory a model of the atom has been developed. Biologists propose

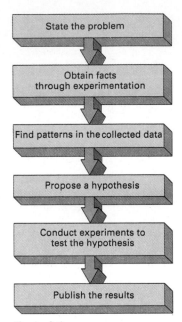

Figure 2.1
Typical steps followed when investigating scientific problems.

Variable quantities do not have a fixed numerical value. The value of the dependent variable depends on the value of the independent variable.

Think of an experiment as a means of discovering a cause and effect relationship.

No amount of experimentation can ever prove me right; a single experiment can prove me wrong.
Albert Einstein

theories to explain the behavior of animals, and physicists propose theories to explain the origin of the universe.

Theories are generally regarded as useful or not useful, rather than right or wrong. Once they are published, theories are subject to scientific scrutiny and criticism. Good theories stand up under the most severe testing; theories that are not so good fall apart under similar conditions. Useful theories have fewer assumptions and exceptions than inferior theories. In chemistry, many phenomena—atoms, molecules, physical states, and chemical changes—are understood through the application of generally accepted theories. It should also be noted that these theories evolve in light of new evidence. Good scientific theories are flexible and can be modified to accommodate new discoveries.

Theories also explain scientific laws. A **scientific law** is a statement as to how a process or event in the universe occurs given a particular set of conditions. There are only a few basic scientific laws; each one explains some consistency of behavior in the natural world. For example, the law of conservation of matter states that matter cannot be created or destroyed under normal conditions. We will discuss this and other laws of conservation in Chap. 4. When applying a scientific law, one is confident of its universality—laws have few, if any, exceptions.

Even though the scientific method has passed the test of time and is based on a sound logical foundation, it has some limitations and flaws. The scientific method assumes that nature acts in a consistent, rational, and understandable way. But in some cases the laws of nature only apply to a subset of matter. For example, some laws work for only normal-size objects; they fail miserably when explaining the behavior of extremely small or large objects. Certain "why" questions regarding the universe are unanswerable, and no matter what problem-solving method is used, they will remain unanswered. Finally, the scientific method is based on a foundation of logic that has inherent limitations; consequently, scientific investigators must be careful not to overstep the limits of logical thought or they can logically fabricate incorrect results with a high level of confidence!

Nature's laws affirm instead of prohibit. If you violate her laws you are your own prosecuting attorney, judge, jury, and hangman.

Luther Burbank

2.1 List the six basic steps of the scientific method.

2.2 Define each term completely: *(a)* fact, *(b)* data, *(c)* hypothesis, *(d)* theory, *(e)* law, *(f)* variable quantity, *(g)* constant quantity.

2.3 What are the limitations of the scientific method?

REVIEW EXERCISES

T o solve chemistry problems successfully, you must employ some of the same procedures used by research scientists. Haphazard attempts to solve problems are generally unproductive, whether they are undertaken in the research laboratory or in the classroom. The "key" which unlocks the mysteries of how to succeed in chemistry is a systematic procedure for approaching and solving chemistry problems.

2.2 STUDENT GUIDE TO PROBLEM SOLVING IN CHEMISTRY

Before we plunge into the details of learning how to solve chemistry problems, let us consider a simple problem faced by most people everyday: How does one go from one place to another? Solving this problem is easy: (1) Determine where you are located initially; (2) pinpoint the exact location of your destination; (3) using a map or prior knowledge, plan a pathway to follow; and (4) go! Few people would jump into their cars and drive endlessly until they found their destination. Instead, they follow a logical plan.

If you think carefully about this simple problem, it is evident that the problem cannot be solved if you do not know either your location or your destination. How can you go to an unknown place? How can you plan a trip if you do not know where you are starting from? Even if you know where you are located and where you are headed, it is impossible to plan a trip if you lack appropriate knowledge of what lies between the two points. If the destination is a familiar place, then you search your brain for the route that has the smallest number of obstacles such as traffic jams, traffic lights, and stop signs. If you are unfamiliar with the destination, you will probably use a map to plan an expedient route. When reading a map, most people usually look for major highways and the shortest path possible, although on some trips one must use back roads and alternative routes. The situation dictates the pathway.

Thinking about the way you solve the problem of going from one place to another will help you understand how to solve chemistry problems. Chemistry students must also know (1) where they are, (2) where they are going, and (3) what pathway they intend to follow before they can solve a chemistry problem. Too often, in a rush to solve a problem, students forget about following a logical route and, in effect, jump into their cars without a map, to search endlessly for their destinations. Random problem-solving methods usually end in frustration.

Chemistry problem solving begins with a careful reading of the problem to determine what is unknown. **What are you trying to find?** If the answer to this question is not apparent, reread the problem and list any unfamiliar words or terms. Find the exact meanings of the unfamiliar terms by referring to the chapter or glossary in the textbook. If you still cannot decide what is unknown, go back to the chapter and reread appropriate sections, paying attention to the example problems. After you have figured out what is unknown, write it down. Now you know where you are headed.

First determine what is unknown.

Continue the chemistry-problem-solving process by listing all relevant information (data) given or known. In other words, answer the question: **What is known?** Write numbers with their labels, or units. A number is meaningless without a label (unless it's unitless). **Units,** or labels, are words that describe the number. For example, 6 dogs, 10 houses, 2 days, and 5 seconds are examples of labeled numbers. In these examples, dogs, houses, days, and seconds describe the numbers to which they are attached. In some problems, besides the data and conversion factors found in the problem, additional information may be required. Such information is found in tables or charts located in the chap-

List all given information.

ter and appendixes. After completing this second step in chemistry problem solving, you know where you are located.

Now you must develop a logical plan for traveling from the location to the destinaton. The logical plan comes from an understanding of the chemistry principles that pertain to the problem. Ask yourself: What is the connection between what is known and what is unknown? One of the most important procedures applied to the solution of chemistry problems is the **factor-label method** (sometimes called the **unit-conversion method**).

Develop a logical plan for solving the problem and gather any other relevant information.

The factor-label method is an orderly procedure in which known, labeled numbers are converted to new numbers with new labels. For example, 14 days can be changed to 2 weeks once we know that 1 week equals 7 days, or 3 dozen doughnuts can be changed to 36 doughnuts using our knowledge that there are 12 doughnuts in a dozen. The specifics of the factor-label method are discussed and illustrated in the next section. At this point, it is important for you to realize that some systematic, logical procedure such as the factor-label method is required if you are to solve chemistry problems. A systematic procedure is the "map" that guides you to your destination.

A chemistry problem is finally solved when you **perform the indicated mathematical operations.** This final step is purely mechanical. Using the correct setup, all numbers and their associated units are added, subtracted, multiplied, or divided to yield the final numerical answer with its units. After you have obtained an answer always ask yourself: **Is this a reasonable answer to the problem?** If the answer is not reasonable, go back and look for mistakes that you may have made. Too often students submit unreasonable answers because they blindly accept an answer without giving thought to its validity.

Perform the indicated mathematical operations.

When possible, check your answer with the correct answer located in the Appendix of selected answers to the exercises. If you have successfully solved the problem, your answer should agree with the correct answer. If your answer does not agree, check to see if you made an arithmetic error. If no arithmetic error is detected, assume you have made an error in logic. Analyze the reasoning that you used originally, or attempt to work back from the correct answer. A word of caution is in order: Do not rely totally on this last method; the correct answer is not available during quizzes and exams!

In summary, the four steps most commonly followed when solving chemistry problems are:

1. Carefully read the problem and determine what it asks. On paper, write down exactly what you are trying to find—the unknown quantity.

2. Extract from the problem all information that is given, and obtain any other information that is necessary to solve the problem.

3. Apply appropriate chemistry principles, logic, and the factor-label method to convert the known information to what is desired.

4. Perform the indicated mathematical operations and find the answer to the problem. If possible, check to see that the answer is correct. If the answer is incorrect, repeat any steps as required.

Before proceeding, it is extremely important to understand and learn all four chemistry-problem-solving steps. **Possibly the most important thing learned in a beginning chemistry course is how to approach and solve problems.**

2.4 List all things that must be known before a problem can be solved.

2.5 What are the four principal steps used to solve chemistry problems?

2.3 THE FACTOR-LABEL METHOD

Numerous problems encountered in chemistry are conveniently solved using the factor-label method. In the factor-label method, one or more conversion factors are used to change the given units to the desired units. A **conversion factor** is an exact relationship between two quantities expressed as a fraction. For example, one dozen objects is defined as 12 objects.

$$1 \text{ dozen objects} = 12 \text{ objects}$$

The correct way to express this equality as a conversion factor is

$$\frac{12 \text{ objects}}{1 \text{ dozen objects}} \quad \text{or} \quad \frac{1 \text{ dozen objects}}{12 \text{ objects}}$$

In a conversion factor, the fraction line is read as "per." So the above pair of expressions is read as "12 objects per 1 dozen objects or 1 dozen objects per 12 objects."

Mathematically, conversion factors are obtained by dividing both sides of the equality by one of the quantities. Dividing both sides of the equality 1 dozen objects = 12 objects by 1 dozen objects yields

$$\frac{\cancel{1 \text{ dozen objects}}}{\cancel{1 \text{ dozen objects}}} = \frac{12 \text{ objects}}{1 \text{ dozen objects}}$$

and canceling the 1 dozen objects gives us the conversion factor

$$1 = \frac{12 \text{ objects}}{1 \text{ dozen objects}}$$

To obtain the inverted form of the conversion factor, divide the equality by "12 objects,"

$$\frac{1 \text{ dozen objects}}{12 \text{ objects}} = \frac{\cancel{12 \text{ objects}}}{\cancel{12 \text{ objects}}}$$

and then cancel the "12 objects."

$$\frac{1 \text{ dozen objects}}{12 \text{ objects}} = 1$$

Note that a conversion factor always equals 1. Therefore, if a quantity is multiplied by a conversion factor, the value of the quantity is unchanged, even though the number and units change. Multiplying 1 times any number does not alter the value of the number.

Other examples of conversion factors are

The value of a number is unchanged when it is multiplied by 1. Consider the following examples: 5 × 1 = 5, 10 × 1 = 10, 85 × 1 = 85.

$$\frac{60 \text{ seconds}}{1 \text{ minute}} \qquad \frac{12 \text{ inches}}{1 \text{ foot}} \qquad \frac{4 \text{ quarts}}{1 \text{ gallon}}$$

Each of these conversion factors was obtained from exact relationships between the two units. For example, by definition 60 seconds elapse per 1 minute (60 seconds = 1 minute). After dividing both sides by 1 minute, the above conversion factor is obtained. If we had divided by 60 seconds, the reciprocal of this factor would have been obtained: 1 minute/60 seconds. Similarly, the reciprocals of the other two conversion factors are 1 foot/12 inches and 1 gallon/4 quarts.

Conversion factors are used to change the units associated with a number to another set of units. This is accomplished by multiplying the conversion factor times the given quantity, so that the given unit cancels and yields the desired unit.

$$\cancel{\text{Given unit}} \times \underbrace{\frac{\text{desired unit}}{\cancel{\text{given unit}}}}_{\text{Conversion factor}} = \text{desired unit}$$

For example: How many dozen eggs is 120 eggs?

$$120 \cancel{\text{ eggs}} \times \frac{1 \text{ dozen eggs}}{12 \cancel{\text{ eggs}}} = 10 \text{ dozen eggs}$$

In this case, we see that the given unit, eggs, is canceled by the unit eggs in the denominator of the conversion factor, yielding dozen eggs, the unit in the numerator of the conversion factor. Thus, 120 eggs equals 10 dozen eggs.

Study the following examples of simple factor-label conversions used to solve nonscientific problems. The purpose of these problems is to show the mechanics of the factor-label method.

──────────── **Example Problem 2.1** ────────────

A builder constructs 50 houses of 10 rooms each. How many rooms are there in all 50 houses?

──────────── **Solution** ────────────

1. What is unknown?   Number of rooms in 50 houses
2. What is known?   10 rooms/1 house; 50 houses
3. Apply the factor-label method.

Because the problem asks for the number of rooms, and we know the number of houses and the number of rooms per house, we write the conversion factor with rooms in the numerator (top) and houses in the denominator (bottom). When houses are multiplied by rooms per house, the houses cancel, leaving the number of rooms.

$$\text{Houses} \times \frac{10 \text{ rooms}}{1 \text{ house}} = \text{? rooms}$$

4. Perform the indicated math operations.

$$50 \text{ houses} \times \frac{10 \text{ rooms}}{1 \text{ house}} = \textbf{500 rooms}$$

In this example, we have applied the basic principles of problem solving: (1) identifying what is unknown, (2) identifying what is known, (3) applying the factor-label method to find the desired units, and (4) performing the indicated arithmetic operations.

──────────── **Example Problem 2.2** ────────────

A soup company packages 30 cans of soup per box. How many boxes are needed to hold 8700 soup cans?

──────────── **Solution** ────────────

1. What is unknown?   Number of boxes that hold 8700 soup cans
2. What is known?   30 soup cans/1 box; 8700 soup cans
3. Apply the factor-label method.

$$\text{Soup cans} \times \frac{1 \text{ box}}{30 \text{ soup cans}} = \text{? boxes}$$

To obtain the number of boxes and cancel soup cans, we invert the conversion factor, placing the desired unit, boxes, in the numerator and the unit soup cans in the denominator.

4. Perform the indicated math operations.

$$8700 \; \cancel{\text{soup cans}} \times \frac{1 \text{ box}}{30 \; \cancel{\text{soup cans}}} = \textbf{290 boxes}$$

The factor-label setup shows that 30 is divided into 8700 to obtain the correct answer, 290 boxes.

Example Problems 2.1 and 2.2 are simple, almost trivial, nonchemical examples that illustrate the general procedure for solving chemistry problems. Most chemistry problems are as easy to solve as these, once the techniques of problem solving and the factor-label method are learned. Use the factor-label method to solve the following Review Exercises before going on to the next section.

REVIEW EXERCISES

2.6 An orange crate holds 66 oranges. *(a)* How many crates are needed to hold 2310 oranges? *(b)* How many oranges are contained in 175 full crates?

2.7 One brand of gasoline costs $1.15 per gallon. *(a)* How many gallons can be purchased with $16.10? *(b)* What is the cost of exactly 11 gallons of this gasoline? *(c)* How many gallons can be purchased with $1.00?

In many cases, more than one conversion factor is utilized to solve a problem. Let's consider the problem of converting a given number of years to hours. Most people do not know an exact relationship between years and hours, and such a relationship is generally not found in a table. But application of the rules regarding problem solving and conversion factors efficiently gives us the correct answer.

Knowing that we want to find the number of hours, given the number of years, and that there are 365 days/year and 24 hours/day we write

$$\cancel{\text{Years}} \times \frac{365 \; \cancel{\text{days}}}{1 \; \cancel{\text{year}}} \times \frac{24 \text{ hours}}{1 \; \cancel{\text{day}}} = \text{hours}$$

The number of years is first multiplied by the conversion factor that relates days to years. The years cancel, giving the number of days. Days are then converted to hours by multiplying by the conversion factor that equates hours and days. In a similar manner the days cancel, yielding the number of hours.

Figure 2.2 illustrates the pathway followed to find the number of hours in a given number of years. Because a direct path is not available, an indirect route, obtaining the number of days, is followed.

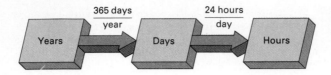

Figure 2.2
To convert years to hours, multiply the given number of years by the conversion factor that equates years to days, and then multiply the result by the conversion factor that equates days to hours.

Conversion factors are introduced successively until the desired unit is found. Any number of conversion factors may be used to solve a problem. Example Problems 2.3 and 2.4 illustrate time conversions that require more than one conversion factor.

─────────────── **Example Problem 2.3** ───────────────

How many seconds elapse in exactly 4 hours?

─────────────── **Solution** ───────────────

1. What is unknown?   Number of seconds
2. What is known?   4 hours; 60 minutes/hour; 60 seconds/minute
3. Apply the factor-label method.

$$\text{Hours} \times \frac{60 \text{ minutes}}{1 \text{ hour}} \times \frac{60 \text{ seconds}}{1 \text{ minute}} = ? \text{ seconds}$$

Hours are first converted to minutes, using the conversion factor 60 minutes per 1 hour. In a similar manner, minutes are converted to seconds with the conversion factor 60 seconds per 1 minute. The first conversion factor cancels hours, yielding minutes; the second conversion factor cancels minutes, giving seconds, the units of the answer. This pathway is represented diagrammatically in Fig. 2.3.

4. Perform the indicated math operations.

$$4 \text{ hours} \times \frac{60 \text{ minutes}}{1 \text{ hour}} \times \frac{60 \text{ seconds}}{1 \text{ minute}} = \textbf{14,400 seconds}$$

─────────────── **Example Problem 2.4** ───────────────

Convert exactly 3600 minutes to days.

─────────────── **Solution** ───────────────

1. What is unknown?   Number of days
2. What is known?   3600 minutes; 60 minutes/hour; 24 hours/day
3. Apply the factor-label method.

$$\text{Minutes} \times \frac{1 \text{ hour}}{60 \text{ minutes}} \times \frac{1 \text{ day}}{24 \text{ hours}} = ? \text{ days}$$

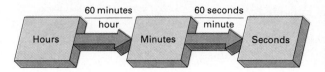

Figure 2.3
To change hours to seconds, multiply the given number of hours by the conversion factor that equates hours to minutes, and then multiply the result by the conversion factor that equates minutes to seconds.

Because we are starting with minutes, we must invert the conversion factor that equates minutes to hours so that minutes cancel when we multiply. Hours are converted to days by inverting the conversion factor that equates hours and days, and canceling hours when we multiply.

4. Perform the indicated math operations.

$$3600 \text{ } \cancel{\text{minutes}} \times \frac{1 \text{ } \cancel{\text{hour}}}{60 \text{ } \cancel{\text{minutes}}} \times \frac{1 \text{ day}}{24 \text{ } \cancel{\text{hours}}} = \textbf{2.5 days}$$

A common error when performing arithmetic operations of this type is to forget that both conversion factors require division—both 60 minutes and 24 hours are divided into 3600 minutes.

After completing a conversion using the factor-label method, look at the answer and ask yourself if it is reasonable or not. When you convert units small in size to units of larger size, the numerical value decreases. Conversely, when you change large units to smaller units, the numerical value increases. For example, if you change years, a larger time unit, to seconds, a smaller time unit, the numerical value for the number of seconds will be significantly larger. As we saw in Example Problem 2.4, 3600 minutes elapse in 2.5 days. Thus, changing a unit of smaller magnitude to one of larger magnitude results in a decrease in the size of the number. Do not let your use of the factor-label method become a mechanical operation. **Apply the factor-label method prudently, considering where you are coming from and where you are heading.**

On some occasions you may need to calculate the value of a conversion factor required to solve a problem. For example, if we know the number of hours per day and the number of days per week, we can calculate the number of hours per week by multiplying the conversion factors in such a way as to cancel the number of days. This is accomplished in the following manner:

$$\frac{24 \text{ hours}}{1 \text{ } \cancel{\text{day}}} \times \frac{7 \text{ } \cancel{\text{days}}}{1 \text{ week}} = \frac{168 \text{ hours}}{1 \text{ week}}$$

We find that 168 hours elapse during 1 week. Example Problem 2.5 gives another example of the calculation of a new conversion factor.

Example Problem 2.5

One mile is 5280 feet, and the surveyor's unit called a link is exactly 7.92 inches. Calculate the number of links per mile.

Solution
---

1. What is unknown? Number of links per mile

2. What is known? 5280 feet/mile; 7.92 inches/link; 12 inches/foot We must know the relationship between inches and feet because no common units are found in the first two conversion factors (the ones given in the problem).

3. Apply the factor-label method.

$$\frac{5280 \text{ feet}}{1 \text{ mile}} \times \frac{12 \text{ inches}}{1 \text{ foot}} \times \frac{1 \text{ link}}{7.92 \text{ inches}} = \frac{? \text{ links}}{\text{mile}}$$

In the factor-label setup the number of feet per mile is first converted to inches per mile, and then it is converted to links per mile, the desired conversion factor.

4. Perform the indicated math operations.

$$\frac{5280 \text{ feet}}{\text{mile}} \times \frac{12 \text{ inches}}{\text{foot}} \times \frac{1 \text{ link}}{7.92 \text{ inches}} = \frac{\textbf{8000 links}}{\textbf{mile}}$$

We find that 1 mile is exactly 8000 links.

REVIEW EXERCISES

2.8 Convert exactly 8 days to: (*a*) hours, (*b*) minutes, (*c*) seconds, (*d*) years.

2.9 Convert exactly 12 centuries to decades (100 years = 1 century; 10 years = 1 decade).

2.10 Calculate the numerical value for each of the following conversion factors: (*a*) seconds/year, (*b*) centuries/day, (*c*) days/millennium (1 millennium = 1000 years).

SUMMARY

A systematic set of procedures, called the **scientific method,** is more or less followed in modern scientific research. The heart of the scientific method is the collection of facts and data in order to propose a **hypothesis,** a suggested explanation of the problem. A **controlled experiment** is then conducted to determine if the hypothesis is valid or not. Experiments provide evidence to support hypotheses.

The best way to study chemistry is to follow a systematic plan. Solving problems and answering questions is the most effective way to learn chemistry. You must know what you are looking for and what infor-

mation is available before mapping out a strategy for solving a problem. Once the problem is solved, check to see if you have obtained the correct answer. If not, go back and try again.

Many chemistry problems are solved using the **factor-label method.** This is a systematic procedure for converting a number with one unit to a new number with another unit using conversion factors, i.e., fractions that express an equality. Conversion factors are multiplied in such a way that the given unit is canceled and the desired unit is retained.

EXERCISES*

Scientific Method

2.11 List reasons why it is useless to conduct an uncontrolled experiment.

2.12 Why is it important to spend a significant amount of time in stating precisely the problem to be solved?

2.13 (a) How is a fact different from a law? (b) How is a theory different from a law? (c) How is a hypothesis different from a theory?

2.14 Give three examples of theories from everyday life.

2.15 Why are theories labeled valid and invalid, not right and wrong?

2.16 If you were conducting the following experiments, what variables should you control to produce the most valid results? (a) Which of two insecticides is most effective in killing roaches? (b) Which of two automobiles is most fuel efficient? (c) What pain reliever is most effective at controlling headaches?

2.17 Classify each of the following as a fact, theory, or law:
(a) Grass is green.
(b) Whatever goes up must come down.
(c) Human beings evolved from apes.
(d) The earth revolves around the sun.
(e) Matter is composed of atoms.
(f) Diamonds are forever.
(g) The harder an object is pushed, the faster it accelerates.

2.18 Why can't an experiment be conducted to measure "love"?

2.19 Outline a complete experiment (all six steps) that might be used to determine the effectiveness of a newly synthesized antimicrobial agent, i.e., a drug that fights disease.

Factor-Label Method

2.20 A small economy car, when cruising on a highway, gets exactly 47 miles per gallon. (a) How many miles can the economy car travel on exactly 12 gallons of gasoline? (b) How many gallons of gas are consumed in this car during a trip of 164.5 miles?

2.21 Perform the following conversions:
(a) 8 minutes = ? seconds
(b) 253 days = ? years
(c) 9 decades = ? centuries
(d) 5184 apples = ? dozen apples

(e) 14,500 hours = ? days
(f) 5616 pencils = ? gross (1 gross = 144 items)

2.22 (a) How many hours would it take to travel 462 miles if a constant speed of exactly 55 miles per hour were maintained? (b) How far can you travel at 55 miles per hour in 16.4 minutes?

2.23 How many seconds elapse during each of the following time intervals: (a) 5.00 minutes, (b) 0.850 minute, (c) 1.25 hours, (d) 3.1 days?

2.24 What is the age of a 25-year-old person in (a) days, (b) hours, (c) minutes, (d) seconds?

2.25 Perform each of the following time conversions:
(a) 9641 hours = ? years
(b) 6505 days = ? decades
(c) 12 centuries = ? days
(d) 915,000 seconds = ? days
(e) 100,000 days = ? millennia (1 millennium = 1000 years)

2.26 In typography, one pica equals twelve points, and one point equals 0.013837 inch exactly. (a) How many inches are there per pica? (b) How many picas per inch?

2.27 In the avoirdupois system of weights, 16 drams equal 437.5 grains, 16 ounces equal one pound, and 16 drams equal one ounce. (a) How many drams are there per pound? (b) How many grains per pound? (c) How many drams per grain?

2.28 Excellent long distance runners run 25 miles in 2 hours. (a) Assuming they run at the same pace throughout the 2 hours, how many miles will they travel in 0.75 hour? (b) How many feet are traveled in 0.75 hour? (5280 feet = 1 mile) (c) How long would it take a runner to travel 10,000 feet at this pace?

2.29 Nine bananas weigh 2.9 pounds, and bananas cost $0.35 per pound. (a) What is the cost of six bananas? (b) How many dozen of bananas can be bought for $12.40? (c) What is the cost of one banana?

2.30 On a trip from New York to Miami, 1317 miles, an automobile consumed exactly 49 gallons of gasoline costing $1.05 per gallon. (a) What was the average number of miles traveled per gallon? (b) How many gallons of gasoline were consumed per mile? (c) What was the cost of gaso-

*For exercise numbers printed in color, answers can be found at the end of the book.

line per 100 miles? *(d)* What was the total cost of gasoline for the trip?

2.31 A box that contains 100 paper clips costs $0.79; each paper clip weighs 0.0013 pounds. *(a)* What is the cost of 34 paper clips? *(b)* What is the cost of a single paper clip? *(c)* What is the cost of paper clips per pound? *(d)* How many boxes must be emptied to produce 8 pounds of paper clips? *(e)* How many pounds of paper clips can be purchased with $54?

2.32 Convert exactly 50,000 seconds to *(a)* centuries, *(b)* days, *(c)* decades.

2.33 The approximate life-span of an American is 75 years. Convert the life-span to *(a)* seconds, *(b)* decades, *(c)* days.

2.34 A furlong is a unit of length equal to 660 feet. Another unit of length, the rod, is equal to 198 inches. *(a)* How many rods are there per furlong. *(b)* If a centimeter is 0.394 inch, how many furlongs are there in 1 million centimeters? *(c)* A fathom is exactly 6 feet. How many rods are there in a fathom? *(d)* One chain equals exactly 792 inches. If a link equals 0.04 rod, how many yards are there in a link?

Additional Exercises

2.35 Explain specifically how you might solve one of your personal problems making use of the scientific method.

2.36 *(a)* Propose a means by which a study could be conducted to determine if one brand of toothpaste with fluoride is more effective in preventing the development of dental caries (cavities) than a second brand that is essentially the same but does not contain fluoride. *(b)* What are the dependent and independent variables? *(c)* What hypothesis might be stated for such a study? *(d)* What variables must be held constant in such a study to give validity to the results?

2.37 Light travels at 186,000 miles per second. Light from the sun takes about 0.23 days to travel to the planet Pluto. *(a)* What is the distance from the sun to Pluto? *(b)* How far does light travel in 1 year?

2.38 Two ancient Roman units of length were the cubit and the stadium. One cubit is exactly 17.5 inches, and one stadium is 202 yards. *(a)* How many cubits are there per stadium? *(b)* How many stadia (plural of *stadium*) are there per cubit?

2.39 Alaska has an area of approximately 571,000 square miles. An acre is equal to 43,560 square feet. *(a)* If one mile is 5280 feet, calculate the area of Alaska in acres. *(b)* Calculate the area of Alaska in square feet. *(c)* Calculate the area of Alaska in square inches.

CHAPTER
Three

Chemical Measurement

STUDY GUIDELINES

After completing Chapter 3, you should be able to

1. Distinguish between precision and accuracy of measurements

2. Distinguish between systematic and random errors in measurements

3. Give a definition of significant figures

4. Determine if a zero in a measured quantity is significant

5. Add, subtract, multiply, and divide measured quantities and express the answer to the correct number of significant figures with the correct units

6. List the seven base SI units and explain what they measure

7. List common prefixes used in the SI and state their meaning

8. Explain the difference between base and derived SI units

9. Convert base SI units to common multiples

10. Distinguish between mass and weight

11. Define temperature and briefly explain how it is different from heat

12. Perform temperature conversions given Kelvin, Celsius, or Fahrenheit temperatures

13. Convert units of volume to multiples and to other units

14. Calculate the density of a substance given its mass and volume

15. Calculate either the mass or volume of an object given its density

16. Explain the relationship of specific gravity to density

When chemists conduct experiments they usually make many different measurements. The taking of a **measurement** is a procedure in which an unknown quantity is compared to a known quantity. Usually, the known quantity is obtained from a device calibrated to display the correct value. For example, when you measure your weight, you stand on a scale constructed to display the weight of a person in pounds. As we have discussed previously, each measurement consists of two parts, a *number* and a *unit.* Because all measurements are uncertain to some degree, chemists making measurements in the laboratory are also concerned with the *reliability* of the measurements. To show how reliable a measurement is, chemists express it to the correct number of significant figures.

In this chapter, we will first discuss uncertainty in measurements and the rules that govern the use of significant figures, and then move to the International System of Units, which is the measurement system used by scientists throughout the world.

3.1 UNCERTAINTY IN MEASUREMENTS

Whenever chemical measurements are made, both precision and accuracy are considered. The **accuracy** of a measurement is how close the obtained value is to a standard, or "true," value. A more accurate measurement is one closer to the standard value than a less accurate measurement. Accuracy is measured in terms of the deviation of the measurement (called the error) from the "true" value.

Precision, on the other hand, refers to how closely repeated measures are grouped, i.e., how reproducible the measurements are. The smaller the range of values obtained when measuring the same quantity, the greater the precision. Most frequently, good precision is an indication of high accuracy, but not always, as we shall see.

The game of darts provides an analogy that is very helpful in understanding the terms "accuracy" and "precision." A dart player attempts to "hit the bull's-eye," as a chemist tries to "hit" the true value when measuring matter. Consider Fig. 3.1*a*; the first dart landed far from the bull's-eye—not an accurate throw. In Fig. 3.1*b*, the dart is inside the bull's-eye—a more accurate throw. Fig. 3.1*c* represents a dart board after six throws; notice that the six attempts are grouped close together, which is an indication of good precision, and in the bull's-eye, which shows good accuracy. In contrast, the dart distribution in Fig. 3.1*d* indicates good precision but poor accuracy. Finally, Fig. 3.1*e* shows poor accuracy and poor precision.

Measurement errors account for the range of different values that are obtained when making the same measurement repeatedly. Two types of errors are generally found in chemical measurements: systematic errors and random errors.

Systematic errors result from (1) poor procedures and methods, (2) malfunctioning and uncalibrated instruments, (3) human error, (4) impure samples, and (5) some unrecognized factors that influence the results. For example, suppose that you measure the weight of an object on a scale but fail to adjust the scale to the zero point prior to weighing the object. The measurement will be either high or low, depending on the initial incorrect setting of the scale. Systematic errors are reduced by finding their causes and eliminating them.

Systematic errors can usually be eliminated if they are identified.

Random errors occur in all chemical measurements. Even if every precaution is taken to avoid systematic errors, small deviations, or random errors, arise that are unavoidable and not identifiable. Random errors, by definition, are impossible to illustrate. If random errors could

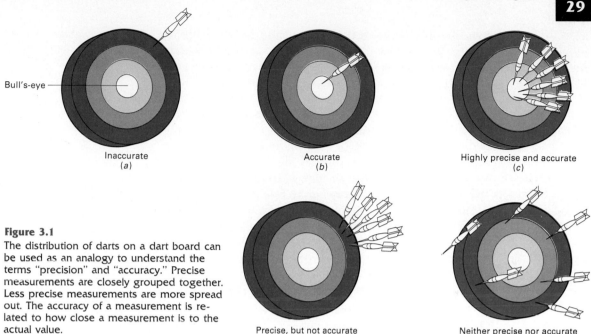

Figure 3.1
The distribution of darts on a dart board can be used as an analogy to understand the terms "precision" and "accuracy." Precise measurements are closely grouped together. Less precise measurements are more spread out. The accuracy of a measurement is related to how close a measurement is to the actual value.

be identified, they would be corrected, and thus would not be considered random errors.

Collectively, systematic and random errors introduce uncertainty—or lack of confidence—into all measured values. Thus, for all reported measurements, the scientist should indicate the degree of certainty and uncertainty in the measurement. In chemistry, this is most frequently accomplished through the use of significant figures, which is described in the following section.

Random errors tend to cancel out and thus do not seriously affect the uncertainty of measurements.

REVIEW EXERCISES

3.1 *(a)* Write definitions for precision and accuracy. *(b)* Can accurate measurements be made without precision? Explain.

3.2 *(a)* What can be said about 10 measurements known to be highly precise? *(b)* Does this mean that these measurements are accurate? Explain.

3.3 The true value for the length of an object is 2.91 meters. What statement may be made about the precision of the following length measurements of the object: 2.02 meters, 3.07 meters, 2.87 meters, and 3.17 meters?

3.4 *(a)* Explain the difference between systematic and random measurement errors. *(b)* Give two examples of systematic errors. *(c)* Why is it impossible to give an example of a random error?

3.2 SIGNIFICANT FIGURES

Significant figures are the measured digits in a number that are known with certainty plus one uncertain digit. Stated differently, the significant figures are all known digits plus the first doubtful or estimated digit.

Before we begin our discussion, it is important to note that significant figures apply only to *measured* values, not to exact numbers. For example, if you correctly count the number of students in a small classroom, the value obtained is absolutely accurate (an exact number) with no uncertainty. When we discuss mass in Sec. 3.4, we will define the kilogram as 1000 grams (1 kg = 1000 g), which is an exact relationship; consequently, significant figures do not apply. Significant figures apply only to measurements that are to some degree uncertain.

To illustrate significant figures, let's consider Fig. 3.2*a*, which shows a liquid in a graduated cylinder (a type of glassware used to measure volume). If you measure the level of the liquid relative to the scale etched on the cylinder, the volume of the liquid may be found. The volume indicated is 35.5 milliliters, because the liquid level (always read the bottom of the meniscus, the concave upper surface of a liquid in a small container) is about halfway between the 35-milliliter line and the 36-milliliter line. This measurement, 35.5 milliliters, represents three significant figures—the 35 is directly measured (these are the digits that are certain), and the .5 is a good estimate from a careful reading of the graduated cylinder (it is the first uncertain digit).

Usually, the last significant figure is considered to be uncertain by ±1. In the above example, by stating the volume as 35.5 milliliters, we indicate that the measured volume is at most 35.6 (+0.1) and at least 35.4 (−0.1). If the same liquid is totally transferred to a more precise volumetric instrument—let's say one that has a scale with 0.1-milliliter marks etched accurately on the side—it is possible to obtain an additional significant figure. In Fig. 3.2*b* we can read the volume to one-hundredth of a milliliter, giving us the value 35.58 milliliters. The first three digits (35.5) are certain, and the last digit (8) is uncertain. Once again, the range of uncertainty is ±1 in the last significant figure, giving a range for our measurement of 35.59 to 35.57 milliliters.

If, using the same graduated cylinder, you read the volume as 35.5872389 milliliters, you are exceeding the limits of this measuring device. There is no way to be certain of the second decimal place, and it is impossible to measure the third or fourth decimal places—they are meaningless numbers, especially if you consider that the uncertainty is ±0.01 milliliter. **Never report more significant figures than the measuring device is capable of providing.**

Whenever you encounter a measurement, always remember that besides the numerical value and units, the number also indicates the precision with which the measurement was made, or, stated differently, the number of significant figures in the measurement. Consider the following measurements and the number of significant figures indicated by each one.

Significant figures include all measured digits plus one uncertain, or estimated, digit.

Significant figures do not pertain to exact numbers because they are not uncertain.

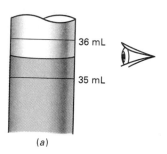

(a)

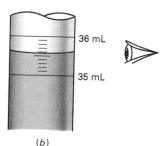

(b)

Figure 3.2
(a) With 1-mL graduations, the volume of a liquid can be estimated to within ±0.1 mL. In this case the volume is reported as 35.5 mL; this is three significant figures.
(b) With 0.1-mL graduations, the volume can be estimated to within ±0.01 mL. Using this more precisely calibrated graduated cylinder, we report the volume as 35.58 mL; this is four significant figures.

1.25 meters indicates three significant figures (1, 2, 5)
(range: 1.26–1.24 meters)

434.56 Kelvins indicates five significant figures (4, 3, 4, 5, 6)
  (range: 434.57–434.55 Kelvins)

3 grams indicates one significant figure (3)
  (range: 4–2 grams)

8.913477 centimeters indicates seven significant figures
    (8, 9, 1, 3, 4, 7, 7)
  (range: 8.913478–8.913476 centimeters)

When writing measured quantities, it is important always to ask yourself: Did I report the correct number of significant figures?

All nonzero numbers in measurements are always significant, but zeros pose a special problem because a zero that acts as a placeholder in a number is *not* significant. Placeholders are not measured quantities; therefore, by definition, they are not significant figures.

The following rules summarize and illustrate all possible cases in which zeros are found in measurements.

Rule 1

Zeros located in the middle of a number. In all cases, zeros in the middle of a number are significant. In each of the following, the zero is a significant figure.

10.004 grams indicates five significant figures (1, 0, 0, 0, 4)

47,000.15 milliliters indicates seven significant figures
    (4, 7, 0, 0, 0, 1, 5)

103 square centimeters indicates three significant figures (1, 0, 3)

Zeros in the middle of measured quantities are measured digits and are not placeholders; accordingly, they are significant in all cases.

Rule 2

Zeros located in front of a number. Zeros in front of numbers are usually to the right of the decimal point. These zeros act as placeholders (they are not measured), so they are not significant figures. Consider each of the following examples of such cases.

.0005 liter indicates one significant figure (5)

.0000477 meter indicates three significant figures (4, 7, 7)

.000000091 kilogram indicates two significant figures (9, 1)

If these examples are converted to scientific notation, you can readily see the insignificance of the zeros.

5×10^{-4} liter 4.77×10^{-5} meter 9.1×10^{-8} kilogram

In each case the zeros are dropped because they act only as placeholders. Sometimes a zero is placed in front of the decimal point to show that no other digit is present. Similarly, this zero is not significant. The following examples serve to illustrate this point.

0.00831 millimeter indicates three significant figures (8, 3, 1)

0.1 milliliter indicates one significant figure (1)

Rule 3

Zeros located after a number to the right of the decimal point. For this specific case, the zero is either a measured quantity (certain) or a good estimate (the first uncertain digit); consequently, zeros after a number and to the right of the decimal point are all significant. Three examples of this case are as follows:

.650 decimeter indicates three significant figures (6, 5, 0)

.178300 millisecond indicates six significant figures (1, 7, 8, 3, 0, 0)

.3550000 kilometer indicates seven significant figures
(3, 5, 5, 0, 0, 0, 0)

For each of the above measurements, the zeros were measured by some instrument. All of the zeros are certain except the last zero, which is uncertain but still significant.

Rule 4

Zeros located after a number to the left of the decimal point. Zeros found after a number and to the left of the decimal point are significant if they are measured, and are not significant if they are placeholders. Suppose you see a statement that an object has a measured mass of 800 grams. The measurement 800 grams here has a questionable number of significant figures; more information is required to determine what the correct number of significant figures is. The measurement, 800 grams, has three significant figures only if the second zero (units' place) is the first uncertain figure. It has two significant figures if the first zero (tens' place) is the first uncertain figure. There is a third possibility: the 8 could be the uncertain digit; if this is the case, the measurement has only one significant figure.

To avoid the confusion generated by the ambiguous nature of zeros to the left of decimal points, chemists commonly express such measurements in **scientific notation,** in which the decimal factor represents the correct number of significant figures. Thus, 800 grams is expressed as 8×10^2 grams (one significant figure) or 8.0×10^2 grams (two significant figures) or 8.00×10^2 grams (three significant figures), depending on what the actual number of significant figures is.

Let's look at a second example. The measurement 45,000 millimeters is expressed as 2, 3, 4, or 5 significant figures as follows:

4.5×10^4 millimeters represents two significant figures (4, 5)

4.50×10^4 millimeters represents three significant figures (4, 5, 0)

4.500×10^4 millimeters represents four significant figures
(4, 5, 0, 0)

4.5000×10^4 millimeters represents five significant figures
(4, 5, 0, 0, 0)

3.5 What is the purpose of expressing measurements to the correct number of significant figures?

3.6 What is the number of significant figures indicated by each of the following measurements: *(a)* 219,977 grams, *(b)* 0.2198 centimeter, *(c)* 5000.020 liters, *(d)* 0.0005000 kilogram, *(e)* 5.0110×10^{12} milligrams, *(f)* 0.0000000070 meter, *(g)* 27,333,000.00080000 nanometers, *(h)* 0.2 second?

Normally, after measurements are obtained, they are used in subsequent calculations. Specific rules are followed so that results of the calculations also have the proper number of significant figures. Two different rules are followed, depending on the arithmetic operation performed. The rules for handling significant figures when adding and subtracting are different from those used when multiplying and dividing. Attempt to learn each rule, and try not to confuse one with the other.

When measured quantities are added and subtracted, the answer can have no more digits to the right of the decimal point than does the measured quantity with the least number of decimal places. If, for instance, we add the masses 2.0965 grams and 1.41 grams, the answer can only have two decimal places. First calculate the sum of the two numbers:

Addition and Subtraction of Significant Figures

$$
\begin{array}{r}
2.0965 \text{ grams} \\
+ \ 1.41 \quad\ \text{ grams} \\
\hline
3.5065 \text{ grams}
\end{array}
$$

Then round off to the correct number of decimal places. Because the second mass was measured only to two decimal places, the answer cannot have more than two decimal places.

When you round off, look at the first nonsignificant figure (6 in our example). The first nonsignificant figure is the figure one place to the right of the least significant figure (0, in our example). The *least significant figure* is the last digit in the number retained when rounding off. Then apply the following three rules for rounding off.

Rule 1

If the value of the first nonsignificant figure is greater than 5, add 1 to the least significant figure and drop all nonsignificant digits.

Rule 2

If the value of the first nonsignificant figure is less than 5, retain the least significant figure and drop all nonsignificant digits.

Rule 3

If the first nonsignificant figure is 5 and it is followed by nonzero digits, increase the value of the least significant figure by 1 and drop all nonsignificant digits. If the first nonsignificant figure is 5 and it is followed by zeros or nothing, add 1 to the least significant figure if it is an odd number and drop all nonsignificant digits; however, if the least significant figure is an even number, retain the least significant figure and drop all nonsignificant figures.

Many people use this simplified version of the rounding rules: "Round up" if the first nonsignificant figure is 5 or greater, and "round down" if the first nonsignificant digit is less than 5.

In our mass example above, the first nonsignificant figure is 6, so 1 is added to 0, giving 1 as the second decimal place. The final answer is expressed as 3.51 grams. Any other answer is incorrect.

$$3.5065 \text{ grams}$$

Least significant figure ——First nonsignificant figure

$$3.5065 \text{ grams}$$

Add 1 —— Drop

Example Problem 3.1 is another illustration of how to apply the rules for adding measurements and expressing the answer to the correct number of significant figures.

--------------------- **Example Problem 3.1** ---------------------

What is the sum of 10.0043 milliliters + 5.5 milliliters + 9.250 milliliters?

--------------------- **Solution** ---------------------

10.0043	milliliters	(Four decimal places)
5.5	milliliters	(One decimal place)
+ 9.250	milliliters	(Three decimal places)
24.7543	milliliters	(Round to one decimal place)

Least significant figure —— First nonsignificant figure

We must round off our answer, 24.7543 milliliters, to one decimal place because the second measured quantity, 5.5 milliliters, has only one decimal place. Because the first nonsignificant figure is a 5 followed by nonzero digits we add 1 to the least significant figure, 7, and drop all nonsignificant digits. This gives a final answer of **24.8 milliliters.**

You may wonder why such a rule is employed when measurements are added and subtracted. Always, when dealing with significant figures, the last digit in a measured quantity is uncertain; thus, if it is added to or subtracted from a figure that is certain, the resultant quantity is rendered uncertain. For example, when 3.42 meters is added to 1.9 meters, the 2 is uncertain in 3.42 meters and the 9 is uncertain in 1.9 meters. Thus, in the answer, 5.32 meters, both the 3 and the 2 are uncertain.

$$3.4\underline{2} \text{ meters} \quad \text{Uncertain}$$
$$+ \ 1.\underline{9} \quad \text{meters} \quad \text{Uncertain}$$
$$5.3\underline{2} \text{ meters} \quad \text{Uncertain}$$

Least significant figure ——

After adding, we see that the new least significant figure is 3; consequently, the number must be rounded off to 5.3 m because the first nonsignificant figure is less than 5 (Rule 2).

When measurements are multiplied and divided, the answer cannot have more significant figures than the measurement with the least number of significant figures. If two numbers are multiplied, one with six and the other with three significant figures, the answer can have only three significant figures. The following example is an illustration of such a case.

$$5.82131 \text{ centimeters} \quad \text{(Six significant figures)}$$
$$\times \ 4.11 \text{ centimeters} \quad \text{(Three significant figures)}$$
$$23.9\underline{255841} \text{ square centimeters}$$
$$\text{Drop} \nearrow$$

The first nonsignificant digit in the answer is 2, which is less than 5; thus, Rule 2 is applied, and the answer is rounded off to 23.9 square centimeters (three significant figures).

The multiplication and division rule results from the fact that, when an uncertain figure—the last figure—is multiplied or divided, it produces uncertain numbers. An answer can have only one certain figure, so all other uncertain figures are dropped. Let's consider another multiplication problem:

$$2.41 \text{ meters} \times 2.1 \text{ meters} = ? \text{ square meters}$$

$$2.4\underline{1} \text{ meters} \quad\quad\quad \text{Uncertain}$$
$$\times \ 2.\underline{1} \text{ meters} \quad\quad\quad \text{Uncertain}$$
$$24\underline{1} \text{ meters} \quad\quad\quad \text{Uncertain}$$
$$48\underline{2} \text{ meters} \quad\quad\quad \text{Uncertain}$$
$$5.0\underline{61} \text{ square meters} \quad \text{Uncertain}$$

Least significant figure ↗

When you are doing calculations with a calculator, remember that it does not keep track of the number of significant figures. It is your job to round the numbers on calculator displays to the correct number of significant figures.

When 2.41 is multiplied by the 1 (an uncertain figure) in 2.1, all resulting numbers are rendered uncertain. When these uncertain figures are added to the product of 2.41 and 2, the least significant figure becomes the first decimal place; thus, the final answer can have only two significant figures. After rounding off, the answer is reported as 5.1 square meters. Example Problem 3.2 shows another example of how to multiply and divide measured quantities.

——————— **Example Problem 3.2** ———————

Perform the indicated arithmetic operations and express the answer to the correct number of significant figures.

$$\frac{7.290 \text{ meters} \times 2.0400 \text{ meters}}{0.95 \text{ meters}} =$$

Solution

Notice that the denominator contains a measurement with only two significant figures; this limits the answer to two significant figures. Perform the indicated math operations and round off the resulting answer to two significant figures.

$$\frac{7.290 \text{ meters} \times 2.0400 \text{ meters}}{0.95 \text{ meters}} = 15.65431579 \text{ meters} = \textbf{16 meters}$$

Least significant figure ⬏ ⬑ First nonsignificant figure

The first nonsignificant figure is 6; hence, the answer is rounded off by adding 1 to 5 and dropping the nonsignificant figures, which leaves 16 meters as the answer.

It is not uncommon to perform calculations in which both addition and multiplication are required. Example Problem 3.3 shows how to solve such problems.

Example Problem 3.3

Perform the indicated arithmetic operations and express the answer to the correct number of significant figures.

(11.2050 millimeters − 10.322 millimeters) × 6.030000 millimeters =

Solution

Both rules regarding significant figures apply in this example; first, after subtracting we can only have three decimal places in the answer because 10.322 millimeters has only three decimal places.

11.2050 millimeters − 10.322 millimeters = 0.8830 millimeter
= 0.883 millimeter

Apply the multiplication rule when you multiply 0.883 millimeter (three significant figures) by 6.030000 millimeters (seven significant figures). Three significant figures is the maximum number allowed in the answer.

0.883 millimeter × 6.030000 millimeters = 5.32449 square millimeters
= **5.32 square millimeters**

Subtracting values that have magnitudes close to each other usually results in a loss of significant figures. In this problem, a measurement with five significant figures is subtracted from a measurement with six significant figures and gives an answer with only three significant figures.

3.7 Perform the indicated additions and subtractions and express the answers to the correct number of significant figures:
(a) 32.55 grams − 1.9889 grams =
(b) 0.02 milliliter + 0.183 milliliter =
(c) 554.1864 meters + 42.94 meters =
(d) 34.0000 seconds − 11.01108 seconds + 55.3458702 seconds =
(e) 3.0977×10^{23} atoms + 9.2112×10^{22} atoms =

3.8 Perform the indicated multiplications and divisions and express the answers to the correct number of significant figures:
(a) 83.22 grams/5.4 milliliters =
(b) 123.001 meters × 5.0 meters =
(c) 0.39 centimeter × 6.388 centimeters × 16.5495 centimeters =
(d) 706.0 liters/0.00017613 liter =
(e) 51,087.00 milligrams × 12.7 milligrams =

3.9 Perform the indicated arithmetic operations and express the answers to the correct number of significant figures:
(a) (13.983 meters − 12.98551 meters) × 8.319 meters =
(b) (28.08 grams − 28.0694 grams)/(5.623 seconds + 25.1343 seconds) −
(c) [(6.200 × 10^5 centimeters) × (2.5 × 10^{-6} centimeter)]/[(3.00000 × 10^4 centimeters) × (8.3901 × 10^{-15} centimeter)] =

3.3 INTERNATIONAL SYSTEM OF UNITS (MODERNIZED METRIC SYSTEM)

The measurement system utilized by scientists and virtually all countries in the world is the metric system. It was first developed by the French Academy of Sciences in 1790 in response to a request by the French National Assembly for a simple and organized system of weights and measures.

The metric system has evolved since 1790, but its basic structure has remained the same. The metric system is a decimal system, one that requires only the movement of the decimal point to change larger to smaller units or vice versa.

A conference held in 1960 made significant modifications in the metric system; the changes were significant enough that the name "metric system" was dropped. The revised system is called Le Système International d'Unites or **International System of Units (SI).** Even though we commonly speak of the metric system, in actuality we are usually referring to the International System.

Throughout this book, SI units are used along with some non-SI units still employed frequently by the scientific community.

The measurement system used in the United States is called the U.S. Customary System of Units (USCS). It was developed from the old English system of measurement.

CHEM TOPIC: Metrification in the United States

Of all of the industrialized countries of the world, the United States is the only one that has not adopted a metric system of weights and measures. However, very slowly the United States is moving toward metrification. In 1975 Congress passed the Metric Conversion Act, but it stipulated that metrification be voluntary. Unfortunately, the 17-member United States Metric Board established by the Metric Conversion Act was abolished in 1982 because of federal budget cuts.

In spite of the federal government's attitude toward metrification, many industries have metrified. The automobile industry is one; it has metrified because it uses parts from automobile companies from around the world. You see the results of metrification when considering the displacement volume of an engine, which is expressed in liters instead of cubic inches. If you work on autos, you know that you must use your metric wrenches. When you buy tires for your car, you now must specify the metric dimensions.

As you shop in the supermarket or grocery store, you also see many metric measurements. Volumes of soft-drink containers are given in liters, and the masses of a variety of food packages are given in grams. On reaching the pharmacy section, you find that the contents of pills are given in milligrams, and the volume of liquid medicines is shown in milliliters. For many years, the alcoholic beverage industry has used metric volume measurements to replace measurements such as the fifth.

At the present time, it seems as if the United States is "on hold" with respect to metrification, but let us hope that in the upcoming years we will adopt a system consistent with that used by the rest of the world.

The seven base units of the International System are

1. **Meter** (m)—unit of length
2. **Kilogram** (kg)—unit of mass
3. **Second** (s)—unit of time
4. **Kelvin** (K)—unit of temperature
5. **Mole** (mol)—unit of amount of substance
6. **Ampere** (A)—unit of electric current
7. **Candela** (cd)—unit of luminous intensity

All other SI units are created from combinations of these base units. Examples of derived SI units include the square meter (area), the cubic meter (volume), kilograms per cubic meter (density), and meters per second (velocity).

Prefixes are placed in front of the base SI unit to change the size of the unit. We will use the meter as an example; the meter is a unit of length approximately equal to 39.4 inches. While it is convenient to measure the dimensions of a room in meters, it is unwieldy to express distances between cities in meters. To magnify the meter 1000 times, it is only necessary to add the prefix *kilo*, meaning "1000 times," in front of *meter*, which produces the unit *kilometer*.

1 mile = 1.609 kilometer

1 kilometer = 1000 meters

For measuring small distances, the meter is an awkward unit. Thus, a prefix is added to specify a unit smaller than a meter. The two most commonly used prefixes are *centi,* "one-hundredth," and *milli,* "one-thousandth." When we add these to the unit *meter,* we get two small units of length, the centimeter and millimeter. One meter contains 100 centimeters:

$$\tfrac{1}{100} \text{ meter} = 1 \text{ centimeter}$$

$$100 \times \tfrac{1}{100} \text{ meter} = 100 \times 1 \text{ centimeter}$$

$$1 \text{ meter} - 100 \text{ centimeters}$$

1 inch = 2.54 centimeters

and one meter contains 1000 millimeters:

$$\tfrac{1}{1000} \text{ meter} = 1 \text{ millimeter}$$

$$1000 \times \tfrac{1}{1000} \text{ meter} = 1000 \times 1 \text{ millimeter}$$

$$1 \text{ meter} = 1000 \text{ millimeter}$$

Common prefixes encountered in chemistry are

$$\text{mega (M)} = 1{,}000{,}000 = 10^6$$

$$\text{kilo (k)} = 1000 = 10^3$$

$$\text{deci (d)} = 0.1 = 10^{-1}$$

$$\text{centi (c)} = 0.01 = 10^{-2}$$

$$\text{milli (m)} = 0.001 = 10^{-3}$$

$$\text{micro } (\mu) = 0.000001 = 10^{-6}$$

$$\text{nano (n)} = 0.000000001 = 10^{-9}$$

TABLE 3.1 SI PREFIXES

Prefix	Symbol	Meaning
exa	E	10^{18}
peta	P	10^{15}
tera	T	10^{12}
giga	G	10^{9}
mega	**M**	$\mathbf{10^6}$
kilo	**k**	$\mathbf{10^3}$
hecto	h	10^{2}
deca	da	10^{1}
—	—	10^{0} (= 1)
deci	**d**	$\mathbf{10^{-1}}$
centi	**c**	$\mathbf{10^{-2}}$
milli	**m**	$\mathbf{10^{-3}}$
micro	**μ**	$\mathbf{10^{-6}}$
nano	**n**	$\mathbf{10^{-9}}$
pico	p	10^{-12}
femto	f	10^{-15}
atto	a	10^{-18}

Because these prefixes are used constantly in chemistry, you should learn them. Table 3.1 provides a complete listing of SI prefixes.

REVIEW EXERCISES

3.10 List the seven base SI units and what they measure.

3.11 What is the difference between base and derived SI units? Give an example of each.

3.12 What is the meaning of each of the following prefixes: *(a)* milli, *(b)* mega, *(c)* micro, *(d)* nano, *(e)* kilo?

3.13. What prefix indicates each of the following: *(a)* 10^{-12}, *(b)* 10^{12}, *(c)* 10^3, *(d)* 10^{-9}, *(e)* 10^6, *(f)* 10^9, *(g)* 10^{-3}?

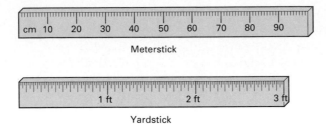

Figure 3.3
One meter is 1.093 yards, or
3.280 feet, or 39.37 inches.

The base SI unit of length is the **meter.** Before 1960, the meter was defined as the distance between two marks on a certain platinum-iridium rod located in France. These etched marks were separated by one ten-millionth the distance from the equator to the north pole on a line going through Sevres, France. After 1960, the meter was redefined in terms of a more universal standard. Today, one meter is 1,650,763.73 wavelengths of a certain light emitted by the gaseous element krypton, ^{86}Kr. Just why the meter is defined in terms of light emitted from krypton gas is beyond the scope of this discussion, but note that this standard is more reproducible than the old standard. Light released by krypton is not affected by temperature or pressure changes, and, with the proper equipment, its wavelength can be measured anywhere.

A meter equals 39.3701 inches, which makes it slightly longer (by 9 percent) than a yard (one yard equals 36 inches). Fig. 3.3 shows the relationship between a meter and a yard. While the meter is a convenient day-to-day measurement, it is generally too large to use for length measurements in chemistry because chemists deal with infinitesimal (minute) objects like atoms and molecules. Therefore, various prefixes are added to the meter to produce more useful units of length.

A centimeter is 0.01 meter and a millimeter is 0.001 meter. Even these units are gigantic when one is measuring dimensions of objects as small as atoms. A prefix commonly added to *meter* when considering atomic dimensions is *nano*, or 10^{-9}. One nanometer equals 0.000000001 meter. To give you an idea of how small a nanometer is, we can apply the factor-label method to calculate what fraction of an inch is 1.000 nm. We know that a nanometer is 10^{-9} meter and a meter is 39.37 inches:

$$1.000 \text{ nm} \times \frac{10^{-9} \text{ m}}{1 \text{ nm}} \times \frac{39.37 \text{ in}}{1 \text{ m}} = \mathbf{3.937 \times 10^{-8} \text{ in}}$$

We find that 1.000 nm is only 3.937×10^{-8} in. Additional multiples of the meter are shown in Fig. 3.4 along with the approximate sizes of objects in our universe.

The **kilogram** is the SI unit of mass. **Mass** is the quantity of matter contained in an object. A more massive object, as a result of containing more matter, has a smaller acceleration than a less massive object when both are pushed by equal forces (Fig. 3.5).

The terms "mass" and "weight" are frequently confused. **Weight** is the gravitational force of attraction that acts on a mass. The mass of a

3.4 BASE SI UNITS: METER, KILOGRAM, AND KELVIN

Length

The spelling of the word "meter" is "metre" in many other countries.

1 m = 39.37 in

Mass

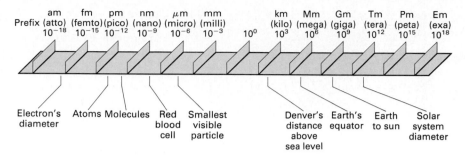

Figure 3.4
Multiples of the meter.

body is constant no matter where it is located, whereas weight is a variable quantity that depends on where the object is situated in the universe. For example, if people travel to the moon, their masses remain the same but their weights are approximately one-sixth of their weights on earth because the gravitational attraction on the moon is one-sixth that on earth. Even on the earth, there is a nonuniform gravitational field in which the gravitational force is greater at the poles than at the equator. Thus, the weight of a person changes slightly as he or she travels from the equator to the north pole; but the person's mass remains constant.

Mass and weight measurements require different instruments. Mass is measured with a balance, and weight with a scale. Because a balance operates in a gravitational field, there must be a way to cancel the effects of gravity. Consider a double-pan balance (Fig. 3.6a): An unknown mass is placed on one pan, and known masses are successively added to the other until the pointer returns to its original setting. Gravity is canceled because it is exerted equally on both sides of the balance. Thus, the unknown mass is equal to the sum of the known masses.

In contrast, a scale—the device for measuring weight—is sensitive to gravitational changes. Consider the common spring scale in Fig. 3.6b. A spring scale has a hook attached to a spring, and the spring has a pointer affixed to it that indicates the weight. When an object is hung on the hook, the spring expands and the pointer indicates how much the spring has expanded. Because the effects of gravity are not canceled, as they are in a balance, a spring scale gives the weight of the object.

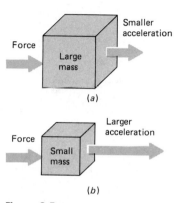

Figure 3.5
If forces of equal magnitude are applied to a large mass (a) and a small mass (b), the acceleration of the smaller mass is greater than the acceleration of the larger mass. The mass of an object is inversely proportional to its acceleration for a constant applied force.

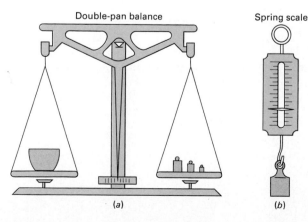

Figure 3.6
(a) A double-pan balance is an instrument for measuring the mass of an object by comparing with known masses. (b) A spring scale is an instrument for measuring the weight of an object.

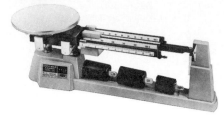

Triple beam balance

Electronic balance

Figure 3.7
Triple beam balances (left) and electronic balances (right) are most commonly used in chemistry laboratories. Most triple beam balances measure the mass of objects to within ±0.01 g, and electronic balances measure the mass of objects to within ±0.001 or ±0.0001 g. *(Fischer Scientific Company.)*

The standard kilogram, a block of platinum-iridium alloy from which all mass measurements are compared, is located in France. A kilogram is equivalent to 2.2046 pounds. The kilogram is too massive for most routine chemical laboratory measurements; consequently, smaller multiples of the kilogram are used. **Grams, milligrams,** and **micrograms** are the most commonly used units of mass in chemistry labs. Inexpensive triple beam balances are used to measure masses to within 0.01 g, or 1 cg, and more expensive analytical balances easily allow chemists to determine masses to the nearest 0.0001 g, or 0.1 mg (100 μg) (Fig. 3.7).

1 kg = 2.2046 lb

Even with expensive balances, the small mass of an individual atom or molecule is undetectable. A staggering quantity of atoms is needed to give a measurable mass on a sensitive analytical balance. For example, it can be calculated that 3.0×10^{17} gold atoms (rather large atoms, as atoms go) is the smallest number of gold atoms detectable on a balance that has a sensitivity of 0.1 mg.

Most mass measurements taken in the laboratory use the unit grams, but on many occasions grams must be changed to some other unit. Example Problem 3.4 shows how to do a mass-conversion problem.

──────────── **Example Problem 3.4** ────────────

A chemistry student performs an experiment in which the mass of a sample in milligrams must be determined. The student first measures the mass of an empty beaker and finds that it is 54.389 g. The combined mass of the sample and beaker is then measured and found to be 55.147 g. What is the mass of the sample in milligrams?

──────────── **Solution** ────────────

1. What is unknown? mg
2. What is known? Mass of beaker and sample = 55.147 g; mass of beaker = 54.389 g; 1000 mg/g
To find the mass of the sample in g, the mass of the beaker is subtracted from the mass of the beaker and sample.

Mass of sample = mass of beaker and sample − mass of beaker

= 55.147 g − 54.389 g

= 0.758 g

3. Apply the factor-label method.

$$0.758 \; \cancel{\text{g sample}} \times \frac{1000 \text{ mg sample}}{1 \; \cancel{\text{g sample}}} = ? \text{ mg sample}$$

4. Perform the indicated math operations.

$$0.758 \; \cancel{\text{g sample}} \times \frac{1000 \text{ mg sample}}{1 \; \cancel{\text{g sample}}} = \textbf{758 mg sample}$$

Temperature is a measure of the degree of hotness of matter. If an object with a higher temperature comes in contact with an object at a lower temperature, the temperature of the hot object decreases and the temperature of the colder object increases until the temperatures of both objects become equal. We say that heat flows from the hotter object to the colder object. **Heat** is a form of energy that is detected only when objects of different temperatures are in contact with each other. Figure 3.8 shows the relationship between temperature and heat. We will discuss heat in more detail in Chap. 4.

Chemists use both the **Celsius** and **Kelvin** temperature scales. In 1742, Anders Celsius, a Swedish astronomer, developed the Celsius scale. He constructed his scale so that the freezing point of water was assigned the value 0.0 degrees, and the normal boiling point of water was assigned the value of 100 degrees; thus, 100 divisions, or degrees, separate the freezing and boiling points of water. On the Celsius scale, room temperature is 25 degrees, and average human body temperature is approximately 37 degrees.

The SI unit of temperature was named after Lord Kelvin, the title given to William Thomson (1824–1907), a brilliant British physicist (Fig. 3.9). The Kelvin temperature scale employs units called *kelvins* that have the same magnitude as Celsius degrees but are displaced by 273 degrees (actually 273.15 degrees) along the Celsius scale. Thus, the zero point on the Kelvin scale, called absolute zero, is equivalent to −273.15°C. Because absolute zero is the lowest possible temperature, no negative values are found in the Kelvin scale.

To change Celsius degrees to kelvins, it is only necessary to add 273.15 to the Celsius temperature:

$$K = °C + 273.15$$

On the Kelvin scale, the freezing and boiling points of water are 273.15 K and 373.15 K, respectively.

$$\text{Freezing point}_{H_2O} = 0°C + 273.15 = 273.15 \text{ K}$$

$$\text{Boiling point}_{H_2O} = 100°C + 273.15 = 373.15 \text{ K}$$

Temperature

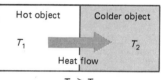

$$T_1 > T_2$$

Figure 3.8
Heat always flows spontaneously from hotter to colder objects. Heat flow continues until T_1 equals T_2 when the two objects are in thermal equilibrium (at the same temperature).

	Temperature (°C)
Star interior	40,000,000
Mercury lamp	14,000
Tungsten, melting point	3,410
Hottest climate (Tripoli)	58
Coldest climate (Alaska)	−63
Liquid air	−195
Helium, boiling point	−269
Absolute zero	−273

While the Celsius and Kelvin scales are universally used in science, in the United States the Fahrenheit scale predominates among nonscientists. The Fahrenheit temperature scale was proposed by the German scientist Gabriel Fahrenheit in 1714. He decided on 180 divisions between the freezing and boiling points of water, and placed the zero point at the coldest temperature that he could attain his laboratory using salt solutions (salts lower the freezing point of water). The resulting scale fixed the freezing point of pure water at 32 degrees and the boiling point at 212 degrees (exactly 180 degrees above the freezing point).

Figure 3.10 shows the relative magnitudes of the freezing and boiling points of common substances on the three temperature scales: Kelvin, Celsius, and Fahrenheit.

Thermometers are used for most temperature measurements in the laboratory. While many different types of thermometers exist, the mercury thermometer is the laboratory "workhorse." A mercury thermometer contains a glass bulb which houses the mercury, a liquid metal. Attached to the bulb is a thin glass tube from which the air has been removed (Fig. 3.11). As the bulb is heated, the mercury expands into the tube. One reason mercury is frequently used in thermometers is that it expands nearly uniformly when heated.

Sometimes it is necessary to convert temperatures from Fahrenheit to Celsius and then to Kelvin or vice versa. Temperature conversions are efficiently made using algebraic conversion formulas. To convert a Fahrenheit temperature to a Celsius temperature, the following formula is used:

$$°C = \tfrac{5}{9}(°F - 32)$$

Figure 3.9
William Thomson, Lord Kelvin. *(Burndy Library.)*

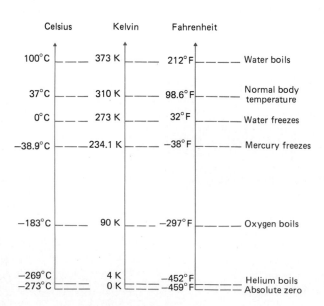

Figure 3.10
Comparisons of the Celsius, Kelvin, and Fahrenheit temperature scales are shown.

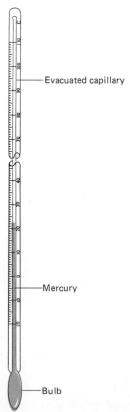

Figure 3.11
Mercury thermometers are the most commonly used temperature-measuring devices in chemistry laboratories. When the bulb containing mercury is heated, the mercury expands and moves up in the evacuated capillary tube.

That is, given a Fahrenheit temperature you would first subtract 32 from it and then multiply by $\frac{5}{9}$ to obtain the equivalent Celsius temperature. For example, when 212°F is substituted into the equation, 100°C is obtained; when 32°F is substituted, 0°C results.

$$°C = \tfrac{5}{9}(212°F - 32) = 100°C$$

$$°C = \tfrac{5}{9}(32°F - 32) = 0°C$$

If a Celsius temperature is being changed to Fahrenheit, the equation is rearranged and solved for °F. First multiply both sides by $\frac{9}{5}$ to eliminate $\frac{5}{9}$ from the right side of the equation, giving

$$\tfrac{9}{5}°C = \tfrac{5}{9}(°F - 32)\tfrac{9}{5}$$

Add 32 to both sides to isolate °F on the right side:

$$32 + \tfrac{9}{5}°C = °F - 32 + 32$$

This yields

$$\tfrac{9}{5}°C + 32 = °F \quad\text{or}\quad °F = \tfrac{9}{5}°C + 32$$

The three temperature conversion equations,

$$K = °C + 273$$

$$°C = \tfrac{5}{9}(°F - 32)$$

$$°F = \tfrac{9}{5}°C + 32$$

are used to convert any given temperature in one scale to a temperature on any other scale. Example Problem 3.5 is an illustration of a temperature-conversion problem.

The conversion formula from Fahrenheit to Celsius results because the zero point on the Celsius scale is 32 degrees below that on the Fahrenheit scale, and the size of a Celsius degree is $\frac{5}{9}$ (reduced from $\frac{100}{180}$) that of a Fahrenheit degree.

Always remember that the 273 in the Celsius to Kelvin formula is not an exact number, but the 32 in the Celsius and Fahrenheit formulas is exact.

───── **Example Problem 3.5** ─────

Convert 300.0 K to Fahrenheit.

───── **Solution** ─────

1. What is unknown?   °F

2. What is known?   300.0 K

3. Apply the appropriate temperature conversion formulas. First convert K to °C, and then convert to °F.

(a) Convert 300.0 K to °C.

To find the °C, first rearrange the equation and substitute the Kelvin temperature in the equation.

$$K = °C + 273.15$$

$$°C = K - 273.15$$

$$°C = 300.0 - 273.15 = 26.8°C$$

The answer, 26.8°C, must be rounded to one decimal place because the temperature was measured to one decimal place.

(b) Convert 26.8°C to °F.

$$°F = \tfrac{9}{5}°C + 32$$

$$°F = \tfrac{9}{5}(26.8°C) + 32$$

$$= \mathbf{80.2°F}$$

The answer, 80.2°F, is equivalent to 300.0 K.

REVIEW EXERCISES

3.14 What are the base SI units for length, mass, and temperature?

3.15 What standards are used for the units of length and mass?

3.16 Use the factor-label method and scientific notation to perform the following conversions:
(a) 9150 m = ? mm
(b) 0.1005 ft = ? cm
(c) 805 g = ? lb
(d) 337.6 kg = ? ng

3.17 (a) What is the difference between mass and weight? (b) What is the difference between temperature and heat?

3.18 Perform each of the following temperature conversions:
(a) 55°C = ? K
(b) −91.4°C = ? K
(c) 2010°F = ?°C
(d) −0.287°C = ?°F

3.5 DERIVED SI UNITS OF MEASUREMENT: VOLUME AND DENSITY

Derived SI units are combinations of the base SI units. While many derived units are in use, only volume and density units are discussed here because of their importance in measuring and describing matter. In addition, we will consider specific gravity, which is closely related to density.

Volume

Volume is the amount of space occupied by matter. All matter takes up some space and therefore has volume. The amount of space taken up by a given amount of matter is measured in cubic units—a cubic unit being a unit of length raised to the third power. For example, the volume V of a rectangular object is calculated by multiplying the object's length l times its width w times its height h (Fig. 3.12).

A volume measurement tells the amount of space occupied by an object.

$$V = lwh$$

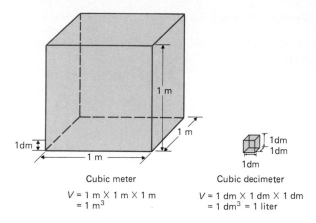

Cubic meter

$V = 1\ m \times 1\ m \times 1\ m$
$= 1\ m^3$

Cubic decimeter

$V = 1\ dm \times 1\ dm \times 1\ dm$
$= 1\ dm^3 = 1\ liter$

Figure 3.12
One cubic meter (1 m³) is a very large unit of volume; it contains 1000 dm³. One cubic decimeter is equal to the non-SI volume unit called the liter.

The base unit of length is the meter; thus, one SI unit of volume is the cubic meter: 1 m × 1 m × 1 m, or 1 m³. One cubic meter contains approximately 264 gallons, and for most chemical measurements the cubic meter is much too large.

Instead of working with such an enormous unit as the cubic meter, chemists frequently use a small fraction of this unit, the cubic decimeter. *Deci* is the prefix that means $\frac{1}{10}$; thus a decimeter is 0.1 meter (3.937 inches). The cubic decimeter is exactly equivalent to a non-SI unit, the liter. Both the cubic decimeter and the liter are equivalent to 0.001 cubic meter. The cubic decimeter is not frequently encountered because of the popularity of the liter. Therefore, the liter is most frequently used throughout this book. Just remember that 1 L and 1 dm³ are different expressions for the same volume, 0.001 m³.

One cubic decimeter is the same volume as one liter.

A liter is approximately equal in volume to a quart (Fig. 3.13). One liter is 1.057 quarts, which means it is slightly larger than a quart. Example Problem 3.6 shows a volume conversion to liters.

1 L = 1.057 qt
1 qt = 946 mL

Figure 3.13
One quart is 0.940 liter, and one liter is 1.06 quart.

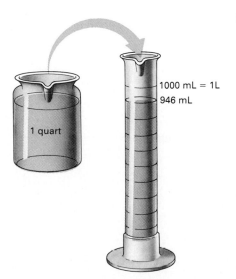

1 quart

1000 mL = 1L
946 mL

Example Problem 3.6

Some stations sell gasoline by the liter. How many liters of gasoline must be purchased to exactly fill a 15.0-gal gasoline tank?

Solution

1. What is unknown?   L

2. What is known?   15.0 gal; 1.057 qt/L; 4 qt/gal

3. Apply the factor-label method.

$$15.0 \; \cancel{gal} \times \frac{4 \; \cancel{qt}}{1 \; \cancel{gal}} \times \frac{1 \; L}{1.057 \; \cancel{qt}} = ? \; L$$

4. Perform the indicated math operations.

$$15.0 \; \cancel{gal} \times \frac{4 \; \cancel{qt}}{1 \; \cancel{gal}} \times \frac{1 \; L}{1.057 \; \cancel{qt}} = \textbf{56.8 L}$$

A 15.0-gal tank holds a maximum of 56.8 L of gasoline.

Even the liter is too large for most routine laboratory volume measurements so the milliliter or the cubic centimeter (cm^3, also written cc), is regularly used in chemistry. The milliliter and the cubic centimeter are measurements for the same volume, 0.001 L.

$$1 \; mL = 1 \; cm^3 = 1 \; cc = 0.001 \; L$$

In some instances, very small volumes must be measured; in such cases the microliter is used. Example Problem 3.7 shows how to find the number of microliters in 1.00 m^3.

Example Problem 3.7

How many microliters are contained in 1.00 m^3?

Solution

1. What is unknown? μL

2. What is known?   1.00 m^3; 1 dm^3/0.001 m^3; 1 dm^3/L; 1 μL/1 $\times 10^{-6}$ L

3. Apply the factor-label method.

$$1.00 \; \cancel{m^3} \times \frac{1 \; \cancel{dm^3}}{0.001 \; \cancel{m^3}} \times \frac{1 \; \cancel{L}}{1 \; \cancel{dm^3}} \times \frac{1 \; \mu L}{1 \times 10^{-6} \; \cancel{L}} = ? \; \mu L$$

4. Perform the indicated math operations.

$$1.00 \, m^3 \times \frac{1 \, dm^3}{0.001 \, m^3} \times \frac{1 \, L}{1 \, dm^3} \times \frac{1 \, \mu L}{1 \times 10^{-6} \, L} = 1 \times 10^9 \, \mu L$$

A common problem encountered when dealing with volumes is calculating a volume unit from a given length unit. For example, if we wanted to know the relationship between cubic millimeters and cubic centimeters, we would start by comparing the relationships between meters and millimeters, and meters and centimeters.

Pipets, burets, graduated cylinders, graduated beakers, volumetric flasks, and syringes are used to measure volumes in the laboratory.

$$1 \, m = 1000 \, mm = 10^3 \, mm$$

$$1 \, m = 100 \, cm = 10^2 \, cm$$

If we cube both sides of the equations, we obtain the respective volume units associated with 1 mm and 1 cm relative to 1 m^3.

$$1 \, m^3 = (10^3 \, mm)^3 = 10^9 \, mm^3$$

$$1 \, m^3 = (10^2 \, cm)^3 = 10^6 \, cm^3$$

Knowing the relation of both 1 mm^3 and 1 cm^3 to 1 m^3, we find after dividing these two equalities, that 1 cm^3 contains 10^3 mm^3.

$$\frac{1 \, m^3}{1 \, m^3} = \frac{10^9 \, mm^3}{10^6 \, cm^3}$$

$$1 = 10^3 \, mm^3/cm^3$$

Example Problem 3.8 presents another case in which a length unit is converted to a volume unit.

—————————— **Example Problem 3.8** ——————————

An automobile engine has a displacement of 3.80 L. What is its displacement in cubic inches?

—————————— **Solution** ——————————

1. What is unknown? in^3

2. What is known?   3.80 L; 1 L/dm^3; 10 dm/m; 39.37 in/m
To convert L to in^3 we must know the conversion factors for L and dm^3, dm and m, and in and m. To find the relationships between dm^3 and m^3,

and in^3 and m^3, we must cube the conversion factors 10 dm/m and 39.37 in/m as follows:

$$\frac{(10 \text{ dm})^3}{(1 \text{ m})^3} = \frac{10^3 \text{ dm}^3}{1 \text{ m}^3}$$

$$\frac{(39.37 \text{ in})^3}{(1 \text{ m})^3} = \frac{6.102 \times 10^4 \text{ in}^3}{1 \text{ m}^3}$$

3. Apply the factor-label method.

$$3.80 \text{ L} \times \frac{1 \text{ dm}^3}{1 \text{ L}} \times \frac{1 \text{ m}^3}{10^3 \text{ dm}^3} \times \frac{6.102 \times 10^4 \text{ in}^3}{1 \text{ m}^3} = ? \text{ in}^3$$

4. Perform the indicated math operations.

$$3.80 \text{ L} \times \frac{1 \text{ dm}^3}{1 \text{ L}} \times \frac{1 \text{ m}^3}{10^3 \text{ dm}^3} \times \frac{6.102 \times 10^4 \text{ in}^3}{1 \text{ m}^3} = \textbf{232 in}^3$$

We find that a 3.80-L engine displacement is equivalent to a displacement of 232 in^3.

3.19 (a) What is the SI unit of volume? (b) What other volume units are used in chemistry?

3.20 Perform each of the following volume conversions:
(a) 2.59 L = ? mL = ? cm^3 = ? dm^3 = ? m^3
(b) $8.16 \times 10^3 \ \mu L$ = ? m^3 = ? L = ? mL = ? mm^3

3.21 A barrel is a unit of volume used to measure crude oil. One barrel contains 44 gallons. (a) How many liters of oil are there in a barrel? (b) How many cubic centimeters of oil are there in a barrel? (c) How many barrels of oil are required to fill a tank with a volume of 8.77×10^6 mL?

3.22 (a) Calculate the displacement of an engine in liters if it has a displacement of 425 in^3. (b) What is the displacement of an engine in cubic inches if it is a 1.8-L engine?

Density

Density is an important property of matter. The **density** of a substance describes how much mass is contained in a unit volume.

$$\text{Density} = \frac{\text{mass}}{\text{volume}}$$

Substances with higher densities have more mass packed into equivalent volumes than those with lower densities. For example, 1.0 mL of gold has a mass of 19 g; in contrast, 1.0 mL of water has a mass of only 1.0 g. If you compare equal masses of gold and water, the gold occupies one-nineteenth the volume of the water.

A density can be thought of as a conversion factor that equates the mass and volume of a substance.

Densities of substances are measured by finding the mass of a known volume of the substance. Mass is most commonly measured in grams, and volume is measured in milliliters or cubic centimeters and in liters. For the more compact forms of matter, liquids and solids, densities are expressed in grams per cubic centimeter or grams per milliliter. For less dense forms of matter, mainly gases, densities are expressed in grams per liter and grams per cubic decimeter.

Each substance has a characteristic density; for example, the density of water at 4°C is 1.00 g/cm³, which means that 1.00 g of water occupies a volume of 1.00 cm³ at 4°C. Table 3.2 lists the densities of common materials.

TABLE 3.2 DENSITIES OF SELECTED MATERIALS

Liquids and solids	Density, g/cm³ (25°C)	Gases	Density, g/dm³ (1 atm, 0°C)
Aluminum	2.70	Air	1.29
Battery acid (38% sulfuric acid)	1.285	Carbon dioxide, CO_2	1.977
Benzene	0.860	Chlorine, Cl_2	3.21
Ethanol (grain alcohol)	0.785	Hydrogen, H_2	0.090
Gold	19.3	Krypton, Kr	3.71
Iron	7.86	Nitrogen, N_2	1.251
Lead	11.4	Oxygen, O_2	1.43
Mercury	13.6	Xenon, Xe	5.85
Osmium	22.57		
Potassium	0.86		
Uranium	18.9		

When the temperature of matter changes, its volume also changes; consequently, the density of a substance depends on its temperature. The density of water is 1.00 g/cm³ only at 4°C. At room temperature, 25°C (298 K), the density of water decreases to 0.997 g/cm³. When the temperature is raised to 80°C, the density of water decreases to 0.971 g/cm³. Therefore, the temperature must be stated whenever a density is given.

Density varies with temperature because substances expand and contract when heated and cooled.

Densities may be used to compare masses of substances that have the same volume. It is incorrect to say, "Gold is a heavier metal than iron." Properly, one says, "Gold is more dense than iron." Table 3.2 shows that the density of gold is 19.3 g/cm³, which is higher than the density of iron, 7.86 g/cm³. Relative volumes of substances with the same mass are shown in Fig. 3.14.

In the laboratory, densities are measured by finding the mass of a known volume. Liquid densities are easily obtained. A volumetric instrument such as a graduated cylinder or volumetric flask is weighed, filled to a specific volume level, and reweighed. If we subtract the mass of the container from the mass of the container plus the liquid, we get the mass of the liquid (Fig. 3.15).

Mass of liquid = mass of container and liquid − mass of container

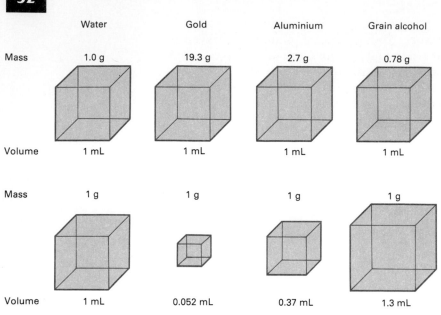

	Water	Gold	Aluminium	Grain alcohol
Mass	1.0 g	19.3 g	2.7 g	0.78 g
Volume	1 mL	1 mL	1 mL	1 mL

Mass	1 g	1 g	1 g	1 g
Volume	1 mL	0.052 mL	0.37 mL	1.3 mL

Figure 3.14
If you consider equal volumes of matter, substances with higher densities have greater masses than those with smaller densities. If you consider equal masses of matter, substances with higher densities have smaller volumes than those with smaller densities.

To find the density of the liquid, its volume is divided into the mass. A typical density determination for a liquid is found in Example Problem 3.9.

─────────────── **Example Problem 3.9** ───────────────

An empty 10-mL graduated cylinder has a mass of 53.24 g. When 10.0 mL of an unknown liquid is added to the cylinder, the total mass is 63.12 g. Calculate the density of the unknown liquid.

─────────────── **Solution** ───────────────

1. What is unknown?   Density of the unknown liquid, g/mL

2. What is known?   Volume = 10.0 mL; mass of empty graduated cylinder = 53.24 g; mass of liquid plus graduated cylinder = 63.12 g

$$d_{liq} = \frac{\text{mass of liquid}}{\text{volume of liquid}}$$

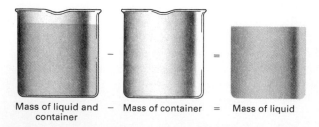

Mass of liquid and   −   Mass of container   =   Mass of liquid
container

Figure 3.15
To find the mass of a liquid, measure the mass of the container that holds the liquid and then measure the mass of the container plus the liquid. The mass of the liquid is the difference between these two masses.

3. To find the density of a substance, it is necessary to measure the mass of a known volume of the substance. The mass of the unknown liquid is found by subtracting the mass of the graduated cylinder from the mass of the graduated cylinder and liquid.

$$\text{Mass of liquid} = 63.12 \text{ g} - 53.24 \text{ g} = 9.88 \text{ g}$$

4. Divide the mass of the liquid by its volume.

$$d = \frac{\text{mass}}{\text{volume}} = \frac{9.88 \text{ g}}{10.0 \text{ mL}}$$

$$d = \textbf{0.988 g/mL}$$

The density of the unknown liquid is 0.988 g/mL.

Densities are helpful quantities when working in the laboratory. Sometimes it is inconvenient to measure the mass of a substance directly. If the volume and density of a substance are known, the mass can be calculated. Think of the density of an object as another conversion factor, mass/volume. Example Problem 3.10 shows how density is applied as a conversion factor.

―――――――――― **Example Problem 3.10** ――――――――――

A student places 25.0 cm^3 of decane in a graduated cylinder. If the density of decane is 0.730 g/cm^3, what is the mass of decane in the cylinder?

―――――――――― **Solution** ――――――――――

1. What is unknown?   Mass of decane in g
2. What is known?   Volume of decane = 25.0 cm^3; density of decane = 0.730 g/cm^3
3. Apply the factor-label method.

$$25.0 \text{ cm}^3 \text{ decane} \times \frac{0.730 \text{ g decane}}{1 \text{ cm}^3 \text{ decane}} = ? \text{ g decane}$$

4. Perform the indicated math operations.

$$25.0 \text{ cm}^3 \text{ decane} \times \frac{0.730 \text{ g decane}}{1 \text{ cm}^3 \text{ decane}} = \textbf{18.3 g decane}$$

Specific gravity is a measurement closely related to density. **Specific gravity** is defined as the ratio of the density of a substance to the density of water. Stated another way, specific gravity is the ratio of the mass of a substance to the mass of an equal volume of water.

$$\text{Specific gravity} = \frac{d_{\text{substance}}}{d_{\text{water}}}$$

$$= \frac{(g_{\text{substance}}/\text{cm}^3)}{(g_{\text{water}}/\text{cm}^3)}$$

$$= \frac{g_{\text{substance}}}{g_{\text{water}}}$$

Because the density of water is 1.0 g/cm^3 at 25°C, the numerical values for the specific gravity and density of a substance are the same. However, specific gravity is a unitless quantity because the units cancel when the ratio is calculated. For example, the density of mercury is 13.6 g/cm^3, and the specific gravity of mercury is 13.6. If equal volumes of mercury and water are compared, the sample of mercury has 13.6 times the mass of water.

In medical and technical labs, researchers usually make specific gravity determinations rather than density determinations, because specific gravity is easily measured with a hydrometer. A *hydrometer* is usually a glass tube designed to measure specific gravity (Fig. 3.16); the instrument floats in the liquid; its side is calibrated so that the specific gravity of the liquid can be read at the surface of the liquid. In medical laboratories, a special hydrometer, called a *urinometer,* is used to measure the specific gravity of urine. Knowing the specific gravity of a patient's urine helps physicians to diagnose abnormal conditions such as kidney disease or diabetes.

Body fluid	Specific gravity
Amniotic fluid	1.006–1.008
Blood plasma	1.024–1.030
Saliva	1.010–1.020
Urine	1.002–1.040
Whole blood	1.052–1.064

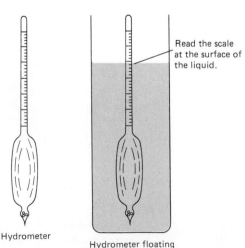

Read the scale at the surface of the liquid.

Hydrometer

Hydrometer floating in a liquid

Figure 3.16
A hydrometer is constructed to float in liquids in such a way that the specific gravity of the liquid is read from the scale at the surface of the liquid.

Many other derived SI units of measurement are used in chemistry, including measures of (1) energy, (2) force, (3) power, (4) electric charge, and (5) pressure. We will investigate some of these in later chapters.

3.23 *(a)* An unknown solid is found to have a mass of 133.1 g and a volume of 11.68 cm^3. What is the density of the solid? *(b)* If the unknown solid is one of those listed in Table 3.2, what is the unknown solid?

3.24 *(a)* What is the volume of 937 g of aluminum? *(b)* What is the mass of 3.91 L of aluminum? (The density of aluminum is found in Table 3.2.)

3.25 An empty 25.0-cm^3 graduated cylinder has a mass of 98.24 g. *(a)* If 25.0 cm^3 of an unknown liquid and the graduated cylinder have a total mass of 121.44 g, what is the density of the liquid? *(b)* What is the mass of 75.0 mL of the liquid?

3.26 *(a)* What is specific gravity? *(b)* How is specific gravity different from density? *(c)* What instrument is used to measure the specific gravity of a liquid?

SUMMARY

All measurements are to some degree uncertain because of measurement errors. Systematic and random errors are the two general types of errors found in measurements. **Systematic errors** result from improper techniques, uncalibrated equipment, or human error: they may be corrected. **Random errors** are those that cannot be identified but are present: they cannot be corrected.

Chemists use significant figures to express the degree of certainty of a measurement. **Significant figures** are the measured digits plus one estimated digit. Measurements with a greater number of significant figures are more certain than those with less. When we do calculations involving measurements, we must be careful to follow the rules regarding significant figures so that the results of the calculations have only one uncertain digit.

The **International System of Units** (SI) is the system of measurement used by scientists. All SI units of measurement are either base or derived units. The seven base units SI are the meter, kilogram, kelvin, second, mole, ampere, and candela. All other units are derived from these seven base units; thus, they are called derived SI units. To increase or decrease the magnitude of SI units, it is only necessary to place the appropriate prefix in front of the unit. Frequently encountered SI prefixes are kilo (1000 ×), centi ($\frac{1}{100}$ ×), milli ($\frac{1}{1000}$ ×), and micro (10^{-6} ×).

Mass is the quantity of matter in an object and is usually measured in kilograms, grams, or milligrams. No matter where an object is located, mass remains constant. **Weight** is the force of gravity on a mass. The weight of an object varies, depending on its location in the universe. **Temperature** is a measure of the degree of hotness of matter. The **kelvin** (K) is the SI unit of temperature. The zero point on the Kelvin scale is absolute zero, the lowest possible temperature. One Kelvin has the same magnitude as one Celsius degree, the non-SI unit of temperature.

Volume is measured in cubic meters, cubic decimeters, or cubic centimeters in SI. These measures of volume are derived from the base units of length—meters, decimeters, and centimeters. Liters and milliliters, two non-SI volume units, are commonly used in chemistry. **Density** is the ratio of mass to volume and is most frequently measured in grams per cubic centimeter or grams per liter. **Specific gravity** is the ratio of the density of a substance to the density of water.

KEY TERMS
---

absolute zero	derived unit	mass	random error	systematic error
balance	International System	measurement	scale	temperature
base unit	of Units (SI)	uncertainty	significant figure	volume
density	kilogram	metric system	specific gravity	weight

EXERCISES*

3.27 Define each of the following terms: SI, metric system, base unit, derived unit, kilogram, mass, weight, balance, scale, absolute zero, volume, density, temperature, systematic error, random error, measurement uncertainty, significant figure, specific gravity.

Uncertainty in Measurements

3.28 By what means is the accuracy of a measurement determined?

3.29 What factors decrease the accuracy of the measurement you take of your weight on a bathroom scale?

3.30 What could be said about the precision of the following group of volume measurements for the same object: 144 mL, 163 mL, 155 mL, 182 mL, 159 mL, and 149 mL?

3.31 Select the more precise measuring device in each of the following pairs: (a) centigram balance or analytical balance; (b) pipet or calibrated beaker; (c) large graduated cylinder or small graduated cylinder; (d) volumetric flask or graduated cylinder.

3.32 What systematic errors might influence the measurement of air temperature on an outdoor thermometer?

3.33 Why can't random errors be eliminated from measurements?

Significant Figures

3.34 How many significant figures are there in each of the following measured quantities?
(a) 24.084   (b) 128.0700
(c) 0.098   (d) 0.0012000
(e) 500.0   (f) 0.00000000191010
(g) 0.00133490 (h) 0.0002
(i) 8.90×10^1 (j) 5×10^{-8}

3.35 For each measurement listed in exercise 3.34, write the least significant figure.

3.36 Express each measurement listed in exercise 3.34 with one significant figure.

3.37 Round off each of the following to four significant figures:
(a) 194.645   (b) 10.998
(c) 962.1539   (d) 0.659119999
(e) 999,650,004,000

3.38 Use scientific notation to express 90,000 with (a) one, (b) two, (c) three, (d) four, and (e) five significant figures.

3.39 Add each of the following and express the answer to the correct number of significant figures:
(a) 2.345 g + 2.5 g =
(b) 32.0030 mL + 11.87 mL =
(c) 4.5666 g + 3.2388 g + 10.382 g =
(d) 494.14320 m + 171.579 m + 91.15 m =

3.40 Subtract each of the following and express the answer to the correct number of significant figures:
(a) 154.236 mL − 56.9 mL =
(b) 1.00666 cm − 0.839 cm =
(c) 40.311 g − 40.277 g =
(d) 900.0088 mm − 65.3 mm − 138.5454 mm =

3.41 Multiply each of the following and express the answer to the correct number of significant figures:
(a) 0.077 g × 6 g =
(b) 732.2 s × 8.1 s =
(c) 5.5050 cm × 9.9 cm =
(d) 0.009470 kg × 472.0 kg =

3.42 Divide each of the following and express the answer to the correct number of significant figures with the correct units:
(a) 9.233 g/1.4 mL =
(b) 89.08 m/0.509514 m =
(c) $(5.83 \times 10^{14} \text{ m})/(4 \times 10^{-6} \text{ s}) =$
(d) $0.00071000 \text{ g}/0.93 \text{ cm}^3 =$

3.43 Perform the indicated arithmetic operations and express the answer to the correct number of significant figures:
(a) (133 × 534.00)/(9.1 + 0.4543) =
(b) (154.7325 − 154.7036) × 9.89892 =
(c) [(0.000100/0.0027711) − 0.0000522 + 0.00046]/5.198 =
(d) [459.541/399.06 + (2.0477 × 3.91)]/[(0.00355 + 0.0019955) × 4050.7] =

Base SI Units

3.44 Write the base SI unit that corresponds to the following measurements: (a) time, (b) quantity

---

*For exercise numbers printed in color, answers can be found at the back of the book.

of substance, *(c)* length, *(d)* electric current, *(e)* mass, *(f)* temperature.

3.45 *(a)* Use a dictionary to find the meaning of the following English measurement units: dram, scruple, minim, and slug. *(b)* What difficulties are encountered when using such units that are not encountered when using the International System?

3.46 What prefix is used for each of the following: *(a)* 0.01 ×, *(b)* 1,000,000 ×, *(c)* 10^{-6} ×, *(d)* 0.1 ×, *(e)* 1000 ×, *(f)* 0.001 ×, *(g)* 10^{-9} ×?

3.47 What is the meaning of each of the following prefixes: *(a)* milli, *(b)* deca, *(c)* giga, *(d)* pico, *(e)* nano, *(f)* deci, *(g)* kilo, *(h)* centi, *(i)* micro, *(j)* hecto?

Length

3.48 Perform each of the following length conversions:
(a) 91 m = ? mm
(b) 1669 cm = ? m
(c) 4.3 km = ? cm
(d) 6.8 km = ? μm
(e) 358.0 mm = ? nm
(f) 74.6 Mm = ? km

3.49 If one mile equals 5280 feet, change each quantity that follows to the designated unit of length:
(a) 1 mi = ? m
(b) 23 mi = ? in
(c) 6.1 mi = ? cm
(d) 56 mi = ? mm
(e) 705 mi = ? km
(f) 486.0 mi = ? Mm

3.50 Change each to the designated SI unit:
(a) 19 in = ? km
(b) 36.9 ft = ? mm
(c) 5 ft 10 in = ? m
(d) 136.0 yd = ? m
(e) 0.0015 in = ? nm
(f) 35.156 mi = ? mm

3.51 The distance from the earth to the sun is 9.3×10^7 mi. Express this distance in *(a)* kilometers, *(b)* meters, *(c)* millimeters, and *(d)* picometers.

3.52 *(a)* What is the modern standard of length in the SI? *(b)* Why did it replace the old standard?

3.53 A person's body measurements are as follows: weight = 130 lbs; height = 5 ft 7 in; chest = 37 in; waist = 30 in; hips = 35 in. Convert the weight measurement to kilograms and the length measurements to centimeters.

Mass

3.54 Assuming that you could travel to Jupiter and make a quick weight measurement, how would your weight there compare to your weight here on earth? (Jupiter is the largest planet in our solar system, with a mass many times that of earth.)

3.55 Perform the following mass conversions:
(a) 58.73 kg = ? mg
(b) 74 mg = ? g
(c) 9.7115×10^{12} μg = ? kg
(d) 12.93 kg = ? pg
(e) 43.8×10^{-4} g = ? μg
(f) 45.2×10^{19} mg = ? Mg
(g) 0.000022 kg = ? dg
(h) 8.395×10^8 g = ? cg

3.56 If there are exactly 16 ounces in a pound and 2000 pounds constitute one ton, change each of the following to the designated SI unit:
(a) 52 lb = ? g
(b) 6186 oz = ? mg
(c) 1.088 tons = ? Mg
(d) 12.033 lb = ? cg
(e) 8.43×10^{-3} lb = ? ng
(f) 3.94×10^{13} oz = ? dg

Temperature

3.57 Change each Celsius temperature to Kelvin: *(a)* 11°C, *(b)* −61°C, *(c)* −235.6°C, *(d)* 2.04°C.

3.58 Change each Kelvin temperature to Celsius: *(a)* 88 K, *(b)* 314.7 K, *(c)* 0.388 K, *(d)* 3221 K.

3.59 Change each Fahrenheit temperature to Celsius: *(a)* 77.7°F, *(b)* −77.7°F, *(c)* 1466°F, *(d)* −459°F.

3.60 Change each Celsius temperature to Fahrenheit: *(a)* 77.7°C, *(b)* −77.7°C, *(c)* 1466°C, *(d)* −140.0°C.

3.61 Convert each Kelvin temperature to Fahrenheit: *(a)* 77.7 K, *(b)* 98.6 K, *(c)* 1111.11 K, *(d)* 0.89 K.

3.62 One temperature has exactly the same numerical value on both the Celsius and Fahrenheit temperature scales. Use algebra to find this temperature.

3.63 Cesium is a highly reactive metal that melts at 28.7°C. *(a)* What is the melting point of cesium in degrees Fahrenheit? (b) What is the melting point on the Kelvin scale? (c) If you could hold cesium in your hand, what would happen to it?

3.64 The melting and boiling points of hydrogen are $-269.7°C$ and $-268.9°C$, respectively. (a) Calculate the melting and boiling points of hydrogen in kelvins. (b) Find the melting and boiling points of hydrogen in Fahrenheit.

Volume

3.65 A block of wood is 2.1 cm long, 0.54 cm wide, and 934 mm high. What is the volume of the block in (a) cubic centimeters, (b) cubic decimeters, (c) milliliters, (d) liters?

3.66 Calculate the volume in liters of a metal block that is 4503 mm long, 439 cm wide, and 0.0229 m high.

3.67 Perform the indicated volume conversions:
(a) $3.98 \text{ dm}^3 = ? \text{ cm}^3$
(b) $12 \text{ mL} = ? \text{ m}^3$
(c) $73.570 \text{ mL} = ? \mu\text{L}$
(d) $1.14 \text{ m}^3 = ? \text{ mL}$
(e) $508.2 \text{ L} = ? \text{ m}^3$
(f) $1.2 \times 10^7 \mu\text{L} = ? \text{ m}^3$

3.68 Change each to the indicated unit:
(a) $821.0 \text{ mL} = ? \text{ qt}$
(b) $1.000 \text{ m}^3 = ? \text{ mm}^3$
(c) $199 \text{ mL} = ? \text{ qt}$
(d) $3.501 \text{ cm}^3 = ? \text{ ft}^3$
(e) $8.1 \text{ ft}^3 = ? \text{ cc}$

3.69 Change each to the indicated SI unit:
(a) $177 \text{ gal} = ? \text{ m}^3$
(b) $50.45 \text{ pints} = ? \text{ cm}^3$
(c) $3.04 \times 10^{-5} \text{ qt} = ? \text{ dm}^3$
(d) $0.0011 \text{ gallons} = ? \text{ mm}^3$

Density

3.70 Calculate the density of each substance in grams per milliliter, given mass and volume:
(a) Mass = 651 g; volume = 49.9 mL
(b) Mass = 8.32 g; volume = 9.67 mL
(c) Mass = 1.58 kg; volume = 1.677 L
(d) Mass = 7.24 lb; volume = 0.03119 ft^3

3.71 Vanadium, an element, has a density of 6.11 g/mL. Calculate the mass of vanadium contained in the following volumes: (a) 33 mL, (b) 544 mL, (c) 211.5 L, (d) 0.289 cm^3, (e) 1 qt, (f) 0.000671 m^3.

3.72 The density of pure silicon is 2.33 g/mL. Cal-

culate the volume of the following masses of silicon: (a) 834.6 g, (b) 900.5 mg, (c) 3.330 kg, (d) 71.9 lb, (e) 9.00 t (1 t = 1000 kg).

3.73 Air has a density of 1.29 g/L. Express the density of air in each of the following units: (a) grams per cubic foot, (b) kilograms per cubic meter, (c) pounds per cubic foot, (d) milligrams per cubic millimeter, (e) kilograms per cubic centimeter.

3.74 The metal platinum has a density of 21.4 g/mL. If you are given a metal sample weighing 9.29 g that occupies 0.434 mL, how could you quickly determine if the sample is pure platinum or not? Is it?

3.75 A graduated beaker is used to determine the density of an unknown liquid. The mass of the beaker is 40.1 g. What is the density of the unknown liquid if the beaker and 18.6 mL of the unknown liquid together weigh 69.7 g?

3.76 A 1.28-qt bottle is used to store liquid mercury (density = 13.6 g/cc). What is the mass of the mercury in the bottle in (a) grams, (b) kilograms, and (c) milligrams?

3.77 A 45.045-g graduated cylinder is used to measure the density of a liquid. After 15.0 mL of this liquid is poured into the cylinder, the combined mass of the cylinder and liquid is 69.731 g. Calculate the density of the liquid.

3.78 What is the mass of air in a room that measures $6.1 \text{ m} \times 5.5 \text{ m} \times 4.3 \text{ m}$?

3.79 A bar of osmium metal has the following dimensions: $0.35 \text{ m} \times 2.43 \text{ cm} \times 453.9 \text{ mm}$. What is the mass of the osmium bar in kilograms?

Specific Gravity

3.80 Ethyl alcohol is also called grain or drinking alcohol. Calculate the specific gravity of ethyl alcohol if the density of ethyl alcohol is 0.789 g/cm^3 and the density of water is 1.00 g/cm^3.

3.81 The specific gravity of blood is 1.05. (a) What is the mass in grams of 1 pt of blood? (b) If a person has 5.0 L of blood, what is the mass of the person's blood in kilograms? (c) What is the volume of 498 g of blood?

3.82 Calculate the specific gravity of each of the following: (a) uranium, (b) benzene, (c) grain alcohol, (d) mercury. (Table 3.2 lists the densities of these substances.)

3.83 Explain why it is incorrect to say that mercury is a heavier liquid than water.

3.84 (a) What is the name of the instrument used to measure the specific gravity of urine? (b) What is the purpose of measuring the specific gravity of urine?

Additional Exercises

3.85 Perform the following arithmetic operations and express the answers to the correct number of significant figures with the correct units:

(a) $\dfrac{(0.3441 \text{ g} - 0.3294 \text{ g})}{(8.03 \text{ g} \times 0.0233 \text{ g})} =$

(b) $(10.004 \text{ m} \times 7.322 \text{ m}) - 67.9 \text{ m}^2 + (4.558 \text{ m} \times 81.2 \text{ m} \times 16.03 \text{ m})/81.04 \text{ m} =$

(c) $(5.79 \text{ cm}^3/4.220 \text{ cm}^3) \times (203.1 \text{ cm}^3 + 821.9 \text{ cm}^3)/0.00345 \text{ m}^3 =$

3.86 Perform each of the following conversions:

(a) $9.32 \text{ mg} = ? \text{ Mg}$

(b) $0.02445 \text{ cm}^3 = ? \text{ m}^3$

(c) $3.22 \text{ cm} = ? \text{ pm}$

(d) $81.4°C = ?°F$

(e) $9561 \text{ g} = ? \text{ cg}$

(f) $1.63 \text{ g/mL} = ? \text{ kg/dm}^3$

(g) $-248°C = ? \text{ K}$

(h) $451 \text{ nL} = ? \text{ m}^3$

3.87 One pennyweight equals exactly 24 grains, and one grain equals 15.432 grams. (a) How many grams are there in 1 pennyweight? (b) How many pennyweights are there in 3.699 kg? (c) How many milligrams are there in 9.872 grains?

3.88 A gill is a unit of volume equal to 7.219 cubic inches. (a) What is the volume of a gill in cubic centimeters? (b) The density of xenon is 5.897 kg/m³. What is the mass of 25 gills of xenon?

3.89 The hectare is a unit of area equal to exactly 10,000 square meters. (a) How many square centimeters are there in 1.00 hectare? (b) How many hectares are there in 1.00 km²? (c) How many square miles are there in 1.00 hectare? (One mile equals 5280 feet.)

3.90 (a) One nautical mile (nmi) is equivalent to 6076.1 feet. If a ship travels at 23.0 nmi/hr, how many kilometers will the ship travel in 1 day? (b) One statute mile is equivalent to exactly 5280 feet. How fast in statute miles per hour would the same ship have to travel to traverse the distance obtained in part a?

3.91 Commercial airplanes regularly travel in excess of 555 mi/hr. Convert this speed to: (a) kilometers per hour, (b) meters per second, (c) millimeters per nanosecond.

3.92 A cylinder is measured and is found to weigh 124.54 g. Its height is 3.22 cm, and its circular diameter is 1.88 cm. The formula for the volume of a cylinder is $V = \pi r^2 h$. Calculate the density of the cylinder.

3.93 What volume of uranium has the same mass as 93.4 cm³ of mercury? (The densities of uranium and mercury are found in Table 3.2.)

3.94 What size cube—length of each side—can be formed from 961 g of chromium metal? The density of chromium is 7.20 g/cm³. (*Hint:* An equation to find the volume of a cube is $V_{cube} = l^3$, in which l is the length of each side.)

3.95 Automakers express engine displacement in liters. What is the displacement of a 2.0-L engine in (a) cubic inches (b) cubic meters?

3.96 The densities of ethanol and water at 20°C are 0.7894 g/cm³ and 0.9982 g/cm³, respectively. (a) Calculate the density of an ethanol-water mixture that contains 31.24 g of ethanol and 43.11 g of water. (b) Calculate the density of an ethanol-water mixture that contains 472.1 cm³ of ethanol and 93.77 cm³ of water.

3.97 (a) What Fahrenheit temperature is numerically equal to twice the Celsius temperature? (b) What Celsius temperature is numerically equal to twice the Fahrenheit temperature?

3.98 A 32.326-g container is completely filled with water. The mass of the container with the water is found to be 57.205 g. The water is removed, and the container is filled with an unknown liquid. The mass of the container with the unknown liquid in it is 55.223 g. Assume the density of water is 1.000 g/mL and calculate the density of the unknown liquid.

3.99 A small particle is found to have a volume of 8.22 μm³ and a density of 2.39 g/cm³. How many particles with the same volume and density are contained in a 1.33-kg sample?

3.100 The density of mercury is 13.546 g/cm³ at 20°C and is 13.521 g/cm³ at 30°C. What change in volume occurs when 1.000 g of mercury is heated from 20°C to 30°C, assuming that none of the mercury evaporates?

3.101 The density of thorium, Th, is 11.71 g/cm³. If a sphere of pure Th is found to have a diameter of 4.551 cm, what is the mass of the sphere?

CHAPTER
Four

Matter
and Energy

After completing Chapter 4, you should be able to

1. Distinguish between the composition and structure of matter

2. Define and give examples of physical and chemical properties, and physical and chemical changes

3. List the fundamental properties of solids, liquids, and gases

4. Describe what changes of state occur during melting, freezing, boiling, and subliming

5. List and explain the criteria used to classify pure substances and mixtures

6. Distinguish between elements and compounds

7. Write symbols and names for the elements

8. State the number of atoms of each type indicated by chemical formula

9. Give examples of homogeneous and heterogeneous mixtures and explain how they can be separated into their components

10. Define energy in terms of work

11. Describe different forms of potential and kinetic energy

12. Explain the direction of heat flow in terms of the temperature of objects in contact with each other

13. Explain the difference between heat and temperature

14. Identify and define the two units most commonly used to measure heat energy

15. Convert joules to calories and vice versa

16. Define specific heat and calculate the specific heat of a substance

17. Calculate the heat gain or loss by a substance given the specific heat, mass, and temperature change

18. State the laws of conservation of matter and energy and describe their importance to chemistry

---

As we discussed in Chap. 3, matter is anything that has mass and occupies space. One of the principal concerns of chemists is to study the composition and structure of matter. **Composition** refers to the identity and quantity of the components (ingredients) of matter, and

4.1 PROPERTIES OF MATTER

structure refers to the physical arrangement of the components within matter. Collectively, the composition and structure determine the **properties,** or characteristic traits, of matter.

Each type of matter has its own unique set of properties. Thus, different types of matter are distinguished by their properties, just as people are distinguished by their physical appearance and personality traits. Properties are classified as being either physical or chemical. We will begin our study of matter by discussing these two general groups of properties.

A **physical property** is a characteristic of an individual substance that can be determined without changing the composition of that substance. A physical property that we have already discussed thoroughly is density. Density is the mass of a unit volume. When density is measured, it is only necessary to find the mass and volume of a sample of the substance—no change in composition occurs when making this measurement.

Examples of physical properties are color, hardness, electrical conductivity, heat conductivity, physical state, melting point, boiling point, and tensile strength. Physical properties are measured by observing what happens when matter interacts with heat, light, electricity, and other forms of energy, or when matter is subjected to various stresses and forces.

A substance has a unique set of physical properties that distinguishes it from all other substances. If two substances have exactly the same set of physical properties, the most plausible conclusion is they are the same substance with the same composition and structure.

To illustrate, let's investigate the physical properties of pure gold. Gold is a bright-yellow metal that melts at 1063°C and boils at 2808°C. Its density is 19.3 g/mL, which is a very high density for a metal. On a hardness scale of 1 to 10, gold has a value of 2.8, which means it is very soft. It shares the common physical property of metals called malleability, i.e., the ability of a substance to be hammered into different shapes or thin foils (called leaf). Gold leaf, 1.3×10^{-5} cm thick, is used in expensive decorations. Gold is one of the best conductors of electricity and is an excellent heat conductor. Table 4.1 lists the physical properties of other selected substances.

Physical Properties and Changes

Tensile strength is the resistance to pulling, and is measured by finding the breaking stress.

TABLE 4.1 PHYSICAL PROPERTIES OF SELECTED SUBSTANCES

Substance	Color	Physical state (25°C)	Density	Melting point, °C
Aluminum, Al	Grayish	Solid	2.70 g/cm³	660
Bromine, Br_2	Red brown	Liquid	3.10 g/cm³	−7.3
Chlorine, Cl_2	Pale green	Gas	2.98 g/L	−101
Copper, Cu	Red brown	Solid	8.92 g/cm³	1085
Helium, He	Colorless	Gas	0.178 g/L	—
Water, H_2O	Colorless	Liquid	1.00 g/cm³	0.0

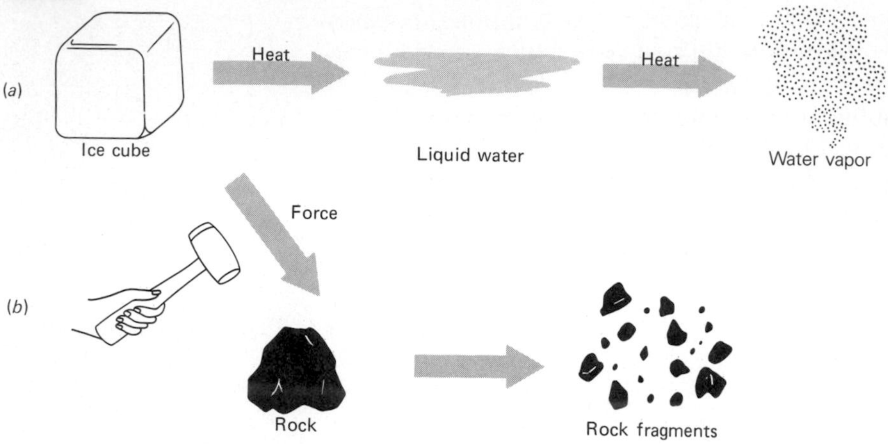

Figure 4.1
(a) After undergoing a physical change, the substance's composition remains the same. Addition of heat to ice changes it to liquid water. Addition of heat to liquid water changes it to water vapor. *(b)* Only the particle size of the rock changes after being broken into pieces. A change in shape is a physical change.

When the physical properties of a substance are altered but the composition remains the same, we say that a **physical change** has occurred. *No new substance is formed when a physical change takes place.* Examples of physical changes are changes in state, density, shape, magnetic properties, and conductivity. After a physical change, the starting substance is still present but in a modified state (Fig. 4.1). For example, after a rock is crushed, it still has the same composition; only the particle size has changed. When ice melts, it changes from the solid state to the liquid state. When iron is magnetized, it is still iron.

The composition of a substance remains the same when a physical change occurs.

Chemical properties describe how the composition of a substance changes or does not change when the substance interacts with other substances or energy forms. Terms used to describe some chemical properties are "reactive," "inert," "unstable," and "combustible." Chemical properties are observed when a substance changes composition. Paper burns in air, iron rusts, silver tarnishes, TNT explodes—these are all illustrations of chemical properties. In each case a new substance is formed, and a **chemical change** occurs.

Chemical Properties and Changes

Chemical changes or *chemical reactions* are the result of the chemical properties of matter. After a chemical change, the composition of the substance that underwent the change is no longer the same. For each chemical change, we can write a chemical equation that indicates what the original substance or substances are ultimately changed to. The starting materials, called **reactants,** undergo a chemical change and produce the **products.**

$$\text{Reactants} \longrightarrow \text{products}$$

The arrow that separates the reactants from the products is a symbol meaning "yield" or "produce." In chemical reactions, the reactants combine to *yield* the products. If only one reactant is present initially, we usually say the reactant "decomposes" or "rearranges" to yield the products.

To monitor chemical changes, we first observe the physical properties of the reactants, and then consider the physical properties of the products. If a chemical change has truly taken place, some or all of the physical properties of the products will be different from those of the reactants.

Examples of everyday chemical changes include the rusting of iron, the cooking of foods, the ignition of gasoline, the explosion of dynamite (Fig. 4.2), and the burning of wood. Each of these examples involves a fairly complex reaction, requiring elaborate equations to explain what happens to the reactants. Let's consider a less complex chemical change, the decomposition, or breakdown, of water using electricity.

When an electric current is passed through water (Fig. 4.3), the water is changed to new substances, hydrogen and oxygen. A word equation for the decomposition of water reads as follows:

$$\text{Liquid water} \xrightarrow{\text{electricity}} \text{hydrogen gas} + \text{oxygen gas}$$

This equation states: "Liquid water is decomposed by electricity to yield the gases hydrogen and oxygen." How do we know that a chemical change rather than a physical change has taken place? First we must consider the physical properties of water: (1) A liquid; (2) normal boiling point is 100°C; (3) melting point is 0°C; (4) flows easily; (5) has a high specific heat; (6) is a poor heat conductor; (7) density equals 1 g/mL. These hardly resemble the physical properties of the products, hydro-

Figure 4.2
An explosion is an example of a chemical change. Explosives such as nitroglycerine or TNT are rapidly converted to gaseous products and a large amount of energy.

The decomposition of water by electricity is called electrolysis. We will discuss electrolysis in Chap. 18.

Figure 4.3
Electrolysis occurs when water is decomposed by a direct electric current. During electrolysis, water is changed chemically to hydrogen gas and oxygen gas.

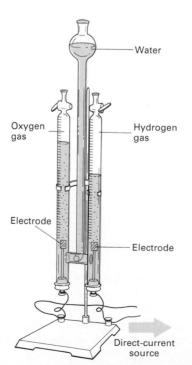

Water

Oxygen gas

Hydrogen gas

Electrode

Electrode

Direct-current source

gen and oxygen. Both of these are gases with boiling and melting points well below 0°C, and their densities are quite small relative to the density of water.

Example Problem 4.1 shows how to predict whether a property of matter is classified as physical or chemical. Study each of the examples given.

─────────────────── **Example Problem 4.1** ───────────────────

For each of the following, determine if they are chemical or physical properties. *(a)* When limestone is heated, carbon dioxide and calcium oxide are produced. *(b)* Ice floats on water. *(c)* Copper can be drawn into thin wires. *(d)* Milk sours if not refrigerated.

─────────────────── **Solution** ───────────────────

(a) "When limestone is heated, carbon dioxide and calcium oxide are produced" describes a chemical property of limestone because two new substances are formed after heating. If a physical change had occurred, limestone would have been present after heating.

(b) "Ice floats on water" describes a physical property. Floating involves no change in physical properties, and thus immediately eliminates the possibility of a chemical change. Ice floats on water because ice has a lower density than water. Density is a physical property. A substance always floats on the surface of a fluid of higher density (assuming it does not mix with the fluid).

(c) "Copper can be drawn into thin wires" describes a physical property. Copper's shape is changed when it is drawn into wires; its composition does not change.

(d) "Milk sours" describes a chemical property. A change in the taste of the milk indicates that one or more new substances are present in the sour milk that were not present in the fresh milk. The increase in the concentration of acid produced by microorganisms causes milk to sour.

When classifying properties as physical or chemical, ask yourself: Has the composition changed? Can an equation be written that shows the reactants and products? If the answer is "yes" to both of these questions, the property being described is chemical. Solving problems and answering questions by eliminating possibilities is an effective technique to use. Sort through all given information and discard what you can.

Example Problem 4.2 shows you how to classify chemical and physical changes.

─────────────── **Example Problem 4.2** ───────────────

For each of the following, determine if a physical or chemical property is described. *(a)* Liquid water evaporates to produce water vapor. *(b)* Fermenting grapes produce ethanol. *(c)* A candle is burned.

─────────────── **Solution** ───────────────

(a) "Liquid water evaporates to produce water vapor" describes a physical change. When water evaporates, the composition remains constant. Water is present in two different states; thus, a physical change has occurred.

(b) "Fermenting grapes" is an example of a chemical change. Fermentation is the process in which yeast is added to crushed fruits or grains to produce ethanol, the drinking variety. Ethanol, a new substance, is formed by fermentation, indicating that a chemical change has taken place.

(c) "A candle is burned" illustrates a chemical change. Whenever a substance is burned, the initial substance, a candle in this example, is changed to new substances with new compositions; accordingly, the change is chemical.

REVIEW EXERCISES

4.1 For each of the following, determine if a chemical or physical property is described. *(a)* Sulfur is bright yellow. *(b)* Silicon is a relatively hard substance. *(c)* At low temperatures mercury exists in the solid state. *(d)* Cadmium is corroded by acids. *(e)* Hydrogen explodes when ignited in the air.

4.2 For each of the following, determine if a chemical or physical change is described. *(a)* Paper ignites when placed in a flame. *(b)* Sand and water are separated when passed through a filter. *(c)* Charcoal burns, leaving ashes. *(d)* Dry ice forms a vapor "cloud" upon heating. *(e)* Sugar dissolves in water.

4.2 STATES OF MATTER

All matter on earth exists in three physical states: **solid, liquid,** and **gaseous.** Various physical properties distinguish the three states of matter. The properties most often considered are shape, volume, average density, structure, viscosity, and compressibility. Shape, volume, and density have been discussed previously; the last two properties require some explanation.

Viscosity is a measure of the resistance to flow. Substances with high viscosities do not flow readily, whereas substances with low viscosities flow more readily. If we are told that water is more fluid than motor oil, we know that the viscosity of water is less than that of motor oil. **Compressibility** is the measure of the decrease in volume of a substance with an applied pressure. A substance is deemed compressible if a force

A fourth state of matter known as the plasma state also exists. Only matter at temperatures that exceed 100 million degrees is found in the plasma state.

exerted on its surface (a pressure) results in a compacting of the sub-stance.

Let's consider each physical state individually, starting with the solid state and proceeding to the liquid and gaseous states. The physical state of a substance depends on its temperature and pressure. Unless other-wise noted, room conditions of 25°C (298 K) and normal atmospheric pressure are assumed. Atmospheric pressure is measured in atmos-pheres, the atmosphere being a unit of gas pressure. Normal atmos-pheric pressure is equivalent to one atmosphere.

Pressure results when a force is exerted on an area. One of the most common units of pressure is the atmosphere.

Solids

Solids have fixed shapes that are independent of their container. The volume of a solid is also fixed, and does not change when a pressure is exerted. Solids are almost completely incompressible. Those that seem to be compressible, such as foams or corrugated paper, actually are sol-ids that contain holes, or empty regions, throughout their volume. When these are "compressed," the solid structure fills into the empty regions: the solid itself is not compressed.

Of the three states of matter, solids have the highest average den-sity. Densities in excess of 1 g/cm^3 are the norm for solids. Such is not the case for most liquid and gases. A high average density reflects the fact that the particles within solids are usually packed closer than those in liquids or gases. The tightly packed particles of solids are also highly organized (Fig. 4.4). The regular patterns of particles found in solids are not detected in either liquids or gases.

Solids have practically no ability to flow because the particles that compose a solid are very tightly bonded. Stated in another way: Solids have very high viscosities.

Liquids

Liquids are quite different from solids in many respects, but the two share some characteristics. Like solids, liquids are essentially incompress-

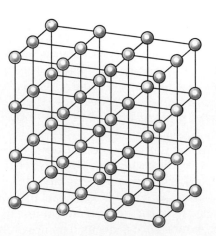

Figure 4.4
Most solids are composed of a regular array of closely packed particles. Particles within solids are usually more organized and packed more tightly than are the particles within liquids and gases.

ible; pressure exerted on liquids generally produces little, if any, change in their volumes. When placed in a container, liquids assume the shape of the container to the level they fill (Fig. 4.5).

As previously mentioned, the average density of liquids is less than that of solids but greater than that of gases. Liquid particles are not bonded as strongly as those in solids, and they are less orderly—more randomly distributed. Both of these factors tend to increase the average volume of liquids relative to that of solids. Thus, for equal masses of an average solid and liquid, the volume of the liquid is usually larger than that of the solid, which results in a lower density.

Viscosities of liquids vary over a broad range. Liquids have much lower viscosities than solids; i.e., they are significantly more fluid than solids. However, the viscosities of liquids are greater than those of gases. The gaseous state is the most fluid state of matter.

Gases bear little resemblance to the more-condensed states of matter, solids and liquids. To a degree, the properties of gases are the opposite of those of solids. Gases completely fill the volume of their containers, are compressible (Fig. 4.6), have a completely disorganized structure, possess the lowest average density of the three states, and have the lowest viscosities. Table 4.2 summarizes the properties of the physical states, and Table 4.3 lists examples of common solids, liquids, and gases.

Gases

Figure 4.5
A liquid completely fills and takes the shape of the bottom of its container.

TABLE 4.2 PROPERTIES OF THE THREE PHYSICAL STATES

Property	Solid	Liquid	Gas
Shape	Constant	Variable	Variable
Volume	Constant	Constant	Variable
Density	Highest	Moderately high	Lowest
Structure	Organized	Semiorganized	Random
Viscosity	Highest	High to low	Lowest
Compressibility	Incompressible	Incompressible	Compressible
Particles	Closely packed	Less closely packed	Widely separated

TABLE 4.3 EXAMPLES OF SOLIDS, LIQUIDS, AND GASES (25°C, 1 atm)

Solids	Liquids	Gases
Carbon	Alcohol	Carbon dioxide
Ice	Ether	Helium
Iron	Kerosene	Methane
Salt	Mercury	Neon
Steel	Oils	Nitrogen
Sugar	Water	Water vapor

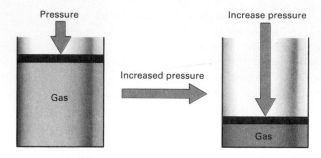

Pressure

Increased pressure

Gas

Increase pressure

Gas

Figure 4.6
Gases are compressible. If the external pressure increases on a gas sample, its volume decreases. Liquids and solids are almost incompressible.

Changes of State

Matter can change from one physical state to another. For example, solids, when heated, change to liquids. The characteristic temperature at which a particular solid changes to a liquid is called its **melting point.** At the melting point, the solid and liquid states of the substance coexist. Liquids, in turn, change to solids as they are cooled. The temperature at which a liquid becomes a solid is called the **freezing point.** Freezing and melting occur at the same temperature. In one case, the solid changes into a liquid—it melts. Moving in the other direction, a liquid changes into a solid—it freezes.

$$\text{Solid} \underset{\text{freezing}}{\overset{\text{melting}}{\rightleftharpoons}} \text{liquid}$$

For example, water freezes or melts at 0.0°C. Table 4.4 lists the melting (freezing) points of some common substances.

The solid and liquid states are in equilibrium at the melting point.

TABLE 4.4 NORMAL MELTING AND BOILING POINTS OF COMMON SUBSTANCES

Substance	Melting point, °C	Boiling point, °C
Aluminum, Al	660	2450
Carbon dioxide, CO_2	−56.2	−78.4
Helium, He	−269.7	−268.9
Hydrogen, H_2	−259.2	−252.7
Iron, Fe	1536	3000
Mercury, Hg	−38.4	357
Oxygen, O_2	−218.8	−183
Sodium, Na	97.8	892
Water, H_2O	0.0	100

Liquids, when heated, change to their vapors. A vapor is the gaseous phase of a substance. The transition temperature for this change is termed the **boiling point.** At the boiling point the liquid and vapor states coexist. When a substance boils, vapor bubbles can be seen throughout the liquid phase. The change in the opposite direction, from vapor to liquid, is called **condensation.**

$$\text{Liquid} \xrightleftharpoons[\text{condensation}]{\text{boiling}} \text{liquid}$$

Once again, refer to Table 4.4 for examples of boiling points of common substances. A pressure of 1 atm is specified for the boiling points listed in the table because pressure significantly affects the boiling points of liquids. We will discuss the effect of pressure on boiling points in Chap. 13.

The liquid and vapor states are in equilibrium at the boiling point.

Numerous solids change directly to their vapors without going through the liquid state. This state change is called **sublimation.** At the temperature and pressure at which a substance sublimes, the solid and vapor states coexist.

$$\text{Solid} \xrightarrow{\text{sublimation}} \text{vapor}$$

A good example of a solid that sublimes is "dry ice," or solid carbon dioxide. At $-78°C$ (195 K), solid carbon dioxide and gaseous carbon dioxide coexist (Fig. 4.7).

A solid that changes to a vapor is said to sublime.

4.3 *(a)* What is viscosity? *(b)* What trend exists in viscosities when considering the solid, liquid, and gaseous states?

REVIEW EXERCISES

4.4 What properties distinguish an average solid from an average liquid? Give specific examples.

4.5 *(a)* What physical state has the most unorganized structure? *(b)* What accounts for the unorganized structure?

4.6 What physical states coexist at the following transition points: *(a)* melting point, *(b)* subliming point, *(c)* freezing point, *(d)* boiling point?

Figure 4.7
Solid carbon dioxide, commonly called "dry ice," sublimes at room temperature and pressure. When the cold carbon dioxide vapor contacts moist air, it produces the effect shown here. *(Direct Positive Imagery.)*

Matter is found in many different forms throughout the earth, and every year thousands of new types of matter are synthesized. When dealing with such a gigantic variety of substances, it is best to divide this enormous group into smaller categories of similar matter forms.

Matter may be grouped into two major classes: **pure substances** and **mixtures.** We will discuss pure substances before we consider the mixtures, which are more complex. Figure 4.8 shows a complete classification of matter.

A substance is classified as pure if it meets the following three criteria:

1. It has the same composition throughout the sample.
2. Its components are inseparable using physical methods.
3. Changes of state occur at a constant temperature.

Analysis of a pure substance reveals the same composition throughout the sample. All parts of a pure substance contain the same percent of each component. For example, water is a pure substance—all water samples are composed of 11% hydrogen and 89% oxygen by mass. If some other percents are found for hydrogen and oxygen, then the substance is impure water or it is not water at all.

The components of pure substances cannot be separated by physical methods. If the components of pure substances are capable of being separated (not all are), chemical means are required. For example, one way of separating water into its component elements is to pass an electric current through it. This results in the production of hydrogen gas and oxygen gas. Physical methods, such as filtering or heating at normal temperatures, have no effect on water with respect to separating it into its components. Water passes through filters, and heat changes liquid water to water vapor.

All pure substances undergo state changes at constant temperatures. In contrast, state changes of mixtures occur over broader temper-

4.3 CLASSIFICATION OF MATTER

Pure Substances

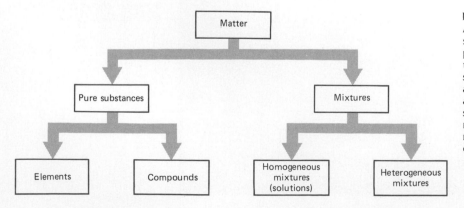

Figure 4.8
All forms of matter may be subdivided into two classes: pure substances and mixtures. Two classes of pure substances exist: elements and compounds. Mixtures are combinations of pure substances. Two classes of mixtures exist: homogeneous mixtures (solutions) and heterogeneous mixtures.

ature ranges. Chemists can use this property to assess the purity of a substance; a sharper melting point (smaller observable temperature range) indicates a higher level of purity. A fairly broad range in the melting temperature is characteristic of an impure sample, a mixture.

Pure substances are subdivided into two groups: elements and compounds (Fig. 4.8). Elements are pure substances that cannot be decomposed by chemical changes. Compounds are pure substances that may be chemically decomposed to elements.

Elements

Elements are the basic units of matter. All of the other types of matter contain elements. Today, we know of approximately 110 different elements. Whenever the number of elements is stated, it is necessary to introduce some doubt. New elements are occasionally discovered, and they are added to the list. About 92 elements occur in nature, and the remaining are synthetic. At 25°C, 97 elements are solids, 2 are liquids, and 11 are gases.

The forerunner to our modern periodic table was developed at the end of the nineteenth century by the famous Russian scientist Dmitri Mendeleev.

All of the known elements are listed, using their symbols, in the **periodic table,** which is probably the most important table in chemistry (Fig. 4.9). Each element is located in a horizontal row called a **period,** and in a vertical column called a **group** (sometimes called a *family*). Each period is numbered consecutively from 1 to 7; each group of elements is assigned a Roman numeral and a letter. Presently, the numbering of the groups of the periodic table is undergoing a change. It has been recom-

Figure 4.9
All the elements are listed on the periodic table. Each element is a member of a chemical group, a vertical column of elements, and a period, a horizontal row of elements. In addition, two numbers are found next to each symbol. These numbers are discussed fully in Chap. 5.

PERIODIC TABLE OF THE ELEMENTS

Periods \ Groups	IA(1)	IIA(2)	IIIB (3)	IVB (4)	VB (5)	VIB (6)	VIIB (7)	VIIIB (8)	VIIIB (9)	VIIIB (10)	IB (11)	IIB (12)	IIIA (13)	IVA (14)	VA (15)	VIA (16)	VIIA (17)	VIIIA (18)
1	1 H 1.0079																	2 He 4.00260
2	3 Li 6.941	4 Be 9.01218											5 B 10.81	6 C 12.011	7 N 14.0067	8 O 15.9994	9 F 18.998403	10 Ne 20.179
3	11 Na 22.98977	12 Mg 24.305											13 Al 26.98154	14 Si 28.0855	15 P 30.97376	16 S 32.06	17 Cl 35.453	18 Ar 39.948
4	19 K 39.0983	20 Ca 40.08	21 Sc 44.9559	22 Ti 47.90	23 V 50.9415	24 Cr 51.996	25 Mn 54.9380	26 Fe 55.847	27 Co 58.9332	28 Ni 58.70	29 Cu 63.546	30 Zn 65.38	31 Ga 69.72	32 Ge 72.59	33 As 74.9216	34 Se 78.96	35 Br 79.904	36 Kr 83.80
5	37 Rb 85.4678	38 Sr 87.62	39 Y 88.9059	40 Zr 91.22	41 Nb 92.9064	42 Mo 95.94	43 Tc (98)	44 Ru 101.07	45 Rh 102.9055	46 Pd 106.4	47 Ag 107.868	48 Cd 112.41	49 In 114.82	50 Sn 118.69	51 Sb 121.75	52 Te 127.60	53 I 126.9045	54 Xe 131.30
6	55 Cs 132.9054	56 Ba 137.33	57 La 138.9055 *	72 Hf 178.49	73 Ta 180.9479	74 W 183.85	75 Re 186.207	76 Os 190.2	77 Ir 192.22	78 Pt 195.09	79 Au 196.9665	80 Hg 200.59	81 Tl 204.37	82 Pb 207.2	83 Bi 208.9804	84 Po (209)	85 At (210)	86 Rn (222)
7	87 Fr (223)	88 Ra 226.0254	89 Ac 227.0278 †	104 Unq (261)	105 Unp (262)	106 Unh (263)	107 Uns (262)	108	109 Une (266)									

* Lanthanide series

58 Ce 140.12	59 Pr 140.9077	60 Nd 144.24	61 Pm (145)	62 Sm 150.4	63 Eu 151.96	64 Gd 157.25	65 Tb 158.9254	66 Dy 162.50	67 Ho 164.9304	68 Er 167.26	69 Tm 168.9342	70 Yb 173.04	71 Lu 174.967

† Actinide series

90 Th 232.0381	91 Pa 231.0359	92 U 238.029	93 Np 237.0482	94 Pu (244)	95 Am (243)	96 Cm (247)	97 Bk (247)	98 Cf (251)	99 Es (252)	100 Fm (257)	101 Md (258)	102 No (259)	103 Lr (260)

mended that the groups be numbered consecutively from 1 to 18. In this textbook we will use both the old and the new numbering systems. A periodic table and an alphabetical listing of the elements are found on the inside of the front cover.

Chemists use symbols to represent elements. The use of symbols to represent elements dates back to the ancient Greeks, who originally suggested that matter was composed of elements. Table 4.5 lists four of the ancient Greek symbols, those for the elements sulfur, gold, copper, and lead. In addition, the symbols used by the alchemists and by John Dalton, the developer of the first scientific theory of the atom, are given in this table.

TABLE 4.5 ANCIENT SYMBOLS FOR THE ELEMENTS

Source	Sulfur	Gold	Copper	Lead
Ancient Greeks	ω	♂	♀	♄
Alchemists	♁	⊙	♀	♃
John Dalton (1808)	⊕	Ⓖ	Ⓒ	Ⓛ
Modern	S	Au	Cu	Pb

Modern symbols are usually derived from the first one or two letters of the name of the element. Twelve of the elements have one-letter symbols which correspond to the first letter of the element's name. They are hydrogen, H; boron, B; carbon, C; nitrogen, N; oxygen, O; fluorine, F; phosphorus, P; sulfur, S; vanadium, V; yttrium, Y; iodine, I; and uranium, U.

Two other elements are designated by one-letter symbols, but these symbols are not the first letter of the elements' English names. K is the symbol for the element potassium. Why K? K is the first letter of the old Latin name for potassium, *kalium*, which means ashes. It is convenient to use K instead of P or Po because other elements have these symbols. Similarly, the symbol for tungsten, W, is derived from its old name, "wolfram." Today, many scientists use this old name and no longer refer to W as tungsten.

Most of the remaining elements are assigned two-letter symbols. The first letter is always an uppercase letter, and the second is a lowercase letter. The symbol for cobalt is Co, not CO. CO is not a symbol; it is the formula of a compound called carbon monoxide. Some symbols are made up of the first two letters of the English name, others are two letters of an old name, and the remainder combine the first letter with some other letter in the name.

Origins of the modern names of elements are interesting. Some elements are named after geographical locations; others have names

Ytterby, a town in Sweden, has four elements named after it: yttrium, Y; terbium, Tb; erbium, Er; and ytterbium, Yb.

——— **CHEM TOPIC: Names of the Transuranium Elements** ———

Elements with higher atomic numbers than uranium for the most part are synthetic and have been produced by scientists in sophisticated machines called particle accelerators. The scientists who discovered these new elements were allowed to assign their names. However, when element 104 was reached, two names were assigned to it because of a dispute between Soviet and American scientists. The Soviet name, assigned in 1964, was kurchatovium after an important Soviet nuclear scientist. A few years later, American scientists assigned the name rutherfordium, after the famous scientist whose discovery of the nuclei of atoms was among his many significant scientific contributions. A similar problem arose with element 105, which was given the name neilsbohrium by the Soviets and hahnium by the Americans. The Soviets honored the Danish physicist Neils Bohr, while the Americans recognized the German nuclear scientist Otto Hahn.

The International Union of Pure and Applied Chemistry (IUPAC), an international organization of chemists, is responsible for establishing the rules that govern the naming of elements and compounds. They rectified the problem by devising a system for assigning names to elements. Element 104 was given the name unnilquadium and the symbol Unq. If we break the word into its component parts, we get *un, nil, quad,* which means 104 (*un* = 1, *nil* = 0, *quad* = 4). Element 105 is given the name unnilpentium (and the symbol Unp), which literally means 105 (*un* = 1, *nil* = 0, *pent* = 5). Unnilhexium, Unh, and unnilseptium, Une, are the next two elements—106 and 107. The most important advantage of such a system is that it unambiguously gives the names and symbols for the elements up to 999.

---

derived from those of great scientists, while some are named for mythological gods and astronomical bodies. Table 4.6 lists the origins of some of the elements' names.

Atoms are the smallest particles that retain the chemical properties of elements. Atoms are extremely small; 1 g of carbon, C, contains 5×10^{22} C atoms. Placed end to end, approximately 2×10^8 atoms of C are needed to span an inch. Each element is composed of similar atoms. Both the chemical and physical properties of an element are directly related to the composition and properties of its atoms. In Chap. 5 we will investigate this central topic in chemistry.

An atom is the smallest unit of an element.

Table 4.7 lists the abundances, in percent by mass, of the elements found in the earth's crust and in the entire earth. The crust of the earth is the thin outer layer that surrounds the mantle and core of the earth. Outside of the crust is the atmosphere. In the earth's crust, O and Si are the most abundant elements, and they comprise 74.3% of the mass of the crust. Al, Fe, Ca, Na, K, and Mg are the next most abundant elements. Together these six elements compose 24.2% of the mass of the earth's crust. If the whole earth is considered, a different distribution of elements is found. Iron is the most abundant element on earth because it makes up the core of the earth. Iron is followed by O, Si, and Mg; these elements account for 92.0% of the mass of the earth.

TABLE 4.6 ORIGINS OF THE NAMES OF ELEMENTS

Symbol and name	Origin
	Location
Am, americium	America
Bk, berkelium	Berkeley, California
Cf, californium	California
Eu, europium	Europe
Fr, francium	France
Ge, germanium	Germany
Po, polonium	Poland
Sr, strontium	Strontia, Scotland
	Scientist
Cm, curium	Marie and Pierre Curie
Es, einsteinium	Albert Einstein
Fm, fermium	Enrico Fermi
Lr, lawrencium	Ernest O. Lawrence
Md, mendelevium	Dmitri Mendeleev
No, nobelium	Alfred Nobel
	God or astronomical body
He, helium	Greek, *helios*, "sun"
Nb, niobium	Niobe, daughter of Tantalus
Np, neptunium	Neptune
Pd, palladium	an asteroid called Pallas
Pu, plutonium	Pluto
Se, selenium	Greek, *Selene*, "moon"
Th, thorium	Thor
U, uranium	Uranus

REVIEW EXERCISES

4.7 What are the three criteria used to identify pure substances?

4.8 *(a)* What are the two classes of pure substances? *(b)* Give an example of each.

4.9 Write the names of each of the following elements: *(a)* B, *(b)* Be, *(c)* Ba, *(d)* Br, *(e)* Bi.

4.10 Write the symbols and names for 10 elements whose names begin with C.

4.11 List five elements in which the first letter of the symbol is different than the first letter of the name.

Compounds

Compounds make up the other class of pure substances. They are more complex than elements. Elements undergo chemical reactions to form compounds; thus, **compounds** are chemical combinations of elements. Consequently, compounds can only be separated into their component elements by chemical means. The smallest subdivision of a compound is a **molecule,** which is a chemical combination of atoms.

TABLE 4.7 ABUNDANCE OF ELEMENTS IN THE EARTH'S CRUST AND THE WHOLE EARTH

Earth's crust		Whole earth	
Element	Abundance, % by mass	Element	Abundance, % by mass
O	46.6	Fe	34.6
Si	27.7	O	29.5
Al	8.1	Si	15.2
Fe	5.0	Mg	12.7
Ca	3.6	Ni	1.6
Na	2.8	S	1.9
K	2.6	Ca	1.1
Mg	2.1	Al	1.1
Ti	0.44	Na	0.57
H	0.14	Cr	0.26
P	0.11	Mn	0.22
Mn	0.095	Co	0.13
F	0.063	P	0.10

Formulas are used to represent compounds. Each formula shows the specific composition of a compound. Most people are familiar with the chemical formula of water, H_2O. What information is conveyed by the formula of water? It indicates that all water is composed of two parts hydrogen and one part oxygen. If water is decomposed by electrolysis, two volumes of hydrogen are formed for each volume of oxygen.

$$\text{Water} \longrightarrow 2 \text{ hydrogen} + 1 \text{ oxygen}$$

Additionally, the formula of a compound gives the ratio of the different types of atoms that make up one molecule of that compound. For example, water molecules are particles that contain two atoms of hydrogen and one atom of oxygen.

In each chemical formula the symbol of each atom is listed with a subscript—a number written to the right of and below the symbol—which indicates the total number of that type of atom in the molecule. If only one atom of a given element is found in the molecule, the subscript is not written; it is understood to be 1. In the formula of the water molecule, H_2O, the subscript 2 is placed next to hydrogen, but no subscript is written next to oxygen because there is only one oxygen atom per molecule. As an additional example, let's consider carbon dioxide.

$$CO_2 \text{ (carbon dioxide)}$$
$$\text{1 carbon atom} \overline{}\!\!\lceil\;\;\rceil\!\!\underline{} \text{2 oxygen atoms}$$

A CO_2 molecule contains one carbon atom and two oxygen atoms. When reading the formula, we say "see-oh-two."

Other examples of compounds are NaCl, sodium chloride; NH_3, ammonia; CH_4, methane; and $C_{12}H_{22}O_{11}$, sucrose. NaCl is common table salt. All living systems maintain a balance between the dissolved NaCl and fluids within their cells. An excess amount of NaCl in your diet may lead to problems such as hypertension. NH_3, ammonia, is an ingredient of many common household cleaners. Ammonia is also found in some fertilizers and is used to synthesize many important chemicals. CH_4, methane, is the major component of natural gas, which is an important fuel. Sucrose, $C_{12}H_{22}O_{11}$, also known as table sugar, is the sweetening agent used by most people in the world.

In more complex formulas, parentheses are used to group repeating units. For example, the formula of calcium phosphate is $Ca_3(PO_4)_2$. Calcium phosphate contains two PO_4 (phosphate, which is actually $PO_4{}^{3-}$) groups. Instead of writing PO_4 twice, we enclose it in parentheses and write 2 as a subscript to the right. The formula for calcium phosphate indicates that three calcium atoms, two phosphorus atoms, and eight oxygen atoms make up one molecule of calcium phosphate.

Calcium phosphate is a white powder that melts at 1730°C.

4.12 What information can be obtained from the formula of a compound?

4.13 For each of the following, write the name of each atom and tell how many atoms are contained in the formula: (a) KBr, (b) BeF_2, (c) NiO, (d) $Fe(OH)_3$, (e) PCl_3, (f) $Ba(NO_3)_2$, (g) $RbMnO_4$, (h) $(NH_4)_2C_2O_4$, (i) $Ca(H_2PO_4)_2$.

REVIEW EXERCISES

Mixtures make up the second major division of matter. They are more complex than pure substances because **mixtures** are composed of two or more pure substances that are physically associated. Three criteria are used to classify a mixture:

Mixtures

1. Its composition is variable.
2. Its components may be separated by physical methods.
3. Changes of state occur over a range of temperatures.

Many possibilities exist when a mixture is prepared. Two substances can be mixed in virtually any proportion. For example, a mixture of sugar and water might contain 1 g of sugar in 100 mL of water, or 15 g of sugar in 80 mL of water, or many other possible combinations.

Unlike pure substances, mixtures may be separated into their components by physical methods. For example, a mixture of sugar and water is separated by evaporating the water, leaving the sugar behind. A mixture of iron and aluminum can be separated with a magnet. Iron is attracted to the magnet, leaving the aluminum behind (Fig. 4.10).

When a mixture undergoes a change of state, the observed temperature at which the change takes place is not constant. Usually, a fairly broad range of temperatures is observed from the time when the mixture begins to change state until it is entirely in the new state. Because

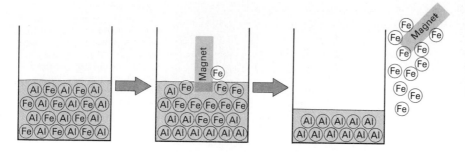

Figure 4.10
Mixtures are separated by physical methods. A mixture of iron and aluminum is separated by placing a magnet into the mixture. Aluminum is not attracted by a magnet, but iron is strongly attracted.

mixtures are composed of two or more pure substances, each component changes state at a different temperature; this is what produces the wide range.

Milk, gasoline, asphalt, ocean water, granite, and air are some examples of mixtures. The components of milk include such compounds as water, proteins, fats, and vitamins. Gasoline is a mixture of organic compounds that are called hydrocarbons because they are chiefly carbon-hydrogen compounds. Asphalt contains tarry substances blended with sand, gravel, glass, and stones. Ocean water is primarily water with a large number of dissolved substances, which include minerals, salts, and gases. Granite is a rock composed of quartz, feldspar, and mica—three common minerals (Fig. 4.11). Finally, air is a mixture of gases, mainly nitrogen and oxygen, with smaller amounts of carbon dioxide, water vapor, argon, and others.

Mixtures are divided into two classes, homogeneous mixtures and heterogeneous mixtures. "Homogeneous" is a word derived from the Greek words *homos,* which means "the same" or "equal," and *genus,* which means "kind" or "structure." "Hetero" is a prefix that means "different."

A **heterogeneous mixture** is one that exhibits more than one phase.

Homogeneous and Heterogeneous Mixtures

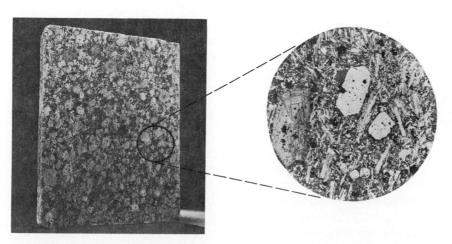

Figure 4.11
Granite is an example of a heterogeneous mixture. It is composed of quartz, feldspars, and mica. A close look at granite reveals these different minerals within the rock. *(Photograph by Ron Testa, Field Museum of Natural History, Chicago.)*

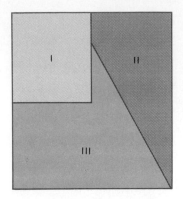

Figure 4.12
In a heterogeneous mixture, two or more phases can be seen. A phase is an observable region of matter with a composition different from that of the surrounding regions. This diagram shows a three-phase system. An interface (boundary) is found between one phase and another.

A phase is an observable region of matter with a composition different from the surrounding regions. The phases can be distinguished from each other by their properties (Fig. 4.12). For example, when sand is added to water, the sand does not dissolve; it just falls to the bottom of the water. When observing sand and water, you see the solid sand phase and the liquid water phase. Oil and water, salt and sand, and granite are all examples of heterogeneous mixtures.

Homogeneous mixtures are also called **solutions.** Only one phase is found in homogeneous mixtures. For example, let's consider a sugar water solution. It is prepared by mixing solid sugar and liquid water. After the sugar dissolves, a homogeneous mixture results. When looking at sugar water, you cannot tell if it is pure water or a solution.

Other examples of solutions are alcohol and water, air, and most alloys. Ethanol (grain alcohol) and water is a popular beverage throughout the world. The alcohol content of drinks can range from a few percent to as high as 60% alcohol by volume. Air is a gaseous solution. All gaseous mixtures exhibit only one phase; therefore, they are classified as solutions. Most alloys are solutions of metals. Examples of common alloys include sterling silver, which is 92.5% Ag (silver) and 7.5% Cu (copper), and brass, which is 67% Cu (copper) and 33% Zn (zinc).

Mixtures, whether homogeneous or heterogeneous, may be separated into their components by physical methods. A solution of salt and water can be separated by heating, for example. During heating, water changes to a vapor, which is then condensed, leaving the salt behind. This is the basis of the process called **distillation.** Figure 4.13 illustrates a laboratory distillation setup. After crude oil is pumped out of the ground, it is separated into its components by distillation.

Filtration is a method used to separate some heterogeneous mixtures. Such mixtures are poured into a funnel that contains a filter paper, which is paper containing small, uniform openings or pores. The pores block large particles and allow small particles to pass through. Sand and water are separated by pouring the mixture into a filter. The small water molecules easily pass through the filter paper, but the larger particles of sand are blocked by the filter paper (Fig. 4.14).

Wine is filtered to remove particulate matter before it is bottled.

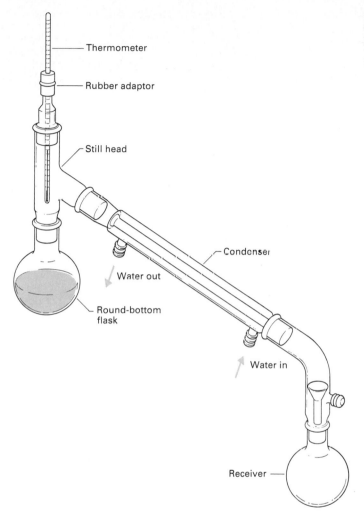

Thermometer

Rubber adaptor

Still head

Condenser

Water out

Round-bottom flask

Water in

Receiver

Figure 4.13
Some mixtures can be separated by distillation. A mixture is placed in a round-bottom flask and is then carefully heated to vaporize the component with the lowest boiling point. The vapors are condensed in a water-cooled condenser, and the liquid drips into a receiving flask.

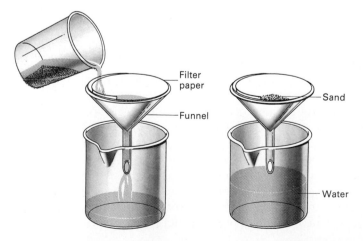

Sand and water

Filter paper

Funnel

Sand

Water

Figure 4.14
A sand-and-water mixture may be separated by gravity filtration. The mixture is poured into a funnel containing a filter paper cone. Water passes through the pores in the filter paper, and the sand particles are trapped in the filter paper.

Example Problem 4.3 shows how another simple mixture can be separated.

———————————— **Example Problem 4.3** ————————————

How can a mixture of sand and salt be separated?

———————————————— **Solution** ————————————————

A combination of sand and salt is a heterogeneous mixture. One contrasting property, the different abilities of the two substances to dissolve in water, provides for a convenient method of separation. Salt dissolves in water, whereas sand does not. Thus, if water is added to the mixture, the salt dissolves, leaving the sand undissolved. The salt water solution that results is filtered. Sand is trapped by the filter paper, while the salt water passes through. To recover the salt, the water is evaporated. In summary:

1. Add water to dissolve salt. The sand is unaffected.
2. Filter the resulting salt water solution from the sand.
3. Evaporate water to recover the salt.

4.14 *(a)* What criteria are used to classify mixtures? *(b)* How do these criteria differ from the criteria used to classify pure substances?

4.15 *(a)* Distinguish between heterogeneous and homogeneous mixtures. *(b)* Give an example of each.

4.16 Give two examples of solutions commonly used around the house.

4.17 How are the following mixtures separated: *(a)* oil and water, *(b)* carbonated water (carbon dioxide, a gas, dissolved in water), *(c)* crushed rock and salt?

REVIEW EXERCISES

Energy is defined as the capacity to do work. What does this mean? Work is done when matter is moved by applying a force—a push or pull. To lift a book off a table or push a stalled car requires work. In science, something must be moved in order for us to say that work has been done. Energy, therefore, is the capacity to move matter or to effect changes in matter.

Two general classes of energy exist, potential and kinetic energy. **Potential energy** is stored energy. This potential or stored energy results from the position, condition, or composition of a body. A boulder moved from the ground to the top of a cliff has potential energy of position with respect to the ground. The boulder can fall off the cliff and crush objects below. A compressed spring has potential energy of condition; spontaneously, the spring can expand and do work by pushing something. A vial of nitroglycerine possesses potential energy of compo-

4.4 ENERGY

Work can be done in many different ways; e.g., mechanical, expansion, electrical, and gravitational work.

Potential energy of position depends on mass, height, and acceleration due to gravity.

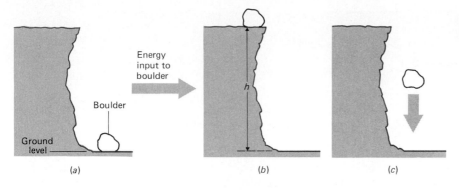

Figure 4.15
(a) A boulder on the ground has no kinetic energy because it is not moving, and has no potential energy with respect to the ground.
(b) Energy is required to move the boulder from the ground to the top of the cliff. On top of the cliff the boulder has no kinetic energy, but it has potential energy with respect to the ground level. The greater the distance above the ground *(h)*, the greater its potential energy. *(c)* If the boulder falls off the cliff, it possesses both kinetic energy and potential energy until it reaches the ground.

sition, or chemical potential energy. You do not want to drop a bottle of nitroglycerine and release this stored energy!

Kinetic energy is the energy associated with matter in motion. Whenever an object is moving, it possesses kinetic energy. The following formula is used to calculate kinetic energy:

$$KE = \tfrac{1}{2} mv^2$$

in which KE is the kinetic energy of an object, m is its mass, and v is its velocity or speed. This equation shows that the kinetic energy of a body is proportional to its mass and to the square of its velocity. The more massive an object and the faster the object is moving, the greater the kinetic energy it possesses.

Potential energy may be converted to kinetic energy. For example, a boulder on top of a cliff has potential energy only (Fig. 4.15). But as soon as the boulder falls from the cliff, its potential energy is changed to energy of motion, or kinetic energy. When the boulder hits the ground, it stops moving and thus no longer possesses kinetic energy.

The velocity of an object is its speed in a particular direction.

Figure 4.16
Lightning is an electric discharge that produces light, heat, and sound.

Figure 4.17
A nuclear power plant differs from a conventional power plant primarily in the way it produces heat. In a nuclear power plant, energy released from splitting atoms (nuclear energy) is converted into heat energy. The liberated heat changes liquid water into high-pressure steam. The steam turns a turbine that is connected to an electric generator. *(Atomic Industrial Forum.)*

Energy is encountered in a wide variety of forms: (1) mechanical energy, (2) electric energy, (3) nuclear energy, (4) light, (5) heat, and (6) sound (Fig. 4.16). All of these are energy forms because each has the capacity to produce changes in matter. Energy can be converted from one form to another. For example, the energy released when gasoline is burned in the engine of an automobile is transformed into mechanical energy. The chemical potential energy of the gasoline is released, causing a piston to move, which is mechanically linked to the wheels of the automobile. Along with the mechanical energy, heat is also released, an unavoidable transformation. Electric potential energy stored in a battery is transformed into light (radiant) energy inside a flashlight or into sound inside a radio.

Most power plants burn fuels to release heat. This heat is transferred to water, which is converted to high-pressure steam, which turns a turbine. A turbine is connected to an electric generator that changes the mechanical energy of the turbine into electricity. Finally, the electricity is sent to consumers who use the electricity to perform innumerable tasks. In this example, chemical potential energy is converted to heat, mechanical energy, and finally electric energy. In nuclear power plants (Fig. 4.17), the heat is generated by nuclear reactions. This is discussed in detail in Chap. 19.

REVIEW EXERCISES

4.18 What is the effect of work on an object? Give an example.

4.19 Distinguish between kinetic and potential energy. Give examples of each.

4.20 List four forms of energy.

4.21 What energy transformations take place within a television set?

Heat, or thermal energy, is especially important to chemists. Why? All other forms of energy can be transformed into heat, and chemical reactions either release or absorb heat. **Heat** is a form of kinetic energy; it can never be classified as potential energy. As we have previously discussed, heat brings about both chemical and physical changes. Solids melt, liquids evaporate, and some substances decompose or undergo chemical changes when heated.

When heat is transferred to matter, the heat increases the matter's potential energy, or its kinetic energy, or both. What happens when a substance is heated? It becomes hotter, you may say. In some cases you are correct. Substances become hotter because the heat transferred increases the average motion of the molecules—an average kinetic energy increase. We observe this increase in kinetic energy by measuring the temperature of the substance. **Temperature** is a measure of the average velocity of atoms and molecules. Higher temperatures of pure substances indicate faster-moving molecules, on average.

But heat transferred to ice at 0°C gives a different result. As long as ice is present, added heat does not increase its temperature. State changes of pure substances occur at a constant temperature. Added heat increases the potential energy of the solid ice molecules to the point where they become a liquid. Only after all the ice has melted does the temperature increase; this temperature change indicates an increase in the average kinetic energy of the molecules.

Heat is detected only when it moves from one body to another. In our world heat travels down a one-way street. Heat is always transferred from a hotter object to a colder object; the reverse never occurs spontaneously (Fig. 4.18).

When two objects at different temperatures come in contact with each other, heat is transferred from the hotter object to the colder object. Faster-moving molecules in the hotter object collide with the slower-moving molecules in the colder object. During the molecular collisions energy is transferred, decreasing the average kinetic energy of the faster-moving molecules and increasing that of the slower-moving molecules. When the average energies of the molecules in both objects are equal, heat flow stops.

Heat

A range of velocities is found for the particles in any sample of matter.

Heat can only be detected when it is in transit. Bodies cannot possess heat.

Figure 4.18
(a) If two objects at different temperatures $(T_1 > T_2)$ contact each other, heat flows spontaneously from the hot object at T_1 to the colder object at T_2. Initially T_1 decreases and T_2 increases until T_1 equals T_2; at that time, heat flow ceases. *(b)* If two objects at the same temperature $(T_1 = T_2)$ contact each other, no heat is transferred between the objects.

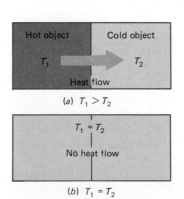

Consider what happens when a hot cup of coffee is allowed to sit for a period of time in a room. As time passes, the temperature of the coffee drops. If the coffee is not consumed, its temperature continues to drop until it equals the temperature of the room. What happens to the heat? Heat flows from the coffee and heats the room. But, compared with the total amount of heat in the room, the small quantity of heat released by the coffee is insignificant.

To summarize, heat is a form of energy that always flows spontaneously from warmer to colder objects. Temperature is a measure of how fast the particles that compose matter are moving. In other words, temperature is a measure of the average kinetic energy of atoms and molecules. Knowing the temperature of two objects in contact with each other allows us to predict the direction of heat flow.

4.22 Describe the difference between heat and temperature.

4.23 (a) Explain what happens to the particles within a hot object when it contacts a colder object. (b) How is the motion of the particles in a body related to its temperature?

4.24 Describe the heat flow and temperature changes when a hot metal at 200°C is placed in a container of water at 25°C.

4.5 MEASUREMENT OF ENERGY

Joules and Calories

The joule is the SI unit of energy. It was named after James P. Joule (1818–1889), an English scientist who performed significant early experiments that related energy forms (Fig. 4.19). Prior to the adoption of the SI, the joule was used exclusively to express mechanical equivalents of energy. A second unit, the calorie, was employed to express heat energy. Today, the joule is the principal unit of energy used in chemistry, but in many other areas of science the calorie, a non-SI unit, is still used. In the past, a calorie was defined as the amount of heat necessary to raise the temperature of exactly one gram of pure water from 14.5°C to 15.5°C. Because the calorie is a rather small unit of heat, the kilocalorie is the unit most commonly encountered. A kilocalorie is equal to 1000 calories. If 1 kcal of heat is transferred to 1 kg of water at 14.5°C, the added heat increases the temperature to 15.5°C.

How does the magnitude of the calorie compare with that of the joule? The calorie is more than four times larger than the joule. By definition, one calorie is exactly equal to 4.184 joules.

$$1 \text{ cal} = 4.184 \text{ J}$$

As you can see, the joule is a relatively small unit of energy. Because the joule is small, it is the kilojoule that is most frequently encountered in chemistry. One kilocalorie is exactly equivalent to 4.184 kilojoules.

$$1 \text{ kcal} = 4.184 \text{ kJ}$$

Example Problem 4.4 shows how kilocalories are converted to joules.

Joule, self-educated, constructed his laboratory in his house. Joule was so dedicated to his work that he took a thermometer with him to make measurements during his honeymoon.

Figure 4.19
James P. Joule. *(National Portrait Gallery, Smithsonian Institution, Washington, D.C.)*

Example Problem 4.4

Convert 34.5 kcal to joules and kilojoules.

Solution

1. What is unknown? J and kJ

2. What is known?   34.5 kcal; 1000 cal/kcal; 1 cal/4.184 J; 1 kJ/1000 J

3. Apply the factor-label method.

$$34.5 \text{ kcal} \times \frac{1000 \text{ cal}}{1 \text{ kcal}} \times \frac{4.184 \text{ J}}{1 \text{ cal}} = ? \text{ J}$$

$$34.5 \text{ kcal} \times \frac{1000 \text{ cal}}{1 \text{ kcal}} \times \frac{4.184 \text{ J}}{1 \text{ cal}} \times \frac{1 \text{ kJ}}{1000 \text{ J}} = ? \text{ kJ}$$

4. Perform the indicated math operations.

$$34.5 \text{ kcal} \times \frac{1000 \text{ cal}}{1 \text{ kcal}} \times \frac{4.184 \text{ J}}{1 \text{ cal}} = \mathbf{1.44 \times 10^5 \text{ J}}$$

$$34.5 \text{ kcal} \times \frac{1000 \text{ cal}}{1 \text{ kcal}} \times \frac{4.184 \text{ J}}{1 \text{ cal}} \times \frac{1 \text{ kJ}}{1000 \text{ J}} = \mathbf{1.44 \times 10^2 \text{ kJ}}$$

CALORIC VALUE OF SELECTED FOODS

Food	Food energy (kcal)
Apple	55
Egg	75
10 potato chips	100
Orange juice, $\frac{1}{2}$ glass	50
10 peanuts	50
1 soda	325
Coffee, plain	0
Beer	200
Cheesecake, 1 slice	275
Chocolate bar with nuts	250

Specific Heat

When substances that are not undergoing state changes are heated, they increase in temperature. The **specific heat** (c) of a substance is the amount of heat required to increase the temperature of one gram of the substance by one degree Celsius. Each substance has its own characteristic specific heat. For example, the specific heat of water is 4.184 J/(g·°C) or 1.00 cal/(g·°C), which means that 4.184 J or 1.00 cal is required to increase the temperature of 1 g of water by 1.00°C. In comparison, the specific heat of copper is 0.38 J/(g·°C), or 0.092 cal/(g·°C). Only 0.38 J or 0.092 cal is required to increase 1 g Cu by 1°C. Less heat is needed to raise the temperature of a quantity of Cu by 1°C than is needed to raise the temperature of an equal quantity of water by the same amount. Substances with higher specific heats require more heat to increase the temperature of equal-mass samples by a fixed amount than do substances with lower specific heats.

The high specific heat of water makes it an excellent coolant.

Copper, a metal with a low specific heat, is placed on the bottom of some cookware to maximize the heat transfer from the heat source to the food.

The specific heat of a substance is calculated using the following equation:

$$c = \frac{q}{g \, \Delta T}$$

in which c is specific heat in joules or calories per gram degree Celsius, q is the heat in joules or calories, g is mass of the substance in grams, and ΔT is the change in temperature $(T_2 - T_1)$, where T_2 is the final temperature and T_1 is the initial temperature.

You should add specific heat to your ever-growing list of conversion factors. Think of specific heat as the conversion factor that equates the quantity of heat transferred to raise 1 g of substance by 1°C or 1 K.

Table 4.8 lists the specific heats of some common substances. They are determined by applying a known quantity of heat to a known mass, and then measuring the increase in temperature. Example Problem 4.5 illustrates the calculations required to find the specific heat of a solid.

TABLE 4.8 SPECIFIC HEAT OF SELECTED SUBSTANCES

Substance	Specific heat	
	J/(g·°C)	cal/(g·°C)
Boron(s)	1.29	0.309
Copper(s)	0.38	0.092
Gold(s)	0.13	0.031
Helium(g)	5.23	1.25
Hydrogen(g)	14.4	3.45
Mercury(l)	0.138	0.331
Sand(s)	0.80	0.19
Silver(s)	0.23	0.056
Sugar(s)	1.3	0.30
Water(l)	4.184	1.000

─────── **Example Problem 4.5** ───────

1.448 kJ of heat is required to raise the temperature of a 215-g sample of an unknown solid from 25.3°C to 44.9°C. Calculate the specific heat of the solid in J/(g·°C).

─────── **Solution** ───────

1. What is unknown? c, the specific heat in J/(g·°C)

2. What is known? $q = 1.448$ kJ; $g = 215$ g; $T_1 = 25.3°C$; $T_2 = 44.9°C$

3. Apply the appropriate equation.

$$c = \frac{q}{g\Delta T} = \frac{q}{g(T_2 - T_1)}$$

$$c = \frac{1.448 \text{ kJ}}{215 \text{ g} \times (44.9°C - 25.3°C)}$$

4. Perform the indicated math operations.

$$c = \frac{1.448 \text{ kJ}}{215 \text{ g} \times 19.6°C}$$

$$= 3.44 \times 10^{-4} \text{ kJ/(g·°C)}$$

Our answer has the units kJ/(g·°C), so we must convert kJ to J to obtain the answer in the desired units.

5. Convert kJ/(g·°C) to J/(g·°C) using the conversion factor 1000 J/kJ.

$$\frac{3.44 \times 10^{-4} \text{ kJ}}{\text{g·°C}} \times \frac{1000 \text{ J}}{1 \text{ kJ}} = \mathbf{0.344 \text{ J/(g·°C)}}$$

The specific heat of the unknown solid is 0.344 J/(g·°C).

The amount of heat q released or absorbed by a substance can be calculated from the specific heat, mass, and change in temperature of the substance. To find the value of q, the specific heat equation is rearranged a follows:

$$c = \frac{q}{g\Delta T}$$

Multiply both sides of the equation by $g\Delta T$ to obtain

$$(g\Delta T) \times c = (g\Delta T)\frac{q}{(g\Delta T)}$$

$$q = g\Delta Tc$$

Thus, to find the amount of heat transferred, the mass of substance is multiplied by its change in temperature and its specific heat. Example Problems 4.6 and 4.7 show how to apply this equation.

─────────── **Example Problem 4.6** ───────────

How many joules of heat are required to increase the temperature of 1.0 kg of silver, Ag, from 22.1°C to 71.9°C?

——————————— **Solution** ———————————

1. What is unknown? J

2. What is known? $g = 1.0$ kg Ag; 1000 g/kg; $\Delta T = 71.9°C - 22.1°C$; $c_{Ag} = 0.23$ J/(g·°C) (from Table 4.8)

3. Apply the equation.

$q = g \times \Delta T \times c$

$= 1.0 \, \cancel{\text{kg Ag}} \times \dfrac{1000 \, \text{g Ag}}{1 \, \cancel{\text{kg Ag}}} \times (71.9°\cancel{C} - 22.1°\cancel{C}) \times \dfrac{0.23 \, \text{J}}{\cancel{\text{g·°C}}}$

The symbol for silver is Ag because the Latin name for silver was argentum. Silver was known to the ancient peoples and is mentioned in the book of Genesis.

The mass of Ag in kg is converted to g so the units will cancel when dividing by g in the specific heat.

4. Perform the indicated math operations.

$q = 1.0 \text{ kg} \times 1000 \, \dfrac{\cancel{\text{g}}}{1 \text{ kg}} \times 49.8°\cancel{C} \times \dfrac{0.23 \, \text{J}}{\cancel{\text{g·°C}}}$

$= 11,454 \text{ J, rounded to } \mathbf{1.1 \times 10^4 \, J}$

The answer, 1.1×10^4 J, is expressed to two significant figures because the c_{Ag} and mass are only known to two significant figures. When multiplying, the answer can have no more significant figures than the measured quantity with the smallest number of significant figures.

——————————— **Example Problem 4.7** ———————————

When 3.55 kJ is added to 375 g sand that is originally at 29.4°C, what is the final temperature of the sand?

——————————— **Solution** ———————————

1. What is unknown?   Final temperature T_2

2. What is known? $q = 3.55$ kJ; 1000 J/kJ; $g = 375$ g; $T_1 = 29.4°C$; $c_{sand} = 0.80$ J/(g·°C) (from Table 4.8)

3. Apply the equation.

$q = g \times (T_2 - T_1) \times c_{sand}$

$3.55 \text{ kJ} \times \dfrac{1000 \text{ J}}{1 \text{ kJ}} = 375 \text{ g} \times (T_2 - 29.4°C) \times \dfrac{0.80 \text{ J}}{\text{g·°C}}$

4. Perform the indicated math operations.

First divide both sides of the equation by 375 g × 0.80 J/(g·°C), and then add 29.4°C to both sides to obtain the final answer.

$$12°C = T_2 - 29.4°C$$

$$T_2 = \mathbf{41°C}$$

The sand reaches a final temperature T_2 of 41°C.

4.25 (a) What is the SI unit of energy? (b) What other energy units are commonly used in chemistry?

4.26 Perform the following conversions:
   (a) 5.09×10^3 J to calories
   (b) 8.9 cal to joules
   (c) 3.087×10^5 J to kilocalories

4.27 What does the specific heat tell you about a substance?

4.28 (a) Calculate the specific heat of a substance that requires 659 J to raise 93.4 g from 15.9°C to 34.6°C. (b) What is the specific heat of the substance expressed in cal/(g·°C)?

4.29 Calculate the amount of heat in joules required to raise 0.984 kg of water from 19.3°C to 84.5°C.

4.30 If 0.312 kJ of heat is added to 543 g of mercury at 0.0°C, what is the final temperature of the mercury? (Table 4.8 gives the specific heat of mercury.)

4.6 CONSERVATION OF MATTER AND ENERGY

One of the most fundamental laws of nature is applied when dealing with matter and energy. This is called the **law of conservation of mass/energy.**

At one time this was stated separately as two laws: the law of conservation of mass and the law of conservation of energy. However, these laws were merged by Albert Einstein (Fig. 4.20), who proposed that matter and energy are equivalent. He showed that this relationship may be expressed by the following equation:

$$\Delta E = \Delta mc^2$$

in which ΔE is the change in energy in joules, Δm is the change in mass in kilograms, and c is the velocity of light, 3.00×10^8 m/s.

The Einstein equation tells us that matter and energy are interconvertible. A tiny amount of matter is converted into a huge quantity of energy in nuclear bombs or nuclear power plants. For example, 1 g of uranium, U, is equivalent to approximately 9×10^{13} J, or 2×10^{13} cal. Looking at this from a different point of view, 1 g U is equivalent to the energy needed to heat 2×10^5 tons of water from 0°C to 100°C.

For most chemical reactions, which are nonnuclear changes, the amount of matter converted to energy is too small to measure on a bal-

In 1905, while working in a patent office in Switzerland, Albert Einstein published three papers. Each paper was worthy of a Nobel prize in physics. Later, in 1921, he was awarded the Nobel prize for one of these works.

ance, and it is safe to say that both matter and energy are conserved. Chemists are concerned with matter-energy conversions principally in nuclear reactions (see Chap. 19).

The **law of conservation of mass** states that there is no detectable change in mass in a chemical reaction. Stated another way: The total mass of the reactants equals the total mass of the products. Let's illustrate by considering what happens when a log is burned in a fireplace. The mass of all gaseous products, soot, dust particles, and ash equals the initial mass of the log burned plus the mass of oxygen required for combustion. If for some reason we find that the masses are not equal, we can be perfectly sure we overlooked something. Matter cannot vanish. If matter is "lost" in one part of the universe, then it is residing somewhere else; it is not really lost.

The **law of conservation of energy** states that energy cannot be created or destroyed in chemical reactions. As we have previously discussed, energy may be converted from one form to another. During these conversions no energy is lost. If there is an apparent energy loss, we have not looked hard enough; the energy is somewhere else in the universe. Generally, energy that cannot be accounted for has escaped as heat.

Heat is unavoidably released in energy transformations. For example, only a small quantity of the electricity pumped into the filament of a light bulb is changed into light; most is released as heat. An efficient engine only converts 10 to 20 percent of the energy stored in the fuel into mechanical energy. Most of the energy released heats the engine and surrounding areas.

Figure 4.20
Albert Einstein. *(The American Friends of the Hebrew University.)*

4.31 How many joules of energy are equivalent to 0.003 mg of matter?

4.32 *(a)* State the law of conservation of matter. *(b)* State the law of conservation of energy.

REVIEW EXERCISES

SUMMARY

Our universe is composed entirely of matter and energy. **Matter** is anything that has mass and occupies space. **Energy** is the capacity to do work.

Physical properties are characteristics of individual substances that can be measured without changing the composition of the substance. Examples of physical properties are melting point, tensile strength, color, and shape. **Chemical properties** describe how the composition of a substance changes when it interacts with other substances or energy forms. Inertness, reactivity, and combustibility are terms employed to describe chemical properties.

The solid, liquid, and gaseous states are the three physical states. **Solids** are the most dense and most viscous. In contrast, **gases** are least dense and least viscous. Solids have a fixed shape and volume; gases expand and take the shape of their containers, and they have a variable volume. The structure of **liquids** more closely resembles that of solids than that of gases. Liquids have a relatively high average density and are incompressible.

Matter is subdivided into two general classes, pure substances and mixtures. **Pure substances** have a constant composition, cannot be separated by physical means, and undergo state changes at a constant temperature. **Mixtures** have a variable composition, can be separated using physical means, and undergo state changes over a wide temperature range.

Pure substances are subdivided into elements and compounds. **Elements** are the most fundamental units of matter. Approximately 110 elements are known. **Compounds** are produced when elements are chemically combined. Elements are composed of small particles called atoms, and compounds are made up of molecules, which are chemical combinations of atoms.

Mixtures can either be homogeneous or heterogeneous. A **homogeneous mixture** is a combination of pure substances that can be separated by physical means, has a variable composition, undergoes state changes over a temperature range, and exhibits only one phase. Homogeneous mixtures are called **solutions. Heterogeneous mixtures** differ from homogeneous mixtures in that they exhibit two or more phases.

Potential energy is stored energy, and results from the position, condition, or chemical composition of an object. **Kinetic energy** is the energy of motion. All things that move possess kinetic energy. The kinetic energy of an object is proportional to its mass and to the square of its velocity.

Energy can be interconverted from one form to another. Whenever energy is interconverted, some of the energy is usually lost as heat. The SI unit of energy is the **joule,** and the non-SI unit is the **calorie.** One calorie is equivalent to 4.184 joules. The quantity of heat flow depends on the difference in temperature of two objects that contact each other. Heat always flows from a hotter object to a cooler one.

Matter and energy are interconvertible—matter can be changed to energy or vice versa. However, in normal chemical changes, both matter and energy are conserved. The same quantity of matter is present after a chemical reaction as was originally present. Matter cannot be created or destroyed. Likewise, energy cannot be created or destroyed.

KEY TERMS

alloy	chemical property	homogeneous	physical property	solution
boiling point	compound	joule	physical change	specific heat
calorie	decomposition	kinetic energy	potential energy	subliming point
change of state	element	matter	product	viscosity
chemical formula	energy	melting point	property of matter	work
chemical symbol	freezing point	mixture	pure substance	
chemical change	heterogeneous	physical state	reactant	

EXERCISES*

4.33 Define each of the following terms: matter, energy, property of matter, physical property, chemical property, reactant, product, physical change, chemical change, decomposition, physical state, viscosity, change of state, melting point, boiling point, freezing point, subliming point, pure substance, mixture, element, compound, chemical symbol, chemical formula, homogeneous, heterogeneous, solution, work, kinetic energy, potential energy, calorie, joule, specific heat.

Physical and Chemical Properties

4.34 Classify each as either a physical or chemical property: *(a)* existence in the solid state, *(b)* magnetic properties, *(c)* explosiveness, *(d)* combustibility, *(e)* flammability, *(f)* boiling point, *(g)* rusting, *(h)* density, *(i)* specific heat, *(j)* undergoes decay, *(k)* viscosity, *(l)* hardness, *(m)* tensile strength, *(n)* reactivity, *(o)* inertness.

4.35 Classify each as a physical or chemical change: *(a)* formation of an ice cube from liquid water, *(b)* frying of an egg, *(c)* fizzing of an Alka-Seltzer tablet, *(d)* gasoline evaporating, *(e)* distillation of alcohol, *(f)* cutting a piece of paper, *(g)* digestion of food, *(h)* corrosion of a metal, *(i)* shaping of steel, *(j)* explosion of a bomb, *(k)* heating of a metal until it is red-hot.

4.36 Consider the following properties of diamond (a pure form of C): *(a)* Good conductor of heat; *(b)* electric insulator; *(c)* density = 3.51 g/cm^3; *(d)* chemically inert; *(e)* extremely hard; *(f)* burns in oxygen to produce carbon dioxide. Classify each of the listed properties of diamond as physical or chemical.

*For exercise numbers printed in color, answers can be found at the back of the book.

4.37 Sulfur is a yellow solid that burns in air to yield poisonous sulfur oxides. On heating, sulfur discolors and turns dark brown at 180°C. Sulfur melts at 115°C and boils at 445°C. Identify all stated properties of sulfur as physical or chemical.

4.38 Cesium, Cs, is a reactive metal that melts at 28°C and has a density of 1.87 g/cm^3. It reacts violently with cold water to produce cesium hydroxide, which is one of the strongest bases known. Cs is a soft, silvery-white solid that is a good conductor of electricity. It combines with chlorine to produce cesium chloride, and combines with oxygen to produce cesium oxide. Classify all of the stated properties of cesium as either physical or chemical. Explain your reasoning in each case.

Physical States

4.39 List four general properties for each of the following: (a) solids, (b) liquids, (c) gases.

4.40 For each of the following pairs, determine which substance has the higher viscosity: (a) water or vegetable oil; (b) motor oil or antifreeze; (c) pudding or soft drink; (d) shaving cream or molasses.

4.41 Identify one or more of the states of matter with each of the following properties: (a) highest average density, (b) lowest viscosity, (c) intermediate densities, (d) constant volume, (e) takes the shape of the bottom of its container, (f) most orderly structure of particles, (g) strongest forces among particles, (h) highest viscosity, (i) random structure of particles, (j) highest average specific heat.

4.42 What accounts for the fact that some substances cannot exist in all three states?

4.43 What physical state is most commonly found under each of the following conditions: (a) very high temperatures and low pressures, (b) very low temperatures and high pressures?

4.44 What type(s) of matter possess the following properties: (a) Has a variable composition with one phase; (b) is inseparable by chemical means; (c) exhibits two or more phases; (d) changes state at constant temperature, and its components can be separated chemically; (e) is composed of uncombined atoms?

Classification of Matter

4.45 Classify each of the following as a pure substance or a mixture: (a) wine, (b) beef, (c) gold bars at Fort Knox, (d) tap water, (e) charcoal, (f) baking soda, (g) sugar cube, (h) paint, (i) water vapor, (j) air, (k) cola drink.

4.46 Write the name for each of the following elements: (a) He, (b) Fe, (c) Li, (d) Se, (e) Ne, (f) Zr, (g) Mg, (h) Be, (i) F, (j) Ce, (k) C, (l) Ca.

4.47 Write the name for each of the following elements: (a) Hg, (b) Zn, (c) W, (d) Xe, (e) Sr, (f) Al, (g) Ge, (h) Kr, (i) Lu, (j) Hf.

4.48 Give the symbols for each of the following elements: (a) nickel, (b) nitrogen, (c) neodymium, (d) neon, (e) niobium, (f) nobelium, (g) neptunium.

4.49 Give the symbols for each of the following elements: (a) indium, (b) silicon, (c) chlorine, (d) potassium, (e) manganese, (f) beryllium, (g) platinum, (h) rubidium.

4.50 Write the names and symbols for all seven elements whose names begin with A.

4.51 Write the names and symbols for all eight elements in the second period of the periodic table.

4.52 (a) What are the four most abundant elements in the earth's crust? What percent of the earth's crust do they compose? (b) What are the four most abundant elements in the whole earth? What percent of the whole earth do they compose?

4.53 Write the names and symbols for all elements in the second chemical group (IIA) of the periodic table.

4.54 What is the difference between a chemical formula and a chemical symbol? Give an example of each.

4.55 State the name and number of each atom in the following formulas: (a) Na_2O, (b) SO_3, (c) N_2O_5, (d) Li_2CO_3, (e) $NaNO_3$, (f) RbH_2PO_4, (g) $Al(OH)_3$, (h) $(NH_4)_2C_2O_4$, (i) $XePtCl_6$, (j) CCl_2Br_2.

4.56 Classify each of the following as an element or a compound from the given information: (a) a substance that melts at 120°C, boils at 228°C, and decomposes to Si and I; (b) a white solid that melts at 44°C and combines with oxygen to form P_2O_5; (c) a soft, silvery metal that reacts violently with water.

4.57 Explain the following statement: "Mixtures have variable compositions."

4.58 Classify the following as homogeneous or heterogeneous mixtures: *(a)* brass, *(b)* coffee, *(c)* cement, *(d)* motor oil, *(e)* cotton, *(f)* paper, *(g)* oil and vinegar, *(h)* smog.

4.59 Explain each step that is required to separate the following mixtures: *(a)* sand and salt; *(b)* alcohol and water; *(c)* sand and sugar; *(d)* oil and water.

4.60 How many phases are observed in each of the following (not including the container and air): *(a)* glass of iced tea; *(b)* bottle of seawater and oil residue; *(c)* aquarium with four different colors of sand; *(d)* glass of soda water with ice cubes?

Energy

4.61 What are the three ways that potential energy is stored? Give an example for each.

4.62 *(a)* What two factors are directly related to an object's kinetic energy? *(b)* How is the kinetic energy of an object calculated?

4.63 Kinetic energy in joules is calculated from the mass of a body in kilograms and its velocity in meters per second. *(a)* If a body has a mass of 5.23 kg and a velocity of 2.25 m/s, what is the kinetic energy of the body? *(b)* What is the kinetic energy of the body if its velocity decreases to 1.13 m/s? *(c)* What is the kinetic energy of the body if its velocity is 5.00 m/s?

4.64 Calculate the velocity of a body that has a mass of 8.23×10^3 g and a kinetic energy of 9.97 kJ.

4.65 What type(s) of potential energy are possessed by each of the following: *(a)* TNT, *(b)* apple on a tree, *(c)* mainspring of a watch that has just been wound, *(d)* stretched rubber band, *(e)* water at the top of a dam, *(f)* hamburger?

4.66 For each of the following pairs, determine which can transfer the largest quantity of heat: *(a)* match flame or bunsen burner flame; *(b)* cup of water at 90°C or bathtub filled with 90°C water; *(c)* ice cube at 0°C or a large block of ice at 0°C; *(d)* teaspoon of boiling water or gallon of water at 50°C; *(e)* two identical wooden blocks in contact with each other.

4.67 Perform the indicated energy conversions:
(a) 9.21 cal = ? J
(b) 7.09 J = ? cal
(c) 2.168×10^4 J = ? kcal
(d) 8.1×10^7 cal = ? kJ
(e) 0.00535 cal = ? kJ

Specific Heat

4.68 Calculate the specific heat of a substance if 812 J is required to raise the temperature of a 45.8-g sample by 2.11°C.

4.69 Calculate the specific heat of a substance if the addition of 521 J increases the temperature of a 217-g sample from 8.32°C to 15.9°C.

4.70 How much heat in joules is required to raise the temperature of 8.43 kg of water from 29.8°C to 54.1°C?

4.71 How many calories of heat are released when 8.19 kg of water at 75.0°C cools to 22.7°C?

4.72 *(a)* How many joules and how many calories are required to increase a 9.23-kg sample of hydrogen gas from −52.9°C to 25.0°C? *(b)* How many joules and how many calories are required to raise the temperature of a 9.23-kg sample of helium gas from −52.9°C to 25.0°C?

4.73 Calculate and compare the amount of heat released in joules when 1.00 kg of water and when 1.00 g of water, both at 100.0°C, cool to 25.0°C.

4.74 The specific heat of gold, Au, is 0.13 J/(g·°C). *(a)* How much heat is required to increase 175 g Au from 15.0°C to 40.0°C? *(b)* Compare this quantity of heat with the amount of heat that would be required to increase the temperature of an equal mass of water through the same temperature range.

4.75 The initial temperature of a 29.3-g sample of copper is 15.4°C. What is the final temperature after 9.21×10^3 J of heat is transferred to it?

4.76 What is the final temperature of a 1.25-kg sample of helium initially at 0.0°C, if 3.34 kJ is transferred to it?

Conservation Laws

4.77 *(a)* If 12 g of carbon is exactly combined with 32 g of oxygen, how many grams of carbon dioxide form? *(b)* What law does this illustrate?

4.78 What energy transformations occur when electricity is produced through hydroelectric generation?

4.79 What mass in milligrams is equivalent to each of the following energies: *(a)* 6.0×10^9 J, *(b)* 5.5×10^{35} J, *(c)* 1.0 J?

4.80 What energy in joules is equivalent to each of the following masses: *(a)* 0.0015 g, *(b)* 12.9 mg, *(c)* 1.50 g?

Additional Exercises

4.81 An object has a kinetic energy of 5.78×10^4 J and a mass of 0.985 kg. What is the velocity of the object?

4.82 How many atoms of each type are indicated by each of the following formulas: (a) $Ca_3(PO_4)_2$, (b) $Al(C_2H_3O_2)_3$, (c) $(NH_4)_2Cr_2O_7$?

4.83 Write the names and symbols for elements 57 to 81.

4.84 (a) If all of the energy from the food that we consume in a day could be transferred to water and if a person eats a total of 3000 kcal per day, what mass of cold water at 0.0°C would be heated to 50.0°C by one day's food? (b) How many liters of water would this be? (Assume the density of water is 1.0 g/cm³.)

4.85 Describe the temperature change and heat flow when liquid oxygen, $O_2(l)$, at −183°C comes in contact with an equal mass of liquid nitrogen, $N_2(l)$ at −196°C.

4.86 A 30.0-g sample of an alloy was heated to 100.0°C and then placed in a beaker that contained 123.8 g of water. The temperature of the water increased from 23.19°C to 26.88°C. If all of the heat from the metal was transferred to the water, what is the specific heat of the alloy?

4.87 Freon is a gas used as a heat transfer agent in cooling systems. The specific heat of freon is 0.543 J/(g·°C). What mass of freon would absorb 34.5 kJ while increasing in temperature from −10.0°C to 30.0°C?

4.88 Closely related to specific heat is the heat capacity of a substance. Heat capacity is defined as the amount of heat required to raise the temperature of a given mass of a substance by one degree Celsius. What is the heat capacity in joules per degree Celsius of each of the following: (a) 449 g of water, (b) 9.22 kg of mercury, (c) 1.99 g of boron?

4.89 How much heat is required to increase the temperature of a cube of Pt that has an edge length of 3.28 cm from 11.7°C to 65.8°C? [The density of Pt is 21.45 g/cm³, and its specific heat is 0.133 J/(g·°C).]

4.90 A swimming pool is 3.25 m wide, 6.50 m long, and has an average depth of 2.50 m. Calculate the number of joules of heat required to increase the temperature of the water from 18.5°C to 30.4°C, assuming that the pool is completely filled and the density of the water is 1.00 g/cm³.

4.91 Sterling silver is 92.5% Ag and 7.5% Cu. Calculate the amount of heat in joules required to increase the temperature of 0.438 kg of sterling silver from 24.9°C to 100.0°C.

CHAPTER
Five

Atoms

STUDY GUIDELINES

After completing Chapter 5, you should be able to

1. List the main principles of Dalton's atomic theory, and explain how the Dalton model of the atom differs from the quantum mechanical model

2. Discuss how J. J. Thomson and Ernest Rutherford changed the model of the atom

3. List the properties of the fundamental components of an atom

4. Determine the composition of an atom given the atomic and mass numbers

5. Write the complete symbol of an atom given the number of protons and neutrons

6. Identify isotopes of an element

7. Calculate the atomic mass of an element given the masses of its isotopes and their natural abundances

8. Explain the importance of the unified atomic mass scale

9. Describe the Bohr model of the H atom

10. Discuss the quantum mechanical model of the atom

11. Describe what is meant by an electron orbital

12. Distinguish among electron energy levels, sublevels, and orbitals

13. List the maximum number of electrons found in energy levels, sublevels, and orbitals

14. Write the electronic configuration, Lewis symbol, and orbital diagram for a given atom

15. Write the outer-level (valence) electron configuration for an atom

O ur contemporary theory regarding atoms has evolved over the last 200 years. Theory development is a dynamic process. Whenever new, solid evidence is uncovered that contradicts a theory, the theory must be flexible enough to incorporate the new data. Theories are not fact; they are the best explanation we can offer in light of known information.

The historical development of the modern atomic theory is an unusually interesting story which begins with **John Dalton** (1766–1844), an

5.1 DEVELOPMENT OF THE MODERN ATOMIC THEORY

English chemist (Fig. 5.1). He is credited with developing the first scientific theory of atoms. Dalton, unlike the Greek philosophers (Sec. 1.2), based his theory on approximately 150 years of investigations by scientists.

Dalton's atomic theory stated that

1. Matter is composed of small, solid, spherical particles called atoms.

2. Atoms of one element have the same properties, but differ from atoms of all other elements.

3. Atoms cannot be subdivided or changed to other atoms.

4. Atoms combine chemically in simple, whole-number ratios.

5. Chemical changes involve the linkage and separation of atoms.

Dalton's model of the atom was accepted for 100 years without a serious challenge. The longevity of Dalton's atomic theory is attributed to its ability to support two fundamental laws of nature; the law of conservation of matter—matter cannot be created or destroyed—and the law of constant composition, which states that all compounds are composed of elements in a fixed proportion by mass. In spite of this, scientific research during the late nineteenth and early twentieth centuries significantly changed our ideas about the structure of atoms.

Research into the nature of electricity showed that the atom was not just a solid sphere but was composed of even smaller particles. The first particles to be discovered were the proton and the electron. As we will soon discuss in detail, the proton is a positively charged body and the electron is a negatively charged body. Studies concerning the properties of these particles were conducted in discharge tubes. Figure 5.2 shows a diagram of a discharge tube. Scientists found that if they first pumped out much of the air in the tube, the tube glowed when a voltage was applied. Later it was found that negative particles, which were called cathode rays (electrons), passed from the negative electrode, the cathode, to the positive electrode, the anode. This and similar studies led

Figure 5.1
John Dalton. *(National Portrait Gallery, Smithsonian Institution, Washington, D.C.)*

Dalton was also a meteorologist, and he wrote a book entitled Meteorological Observations and Essays. *Dalton also discovered color blindness, an affliction he had.*

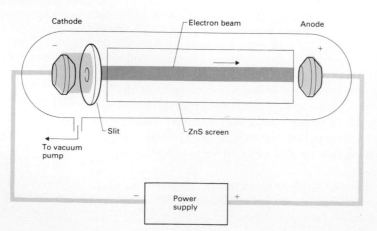

Figure 5.2
A discharge tube is composed of an evacuated glass container that has two electrodes. The negative electrode is called the cathode and the positive electrode is called the anode. Electrons travel from the cathode to the anode. The electron beam can be seen when it strikes the zinc sulfide screen (ZnS), causing it to glow.

English physicist **J. J. Thomson** (1856–1940) to propose in 1898 what is now known as the **"plum pudding"** model of the atom. Thomson envisioned a structure in which negatively charged electrons were embedded in an atom that had a uniform positive charge from the protons (Fig. 5.3).

By 1911 evidence was collected that allowed scientists to modify the Thomson model of the atom. **Ernest Rutherford** (1871–1937) performed one of the most classic experiments of scientific history. He bombarded different types of matter with high-energy, positively charged alpha particles. We will discuss the properties of these particles in the next section. Based on the Thomson model of the atom, Rutherford hypothesized that the alpha particles should go through very thin metal foils undeflected. However, after performing an experiment on gold leaf, which is very thin gold foil, he found that a small but significant number of the alpha particles were deflected through large angles by the gold atoms (Fig. 5.4a). These results showed Rutherford that the Thomson model of the atom was not valid. A uniform positive charge with embedded negatively charged electrons would not interact with the alpha particles in such a manner. Rutherford proposed that the atom has a small, positively charged nucleus which repels the alpha particles and produces the large deflected angles (Fig. 5.4b). In other words, Rutherford gave us the **nuclear model** of the atom.

While Rutherford showed us that the atom has a nucleus, he did not know how the electrons were arranged outside the nucleus. In 1913, the Danish physicist Neils Bohr (1885–1962) proposed an atomic model in which the electrons traveled around the nucleus in circular paths called orbits. His model was replaced with the current quantum mechanical model of the atom proposed by Erwin Schrödinger. We will discuss both of these models of the atom when we consider the electron configuration of atoms in Sec. 5.3. The discovery of a third particle in the atom did not occur until 1932. In that year, James Chadwick, an English scientist, discovered the neutral (uncharged) neutron.

Most work since 1930 has been directed toward elucidating the composition of the subatomic particles. Little has changed in the minds of scientists regarding overall atomic structure since the 1930s, a fact which suggests that our current atomic model is a valid one. Figure 5.5 shows the evolution of atomic models from Dalton to the present. In the next section, we will consider the properties of protons, neutrons, and electrons, as well as the composition of atoms.

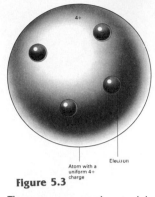

Figure 5.3

Thomson proposed a model of the atom in which electrons were embedded in a matrix of positive charge. This figure shows a Be atom which has four protons and four electrons.

The famous experiment in which Ernest Rutherford, a New Zealander who lived in England, discovered the nucleus is called the gold leaf experiment. His assistant for this experiment, Hans Geiger, invented the radiation-detecting device that bears his name.

REVIEW EXERCISES

5.1 Why is an atomic theory required to describe atoms?

5.2 List the most significant points of Dalton's atomic theory.

5.3 *(a)* Describe the Thomson model of the atom. *(b)* Describe the Rutherford model of the atom. *(c)* On what evidence did Rutherford base his model of the atom?

5.4 What are the names of the three particles that compose an atom?

Atoms

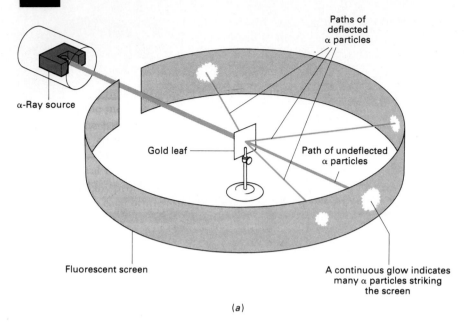

Paths of deflected α particles

α-Ray source

Gold leaf

Path of undeflected α particles

Fluorescent screen

A continuous glow indicates many α particles striking the screen

(a)

Gold nucleus

X-Ray beam

(b)

Figure 5.4

(a) Rutherford detected the alpha rays with a fluorescent screen that glowed when struck by radiation. Most of the alpha rays went through the gold leaf undeflected, but a significant number were deflected through large angles. (b) Only the alpha particles that passed close to the gold nuclei were deflected through large angles.

**Quantum mechanical model
(current model)**

In 1923, Schrödinger proposed a wave equation from which atomic orbitals are derived. In 1932, Chadwick discovered the neutron, a second major particle in the nucleus.

Figure 5.5
A summary of the development of the atomic theory.

Bohr-Sommerfeld (1916)

Modification of the Bohr model, placing electrons in elliptical orbits.

Bohr model (1913)

First quantum model of the atom with the electrons following circular orbits around the nucleus.

Nuclear model (1911)

Rutherford discovered that the atom possessed a small dense core called the *nucleus*.

Thomson model (1898)

"Plum pudding" model of the atom, with electrons as the "plums" in a matter "pudding."

Dalton model (1803)

Atoms as solid indestructible spheres.

Atoms are composed of three fundamental particles—protons, neutrons, and electrons. These particles are characterized by their mass and electric charge (Table 5.1). Protons and neutrons have approximately the same mass, about 1.67×10^{-24} g. The mass of an electron is only 9.11×19^{-28} g, about $\frac{1}{1837}$ that of a proton (or neutron); in other words, 1837 electrons are needed to equal the mass of one proton.

5.2 ATOMIC SUBSTRUCTURE

Protons, Neutrons, and Electrons

TABLE 5.1 PROPERTIES OF SUBATOMIC PARTICLES

Particle	Symbol	Mass, g	Mass, u	Relative charge
Proton	p^+	1.6726×10^{-24}	1.007276	1+
Neutron	n^0	1.6749×10^{-24}	1.008666	0
Electron	e^-	9.1096×10^{-28}	0.0005486	1−

Masses of subatomic particles are frequently expressed using a relative unit, termed a **unified atomic mass unit,** u. Later, we shall discuss how this unit is derived. For now, we can say that one unified atomic mass unit is approximately the mass of a proton (or neutron). The mass of an electron on this scale is only 0.000549 u.

$1\,u = 1.6606 \times 10^{-24}g$

Bodies can have a positive (+), negative (−), or no net charge (neutral). Electrons and protons have the smallest elementary unit of charge found in matter. Electrons are negatively charged (1−); protons possess the same magnitude of charge as electrons, but their charge is positive

(1+). Neutrons, as the name implies, are electrically neutral particles.

From physics, we find that particles with the same electric charge repel each other. Objects with unlike charges attract each other (Fig. 5.6). Two electrons or two protons in close proximity repel each other, or push each other apart. Unlike charged particles (+ and −), when brought close together, attract each other. The force of electric attraction or repulsion is inversely related to the square of the distance that separates the particles. This relationship is known as *Coulomb's Law* and is expressed as follows:

$$F = k\frac{q_1 q_2}{r^2}$$

In this equation q_1 and q_2 are the magnitudes of the charges, k is Coulomb's constant, r is the distance that separates the particles, and F is the force either of attraction or repulsion. Neutral particles do not interact with charged particles.

Protons and neutrons are located in a very small region of the atom called the **nucleus** (plural, *nuclei*). Most nuclei have diameters of roughly 10^{-6} nm (1 nm = 1×10^{-9} m). Diameters of whole atoms are many times ($100,000\times$) larger than those of nuclei, ranging from 0.1 to 0.5 nm. The electrons are found in the relatively vast space outside the nucleus.

Nuclear density is incredibly large; virtually all of the mass of an atom (protons and neutrons) is concentrated in the infinitesimal volume of the nucleus. Nuclear densities, roughly, are 100,000,000 tons/mL!

To summarize atomic structure: The atom consists of a small, dense core, the nucleus, surrounded by minute particles, electrons, which occupy an immense, mainly empty, region of space. Atoms are not solid forms of matter; they are sparsely populated with matter.

Chemists are concerned with the number of protons and neutrons located in the nucleus. Two values are needed to determine the composition of a nucleus. One is called the atomic number, and the other is the mass number. The **atomic number** of an atom equals the number of positive charges found in the nucleus of that atom, or stated another way, the number of protons in the nucleus.

Atomic number = number of protons in the nucleus of an atom

Hydrogen, the simplest atom, has an atomic number of 1. Given its atomic number, we know that a hydrogen atom has one proton in its nucleus. The atomic number of helium is 2; thus, two protons are found in the nucleus of a helium atom. Lithium nuclei each possess three protons; accordingly, the atomic number of lithium atoms is 3. If you look at the periodic table located inside the front cover, you will find that the integer (whole number) next to the symbol of each element is the atomic

Repulsion of like charges

Attraction of unlike charges

Figure 5.6
Two particles with the same electric charge repel each other while two particles with unlike electric charges attract each other.

The Nucleus

Electric charge is measured in coulombs.

Nuclear Properties of Atoms

H. G. J. Moseley (1887–1915), a student of Rutherford, was the first to measure the charge on the nucleus—its atomic number. He accomplished this by reflecting x-rays off the surfaces of metals.

number. Atoms are arranged in the periodic table in order of increasing atomic number.

In addition to the atomic number, the **mass number** of an atom is required to find the composition of the nucleus of an atom. The mass number of an atom equals the total number of protons and neutrons in the nucleus.

$$\text{Mass number} = \text{number of protons} + \text{number of neutrons}$$

You should note that mass numbers are just numbers and are not masses.

To express the atomic number and mass number of an atom, atomic symbols are written as follows:

$$^{A}_{Z}X$$

in which X is the symbol of the atom, A is the mass number, and Z is the atomic number.

To illustrate the writing of atomic symbols let's write the atomic symbol for a krypton atom that has 36 protons and 48 neutrons in its nucleus. The atomic number of Kr is 36 because it has 36 protons. Its mass number is 84 because the sum of the protons plus neutrons is 84 ($36 \text{ p}^+ + 48 \text{ n}^0$). After writing the symbol of krypton (Kr), we write 36 as a subscript to the left of the symbol and 84 as a superscript, also to the left of the symbol.

$$^{84}_{36}\text{Kr}$$

To find the number of neutrons in a nucleus, given the mass number and atomic number, we subtract the atomic number (number of p^+) from the mass number (number of p^+ plus number of n^0).

$$\text{Number of n}^0 = \text{mass number} - \text{atomic number}$$
$$= (\text{number of p}^+ + \text{number of n}^0) - \text{number of p}^+$$

Mass numbers are not found in the periodic table, so they will be given whenever they are needed.

Let's consider the hydrogen atom, and find the complete composition of its nucleus. Most hydrogen atoms have a mass number equal to 1 and an atomic number also equal to 1; they are written $^{1}_{1}\text{H}$.

$$\text{Number of n}^0 = \text{mass number} - \text{atomic number}$$
$$\text{Number of n}^0 \text{ in } ^{1}_{1}\text{H} = 1 - 1 = 0$$

Consequently, $^{1}_{1}\text{H}$ atoms have one proton and zero neutrons in their nuclei. The $^{1}_{1}\text{H}$ atom is the only atom that does not have a neutron in the nucleus.

What is the nuclear condition of $_{26}^{56}\text{Fe}$ atoms? The symbol tells us that the atomic number is 26; hence, 26 protons are found in this atom. To find the number of neutrons, we subtract the atomic number, 26, from the mass number, 56, to obtain 30 neutrons.

$$\text{Number of } n^0 = \text{mass number} - \text{atomic number}$$

$$= 56 - 26$$

$$= 30$$

$_{26}^{56}\text{Fe}$ atoms have 26 protons and 30 neutrons in their nuclei. Example Problem 5.1 is another illustration of how to find the composition of a nucleus.

Example Problem 5.1

Find the composition of the nucleus of the following atom.

$$_{88}^{226}\text{Ra}$$

Solution

In our example, the mass number is 226, and the atomic number is 88. The mass number is the number of protons and neutrons, and the atomic number is the number of protons. Therefore, there are **88 protons** in the nucleus. The number of neutrons is calculated by subtracting the atomic number from the mass number.

$$\text{Number of } n^0 = \text{mass number} - \text{atomic number}$$

$$= 226 - 88$$

$$= \textbf{138}$$

^{226}Ra contains 88 protons and 138 neutrons in its nucleus.

Marie Curie first isolated Ra in 1911. Ra causes severe biological effects if inhaled or ingested because of the intense radiation that it releases.

REVIEW EXERCISES

5.5 What happens when particles with (a) the same and (b) different charges interact with each other?

5.6 What can be determined about an atom given its (a) atomic number, (b) mass number, (c) atomic mass, (d) atomic number and mass number?

5.7 What are the atomic numbers of each of the following elements: (a) Ca, (b) Ag, (c) U, (d) Lr?

5.8 How many protons and neutrons are found in the nucleus of (a) ^{48}Ti, (b) ^{85}Rb, (c) ^{227}Ac, (d) ^{195}Pt?

Whenever atoms have the same atomic number but different mass numbers, they are referred to as **isotopes.** Stated differently: Isotopes are atoms with the same number of protons but different numbers of neutrons in their nuclei.

A small percentage of naturally occurring hydrogen atoms are not $_1^1H$ atoms; they have a different nuclear composition. These hydrogen atoms have a mass number of 2 and an atomic number of 1.

$$\text{Number of neutrons in } {}_1^2H = 2 - 1 = 1n^0$$

Each of these atoms has one proton and one neutron in its nucleus. A third hydrogen isotope exists, with a mass number of 3, $_1^3H$. In these atoms two neutrons and a single proton are found in the nucleus. Consequently, $_1^1H$, $_1^2H$, and $_1^3H$ are all said to be isotopes of hydrogen.

The three isotopes of hydrogen are expressed as follows:

$$_1^1H \qquad\qquad _1^2H \qquad\qquad _1^3H$$
$$\text{Protium} \qquad \text{Deuterium} \qquad \text{Tritium}$$

$_1^1H$ is called regular hydrogen or protium and is the most abundant type of hydrogen. $_1^2H$ is called deuterium or heavy hydrogen. Deuterium has a mass approximately twice that of regular hydrogen because it has both a proton and a neutron in its nucleus. Tritium, $_1^3H$, the heaviest form of hydrogen, is not found in detectable amounts in naturally occurring samples of hydrogen. Figure 5.7 shows the nuclear composition of the three hydrogen isotopes.

A large percent of the elements are composed of mixtures of different isotopes. For example, three isotopes are found in naturally occurring samples of uranium: $_{92}^{234}U$, $_{92}^{235}U$, and $_{92}^{238}U$. Each of these uranium isotopes contains 92 protons in the nucleus. They differ with respect to the number of neutrons in the nucleus. $_{92}^{234}U$ nuclei contain 142 neutrons, $_{92}^{235}U$ nuclei contain 143 neutrons, and $_{92}^{238}U$ nuclei contain 146 neutrons. Of the three, $_{92}^{238}U$ is most abundant in natural samples, repre-

Isotopes

Iso is a prefix that means "the same."

All of the isotopes of uranium are radioactive. Each releases alpha radiation, and some undergo spontaneous fission—the nucleus splits.

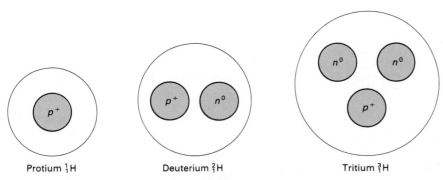

Figure 5.7
Nuclei of regular hydrogen atoms (^1H) contain one proton and no neutrons. Nuclei of deuterium atoms (^2H) contain one proton and one neutron, and tritium atoms (^3H), the most massive isotope of hydrogen, contain nuclei with one proton and two neutrons.

Protium $_1^1H$ Deuterium $_1^2H$ Tritium $_1^3H$

Hydrogen nuclei

TABLE 5.2   NATURAL ISOTOPIC COMPOSITION OF SELECTED ELEMENTS

Atomic number	Isotope	Number of protons	Number of neutrons	Natural abundance, %	Mass, u
6	$^{12}_{6}C$	6	6	98.89	12.000
	$^{13}_{6}C$	6	7	1.11	13.003
8	$^{16}_{8}O$	8	8	99.76	15.9949
	$^{17}_{8}O$	8	9	0.04	16.9991
	$^{18}_{8}O$	8	10	0.20	17.9992
29	$^{63}_{29}Cu$	29	34	69.09	62.9298
	$^{65}_{29}Cu$	29	36	30.91	64.9278
32	$^{70}_{32}Ge$	32	38	20.51	69.9243
	$^{72}_{32}Ge$	32	40	27.43	71.9217
	$^{73}_{32}Ge$	32	41	7.76	72.9234
	$^{74}_{32}Ge$	32	42	36.54	73.9219
	$^{76}_{32}Ge$	32	44	7.76	75.9214

senting 99 percent of the total number of atoms. Consequently, the other two isotopes represent less than 1 percent of the atoms. $^{235}_{92}U$ is used as the fuel in nuclear power plants (Chap. 19). Table 5.2 gives more examples of natural isotopic mixtures.

Masses of Individual Atoms

An atom has a very small mass. A hydrogen atom, for example, has a mass of only 1.67×10^{-24} g. To avoid the inconvenience of working with such small numbers, chemists use a relative scale for the masses of individual atoms. On this scale, masses of all atoms are expressed relative to the mass of one $^{12}_{6}C$ atom.

By definition, the mass of one $^{12}_{6}C$ atom is equal to exactly 12 unified atomic mass units. Thus, **one unified atomic mass unit** is one-twelfth the mass of the $^{12}_{6}C$ atom. Because $^{12}_{6}C$ is composed of six protons and six neutrons, 1 u is about the average mass of a proton or a neutron.

The masses of other atoms are determined relative to the standard, $^{12}_{6}C$. For example, if an atom is found to have a mass three times that of $^{12}_{6}C$, its relative mass is 36 u, 3 times 12 u. An atom with a mass one-fourth the mass of $^{12}_{6}C$ is assigned a mass of 3 u, one-fourth of 12 u. When the mass of an individual atom is given, it should be thought of relative to the mass of a $^{12}_{6}C$ atom.

Atomic Mass (Atomic Weight)

Atomic masses are the numbers listed in each block in the periodic table along with the symbol and atomic number of an element. Notice that the atomic masses of the elements are decimal numbers (e.g., Al, 26.9815; S, 32.06; V, 50.942); none are integers like the atomic number. Why is this?

A large percent of the naturally occurring elements exist as a mixture of isotopes, each isotope with a different mass. Because chemists work with a large quantity of atoms that have different masses, they usually find it convenient to use the average mass of an element's isotopes, which is called the atomic mass. The **atomic mass** of an element is the average mass of its naturally occurring isotopes relative to the mass of $^{12}_6C$.

Consider the atomic mass of carbon, which is 12.011 u. If you refer to Table 5.2, you will find that carbon is composed principally of two isotopes, $^{12}_6C$, the one we just referred to as the standard for the atomic mass scale, and $^{13}_6C$. In nature, approximately 9889 out of 10,000 carbon atoms are $^{12}_6C$, and 111 are the heavier $^{13}_6C$ isotope. If we average 9889 particles with a mass of 12.00 u and 111 particles with a mass of 13.00 u, the average mass is 12.01 u. How is this number calculated? First we must consider how a *weighted average* is calculated.

Weighted averages are calculated in a similar way to any other average: We add each of the values and divide by the total number of values. For example, if a test is given to 100 students, and 60 students get a score of 80 points while the remaining 40 students score 95 points, the average score is 86 points. The weighted average of scores is obtained in the following way. First, multiply 60, the number of students with 80 points, times their score, 80 points. Second, multiply 40, the number of students with 95 points, times their score, 95. Finally, add these two products and divide by the total number of students.

$$\text{Average} = \frac{(60 \text{ students} \times 80 \text{ pts}) + (40 \text{ students} \times 95 \text{ pts})}{100 \text{ students}}$$

$$= 86 \text{ pts}$$

A similar calculation allows us to find the atomic mass of chlorine once we know the natural abundance of the two principal isotopes, $^{35}_{17}Cl$ and $^{37}_{17}Cl$. In natural samples of the element chlorine, 75.53% is $^{35}_{17}Cl$ with a mass of 34.969 u. The remaining 24.47% is $^{37}_{17}Cl$, which has a mass of 36.966 u. To obtain the atomic mass of chlorine, the weighted average of the masses of the two isotopes, assume that you have 100 atoms. Of these 100 total atoms, 75.53 have a mass of 34.969 u, and 24.47 have a mass of 36.966 u. Therefore, multiply 75.53 times 34.969 u, add that to the product of 24.47 times 36.966 u, and divide by the total number of atoms, 100.

$$\text{Atomic mass of Cl} = \frac{(75.53 \times 34.969 \text{ u}) + (24.47 \times 36.966 \text{ u})}{100 \text{ atoms}}$$

(handwritten annotations: Natural Abundance, mass, Natural Abundance, mass, atoms)

$$= 35.46 \text{ u}$$

Thus the atomic mass of chlorine is 35.46 u.

Example Problem 5.2 is a second illustration of how to calculate the atomic mass of an element.

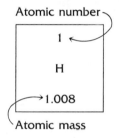

Atomic number

1

H

1.008

Atomic mass

Example Problem 5.2

Calculate the atomic mass of boron, using the following data:

Isotope	Relative mass, u	Percent abundance
$^{10}_{5}B$	10.013	19.70
$^{11}_{5}B$	11.009	80.30

Solution

The atomic mass of B is obtained by calculating the average mass of its naturally occurring isotopes:

$$\text{Atomic mass of B} = \frac{(10.013 \text{ u} \times 19.70) + (11.009 \text{ u} \times 80.30)}{100}$$

$$= \textbf{10.81 u}$$

Percents indicate the number of parts per 100 total parts. In this calculation, for every 100 B atoms, 19.70 have a mass of 10.013 u, and 80.3 have a mass of 11.009 u. Thus, 19.70 is multiplied by 10.013 u, and 80.3 is multiplied by 11.009 u. The two resultant quantities are added, and the sum is divided by 100, giving the average, 10.81 u.

REVIEW EXERCISES

5.9 (a) What are isotopes? (b) List the naturally occurring isotopes of germanium, Ge (Table 5.2). (c) Which isotopes of germanium are the most and least abundant in nature?

5.10 (a) Define atomic mass. (b) How is the atomic mass of an element calculated?

5.11 Calculate the atomic mass of thallium using 202.97 as the mass of $^{203}_{81}Tl$ and 204.97 as the mass of $^{205}_{81}Tl$ given that the natural abundances of $^{203}_{81}Tl$ and $^{205}_{81}Tl$ are 29.50% and 70.50%, respectively.

Radioactivity

Nuclei of certain atoms are unstable and undergo spontaneous changes that result in a particle or ray being emitted from the nucleus at high speed. Such nuclei are said to be **radioactive** because they emit radiation. The three principal types of nuclear emissions are alpha (α) rays, beta (β) rays, and gamma (γ) rays.

Alpha rays are composed of the most massive particles of the three basic forms of radiation. An alpha particle has a mass of 4 u and a charge of 2+. Beta rays are composed of electrons that are ejected from the nucleus at high velocities. Gamma rays are one type of electromagnetic radiation. Other examples of electromagnetic radiations include x-rays, ultraviolet light, visible light, and infrared. These types of radiation exhibit properties of waves and have no measurable mass or charge.

Henri Becquerel accidentally discovered radioactivity in 1896 when he found that a uranium salt emitted rays that fogged his photographic plates.

Table 5.3 summarizes the properties of these three forms of radioactive emission.

TABLE 5.3 PROPERTIES OF THREE TYPES OF RADIATION

Name	Symbol	Mass, u	Relative charge	Penetration power
Alpha	α	4	2+	Low
Beta	β	$\frac{1}{1837}$	1−	Intermediate
Gamma	γ	0	0	High

Types of electromagnetic radiation include radio waves, microwaves, infrared, visible light, ultraviolet, x-rays, and gamma rays.

Another distinguishing characteristic of the three types of radiation is their *penetration power,* their ability to penetrate and travel through matter. Gamma rays have the largest capacity to penetrate matter; in other words, they have the highest penetration power. Most gamma rays easily pass through wood and various thin metals. About 100 mm of lead or other dense type of matter is needed to provide adequate shielding from gamma rays (Fig. 5.8). Beta rays are less penetrating than gamma rays. A 1 mm thick metal barrier will generally stop most beta rays. Alpha particles have the lowest penetration power. Most alpha rays are shielded by a 0.01 mm thickness of metal. In many cases a piece of paper, clothing, or your skin is an adequate shield against alpha rays.

The greater the mass and charge of a radioactive particle, the more it interacts with matter and the lower its penetration power.

5.12 *(a)* What are the three principal types of radioactivity? *(b)* What is the mass and charge of each of these forms of radioactivity?

5.13 *(a)* What is meant by penetration power? *(b)* Which form of radiation has the greatest capacity to penetrate matter?

REVIEW EXERCISES

So far we have learned that electrons have a very small mass with respect to neutrons and protons, have a negative charge, and are found somewhere outside the nucleus. Because all atoms are electrically neutral, the number of electrons in an atom equals the number of protons (atomic number).

5.3 ELECTRONIC CONFIGURATIONS OF ATOMS

Electrons

$$\text{Number of } e^- = \text{number of } p^+ = \text{atomic number}$$

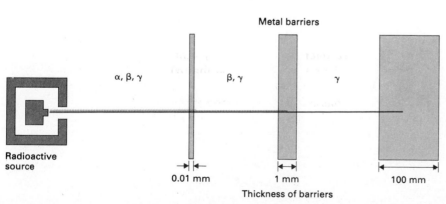

Figure 5.8
Alpha rays have the lowest penetration power of the three principal types of radiation. They are effectively blocked by a 0.01-mm barrier of metal. Beta rays have a higher penetration power and can penetrate the barrier that blocks alpha rays. Most beta rays are stopped by a 1-mm thickness of metal. Gamma rays are the most penetrating. A barrier of about 100 mm of metal is required to provide adequate shielding from gamma rays.

It is only necessary to look at the periodic table to find the number of electrons in an atom: e.g., H, 1 e⁻; C, 6 e⁻; Ne, 10 e⁻; and U, 92 e⁻.

Neils Bohr (1885–1962) (Fig. 5.9), a Danish physicist, proposed a model of the atom in 1913 that was the forerunner of the current atomic model. He was principally concerned with the structure of the hydrogen atom, the simplest of all atoms, and how it related to the light released when the atom was heated to high temperatures. He visualized the H atom as having a nucleus with a series of circular orbits in which the electron traveled. Figure 5.10 shows a picture of the Bohr model of the atom. Bohr theorized that the electron in a H atom could only travel around the nucleus in these orbits and could never be found in the regions between the orbits.

The Bohr model of the atom was a very significant advance in the understanding of the structure of atoms. It incorporated the ideas of Max Planck (1858–1947), who proposed in 1900 that energy is not emitted in a continuous manner, but is released in discrete packets called *quanta*. Planck's proposal provided the foundation for one of the central theories of present-day science—**quantum theory.** As an analogy to the quantum theory let us compare a person who walks up stairs to a person who walks up a ramp (Fig. 5.11). A person who walks up stairs can move only in fixed intervals as he climbs the stairs and can stop only at certain levels, corresponding to the steps, above the ground. In contrast a person who walks up the ramp can move up in any desired interval and can stop at any point above the ground.

The Bohr model could explain the spectrum of hydrogen. When samples of hydrogen gas or other gases are subjected to either high voltages or high temperatures, they release colored lights. For example,

Bohr Model of the Atom

Figure 5.9
Neils Bohr. (Bettmann Archive.)

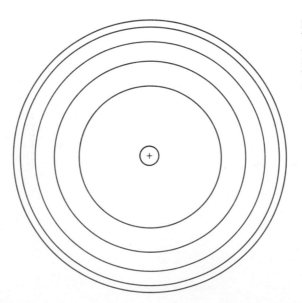

Figure 5.10
The Bohr model of the H atom has a nucleus that is surrounded by circular orbits where its one electron is located.

you have seen the characteristic intense red light released by a neon sign in a storefront. If the light released by gases is passed through a narrow slit and then through a prism, it is broken up into a series of bright lines (Fig. 5.12). Such an instrument for analyzing light is called a **spectroscope,** and the resulting colored lines at specific frequencies are called a **spectrum.** Figure 5.13 shows the spectrum produced by hydrogen. Each different element has its own characteristic spectrum.

Bohr used the quantum theory to explain the lines in the spectrum of hydrogen. He suggested that each orbit in the H atom corresponded to an energy level (Fig. 5.14). The first orbit, the one closest to the nucleus, was the lowest energy level; he called it the *ground-state* energy level. The second orbit was the second energy level, and each succeeding orbit was a higher energy level. He assigned a number, called the **principal quantum number** n to each energy level. The first energy level has the principal quantum number equal to 1 ($n = 1$), the second level has a principal quantum number equal to 2 ($n = 2$), and so forth. Most of the time the electron in hydrogen occupies the lowest energy state, the ground state ($n = 1$). When energy is transferred to the atom, the electron is moved, or excited, to a higher energy level (Fig. 5.15). Ultimately, the excited electron drops to lower energy levels and finally to the ground state. When electrons fall back, they release their energy in the form of quanta, discrete packets of energy with specific frequencies. Calculations by Bohr showed that his proposed energy levels could account for the frequencies of the lines in the H spectrum.

Bohr's theory showed that electrons are found in quantized energy levels. He used the model to explain the spectrum of H and also calculated the velocity of the electron and the distance each energy level was from the nucleus. Nonetheless, the Bohr model had many limitations.

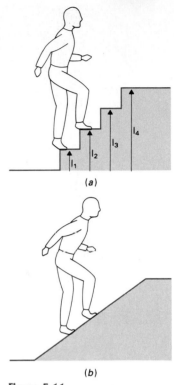

(a)

(b)

Figure 5.11
(a) A person who walks up stairs can only be at fixed distances from the ground (1_1, 1_2, 1_3, or 1_4) because he must stop at one of the stairs. *(b)* A person who walks up a ramp can stop at many different heights above the ground.

Figure 5.12
Light emitted from a gas that is excited by a high voltage is passed through a narrow slit and a prism. The prism diffracts (scatters) the light and separates it into its component frequencies.

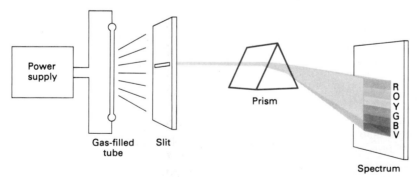

Power supply

Prism

Gas-filled tube

Slit

R O Y G B V

Spectrum

Figure 5.13
The visible spectrum of hydrogen shows red, green, blue, and violet lines. Bohr showed that the frequencies of these lines are related to differences in energies of the orbits in the H atom.

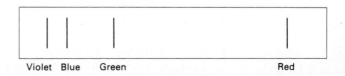

Violet Blue   Green   Red

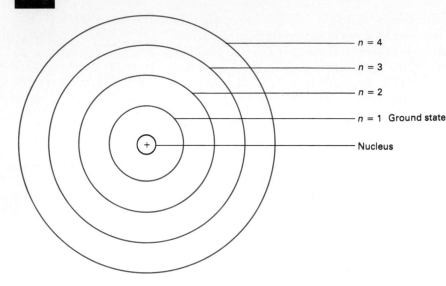

$n = 4$

$n = 3$

$n = 2$

$n = 1$ Ground state

Nucleus

Figure 5.14
In the Bohr model of the atom, the lowest energy level or orbit is called the ground state ($n = 1$). The first excited state is the second energy level. Each orbit is at a fixed distance from the nucleus.

Figure 5.15
Energy is required to excite an electron from the first to the second energy level (1). When the electron falls back (2) to the first level, it releases that energy in the form a quantum, a packet of energy with a specific frequency. An electron excited from the first to the third energy level (3) can either release two quanta, falling from the third to the second (4) and then from the second to the first (5), or release one quantum if it falls directly from the third to the first level (6).

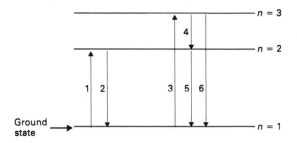

$n = 3$

$n = 2$

1 2   3 5 6

Ground state

$n = 1$

The most serious limitation was that it could only explain the spectrum of H and hydrogenlike ions. It could not explain the structure of atoms that had two or more electrons.

The Bohr model of the atom was replaced with the **quantum mechanical** model of the atom in the mid-1920s, when men such as Louis de Broglie, Werner Heisenberg, and Erwin Schrödinger proposed new ideas concerning the nature and properties of electrons. The quantum mechanical model of the atom places electrons in regions of space called **orbitals.** While we no longer believe that electrons travel around the nucleus in circular orbits as suggested by Bohr, the idea of energy levels is carried over to the quantum mechanical model.

An **orbital** is a volume of space where there is a specific probability of finding electrons. An orbital can contain no more than two electrons: orbitals can be empty (no electrons), half-filled (one electron), or filled (two electrons).

Electrons are elusive particles, ones that cannot be directly observed. A German physicist, Werner Heisenberg, first proposed what is

Orbitals

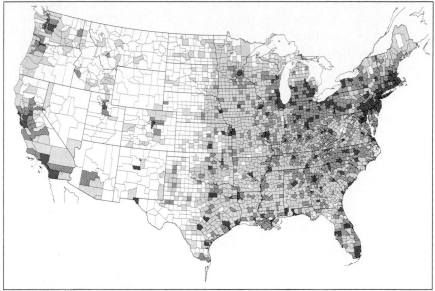

Figure 5.16
A population density map of the United States shows the regions where there is a high probability of locating people (high population density). Regions that are shaded darkly indicate a higher probability of locating a person. Unshaded regions are places where there is a low probability of locating a person (low population density).

U.S. DEPARTMENT OF COMMERCE

now called the **uncertainty principle,** which states that it is impossible to accurately determine simultaneously the exact position and velocity of an electron. In simpler terms, the position of an electron in space cannot be pinpointed at a particular instant. Consequently, scientists no longer are concerned about identifying what path (the orbit) an electron takes as it travels around the nucleus. Instead, they identify the regions, *or orbitals*, where electrons are most likely to be found.

A chemist thinks of orbitals as a geographer thinks of a population density map. A population density map shows an area such as the United States with regions shaded where a high probability of finding a person exists (Fig. 5.16). Darkly shaded areas indicate regions where there is a good chance of finding a person (a high population density), and lightly shaded areas indicate regions where there is a lower probability (a low population density). Actually going to a darkly shaded region does not ensure that you will find a person; there is just a better chance of finding a person in this region.

Within an orbital, electrons behave as if they are spinning on an axis. An electron can spin in either one direction or the other, in a manner similar to a spinning top. If there are two electrons in an orbital, one behaves as if it spins in one direction and the other behaves as if it spins in the opposite direction (Fig. 5.17). This idea that two electrons in an orbital have opposite spins was proposed by Wolfgang Pauli (1900–1958) and is referred to as **Pauli's exclusion principle.**

A collection of orbitals at approximately the same average distance from the nucleus is referred to as an **energy level.** Electrons closer to the nucleus are in lower energy levels, and electrons farther from the nu-

Werner Heisenberg (1901–1975), a student of Bohr, proposed the uncertainty principle in 1927. This radical idea and its implications were ultimately accepted by the scientific community, but it was never totally acceptable to Albert Einstein.

Figure 5.17
Electrons behave as if they were spinning on an axis. An electron can spin either in one direction or the opposite direction.

Electron Energy Levels

cleus are in higher energy levels. An electron in the first energy level is, on average, closer to the nucleus than an electron that occupies the second energy level. The average distance from the nucleus of an electron in the third energy level is greater than that of one in the second energy level (Fig. 5.18). Each energy level is denoted by the **principal quantum number** n. The principal quantum number for the first energy level is 1, the principal quantum number for the second energy level is 2, etc.

Each energy level contains a maximum number of orbitals and electrons. Lower energy levels are closer to the nucleus, where less volume is available for the electrons to occupy. Mutual repulsion of electrons limits the number of electrons in a given region. Succeedingly higher energy levels have greater volume for electrons to populate, diminishing the repulsion of the electrons. An energy level n can contain a maximum of n^2 orbitals and $2n^2$ electrons. Table 5.4 presents the theoretical maximum number of orbitals and electrons for each energy level.

In his teens, Wolfgang Pauli (1900–1958) published articles on relativity that were accepted by the scientific community. Pauli was an extremely productive theoretical physicist, but was very clumsy in the laboratory.

TABLE 5.4 POPULATION OF ELECTRON ENERGY LEVELS

Energy level, n	Orbitals, n^{2*}	Electrons, $2n^{2*}$
1	1	2
2	4	8
3	9	18
4	16	32
5	25	50

*Theoretical maximum number

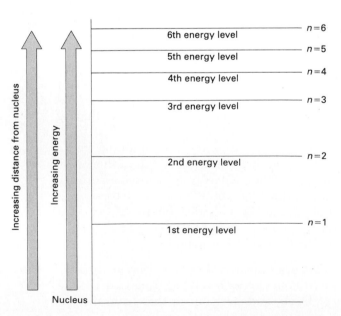

Figure 5.18
The first electron energy level is the region closest to the nucleus. Higher energy levels are regions that are farther from the nucleus.

All energy levels beyond the first level are divided into two or more sublevels. A **sublevel** is composed of one or more orbitals within an energy level that have similar characteristics. The number of sublevels within an energy level corresponds to the principal quantum number n of the energy level. Thus, the first energy level ($n = 1$) contains only one sublevel—they are one and the same. The second energy level ($n = 2$) is composed of two sublevels, the third energy level ($n = 3$) has three sublevels, and so on. We will consider the four lowest energy sublevels. Sublevels are denoted by a letter: s for the lowest-energy sublevel; p, next higher energy sublevel; d, higher still; and f, the highest-energy sublevel of the four.

Each sublevel contains a maximum population of orbitals and electrons that it can hold. Table 5.5 lists the maximum number of orbitals and electrons in each sublevel. An s sublevel can hold a maximum of two electrons because the s sublevel is composed of only a single orbital. The p sublevel is composed of three orbitals; thus six electrons are the maximum number of electrons in a p sublevel. Five and seven orbitals are found in the d and f sublevels, which hold 10 and 14 electrons, respectively.

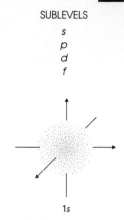

SUBLEVELS

s
p
d
f

1s

Figure 5.19
The 1s sublevel is the lowest energy sublevel in an energy level. Only one orbital makes up an s sublevel; therefore, the 1s sublevel and 1s orbital are the same region of space. An s orbital has the shape of a sphere. The highest probability for finding an electron in an 1s orbital is close to the nucleus.

TABLE 5.5 ELECTRON SUBLEVEL POPULATIONS

Sublevel	Number of orbitals	Maximum number of electrons
s	1	2
p	3	6
d	5	10
f	7	14

Sublevels are distinguished by the shapes of the orbitals of which they are composed. For example, the shape of the s sublevel (or orbital) is shown in Fig. 5.19. You should notice that the shading is darkest near the nucleus and becomes lighter farther from the nucleus, indicating that the highest probability of finding an electron in an s orbital occurs closer to the nucleus rather than farther away. The overall shape of an s orbital is that of a sphere.

A more complex distribution is found for the p sublevel. Each of the three p orbitals is shown in Fig. 5.20a. One of the p orbitals is aligned on the x axis, another along the y axis, and the third along the z axis. Note that the shape of a p orbital is much different from that of the s orbital. The highest probability of finding electrons in p orbitals occurs along the axes. Figure 5.20b shows the three p orbitals superimposed on each other. Both d and f orbitals have even more complex distributions, and will not be considered in our discussion.

5.14 How many total electrons are found in each of the following atoms: (a) B, (b) K, (c) Mg, (d) Co, (e) Kr?

5.15 Explain how the Bohr description of the placement of electrons in atoms is different from the quantum mechanical description.

REVIEW EXERCISES

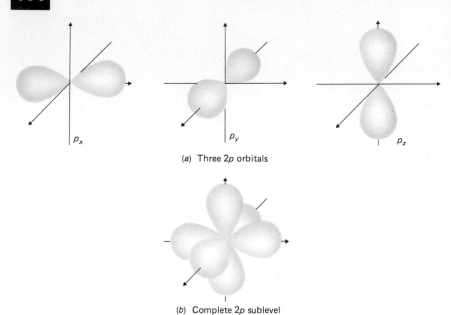

(a) Three 2p orbitals

(b) Complete 2p sublevel

Figure 5.20
The 2p sublevel is composed of three 2p orbitals. Each 2p orbital is located on a different axis and has a dumbbell shape. They are referred to as the p_x, p_y, and p_z orbitals.

5.16 *(a)* Where is the location of the fifth energy level in relation to the fourth energy level? *(b)* Explain why electrons are identified according to the energy level in which they reside.

5.17 What is the maximum number of electrons that can populate each of the following: *(a)* second energy level, *(b)* third energy level, *(c)* an *s* sublevel, *(d)* a *p* sublevel, *(e)* an orbital?

Electron Arrangements

So far, we have found that each energy level is divided into smaller regions called sublevels, and each sublevel is divided into orbitals. Each orbital contains a maximum of two electrons. Figure 5.21 shows the first through the sixth energy levels with their sublevels and orbitals. Electrons fill these orbitals starting from the lowest-energy orbital (first energy level), and proceed, one electron at a time, filling each lower-energy orbital before filling a higher-energy orbital. The filling of lower-energy orbitals before higher-energy orbitals is known as the *aufbau principle.*

We will now begin our study of the arrangement of electrons in atoms by considering where hydrogen's one electron is located. Then we will proceed to helium, which has two electrons. After helium, we will add another electron to the two electrons of helium to find the electronic configuration of lithium. In a similar manner, we will follow the order of the elements on the periodic table, adding one more electron each time to the number found in the previous atom.

In all atoms, the lowest-energy orbital, the one closest to the nucleus, is the 1s orbital.

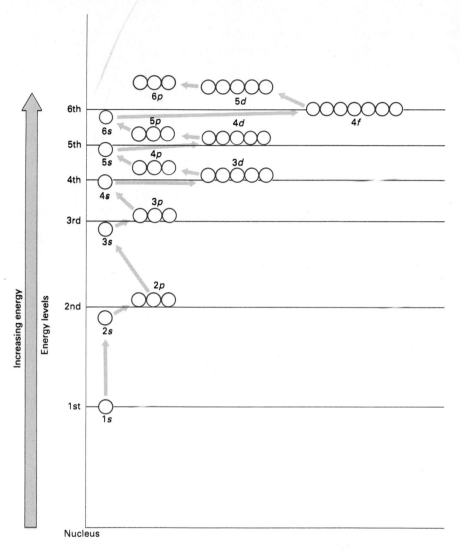

Figure 5.21
Each orbital is represented by a circle. Two electrons fill each orbital, starting with the lowest energy orbital and proceeding to the next higher energy orbital until all the electrons of an atom are accounted for.

Sublevels

S = lowest energy

P = next highest

d = higher

f = highest

$$\text{Energy level} \overset{\displaystyle 1s}{\underset{\displaystyle \uparrow\uparrow}{}}\text{Sublevel}$$

The first number, 1 (the principal quantum number), refers to the energy level where the electron is located, and the *s* identifies the sublevel.

Hydrogen is the simplest atom, containing only one electron. We symbolize hydrogen's electronic configuration (the representation of its occupied orbitals) as follows:

$$\overset{\displaystyle \text{Number of } e^- \text{ in the sublevel}}{1s^1}$$

Hydrogen's electron is found in the s orbital of the lowest energy level.

Helium atoms contain two electrons (Z = 2). Because space is available for another electron in the $1s$ orbital, both electrons populate this orbital. The electronic configuration of helium is

$$\text{He} \qquad 1s^2$$

A 2 is written as a superscript above the s to show that there are two electrons in the $1s$ orbital; the orbital is now filled. Because two is the maximum number of electrons that can occupy the first energy level, the first energy level is also filled.

Lithium (Z = 3) is the first atom to have an electron in the second energy level. Three electrons are found in Li atoms; the first two electrons occupy the lower energy level, $1s$, and the remaining electron is found in the next higher energy $2s$ orbital. We write the electronic configuration of lithium as

Lithium is a reactive metal with a very low density, 0.53 g/cm³. Because of its reactivity, it is not found uncombined in nature.

$$\text{Li} \qquad 1s^2 2s^1$$

Even though the second energy level contains two sublevels, s and p, the s sublevel is lower in energy than the p (Fig. 5.22). Accordingly, the $2s$ sublevel fills before an electron enters the $2p$ sublevel.

The two lowest-energy electrons in beryllium (Z = 4) occupy the $1s$ orbital, the He configuration. The two outer electrons are located in the $2s$ sublevel, filling it. The configuration of Be is

$$\text{Be} \qquad 1s^2 2s^2$$

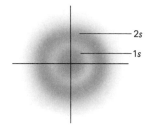

Figure 5.22
Electrons that occupy the $1s$ orbital are on an average closer to the nucleus than are electrons in the $2s$ orbital. The $2s$ orbital distribution is similar to the $1s$ except for being farther from the nucleus.

In boron (Z = 5) the first four electrons occupy the same orbitals as the four electrons in beryllium, $1s^2$ and $2s^2$. The fifth electron enters the higher-energy $2p$ sublevel. Boron is the first element to possess an electron in the $2p$ sublevel. The electronic configuration of boron is

$$\text{B} \qquad 1s^2 2s^2 2p^1$$

Because the p sublevel has three orbitals and has the capacity to hold six electrons (Table 5.5), the next five atoms on the periodic table, C through Ne, fill the $2p$ sublevel. Carbon (Z = 6) has one more electron than boron; therefore, the electronic configuration of carbon is

$$\text{C} \qquad 1s^2 2s^2 2p^2$$

In nitrogen (Z = 7), the next element on the periodic table, the $2p$ sublevel becomes half-filled.

$$\text{N} \qquad 1s^2 2s^2 2p^3$$

In the atoms O, F, and Ne, the $2p$ sublevel fills as follows:

$$
\begin{array}{ll}
\text{O} & 1s^2 2s^2 2p^4 \\
\text{F} & 1s^2 2s^2 2p^5 \\
\text{Ne} & 1s^2 2s^2 2p^6
\end{array}
$$

The $2p$ sublevel of Ne (Z = 10) contains six electrons which is the maximum that the $2p$ can hold. Therefore, a neon atom has a completely filled second energy level. All atoms beyond Ne on the periodic table have more than 10 electrons, and thus have outer electrons in higher energy levels.

Sodium (Z = 11) has the same electronic configuration as neon for its inner level electrons ($1s^2 2s^2 2p^6$); the remaining electron occupies the lowest-energy sublevel in the third energy level: $3s^1$. The electronic configuration for the Na atom is

$$
\text{Na} \qquad 1s^2 2s^2 2p^6 3s^1
$$

You should note that the outer electronic configuration of Na is the same as that of H and Li, except that the electron is in a higher energy level. Each of these atoms has one outer electron in an s sublevel.

$$
\begin{array}{ll}
\text{H} & 1s^1 \\
\text{Li} & 2s^1 \\
\text{Na} & 3s^1
\end{array}
$$

All group IA atoms have one outer-level s electron (s^1).

The outer electronic configuration of an atom is also called the **valence** electronic configuration. Both of these terms refer to the highest occupied electron energy level in an atom.

Magnesium (Z = 12), the next atom after Na on the periodic table, has the electronic configuration

$$
\text{Mg} \qquad 1s^2 2s^2 2p^6 3s^2
$$

All atoms in group IIA have two outer electrons in the s sublevel.

$$
\begin{array}{ll}
\text{Be} & 2s^2 \\
\text{Mg} & 3s^2 \\
\text{Ca} & 4s^2 \\
\text{Sr} & 5s^2 \\
\text{Ba} & 6s^2 \\
\text{Ra} & 7s^2
\end{array}
$$

In all cases, atoms in the same group on the periodic table have the same number of outer-level electrons. Table 5.6 lists the outer electronic configurations of the **representative elements,** those in groups IA (1) through VIIIA (18). From Table 5.6, we see that Al, an atom in group

TABLE 5.6 OUTER ELECTRONIC CONFIGURATIONS OF THE REPRESENTATIVE ELEMENTS

Group number in periodic table	Number of outer electrons	Outer electronic configuration
IA (1)	1	s^1
IIA (2)	2	s^2
IIIA (13)	3	s^2p^1
IVA (14)	4	s^2p^2
VA (15)	5	s^2p^3
VIA (16)	6	s^2p^4
VIIA (17)	7	s^2p^5
VIIIA (18)	8	s^2p^{6*}

*Except He, which is s^2 only.

IIIA (13) with 13 electrons, has three outer-level electrons, two electrons in the $3s$ sublevel and one in the $3p$ sublevel. The complete electronic configuration of Al is

$$\text{Al} \qquad 1s^22s^22p^63s^23p^1$$

Each succeeding atom after aluminum has one more electron in the $3p$ sublevel, ending with Ar with a completed $3p$ sublevel.

$$\text{Si} \qquad 1s^22s^22p^63s^23p^2$$
$$\text{P} \qquad 1s^22s^22p^63s^23p^3$$
$$\text{S} \qquad 1s^22s^22p^63s^23p^4$$
$$\text{Cl} \qquad 1s^22s^22p^63s^23p^5$$
$$\text{Ar} \qquad 1s^22s^22p^63s^23p^6$$

Up to this point, the electrons have filled the orbitals in an orderly fashion. Despite the fact that the third energy level has three sublevels, s, p, and d, the $3d$ sublevel is slightly higher in energy than the $4s$ sublevel. Hence, the $4s$ sublevel fills before the $3d$. Thus, the electronic configurations for potassium, K, and calcium, Ca, are

$$\text{K} \qquad 1s^22s^22p^63s^23p^64s^1$$
$$\text{Ca} \qquad 1s^22s^22p^63s^23p^64s^2$$

Scandium, Sc, is the first atom to have an electron in the $3d$ orbital.

$$\text{Sc} \qquad 1s^22s^22p^63s^23p^63d^14s^2$$

From Sc to Zn, the $3d$ sublevel fills somewhat irregularly. Zinc is the first atom to have a complete $3d$ sublevel. Its electronic configuration is

$$\text{Zn} \qquad 1s^22s^22p^63s^23p^63d^{10}4s^2$$

Scandium is a metal named for the Scandinavian countries where it is almost exclusively found on earth. It is also found in rather large amounts in the sun.

After the $3d$ is full, the $4p$ fills. Gallium, Ga, is the first element with a $4p$ electron.

$$\text{Ga} \qquad 1s^2 2s^2 2p^6 3s^2 3p^6 3d^{10} 4s^2 4p^1$$

Table 5.7 lists the electronic configurations for all the elements.

TABLE 5.7 GROUND STATE ELECTRONIC CONFIGURATIONS OF ATOMS

I. Atoms with atomic numbers 1 to 54

Atomic number	Symbol	1s	2s	2p	3s	3p	3d	4s	4p	4d	4f	5s	5p
1	H	1											
2	He	2											
3	Li	2	1										
4	Be	2	2										
5	B	2	2	1									
6	C	2	2	2									
7	N	2	2	3									
8	O	2	2	4									
9	F	2	2	5									
10	Ne	2	2	6									
11	Na	2	2	6	1								
12	Mg	2	2	6	2								
13	Al	2	2	6	2	1							
14	Si	2	2	6	2	2							
15	P	2	2	6	2	3							
16	S	2	2	6	2	4							
17	Cl	2	2	6	2	5							
18	Ar	2	2	6	2	6							
19	K	2	2	6	2	6		1					
20	Ca	2	2	6	2	6		2					
21	Sc	2	2	6	2	6	1	2					
22	Ti	2	2	6	2	6	2	2					
23	V	2	2	6	2	6	3	2					
24	Cr	2	2	6	2	6	5	1					
25	Mn	2	2	6	2	6	5	2					
26	Fe	2	2	6	2	6	6	2					
27	Co	2	2	6	2	6	7	2					
28	Ni	2	2	6	2	6	8	2					
29	Cu	2	2	6	2	6	10	1					
30	Zn	2	2	6	2	6	10	2					
31	Ga	2	2	6	2	6	10	2	1				
32	Ge	2	2	6	2	6	10	2	2				
33	As	2	2	6	2	6	10	2	3				
34	Se	2	2	6	2	6	10	2	4				
35	Br	2	2	6	2	6	10	2	5				
36	Kr	2	2	6	2	6	10	2	6				
37	Rb	2	2	6	2	6	10	2	6			1	
38	Sr	2	2	6	2	6	10	2	6			2	
39	Y	2	2	6	2	6	10	2	6	1		2	
40	Zr	2	2	6	2	6	10	2	6	2		2	
41	Nb	2	2	6	2	6	10	2	6	4		1	

TABLE 5.7   GROUND STATE ELECTRONIC CONFIGURATIONS OF ATOMS *Continued*

I. Atoms with atomic numbers 1 to 54

Atomic number	Symbol	1s	2s	2p	3s	3p	3d	4s	4p	4d	4f	5s	5p
42	Mo	2	2	6	2	6	10	2	6	5		1	
43	Tc	2	2	6	2	6	10	2	6	6		1	
44	Ru	2	2	6	2	6	10	2	6	7		1	
45	Rh	2	2	6	2	6	10	2	6	8		1	
46	Pd	2	2	6	2	6	10	2	6	10			
47	Ag	2	2	6	2	6	10	2	6	10		1	
48	Cd	2	2	6	2	6	10	2	6	10		2	
49	In	2	2	6	2	6	10	2	6	10		2	1
50	Sn	2	2	6	2	6	10	2	6	10		2	2
51	Sb	2	2	6	2	6	10	2	6	10		2	3
52	Te	2	2	6	2	6	10	2	6	10		2	4
53	I	2	2	6	2	6	10	2	6	10		2	5
54	Xe	2	2	6	2	6	10	2	6	10		2	6

It is not necessary to memorize the order of the filling of sublevels. Instead recognize that the periodic table is organized according to the electronic configurations of atoms. In Fig. 5.23, the periodic table is marked to indicate the order in which electrons fill sublevels. All elements in groups IA and IIA have s electrons in their outer energy level. Elements in groups IIIA through VIIIA have both s and p electrons in their outer levels. Elements in groups IB through VIIIB, the transition elements, have electrons that fill the d and f sublevels.

To write electronic configurations, follow the periodic table in order of increasing atomic number. Numbers that denote periods correspond to electron energy levels, and the number at the top of each vertical column helps give the outer-level configuration. Example Problem 5.3 illustrates how the periodic table may be used to help you write electronic configurations.

Example Problem 5.3

Use the periodic table to write the complete electronic configuration for strontium, Sr.

Solution
---

1. Find Sr on the periodic table.
Sr has 38 electrons, is in the fifth period, and is a member of group IIA. All elements in group IIA have an outer electronic configuration of two electrons in the s orbital, s^2.

2. Follow the periodic table in order of increasing atomic number.

Strontium is a metal that was discovered by Sir Humphry Davy in 1808. Sr is found in nature as a mixture of four isotopes. It is a reactive metal that will spontaneously combust if very finely divided.

II. Atoms with atomic numbers 55 to 104

Atomic number	Symbol	[Xe]*	4f	5d	5f	6s	6p	6d	6f	7s
55	Cs					1				
56	Ba					2				
57	La			1		2				
58	Ce		2			2				
59	Pr		3			2				
60	Nd		4			2				
61	Pm		5			2				
62	Sm		6			2				
63	Eu		7			2				
64	Gd		7	1		2				
65	Tb		9			2				
66	Dy		10			2				
67	Ho		11			2				
68	Er		12			2				
69	Tm		13			2				
70	Yb		14			2				
71	Lu		14	1		2				
72	Hf		14	2		2				
73	Ta		14	3		2				
74	W		14	4		2				
75	Re		14	5		2				
76	Os		14	6		2				
77	Ir		14	7		2				
78	Pt		14	9		1				
79	Au		14	10		1				
80	Hg		14	10		2				
81	Tl		14	10		2	1			
82	Pb		14	10		2	2			
83	Bi		14	10		2	3			
84	Po		14	10		2	4			
85	At		14	10		2	5			
86	Rn		14	10		2	6			
87	Fr		14	10		2	6			1
88	Ra		14	10		2	6			2
89	Ac		14	10		2	6	1		2
90	Th		14	10		2	6	2		2
91	Pa		14	10	2	2	6	1		2
92	U		14	10	3	2	6	1		2
93	Np		14	10	4	2	6	1		2
94	Pu		14	10	6	2	6			2
95	Am		14	10	7	2	6			2
96	Cm		14	10	7	2	6	1		2
97	Bk		14	10	9	2	6			2
98	Cf		14	10	10	2	6			2
99	Es		14	10	11	2	6			2
100	Fm		14	10	12	2	6			2
101	Md		14	10	13	2	6			2
102	No		14	10	14	2	6			2
103	Lr		14	10	14	2	6	1		2
104	Unq		14	10	14	2	6	2		2

*Elements 55 to 104 have the inner electronic configuration of Xe.

Period	IA	IIA	IIIB	IVB	VB	VIB	VIIB	VIIIB	VIIIB	VIIIB	IB	IIB	IIIA	IVA	VA	VIA	VIIA	VIIIA
	s	s	d	d	d	d	d	d	d	d	d	d	p	p	p	p	p	p
1	1 H $1s^1$																	2 He $1s^2$
2	3 Li $2s^1$	4 Be $2s^2$											5 B $2s^22p^1$	6 C $2s^22p^2$	7 N $2s^22p^3$	8 O $2s^22p^4$	9 F $2s^22p^5$	10 Ne $2s^22p^6$
3	11 Na $3s^1$	12 Mg $3s^2$											13 Al $3s^23p^1$	14 Si $3s^23p^2$	15 P $3s^23p^3$	16 S $3s^23p^4$	17 Cl $3s^23p^5$	18 Ar $3s^23p^6$
4	19 K $4s^1$	20 Ca $4s^2$	21 Sc	22 Ti	23 V	24 Cr	25 Mn	26 Fe	27 Co	28 Ni	29 Cu	30 Zn	31 Ga $4s^24p^1$	32 Ge $4s^24p^2$	33 As $4s^24p^3$	34 Se $4s^24p^4$	35 Br $4s^24p^5$	36 Kr $4s^24p^6$
5	37 Rb $5s^1$	38 Sr $5s^2$	39 Y	40 Zr	41 Nb	42 Mo	43 Tc	44 Ru	45 Rh	46 Pd	47 Ag	48 Cd	49 In $5s^25p^1$	50 Sn $5s^25p^2$	51 Sb $5s^25p^3$	52 Te $5s^25p^4$	53 I $5s^25p^5$	54 Xe $5s^25p^6$
6	55 Cs $6s^1$	56 Ba $6s^2$	57 La*	72 Hf	73 Ta	74 W	75 Re	76 Os	77 Ir	78 Pt	79 Au	80 Hg	81 Tl $6s^26p^1$	82 Pb $6s^26p^2$	83 Bi $6s^26p^3$	84 Po $6s^26p^4$	85 At $6s^26p^5$	86 Rn $6s^26p^6$
7	87 Fr $7s^1$	88 Ra $7s^2$	89 Act†	104 Unq	105 Unp	106 Unh	107	108										

	f													
*	58 Ce	59 Pr	60 Nd	61 Pm	62 Sm	63 Eu	64 Gd	65 Tb	66 Dy	67 Ho	68 Er	69 Tm	70 Yb	71 Lu
	4f													
†	90 Th	91 Pa	92 U	93 Np	94 Pu	95 Am	96 Cm	97 Bk	98 Cf	99 Es	100 Fm	101 Md	102 No	103 Lw
	5f													

Figure 5.23
Elements are listed on the periodic table according to the electronic configurations of their atoms. Atoms in the same group have the same outer-level sublevel configurations. Following the periodic table in order of increasing atomic number is an easy way to write the electronic configuration of an atom.

Considering the periodic table in Fig. 5.24, we see that the inner electronic configuration of Sr is the same as that of Kr (Z = 36). The electronic configuration of Kr is

$$\text{Kr} \quad 1s^22s^22p^63s^23p^63d^{10}4s^24p^6$$

To this we add the two outer-level electrons, $5s^2$, to give the complete configuration for Sr:

$$\textbf{Sr} \quad \mathbf{1s^22s^22p^63s^23p^63d^{10}4s^24p^65s^2}$$

Always check to see that the total number of electrons in the electronic configuration equals the atomic number. For this example, the sum of the superscripts equals 38, which is the atomic number of Sr.

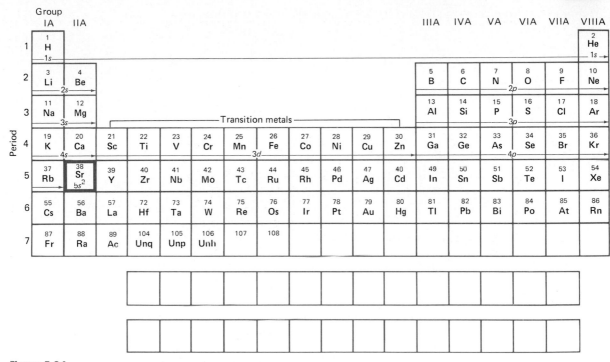

Figure 5.24
Start with H and follow along the periodic table until you reach Sr. As you proceed toward Sr write the electrons that fill each sublevel in Sr: $1s^2 2s^2 2p^6 3s^2 3p^6 4s^2 3d^{10} 4p^6 5s^2$.

5.18 Write the complete electronic configuration for each of the following atoms: *(a)* Li, *(b)* N, *(c)* K, *(d)* Ga, *(e)* Br, *(f)* Cs.

The outer electronic configurations of atoms for the most part determine the properties of elements, especially their chemical properties. Chemists frequently draw Lewis symbols (also called dot formulas) as a means of conveniently expressing the outer electronic configurations of atoms.

Use the following two rules to write the Lewis symbol for an atom:

1. Write the symbol of the atom.

2. Place one dot around the symbol for each electron in the **outermost** energy level of the atom.

Always remember that inner-level electrons are never shown in Lewis symbols.

Because a hydrogen atom only has one electron ($1s^1$), its Lewis symbol has only one dot next to its atomic symbol.

H ·

REVIEW EXERCISE

Lewis Symbols

G. N. Lewis (1875–1946) was an American chemist who obtained his Ph.D. from Harvard University. In 1933, he prepared a sample of heavy water, D_2O. Heavy water has two deuterium atoms, $_1^2H$, in place of regular hydrogen atoms, $_1^1H$.

Helium has two electrons ($1s^2$), so its Lewis symbol shows two dots next to its symbol.

$$\text{He}\,\text{:}$$

You should note that the two dots in the Lewis symbol for He are written together to show that the electrons are located in the same orbital.

Lithium has an electronic configuration of $1s^2 2s^1$. Only the $2s^1$ is in the outer energy level, so only one dot is placed next to the symbol of Li (don't count the dot in the i of Li).

$$\text{Li}\,\cdot$$

The outer electronic configuration of Be is $2s^2$; therefore, two dots are placed next to the Be in the Lewis symbol.

$$\text{Be}\,\text{:}$$

If the atom has p electrons in the outer energy level, a number of different arrangements of dots around the symbol can be written. For example, the Lewis symbol for carbon ($1s^2 2s^2 2p^2$) can be written as follows:

$$\cdot\,\overset{\textstyle\cdot}{\underset{\textstyle\cdot}{\text{C}}}\,\cdot \quad \text{or} \quad \text{:}\,\overset{\textstyle\cdot}{\text{C}}\,\cdot$$

In the first Lewis symbol, the dots representing electrons are written symmetrically around the symbol. However, some chemists prefer to write the dots in a way corresponding to the way the electrons are paired in orbitals. In carbon, two electrons are paired in the $2s$; but each of the two electrons in the $2p$ occupies a different p orbital.

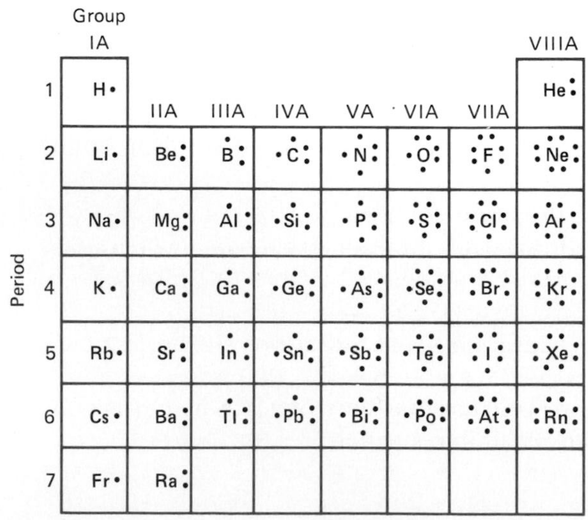

Figure 5.25
Lewis symbols of the representative elements.

All atoms in a chemical group have the same number of dots around their Lewis symbols because all have the same number of outer electrons. For example, atoms in group VIIA all possess seven outer-level electrons. Thus, the general Lewis symbol for these atoms is

$$: \overset{\displaystyle ..}{\underset{\displaystyle .}{X}} :$$

in which X is either F, Cl, Br, I, or At.

Figure 5.25 illustrates the Lewis symbols for all atoms in groups IA through VIIIA on the periodic table, and Example Problem 5.4 shows how the Lewis symbols are written for three different atoms.

--- **Example Problem 5.4** ---

Draw the Lewis symbols for *(a)* Si, *(b)* Rb, and *(c)* Te.

--- **Solution** ---

(a) Silicon belongs to group IVA and has an outer electronic configuration of $3s^2 3p^2$. Each member of group IVA has four outer electrons ($s^2 p^2$); therefore, the Lewis symbol for Si is

$$\cdot \underset{\displaystyle .}{Si} :$$

(b) Rubidium belongs to group IA and has an outer electronic configuration of $5s^1$, one outer electron. Consequently, the Lewis symbol for Rb is

$$Rb \cdot$$

(c) Tellurium belongs to group VIA and has an outer electronic configuration of $5s^2 5p^4$; thus, six dots are placed around its symbol:

$$: \underset{\displaystyle .}{Te} :$$

Orbital Diagrams

Another way that chemists represent the population of electron orbitals is to draw **orbital diagrams.** In these diagrams, orbitals are frequently represented by circles or underlines and electrons by arrows. Arrows are used to represent the apparent spin of electrons. *In a filled orbital the two electrons spin in opposite directions;* thus, the arrows that indicate the two electrons in a filled orbital point in opposite directions. Two electrons in one orbital are said to have **paired spins.** Figure 5.26 shows the orbital diagrams for the first 12 elements.

An H atom has one electron in the 1s orbital; therefore, one arrow is placed in the circle. It does not matter in which direction the arrow

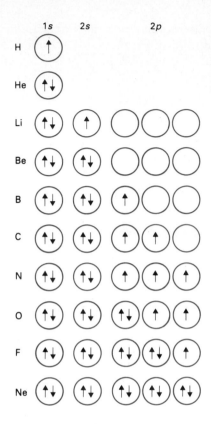

Figure 5.26
Orbital diagrams show each orbital as a circle and each electron as an arrow. Two electrons within an orbital always have opposite spins. If more than one orbital is at exactly the same energy, electrons fill the empty orbitals before occupied orbitals.

points in this diagram because the spin of the H atom can be in either direction. Because a He atom has two electrons in the 1s orbital, two electrons are shown by arrows pointing in opposite directions; the arrows can never point in the same direction. After the 2s fills, the 2p begins to fill. B has one electron in the first p orbital. Two possibilities exist for placing the two electrons in the 2p sublevel of a C atom. Either both electrons are in one 2p orbital or they are in different 2p orbitals. Because of the mutual repulsion of the negatively charged electrons, electrons fill empty orbitals within a sublevel before filling occupied orbitals. This is a statement of *Hund's rule*. Therefore, the orbital diagram of C shows two electrons with the same spin characteristics in different 2p orbitals. Nitrogen has three 2p electrons, each of which is in a different 2p orbital. Oxygen is the first atom in which electrons are paired in the 2p sublevel. Two of the 2p orbitals contain paired electrons in F atoms, and all of the orbitals contain paired electrons in Ne atoms.

REVIEW EXERCISES

5.19 Draw the Lewis symbols for each of the following atoms: *(a)* Mg, *(b)* S, *(c)* Kr, *(d)* Ra, *(e)* As.

5.20 Draw orbital diagrams for each of the following atoms: *(a)* Be, *(b)* F, *(c)* Si, *(d)* Ar.

SUMMARY

John Dalton proposed the first true scientific theory of atoms in the beginning of the nineteenth century. He thought that atoms were very small solid spheres. After the discovery of protons and electrons, a new model of the atom was proposed by **J. J. Thomson.** The Thomson model of the atom is called the "plum pudding" model because he pictured a structure that had electrons embedded in an atom that had a uniform positive charge from the protons. **Rutherford** changed our thoughts about atoms when he discovered that the atom has a nucleus.

According to modern atomic theory, the atom is a small particle composed of a very dense nucleus that contains protons and neutrons, with outer regions sparsely populated by electrons. **Protons** and neutrons have approximately the same mass, 1 u (unified atomic mass unit). **Electrons** have an extremely tiny mass, only $\frac{1}{1837}$ u. However, electrons possess a full negative charge, equal in magnitude but opposite in sign to that of the proton. **Neutrons** have no electric charge.

The **atomic number** of an atom equals the number of protons in the nucleus of the atom, and the **mass number** equals the total number of protons and neutrons in the nucleus. If atoms have the same atomic number but different mass numbers, they are called **isotopes.** When dealing with elements composed of a mixture of isotopes, chemists use **atomic mass** (atomic weight), which is the average mass of the isotopes of an element compared to the mass of the $^{12}_{6}C$ isotope.

Electrons are located in regions of space called **orbitals.** Orbitals are regions around the nucleus where there is a high probability of finding electrons. An orbital holds a maximum of two electrons. A set of orbitals with similar characteristics and nearly the same energy is called a **sublevel.** Electrons in their lowest energy states are found in four different sublevels: s, p, d, and f. Sublevels with similar energies are grouped into **energy levels.**

Electrons fill orbitals in atoms, starting with the lowest-energy orbital, and proceeding to higher-energy orbitals. Each different atom has its own specific **electronic configuration.** Electronic configurations are represented by writing the number that corresponds to the energy level next to the letter that designates the sublevel, then writing the number of electrons that occupy the sublevel as a superscript above the letter. For example, the electronic configuration of a hydrogen atom is $1s^1$.

Atoms in the same group on the periodic table have the same outer-level sublevel configuration. The number at the top of each group corresponds to the total number of electrons in the outermost energy level. For example, each element in group IA has one outer-level electron. Outer electronic configurations are represented by writing **Lewis symbols** or **orbital diagrams.**

KEY TERMS

atomic number	energy level	nucleus	radiation	sublevel
atomic mass	isotope	orbit	radioactivity	unified atomic mass unit
electric charge	Lewis symbols	orbital diagrams	spectroscopy	
electron	mass number	orbital	spectrum	

EXERCISES*

5.21 Define the following terms: unified atomic mass unit, electric charge, nucleus, radioactivity, atomic number, mass number, isotope, atomic mass, orbit, orbital, electron energy level, sublevel, Lewis symbol, orbital diagram, spectrum.

Development of the Modern Atomic Theory

5.22 Why is Dalton credited with the discovery of the first scientific atomic model when Democritus proposed an atomic theory many years before?

*For exercise numbers printed in color, answers can be found at the back of the book.

5.23 (a) What parts of Dalton's atomic theory are accepted today? (b) What parts of Dalton's atomic theory are no longer accepted?

5.24 Explain how J. J. Thomson changed the model of the atom.

5.25 (a) How did Rutherford change the model of the atom? (b) Describe the experiment he performed.

5.26 (a) What type of radiation did Rutherford use in the experiment in which he discovered the nucleus? (b) Why did Rutherford hypothesize that the radiation would go through the gold leaf undeflected? (c) Why was only a small percent of the radiation deflected through large angles?

5.27 (a) Describe the Bohr model of the atom. (b) How was the Bohr model different from the Rutherford model of the atom? (c) How does the Bohr model of the atom differ from our modern concept of the atom?

5.28 How does the Bohr model of the atom explain the spectrum of hydrogen? Use a diagram in your explanation.

5.29 Propose a reason why the neutron was not discovered until 1932, approximately 50 years after the discovery of the proton and electron.

Structure of Atoms

5.30 What is the mass in unified atomic mass units and the charge of (a) a proton, (b) an electron, (c) a neutron?

5.31 (a) Derive a conversion factor that equates unified atomic mass units and grams. (b) Calculate the mass in grams of a proton from its mass in unified atomic mass units.

5.32 Describe what happens when the following charged particles are brought close to each other: (a) two protons, (b) proton and electron, and (c) neutron and electron.

5.33 Complete the following table by providing all missing information:

Symbol	Atomic number	Mass number	Number of protons	Number of neutrons	Number of electrons
(a) $^3_1$H	1	3	1	2	1
(b) $^{18}_8$O	8	16	8	10	8
(c) $^{23}_{11}$Na					
(d) ———	15	31			
(e) ———	18	40			
(f) ———	23	51			
(g) ———			27	32	
(h) ———			35	46	
(i) ———				62	46
(j) Pt	78	195	78	117	78

5.34 How many protons, neutrons, and electrons are there in each of the following atoms: (a) $^{108}_{46}$Pd, (b) $^{70}_{32}$Ge, (c) $^{48}_{22}$Ti, (d) $^{210}_{85}$At?

5.35 Write the symbols for the atoms that have the following number of protons and neutrons:
(a) $p^+ = 65$; $n^0 = 94$
(b) $p^+ = 53$; $n^0 = 74$
(c) $p^+ = 44$; $n^0 = 58$

5.36 What types of radiation have the following properties: (a) Penetrates relatively dense matter; (b) is an electron that travels at high speed; (c) resembles energy more than matter; (d) most massive and has a low penetration power?

Isotopes and Atomic Mass

5.37 (a) What are the three principal isotopes found in a natural sample of uranium? (b) How do the isotopes differ?

5.38 (a) What atom is used as the standard for the atomic mass scale? (b) Could another atom be used as the standard? If so, how would this be accomplished?

5.39 How does atomic mass differ from mass number? Give an example.

5.40 Europium, Eu, is composed of two isotopes: $^{151}_{63}$Eu (mass = 150.92 u) and $^{153}_{63}$Eu (mass = 152.92 u). If $^{151}_{63}$Eu and $^{153}_{63}$Eu natural abundances are 47.82% and 52.18%, respectively, calculate the atomic mass of Eu.

5.41 Rubidium is composed of two isotopes: $^{85}_{37}Rb$ and $^{87}_{37}Rb$. The mass of $^{85}_{37}Rb$ is 84.9117 u, and its natural abundance is 72.15%; the mass of $^{87}_{37}Rb$ is 86.909 u, and its natural abundance is 27.85%. What is the atomic mass of rubidium?

5.42 Use the following mass and natural abundance data to calculate the atomic mass of thallium, Tl.

Isotope	Mass, u	Percent abundance
$^{203}_{81}Tl$	202.97	29.50
$^{205}_{81}Tl$	204.97	70.50

5.43 (a) Calculate the atomic mass of copper, Cu, from the information given in Table 5.2. (b) Calculate the atomic mass of germanium, Ge, from the data in Table 5.2.

Electronic Configurations

5.44 (a) Draw a figure that shows the Bohr model of the hydrogen atom. (b) Label the first four energy levels. (c) What is the ground state orbit in the Bohr atom?

5.45 (a) Who first proposed the quantum theory? (b) What is the main premise of quantum theory?

5.46 Explain how Bohr incorporated quantum theory into his model of the atom.

5.47 (a) How is the spectrum of an element obtained? (b) What is the name of the instrument used to obtain the spectrum of an element?

5.48 How does the Bohr model of the atom explain the existence of the lines in the spectrum of H?

5.49 (a) What is the importance of the principal quantum number? (b) What is the principal quantum number for the ground state electron in a H atom?

5.50 What were the main limitations of the Bohr model of the atom?

5.51 (a) What is an orbital? (b) How are orbitals designated?

5.52 Write the maximum number of electrons that can be located in the ground state of each of the following: (a) an orbital, (b) the d sublevel, (c) a Be atom, (d) third energy level, (e) the f sublevel, (f) a Ca atom, (g) 4p sublevel of As, (h) 3d sublevel of Cr, (i) 5s sublevel of Sr, (j) fifth energy level of Br, (k) 4d sublevel of Cd.

5.53 How does the quantum mechanical model describe the location of electrons in atoms?

5.54 For each of the following, identify the region closest to the nucleus: (a) fifth, sixth, or seventh energy level, (b) 4s, 4p, or 4d sublevel, (c) 3s, 4s, or 5s sublevel.

5.55 Write the complete electronic configuration for each of the following atoms: (a) Li, (b) C, (c) Na, (d) Al, (e) K, (f) Zn, (g) Kr, (h) Sc, (i) Ge, (j) Se.

5.56 Write the complete electronic configuration for each of the following atoms: (a) Ba, (b) Fr, (c) Te, (d) Pb, (e) Rn.

5.57 Write the outer-level electronic configuration for each of the following atoms: (a) Rb, (b) Ca, (c) Ga, (d) Se, (e) Br, (f) Sn, (g) Sb, (h) Ra, (i) At, (j) Fr.

5.58 Write the symbols for the elements with the following electronic configurations:
(a) $1s^2 2s^2 2p^3$
(b) $1s^2 2s^2 2p^6 3s^2$
(c) $1s^2 2s^2 2p^6 3s^2 3p^3$
(d) $1s^2 2s^2 2p^6 3s^2 3p^6 4s^2 3d^{10} 4p^2$
(e) $1s^2 2s^2 2p^6 3s^2 3p^4$
(f) $1s^2 2s^2 2p^6 3s^2 3p^6 4s^2 3d^{10} 4p^5$
(g) $1s^2 2s^2 2p^6 3s^2 3p^6 4s^2 3d^{10} 4p^6 5s^1$
(h) $1s^2 2s^2 2p^6 3s^2 3p^6 4s^2 3d^{10} 4p^6 5s^2 4d^{10} 5p^6$

5.59 Draw the Lewis symbols for each of the following: (a) N, (b) Rb, (c) Sb, (d) Br, (e) Rn, (f) P, (g) Ge, (h) Se, (i) Si, (j) Ba.

5.60 Draw the Lewis symbols for each of the following atoms: (a) Al, (b) O, (c) Mg, (d) Pb, (e) Xe, (f) K, (g) S, (h) Ra, (i) Cs, (j) As.

5.61 Identify the atom or atoms with the following characteristics: (a) First group IIA atom to have a 3s outer electron; (b) group VIA atom(s) with 4p electrons; (c) atom(s) with atomic number less than 43 that have electrons in the 4d sublevel; (d) group IIA atom(s) without occupied f orbitals; (e) period 4 atoms with more than five outer electrons; (f) atom(s) that only have s electrons.

5.62 What electronic configuration would each of the following achieve if they lost or gained the stated number of electrons: (a) Mg, loses 2 electrons; (b) N, gains 3 electrons; (c) F, gains 1 electron; (d) I, gains one electron?

5.63 Explain the importance of Hund's rule.

5.64 Draw orbital diagrams for each of the following atoms; (a) B, (b) F, (c) Mg, (d) P, (e) Cl.

5.65 How many unpaired electrons are found in each of the following atoms: *(a)* He, *(b)* Si, *(c)* S, *(d)* Ca, *(e)* As?

Additional Exercises

5.66 How did the Dalton model of the atom differ from the model proposed by Democritus?

5.67 Which scientist gave us each of the following models of the atom: *(a)* quantum mechanical model, *(b)* nuclear model, *(c)* electrons orbiting a central nucleus, *(d)* "plum pudding" model?

5.68 How many protons, neutrons, and electrons are found in each of the following atoms: *(a)* $^{208}_{84}$Po, *(b)* $^{235}_{92}$U, *(c)* $^{180}_{72}$Hf, *(d)* $^{205}_{81}$Tl?

5.69 Write the symbols for at least two isotopes of each of the following atoms: *(a)* $^{131}_{54}$Xe, *(b)* $^{16}_{8}$O, *(c)* $^{24}_{12}$Mg, *(d)* $^{139}_{57}$La.

5.70 Write the electronic configurations for the lowest-energy excited states for the following atoms: *(a)* Na, *(b)* Ne, *(c)* Xe.

5.71 *(a)* What is the electronic configuration for Mg if it loses two electrons? *(b)* What atom has the same electronic configuration as Mg has after it loses two electrons? *(c)* What would you have to do in order to remove two electrons from Mg?

5.72 *(a)* State the Pauli exclusion principle. *(b)* Why must we talk of the apparent spin of electrons instead of being more definite with regard to their spin properties?

5.73 If the diameter of the nucleus of an atom could be increased in size to 1 cm, what would be the approximate diameter of the atom?

5.74 By what degree would the attractive force between a proton and an electron decrease if the distance between the two was tripled?

5.75 Use the information in the following table to calculate the atomic mass of lead, Pb.

Isotope	Mass, u	Percent abundance
$^{204}_{82}$Pb	203.9731	1.50
$^{206}_{82}$Pb	205.9745	23.6
$^{207}_{82}$Pb	206.9759	22.6
$^{208}_{82}$Pb	207.9766	52.3

5.76 The atomic mass of silicon, Si, is 28.086. $^{28}_{14}$Si, $^{29}_{14}$Si, and $^{30}_{14}$Si are the three isotopes that compose Si. $^{28}_{14}$Si has a mass of 27.9769 u and natural abundance of 92.21%. $^{29}_{14}$Si has a mass of 28.9765 u and a natural abundance of 4.70%. If the percent abundance is 3.09, calculate the mass of $^{30}_{14}$Si.

5.77 Manganese is composed of only one isotope. What is the mass number of this isotope?

5.78 How did Heisenberg's uncertainty principle change our thoughts about the nature of the atom?

5.79 Use Table 5.7 to help you describe the "irregular" filling of the $3d$ sublevel.

5.80 Bohr proposed the following equation to calculate the energies of the orbits in the H atom:

$$E_n = \frac{-2.18 \times 10^{-18} \text{ J}}{n^2}$$

in which E_n is the energy of the nth orbit and n is the principal quantum number. Calculate the energies associated with the first four orbits of the H atom.

5.81 Describe how the shapes of each of the following pairs of orbitals differ: *(a)* $1s$ and $2s$, *(b)* $1s$ and $2p$, *(c)* $2p$ and $3p$.

5.82 Draw the orbital diagram for the outermost energy level of each of the following: *(a)* In, *(b)* Sb, *(c)* Po, *(d)* Ba.

5.83 Write the electronic configurations for the two elements that have completely filled outer energy levels.

5.84 Use three orbital diagrams to illustrate Hund's rule.

5.85 $^{6}_{3}$Li and $^{7}_{3}$Li are the two isotopes of Li that exist in nature. $^{6}_{3}$Li has a mass of 6.015 u, and $^{7}_{3}$Li has a mass of 7.016 u. If the atomic mass of Li is 6.941, calculate the percent abundance of each isotope.

5.86 The approximate diameter of the nucleus of an Al atom is 286 pm. Calculate the volume of the nucleus of Al in cubic centimeters, using the formula $V = \frac{4}{3}\pi r^3$, in which V is the volume, π is 3.14, and r is the radius of the nucleus.

CHAPTER
Six

Periodic Properties

STUDY GUIDELINES

After completing Chapter 6, you should be able to

1. State the periodic law and use it to predict the properties of elements

2. State the names of representative groups of elements on the periodic table

3. Distinguish between the properties of metals and nonmetals

4. Define and give examples of cations and anions

5. Predict the charge on ions from their location in the periodic table

6. Predict a property of an element given the properties of related elements

7. Predict trends in ionization energy and atomic size

8. Discuss properties, characteristics, and uses of Na, Cl, and Si as representative of metals, nonmetals, and metalloids, respectively

9. Explain why hydrogen could be placed in a group all by itself

---

B efore the electronic configurations of atoms were determined, scientists such as Lothar Meyer and Dmitri Mendeleev (Fig. 6.1) recognized that if elements were placed in order of atomic mass (they did not know about atomic numbers) certain properties of atoms recurred at regular intervals. Both Meyer and Mendeleev proposed periodic tables that placed elements with similar properties in the same group (Fig. 6.2). Today, we know that these recurring properties result from the fact that elements in the same chemical group have the same outer electronic configuration; only the energy level occupied by the electrons differs.

In the late nineteenth century, Mendeleev gained the attention of the scientific world by using his periodic table to predict the properties of elements that had not yet been discovered. Table 6.1 lists Mendeleev's predictions for the undiscovered element that he called "ekasilicon" (lit-

6.1 INTRODUCTION

Mendeleev (1834–1907) published his first periodic table in 1869. In Russia, scientific papers were rarely translated into western languages, so the western world usually had to rediscover such findings. Because of his fame as a renowned scientist, his table and explanations were translated immediately into German.

erally, "under Si" on his table). After its discovery ekasilicon was named germanium, Ge.

TABLE 6.1   MENDELEEV'S PREDICTIONS FOR EKASILICON (GERMANIUM)

Property	Modern values	Predicted values (1871)*
Atomic mass	72.6	72
Color	Grayish white	Dark gray
Density	5.4 g/cm³	5.5 g/cm³
Specific heat	0.074 cal/g·°C	0.073 cal/g·°C
Formula of oxide	GeO_2	ESO_2

*Mendeleev's predictions were made 15 years prior to the discovery of Ge.

Over the years since Mendeleev developed the modern form of the periodic table, groups of elements have been given "family" names, which are summarized in Fig. 6.3. Elements in group IA (1), except H,

Figure 6.1
Dmitri Mendeleev. *(Bettman Archives.)*

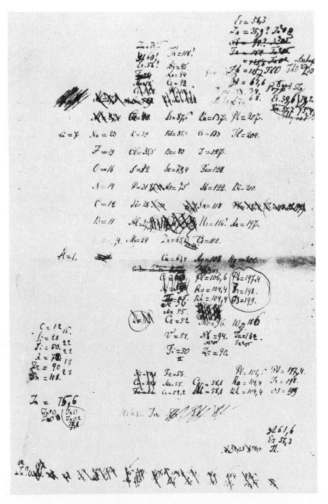

Figure 6.2
Mendeleev placed elements in order of their atomic masses and grouped elements with similar properties. This figure shows an early draft of his periodic table that he published in 1869. *(Burndy Library.)*

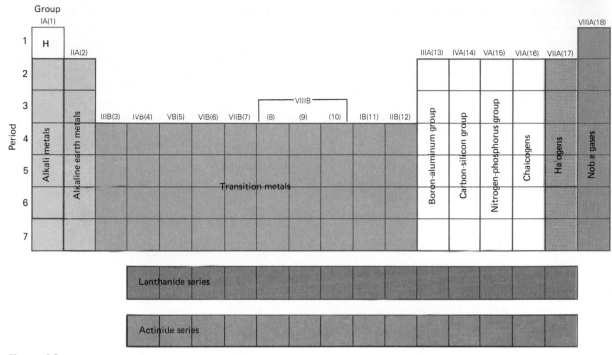

Figure 6.3
Periodic table with the names of the chemical groups.

are called **alkali metals,** while elements in group IIA (2) are called **alkaline earth metals** (the name used by the alchemists). The next 10 columns on the periodic table (group B elements, 3 to 12) are the **transition elements** or **transition metals.** This group contains the most common metallic elements. Two sets of 14 elements written below the main body of the periodic table are also transition elements: elements with atomic numbers from 58 to 71 are members of the **lanthanide series,** and elements with atomic numbers from 90 to 103 are members of the **actinide series.** All nontransition elements, group A elements including groups IA through VIIIA, are called **representative elements.**

Group IIIA (13) has no unique name; it is often called the **aluminum** or **boron-aluminum group.** Similarly, groups IVA (14) and VA (15) are designated the **carbon** and **nitrogen groups,** respectively. An old name for group VIA (16), the **chalcogens,** is presently being used. Finally, group VIIA (17) are the **halogens,** and VIIIA (18) are the **noble gases.**

REVIEW EXERCISES

6.1 How did Mendeleev show that his periodic table was a valid grouping of elements?

6.2 Write the names of the groups on the periodic table with the following Roman numeral and letter designations: *(a)* IA, *(b)* IIA, *(c)* IVA, *(d)* VIA, *(e)* VIIA, *(f)* VIIIA, *(g)* all group B elements.

Let's turn our attention to what is now called the periodic law. The **periodic law** states that the properties of the elements are periodic functions of their atomic number. In other words, if the elements are listed in order of their atomic numbers, a regular pattern of chemical and physical properties is found. To illustrate the periodic law, we will consider the two main classes of elements—metals and nonmetals.

Metals and Nonmetals

Metals are usually solid elements with a silvery gray color; their melting and boiling points are usually quite high. Metals share a set of common properties: (1) They have a high average density; (2) they are excellent conductors of heat and electricity; and (3) they are malleable—i.e., they can be hammered into various shapes and foils.

The largest percentage of elements, over 80, are metals.

Nonmetals possess properties that, in many cases, are opposite to those of metals. A large percent of the nonmetals are liquids and gases, not solids. On average, the melting points, boiling points, densities, and electric and heat conductivities of nonmetals are lower than those of metals.

As an example, let us consider the second-period elements. Li and Be are metals, and C, N, O, F, and Ne are nonmetals. B has intermediate properties, not exactly metallic or nonmetallic, and is classified as a **metalloid** (or semimetal). Elements on the right side of the periodic table have a greater degree of nonmetallic character than those on the left side.

If we look beyond Ne to the third period of the periodic table, a similar trend in metallic and nonmetallic properties is found: Na, Mg, and A1 possess metallic properties; Si is a metalloid; and P, S, C1, and Ar are nonmetals. Table 6.2 presents selected physical properties of the elements in the third period of the periodic table.

TABLE 6.2   SELECTED PROPERTIES OF THIRD-ROW ELEMENTS

Element	Atomic number	Physical state (25°C)	Melting point, °C	Density, g/cm^3	Type
Na	11	Silvery solid	97.8	0.97	Metal
Mg	12	Gray solid	650	1.74	Metal
A1	13	Light-gray solid	660	2.70	Metal
Si	14	Bluish solid	1414	2.33	Metalloid
P	15	White or red solid	44	1.83	Nonmetal
S	16	Pale-yellow solid	112	2.07	Nonmetal
C1	17	Green gas	−101	0.00321	Nonmetal
Ar	18	Colorless gas	−189	0.00178	Nonmetal

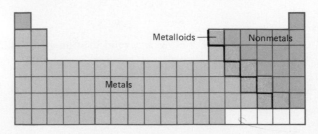

Figure 6.4
Metals are the elements located on the left side of the periodic table. Nonmetals are listed on the right side of the periodic table. Metalloids are located along a zigzag line separating the metals from the nonmetals.

CHEM TOPIC: Metals from Ores

Only a few metals exist free in nature—e.g., gold, silver, copper, platinum, and palladium. Most metals are found in **ores,** which are complex mixtures of different minerals. For example, Al is found as aluminum oxide, Al_2O_3, in a mineral called bauxite. Along with aluminum oxide in bauxite is silica, SiO_2, and hematite, Fe_2O_3.

Three major steps are required to extract metals from their ores. After the ore is mined, either it is treated to concentrate the desired mineral or it is converted to a form that is more easily extracted. Then the metallic compound is converted to the metal. This process is known as reduction. **Reduction** is accomplished by heating the compound to a high temperature (smelting) or through the use of electricity (electrolysis). Finally, the free metal is **refined** to remove impurities; in many cases, these are recovered and sold.

Figure 6.4 shows a periodic table that classifies elements as metals, nonmetals, and metalloids. You should note that the eight metalloids border a zigzag line that begins to the left of B and ends between Po and At. This line separates metals from nonmetals.

Specifically, if a physical property such as melting point is plotted on a graph, a better view of the periodic nature of properties of metals and nonmetals is obtained. Figure 6.5 presents a graph of melting point versus atomic number for the first 36 elements. Note the somewhat regular pattern of increasing and decreasing melting points. With the exception of carbon, metals and metalloids are found at the maximum values, the peaks, while the nonmetals occupy the valleys, the minimum values. This repeating pattern of peaks and valleys as atomic number increases is an illustration of the periodic law.

The chemical properties of metals differ from those of nonmetals. Metals have a small number of outer-level electrons, and they tend to lose these electrons during chemical changes. In contrast, nonmetals have more complete outer energy levels and tend to gain electrons (except for the noble gases). When atoms either lose or gain electrons, they form ions (Fig. 6.6). An **ion** is a charged atom, or as we will see in a later chapter, a charged group of atoms. Two different types of ions exist:

The melting point of a substance is the temperature at which the solid and liquid states coexist in equilibrium. Higher melting points indicate stronger forces of attraction among atoms in the solid state.

An easy way to remember that anions are negative ions is to break the word anion *into **a n (negative) ion.** Mnemonics (memory aids) such as this can simplify the task of learning chemistry.*

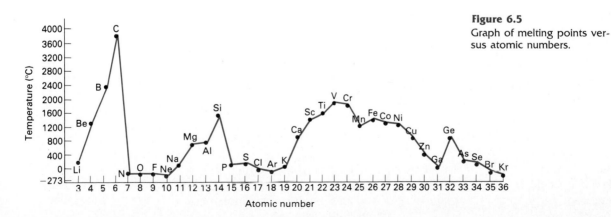

Figure 6.5
Graph of melting points versus atomic numbers.

positive ions, called **cations,** and negative ions, called **anions.** Cations are produced when metals release one or more electrons. For example, if a metal (M) loses one electron, it becomes an ion with a charge of 1+:

$$M \cdot \longrightarrow M^+ + e^-$$

However, if it loses two electrons it becomes an ion with a charge of 2+:

$$M : \longrightarrow M^{2+} + 2e^-$$

Anions form when nonmetals (X) accept electrons. A 1− ion results if one electron is accepted and a 2− ion results if two electrons are taken in.

$$: \overset{..}{\underset{..}{X}} \cdot + e^- \longrightarrow [: \overset{..}{\underset{..}{X}} :]^-$$
$$: \overset{..}{\underset{..}{X}} + 2e^- \longrightarrow [: \overset{..}{\underset{..}{X}} :]^{2-}$$

Figure 6.6
Metal atoms lose one or more electrons, forming cations. The resulting cation is smaller than the metal atom because the outer-level electrons are removed. Nonmetal atoms gain one or more electrons, forming anions. The resulting anion is larger than the nonmetal atom because there are more electrons than protons.

Charges on ions are written as superscripts to the right of symbols, with the magnitude preceding the + or the − .

Metals give up their outer-level electrons and form cations that have noble gas electronic configurations. Nonmetals often take in enough electrons to complete their outer levels, also producing a noble gas configuration. As we will discuss in Chap. 8, the most stable electronic configuration is that of the noble gases. Ions that have the same electronic configurations as noble gases are said to be **isoelectronic** to noble gases.

Metals in group IA, the alkali metals, tend to lose one electron when they combine with other substances, producing 1+ cations. For example, Na and K each lose one electron to form 1+ cations.

$$Na \longrightarrow Na^+ + e^-$$
$$K \longrightarrow K^+ + e^-$$

The electronic configuration for a Na atom is $1s^2 2s^2 2p^6 3s^1$. After the atom loses an electron, the electronic configuration of the sodium ion, Na^+, is $1s^2 2s^2 2p^6 3s^0$, making the sodium ion isoelectronic to Ne.

A sodium ion, Na^+, becomes isoelectronic to Ne, and a potassium ion, K^+, becomes isoelectronic to Ar. Metals in group IIA, alkaline earth metals, tend to lose two electrons when they combine with other elements, yielding 2+ cations. Ca and Ba each lose two electrons to form 2+ cations.

$$Ca \longrightarrow Ca^{2+} + 2e^-$$
$$Ba \longrightarrow Ba^{2+} + 2e^-$$

Calcium ions, Ca^{2+}, are isoelectronic to Ar, and barium ions, Ba^{2+}, become isoelectronic to Xe.

Nonmetals, except noble gases, accept electrons and produce anions that are isoelectronic to a noble gas. The elements in group VIIA, the halogens, most frequently gain one electron and produce 1− anions. For

example, Cl and Br atoms each accept one electron and produce 1−
halide ions.

$$Cl + e^- \longrightarrow Cl^-$$

$$Br + e^- \longrightarrow Br^-$$

The chloride ion, Cl^-, is isoelectronic to Ar, and the bromide ion, Br^-, is isoelectronic to Kr. Similarly, group VIA elements, chalcogens, accept two electrons and form 2− anions. O and S each take in two electrons to form 2− ions.

$$O + 2e^- \longrightarrow O^{2-}$$

$$S + 2e^- \longrightarrow S^{2-}$$

The oxide ion, O^{2-}, is isoelectronic to Ne, and the sulfide ion, S^{2-} is isoelectronic to Ar. A periodic table with selected ions is presented in Fig. 6.7.

The electronic configuration of a Cl atom is $1s^2 2s^2 2p^6 3s^2 3p^5$. After the atom gains an electron, the electronic configuration of the chloride ion, Cl^-, is $1s^2 2s^2 2p^6 3s^2 3p^6$; the Cl^- ion is isoelectronic to Ar.

REVIEW EXERCISES

6.3 State the periodic law and give an example that demonstrates it.

6.4 List three properties that distinguish metals from nonmetals.

6.5 What periodic trend is found for densities within the second period of the periodic table?

6.6 (a) How does a cation differ from an anion? (b) Describe how cations and anions are formed.

Periodic trends are found when the properties of individual atoms are studied. An important property of atoms that helps explain ion formation is called ionization energy. **Ionization energy** is the minimum

6.3 PERIODIC PROPERTIES OF ATOMS

Figure 6.7
A periodic table is shown with selected ions. Many blank spaces are found in the table because some atoms do not readily form ions and some can form two or more ions with different charges.

Period	IA	IIA	IIIA	IVA	VA	VIA	VIIA
1	H^+						
2	Li^+	Be^{2+}			N^{3-}	O^{2-}	F^-
3	Na^+	Mg^{2+}	Al^{3+}		P^{3-}	S^{2-}	Cl^-
4	K^+	Ca^{2+}	Ga^{3+}				Br^-
5	Rb^+	Sr^{2+}					I^-
6	Cs^+	Ba^{2+}					At^-
7	Fr^+	Ra^{2+}					

Group

amount of energy required to remove the most loosely held electron from a neutral gaseous atom.

$$A(g) + \text{ionization energy} \longrightarrow A^+(g) + e^-$$

The ionization energy of an atom is a measure of the degree to which the nucleus attracts the most loosely held outer-level electron, which is the one most distant from the nucleus. A low ionization energy indicates a smaller attractive force, and a high ionization energy indicates a larger attractive force between the nucleus and the most loosely held electron.

A graph of ionization energy versus atomic number for the atoms in the first three periods is presented in Fig. 6.8, and Fig. 6.9 lists the ionization energies of most of the atoms in megajoules per mole. Two distinct trends are evident. First, as we proceed from left to right across a period (increasing atomic number within a period), the ionization energy generally increases. More energy is required, on average, to remove the most loosely held electron from nonmetals than metals. Alkali metals, group IA elements, have the lowest ionization energies, and noble gases, group VIIIA elements, have the highest ionization energies within a period.

Second, within a group like the noble gases (VIIIA), ionization energies decrease with increasing atomic number. Helium has the highest ionization energy (2.37 MJ/mol) of the noble gases, and Rn has the lowest (1.04 MJ/mol). Similar trends are found in all chemical groups.

Overall, elements in the lower left corner of the periodic table have the lowest ionization energies; francium, Fr, is the element with the lowest ionization energy. As we move up and to the right on the periodic

Our definition of ionization energy is actually for the first ionization energy. The second ionization energy is that minimum amount of energy required to remove the most loosely held electron in a monopositive cation, M^+.

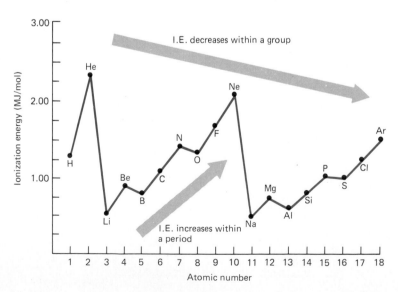

Figure 6.8
A graph of the ionization energies of atoms reveals two trends. Within a period, the ionization energy increases with increasing atomic number. Within a group, the ionization energy decreases with increasing atomic number.

Group

Period	IA(1)	IIA(2)	IIIB(3)	IVB(4)	VB(5)	VIB(6)	VIIB(7)	VIIIB (8)	(9)	(10)	IB(11)	IIB(12)	IIIA(13)	IVA(14)	VA(15)	VIA(16)	VIIA(17)	VIIIA(18)
1	1 H 1.31																	2 He 2.37
2	3 Li 0.520	4 Be 0.900											5 B 0.800	6 C 1.09	7 N 1.40	8 O 1.31	9 F 1.68	10 Ne 2.08
3	11 Na 0.496	12 Mg 0.738											13 Al 0.578	14 Si 0.787	15 P 1.01	16 S 1.00	17 Cl 1.25	18 Ar 1.52
4	19 K 0.418	20 Ca 0.590	21 Sc 0.631	22 Ti 0.658	23 V 0.650	24 Cr 0.653	25 Mn 0.717	26 Fe 0.759	27 Co 0.758	28 Ni 0.737	29 Cu 0.746	30 Zn 0.906	31 Ga 0.579	32 Ge 0.762	33 As 0.944	34 Se 0.941	35 Br 1.14	36 Kr 1.35
5	37 Rb 0.403	38 Sr 0.550	39 Y 0.616	40 Zr 0.660	41 Nb 0.664	42 Mo 0.685	43 Tc 0.702	44 Ru 0.711	45 Rh 0.720	46 Pd 0.805	47 Ag 0.731	48 Cd 0.868	49 In 0.558	50 Sn 0.709	51 Sb 0.832	52 Te 0.869	53 I 1.01	54 Xe 1.17
6	55 Cs 0.376	56 Ba 0.503	57 La 0.538	72 Hf 0.654	73 Ta 0.761	74 W 0.770	75 Re 0.760	76 Os 0.84	77 Ir 0.88	78 Pt 0.87	79 Au 0.890	80 Hg 1.01	81 Tl 0.589	82 Pb 0.716	83 Bi 0.703	84 Po 0.812	85 At 0.916	86 Rn 1.04

Figure 6.9
Periodic table that contains the ionization energies of atoms in MJ/mol.

table, the ionization energies generally increase; He has the highest ionization energy of all atoms (Fig. 6.10).

Two factors account for the trends in ionization energy: (1) attractive forces between the nucleus and the electrons, and (2) the shielding effect of the inner-level electrons. Greater attraction of the electrons by the nucleus results in higher ionization energies. Shielding refers to the blocking effect that inner-level electrons have on the nuclear attraction of the outer electrons. Greater shielding of the nucleus by inner electrons results in lower ionization energies.

The overall increasing trend in ionization energy across a period is explained as follows. The shielding effect of the inner electrons remains constant across a period, while the charge on the nucleus increases. Increasing nuclear charge produces a greater attractive force on the outer electrons, resulting in a higher ionization energy.

Decreasing trends in ionization energy within a chemical group are explained in terms of the location of the outer electrons. With each new energy level, the outer-level electrons are farther from the nucleus with more levels of inner electrons shielding the nuclear charge (Fig. 6.11). Greater distance from the nucleus and the greater shielding effects re-

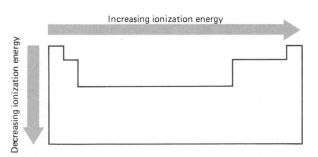

Increasing ionization energy

Decreasing ionization energy

Figure 6.10
Periodic table that shows the trends in ionization energies.

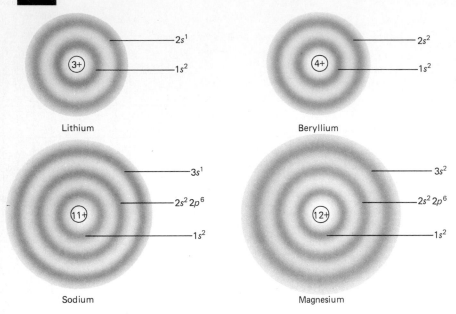

Lithium   Beryllium

Sodium   Magnesium

Figure 6.11
In Li, two of the three nuclear charges are shielded by the inner 1s orbital, leaving only one nuclear charge to hold the outer electron. In Be, two of the four nuclear charges are shielded, leaving two nuclear charges to hold the outer electrons. As a result of having a greater nuclear charge after shielding, Be has a higher ionization energy than Li. A similar argument can be used for Na and Mg.

sult in a smaller attractive force on the outer electrons; hence, they are easier to remove.

The **atomic size** is an estimate of the distance from the nucleus to a point that corresponds to the outermost region of a neutral atom. Atomic sizes for the elements in the second and third periods are plotted in Fig. 6.12, and the relative sizes of the atoms are illustrated as circles in Fig. 6.13.

Once again, two trends can be identified. Moving across a period from left to right, the atomic size decreases, and within a group the atomic size increases. Thus, the largest atoms are located in the bottom left corner of the periodic table, and the smallest atoms are located in the upper right corner (Fig. 6.14).

Is this consistent with the trends in ionization energy? Yes. Across a period, the shielding effects of inner electrons are constant, and the nuclear charge increases with the addition of protons to the nucleus. Consequently, the attractive force of the nucleus for the electrons increases, and the size of the atom decreases as the electrons are attracted closer to the nucleus. Within a group, the attractive forces exerted on the outer-level electrons decrease because the outer electrons in each succeeding group member are farther from the nucleus and are more shielded by the inner electrons. Thus, the atomic size increases.

Example Problems 6.1 and 6.2 show how to predict the properties of elements and atoms from the general trends.

Trends in Atomic Sizes

Sizes of atoms are very difficult to measure. In most cases estimates are obtained from measurements of atoms that are components of molecules.

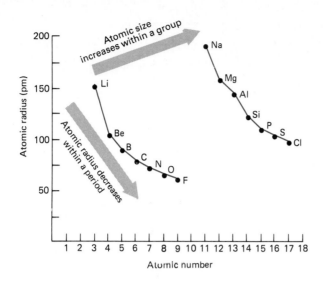

Figure 6.12
The atomic sizes in pm are plotted versus atomic number for the first 18 atoms. Noble gases are not included because a different means is used to measure their size.

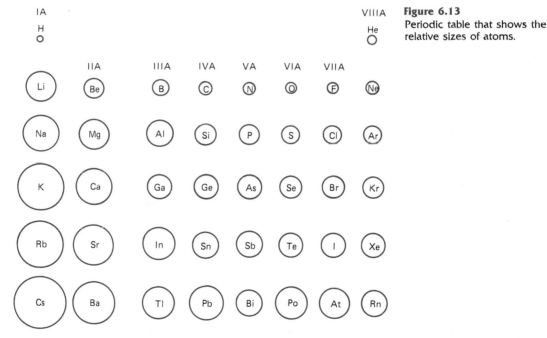

Figure 6.13
Periodic table that shows the relative sizes of atoms.

Figure 6.14
Periodic table that shows the major trends in atomic sizes.

Example Problem 6.1

Of the four atoms, Si, P, S, and Cl, which *(a)* is the most nonmetallic, *(b)* has the highest ionization energy, and *(c)* is the largest atom?

Solution

Si, P, S, and Cl belong to the third period of the periodic table. Thus, to answer this question, it is only necessary to know the periodic trends as the atomic number increases across the periodic table.

(a) Metallic character decreases or nonmetallic character increases as we go from left to right across the periodic table. Accordingly, Cl, a halogen, is the most nonmetallic atom of this group.

(b) Ionization energies increase from left to right across a period; consequently, Cl has the highest ionization energy of the four atoms.

(c) Atomic sizes decrease across a period; hence, Si is the largest atom.

Example Problem 6.2

Predict the *(a)* atomic size and *(b)* boiling point of krypton given the following data.

	Atomic size, nm	Boiling point, °C
Neon, Ne	0.065	−246
Argon, Ar	0.095	−186
Xenon, Xe	0.130	−108

Solution

Kr is a member of the noble gases (group VIIIA). Its atomic mass is greater than that of Ar and less than that of Xe; thus, its atomic size and boiling point should be intermediate between the values for Ar and Xe. To make good predictions, it is necessary to follow the trends within the group.

(a) Atomic size increases within a group, which means that the atomic size of Kr is larger than that of Ar but smaller than that of Xe. We see an increase of 0.030 nm from Ne to Ar; however, we would not expect the same increase from Ar to Kr because such an increase gives a value (0.125 nm) almost identical to that for Xe. A better estimate might be the average value of Ar and Xe, (0.095 + 0.130)/2, which gives **0.113 nm.** The actual value for Kr is 0.110 nm.

(b) Considering the trend in boiling points of the noble gases, we find an increasing trend. Therefore, the boiling point of Kr is larger than that of Ar but smaller than that of Xe. If we once again take the average value of the boiling points of Ar and Xe, we get a reasonable prediction. The value obtained when we average −186°C and −108°C is **−147°C,** which is fairly close to the actual value, −153°C.

6.7 *(a)* Define ionization energy. *(b)* What general trends in ionization energies are found?

6.8 From the set of elements that follows each description, select the element that best fits that description.
(a) Highest ionization energy: P, S, Cl, Ar
(b) Lowest ionization energy: Li, Be, Na, Mg
(c) Largest atomic size: Ga, Ge, As, Se
(d) Smallest atomic size: Br, Se, I, Te
(e) Most metallic: Be, B, C, N

6.4 A REPRESENTATIVE METAL, NONMETAL, AND METALLOID

B efore leaving the topic of periodic properties, let's take a more in-depth look at a representative metal, nonmetal, and metalloid; this will help us gain a better understanding of elements and their properties.

Sodium

Sodium, the most abundant alkali metal, was discovered in 1807 by Sir Humphry Davy (1778–1829). He isolated Na by passing an electric current through a molten sodium compound. What he found was a reactive metal with properties unlike those of frequently encountered metals such as Fe, Ag, Au, or Cu.

Sodium is a soft, silvery-gray solid with a low density, 0.97 g/cm^3 (Fig. 6.15). Most metals have high densities and sink when placed in water, but sodium does not. If a small chunk of sodium is placed in water, it floats. However, this is of secondary importance, for immediatly on contacting water, Na reacts violently, liberating hydrogen gas and a large amount of heat.

Davy discovered K, Na, Ba, Sr, Ca and Mg, but his greatest discovery was the man he selected for his assistant, Michael Faraday. Faraday became one of the world's most productive scientists.

$$2Na(s) + 2H_2O(l) \longrightarrow 2NaOH(aq) + H_2(g) + energy$$

Sodium melts at 97.5°C and boils at 889°C. Both the melting and boiling points are very low for nonalkali metals, but only lithium has higher melting and boiling points among the alkali metals. The remaining alkali metals have even lower melting and boiling points (Table 6.3). On hot days cesium, Cs, is a liquid metal.

TABLE 6.3 NORMAL MELTING AND BOILING POINTS OF THE ALKALI METALS

Symbol	Melting point, °C	Boiling point, °C
Li	180	1326
Na	98	889
K	63	757
Rb	39	679
Cs	29	690

Sodium is much too reactive to exist in nature as a free element. Instead, it is found as sodium ions, Na$^+$, in numerous compounds. Many rocks are composed of sodium compounds that, on weathering, dissolve, and ultimately are carried to the oceans as dissolved ions. Many dis-

Slightly less than 3% by mass of the earth's crust is sodium chemically bonded in compounds.

solved sodium compounds are present in the oceans, most notably sodium chloride (table salt), NaCl.

Sodium metal has many uses. It is a dehydrating agent, a compound that removes water from many liquids. Sodium vapor lamps illuminate highways. In industry, sodium is combined with other metals (Hg, K, Sn, and Sb) to form important alloys (metallic solutions). The petroleum industry utilizes sodium in the production of antiknock compounds for gasoline. Some nuclear power plants use molten (liquid) sodium as a heat transfer agent, i.e., to carry away heat released by radioactive substances.

Chlorine

Among his many discoveries, Karl Wilhelm Scheele (1742–1786) is credited with the discovery of chlorine in 1774. Over 30 years elapsed before it was given its modern name by Davy in 1810. Chlorine is a greenish-yellow gas whose density (3.2 g/L) is greater than that of air. The melting point of chlorine is $-101°C$ and its boiling point is $-34.1°C$. A regular trend in melting and boiling points is found in the halogens. Fluorine and chlorine are gases at room conditions of 25°C and 1 atm, bromine is a liquid, and iodine is a solid (Table 6.4).

TABLE 6.4 NORMAL MELTING AND BOILING POINTS OF THE HALOGENS

Element	Melting point, °C	Boiling Point, °C
Fluorine	−220	−188
Chlorine	−101	−34
Bromine	−7	59
Iodine	114	184

Pure chlorine gas is not composed of individual chlorine atoms. A chlorine atom is so reactive that it combines chemically with another chlorine atom to form a diatomic chlorine molecule, Cl_2. A *diatomic molecule* is composed of two bonded atoms. All of the halogens share this property of chlorine and exist as diatomic molecules. Molecules and their formation are discussed in Chap. 8.

Chlorine is mainly found in nature as chloride ion, Cl^-.

Like sodium, chlorine is very reactive and combines with many substances. Due to its extreme reactivity, chlorine is acutely toxic. It is so toxic that, as a precautionary measure, towns are evacuated when a train that contains chlorine tank cars derails nearby. During World War I, chlorine was used in trench warfare. The toxicity of chlorine and its high density made it an effective killing agent, until gas masks were perfected.

Chlorine gas is used to bleach paper and pulp products.

Chlorine compounds, some resembling the active ingredient in laundry bleach, are added to swimming pools to destroy bacteria and algae. Chlorine gas, under controlled conditions, is pumped into drinking water and then removed during the purification process. Chlorine destroys most microorganisms that live in the water and removes compounds responsible for bad taste and unpleasant odor in water.

Over 1000 chlorine compounds are produced by the chemical industry. They are used in pesticides, refrigerants, anesthetics, cleaning solutions, and plastics.

Like sodium, chlorine can be prepared from sodium chloride, NaCl. Sodium chloride is obtained from salt mines and ocean water. An electric current is passed through a sodium chloride solution; the process produces chlorine gas and another industrially significant substance, sodium hydroxide (lye), NaOH.

Silicon is a representative metalloid. It is a brittle, shiny, black-gray solid that appears to be metallic but is not. Structurally, Si resembles diamond (a pure form of carbon). Si is extremely hard, is capable of scratching glass, melts at 1414°C, and boils at 2327°C. Jöns Jakob Berzelius (1779–1848), a Swedish chemist, discovered silicon in 1823.

Silicon belongs to a diverse group on the periodic table, group IVA. C, the first member of group IVA, is a nonmetal; both Si and Ge are metalloids, and Sn and Pb are metals. The melting point of Si is lower than that of carbon, but is higher than those of the other members of group IVA. Both C and Si have low densities compared to the more metallic elements in the group, Sn and Pb. Selected properties of the group IVA elements are found in Table 6.5.

Silicon

Silicon is not the same as silicone. Silicon is an element; silicone is a complex compound.

TABLE 6.5 SELECTED PROPERTIES OF GROUP IVA ELEMENTS

Symbol	Melting point, °C	Density, g/cm^3
C	3570	2.25
Si	1414	2.33
Ge	937	5.32
Sn	232	7.30
Pb	328	11.4

Figure 6.15
Quartz, SiO_2, is one of the many silicon oxides on earth. Quartz is a hard, colorless, brittle crystalline solid. It is used to control frequencies in electronic devices, and is used in optical instruments. *(Ron Testa/Field Museum of Natural History, Chicago)*

Silicon is the second most abundant element in the earth's crust; oxygen is the most abundant. Approximately 25 percent of the earth's crust is silicon in silicon compounds. Quartz (Fig. 6.16), sand, agate, jasper, and opal are oxides of silicon (i.e., compounds in which silicon is bonded to oxygen). In many compounds, silicon is combined chemically with both oxygen and metals; common examples include talc, mica, asbestos, beryl, and feldspar.

Silicon compounds are commercially important, especially the group of compounds known as silicates (silicon-oxygen compounds). Clay, cement, and glass are all silicates. When Si is combined with C, the resulting compound is silicon carbode, SiC, a very hard compound that has many industrial uses. Very pure Si is used in the production of transistors and integrated circuits.

Silicon dioxide (silica), SiO_2, is found in 20 or more different forms in the earth and its inhabitants. Silica dissolves in water to form silicic acid, $Si(OH)_4$. This acid provides the silica to produce quartz crystals inside geodes, skeletons for diatoms, and petrified wood.

REVIEW EXERCISES

6.9 Which scientist discovered *(a)* sodium, *(b)* chlorine, *(c)* silicon?

6.10 What properties distinguish sodium from metals such as Cu, Fe, or Au?

6.11 Compare the melting-point trends in the alkali metals with that for the halogens.

6.12 *(a)* What are the main physical properties of Si? *(b)* List three silicon compounds.

6.5 HYDROGEN: AN ELEMENT AND A GROUP

Hydrogen is usually written at the top of group IA of the periodic table because it shares one property with the alkali metals: Hydrogen and the alkali metals all possess an s^1 outer-level electronic configuration. Nevertheless, hydrogen is unique in so many other ways that it truly belongs in a group by itself.

Hydrogen was discovered by Henri Cavendish in 1766 and was named by Lavoisier. The name "hydrogen" literally means "water former." Just about all the hydrogen on earth is chemically combined with another element. The largest percent is combined with O in water, H_2O. In the uncombined state, hydrogen exists diatomically as H_2. Free, isolated hydrogen atoms, H, do not exist under normal conditions.

Hydrogen is fairly abundant on earth and is the most abundant element in the universe. Most of the mass of a star is ionized hydrogen gas at very high temperatures.

Hydrogen has the lowest atomic mass and density of all the elements. It must be cooled to $-253°C$ before H will liquefy, and it freezes at $-259°C$, only 6°C below its boiling point! Hydrogen is a poor conductor of heat and electricity. Notice the extreme differences between the physical properties of hydrogen and those of the alkali metals, which are solids, have higher densities, and are good electric conductors. Table 6.6 compares the properties of hydrogen with those of the lowest-mass alkali metal, lithium.

Henry Cavendish (1731–1810) was truly an eccentric scientist. He was a total "loner"; rarely did he speak, and when he did it was only to men, never to women. He wrote notes when forced to communicate with women. Cavendish spent virtually none of an inheritance worth more than $5 million. He devoted 60 years of his life to scientific research.

TABLE 6.6   COMPARISON OF THE PHYSICAL PROPERTIES OF HYDROGEN AND LITHIUM

Property	Hydrogen	Lithium
Outer electronic configuration	$1s^1$	$2s^1$
Melting point, °C	-259.2	180.5
Boiling point, °C	-252.7	1330
Density, g/cm^3	9.0×10^{-5}	0.53
Ionization energy, MJ/mol	1.3	0.52

In terms of specific atomic properties, we have seen that the alkali metals have the lowest ionization energies and largest atomic sizes in their period. In contrast, hydrogen has a high ionization energy and a small atomic size.

Hydrogen is highly reactive. It combines explosively with oxygen, O_2, to form water.

$$2H_2(g) + O_2(g) \longrightarrow 2H_2O(g)$$

Industrially, a large quantity of hydrogen, H_2, is combined with nitrogen, N_2, to produce ammonia, NH_3.

$$3H_2(g) + N_2(g) \longrightarrow 2NH_3(g)$$

In the future, hydrogen might occupy the position that fossil fuels hold today as the world's primary energy source. Stars obtain their energy from hydrogen in a process called nuclear fusion. If scientists could sustain and control nuclear fusion reactions, then the energy released from these reactions could be converted to electricity. One of the technological problems with using hydrogen as a fuel is the enormous temperatures ($>100,000,000$°C) needed to sustain nuclear fusion reactions. Nuclear fusion and other nuclear phenomena are discussed in Chap. 19.

Nuclear fusion occurs when the nuclei of low-mass atoms are combined to produce an atom of higher mass. During nuclear fusion, a large quantity of energy is released.

6.13 What evidence could be presented to show that hydrogen is not an alkali metal?

6.14 All halogens obtain the noble gas configuration by gaining one electron. If H receives one electron, it also obtains the noble gas configuration. What other properties of hydrogen could possibly allow it to be classified as a halogen?

REVIEW EXERCISES

SUMMARY

The **periodic law** states that the properties of the elements are periodic functions of their atomic number. When atoms are arranged in order of increasing atomic number, recurring chemical and physical properties are found. Elements with the same outer electronic configuration are in the same group (vertical column). Group members have many common properties, and within most groups regular trends in properties are found. Periods (horizontal rows) contain elements that have outer electrons in the same energy level. Likewise, somewhat regular trends in properties are observed when progressing from one side of the table to the other.

Elements are classified as metals, nonmetals, or metalloids (semimetals), depending on their properties. **Metals** occupy the left side of the periodic table, and are usually solids with high densities and high melting and boiling points. They are also good conductors of heat and electricity and tend to lose electrons in chemical changes. **Nonmetals,** on the right side of the periodic table, have lower average densities, melting points, boiling points, and conductivities than metals. In chemical reactions, nonmetals tend to gain electrons and form anions. Metalloids resemble metals but have some nonmetallic properties.

Ionization energy is the amount of energy required to remove the most loosely held electron from a neutral gaseous atom. From left to right across a period, ionization energy increases. Within a chemical group, the ionization energy decreases with increasing atomic number. Elements near the bottom left side of the periodic table have the lowest ionization energies, and those near the top right side have the highest ionization energies.

From left to right across a period, the **atomic size** decreases, and within a group the atomic size increases as atomic number increases. Elements at the bottom left side of the periodic table are the largest, and those near the top right are the smallest.

Sodium is a soft metal with low melting and boiling points. It is a good conductor of heat and electricity. Na combines with many other elements; in many cases, it releases an electron to form a 1+ cation.

Chlorine is a nonmetal, and is a gas at room temperature. It possesses the general properties of nonmetals: it has a low density, is a poor conductor of heat and electricity, and forms anions in chemical reactions. Chlorine is a reactive gas that combines with most other elements to produce chlorine compounds.

Hydrogen is unique because its properties are so diverse. It does not readily fit into any group on the periodic table. Hydrogen is best placed in a group by itself. In spite of this, because H has a $1s^1$ electronic configuration, it is usually placed in group IA of the periodic table.

KEY TERMS

anion
atomic size
cation

diatomic molecule
electron shielding
ion

ionization energy
isoelectronic
metal

metalloid
nonmetal
periodic properties

EXERCISES*

6.15 Define the following terms: periodic properties, metal, nonmetal, metalloid, ion, cation, anion, ionization energy, atomic size, diatomic molecule, electron shielding, isoelectronic.

Periodic Properties

6.16 Explain why the periodic table is called by that name.

6.17 Write the name of the group on the periodic table to which each of the following elements belongs: (a) Na, (b) Al, (c) C, (d) As, (e) Cu, (f) I, (g) Be, (h) Fe, (i) Sb, (j) V, (k) Ce.

6.18 (a) How is a representative element distinguished from a transition element, using the periodic table? (b) Give three examples each of representative and transition elements.

6.19 What is the outer-level electronic configuration of each: (a) alkali metals, (b) halogens, (c) chalcogens, (d) noble gases, (e) alkaline earth metals, (f) nitrogen-phosphorus group elements?

6.20 (a) How do the properties of nonmetals differ from those of metals? (b) Give two specific examples.

6.21 Classify each of the following as a metal, nonmetal, or metalloid: (a) Ge, (b) Ar, (c) Zr, (d) N, (e) H, (f) Tl, (g) Li, (h) Sr, (i) Pb, (j) Ga.

6.22 Consider the following properties of hypothetical elements A and B, and classify each as a metal, nonmetal, or metalloid. (a) Element A boils at −195.8°C, has a density of 1.3 g/L, and is a colorless gas at room temperature. (b) Ele-

*For exercise numbers printed in color, answers can be found at the back of the book.

ment B boils at 3200°C, has a density of 10 g/cm³, is a solid at room temperature, and is a good conductor.

6.23 Select the element that is the most metallic in each of the following groups of elements: (a) Be, B, C, N; (b) C, Si, Ge, Sn; (c) As, Se, Sb, Te; (d) Mg, Al, Si, P.

6.24 (a) How many added electrons are required to complete the outer energy level of (1) P, (2) Cl, (3) F, (4) Se, (5) N, and (6) I? (b) What charge does each of the atoms in part a have after the atom gains the indicated number of electrons?

6.25 (a) How many electrons must be removed from each of the following atoms to give it a noble gas electronic configuration: (1) Cs, (2) Be, (3) Al, (4) Sc, (5) Mg, and (6) Rb? (b) What charge does each of the atoms in part a have after the electrons are lost?

6.26 What charge would the most stable ions of the following atoms possess: (a) Sr, (b) P, (c) S, (d) O, (e) Mg, (f) Te, (g) N, (h) Rb, (i) F, (j) Ra?

6.27 Write equations that make use of Lewis symbols to illustrate what happens when the following atoms form ions: (a) N, (b) Cl, (c) Ba, (d) Cs, (e) P, (f) Ca.

6.28 What noble gases are isoelectronic to the following ions: (a) Mg^{2+}, (b) Al^{3+}, (c) N^{3-}, (d) I^-, (e) Rb^+, (f) Se^{2-}?

6.29 Predict the value for the ionization energy of antimony, Sb, given the values of the ionization energies for all other members of group VA. Their ionization energies in kilojoules per mole are: N, 1400; P, 1062; As, 966.5; Bi, 774.0.

6.30 For each of the following groups of atoms predict which atom will have the lowest ionization energy: (a) C, Si, Ge, Sn; (b) As, Se, Br, Kr; (c) K, Ca, Rb, Sr; (d) F, Ne, Cl, Ar.

6.31 For each set of elements in question 6.30, select the element with the smallest atomic size.

6.32 Write the symbol for the element or elements that best fit these descriptions: (a) noble gas with the lowest ionization energy, (b) group IIIA element with the least metallic character, (c) highest density in the chalcogens, (d) highest melting point of the alkali metals, (e) period 2 member that forms 3− ions, (f) a member of the lanthanide series, (g) elements whose atomic size is smaller than that of fluorine.

6.33 The element with atomic number 118 (ununoctium, Uuo) has not been discovered. For each of the listed properties make a prediction about the properties of element 118, and explain your decision: (a) metal, nonmetal, or metalloid; (b) gas, liquid, or solid; (c) good or bad conductor of electricity; (d) colored or colorless; (e) large or small atomic size compared to its group; (f) high or low ionization energy compared to its group; (g) charge on its ions; (h) high or low boiling point.

6.34 Write the symbol for the element that best fits each description: (a) lowest ionization energy on the periodic table, (b) largest atomic size on the periodic table, (c) smallest chalcogen, (d) atom that forms a 2− ion that is isoelectronic to Kr, (e) metalloid that belongs to the chalcogens, (f) group VA element with the highest density, (g) largest alkaline earth metal, (h) member of the actinide series, (i) transition element that has a completely filled 3d sublevel, (j) metallic members of group IIIA.

Representative Metal, Nonmetal, and Metalloid

6.35 Considering the electronic configurations of sodium and chlorine, what could explain the reactivity of these elements?

6.36 (a) What happens when Na contacts water? (b) Write an equation for this reaction. (c) Write the equation for the analogous reaction that occurs when K contacts water.

6.37 How are (a) Na, (b) Cl, and (c) Si used commercially?

6.38 Study the properties of Si and determine whether it has more metallic or nonmetallic character.

6.39 In what types of rocks are silicon compounds found?

6.40 What are some everyday uses of silicates?

Hydrogen

6.41 Write equations for the reaction of hydrogen with oxygen and nitrogen.

6.42 Use data from the chapter to show that the properties of hydrogen do not resemble those of the alkali metals.

6.43 Explain how hydrogen might be used as a fuel.

Additional Exercises

6.44 From the data in Table 6.4 predict the melting and boiling points of astatine, At.

6.45 Use data from Table 6.2 to plot a graph of the density of period 3 elements. Describe density trends in the third period.

6.46 (a) Graph the ionization energies of the period 4 representative elements versus their atomic numbers, using data from Fig. 6.9. (b) Compare and contrast the trends in ionization energy of the third period elements (Fig. 6.9) with those of the fourth period.

6.47 The densities of Ne, Ar, and Kr at 0°C are 0.900, 1.78, and 3.75 g/L, respectively. Predict the density of Xe at the same conditions.

6.48 Predict the atomic size and density of Rb from the following data.

	Atomic size, nm	Density, g/cm^3
Na	0.186	0.968
K	0.227	0.856
Cs	0.265	1.88

6.49 Provide a complete explanation for the fact that helium has the highest ionization energy of all atoms. Consider nuclear charge, shielding, and the size of a helium atom.

6.50 Predict as many properties as possible for the yet to be discovered element 120, unbinilium, Ubn.

6.51 The second ionization energy of an atom is the amount of energy required to remove an electron from the 1+ cation of the atom. What group on the periodic table would be expected to have the highest second ionization energy? Fully explain.

6.52 Propose a reason why the ionization energy for S (1.00 MJ/mol) is slightly less than that for P (1.01 MJ/mol).

6.53 (a) Arrange the following in order of increasing atomic size: Cl, Cl$^-$, Cl$^+$. (b) Explain your reasoning completely.

6.54 (a) Write the complete electronic configuration for Ba. (b) Write the complete electronic configuration for Ba^{2+}. (c) What atom is Ba^{2+} isoelectronic to? (d) Compare the atomic size of Ba^{2+} to that of its isoelectronic species.

6.55 Answer the following questions for the period 2 elements. (a) Which elements are metallic? (b) Which element is composed of the largest atoms? (c) Which elements do not conduct an electric current? (d) Which has the highest ionization energy?

6.56 From what sublevel is an electron removed when the ionization energy is added to each of the following atoms: (a) Al, (b) Sr, (c) I, (d) Ar, (e) Ga?

6.57 Write Lewis symbols and orbital diagrams for each of the following: (a) Mg^{2+}, (b) F$^-$, (c) S^{2-}, (d) K$^+$.

6.58 Use data in the chapter to calculate the melting point of carbon in kelvins and the density of carbon in kilograms per liter.

CHAPTER
Seven

Mole Concept and Calculations

A toms and molecules are very small particles and cannot be handled on an individual basis. Minute samples of matter contain a staggering number of particles. In this chapter we will begin our study of how chemists deal with matter in such a way that macroscopic (large) samples reflect their microscopic (very small) composition.

In Chap. 3 we discovered that the mole is a base SI unit for the amount of substance. What is a mole, and how is it applied in chemistry? The mole is a counting unit, a unit which allows us to keep track of the

7.1 THE MOLE: A COUNTING UNIT

number and mass of atoms, molecules, and ions. While you may not have encountered moles in the past, you are familiar with a common counting unit called the "dozen."

A dozen is the counting unit for 12 objects.

1 dozen objects = 12 objects

Moles and dozens are alike, except that a mole refers to a different number of objects.

Discrete (separate) objects, like oranges or eggs, often are not sold by weight, but by quantity. In such a case, the dozen can represent a mass of oranges or eggs. Counting units can be used to "weigh things by counting."

Frequently, the dozen is too small, so other counting units are employed. Paper is bought by the ream (480 sheets or 20 quires). We say that a gross of pencils are purchased, rather than 144 pencils. Counting units are used whenever it is not easy to deal with objects because of their large number or small size.

Chemists have produced a counting unit that allows them to efficiently keep track of small particles such as atoms, ions, or molecules. This counting unit, the mole, is a fixed number of objects. The number of objects in a **mole** is 6.022×10^{23}, called **Avogadro's number.** Amedeo Avogadro (1776–1856) (Fig. 7.1) was an Italian scientist who performed pioneering experiments on the properties of gases.

What is the importance of Avogadro's number, 6.022×10^{23}? If you multiply Avogadro's number by the mass in *grams* of one atom, the product is the atomic mass in *grams*. For example, the mass of one H atom is approximately 1.67×10^{-24} g, and the atomic mass of H (obtained from the periodic table) is 1.01. What number, when multiplied times 1.66×10^{-24} g, gives 1.01 **grams?** You guessed it, the answer is Avogadro's number, 6.022×10^{23}.

Stated another way, if 6.02×10^{23} atoms of an element are placed on a balance, their mass is the atomic mass in grams. Consider the following 1-mol samples of atoms:

Element	Number of atoms	Mass
1.00 mol He	6.02×10^{23} atoms He	4.00 g
1.00 mol Li	6.02×10^{23} atoms Li	6.94 g
1.00 mol C	6.02×10^{23} atoms C	12.0 g
1.00 mol Ne	6.02×10^{23} atoms Ne	20.2 g
1.00 mol Fe	6.02×10^{23} atoms Fe	55.8 g
1.00 mol U	6.02×10^{23} atoms U	238 g

In all cases, 1 mol of atoms contains Avogadro's number of atoms, which has a mass equal to the atomic mass in grams.

Figure 7.1
Amedeo Avogadro. *(The Bettmann Archive, Inc.)*

Moles

1 mol = 6.022045 $\times$ 10²³ entities

Amedeo Avogadro, Count of Quaregna (1776–1856), was the first person to distinguish between atoms and molecules (a word he coined). Most of his work was overlooked during his lifetime, and it was not until a few years after his death that his discoveries were recognized.

The actual SI definition of the mole is as follows: One mole is the amount of pure substance that contains the same number of particles as there are atoms in exactly 12 grams of ^{12}C.

How large is 1 mol? Well, 1 mol is much too large for any person to conceive. For example, it has been estimated that approximately 1 mol of grains of sand are found on all the beaches on earth. It is best to consider the fact that the magnitude of a mole is beyond comprehension when considering normal-size objects, but is a convenient number of particles when dealing with atoms, ions, and molecules. Avogadro's number of atoms can readily be measured on a balance in the laboratory.

One mole of baseballs would cover the entire earth (land and water) to a depth of about 50 mi.

7.1 What similarities are there between the units dozen and mole?

7.2 State the accepted SI definition of a mole.

7.3 What is Avogadro's number, and why is it important?

7.4 Calculate how many billions there are in exactly 1 mol.

REVIEW EXERCISES

E ach element has a molar mass, the number of grams of that element that contains 6.022×10^{23} atoms (1.000 mol). In all cases, the molar mass is the atomic mass in grams.

The mass of 1.00 mol of hydrogen atoms is 1.00 g, while the mass of 1.00 mol of He atoms is 4.00 g. Figure 7.2 shows a periodic table with the molar masses of the elements. Note that the ratio of molar masses of He to H is 4 to 1, or the same ratio found when comparing one He atom to one H atom. As long as an equal number of particles are compared, the mass ratio remains fixed.

Let us discuss normal-size objects to help develop a mental picture of what is unobservable. Consider two bags of potatoes, the first a 1-kg bag, and the other a 4-kg bag (Fig. 7.3). What is the mass ratio of the two potato bags? It is 1 to 4. If two 1-kg bags are compared to two 4-kg bags, the mass ratio is 2 kg to 8 kg, still a 1-to-4 ratio. Comparing a dozen 1-kg bags of potatoes to a dozen 4-kg bags, we find a mass ratio of 12 kg to 48 kg; a 1-to-4 ratio of masses still exists. If we compare a thousand 1-kg bags to a thousand 4-kg bags, the mass ratio is still 1 to 4. If 1 mol of each is compared, the ratio remains unchanged.

We can use the same reasoning for atoms. For example, a He atom is four times as massive as a H atom. Therefore, 6.022×10^{23} atoms of He (1.000 mol He) have four times the mass of 6.022×10^{23} atoms of H (1.000 mol H).

The molar mass of an element is obtained by writing the numerical value for the atomic mass (from the periodic table), and placing the unit grams after the number (Table 7.1). For example, the atomic mass of Ne is 20; thus, its molar mass is 20 g. The atomic mass of Kr is 40, and its molar mass is 40 g. Both 20 g of Ne and 40 g of Kr contain Avogadro's number of atoms. One mol of Kr is twice as massive as one mol of Ne, because one Kr atom is twice as massive as one Ne atom (Fig. 7.4).

7.2 MOLAR MASS OF ATOMS

Molar masses are sometimes called gram atomic masses.

Group

Period	IA(1)	IIA(2)	IIIB(3)	IVB(4)	VB(5)	VIB(6)	VIIB(7)	(8)	VIIIB (9)	(10)	IB(11)	IIB(12)	IIIA(13)	IVA(14)	VA(15)	VIA(16)	VIIA(17)	VIIIA(18)
1	1 H 1.008 g																	2 He 4.003 g
2	3 Li 6.941 g	4 Be 9.012 g											5 B 10.81 g	6 C 12.01 g	7 N 14.01 g	8 O 16.00 g	9 F 19.00 g	10 Ne 20.18 g
3	11 Na 22.99 g	12 Mg 24.31 g											13 Al 26.98 g	14 Si 28.09 g	15 P 30.97 g	16 S 32.06 g	17 Cl 35.45 g	18 Ar 39.95 g
4	19 K 39.10 g	20 Ca 40.08 g	21 Sc 44.96 g	22 Ti 47.90 g	23 V 50.94 g	24 Cr 52.00 g	25 Mn 54.94 g	26 Fe 55.85 g	27 Co 58.93 g	28 Ni 58.70 g	29 Cu 63.55 g	30 Zn 65.38 g	31 Ga 69.72 g	32 Ge 72.59 g	33 As 74.92 g	34 Se 78.96 g	35 Br 79.90 g	36 Kr 83.80 g
5	37 Rb 85.47 g	38 Sr 87.62 g	39 Y 88.91 g	40 Zr 91.22 g	41 Nb 92.91 g	42 Mo 95.94 g	43 Tc 98 g	44 Ru 101.1 g	45 Rh 102.9 g	46 Pd 106.4 g	47 Ag 107.9 g	48 Cd 112.4 g	49 In 114.8 g	50 Sn 118.7 g	51 Sb 121.8 g	52 Te 127.6 g	53 I 126.9 g	54 Xe 131.3 g
6	55 Cs 132.9 g	56 Ba 137.3 g	57* La 138.9 g	72 Hf 178.5 g	73 Ta 180.9 g	74 W 183.8 g	75 Re 186.2 g	76 Os 190.2 g	77 Ir 192.2 g	78 Pt 195.1 g	79 Au 197.0 g	80 Hg 200.6 g	81 Tl 204.4 g	82 Pb 207.2 g	83 Bi 209.0 g	84 Po 209 g	85 At 210 g	86 Rn 222 g
7	87 Fr 223 g	88 Ra 226.0 g	89† Ac 227.0 g	104 Unq 261 g	105 Unp 262 g	106 Unh 263 g												

*	58 Ce 140.1 g	59 Pr 140.9 g	60 Nd 144.2 g	61 Pm 145 g	62 Sm 150.4 g	63 Eu 152.0 g	64 Gd 157.2 g	65 Tb 158.9 g	66 Dy 162.5 g	67 Ho 164.9 g	68 Er 167.3 g	69 Tm 168.9 g	70 Yb 173.0 g	71 Lu 175.0 g
†	90 Th 232.0 g	91 Pa 231.0 g	92 U 238.0 g	93 Np 237.0 g	94 Pu 244 g	95 Am 243 g	96 Cm 247	97 Bk 247 g	98 Cf 251 g	99 Es 252 g	100 Fm 257 g	101 Md 258 g	102 No 259 g	103 Lr 260 g

Figure 7.2
One mole of like atoms has a mass equal to the atomic mass expressed in *grams*. On this periodic table the molar masses of the atoms are given to four significant figures when possible.

TABLE 7.1   MOLAR MASS OF ATOMS

Element	Atomic mass	Molar mass	Number of atoms
C	12.01	12.01 g	6.022×10^{23}
Na	22.99	22.99 g	6.022×10^{23}
Cl	35.45	35.45 g	6.022×10^{23}
Au	196.9	196.9 g	6.022×10^{23}
Th	232.0	232.0 g	6.022×10^{23}

Mole Calculations: Elements

The beauty of using moles is that after making a simple mass measurement in a laboratory the investigator can readily calculate the number of moles of atoms or the total number of atoms contained in the sample. When you solve mole problems, it is only necessary to apply the factor-label method, employing a procedure similar to the one used when converting one SI unit to another.

You should use the factor-label method to solve mole problems.

Two conversion factors are used in mole calculations that involve elements:

$$\frac{g}{mol} \quad \text{or} \quad \frac{mol}{g}$$

and

$$\frac{6.022 \times 10^{23} \text{ atoms}}{1 \text{ mol atoms}} \quad \text{or} \quad \frac{1 \text{ mol atoms}}{6.022 \times 10^{23} \text{ atoms}}$$

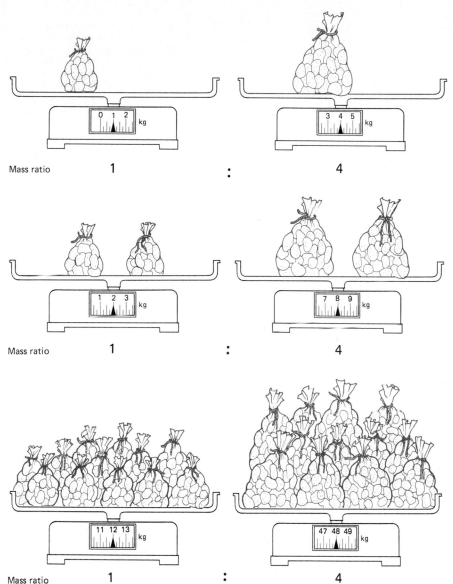

Figure 7.3
If the mass ratio of two different bags of potatoes is 1 to 4 (1 to 4 kg), then the mass ratio of equal quantities of these two different bags of potatoes is always 1 to 4.

For any quantity given—grams, atoms, or moles—the other two quantities can be calculated.

Figure 7.5 shows the pathway to follow when solving most mole problems. Atoms are converted to moles of atoms or moles of atoms are converted to atoms using 6.022×10^{23} atoms/mol. Moles of atoms are converted to grams or grams are converted to moles of atoms using the molar mass. Refer to this diagram when you attempt the problems and exercises.

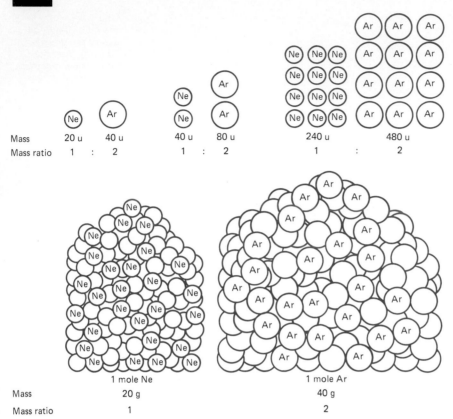

Figure 7.4
The atomic mass of neon is 20, and the atomic mass of argon is 40; an argon atom is therefore twice as massive as a neon atom. The mass of one dozen argon atoms (480 u) is twice as massive as one dozen neon atoms (240 u), and the mass of one mole of argon atoms (40 g) is twice as massive as one mole of neon atoms (20 g).

In the following series of example problems, we will consider different types of mole problems that involve atoms. Study each carefully before proceeding to the next one. Example Problem 7.1 illustrates the procedure for converting the mass of a sample to moles.

──────────── **Example Problem 7.1** ────────────

Chromium, Cr, is a hard, silvery-white solid that is used to plate iron and copper objects such as plumbing fixtures and automobile trim. How many moles of Cr atoms are there in a 1.00-g sample of Cr?

Chromium was discovered in 1797 by Vauquelin. It is used to harden steel and is plated on metal surfaces to give a bright, shiny look.

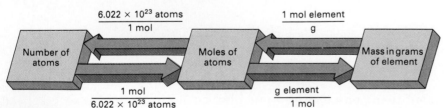

Figure 7.5
To change the number of atoms to moles of atoms, use the conversion factor 1 mol of atoms per 6.022×10^{23} atoms. To change the number of moles of atoms to its mass in grams, use the conversion factor grams of element per 1 mol. To go in the opposite direction, mass to moles, and then to atoms, use the reciprocals of these two conversion factors.

———————————— **Solution** ————————————

1. What is unknown?   Mol Cr

2. What is known?   1.00 g Cr; 1 mol Cr = 52.0 g Cr, or 1 mol Cr/ 52.0 g Cr

Beside the mass, which is given in the problem, we also know the molar mass of Cr, 52.0 g Cr/mol Cr, from the periodic table. In all mole problems, the molar mass is found on the periodic table. Refer to Fig. 7.5, which illustrates the factor-label pathway taken.

3. Apply the factor-label method.

$$1.00 \text{ g Cr} \times \frac{1 \text{ mol Cr}}{52.0 \text{ g Cr}} = ? \text{ mol Cr}$$

Grams are the known units, and moles are the units we want; thus, the conversion factor is written with grams in the denominator and moles (the desired unit) in the numerator. After the grams cancel, we are left with moles.

4. Perform the indicated math operations.

$$1.00 \text{ g Cr} \times \frac{1 \text{ mol Cr}}{52.0 \text{ g Cr}} = \mathbf{0.0192 \text{ mol Cr}}$$

Dividing 52.0 g Cr into 1.0 g Cr, we obtain 0.0192 mol Cr. Note that in the statement of the problem the mass of Cr was 1.00 g, which represents three significant figures; therefore, the answer must also have three significant figures. Never allow the molar mass to determine the number of significant figures in the answer—always use the same or more significant figures to express the molar mass.

————————————————————————————————

Example Problem 7.1 illustrates the general method for performing the majority of mole problems. As in many chemistry problems, careful attention to the units usually results in success in obtaining the correct answer.

If the number of moles of atoms are known, how is the mass of an element found? Study Example Problem 7.2, and see that it is similar to Example Problem 7.1 except for proceeding in the opposite direction with respect to the units.

———————— **Example Problem 7.2** ————————

Aluminum is the most abundant metal in the earth's crust. Al is one of the best reflectors of heat and light and is used to produce low-density alloys. What is the mass of 2.5 mol Al?

Wohler discovered Al in 1827; it was named alumium *by Davy. Its name was subsequently changed to* aluminium. *Many countries still use this name, but in 1925 it was changed to* aluminum *in the United States.*

──────────────────────────── **Solution** ────────────────────────────

1. What is unknown? Mass of Al in g

2. What is known? 2.5 mol Al; 1 mol Al/27.0 g Al (from the periodic table)

A glance at Fig. 7.5 shows us that to solve such a problem we need to know the mass of one mol Al.

3. Apply the factor-label method.

$$2.5 \; \cancel{\text{mol Al}} \times \frac{27.0 \text{ g Al}}{1 \; \cancel{\text{mol Al}}} = ? \text{ g Al}$$

Mol is canceled by g/mol, which leaves g, the desired unit.

4. Perform the indicated math operations.

$$2.5 \; \cancel{\text{mol Al}} \times \frac{27.0 \text{ g Al}}{1 \; \cancel{\text{mol Al}}} = 67.5 \text{ g Al} = \textbf{68 g Al}$$

Our answer is 68 g Al because the initial number of moles was only expressed to two significant figures.

Given a specified quantity of moles of atoms, the number of atoms contained therein is quickly found—somewhat in the same way that the number of eggs is found given how many dozen. Example Problems 7.3 and 7.4 are models for such conversions.

──────────────────── **Example Problem 7.3** ────────────────────

Calcium is a silvery-white metal. Calcium ions, Ca^{2+}, are a major component of the skeletal systems of animals. How many Ca atoms are found in a 1.200-mol sample of Ca?

──────────────────────────── **Solution** ────────────────────────────

1. What is unknown? Atoms of Ca

2. What is known? 1.200 mol Ca; 1 mol Ca/6.022×10^{23} atoms Ca

3. Apply the factor-label method.

$$1.200 \; \cancel{\text{mol Ca}} \times \frac{6.022 \times 10^{23} \text{ atoms Ca}}{1 \; \cancel{\text{mol Ca}}} = ? \text{ atoms Ca}$$

Moles of Ca are converted to atoms using the relationship between Avogadro's number and 1 mol.

Davy and Berzelius discovered calcium metal in 1808, independently of each other. Ca is the fifth most abundant element in the earth's crust. It is always found as part of a compound, never as the free metal. Limestone, gypsum, apatite, and fluorite are the most common minerals that contain Ca.

4. Perform the indicated math operations.

$$1.200 \; \text{mol Ca} \times \frac{6.022 \times 10^{23} \; \text{atoms Ca}}{1 \; \text{mol Ca}} = \textbf{7.226} \times \textbf{10}^{\textbf{23}} \; \textbf{atoms Ca}$$

A 1.200-mol Ca sample contains 7.226×10^{23} atoms Ca. The molar mass is not needed in this problem; no mass is specified, just the number of moles. It should be evident that any element could be substituted for Ca, and the answer would be the same: 1.200 mol of any element contains exactly the same number of atoms (think about dozens, if this statement does not register).

Example Problem 7.4

Gold, silver, and copper are called the coinage metals. They have been used for coins and jewelry for thousands of years. A sample of Au is found to contain 9.7×10^{23} atoms Au. How many moles of Au atoms are there in the sample?

Two forms of gold are found in nature, native gold (1 to 50% silver alloy) and telluride ore ($AuTe_2$). Native gold is found as veins and dust in quartzite rock or in deposits that result after the rock weathers.

Solution

1. What is unknown?   Mol Au

2. What is known? 9.7×10^{23} atoms Au; 6.022×10^{23} atoms Au/mol Au

3. Apply the factor-label method.

$$9.7 \times 10^{23} \; \text{atoms Au} \times \frac{1 \; \text{mol Au}}{6.022 \times 10^{23} \; \text{atoms Au}} = ? \; \text{mol Au}$$

4. Perform the indicated math operations.

$$9.7 \times 10^{23} \; \text{atoms Au} \times \frac{1 \; \text{mol Au}}{6.022 \times 10^{23} \; \text{atoms Au}} = \textbf{1.6 mol Au}$$

When you divide numbers expressed in scientific notation, remember to divide the coefficients and subtract the exponents. Our answer is expressed to two significant figures because two significant figures are given for the initial number of atoms of Au.

In Example Problems 7.1 through 7.4 all conversions were accomplished using one conversion factor. To convert masses of elements to atoms, or find the masses of given numbers of atoms, more than one conversion factor is required. Given the mass of an element, no simple conversion to the number of atoms is known; hence, two conversion factors are utilized: molar mass (grams per mole) and Avogadro's number (atoms per mole) (refer to Fig. 7.5). Consider Example Problems 7.5 and 7.6 as illustrative models for such problems.

──────────────── **Example Problem 7.5** ────────────────

Because lead, Pb, is easily extracted from its ore, it was known and used by the ancient Egyptians. Lead pipes were used to carry the water of the ancient Romans. How many Pb atoms are found in a 100.0-gram sample of pure Pb?

──────────────── **Solution** ────────────────

1. What is unknown?   Number of atoms of Pb

2. What is known?   100.0 g Pb; 1 mol Pb/207.2 g Pb; 6.022×10^{23} atoms Pb/mol Pb

3. Apply the factor-label method.

$$100.0 \text{ g Pb} \times \frac{1 \text{ mol Pb}}{207.2 \text{ g Pb}} \times \frac{6.022 \times 10^{23} \text{ atoms Pb}}{1 \text{ mol Pb}} = ? \text{ atoms Pb}$$

4. Perform the indicated math operations.

$$100.0 \text{ g Pb} \times \frac{1 \text{ mol Pb}}{207.2 \text{ g Pb}} \times \frac{6.022 \times 10^{23} \text{ atoms Pb}}{1 \text{ mol Pb}}$$
$$= \mathbf{2.906 \times 10^{23} \text{ atoms Pb}}$$

Our answer is expressed to four significant figures because the initial mass of Pb is expressed to four significant figures.

A good habit to develop is to ask yourself after performing the indicated math operations if the answer is reasonable. In this problem, the mass, 100.0 grams, is slightly less than half the molar mass, which gives an answer smaller than 0.5 moles, or one-half of Avogadro's number.

──────────────── **Example Problem 7.6** ────────────────

Uranium is a dense, radioactive metal that can burst into flames if finely divided. What is the mass of 5.64×10^{23} atoms of uranium?

──────────────── **Solution** ────────────────

1. What is unknown?   Mass of U in g

2. What is known? 5.64×10^{23} atoms U; 1 mol U/6.022×10^{23} atoms U; 238 g U/mol U

The Latin name for lead is plumbum. From this old name the symbol Pb is derived. Lead is a soft, bluish-white metal that is malleable and has a density of 11.4 g/cm³.

Uranium was discovered by Klaproth in 1789. He found it in an ore called pitchblende. All the isotopes of U are radioactive.

3. Apply the factor-label method.

$$5.64 \times 10^{23} \text{ atoms U} \times \frac{1 \text{ mol U}}{6.022 \times 10^{23} \text{ atoms U}} \times \frac{238 \text{ g U}}{1 \text{ mol U}} = ? \text{ g U}$$

4. Perform the indicated math operations.

$$5.64 \times 10^{23} \text{ atoms U} \times \frac{1 \text{ mol U}}{6.022 \times 10^{23} \text{ atoms U}} \times \frac{238 \text{ g U}}{1 \text{ mol U}} = \mathbf{223 \text{ g U}}$$

Is the answer reasonable? Yes, 5.64×10^{23} atoms U is smaller than Avogadro's number, which is the number of particles in 1 mol. Thus, the answer should be less than the molar mass of U, which is 238 g/mol.

REVIEW EXERCISES

7.5 How many moles of atoms are contained in the following samples: (a) 5.5 g Ne, (b) 3.01 g Fe, (c) 0.087 kg Ar, (d) 22.5 mg Mn?

7.6 Calculate the mass of each of the following samples: (a) 8.19 mol V, (b) 0.0044 mol Ge, (c) 5.03×10^3 mol As, (d) 4.100 mmol Al.

7.7 How many atoms are found in each of the following noble gas samples: (a) 5.00 g He, (b) 5.00 g Ne, (c) 5.00 g Ar, (d) 5.00 g Kr?

7.8 Calculate the number of moles of atoms represented by each of the following:
(a) 8.211×10^{23} atoms B
(b) 4.7×10^{20} atoms Ag
(c) 9.904×10^{26} atoms Cu
(d) 6.29×10^{21} atoms Cs

7.3 MOLAR MASS OF COMPOUNDS

A compound is a chemical combination of atoms. Each compound has a fixed ratio of elements; or on a smaller scale, each molecule is composed of a fixed ratio of atoms. Accordingly, we can apply the mole concept to molecules, as we did to atoms.

Molecular Mass

Molecular mass (traditionally, molecular weight) is the sum of the atomic masses of the atoms within a molecule. Let us use water, H_2O, as an example. Each water molecule contains two H atoms and one O atom. The atomic masses of H and O are 1.0 and 16.0, respectively. To find the molecular mass of water, multiply 2 times 1.0 to obtain the total mass of H, and add that to 16.0; the result, 18, is the molecular mass.

H	2 atoms $\times$ 1.0 =	2.0
O	1 atom $\times$ 16.0 =	+16.0
H_2O	=	18.0

While it appears that water is a simple substance, it is not. Water exhibits "abnormal" properties for such a "simple" compound.

If the molecular mass of water is 18, then its molar mass is 18.0 **g**, and 18.0 g H_2O contains Avogadro's number of molecules, 6.02×10^{23} molecules of H_2O. Example Problem 7.7 shows how to calculate the molecular masses of three compounds.

──────────────── **Example Problem 7.7** ────────────────

Find the molecular masses of (a) CH_4, (b) HNO_3, (c) $B_{10}H_{16}$.

──────────────── **Solution** ────────────────

(a) Molecular mass of CH_4:

CH_4, methane, is the primary component of natural gas, an important fuel.

C	1 atom $\times$ 12.0 =	12.0
H	4 atoms $\times$ 1.0 =	+ 4.0
CH_4	=	16.0

(b) Molecular mass of HNO_3:

HNO_3 is nitric acid, a strong mineral acid of great commercial importance.

H	1 atom $\times$ 1.0	=	1.0
N	1 atom $\times$ 14.0	=	14.0
O	3 atoms $\times$ 16.0	=	+48.0
HNO_3		=	63.0

(c) Molecular mass of $B_{10}H_{16}$:

B	10 atoms $\times$ 10.81 =	108.1
H	16 atoms $\times$ 1.01 =	+ 16.1
$B_{10}H_{16}$	=	124.2

In these molecular mass calculations the number of atoms of each type is multiplied by the element's atomic mass, and then the quantities are added to obtain the sum.

Molar masses of compounds are sometimes called gram molecular masses.

Mole Calculations: Compounds

Mole calculations that involve molecules are the same as those that involve atoms, except that the fundamental particle is a molecule. Therefore, the molar mass of a compound is the molecular mass in grams, and Avogadro's number of molecules are contained within the molar mass of a compound.

What is the molar mass of carbon dioxide, CO_2, and how many particles are contained in that mass? Calculate the molecular mass of CO_2; one C atom, 12, and two O atoms, each 16, give a molecular mass of 44. Adding grams as the unit, we obtain 44 g as the molar mass of CO_2. In 44 g of CO_2, there are 6.0×10^{23} molecules of CO_2.

1 mol CO_2 = 44 g

1 mol CO_2 = 6.022 $\times$ 10^{23} CO_2 molecules

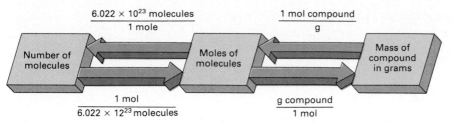

Figure 7.6
To change the number of molecules to moles of molecules, use the conversion factor 1 mol of molecules per 6.022×10^{23} molecules. To change the number of moles of molecules to grams, use the conversion factor grams per 1 mol. To go in the opposite direction, mass to moles, and then to molecules, use the reciprocals of these two conversion factors.

From Example Problem 7.7, we found that the molar masses of CH_4, HNO_3, and $B_{10}H_{16}$ are 16.0 g, 63.0 g, and 124.2 g, respectively. Avogadro's number of CH_4 molecules are found in 16.0 g CH_4. Also, 63.0 g HNO_3 and 124.2 g $B_{10}H_{16}$ each contain 6.02×10^{23} molecules.

As was the case with atoms, most mole calculations for compounds involve interconversions of units—moles, molecules, and grams. Figure 7.6 summarizes mole conversions for compounds.

Carefully go through Example Problems 7.8 through 7.10, using Fig. 7.6 when necessary. Note that the calculations are similar to mole calculations with atoms except that molecular masses are used instead of atomic masses.

Example Problem 7.8

Nitrous oxide, N_2O, is a colorless gas with a pleasant odor. It is also called *laughing gas*. Nitrous oxide has been used as a general anesthetic for medical and dental operations for over 50 years. What is the mass of a 4.05-mol sample of N_2O?

The nickname "laughing gas" for nitrous oxide was coined by Sir Humphry Davy, after he discovered the anesthetic effects of the gas.

Solution

---

1. What is unknown?   Mass of N_2O in g

2. What is known?   4.05 mol N_2O; the molar mass of N_2O, which is calculated from the atomic masses as follows:

$$N \quad 2 \; \cancel{mol \; N} \times \frac{14.0 \; g \; N}{1 \; \cancel{mol \; N}} = \quad 28.0 \; g \; N$$

$$O \quad 1 \; \cancel{mol \; O} \times \frac{16.0 \; g \; O}{1 \; \cancel{mol \; O}} = +16.0 \; g \; O$$

$$N_2O \qquad\qquad\qquad = \quad 44.0 \; g/mol \; N_2O$$

3. Apply the factor-label method.

$$4.05 \; \cancel{mol \; N_2O} \times \frac{44.0 \; g \; N_2O}{1 \; \cancel{mol \; N_2O}} = ? \; g \; N_2O$$

4. Perform the indicated math operations.

$$4.05 \text{ mol N}_2\text{O} \times \frac{44.0 \text{ g N}_2\text{O}}{1 \text{ mol N}_2\text{O}} = \textbf{178 g N}_2\textbf{O}$$

The mass of 4.05 mol N_2O is 178 g.

———————————— **Example Problem 7.9** ————————————

Sulfuric acid, H_2SO_4, is the industrial chemical produced in greatest amount in the United States. It has hundreds of uses in industry. How many sulfuric acid molecules are there in a 0.100-g sample of H_2SO_4?

———————————— **Solution** ————————————

Sulfuric acid is used to make fertilizers, drugs, and dyes. It is an important compound in the refining of crude oil and is used by the steel industry to clean the surfaces of metals.

1. What is unknown?   Molecules of H_2SO_4

2. What is known?   0.100 g H_2SO_4; 1 mol H_2SO_4/6.022 × 10²³ molecules H_2SO_4; the molar mass of H_2SO_4, calculated from its atomic masses:

$$H \qquad 2 \text{ mol H} \times \frac{1.0 \text{ g H}}{1 \text{ mol H}} = \quad 2.0 \text{ g H}$$

$$S \qquad 1 \text{ mol S} \times \frac{32.1 \text{ g S}}{1 \text{ mol S}} = \quad 32.1 \text{ g S}$$

$$O \qquad 4 \text{ mol O} \times \frac{16.0 \text{ g O}}{1 \text{ mol O}} = +64.0 \text{ g O}$$

$$H_2SO_4 \qquad\qquad\qquad\qquad = \quad 98.1 \text{ g/mol } H_2SO_4$$

3. Apply the factor-label method.

$$0.100 \text{ g H}_2\text{SO}_4 \times \frac{1 \text{ mol H}_2\text{SO}_4}{98.1 \text{ g H}_2\text{SO}_4} \times \frac{6.022 \times 10^{23} \text{ molecules H}_2\text{SO}_4}{1 \text{ mol H}_2\text{SO}_4}$$
$$= \text{ ? molecules } H_2SO_4$$

4. Perform the indicated math operations.

$$0.100 \text{ g H}_2\text{SO}_4 \times \frac{1 \text{ mol H}_2\text{SO}_4}{98.1 \text{ g H}_2\text{SO}_4} \times \frac{6.022 \times 10^{23} \text{ molecules H}_2\text{SO}_4}{1 \text{ mol H}_2\text{SO}_4}$$
$$= \textbf{6.14} \times \textbf{10}^{\textbf{20}} \textbf{ molecules H}_2\textbf{SO}_4$$

A 0.100-g sample of H_2SO_4 contains 6.14 × 10²⁰ molecules H_2SO_4.

—————————— **Example Problem 7.10** ——————————

Glucose, $C_6H_{12}O_6$, is also known as blood sugar because it circulates in the blood. It is also found in grapes, honey, and various fruits. What is the mass in grams of 3.39×10^{33} molecules of glucose?

—————————— **Solution** ——————————

1. What is unknown?   Mass of $C_6H_{12}O_6$ in g

2. What is known? 3.39×10^{22} molecules $C_6H_{12}O_6$; 1 mol $C_6H_{12}O_6/6.02 \times 10^{23}$ $C_6H_{12}O_6$ molecules; the molar mass of $C_6H_{12}O_6$, calculated as follows:

$$C \quad 6 \text{ mol C} \times \frac{12.0 \text{ g C}}{1 \text{ mol C}} = \quad 72.0 \text{ g C}$$

$$H \quad 12 \text{ mol H} \times \frac{1.0 \text{ g H}}{1 \text{ mol H}} = \quad 12.0 \text{ g H}$$

$$O \quad 6 \text{ mol O} \times \frac{16.0 \text{ g O}}{1 \text{ mol O}} = +96.0 \text{ g O}$$

$$C_6H_{12}O_6 \qquad\qquad\qquad 180.0 \text{ g/mol } C_6H_{12}O_6$$

3. Apply the factor-label method.

$$3.39 \times 10^{22} \text{ molecules } C_6H_{12}O_6 \times \frac{1 \text{ mol } C_6H_{12}O_6}{6.02 \times 10^{23} \text{ molecules } C_6H_{12}O_6}$$

$$\times \frac{180.0 \text{ g } C_6H_{12}O_6}{1 \text{ mol } C_6H_{12}O_6} = ? \text{ g } C_6H_{12}O_6$$

4. Perform the indicated math operations.

$$3.39 \times 10^{22} \text{ molecules } C_6H_{12}O_6 \times \frac{1 \text{ mol } C_6H_{12}O_6}{6.02 \times 10^{23} \text{ molecules } C_6H_{12}O_6}$$

$$\times \frac{180.0 \text{ g } C_6H_{12}O_6}{1 \text{ mol } C_6H_{12}O_6} = \textbf{10.1 g } \boldsymbol{C_6H_{12}O_6}$$

The mass of 3.39×10^{22} molecules of glucose is 10.1 g.

Moles of Elements in Compounds

In the future, it will sometimes be necessary to calculate the number of moles of atoms or the number of atoms in a specified quantity of a compound. For example, how many moles of H and O are there in 1 mol of water, H_2O? Water molecules contain two atoms of H and one atom of O; consequently, 1 mol H_2O contains 2 mol H (2 g H) and 1 mol O (16 g O).

In 1 mol H_2O, there are 2 mol H and 1 mol O.

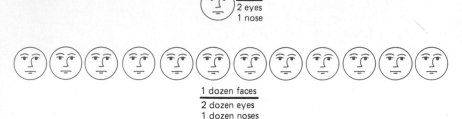

Figure 7.7
One person has two eyes and one nose. One dozen people have two dozen eyes and one dozen noses. One mole of people have 2 mol of eyes and 1 mol of noses.

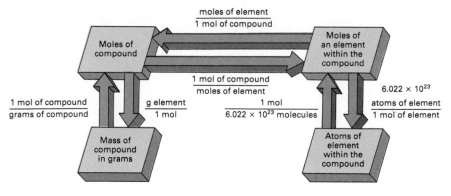

Figure 7.8
To find the number of atoms of a specific type in a given mass of compound, three conversion factors are required. The first is the mass of 1 mol of the compound (1 mol per grams of compound); the second factor is the number of moles of the desired element per 1 mol of compound (moles of element per 1 mol of compound); and the third factor is the number of atoms per 1 mol of the element (6.022 × 10²³ atoms per 1 mol of element).

Using another large-scale analogy, let us consider the face of a person. Each person has one nose and two eyes. One dozen people, among them, possess one dozen noses and two dozen eyes (Fig. 7.7). It stands to reason that 1 mol of people have 1 mol of noses and 2 mol of eyes.

When confronted with such problems, find the number of moles of the compound of interest, and then use the appropriate conversion factor that gives the number of atoms per molecule. Figure 7.8 maps the pathway to be taken, and Example Problem 7.11 shows how such a problem is solved.

Example Problem 7.11

How many moles of C and H atoms are in 9.25 g C_5H_{10}?

Solution

1. What is unknown?   Mol C and mol H

2. What is known?   9.25 g C_5H_{10}; 5 mol C/mol C_5H_{10}; 10 mol H/mol C_5H_{10}; the molar mass of C_5H_{10}:

$$\text{C} \qquad 5 \text{ mol C} \times \frac{12.0 \text{ g C}}{1 \text{ mol C}} = \quad 60.0 \text{ g C}$$

$$H \quad 10 \, \overline{mol \, H} \times \frac{1.0 \, g \, H}{1 \, \overline{mol \, H}} = +10.0 \, g \, H$$

$$C_5H_{10} \qquad\qquad = \quad 70.0 \, g/mol \, C_5H_{10}$$

3. Apply the factor-label method (refer to Fig. 7.8).

$$9.25 \, \overline{g \, C_5H_{10}} \times \frac{1 \, \overline{mol \, C_5H_{10}}}{70.0 \, \overline{g \, C_5H_{10}}} \times \frac{5 \, mol \, C}{1 \, \overline{mol \, C_5H_{10}}} = ? \, mol \, C$$

It is not necessary to write the full setup to find the number of moles of H. The ratio of H to C in C_5H_{10} is 10 to 5, or 2 to 1: 2 mol H/1 mol C. After calculating the number of moles of C, we can use this conversion factor to find the moles of H.

$$\overline{mol \, C} \times \frac{2 \, mol \, H}{1 \, \overline{mol \, C}} = mol \, H$$

4. Perform the indicated math operations.

$$9.25 \, \overline{g \, C_5H_{10}} \times \frac{1 \, \overline{mol \, C_5H_{10}}}{70.0 \, \overline{g \, C_5H_{10}}} \times \frac{5 \, mol \, C}{1 \, \overline{mol \, C_5H_{10}}} = \mathbf{0.661 \, mol \, C}$$

$$0.661 \, \overline{mol \, C} \times \frac{2 \, mol \, H}{1 \, \overline{mol \, C}} = \mathbf{1.32 \, mol \, H}$$

After calculating the number of moles of C in 9.25 g C_5H_{10}, it is only necessary to double the value to obtain the number of moles of H atoms.

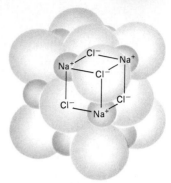

Figure 7.9
Sodium chloride, NaCl, is composed of a cubic network of sodium ions, Na^+, surrounded by chloride ions, Cl^-, (or chloride ions surrounded by sodium ions). Discrete molecules of sodium chloride do not exist in this structure.

Compounds Not Composed of Molecules

When two nonmetals are combined chemically, the resulting compound is usually composed of discrete molecules; examples of compounds such as these are CO, NO, CH_4, and H_2O. Other compounds are not comprised of discrete, or separate, molecules. Instead, their structures are networks of ions chemically bonded together. For these compounds, the term molecular mass has no real meaning because no molecules are present. A more accurate term is applied to these compounds: formula mass. **Formula mass** is used for those substances not composed of discrete molecules in the same way that molecular mass is used for compounds composed of molecules.

An example of a compound not composed of discrete molecules is sodium chloride, NaCl. Sodium chloride is composed of Na^+ and Cl^- ions in a three-dimensional pattern (Fig. 7.9). Each Na^+ cation is surrounded by Cl^- ions, and each Cl^- anion is surrounded by Na^+ ions. In NaCl, for every Na^+ there is one Cl^-: the two elements are found in a 1-to-1 ratio. We speak of this as the **formula unit** or the simplest ratio of atoms within the compound.

Formula mass NaCl = 58.5 g NaCl = 6.022 × 10^{23} formula units NaCl

In Chap. 8, we will investigate the structure of these compounds; for now, we will simply say that most compounds that result from combinations of metals and nonmetals or metals and the charged groups of atoms called polyatomic ions do not exist as discrete molecules. Members of this group of substances are called *ionic compounds*. Examples include calcium fluoride, CaF_2; sodium nitrate, $NaNO_3$; copper sulfate, $CuSO_4$; and many others. Table 7.2 lists some commonly encountered ionic substances.

TABLE 7.2 EXAMPLES OF IONIC COMPOUNDS

Compound	Formula	Formula mass
Ammonium nitrate	NH_4NO_3	80.0
Calcium chloride	$CaCl_2$	111.0
Iron(III) oxide	Fe_2O_3	159.7
Magnesium hydroxide	$Mg(OH)_2$	58.3
Titanium(IV) oxide	TiO_2	79.9

To calculate the formula mass of an ionic compound, you should follow the same procedure you used to find the molecular mass of a substance composed of molecules. In most cases, the formula mass is treated in the same way as molecular mass. For example, NaCl has a formula mass of 58.5 and a molar mass of 58.5 g that is composed of Avogadro's number of formula units.

REVIEW EXERCISES

7.9 Calculate the molar mass of (a) Br_2, (b) SO_3, (c) HNO_3, (d) $Al_2(SO_4)_3$.

7.10 What is the mass in grams of each of the following samples: (a) 5.11 mol PF_3, (b) 8.012×10^{23} molecules $SiCl_4$, (c) 0.0809 mmol P_4O_{10}?

7.11 How many molecules are there in each of the following samples: (a) 10.00 g H_2S, (b) 10.00 g ClF_3, (c) 10.00 g C_5Br_{12}?

7.12 Which of the following contains the largest number of moles of oxygen atoms: (a) 25 g CO_2, (b) 25 g $C_6H_{12}O_6$, or (c) 25 g H_3PO_4?

7.13 Explain why it is incorrect to refer to the "molecular mass" of NaCl.

7.4 MOLES AND CHEMICAL FORMULAS

Armed with the mole concept, chemists can obtain a tremendous amount of information about a compound, solution, or mixture by finding the mass of its components. Moles are used by chemists to calculate (1) formulas of compounds, (2) mass relationships in chemical reactions, and (3) concentrations of solutions and mixtures.

In this section, we will study formula calculations and determinations. We will apply the mole concept to equations and solutions in Chaps. 11 and 15.

Law of Constant Composition

Each compound is composed of elements in a fixed mass ratio. Any sample of H_2O contains by mass 11% H and 89% O. If a sample of a H and O compound is studied and different percents by mass are found, it must be concluded that the substance is not pure water. To generalize:

All samples of a given compound contain the same elements in a fixed mass ratio. This is a statement of the **law of constant composition,** which is also called the **law of definite proportions.**

The law of constant composition was proposed by the French scientist Joseph-Louis Proust (1754–1826). At the time, it was thought that the composition of a substance could vary, depending on the sample. Proust's "radical" new hypothesis took about 10 years to catch on.

A direct result of the law of constant composition is that each element in a compound can be expressed as a mass percent. Collectively, all mass percents of elements in a compound are called the **percent composition** of the substance. The percent composition of water is 11% H and 89% O, and the percent composition of table sugar (sucrose) is 42.1% C, 6.4% H, and 51.5% O. For all compounds, the sum of the mass percents equals 100%.

How is the percent composition of a compound determined? A specific mass of the compound is analyzed to find the masses of each element. Percent by mass of each element is then calculated by dividing the mass of each element by the total mass of the compound, and then multiplying by 100. As with all percent calculations you take the part and divide it by the whole and multiply times 100.

It is most convenient to use the molar mass of a compound in percent composition calculations. When you determine the molar mass, the mass of each element is calculated. It is then only necessary to convert these masses to percents. Example Problem 7.12 shows how the percent composition of nitrogen(V) oxide, N_2O_5, is found.

Proust was the son of an apothecary and was one of the first chemists to study sugars. Proust was an avid balloonist, making one of the first ascensions in 1784.

Percent Composition

A percent is the number of parts per 100 total parts. Percent by mass is the ratio of the mass of a component to 100 g total mass.

Example Problem 7.12

Nitrogen(V) oxide, N_2O_5, is a gas that combines with water to produce nitric acid, HNO_3, which is one of the most important industrial acids. What is the percent composition of N_2O_5?

Solution

---

1. What is unknown?   %N and %O in N_2O_5, in which %N is (g N/g N_2O_5) × 100, and %O is (g O/g N_2O_5) × 100

2. What is known?   Molecular formula of the compound, N_2O_5; atomic masses of N and O, 14.0 g N/mol N and 16.0 g O/mol O

From this information, the molar mass is determined:

$$N \quad 2\ \text{mol N} \times \frac{14.0\ \text{g N}}{1\ \text{mol N}} = \quad 28.0\ \text{g N}$$

$$O \quad 5\ \text{mol O} \times \frac{16.0\ \text{g O}}{1\ \text{mol O}} = +\ 80.0\ \text{g O}$$

$$N_2O_5 \qquad\qquad\qquad = \quad 108.0\ \text{g/mol } N_2O_5$$

3. Calculate the percent by mass of N and O in N_2O_5.

$$\%N = \frac{g\ N}{g\ N_2O_5} \times 100 = \frac{28.0\ g\ N}{108.0\ g\ N_2O_5} \times 100 = \textbf{25.9\% N}$$

$$\%O = \frac{g\ O}{g\ N_2O_5} \times 100 = \frac{80.0\ g\ O}{108.0\ g\ N_2O_5} \times 100 = \textbf{74.1\% O}$$

N_2O_5 has a percent composition of 25.9% N and 74.1% O. Note that the percents always add up to 100%. A common practice is to calculate the percent of one element and then subtract that value from 100 to obtain the percent for the second element. This philosophy is fine as long as the first value is correct. A better strategy is to compute both percents and check to see if they add up to 100%.

Example Problem 7.13 illustrates the calculation of the percent composition of a compound containing more than two elements: ethanol, C_2H_6O, which is also called grain or drinking alcohol.

─────────────── **Example Problem 7.13** ───────────────

Ethanol, C_2H_6O, is produced when grains or fruits are fermented by yeast. Find the percent composition of ethanol.

─────────────── **Solution** ───────────────

1. What is unknown? %C, %H, and %O in C_2H_6O

2. What is known? The molar mass of each element and the molecular formula
The molar mass of ethanol is calculated as follows:

C $\quad 2\ \text{mol C} \times \dfrac{12.0\ g\ C}{1\ \text{mol C}} = \quad 24.0\ g\ C$

H $\quad 6\ \text{mol H} \times \dfrac{1.01\ g\ H}{1\ \text{mol H}} = \quad 6.06\ g\ H$

O $\quad 1\ \text{mol O} \times \dfrac{16.0\ g\ O}{1\ \text{mol O}} = +16.0\ g$

$C_2H_6O \qquad\qquad\qquad\qquad = \quad 46.1\ g/mol\ C_2H_6O$

3. Calculate the percent by mass of C, H, and O in C_2H_6O.

$$\%C = \frac{g\ C}{g\ C_2H_6O} \times 100 = \frac{24.0\ g\ C}{46.1\ g\ C_2H_6O} \times 100 = \textbf{52.1\%}$$

Ingestion of a large quantity of ethanol causes unconsciousness and a comalike state. Death can result if the blood ethanol concentration exceeds 0.5%.

$$\%H = \frac{g\ H}{g\ C_2H_6O} \times 100 = \frac{6.06\ g\ H}{46.1\ g\ C_2H_6O} \times 100 = \mathbf{13.1\%}$$

$$\%O = \frac{g\ O}{g\ C_2H_6O} \times 100 = \frac{16.0\ g\ O}{46.1\ g\ C_2H_6O} \times 100 = \mathbf{34.7\%}$$

Our calculations indicate that the percent composition is 52.1% C, 13.1% H, and 34.7% O. The sum of these three percents 99.9%. Sometimes, due to rounding errors, the percents will not add up to exactly 100%.

Empirical Formula Calculations

In the laboratory, chemists determine formulas of compounds as one way of characterizing compounds. To calculate a compound's empirical formula, it is only necessary to obtain the percent composition or other mass data on the elements in the compound. **Empirical formulas** express the smallest whole-number ratio of atoms within a molecule.

To calculate the empirical formula of a compound, we must know either the percent composition or the mass ratio of the elements. The mass of each element in the compound is converted to moles and the mole ratio is found. The empirical formula is obtained when the simplest mole ratio is calculated. In most cases, the empirical formula is a ratio of whole numbers, such as 3 to 1, or 5 to 2.

Let us consider hydrogen peroxide, which is an antiseptic and bleaching agent. Its molecular formula is H_2O_2. A molecular formula expresses the actual number of atoms in the molecule. One hydrogen peroxide molecule contains two H atoms and two O atoms. Its empirical formula, HO (H_1O_1), is obtained by dividing the subscripts of the molecular formula by 2. Table 7.3 lists molecular and empirical formulas for selected molecules.

A 3% solution of hydrogen peroxide is used as a germicide. In research and industry, 30% solutions are frequently used. Pure H_2O_2 is a pale-blue liquid that must be handled with extreme care because it decomposes violently or readily combines with substances that it contacts.

TABLE 7.3 SIMPLEST AND MOLECULAR FORMULAS

Compound	Molecular formula	Empirical formula
A boron hydrogen	B_4H_{10}	B_2H_5
Benzene	C_6H_6	CH
Glucose	$C_6H_{12}O_6$	CH_2O
Mercury(I) chloride	Hg_2Cl_2	HgCl
Propane	C_3H_8	C_3H_8
Sodium oxalate	$Na_2C_2O_4$	$NaCO_2$

Many compounds have the same molecular and empirical formula—e.g., propane in Table 7.3. Whenever the subscripts of the molecular formula of a compound are not divisible by a common number, the empirical formula is the same as the molecular formula. Other examples of such compounds are potassium nitrate, KNO_3; sulfur trioxide, SO_3; and phosphorus pentachloride, PCl_5.

Example Problem 7.14 shows how the empirical formula of a compound is obtained, given the percent composition. Each step in the calculation is illustrated in Fig. 7.10.

─────────────── **Example Problem 7.14** ───────────────

A nitrogen-fluorine compound contains 26.9% N and 73.1% F. What is the empirical formula of the compound?

─────────────── **Solution** ───────────────

1. What is unknown? Empirical formula, which is the mole ratio of N to F or vice versa

2. What is known? 26.9% N and 73.1% F; the molar masses of N and F, 14.0 g N/mol N and 19.0 g F/mol F

3. Calculate the number of moles of each element.

The percents given are the mass ratios of each element per 100 g of compound; specifically,

$$26.9\% \text{ N} = \frac{26.9 \text{ g N}}{100 \text{ g compound}}$$

$$73.1\% \text{ F} = \frac{73.1 \text{ g F}}{100 \text{ g compound}}$$

Thus, it is convenient to find the number of moles that corresponds to 26.9 g N and 73.1 g F.

$$26.9 \text{ g N} \times \frac{1 \text{ mol N}}{14.0 \text{ g N}} = 1.92 \text{ mol N}$$

$$73.1 \text{ g F} \times \frac{1 \text{ mol F}}{19.0 \text{ g F}} = 3.85 \text{ mol F}$$

4. Calculate the smallest whole-number mole ratio.

Find the element with the smallest number of moles, and then divide this value into the number of moles of each element. In our problem, 1.92 mol N is the smallest quantity of moles; thus, it is divided into itself, giving exactly 1, and then divided into 3.85 mol F, giving 2.01.

$$\frac{1.92 \text{ mol N}}{1.92 \text{ mol N}} = 1.00$$

$$\frac{3.85 \text{ mol F}}{1.92 \text{ mol N}} = \frac{2.01 \text{ mol F}}{1 \text{ mol N}}$$

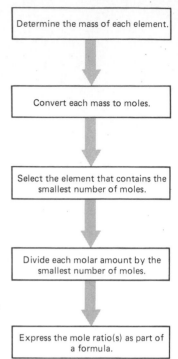

Determine the mass of each element.

Convert each mass to moles.

Select the element that contains the smallest number of moles.

Divide each molar amount by the smallest number of moles.

Express the mole ratio(s) as part of a formula.

Figure 7.10
Use this stepwise procedure to calculate the empirical formula of a compound.

The simplest ratio is 2 mol F to 1 mol N, or, translated into an empirical formula, **NF₂.** If the simplest mole ratio is 1 to 2, this indicates that the ratio of N to F atoms in the molecule is also 1 to 2. With no other information, the molecular formula cannot be calculated. All that we know is the empirical formula; the molecular formula could possibly be NF_2, N_2F_4, N_3F_6, or N_4F_8, all of which have a ratio of 1 N atom to 2 F atoms.

It is not necessary to know the percent composition to calculate the empirical formula of a compound. All that is required is mass data on the elements in a sample of the compound (Fig. 7.10). The mass of each element is converted to moles, and the simplest mole ratio is calculated. Example Problem 7.15 shows such a computation.

--------- **Example Problem 7.15** ---------

Ascorbic acid is the chemical name for vitamin C. Vitamins are essential substances required in the human diet almost every day. Analysis of a 2.642-g sample of ascorbic acid reveals that there is 1.081 g C, 0.121 g H, and the remainder is oxygen. What is the empirical formula of ascorbic acid?

An insufficient amount of ascorbic acid in a person's diet leads to the vitamin deficiency disease called scurvy. A minimum of 60 mg daily of ascorbic acid is required in the human diet to prevent scurvy.

--------- **Solution** ---------

1. What is unknown? Empirical formula of ascorbic acid, which is the simplest mole ratio of C, H, and O

2. What is known? In order to perform this calculation, the masses of all substances must be identified. The problem gives the total mass of ascorbic acid and the masses of C and H. The mass of O can be found by subtracting the sum of the masses of C and H from the total mass.

$$\text{Mass of O} = \text{total mass} - (\text{mass C} + \text{mass H})$$
$$= 2.642 \text{ g total} - (1.081 \text{ g C} + 0.121 \text{ g H})$$
$$= 1.440 \text{ g O}$$

From the periodic table the molar masses of C, H, and O are obtained. They are 12.01 g C/mol C, 1.01 g H/mol H, and 16.00 g O/mol O.

3. Calculate the number of moles of each element.

$$1.081 \text{ g C} \times \frac{1 \text{ mol C}}{12.01 \text{ g C}} = 0.09001 \text{ mol C}$$

$$0.121 \text{ g H} \times \frac{1 \text{ mol H}}{1.01 \text{ g H}} = 0.120 \text{ mol H}$$

$$1.440 \text{ g O} \times \frac{1 \text{ mol O}}{16.00 \text{ g O}} = 0.09000 \text{ mol O}$$

4. Find the simplest mole ratio.

Because 0.09000 mol O is the smallest number of moles, divide 0.09000 mol O into each quantity to find the ratio of whole numbers.

$$\frac{0.09001 \text{ mol C}}{0.09000 \text{ mol O}} = \frac{1.000 \text{ mol C}}{1 \text{ mol O}}$$

$$\frac{0.120 \text{ mol H}}{0.09000 \text{ mol O}} = \frac{1.33 \text{ mol H}}{1 \text{ mol O}}$$

$$\frac{0.09000 \text{ mol O}}{0.09000 \text{ mol O}} = 1.000$$

To eliminate the fraction obtained in the mole ratio of H to O, we must multiply the numbers by small integers to find the ratio of whole numbers. If we multiply by 2, we do not get a ratio of whole numbers, but if we multiply each quantity by 3 we get the following:

$$1.000 \text{ mol C} \times 3 = 3.000 \text{ mol C}$$

$$1.33 \text{ mol H} \times 3 = 3.99 \text{ mol H}$$

$$1.000 \text{ mol O} \times 3 = 3.000 \text{ mol O}$$

Thus, the empirical formula for ascorbic acid is $C_3H_4O_3$. You should always remember that the final calculated ratios will rarely be exact integers, but they should be fairly close, within ± 0.05 mol.

A word of caution: When performing empirical formula calculations, it is imperative to observe all rules regarding significant figures. Failure to observe these rules results in obtaining incorrect ratios.

After calculating the final mole ratio, sometimes whole numbers are not obtained, as was the case in Example Problem 7.15. To eliminate the fraction, you should multiply by succeedingly larger integers until a ratio of whole numbers is obtained. For example, if a mole ratio of 1 to 1.5 results, this indicates a 2-to-3 ratio; a ratio of 1 to 1.33 indicates a 3-to-4 ratio, and 1 to 1.667 is 3-to-5 ratio (see marginal note).

1 × 2 = 2
1.5 × 2 = 3

1 × 3 = 3
1.33 × 3 = 4

1 × 3 = 3
1.667 × 3 = 5

Molecular Formula Calculations

While knowledge of the empirical formula gives the simplest ratio of the various types of atoms in a molecule, the molecular formula expresses the actual ratio.

Once the empirical formula is known, only the molecular mass of a compound is needed to calculate its molecular formula. Because the molecular formula either is identical to the empirical formula or is a higher multiple of the empirical formula, if the mass of the empirical formula unit (empirical formula mass) is divided into the molecular mass, a whole number is obtained. This value gives the number of empirical formula units per molecular formula.

For example, if a substance has an empirical formula of CH and a molecular mass of 104, find the empirical formula mass of CH, and divide it into the molecular mass. The empirical formula mass of CH is $12 + 1 = 13$. Therefore,

$$\frac{\text{Molecular mass}}{\text{Empirical formula mass}} = \frac{104}{13} = 8$$

Eight empirical formula units comprise this molecule, so the molecular formula of the compound is $8 \times \text{CH}$, or C_8H_8. A complete molecular formula calculation is shown in Example Problem 7.16.

―――――――――― **Example Problem 7.16** ――――――――――

The molecular mass of a phosphorus-oxygen compound is 280.4, and a 10.000-g sample contains 4.364 g P. What is the molecular formula of the compound?

―――――――――――― **Solution** ――――――――――――

1. What is unknown? Molecular formula of the phosphorus-oxygen compound (the actual number of P and O atoms in a molecule)

2. What is known? Molecular mass = 280.4, total mass of sample = 10.000 g, mass of P in sample = 4.364 g P
The mass of O in the sample is obtained by subtraction:

$$\text{Mass of O} = \text{total mass} - \text{mass of P}$$

$$= 10.000 \text{ g} - 4.364 \text{ g P}$$

$$= 5.636 \text{ g O}$$

3. Find the mole ratio of P and O.

$$4.364 \text{ g P} \times \frac{1 \text{ mol P}}{30.97 \text{ g P}} = 0.1409 \text{ mol P}$$

$$5.636 \text{ g O} \times \frac{1 \text{ mol O}}{16.00 \text{ g O}} = 0.3523 \text{ mol O}$$

4. Find the simplest mole ratio.

$$\frac{0.1409 \text{ mol P}}{0.1409 \text{ mol P}} = 1.000$$

$$\frac{0.3523 \text{ mol O}}{0.1409 \text{ mol P}} = \frac{2.500 \text{ mol O}}{1 \text{ mol P}}$$

After multiplying 2 times $PO_{2.5}$, we obtain the correct empirical formula, P_2O_5.

5. Calculate the molecular formula by dividing the empirical formula mass into the molecular mass.

Empirical formula mass = $(2 \times$ atomic mass P$) + (5 \times$ atomic mass O$)$

$$= (2 \times 31.0) + (5 \times 16.0)$$

$$= 142$$

$$\frac{\text{Molecular mass}}{\text{Empirical formula mass}} = \frac{284}{142} = 2$$

Two formula units compose the total molecular mass; accordingly, the molecular formula is $2 \times P_2O_5$, or **P_4O_{10}.** This molecule is composed of four P atoms and ten O atoms.

When P_4O_{10} is mixed with water, it produces phosphoric acid, H_3PO_4.

7.14 *(a)* State the law of constant composition. *(b)* Use carbon dioxide, CO_2, as an example to illustrate the law of constant composition.

7.15 Find the percent composition of each of the following compounds: *(a)* OF_2, *(b)* $ZnCl_2$, *(c)* HIO_2, *(d)* $Mg(NO_3)_2$, *(e)* $(NH_4)_2SO_4$.

7.16 Find the empirical formula for compounds with the following percent compositions: *(a)* 75.0% C and 25.0% H; *(b)* 50.0% S and 50.0%; *(c)* 46.7% N and 53.3% O; *(d)* 87.5% Si and 12.5% H; *(e)* 17.98% Li, 26.75% P, and 55.27% O.

7.17 A 10.00-g sample contains 5.99 g Ti, and the remaining mass is oxygen. What is the empirical formula of the compound?

7.18 A compound has a molecular mass of 112 and an empirical formula of CH_2. What is the molecular formula of the compound?

7.19 A boron-hydrogen compound is composed of 78.1% B and 21.9% H and has a molecular mass of 27.7. Calculate the molecular formula of the compound.

REVIEW EXERCISES

SUMMARY

A **mole** is a counting unit that allows chemists to calculate the number of atoms, molecules, or ions contained in a sample by weighing it. One mole contains 6.022×10^{23} particles—**Avogadro's number.**

For each element, the mass of one mole of atoms is the atomic mass in grams. A mole of molecules has a mass equal to the molecular mass in grams; **molecular mass** is the sum of the atomic masses of all the elements in a compound. For those compounds whose structures do not contain identifiable molecules, the **formula mass** is used instead. The formula mass is the mass in grams of one mole of formula units of a compound.

Mole calculations involve the application of the factor-label method to change the given quantity to the desired quantity. Two conversion factors are employed: (1) ratio of moles to grams (either grams per mole or moles per gram) and (2) ratio of particles to moles or moles to particles. For elements, the first ratio (grams per mole) is obtained by finding the atomic mass of the element and adding the unit grams—this is the molar mass. For molecules, grams are the units attached to the molecular mass (molar mass of molecules). The second ratio, particles per mole is the same for all chemical species: it is 6.022×10^{23} particles/mol.

Mole calculations are used to determine chemical formulas, both the empirical formulas and the molecular formulas. The **molecular formula** of a compound gives the actual number of each type of atom in the molecular or formula unit. The **empirical formula** is the simplest ratio of whole numbers of atoms in the molecule. Empirical formulas are calculated from mass data on the elements that compose a compound. This is accomplished by converting the mass of each element to moles and calculating the simplest ratio of whole numbers. Molecular formulas are determined given the empirical formula and molecular mass.

Formulas of compounds are determined as a direct result of a basic law of chemistry called the **law of constant composition.** This law states that all samples of a compound contain the same elements in a fixed mass ratio.

KEY TERMS

Avogadro's number
counting unit
empirical formula
formula mass
formula unit
law of constant composition

molar mass
mole
molecular formula
molecular mass
percent compositon

EXERCISES*

7.20 Define each of the following: counting unit, mole, Avogadro's number, molar mass, molecular mass, formula mass, formula unit, law of constant composition, percent composition, empirical formula, molecular formula.

Moles

7.21 (a) Why is the mole called a counting unit? (b) Give two other examples of counting units.

7.22 If Avogadro's number of like atoms is placed on a balance, what mass is observed?

7.23 What is the mass of 1.000 mol of each of the following atoms: (a) B, (b) Co, (c) Mo, (d) Ge, (e) Ra, (f) Ag, (g) Th?

7.24 Perform a rough calculation to determine how many centuries it would take exactly 1 billion people working 24 hours per day, 365 days per year, to produce exactly 1 mol of doughnuts at a rate of 10 doughnuts per person each second. (*Hint:* First use conversion factors to calculate how many doughnuts could be produced per year.)

Moles and Atoms

7.25 Complete the following table by calculating the missing quantities for each element. For example, in part *a* calculate the number of atoms and the mass of 1.00 mol B.

Element	Number of moles	Number of atoms	Mass, g
(a) B	1.00		
(b) Al		1.99×10^{23}	
(c) Ni			5.87
(d) Zr			182.4
(e) Cs	0.500		
(f) Sn		6.02×10^{24}	
(g)	1.00		101.07
(h)		1.204×10^{24}	80.16
(i) Bi	3.115		
(j) Kr		8.94×10^{22}	

7.26 How many moles of potassium atoms are contained in each of the following: (a) 0.0145 g K, (b) 1.00 g K, (c) 87.2 kg K, (d) 500.00 mg K?

7.27 Calculate the number of moles of atoms in the following: (a) 3.99 g As, (b) 1.414 g Te, (c) 5.66 mg Ni, (d) 3.7 kg Ir.

7.28 How many atoms are contained in each sample listed in 7.27?

7.29 Calculate the mass of each of the following samples of elements: (a) 7.11 mol Ar, (b) 0.0044 mol Li, (c) 9.311 Mmol Na, (d) 2.30 mmol V.

7.30 How many atoms are contained in each of the following samples: (a) 9.97 g Pb, (b) 2.077 g U, (c) 4.62 kg Mo, (d) 23.189 mg Ra?

*For exercise numbers printed in color, answers can be found at the back of the book.

7.31 How many moles of atoms are contained in each of the following samples:
(a) 5.03×10^{23} atoms Xe
(b) 4.1×10^{24} atoms Cu
(c) 5.320×10^{20} atoms Pd
(d) 8.0002×10^{19} atoms Hg

7.32 What is the mass of each of the following:
(a) 3.0×10^{23} atoms Kr
(b) 2.5409×10^{25} atoms Mg
(c) 5.36×10^{20} atoms Ge
(d) 6.413×10^{28} atoms Cd

7.33 What is the mass of each of the following samples of iron atoms: (a) 3×10^{23} atoms Fe, (b) 5.40 billion atoms Fe, (c) 711 atoms Fe, (d) exactly 1 atom Fe?

7.34 Find the unknown quantity:
(a) 5.5 g chromium = ? mol chromium
(b) 7.221×10^{23} atoms osmium = ? mol osmium
(c) 0.000199 mol argon = ? g argon
(d) 4.8×10^{-6} g manganese = ? manganese atoms
(e) 9.374×10^{21} atoms cerium = ? g cerium

7.35 Find the unknown quantity:
(a) 67.5 mg Re = ? mol Re
(b) 2.000 kg Pd = ? atoms Pd
(c) 1 atom Hf = ? mol Hf
(d) 0.001160 mmol Bi = ? mg Bi
(e) 3.66×10^{26} atoms Nb = ? kg Nb

7.36 Arrange the following from highest to lowest mass: (a) 0.651 mol He, (b) 0.750 g He, (c) 0.375 mol Li, (d) 3.95×10^{23} atoms Li.

7.37 Arrange the following from largest to smallest number of atoms: (a) 25.9 mg Pt, (b) 1.51×10^{-2} g Pd, (c) 7.79 mg Ni, (d) 2.62×10^{4} μg Au.

Moles and Molecules

7.38 Calculate the molecular mass to three significant figures for each of the following: (a) IBr, (b) NO, (c) Br_2, (d) CI_4, (e) S_2F_2, (f) XeF_4, (g) $BrCl_5$, (h) $H_2S_2O_8$.

7.39 Calculate the molecular mass to four significant figures for each of the following molecules: (a) P_4O_6, (b) NI_3, (c) OBr_2, (d) $XeOF_4$, (e) N_2O_5, (f) $H_2C_2O_4$, (g) $POCl_3$, (h) $C_{12}H_{22}O_{11}$.

7.40 Complete the following table by calculating the missing quantities for each compound. For example, in part a calculate the number of molecules and the mass of 1.000 mol of SO_2.

Compound	Number of moles	Number of molecules	Mass, g
(a) SO_2	1.000		
(b) BrF	0.34		
(c) NCl_3	0.175		
(d) H_2SO_3		6.02×10^{21}	
(e) N_2O_5		9.641×10^{23}	
(f) $AlCl_3$			34.9
(g) PBr_5			55.721
(h) OF_2	3.030		
(i) SO_3		8.21×10^{25}	
(j) C_5H_{12}		7.54×10^{24}	

7.41 What is the mass of each of the following samples: (a) 0.00344 mol ClO_2, (b) 5.67 mol HBr, (c) 3.4×10^{-4} mol SeO_2, (d) 3 mol H_2SO_4, (e) 1.110 mmol N_2O_4?

7.42 How many molecules are contained in each of the following samples: (a) 6.2 g H_2O_2, (b) 0.04499 g ClF_3, (c) 12.5 g N_2H_4, (d) 7.8 g H_3PO_3, (e) 3.9010 kg ClF?

7.43 How many moles of oxygen atoms are contained in each of the following: (a) 8.207 mol H_2O, (b) 4.5 mol CO_2, (c) 0.00349 g P_2O_5, (d) 8×10^{22} molecules H_5IO_6?

7.44 What is the mass of hydrogen in each of the following: (a) 0.37 mol H_2SO_4, (b) 5.92 mg SiH_4, (c) 5.912×10^{23} molecules C_8H_{16}, (d) 23.775 kg $B_{10}H_{14}$?

7.45 What are the masses of the following quantities of molecules:
(a) 5.1×10^{22} molecules AsH_3
(b) 6.88×10^{21} molecules H_2Te
(c) 1.004×10^{25} molecules $HClO_4$
(d) 4.0×10^{12} molecules UF_6

7.46 Find the unknown quantity:
(a) 3.643 mg SF_6 = ? mol F
(b) 9.99×10^{23} molecules S_2O_3 = ? g S_2O_3
(c) 8.33 mmol C_9H_{20} = ? mol H
(d) 5.00×10^{-6} g H_3PO_4 = ? molecules H_3PO_4

7.47 Arrange the following in order, largest to smallest, of total number of phosphorus atoms: (a) 159 mg PCl_3, (b) 2.96×10^{-5} kg P_4O_{10}, (c) 0.120 mol P_4O_6, (d) 0.120 mol H_3PO_3.

Formula Unit Calculations

7.48 What is the formula mass (expressed to three significant figures) of each of the following: (a) CuCl, (b) CaI_2, (c) $Zn(NO_3)_2$, (d) Li_2SO_4, (e) $Fe_2(SO_4)_3$?

7.49 Calculate the number of moles of compound in each of the following samples: (a) 47 g $NaClO_3$, (b) 700.0 mg $Ca_3(PO_4)_2$, (c) 9.33 kg $RbBrO_2$.

7.50 What is the mass of each of the following: (a) 9.55 mol TiO_2, (b) 7.07 mol K_2SnCl_6, (c) 0.004529 mmol PbC_2O_4?

7.51 How many formula units are contained in each of the following: (a) 9.34 g $MgSiO_3$, (b) 5.8 kg $Hg_2(NO_2)_2$, (c) 234.0 mmol $NH_4C_2H_3O_2$, (d) 8.7 mg Na_2HPO_4?

Percent Composition

7.52 Find the percent composition to three significant figures for each of the following compounds: (a) HI, (b) MgS, (c) Hg_2I_2, (d) Si_3N_4, (e) OsO_5.

7.53 Find the percent composition to four significant figures for each of the following compounds: (a) $KMnO_4$, (b) $KHCO_3$, (c) $Ba(NO_2)_2$, (d) $Ni(CO)_4$, (e) $(NH_4)_2Cr_2O_7$

7.54 For each of the following compounds, calculate the percent by mass of Ag to three significant figures: (a) Ag_2S, (b) $AgIO_3$, (c) Ag_2CrO_4, (d) Ag_3AsS_3.

7.55 For each of the following compounds, calculate the percent by mass of K to three significant figures: (a) $K_2S_2O_6$, (b) K_3AsO_4, (c) K_2PtO_3, (d) $KC_7H_5O_3$.

7.56 Hydrated salts are ionic compounds that have a fixed number of water molecules bonded to them. Calculate the percent by mass of water (to four significant figures) in each of the following hydrated salts:
(a) $CuSO_4 \cdot 5H_2O$
(b) $BaCl_2 \cdot 2H_2O$
(c) $LiClO_4 \cdot 3H_2O$
(d) $Ni(IO_3)_2 \cdot 4H_2O$

7.57 Arrange the following from highest to lowest percent by mass of iron: (a) $FeCl_3$, (b) $Fe(OH)_2$, (c) Fe_3O_4, (d) $Fe_3(PO_4)_2$, (e) $FeCO_3$.

Empirical Formulas

7.58 Calculate the empirical formulas for each of the following compounds, given their percent compositions: (a) 46.55% Fe and 53.45% S; (b) 46.67% N and 53.33% O; (c) 80.0% C and 20.0% H; (d) 5.24% Si and 94.76% I; (e) 11.63% N and 88.37% Cl; (f) 76.62% Ce and 23.38% S.

7.59 Calculate the empirical formulas for each of the following compounds: (a) 60.1% K, 18.4% C, and 21.5% N; (b) 70.2% Pb, 8.1% C, and 21.7% O; (c) 46.54% Cu, 11.72% S, and 41.75% F; (d) 6.90% C, 1.15% H, and 91.95% Br; (e) 18.79% Li, 16.24% C, and 64.97% O.

7.60 A 25.0-g sample of a chromium-oxygen compound contains 13.0 g of chromium, and the remainder is oxygen. What is the empirical formula of the compound?

7.61 A calcium-phosphorus compound is analyzed and is found to contain 0.66 g of calcium and 0.34 g of phosphorus. Calculate the empirical formula of the compound.

7.62 On analysis, a 40.0-g sample was found to contain 16.0 g C, 18.7 g N, and 5.3 g H. Calculate the empirical formula of the compound.

7.63 A 500.0-mg sample contains 64.1 mg C and 152.1 mg F, and the remainder is Cl. What is the empirical formula of this compound?

7.64 Calculate the empirical formula for the compound that contains 28.2% N, 20.8% P, 42.9% O, and 8.1% H.

7.65 Aspirin, acetyl salicylic acid, is the most widely used painkiller in the world. A 0.8164-g aspirin sample contains 0.4898 g C and 0.03657 g H, and the remainder is oxygen. Calculate the empirical formula of aspirin.

Molecular Formulas

7.66 The molecular mass of a compound is 168, and its percent composition is 85.7% C and 14.3% H. Calculate the molecular formula of the compound.

7.67 A 1.000-g sample of a compound contains 0.202 g Al and 0.798 g Cl. The compound's molecular mass is 267. What is its molecular formula?

7.68 Analysis of a compound reveals that it is composed of H, O, and Br. The sample contains 0.64 g H, 10.15 g O, and 50.71 g Br. If the compound's molecular mass is 96.9, calculate its molecular formula.

7.69 Boranes are boron and hydrogen compounds. A borane is analyzed and found to contain 11.843 g B and 0.885 g H, and its molecular mass is 232.4. What is the molecular formula of the compound?

7.70 Calculate the molecular formula of a compound with molecular mass 90.0; a sample of the compound has been analyzed and is composed of 50.00 g C, 66.75 g O, and 8.25 g H.

7.71 Lauric acid is one of the fatty acids contained in living things. A 3.824-g sample of lauric acid contains 2.750 g C and 0.463 g H, and the remainder is O. If the molecular mass of lauric acid is 200, what is its molecular formula?

Additional Exercises

7.72 (a) What mass of sodium phosphate, Na_3PO_4, contains the same number of formula units as are found in 9.971 g KOH? (b) What mass of sodium phosphate contains the same number of Na atoms as 4.506 g of elemental sodium, Na?

7.73 An impure sample of $AgNO_3$ contains 59.5% Ag. Calculate the percent by mass of pure $AgNO_3$ in the sample.

7.74 Various minerals are composed of two or more compounds bonded together. Find the empirical formula of a mineral that contains 60.7% SiO_2, 27.2% MgO, and 12.1% H_2O.

7.75 Penicillin G is a widely used antibiotic. It has a molecular formula of $C_{16}H_{18}N_2O_4S$. (a) Calculate the percent by mass (three significant figures) of C in penicillin. (b) What mass of penicillin contains 1.00 g of carbon? (c) How many carbon atoms are contained in a 1.00-g sample of penicillin? (d) What mass of penicillin contains 4.19×10^{23} atoms of hydrogen?

7.76 Saccharin is a commonly used artificial sweetener that has been found to cause tumors in animals. A 1.000-g sample of saccharin is analyzed and is found to contain 0.459 g C, 0.0275 g H, 0.262 g O, 0.175 g S, and 0.0765 g N. The molecular mass of saccharin is 183.2. Calculate the molecular formula of saccharin.

7.77 Progesterone, $C_{21}H_{30}O_2$, is a steroid hormone. (a) How many moles of progesterone in a 4.44-mg sample? (b) What mass of progesterone contains 25 mmol H? (c) What mass of glucose, $C_6H_{12}O_6$, has the same number of carbon atoms as a 0.0386-g sample of progesterone?

7.78 Iodine pentafluoride, IF_5, is a colorless liquid that has a density of 3.252 g/cm^3. (a) Calculate the volume of 4.31 mol IF_5. (b) How many moles of F atoms are contained in 2.65 L IF_5? (c) What volume of IF_5 contains 4.821×10^{23} atoms of F?

7.79 The psychoactive chemical in marijuana is composed of 71.23% C, 12.95% H, and 15.81% O. Calculate its empirical formula.

7.80 Hemoglobin is the molecule in the blood that transports oxygen to the cells. Its molecular mass is 64,456, and it contains 0.35% iron by mass. (a) What mass of hemoglobin contains 1.00 g of iron? (b) How many iron atoms are found in a 4.33-g sample of hemoglobin? (c) How many moles of Fe are contained in a 9.11-mg sample of hemoglobin? (c) How many iron atoms in one hemoglobin molecule?

7.81 (a) What mass of 24 carat gold (100% pure) could be obtained from exactly 1 lb of 18 carat (75% pure) gold? (b) How many gold atoms are there in this sample?

7.82 Carbon atoms have a diameter of about 1.5×10^{-8} cm. If they are placed in a straight line 10.0 cm long, what is their mass?

7.83 Citric acid is an important compound in cellular metabolism and is a component of citrus fruits. A 4.256-g sample of citric acid contains 1.596 g C and 2.481 g O, and the remainder is H. (a) What is the empirical formula of citric acid? (b) If the molecular mass of citric acid is 192.1, calculate the molecular formula of citric acid.

7.84 Trinitrotoluene (TNT), $C_7H_5N_3O_6$, is a commonly used explosive. (a) What is the percent by mass of hydrogen in TNT? (b) What is the mass of each element in a 3.94-g sample of TNT?

7.85 A compound is found to have the formula XBr_2, in which X is an unknown element. Bromine is found to comprise 71.55% of the mass of the compound. (a) What is the atomic mass of X? (b) What is the name of element X?

7.86 Carbon tetrachloride, CCl_4, is one of the components of dry-cleaning fluids. Its density is 1.587 g/cm^3. (a) What volume of carbon tetrachloride contains 9.56×10^{26} chlorine atoms? (b) How many carbon tetrachloride molecules are there in a 48.3-mm^3 sample? (c) What is the volume of one carbon tetrachloride molecule?

7.87 What is the number of atoms in a pound mole, the atomic mass in pounds?

7.88 Nicotine, $C_{10}H_{14}N_2$, is a mild stimulant found in tobacco products. It is a very toxic compound and was once used as a pesticide. (a) What is the empirical formula of nicotine? (b) If a cigarette has a mass of 1.48 g and it contains 2.1% nicotine by mass, how many nicotine molecules are contained in the cigarette?

CHAPTER
Eight

Molecules Compounds and Chemical Bonding

STUDY GUIDELINES

After completing Chapter 8, you should be able to

1. Contrast the properties of ionic and covalent compounds

2. Write the names of simple binary ionic and covalent compounds

3. Discuss the reason why atoms combine to form molecules

4. Describe the electron transfer that takes place when an ionic bond is formed, given the names of a specific metal and nonmetal that combine

5. Write Lewis structures for ionic substances

6. Write the formula unit of an ionic compound given the chemi-

cal group numbers of the metal and nonmetal on the periodic table

7. Determine if one chemical species is isoelectronic to another

8. Define electronegativity and describe its importance

9. Rank elements according to their electronegativity, using the periodic table

10. Discuss the reason why some nonmetallic elements exist as diatomic molecules instead of as single atoms

11. Apply the steps for writing the Lewis structures of simple covalent molecules

12. Identify and discuss the nature of single, double, and triple bonds

13. Write a Lewis structure for a molecule, and determine if the bonds are polar or nonpolar

14. Identify and draw Lewis structures of simple polyatomic ions

15. Use the valence shell electron pair repulsion method to determine the molecular geometry of a molecule

16. Distinguish among bond order, bond length, bond energy, and bond angle

8.1 COMPOUNDS

Compounds are most frequently classified into two major groups: ionic compounds and covalent compounds. As we will soon learn, the names given to these groups are derived from the two different kinds of chemical bonds that hold the particles in formula units and molecules.

Ionic compounds usually result when metallic elements combine with nonmetallic elements.

$$\text{Metal} + \text{nonmetal} \longrightarrow \text{ionic compound}$$

Before we consider representative ionic compounds, let's learn how the names of simple ionic compounds are assigned. We will begin with **binary ionic compounds,** those that contain two different elements. First write the name of the metal, and then write the name of the nonmetal, replacing its ending with *ide*.

1. Write the name of the metal.

2. Write the name of the nonmetal, replacing its ending with the ending *ide*.

Endings dropped from nonmetals:
*ox**ygen***
*nitr**ogen***
*carb**on***
*sul**fur***
*phosph**orus***
*fluor**ine***
*chlor**ine***
*brom**ine***
*iod**ine***

How do we write the name of NaCl? Notice that Na is an alkali metal and Cl is a halogen, a nonmetal. First, write the name of the metal, *sodium,* and then write the name of the nonmetal, dropping its ending and replacing it with *ide:* Chlorine − *ine* + *ide* = chloride. Sodium chloride is, therefore, the name of NaCl.

Other examples of how to write the names of binary ionic compounds are given in Example Problem 8.1.

--------------- **Example Problem 8.1** ---------------

Write the names for each of the following ionic compounds: *(a)* KF, *(b)* CaO, *(c)* Mg_3N_2.

--------------- **Solution** ---------------

(a) KF
    Metal = potassium
    Nonmetal − ending + *ide* = fluorine − *ine* + *ide*
    Name = **potassium fluoride**

(b) CaO
    Metal = calcium
    Nonmetal − ending + *ide* = oxygen − *ygen* + *ide*
    Name = **calcium oxide**

CaO is also called lime. The expression "being in the limelight" stems from the fact that when lime is heated a brilliant white light is released.

(c) Mg_3N_2
    Metal = magnesium
    Nonmetal − ending + *ide* = nitrogen − *ogen* + *ide*
    Name = **magnesium nitride**

Ionic compounds share many common properties. At 25°C they are all solids with relatively high melting and boiling points (Fig. 8.1). Most are hard, but brittle. Ionic substances are poor conductors of electricity,

TABLE 8.1   PROPERTIES OF SELECTED IONIC COMPOUNDS

Compound	Physical state (25°C)	Color	Melting point, °C	Boiling point, °C	Density, g/cm³	Solubility, g/100 cm³ H₂O
NaCl	Solid	White	801	1413	2.17	35.7 (0°C)
LiF	Solid	White	846	1717	2.64	0.13 (25°C)
CaCl₂	Solid	White	772	1940	2.15	42 (20°C)
Fe₂O₃	Solid	Red brown	1462		5.24	Insoluble

except in the molten (liquid) state, when they are good conductors. Dissolved in water, they break up into ions that help conduct an electric current. Table 8.1 presents the properties of selected ionic compounds.

Covalent compounds result when two or more nonmetals combine chemically.

Covalent Compounds

$$\text{Nonmetal} + \text{nonmetal} \longrightarrow \text{covalent compound}$$

Frequently encountered covalent compounds include water, H_2O; ammonia, NH_3; carbon dioxide, CO_2; and methane, CH_4. In each of these binary covalent compounds two different nonmetallic elements are chemically combined.

A different set of rules is used when writing the names of binary covalent compounds. If there is only one atom per molecule of the nonmetal that appears first in the formula, write that element's name with no change. The name of the second nonmetal is modified in two ways: A prefix is added to it to indicate how many atoms of that element there are in the molecule, and its ending is replaced with *ide*.

Figure 8.1
The particles in an ionic compound are found in regular geometric shapes. In sodium chloride, the particles are arranged in a cubic pattern, as shown by the scanning electromicrograph.
(Dr. Jeremy Burgess/Photo Researchers, Inc.)

TABLE 8.2 PREFIXES USED IN THE NAMES OF COVALENT MOLECULES

Prefix	Number of atoms indicated
mono	1
di	2
tri	3
tetra	4
penta	5
hexa	6
hepta	7
octa	8
nona	9
deca	10

Consider CO_2 as an example for writing the name of a binary covalent compound. Start by writing the name of the nonmetal listed first in the formula, *carbon*. Then modify the name of the second nonmetal, oxygen: Add the prefix *di*, which means "two," drop its ending, *ygen*, and replace it with *ide*. Thus the name of CO_2 is carbon dioxide.

$$CO_2 = \text{carbon } (di + \text{oxygen} - ygen + ide)$$

$$= \text{carbon dioxide}$$

A prefix that indicates the number of atoms is required in the names of covalent compounds because two nonmetals can usually combine in more than one way. For example, CO, carbon monoxide, is a second oxide of carbon. The prefix *mono*, or simply *mon*, is added to indicate that this oxide of carbon contains only one O atom. Table 8.2 lists the most commonly used prefixes in the names of covalent compounds.

In some covalent compounds more than one atom of each nonmetal is found. For example, two N atoms and four O atoms are contained in a molecule of N_2O_4. To write the names of such compounds, a prefix is added to each nonmetal to indicate how many atoms are in the formula. The name of N_2O_4 is dinitrogen tetroxide. The prefix *di* is added to nitrogen to indicate the presence of two N atoms, and the prefix *tetr* is added to oxide to indicate that four O atoms are present. Example Problem 8.2 shows three examples of how to write the names of binary covalent compounds.

Carbon monoxide, CO, is a poisonous gas that kills by not allowing O_2 to reach the hemoglobin in the blood. The affinity of CO for hemoglobin is approximately 200 times greater than that of O_2.

——————— **Example Problem 8.2** ———————

Write the names for each of the following covalent compounds:
(*a*) OCl_2, (*b*) SF_6, (*c*) P_2O_5.

─────────────── **Solution** ───────────────

(a) OCl_2 = oxygen (*di* + chlorine − *ine* + *ide*)
      = **oxygen dichloride**
(b) SF_6 sulfur (*hexa* + fluorine − *ine* + *ide*)
      = **sulfur hexafluoride**
(c) P_2O_5 = (*di* + phosphorus) + (*pent* + oxygen − *ygen* + *ide*)
      = **diphosphorus pentoxide**

    In the last example, *pent* rather than *penta* is added as the prefix for oxide, in order to generate a word that is easier to pronounce. If adding the prefix produces a double vowel, such as *oo* or *ao*, the vowel contributed by the prefix is usually dropped.

    Covalent compounds are markedly different from ionic compounds. Most covalent compounds are either liquids or gases; some are rather soft solids. Compared to average ionic compounds, covalent compounds have lower melting and boiling points and lower densities. Covalent compounds are poor conductors of both heat and electricity. Most do not form ions when dissolved in water. Table 8.3 lists properties of selected covalent compounds.

TABLE 8.3 PROPERTIES OF SELECTED COVALENT COMPOUNDS

Compound	Physical state	Color	Melting point, °C	Boiling point, °C	Density
Methane, CH_4	Gas	Colorless	−182	−162	0.55 g/L
Hydrogen fluoride, HF	Gas	Colorless	−83.1	19.4	0.98 g/L
Ammonia, NH_3	Gas	Colorless	−74.3	−31.1	0.77 g/L
Water, H_2O	Liquid	Colorless	0.0	100	1.0 g/cm³
Carbon tetrachloride, CCl_4	Liquid	Colorless	−23.0	76.8	1.6 g/cm³
Tetraphosphorus decoxide, P_4O_{10}	Solid	White	340	360 (sublimes)	2.3 g/cm³

REVIEW EXERCISES

8.1 What types of elements combine to produce (*a*) ionic compounds, (*b*) covalent compounds?

8.2 Write the name of each of the following ionic compounds: (*a*) KI, (*b*) Li_2O, (*c*) SrO, (*d*) Cs_3N, (*e*) Ca_3P_2.

8.3 (*a*) List four properties of ionic compounds. (*b*) List four properties of covalent compounds.

8.4 Write the name of each of the following covalent compounds: (*a*) OF_2, (*b*) NI_3, (*c*) NO, (*d*) $BrCl_5$, (*e*) P_4O_{10}.

8.5 Compare the properties of NaCl with those of H_2O.

8.2 CHEMICAL BONDS

Chemical bonds are the attractive forces that hold atoms or ions together. In this section we will attempt to understand what drives atoms to combine and produce chemical bonds.

One of the driving forces of nature is the tendency of matter to reach the lowest possible energy state. Generally, a lower energy state implies greater stability. A stable body is more resistant to change than a less stable body.

Elements can be ranked according to their degree of stability. Elements such as sodium, Na, and chlorine, Cl, are ranked as highly reactive (unstable) because they tend to undergo chemical changes and liberate energy. More stable elements remain unaltered, even under extreme conditions. As a group, the noble gases are quite stable. Helium and neon, for example, do not form any compounds.

Certain nonmetallic elements are so unstable that they do not exist in nature as individual atoms, but rather as diatomic molecules (molecules composed of two atoms). Included in this group are hydrogen, H_2; nitrogen, N_2; oxygen, O_2; plus all of the halogens: fluorine, F_2; chlorine, Cl_2; bromine, Br_2; iodine, I_2; and astatine, At_2. More stable elements, such as the noble gases, all exist monatomically ("monatomically" means "as single atoms").

In addition to the diatomic elements, solid sulfur is found as S_8 molecules, and phosphorus exists as P_4 molecules.

Let's consider molecular hydrogen, H_2. All samples of hydrogen gas, at 25°C, are composed of H_2 molecules. How is this explained? One way to answer the question is to consider the stability of two H atoms compared to that of a H_2 molecule. The H_2 molecule is more stable, and the individual H atoms are less stable. When individual H atoms are combined, they release energy:

$$H \cdot + H \cdot \longrightarrow H_2 + 436 \text{ kJ}$$

For each mole of H_2 that forms, 436 kJ of energy is released. This energy is contained initially in the two H atoms but is released when the atoms combine. Thus, one mol of diatomic hydrogen molecules is 436 kJ more stable than two mol of hydrogen atoms (Fig. 8.2).

As an analogy, you should think of a boulder on the ground relative to one on top of a hill. Which one has more potential energy? The boulder on the hill has more energy because kinetic energy was added to the boulder, and stored as potential energy, when it was carried to the top of the hill (Fig. 8.3). A boulder on top of a hill has the capacity to fall to the ground spontaneously, whereas the opposite is not true. You would be quite amazed to see a boulder jump from the ground to the top of a cliff. Think of the two H atoms as being at the top of an energy "cliff" and the H_2 molecule as being at the bottom.

Why don't noble gases form diatomic molecules? Using a similar argument to the one just used, we can say that individual noble gas atoms exist at a lower energy state (at the bottom of the hill because of their stability) compared to diatomic noble gas molecules (at the top of

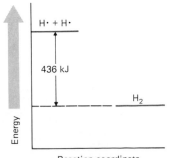

Figure 8.2
When 2 mol of H atoms bond to form 1 mol of H_2, 436 kJ of energy is released. The same quantity of energy, 436 kJ, is required to break the bonds in 1 mol of H_2.

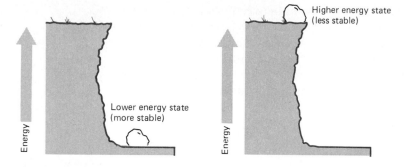

Figure 8.3
A rock on the ground is in a lower energy state and is more stable than a rock on the top of a cliff. Energy must be added to the rock on the ground to move it to the top of the cliff. Thus, at the top of the cliff it is in a higher energy state and is less stable.

the hill) (Fig. 8.4). This is verified by the fact that a large amount of energy is required to produce a diatomic noble gas compound, exactly the opposite of what was found for the formation of diatomic hydrogen, in which energy is released. He_2 does not exit.

Over the course of the twentieth century, chemists and physicists have attempted to develop and refine a theory that explains why some atoms bond and others do not. This bonding theory also tries to explain the degree of stability of compounds, and to account for the arrangement of atoms within molecules.

G. N. Lewis, in 1916, was one of the first scientists to propose that bonding was directly related to the electronic arrangement of atoms. Since that time, a large body of information has been collected to show that chemical bonding is adequately explained in terms of the outer, or valence, electrons. This bonding theory is known as the **valence bond theory.** Due to certain limitations of the valence bond theory, a second bonding theory, known as the **molecular orbital theory,** has also developed. In the molecular orbital theory all the electrons in the atoms are considered in bond formation. As a result of the complexity and the quantitative nature of the molecular orbital theory, it is not included in this beginning discussion of chemical bonds—we will consider only the fundamentals of the valence bond theory.

Throughout the development of the principal concepts of the valence bond theory, never lose sight of the fact that a bonding theory attempts to explain the structure and reactivity of molecules and compounds. A common misunderstanding is that the theoretical arguments of scientists are the facts and the components of the real world are the supporting evidence. If we reach an exception to a rule or guideline, it is not a discrepancy in the natural world, but a flaw in the theory. Bonding theories are nothing more than models produced by scientists to describe molecular systems.

Valence bond theory explains the bonding of atoms in terms of electron "transfers" and electron "sharing." When one or more electrons are transferred from one atom to another in the process of forming a bond, the resultant bond is called an **ionic bond.** If there is no electron

Bonding Theories

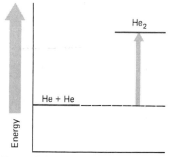

Figure 8.4
Energy must be added to He atoms to produce He_2, if possible. Thus, He_2 is less stable than unbonded He atoms. He_2 does not exist because it is so unstable.

Types of Bonds

transfer, but rather a "sharing" of electrons between two atoms, the resultant bond is classified as a **covalent bond.**

While it is convenient to classify bonds in such a manner, most bonds are not purely ionic or covalent. Just as most occurrences in the world are not "black" or "white," most bonds have varying degrees of ionic and covalent character. Ionic compounds contain bonds with a higher degree of ionic character, and covalent compounds have a greater degree of covalent character.

NaCl, an ionic substance, has 67 percent ionic character and 33 percent covalent character.

REVIEW EXERCISES

8.6 What is the chemical bond?

8.7 List two everyday phenomena that illustrate the tendency of objects to seek spontaneously their lowest energy states.

8.8 What groups of elements in the periodic table are classified as *(a)* more stable and *(b)* less stable? Give specific examples.

8.9 List five elements that exist as diatomic molecules at 25°C.

8.10 Provide an explanation for the fact that diatomic helium molecules do not exist.

8.11 *(a)* What type of bond results when electrons are shared between two atoms? *(b)* What type of bond results when electrons are transferred between atoms?

8.3 IONIC, OR ELECTROVALENT, BONDING

When electrons are transferred from a metal to a nonmetal, an **ionic bond** (sometimes called an electrovalent bond) is produced. In this section, we will consider the formation and nature of ionic bonds. Before we begin our discussion of ionic bonds, we must consider an important property of bonded atoms called electronegativity.

Electronegativity

Electronegativity is a measure of the capacity of an atom to attract electrons in a chemical bond. An element with a high electronegativity has a greater capacity to attract bonded electrons than one with a lower electronegativity. Three or four scales of electronegativity exist, but the original scale developed by Linus Pauling (1901–   ) (Fig. 8.5) is still one of the most popular in beginning chemistry. Pauling collected data on most of the elements, performed various calculations, and produced an electronegativity scale based on the element fluorine, F. Pauling assigned the electronegativity value of 4.0 to fluorine, the most electronegative element.

On the Pauling electronegativity scale (Fig. 8.6) two trends are evident: (1) From left to right across a period, electronegativities increase (excluding the noble gases), and (2) with increasing atomic mass within a chemical group, electronegativities decrease. Francium, Fr, the element located at the bottom left corner of the periodic table, is the least electronegative element. The closer an element is to fluorine on the periodic table, the higher its electronegativity. Oxygen is the second most electronegative element with a value of 3.5.

Figure 8.5
Linus Pauling. *(Linus Pauling Institute of Science and Medicine.)*

Increasing electronegativity →

Decreasing electronegativity ↓

Period	Group IA	IIA											IIIA	IVA	VA	VIA	VIIA	VIIIA
1	H 2.2																	He
2	Li 1.0	Be 1.5											B 2.0	C 2.5	N 3.0	O 3.5	F 4.0	Ne
3	Na 0.9	Mg 1.2											Al 1.5	Si 1.8	P 2.1	S 2.5	Cl 3.0	Ar
4	K 0.8	Ca 1.0	Sc 1.3	Ti 1.5	V 1.6	Cr 1.6	Mn 1.5	Fe 1.8	Co 1.8	Ni 1.8	Cu 1.9	Zn 1.6	Ga 1.6	Ge 1.8	As 2.0	Se 2.4	Br 2.8	Kr
5	Rb 0.8	Sr 1.0	Y 1.2	Zr 1.4	Nb 1.6	Mo 1.8	Tc 1.9	Ru 2.2	Rh 2.2	Pd 2.2	Ag 1.9	Cd 1.7	In 1.7	Sn 1.8	Sb 1.9	Te 2.1	I 2.5	Xe
6	Cs 0.7	Ba 0.9	La 1.1	Hf 1.3	Ta 1.5	W 1.7	Re 1.9	Os 2.2	Ir 2.2	Pt 2.2	Au 2.4	Hg 1.9	Tl 1.8	Pb 1.8	Bi 1.9	Po 2.0	At 2.2	Rn
7	Fr 0.7	Ra 0.9	Ac 1.1															

Figure 8.6
Within a period, the electronegativities of atoms increase with increasing atomic number, excluding the noble gases. Within a group, the electronegativities decrease going from top to bottom.

Trends in electronegativity directly parallel those in ionization energy, and are indirectly related to trends in atomic size. In other words, elements with high ionization energies, those composed of small atoms, are the most electronegative (excluding the noble gases). Elements with low ionization energies, those composed of large atoms, have lower electronegativities.

Ionic Bonding in Sodium Chloride, NaCl

To illustrate ionic bonding, let's consider sodium chloride, NaCl. When sodium combines with chlorine, the ionic salt sodium chloride is the product. Sodium is an alkali metal with one outer-level electron:

$$\text{Na} \quad 1s^2 2s^2 2p^6 3s^1$$

The Lewis symbol for sodium is

$$\text{Na}\cdot$$

Chlorine is a halogen with seven outer electrons:

$$\text{Cl} \quad 1s^2 2s^2 2p^6 3s^2 3p^5$$

The Lewis symbol for chlorine is

$$:\ddot{\underset{\cdot}{\text{Cl}}}:$$

Linus Pauling wrote one of the most significant chemistry books of the twentieth century, entitled The Nature of the Chemical Bond. *He is the recipient of two Nobel prizes. In 1954, he received this award for his work in chemistry, and 9 years later, for his work leading to the banning of atmospheric testing of nuclear bombs.*

-------------------- **CHEM TOPIC: Sodium Chloride** --------------------

When most people ask for salt, they are referring to sodium chloride. Dissolved sodium chloride, NaCl, is a component of each cell in our bodies. We need a constant source of sodium chloride to replenish the salt that we excrete and lose through our sweat glands. NaCl is also required for the production of hydrochloric acid, HCl(aq), which is a component of gastric juice in our stomachs.

In desert regions salt has always been a valuable commodity. Animals will travel great distances to find salt. Salt was an important item in the military baggage of the ancient Roman soldiers. Our word "salary" is derived from the Latin word *salarium,* which originally meant "money for salt."

Today, salt is readily available in a highly refined form. Many people in the United States consume from 5 to 10 g of sodium chloride per day. These amounts are far in excess of the amounts that are nutritionally sound. Recommended dietary allowances indicate that 1 to 3 g of salt are required to maintain maximum health. Too much salt in one's diet leads to high blood pressure (hypertension) and can cause a fluid imbalance. An elevated concentration of sodium ions in the blood is called hypernatremia.

As you may recall, sodium is an extremely reactive metal and chlorine is a reactive nonmetal. Sodium has a low ionization energy (little energy is required to remove an electron) and a lower electronegativity. Chlorine, in contrast, has a relatively high ionization energy and high electronegativity. Sodium is the largest atom in the third period, while chlorine is one of the smallest atoms.

Whenever a sodium atom encounters a chlorine atom, the loosely held outer electron ($3s^1$) of sodium is pulled away by the more compact chlorine atom, creating two ions. In terms of electronegativity, we say that the highly electronegative Cl atom (3.0) attracts the outer-level electron of sodium because of its low electronegativity (0.9). An Na atom loses an electron and becomes a cation (a positive ion), and a Cl atom gains the electron and becomes an anion (a negative ion).

$$\mathrm{Na\cdot} \ + \ \mathrm{:\overset{..}{C}l:} \ \longrightarrow \ \mathrm{Na^+} \ [\mathrm{:\overset{..}{C}l:}]^-$$

$$\mathrm{Na} \quad (1s^2 2s^2 2p^6 3s^1) \ \longrightarrow \ e^- \ + \ \mathrm{Na^+} \ (1s^2 2s^2 2p^6 3s^0)$$

$$\mathrm{Cl} \quad (1s^2 2s^2 2p^6 3s^2 3p^5) + e^- \ \longrightarrow \ \mathrm{Cl^-} \ (1s^2 2s^2 2p^6 3s^2 3p^6)$$

Na^+ and Cl^- have one thing in common; they possess a noble gas electronic configuration. A Na^+ ion possesses 10 electrons and has the same electronic configuration as Ne. In other words, Na^+ is isoelectronic to Ne. Cl^- has 18 electrons and is isoelectronic to Ar.

An ionic bond is the force of attraction between unlike charged ions, in this case Na^+ and Cl^-. Remember that unlike charged particles always attract each other, but it is not the simple attraction of a pair of Na^+ and

Iso is a prefix meaning "the same."

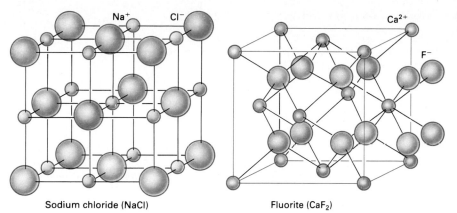

Na⁺ Cl⁻

Ca²⁺

F⁻

Sodium chloride (NaCl)   Fluorite (CaF₂)

Figure 8.7
Ionic solids are composed of regular patterns of alternating anions and cations. This figure shows segments of the crystal lattices of sodium chloride, NaCl, and calcium fluoride, CaF_2.

Cl^- ions that is significant. Ionic compounds like NaCl exist in a *crystal lattice*, a three-dimensional array of Na^+ ions surrounded by Cl^- ions, and vice versa. Figure 8.7 shows the crystal lattice structure of sodium chloride. Each Na^+, except those on the surface, is surrounded and attracted by six Cl^-, and each inner Cl^- is surrounded and attracted by six Na^+.

In Fig. 8.7, note that the Na^+ ions are smaller than the Cl^- ions. After a Na atom loses its outer $3s^1$ electron, the inner, core electrons are held tightly by the nucleus because there is a greater number of protons than electrons. In contrast, Cl^- has one more electron than proton; thus the electrons are not held as tightly as the nucleus. A measure of the size of an ion is its **ionic radius.** Figure 8.8 shows a comparison of the atomic sizes and ionic radii of alkali metals and halide ions.

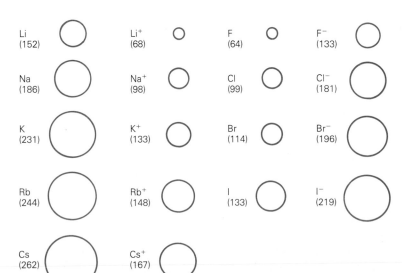

Li (152)   Li⁺ (68)   F (64)   F⁻ (133)

Na (186)   Na⁺ (98)   Cl (99)   Cl⁻ (181)

K (231)   K⁺ (133)   Br (114)   Br⁻ (196)

Rb (244)   Rb⁺ (148)   I (133)   I⁻ (219)

Cs (262)   Cs⁺ (167)

Figure 8.8
Ionic radii of alkali metal cations are smaller than the radii of corresponding alkali metal atoms. The decrease in size is the result of losing the outermost electron. Ionic radii of halide anions are larger than the radii of the corresponding halogen atoms. The increase in size is the result of gaining one electron. The radii are measured in picometers (1 pm = 10^{-12} m).

What drives Na and Cl atoms to combine to form NaCl? When Na and Cl combine they release energy.

$$Na(g) + \tfrac{1}{2}Cl_2(g) \longrightarrow NaCl(g) + 623 \text{ kJ}$$

If energy is released in chemical reactions, the products are more stable (lower-energy) than the reactants. In NaCl, both atoms achieve the stable noble gas configuration—the most stable electronic configuration possible for atoms.

The reaction of Na with Cl is a model for the combination of an alkali metal atom (group IA atom) with a halogen (group VIIA atom). All combinations of alkali metals and halogens yield compounds that contain ionic bonds with formula units of MX, in which M is any alkali metal and X is any halogen. M and X combine in a 1-to-1 ratio because each obtains the stable noble gas electronic configuration after transferring one electron.

Some other ionic compounds that contain an alkali metal and halogen are KI, CsBr, RbCl, and LiF.

A convenient rule to follow is that atoms are most stable when they are isoelectronic to a noble gas. This rule is often called the *octet rule* or *rule of eight*, but is more accurately named the **noble gas rule.** A word of caution when applying the noble gas rule: It is only a generalization which can be applied in many, but not all, cases.

"Octet" refers to eight things, a group of eight electrons in this case.

REVIEW EXERCISES

8.12 What is the electronegativity of an atom?

8.13 *(a)* What is an ionic bond? *(b)* How is it different from a covalent bond?

8.14 Use Lewis symbols to illustrate the electron transfer that occurs when potassium and fluorine atoms combine to produce potassium fluoride.

8.15 Write the formulas for a cation and an anion that are isoelectronic to each of the following noble gas atoms: *(a)* Xe, *(b)* Kr, *(c)* He, *(d)* Ne.

8.16 *(a)* What is the noble gas rule, and how is it applied when considering the formation of ionic bonds? *(b)* What is another name for the noble gas rule?

What happens when Ca atoms bond to F atoms? Calcium belongs to group IIA (2), the alkaline earth metals. Each group IIA element has two loosely held outer electrons. Fluorine is a halogen with seven outer electrons. Because fluorine is the most electronegative element, it can remove an electron from a Ca atom and obtain a noble gas configuration. However, two electrons must be removed from a Ca atom for it to obtain a noble gas configuration. Hence, two F atoms accept one electron each from Ca, allowing the Ca atom to obtain the noble gas configuration of Ar.

Ionic Bonding in Calcium Fluoride, CaF$_2$

Calcium fluoride, commonly called fluorite, is a high-melting solid (1360°C) that has a low water solubility.

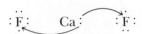

After the two electrons are transferred, Ca is left with a 2+ charge and each F possesses a 1− charge. Both Ca^{2+} and F^- are isoelectronic to noble gases. Ca^{2+} is isoelectronic to Ar, and F^- is isoelectronic to Ne.

$$\text{Ca} \quad (1s^2 2s^2 2p^6 3s^2 3p^6 4s^2) \longrightarrow 2e^- + \text{Ca}^{2+} \ (1s^2 2s^2 2p^6 3s^2 3p^6 4s^0)$$

$$\text{F} \quad (1s^2 2s^2 2p^5) + e^- \longrightarrow \text{F}^- \ (1s^2 2s^2 2p^6)$$

The Lewis structure for calcium fluoride is expressed as follows:

$$\text{Ca}^{2+} \quad 2[\,:\!\ddot{\text{F}}\!:\,]^-$$

A Lewis structure is used to show the bonds in a formula unit or molecule. Figure 8.7 shows the crystal structure of calcium fluoride. Compare its structure to that of sodium chloride.

Once again we can generalize: Alkaline earth metals (except Be, whose compounds have little ionic character) and halogens combine to yield ionic compounds with the formula MX_2.

Magnesium, Mg, belongs to group IIA (2). Oxyen, O, is a member of group VIA (16), the chalcogens. In the formation of magnesium oxide, MgO, an Mg atom must lose two electrons and an O atom must gain two electrons to obtain noble gas configurations. Thus, two electrons are transferred from Mg, which has a low electronegativity, to O, which has a high electronegativity:

Ionic Bonding in Magnesium Oxide, MgO

Magnesium oxide is also called magnesia. It is produced when magnesium carbonate, $MgCO_3$, is heated.

$$\text{Mg}\!:\ +\ \ddot{\text{O}}\!: \ \longrightarrow \ \text{Mg}^{2+}\ [\,:\!\ddot{\text{O}}\!:\,]^{2-}$$

Both elements achieve the noble gas electronic configuration of Ne.

$$\text{Mg} \quad (1s^2 2s^2 2p^6 3s^2) \longrightarrow \text{Mg}^{2+} \ (1s^2 2s^2 2p^6)$$

$$\text{O} \quad (1s^2 2s^2 2p^4) + 2e^- \longrightarrow \text{O}^{2-} \ (1s^2 2s^2 2p^6)$$

The formula unit of magnesium oxide is MgO because one pair of electrons is transferred from the Mg to the O. All members of group IIA combine with nonmetallic members of group VIA in a 1-to-1 ratio.

Potassium, K, an alkali metal atom with one outer-level electron, and sulfur, a chalcogen atom with six outer-level electrons, combine in a 2-to-1 ratio. After a K atom loses one electron, it achieves the noble gas configuration of Ar, but after gaining an electron, the S atom would only have seven outer electrons. Hence, the S atom removes an electron from a second K atom to obtain the noble gas configuration of Ar:

Ionic Bonding in Potassium Sulfide, K_2S

Potassium sulfide is a yellow-brown solid that melts at 840°C.

$$\text{K}\!:\ +\ :\!\ddot{\text{S}}\!:\ +\ \cdot\text{K} \ \longrightarrow \ 2\text{K}^+\ [\,:\!\ddot{\text{S}}\!:\,]^{2-}$$

Table 8.4 lists all possible non-transition metal and nonmetal group combinations that produce ionic compounds. Note that in each case both the metal and nonmetal obtain a noble gas configuration, and each compound is electrically neutral; the sum of positive charges equals the sum of negative charges.

Ionic Compound Summary

TABLE 8.4   SUMMARY OF THE FORMULAS OF IONIC COMPOUNDS

Metal group	Nonmetal group	Formula*	Examples
IA	VIIA	MX (M^+ X^-)	NaBr, KI, CsF, RbI
IA	VIA	M_2X ($2M^+$ X^{2-})	Li_2O, K_2O, Rb_2S, Na_2Se
IA	VA	M_3X ($3M^+$ X^{3-})	Na_3N, K_3P, Cs_3As
IIA	VIIA	MX_2 (M^{2+} $2X^-$)	$MgCl_2$, $SrBr_2$, CaI_2
IIA	VIA	MX (M^{2+} X^{2-})	BaS, SrO, MgS
IIA	VA	M_3X_2 ($3M^{2+}$ $2X^{3-}$)	Ca_3N_2, Mg_3P_2
IIIA	VIIA	MX_3 (M^{3+} $3X^-$)	AlF_3, GaF_3
IIIA	VIA	M_2X_3 ($2M^{3+}$ $3X^{2-}$)	Al_2O_3, In_2O_3
IIIA	VA	MX (M^{3+} X^{3-})	AlN, GaAs

*M = metal; X = nonmetal

As we discussed previously, all ionic substances have a three-dimensional crystal lattice structure composed of alternating cations and anions. Many different crystal lattice patterns exist. Each has an orderly array of cations surrounded by anions, and vice versa. Figure 8.9 shows three different cubic crystal lattice structures.

While the properties of ionic substances are similar, they vary depending on (1) the charge on the ions, (2) the distance between ions, and (3) the pattern of ions within the crystal lattice. For example, more highly charged ions in the crystal lattice generally produce stronger ionic bonds: more energy is needed to break them apart. Ionic bonds in magnesium chloride, $MgCl_2$ (Mg^{2+} $2Cl^-$), are significantly stronger than those in NaCl (Na^+ Cl^-). A greater force of attraction between cations and anions with higher charges creates stronger ionic bonds.

Crystal structures are elucidated through x-ray diffraction analysis.

Figure 8.9
Simple cubic, body-centered cubic, and face-centered cubic crystal lattice patterns are shown. Many other geometric arrangements of ions are found in crystal lattices.

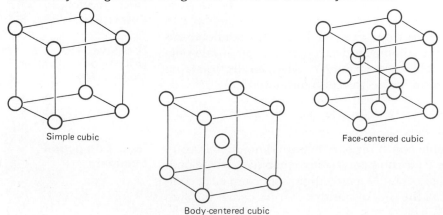

Simple cubic

Body-centered cubic

Face-centered cubic

8.17 Write the complete electronic configurations for *(a)* S^{2-}, *(b)* Rb^+, *(c)* N^{3-}, *(d)* Ba^{2+}.

8.18 Write the Lewis structures for each of the following ionic substances: *(a)* BaS, *(b)* $SrBr_2$, *(c)* Li_2O.

8.19 What is the formula unit for each of the following ionic compounds: *(a)* calcium nitride, *(b)* rubidium oxide, *(c)* magnesium phosphide, *(d)* cesium sulfide?

8.4 INTRODUCTION TO COVALENT BONDING

Many molecules are bonded as a result of "shared" outer-level electrons between two atoms. In this case, the properties of the atoms involved are such that one atom cannot take an electron from the other. A chemical bond that forms without the transfer of electrons is classified as a **covalent bond**.

Covalent Bonding in Hydrogen, H_2

To begin our discussion of covalent bonding, it is easiest to start with the simplest molecule: diatomic hydrogen, H_2. Hydrogen gas is composed of H_2 molecules, rather than discrete H atoms. When one H atom bonds with another H atom, an electron transfer cannot take place because each H atom has the same electronegativity. Instead H atoms must share their electrons in order to reach the noble gas configurations of two electrons (isoelectronic to He).

Hydrogen gas, H_2, is a colorless, highly combustible gas. Its melting point is $-259°C$ and its boiling point is $-253°C$.

$$H \cdot + \cdot H \longrightarrow H{:}H$$

At first glance this might not seem to be an effective means of bonding the two hydrogen atoms. But in fact the covalent bond between two H atoms is relatively strong; 436 kJ/mol is required to break this bond.

Covalent bonds result when the outermost orbital of one atom overlaps with the outermost orbital of another atom. The $1s$ orbital of one H atom overlaps with the $1s$ orbital of the other H atom (Fig. 8.10). Overlapping orbitals are regions between two nuclei where a high probability exists of finding two electrons. Covalent bonds, like ionic bonds, result from the attraction of positive and negative particles. The positively charged nuclei attract the negative region of overlapping orbitals.

Whenever a pair of electrons is located principally in a region of space between two nuclei where orbitals overlap, we say that electrons are shared. In a Lewis structure, shared electrons are illustrated by placing the symbols for the atoms close together and inserting two dots, representing two electrons, between them. Frequently the electron pair is replaced by a dash (—) which is interpreted as a shared electron pair, a single covalent bond.

$$H{-}H = H{:}H$$

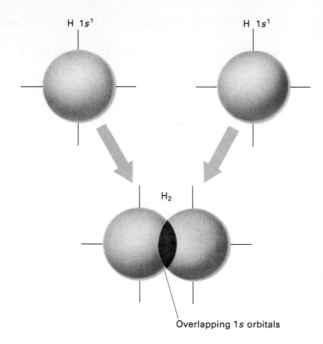

H 1s¹   H 1s¹

H₂

Overlapping 1s orbitals

Figure 8.10
In the formation of a covalent bond between two H atoms, the 1s orbitals from each H atom overlap, producing a negative region between the two nuclei. The force of attraction of the nuclei of the two H atoms for the electrons in the overlapping orbitals is the covalent bond.

Covalent Bonding in Fluorine, F₂

Fluorine gas is composed of diatomic fluorine molecules, F_2. In a manner similar to H_2, F_2 is formed when two atoms with the same electronegativity combine. Neither atom can remove an electron from the other; therefore, the bond between F atoms in F_2 is a covalent bond.

Let's consider the electronic configuration of fluorine:

$$F \qquad 1s^2 2s^2 2p^5$$

If an F atom shares an electronic with another F atom, both F atoms obtain the noble gas configuration.

$$:\overset{..}{F}\cdot \ + \ \cdot \overset{..}{F}: \ \longrightarrow \ :\overset{..}{F}:\overset{..}{F}:$$

After bonding, both F atoms have obtained eight outer-level electrons, making them isoelectronic to Ne. The covalent bond in fluorine is similar to the one that bonds H atoms in H_2, except for the orbitals that overlap. In a F atom, the outermost electrons reside in $2p$ orbitals. Accordingly, a $2p$ orbital that contains one electron from an F atom overlaps with a $2p$ orbital that contains one electron from a second F atom to produce the covalent bond (Fig. 8.11).

$$:\overset{..}{F} \ \underline{\ 2p \ 2p \ } \ \overset{..}{F}:$$

Fluorine, F_2, is extremely reactive. Paper, wood, sulfur, and even powdered metals burst into flames when exposed to F_2.

Overlapping 2p orbitals

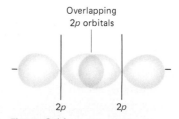

2p   2p

Figure 8.11
In the formation of a covalent bond between two F atoms, the 2p orbitals from each F atom overlap, which produces a negative region between the two nuclei.

Whenever one pair of electrons is shared between two nuclei, as was the case in both F_2 and H_2, the bond is classified as a **single covalent bond.** In molecules in which more than one pair of electrons is shared

Multiple Covalent Bonds—Double and Triple Bonds

between two nuclei, the bonds are classified as **multiple covalent bonds.** In double covalent bonds four electrons are shared; in triple covalent bonds six electrons are shared. A double bond is represented in a Lewis structure by showing four dots or two dashes between two atoms (A).

$$A :: A \quad \text{or} \quad A{=}A$$
Double covalent bond

A triple bond is represented by six dots or three dashes between two atoms (A).

$$A ::: A \quad \text{or} \quad A{\equiv}A$$
Triple covalent bond

As an example of a multiple covalent bond in a diatomic molecule, let's consider the bond between the two N atoms in diatomic nitrogen, N_2. The electronic configuration of a N atom is

$$N \qquad 1s^2 2s^2 2p^3$$

For an N atom to gain the stability of a noble gas configuration, it must share three of its electrons with another N atom. The original five outer-level electrons plus the three shared electrons give each N atom the noble gas configuration of Ne.

$$:N :: N: \qquad :N{\equiv}N:$$

The triple covalent bond in N_2 is a very strong bond.

Double covalent bonds are generally stronger than single bonds between the same atoms. In other words, more energy is required to totally break them. Four negative electrons between two positive nuclei produce a stronger attractive force than two electrons between the same nuclei. Triple bonds are even stronger than double bonds because six electrons are found between the two nuclei. The amount of energy needed to cleave a bond is called the **bond dissociation energy** (bond energies). Most frequently, bond dissociation energies are measured in kilojoules per mole. Table 8.5 lists the bond dissociation energies for selected bonds.

N_2 was discovered in 1772 by Daniel Rutherford (1749–1819), a Scottish chemist. It comprises about 77% of the volume of the atmosphere. N_2 is a relatively inert gas because of its strong triple bond between N atoms.

TABLE 8.5 BOND DISSOCIATION ENERGIES FOR SELECTED BONDS

Single bond	Bond energy, kJ/mol	Multiple bond	Bond energy, kJ/mol
H—H	436	N≡N	946
F—F	159	C≡O	1075
Cl—Cl	243	C≡C	839
Br—Br	192	O=O	498
I—I	151	N=N	418
F—Cl	255	C=C	614
C—C	347	C=N	615

Not all atoms can form multiple bonds. O, N, C, and S are the atoms that most readily produce multiple bonds. Atoms such as H or the halogens obtain a noble gas configuration by sharing one electron with another atom; consequently, they do not form multiple covalent bonds.

In the molecules that we have discussed, both of the bonded atoms were the same, resulting in the equal sharing of electrons. When the bonded atoms are the same, their electronegativities are equal; thus, neither atom has a greater attraction for the shared electron pair. A covalent bond in which both atoms have the same electronegativity is called a **nonpolar covalent bond.**

In most cases, however, the atoms forming covalent bonds have different electronegativities, and, as a result, one atom exerts a greater force of attraction on the electrons than the other. Generally, when two different atoms bond, unequal sharing results. A covalent bond in which the electrons are not shared equally is called a **polar covalent bond.**

The term *polar* implies that a charge separation, or dipole, exists. In other words, one end of the bond is more negative than the other end. As an illustration of a molecule that has a polar covalent bond, let's consider the hydrogen chloride molecule, HCl.

An H atom shares its one electron with one of the electrons of a Cl atom. The electronegativity of chlorine (3.0) is greater than that of hydrogen (2.2). Consequently, whenever a H atom bonds with a Cl atom, a single polar covalent bond results.

$$\text{H} \cdot + \cdot \overset{\cdot\cdot}{\underset{\cdot\cdot}{\text{Cl}}} : \longrightarrow \text{H} - \overset{\cdot\cdot}{\underset{\cdot\cdot}{\text{Cl}}} :$$

Hydrogen becomes isoelectronic to He, and chlorine becomes isoelectronic to Ar. Because the electronegativity of a Cl atom is greater than that of a H atom, the shared pair of electrons are more strongly attracted to the nucleus of the Cl atom (Fig. 8.12). In effect the shared electrons spend a larger percent of the time near the Cl nucleus than near the H nucleus. To show that a dipole exists, we write the symbol delta (δ) followed by a plus or minus to indicate which atom is more positive and which is more negative:

$$\overset{\delta+ \quad \delta-}{\text{H} - \text{Cl}}$$

The delta is read as "partial" or "slightly." So $\delta-$ means that an atom has a partial negative charge, and $\delta+$ indicates a partial positive charge.

Don't confuse a partial charge with the full positive or negative charge that we assigned to ions in ionic compounds. Metals have low electronegativities, and nonmetals have high electronegativities; consequently, a more complete transfer of electrons occurs in ionic compounds.

In polar covalent bonds a transfer of electrons does not occur; in-

Polar Covalent Bonds: Unequal Sharing of Electrons

Covalent Bonding in Hydrogen Chloride, HCl

A water solution of HCl is called hydrochloric acid, HCl(aq).

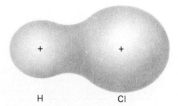

H   Cl

Figure 8.12
In the HCl molecule, the 3p orbital of the Cl atom overlaps with the 1s orbital of the H atom. Because a Cl atom has a higher electronegativity than an H atom, the Cl atom has a stronger force of attraction for the shared pair of electrons.

stead a pair (or pairs) of electrons are unequally shared. Ionic compounds can be thought of as extremely polar covalent compounds, to the point where minimal sharing occurs. The properties of polar covalent compounds are very different from those of ionic compounds.

Bromine monochloride, BrCl, is another example of a molecule with a polar covalent bond. The electronegativity of Cl (3.0) is greater than that of bromine (2.8). Accordingly, the Cl atom has a slightly greater capacity to attract the shared electron pair than does the Br atom. Thus, the electrons spend a slightly larger percent of the time closer to the Cl nucleus. Whenever the Lewis structure of bromine monochloride is written, a $\delta-$ is placed above the Cl and a $\delta+$ is placed above the Br.

Covalent Bonding in Bromine Monochloride, BrCl

$$\overset{\delta^+}{\underset{}{:}}\overset{\delta^-}{\underset{}{}}$$
$$:\overset{..}{\underset{..}{Br}}-\overset{..}{\underset{..}{Cl}}:$$

8.20 Describe how a covalent bond is different from an ionic bond.

8.21 How many electrons are shared in *(a)* a single, *(b)* a double, and *(c)* a triple covalent bond?

8.22 Draw the Lewis structures for *(a)* H_2, *(b)* F_2, *(c)* N_2, *(d)* I_2.

8.23 Explain the difference between a nonpolar and a polar covalent bond, and give examples of each.

8.24 Draw the Lewis structures for each of the following polar molecules, and indicate the partial charges using $\delta+$ and $\delta-$: *(a)* HI, *(b)* IF.

REVIEW EXERCISES

A systematic procedure is usually required to successfully write Lewis structures for larger and more complex covalent molecules. Most Lewis structures for covalent molecules are obtained by following a set of five steps.

8.5 LEWIS STRUCTURES FOR COVALENT MOLECULES

Step 1
Calculate the total number of outer-level electrons in all atoms in the molecule.

Use the periodic table to find the number of outer electrons in each atom, and then add these numbers to obtain the total. For example, CF_4 has a total of 32 outer electrons because C has 4 outer electrons and four F atoms each have 7 outer electrons.

Step 2
Identify the central atom (or atoms), and write the symbols for all other atoms around the symbol for the central atom (Fig. 8.13).

The central atom of a molecule is bonded to two or more other atoms and is the atom that determines the overall shape of small molecules. It is not difficult to identify, and after a little practice this becomes a trivial

Except for F, halogens may be central atoms in molecules when they are bonded to more than one oxygen or another halogen. Cl is the central atom in each of the following molecules: $HClO_3$, ClF_3, ClO_2, and Cl_2O_7.

matter. B, C, Si, N, P, and S are among the most common central atoms encountered.

Hydrogen is never a central atom because it has only one electron to share with another atom, and thus only forms one bond. For the same reason, halogens are usually not central atoms in the simple molecules we will encounter. In binary compounds, oxygen is found as the central atom only when it is bonded to H or halogen atoms. In most other binary compounds, and in compounds with three or more elements, oxygen is rarely a central atom.

It is fairly standard when writing the name or molecular formula of a compound to write the central atom first, except when the compound is an acid (HNO_3, H_2SO_4, H_3PO_4, etc). In each of the following examples, the central atom is written first: CO_2, OF_2, NH_3, and PBr_3.

Step 3

Place a pair of electrons (single bond) between the central atom and each of the other atoms in the molecule. Then, subtract the number of electrons you have written into the Lewis structure so far from the total number of electrons obtained in Step 1. The resulting number is the quantity of electrons that remain to complete the noble gas configurations of each atom.

Step 4

Calculate the number of electrons needed for all atoms to achieve the outer-level noble gas configuration, and compare the result with the number of available electrons (determined in Step 3).

If these numbers are equal, place the appropriate number of dots around each symbol so that every atom has a noble gas configuration. When you write dots into the formula, ask yourself this question: How many outer electrons does the atom have, and how many does it need to obtain a noble gas configuration?

If the number of electrons available is smaller than the number of electrons needed to obtain the outer-level noble gas configuration, then there must be a multiple bond in the molecule. Generally, if the deficit is two electrons, the molecule has a double bond; a shortage of four electrons indicates two double bonds or a triple bond.

Step 5

Check the final Lewis structure in two ways: (1) Count the total number of electrons and confirm that the correct number has been written into the formula, and (2) verify that each atom has a noble gas configuration.

An easy way to check the number of electrons that surround each atom is to circle the electrons around each atom. You should include the electrons that belong to the atom and those that are shared. In all cases, there should be either eight or two electrons. Then and only then are you sure that the Lewis structure is correct (assuming that it is possible for all atoms to achieve a noble gas configuration).

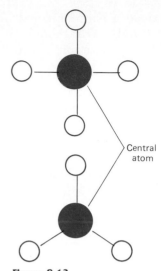

Central atom

Figure 8.13
The central atom in a molecule is bonded to two or more atoms. In binary covalent compounds, the central atom is usually written first in the formula and is the atom with a lower electronegativity.

In summary, when writing the Lewis structure of a covalent compound:

Step 1
Calculate the total number of outer-level electrons.

Step 2
Identify the central atom, and write the symbols for all other atoms around the symbol for the central atom.

Step 3
Place a pair of electrons between the central atom and each of the other atoms. Calculate the quantity of electrons that remain to be written into the structure by subtracting the number of electrons already in the formula from the total (Step 1).

Step 4
Calculate the number of electrons required for all atoms to achieve the outer-level configuration of a noble gas, and compare to the number of electrons available (Step 3).

1. If the number of electrons required to complete the noble gas configuration for all atoms equals the number available, place the electrons around each symbol, completing the octets. The molecule contains only single bonds.

2. If the number of electrons required is greater than the quantity available, there is at least one multiple bond in the molecule. A shortage of two electrons indicates a double bond, and a shortage of four electrons indicates either a triple bond or two double bonds. Locate the atoms that have a multiple bond, write in the correct number of electrons around each of their symbols, and then complete the outer levels of all other atoms.

Step 5
Check the Lewis structure to see that each atom has a noble gas configuration, and that the total number of electrons equals the total in Step 1.

The five rules for writing Lewis structures are general guidelines that may be used for many, but not all, molecules. Nevertheless the Lewis structures of almost all molecules that you will encounter in this book can be successfully written using these rules.

Water is the most important liquid on earth. A significant quantity of the earth's surface is covered with water, and living tissues contain approximately 80% water.

As we will discuss in Chap. 13, the properties of water are unique among low-molecular-mass liquids. The special properties of water cannot be understood until we learn how the water molecule is bonded.

We will use the five steps just outlined to write the Lewis structure of water.

Covalent Bonding in Water, H_2O

Step 1
Calculate the total number of outer-level electrons.

$$2 \text{ atoms H} \times \frac{1 \text{ e}^-}{\text{H atom}} = 2 \text{ e}^-$$

$$1 \text{ atom O} \times \frac{6 \text{ e}^-}{\text{O atom}} = +6 \text{ e}^-$$

$$H_2O = 8 \text{ e}^-$$

Step 2
Write the symbol for the central atom, and place the other atoms' symbols around it. By default the central atom in water is the O atom. H atoms are never central atoms. Therefore write O with the two adjacent H atoms.

$$\text{O H}$$
$$\text{H}$$

It does not matter how the atoms are written around the central atom; Lewis structures do not usually illustrate the spatial arrangement of atoms in a molecule.

Step 3
Place electron pairs between the central and all other atoms, and calculate the number of electrons that remain to be added to the structure.

$$\text{O} \!:\! \text{H}$$
$$\text{H}$$

Four electrons have been written into the Lewis structure, which leaves four electrons.

$$\text{Remaining e}^- = \text{total e}^- - \text{electrons already in formula}$$
$$= 8 - 4 = 4\text{e}^-$$

Step 4
Find the number of electrons needed to give each atom an outer-level noble gas configuration, and compare this number to the number of electrons that remain.

Four more electrons are needed to complete the octet around O, because, by sharing two electrons from each of the two H atoms, the O has four electrons. Each H atom already has a noble gas configuration of two electrons.

Four electrons are available, and four electrons are needed. Thus, place the four electrons around the O to complete its octet.

$$: \overset{..}{\underset{..}{O}} : H$$
$$H$$

Step 5
Check to see the formula has the proper number of electrons.

1. Check to see that each atom has a noble gas configuration: Each H atom has two electrons, and the O atom has eight electrons.

2. Count the total number of electrons in the Lewis structure: There are eight electrons in the Lewis structure, which corresponds to the total number of outer-level electrons in two H atoms and one O atom.

The Lewis structure of water indicates that a water molecule contains two covalent bonds. Each bond is classified as a polar covalent bond, because the electronegativity of oxygen (3.5) is greater than that of hydrogen (2.2). In addition to the electrons in the two bonds, the central oxygen has two pairs of electrons that are not bonded. These electrons are termed either **long pair electrons** or **nonbonded electron pairs.**

Because all members of group VIA (16), the chalcogens, have six outer-level electrons, you might expect that they would bond in a manner similar to oxygen. In simple covalent molecules, this is true. For example, consider the Lewis structure of hydrogen sulfide, H_2S.

$$: \overset{..}{\underset{..}{S}} : H$$
$$H$$

Sulfur is the element directly below oxygen on the periodic table; therefore, the bonding in an H_2S molecule is similar to that in an H_2O molecule. H_2S is a toxic, yellowish gas with the unpleasant odor of rotting eggs. H_2Se and H_2Te also have Lewis structures similar to those of water and hydrogen sulfide.

Ammonia, NH_3, is a gas at room temperature. Water solutions of ammonia are commonly used as household cleaners, and ammonia vapors are used to revive people who have passed out. We will write the Lewis structure of ammonia, using it as an illustration of the covalent bonding of atoms in group VA (15).

Following our rules, we first calculate the total number of outer-level electrons in an ammonia molecule (Step 1):

Covalent Bonding in Ammonia, NH_3

$$3 \ \overline{\text{atoms H}} \times \frac{1 \ e^-}{\overline{\text{H atom}}} = \ 3 \ e^-$$

$$1 \ \overline{\text{atom N}} \times \frac{5 \ e^-}{\overline{\text{N atom}}} = \underline{+5 \ e^-}$$

$$\text{NH}_3 \qquad \qquad = \ 8 \ e^-$$

The central atom of the ammonia molecule is N because H atoms are never central atoms. In Step 2 we write the symbol for N with 3 H atoms placed around it.

H N H
H

After an electron pair is written between each H and the N, six electrons are accounted for out of the eight total electrons, which leaves two electrons to complete the outer-level octet of the N atom.

H : N : H
H

Each H atom has a noble gas configuration, and the N atom needs only two more electrons to complete its octet. Two electrons are required, and two electrons remain to be placed in the structure; consequently, these electrons are placed next to the N atom to complete its outer level.

H : N : H
H

Each atom in a molecule of ammonia has a noble gas configuration, either two or eight, and the total number of outer-level electrons equals eight. The Lewis structure of ammonia reveals a molecule with three polar covalent bonds and one lone pair.

Bonding in phosphine, PH_3, is similar to the bonding in ammonia because P is the atom directly below N on the periodic table. Both have five outer-level electrons. Try to write the Lewis structure of phosphine, following each of the five steps. The Lewis structure of phosphine is

H : P : H
H

All of the group VA hydrides (binary hydrogen compounds) have analogous Lewis structures. These compounds include NH_3, PH_3, AsH_3, and SbH_3.

Nitrogen trifluoride, NF_3, is a colorless gas that is the most stable of the nitrogen halides. Let's draw its Lewis structure.

Covalent Bonding in Nitrogen Trifluoride, NF_3

Step 1
Calculate the total number of outer-level electrons.

$$3 \text{ atoms F} \times \frac{7 \text{ e}^-}{\text{F atom}} = 21 \text{ e}^-$$

$$1 \text{ atom N} \times \frac{5 \text{ e}^-}{\text{N atom}} = +\ 5 \text{ e}^-$$

$$NF_3 = 26 \text{ e}^-$$

Step 2
Identify the central atom, and write the symbols for the other atoms around it. The central atom in NF_3 is N, because F atoms only form one bond and therefore F cannot be a central atom.

F N F
F

Step 3
Write electron pairs between the N atom and the F atoms, and calculate the number of electrons that remain.

F : N : F
..
F

e^- remaining = total number of e^- − e^- written in the Lewis structure

$$= 26e^- - 6e^- = 20e^-$$

Step 4
Calculate the number of electrons needed to complete all octets, and write the remaining electrons into the structure. Each F atom requires six electrons to complete its octet, and the N atom requires two electrons ($8e^- - 6e^- = 2e^-$). Hence, 6×3, or 18, electrons are needed for the F atoms, and adding the 2 electrons for N gives 20 electrons—the number available. Write the 20 available electrons into the Lewis structure to complete the octets of all atoms.

.. .. ..
:F:N:F:
.. .. ..
:F:
..

Step 5
Check the Lewis structure to make sure you have the correct number of electrons. Each F atom has 8 electrons, as does the N, and the total

number of electrons equals 26; consequently, the Lewis structure is written correctly.

Methane, CH_4, commonly called swamp or marsh gas, is the main component of natural gas. Methane is produced when plants are decomposed by various species of microorganisms.

After following the five steps for writing a Lewis structure, we obtain the following structure for methane:

$$\begin{array}{c} \ddot{H} \\ H \!:\! \overset{\displaystyle H}{\underset{\displaystyle H}{C}} \!:\! H \end{array}$$

Covalent Bonding in Methane, CH_4

Verify the Lewis structure of methane by going through each step. Note that this structure shows that methane molecules have four single polar covalent bonds and no lone pair electrons.

Carbon tetrachloride, CCl_4, at one time was used in fire extinguishers (it is a noncombustible liquid that smothers fires) and was also a component of many cleaning fluids and spot removers. However, research studies showed that CCl_4 was toxic and inhalation of its vapors caused liver damage. CCl_4 was also found to be carcinogenic, i.e., to cause cancer. The Lewis structure of carbon tetrachloride is

Covalent Bonding in Carbon Tetrachloride, CCl_4

$$\begin{array}{c} :\ddot{Cl}: \\ :\ddot{Cl}:\overset{\displaystyle :\ddot{Cl}:}{\underset{\displaystyle :\ddot{Cl}:}{C}}:\ddot{Cl}: \end{array}$$

Again verify the Lewis structure by following each of the five steps. Note that the only difference between the Lewis structures for CF_4, CBr_4, and CI_4 and the one shown here for CCl_4 is the symbol for the halogen atom in the molecule.

8.25 List the five steps followed in writing the Lewis structure for a covalent molecule.

8.26 Illustrate the method for writing Lewis structures using each of the following molecules: *(a)* OCl_2, *(b)* NF_3, *(c)* CI_4, *(d)* AsH_3, *(e)* SiH_4, *(f)* H_2Se.

REVIEW EXERCISES

I f the number of electrons that remain after the initial drawing of electron pairs in the structure (Step 3) is less than the number required to complete the octets of all atoms, then there is at least one **multiple bond** in the molecule. We will consider carbon dioxide, CO_2, as an example of such a molecule.

8.6 ADDITIONAL COVALENT BONDING CONSIDERATIONS

Carbon dioxide, CO_2, is a gas at room temperature. It is an important gas because it is a by-product of cellular respiration in animals—the means by which animals produce energy. Carbon dioxide is also a product of the combustion of materials that contain carbon. Carbonated beverages contain dissolved CO_2, and become "flat" when the CO_2 escapes.

Covalent Bonding in Carbon Dioxide, CO_2

To write the Lewis structure of CO_2, we begin by calculating the total number of outer-level electrons. A C atom has 4 electrons, and two O atoms have 12 electrons, which gives a total of 16 electrons. Carbon is the central atom. Rarely will you encounter a compound that has O as its central atom. After placing the pairs of electrons between the O atoms and the C atom we get

$$O : C : O$$

Subtracting 4 electrons from the total of 16 electrons leaves us with 12 electrons. But to complete the octets of one C atom and two O atoms we need 16 electrons—4 for the C atom and 6 for each O atom. The electron deficit is $16 - 12$, or 4, electrons, which indicates that there are either two double bonds or a triple bond in CO_2. In many cases, it is best to select the more symmetrical arrangement of bonds because the more symmetrical arrangements occur more frequently in nature; thus, we would predict that CO_2 has two double bonds and not one triple and one single bond. If we write in two more electrons between the C and each O, and write the remaining eight electrons around the O atoms, four each, the Lewis structure becomes

$$:\overset{..}{\text{O}}::C::\overset{..}{\text{O}}: \quad \text{or} \quad :\overset{..}{\text{O}}=C=\overset{..}{\text{O}}:$$

In the second formula, we use two dashes to indicate a double covalent bond, composed of four shared electrons.

Formaldehyde, H_2CO, a gas at 25°C, is a very important industrial chemical. Formaldehyde molecules also contain multiple bonds. The central atom in formaldehyde is C, and it is bonded to two H atoms and one O atom. Following the five steps for writing a Lewis structure, we find a shortage of two electrons, which indicates that there is a double bond in the molecule. The Lewis structure of formaldehyde is

Covalent Bonding in Formaldehyde, H_2CO

A water solution of formaldehyde is called formalin. Formalin is used to preserve dead biological specimens. Formaldehyde is suspected of being a cancer-causing agent.

$$\begin{array}{c} \quad\quad \overset{..}{} \\ H : C : : \overset{..}{O} \\ \overset{..}{} \\ H \end{array}$$

Verify this structure by going through each step.

We have assumed that in all of the molecules discussed the electrons are localized between two nuclei. However, in some molecules this is not what we find. For example, let's consider the Lewis structure of sulfur

Resonance: Delocalized Bonding

dioxide. Sulfur dioxide gas is released into the atmosphere when sulfur-containing fossil fuels are burned. It is a major contributor to the production of acid rain. Following the five rules, two Lewis structures can be written:

$$:S::O: \quad \text{and} \quad :S:O:$$
$$:O: \qquad\qquad\qquad :O:$$
$$\text{I} \qquad\qquad\qquad \text{II}$$

In structure I the double bond is found between the S atom and the O atom on the right, and in structure II the double bond is found between the S atom and the O atom below. Experimental evidence shows that neither structure I nor structure II can explain the properties of the SO_2 molecule. According to the valence bond theory, when such a situation arises, one of the bonding pairs of electrons is spread out over all three atoms. We say that these electrons are **delocalized.** A better way to represent the bonded electrons in sulfur dioxide is as follows:

Delocalization of electrons occurs only in molecules that have multiple bonds.

$$\overset{S}{\underset{OO}{\diagup\diagdown}}$$

A dashed line indicates a partial bond in which the electron pair is spread out over the entire molecule as opposed to being localized between the nuclei of two atoms. From this structure we see that the bond in SO_2 molecules is neither a single nor a double bond. Instead, the bond is intermediate between a single and double bond.

Molecules such as SO_2 that have delocalized electrons do not have one unique Lewis structure. Instead, these molecules have two or more Lewis structures that differ only with respect to the placement of the electrons. When this situation is encountered, we say that the molecule exhibits **resonance.** Each possible Lewis structure for a molecule is called a **contributing structure.** The arrangement of atoms in each contributing structure is the same; only the electrons are in different places. The actual arrangement of electrons in molecules that exhibit resonance most closely resembles the average of all contributing structures. To symbolize resonance, the contributing structures are separated by a double headed arrow ($\longleftrightarrow$),

If a covalent bond forms in which one atom donates both electrons, as is the case in SO_2, the bond is classified as a coordinate covalent bond. After they form, coordinate bonds have the same properties as other covalent bonds.

Another example of a molecule that exhibits resonance is the nitrous oxide molecule, N_2O. The two contributing structures that best represent N_2O are as follows:

$$:N=N=O: \quad\longleftrightarrow\quad :N\equiv N-O:$$
$$\text{I} \qquad\qquad\qquad\qquad \text{II}$$

Because the distribution of electrons is the average of the contributing structures I and II, we see that the bond between the two N atoms is intermediate between a double and a triple bond, and the bond between the N and O atoms is intermediate between a single and double bond.

The structure of a molecule that exhibits resonance is considered a hybrid (a blend) of all contributing structures.

It is unfortunate that the term "resonance" has a different meaning in physics. Do not confuse the meaning of the term "resonance" as defined in physics with the definition used by chemists. "Resonance" in chemistry refers to delocalization of electrons—these electrons are not resonating back and forth as might be implied from the physics definition.

Molecules that exhibit resonance are more stable than similar molecules that do not exhibit resonance. The greater stability of molecules that exhibit resonance results directly from the delocalization of electrons over more than two atoms. Generally, the more localized the electrons are within a molecule, the more unstable it is.

A **polyatomic ion** is an ion composed of more than one atom. The atoms in polyatomic ions are held together by covalent bonds. Examples of common polyatomic ions include (1) hydroxide, OH^-; (2) nitrate, NO_3^-; (3) sulfate, SO_4^{2-}; (4) phosphate, PO_4^{3-}; and (5) ammonium, NH_4^+. A polyatomic ion bonds ionically to another ion. For example, in sodium hydroxide, NaOH, the negative polyatomic ion hydroxide bonds to the positive sodium ion.

When you write the Lewis structures for polyatomic ions, it is necessary to add or subtract the number of electrons indicated by the charge on the ion in Step 1. For example, phosphate has a 3− charge; it possesses three extra electrons in addition to the outer-level electrons of the P and O atoms. A total of 32 outer-level electrons are found in phosphate.

To illustrate the procedure for writing the Lewis structures of polyatomic ions, let's write the Lewis structure for the nitrate ion, NO_3^-.

Polyatomic Ions

Poly is a prefix that means "many."

Step 1
Calculate the total number of electrons.

$$1 \text{ atom N} \times \frac{5 \text{ e}^-}{\text{N atom}} = 5 \text{ e}^-$$

$$3 \text{ atoms O} \times \frac{6 \text{ e}^-}{\text{O atom}} = 18 \text{ e}^-$$

$$\text{Additional e}^- = + \; 1 \text{ e}^-$$

$$NO_3^- = 24 \text{ e}^-$$

The "extra" electrons in polyatomic ions can come from metals and other polyatomic ions.

Step 2
Place the three O atoms around the central N atom.

$$\begin{array}{cc} & O \\ O & N \\ & O \end{array}$$

Steps 3 and 4

Six electrons are required to bond the three O atoms to the nitrogen atom; thus, 18 electrons remain to be written into the structure. To complete the octets of all atoms, 20 electrons are needed ($3 \times 6e^- = 18\,O\,e^- + 2\,N\,e^- = 20e^-$). A shortage of two electrons is found, indicating a double bond. Consequently, three contributing structures can be written for the nitrate ion, an ion that exhibits resonance. Each of the contributing structures differs with respect to the placement of the double bond.

$$\left[\begin{array}{c} :\!\ddot{O}\!: \\ :\ddot{O}::N \\ :\ddot{O}: \end{array} \right]^{-} \longleftrightarrow \left[\begin{array}{c} :O: \\ :\ddot{O}:N \\ :\ddot{O}: \end{array} \right]^{-} \longleftrightarrow \left[\begin{array}{c} :\ddot{O}: \\ :\ddot{O}:N \\ :\ddot{O}: \end{array} \right]^{-}$$

$$\qquad\text{(I)} \qquad\qquad\qquad \text{(II)} \qquad\qquad\qquad \text{(III)}$$

Step 5

Check to see that there are 24 electrons in each contributing structure, and that each atom has a noble gas electronic configuration.

Lewis structures of polyatomic ions are enclosed in square brackets with their charges as superscripts to indicate that they represent ions that cannot exist alone. Table 8.6 lists some of the more common polyatomic ions and shows their Lewis structures.

Molecules with Atoms That Do Not Achieve the Noble Gas Configuration

The rules we have been using to write Lewis structures are somewhat artificial generalizations that apply to simple molecular systems. Some molecules do not follow these rules. Remember that this is not a fault of the natural world; it is a limitation of using a system of artificial rules.

Some molecules possess a bonded atom with less than eight electrons. Boron trifluoride, BF_3, is an example of such a molecule. Its Lewis structure is

$$\begin{array}{c} :\ddot{F}: \\ | \\ \ddot{B} \\ \diagup \quad \diagdown \\ :\ddot{F} \qquad \ddot{F}: \end{array}$$

The central B atom has only six outer-level electrons. Is BF_3 a very stable molecule? No! Boron trifluoride is reactive and combines with almost any molecule or ion that can donate an electron pair.

Numerous molecules exist with a central atom that has more than eight outer-level electrons. These molecules contain an atom with an *expanded octet*. Only certain elements in the third period and beyond have the capacity to accommodate more than eight outer-level electrons after bonding. Some of these molecules have atoms with 10, 12, and 14 outer-level electrons. Noble gas compounds are good examples of molecules with expanded octets; examples include XeF_2, XeF_4, XeF_6, $XeOF_4$, and XeO_3. The structures of XeF_2 and XeF_4 are best represented as

Xenon difluoride   Xenon tetrafluoride

Noble gas compounds were first synthesized by Neil Bartlett in 1962. At that time it was thought that noble gases could not form stable compounds. Xenon hexafluoroplatinate, $XePtF_6$, was the first noble gas compound synthesized.

TABLE 8.6 POLYATOMIC IONS

Name	Molecular formula	Structural formula	Name	Molecular formula	Structural formula
Borate	BO_3^{3-}		Ammonium	NH_4^+	
Carbonate	CO_3^{2-}		Nitrate	NO_3^-	
			Nitrite	NO_2^-	
Hydrogencarbonate (bicarbonate)	HCO_3^-		Hydroxide	OH^-	
Acetate	$C_2H_3O_2^-$		Phosphate	PO_4^{3-}	
Cyanide	CN^-		Sulfate	SO_4^{2-}	
Chromate	CrO_4^{2-}		Sulfite	SO_3^{2-}	

Still other molecules possess an odd number of outer-level electrons, which precludes the possibility of drawing a structure with eight outer-level electrons in all atoms. Attempt to draw the Lewis structures of nitrogen(II) oxide, NO, or nitrogen(IV) oxide, NO_2, and you will discover that the total number of outer-level electrons is 11 and 17,

The Lewis structure of NO is $:N{=}O:$, and the Lewis structure of NO_2 is $:\ddot{O}{-}\dot{N}{=}\ddot{O}:$

respectively. Also you will find that more than one Lewis structure can be written for each of these oxides of nitrogen: NO and NO_2 exhibit resonance.

8.27 Draw the Lewis structure for carbon monoxide, CO, a molecule with a multiple bond.

8.28 *(a)* What are delocalized electrons? *(b)* Write a definition for resonance. *(c)* Illustrate resonance by drawing the Lewis structures for the three contributing structures of sulfur trioxide, SO_3.

8.29 Draw the Lewis structures for each of the following polyatomic ions: *(a)* OH^-, *(b)* PO_4^{3-}, *(c)* CO_3^{2-}, *(d)* CN^-.

8.30 Which of the following molecules does not obey the octet rule: *(a)* BeF_2, *(b)* PF_3, *(c)* IF_3, *(d)* NO_2?

8.7 PROPERTIES OF MOLECULES

Molecular Geometry

The three-dimensional structure of a molecule is called its **molecular geometry.** More simply, molecular geometry refers to the average shape of a molecule. One way to predict the shape of a simple molecule is to consider the number of bonding and lone pair electrons on its central atom. The best prediction of the molecular geometry of a molecule is the shape that maximizes the distance between the electron pairs on the central atom. Electron pairs, either bonding pairs or lone pairs, are negative regions of space; thus, they repel each other and attempt to be as far as possible from each other. This method of predicting the shape of a molecule is called the **valence shell electron pair repulsion** method, or the **VSEPR** method for short.

To illustrate the VSEPR method, let us first consider the beryllium fluoride, BeF_2, molecule. To use the VSEPR method, first write the Lewis structure of the molecule.

$$: \overset{\cdot\cdot}{\underset{\cdot\cdot}{F}} — Be — \overset{\cdot\cdot}{\underset{\cdot\cdot}{F}} :$$

Then count the number of bonding pairs and lone pairs on the central atom. On the Be atom we find two bonding pairs and no lone pairs. The angle that minimizes the electron repulsions when there are two bonding pairs on the central atoms is 180°, which means that the molecular geometry of BeF_2 is **linear.** Figure 8.14 shows a diagram of a linear molecule. All the atoms in a linear molecule are in a straight line.

As a second example let's consider boron trifluoride, BF_3. After writing the Lewis structure for BF_3, we find that the central B atom has three bonding pairs and no lone pairs.

$$\overset{\cdot\cdot}{\underset{}{: F :}}$$
$$|$$
$$\overset{}{\underset{}{B}}$$
$$: \overset{\cdot\cdot}{F} \qquad \overset{\cdot\cdot}{F} :$$

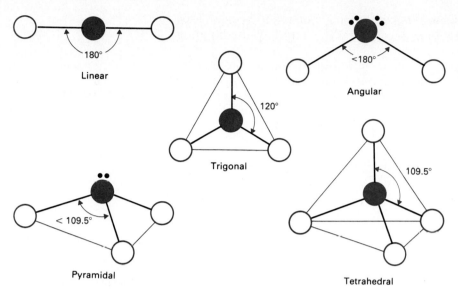

Figure 8.14
In linear molecules, all the atoms are in a straight line (180° bond angle). In angular molecules, three atoms form an angle less tham 180°. Trigonal planar molecules are composed of four atoms that are in the same plane. Each of the atoms bonded to the central atom in a trigonal planar molecule is separated by 120° from the other two atoms. Pyramidal molecules have three bonded atoms and one lone pair around the central atom. Tetrahedral molecules have four atoms bonded to a central atom separated by an angle of 109.5°.

Three bonding pairs can be at a maximum distance from each other when they are separated by 120° and all the atoms are in the same plane. We call this **trigonal planar** geometry (Fig. 8.14). All molecules with three bonding pairs and no lone pairs have trigonal planar geometry.

Three different molecular geometries can result from four electron pairs on the central atom. For example, the Lewis structure of methane, CH_4, shows four bonding pairs and no long pairs on the central C atom.

$$H-\overset{\displaystyle H}{\underset{\displaystyle H}{C}}-H$$

To minimize the repulsions of four bonding pairs of electrons, the maximum angle between H atoms is 109.5°, which gives the molecule a **tetrahedral** geometry (Fig. 8.14). Whenever some combination of four bonding and lone pairs are found on a central atom, the geometry of the *electron pairs* is tetrahedral with approximately a 109.5° angle between them.

If a molecule has three bonding pairs and one lone pair, such as is found in ammonia molecules,

$$H-\overset{\displaystyle \cdot\cdot}{\underset{\displaystyle H}{N}}-H$$

the molecular geometry of the molecule becomes **pyramidal** (pyramid-like) (Fig. 8.14). The angles between the H atoms in ammonia are slightly less than the theoretical maximum angle of 109.5° because lone pairs are diffuse and they tend to compress the angles of the bonding pairs.

Angular, or **V-shaped,** geometry is found for molecules such as water that have two bonding pairs and two lone pairs.

$$\ddot{O} \diagdown_{H}^{H}$$

Because two lone pairs are found on the O atom, the bond angle in water is only 104.5° as a result of the greater force of repulsion by the two lone pairs on the two bonding pairs.

Multiple bonds are treated in the same manner as single bonds when applying the VSEPR method. For example, carbon dioxide, CO_2, which has the Lewis structure

$$\ddot{O}=C=\ddot{O}$$

is a linear molecule because is has no lone pairs to prevent the two double bonds to the C atom from being at a maximum distance from each other. If we reconsider the sulfur dioxide molecule, which has the Lewis structure

$$\ddot{S} \diagup_{O}^{} \diagdown_{O}$$

as one of its contributing structures, we find that it has an angular, or V-shaped, geometry because it has two bonds and one lone pair on the central S atom. Three negative regions are at a maximum distance when they are 120° from each other in the same plane.

Table 8.7 summarizes the principal geometries of molecules that have two, three, and four electron pairs on the central atom.

TABLE 8.7 MOLECULAR GEOMETRIES OF MOLECULES

Number of electron pairs on central atom				
Total	Bonding pairs	Lone pairs	Geometry	Examples
2	2	0	Linear	BeH_2, $BeCl_2$, CO_2
3	3	0	Trigonal planar	BF_3, $AlCl_3$, $CO_3{}^{2-}$
3	2	1	Angular (V-shaped)	SO_2, $SnCl_2$, $NO_2{}^{-}$
4	4	0	Tetrahedral	CH_4, SiF_4, GeH_4
4	3	1	Pyramidal	NH_3, PF_3, AsH_3
4	2	2	Angular (V-shaped)	H_2O, H_2S, OCl_2

Properties of Bonds

Four properties of bonds are commonly considered when studying molecules: (1) bond length, (2) bond order, (3) bond angle, and (4) bond energy (Fig. 8.15).

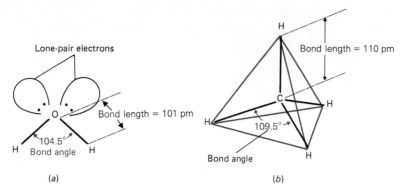

Figure 8.15
(a) Water is composed of angular molecules that have two bonding pairs and two lone pairs on the O atom. The bond angle in water is 104.5° and the O—H bond length is 101 pm. *(b)* Methane is composed of tetrahedral molecules. The bond angle between each pair of H atoms in methane is 109.5°, and the C—H bond length is 110 pm.

Bond length is the average distance between two nuclei of atoms in a covalent bond. For example, carbon–carbon single bonds are approximately 154 pm (1 pm = 1×10^{-12} m). Carbon–carbon double and triple bonds are even shorter: 133 pm and 120 pm, respectively.

Indirectly related to bond length is bond order. **Bond order** is the number of covalent bonds that link two atoms. A single covalent bond has a bond order of 1, a double bond has a bond order of 2, and a triple

C—C
C=C
C≡C

Bond order	Type of bond
1	Single
2	Double
3	Triple

--- **CHEM TOPIC: Molecular Models** ---

Molecular model kits are used to construct three-dimensional models of molecules. A molecular model helps people visualize the spatial relationships of the atoms in the molecule. Many different types of kits are sold. One of the most popular is the *ball-and-stick* model in which atoms are represented by spheres and bonds by sticks and springs. Such a kit contains spheres of different colors with holes drilled into their sides at the same angles as are found between atoms in molecules. For example, black spheres usually have holes separated by approximately 109° to represent the tetrahedral geometry of many atoms. Four sticks are placed in these holes, and four spheres are placed on the ends of the sticks to represent the tetrahedrally bonded atoms. Ball-and-stick models readily show bond angles and overall shapes of molecules, but do not give an accurate picture of the molecule and the nature of the bonds within.

A more realistic representation of molecules is obtained by using *space-filled* models. Figure 8.17 shows that space-filled models of atoms are spheres that have been cut in such a way as to show the overlap of orbitals. "Atoms" in space-filled models snap together to produce an overall model of the regions of space occupied by atoms in a molecule. These models give a better view of the actual size and shape of molecules.

Figure 8.16
The figure shows the ball-and-stick model for butane, C_4H_{10}. The darker spheres represent the C atoms and the lighter ones represent the H atoms. This model clearly shows the tetrahedral geometry around each C atom.

Ball-and-stick model of butane, C_4H_{10}

Figure 8.17
A space-filled model of butane better shows how the orbitals overlap and gives a more accurate picture of the shape and size of the molecule.

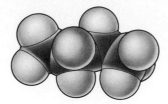

Space-filled model of butane, C_4H_{10}

bond has a bond order of 3. In resonance structures, fractional bond orders are found; e.g., 1.5 and 1.33 are encountered.

As we have already seen, a **bond angle** is the angle between two imaginary lines passing through the nucleus of the central atom and the nuclei of two atoms bonded to the central atom. For example, the bond angle (H—O—H) between H atoms in the water molecule is 104.5°, and the bond angle between H atoms in the methane molecule is 109.5°.

We have already briefly mentioned **bond energy** as being the amount of energy required to break 1 mol of a specific bond in the gas phase. Bond energies are related to how stable a bond is. In general, bond energies for triple bonds are higher than those for double bonds, which are higher than those for single bonds. Carbon–carbon triple bonds have bond energies in excess of 800 kJ/mol. Whereas most carbon–carbon double bond energies are near 600 kJ/mol, only 350 kJ/mol is required to break carbon–carbon single bonds.

REVIEW EXERCISES

8.31 What are the most common molecular geometries exhibited by simple molecules?

8.32 Use the VSEPR method to predict the shapes of the following molecules: (a) NCl_3, (b) CCl_4, (c) OF_2, (d) CS_2.

8.33 Compare single, double, and triple bonds between two carbon atoms with respect to (a) bond order, (b) bond length, (c) bond energy.

8.34 What bond angles are found in each of the following molecular geometries: (a) tetrahedral, (b) linear, (c) trigonal planar?

SUMMARY

Compounds are divided into two groups: (1) ionic and (2) covalent. **Ionic compounds** form when a metal combines with a nonmetal or a polyatomic ion. **Covalent compounds** result when two or more nonmetals combine.

The properties of ionic compounds differ from those of covalent compounds. Generally, ionic com-

pounds have higher melting points, boiling points, and densities than covalent compounds. All ionic compounds are solids at room temperature, while covalent compounds are more frequently liquids and gases. Both groups are nonconductors of electricity, although ionic compounds do conduct an electric current in the liquid state.

A **chemical bond** is the force of attraction that results when atoms either transfer or share electrons. If atoms transfer electrons in the formation of a chemical bond, the bond is classified as an **ionic bond,** and if electrons are shared, the bond is classified as a **covalent bond.**

After one or more electrons are transferred from a metal to a nonmetal, two ions result. The metal atom loses electrons and becomes a cation, and the nonmetal atom gains electrons and becomes an anion. Both ions achieve the configuration of a noble gas. Cations attract anions and vice versa, creating the network of ionic bonds in ionic compounds.

Covalent compounds share electrons through overlapping outer-level orbitals in order to obtain the stable electronic configuration of a noble gas. A covalent bond in which the electrons are equally shared is called a **nonpolar bond.** If the electrons are shared unequally, the bond is classified as a **polar covalent bond.** The polarity of a bond is predicted from the electronegativities of the atoms that make up the bond. **Electronegativity** is the capacity of atoms to attract electrons in chemical bonds.

To identify the bonds in a molecule, we write its Lewis structure. A **Lewis structure** accounts for all outer-level electrons in a molecule. Most frequently, all the atoms in a Lewis structure achieve the noble gas electronic configuration. A series of five steps is usually required to successfully write Lewis structures for molecules.

Covalent molecules contain single, double, and triple bonds. A **single covalent bond** has one shared electron pair. **Double bonds** have two pairs of shared electrons, and **triple bond** have three electron pairs shared. Molecules can have any combination of single, double, or triple covalent bonds.

To predict the shapes of molecules, chemists use the **valence shell electron pair repulsion** method. The Lewis structure is written, and the predicted molecular geometry is the one in which the electron pairs on the central atom are at a maximum distance from each other. When studying molecules, chemists are most concerned with bond lengths, bond orders, bond energies, and bond angles.

KEY TERMS

bond order	contributing structure	electronegativity	molecular geometry	polar covalent bond
bond angle	covalent compound	electrovalent bond	multiple bond	polyatomic ion
bond length	crystal lattice	ionic radius	nonpolar covalent bond	resonance
chemical bond	delocalized electrons	ionic compound	outer-level electron	stability

---

EXERCISES*

8.35 Define each of the following: ionic compound, covalent compound, chemical bond, stability, outer-level electron, electrovalent bond, ionic radius, crystal lattice, electronegativity, nonpolar covalent bond, polar covalent bond, multiple bond, delocalized electrons, resonance, contributing structure, polyatomic ion, bond order, bond length, bond angle, molecular geometry.

Compounds

8.36 How do the rules for writing names of ionic compounds differ from those used to write names of covalent compounds? Give an example.

8.37 Write the names of the following ionic compounds: *(a)* NaBr, *(b)* MgO, *(c)* RbI, *(d)* Ba_3P_2, *(e)* SrF_2, *(f)* PbS, *(g)* AgBr, *(h)* GaP.

8.38 How do the properties of ionic compounds differ from those of covalent compounds? Give specific examples.

8.39 What quantity of atoms do each of the following prefixes refer to: *(a)* tetra, *(b)* hexa, *(c)* di, *(d)* octa, *(e)* tri, *(f)* mono, *(g)* penta?

8.40 Write the name of each of the following covalent compounds: *(a)* N_2O, *(b)* PF_3, *(c)* N_2O_4, *(d)* XeF_4, *(e)* S_2Cl_2, *(f)* AsF_5, *(g)* IF_3, *(h)* $SeBr_4$.

8.41 Write the name of each of the following compounds: *(a)* I_2O_4, *(b)* K_2S, *(c)* Cl_2O_7, *(d)* I_4O_9, *(e)* Al_2O_3, *(f)* CaI_2, *(g)* Te_2F_{10}, *(h)* SeF_6.

8.42 Predict whether an ionic or a covalent compound has, on average, a higher *(a)* melting point, *(b)* boiling point, *(c)* density.

8.43 Explain why ionic substances conduct an electric current in the molten state but are nonconductors when in the solid state.

*For exercise numbers printed in color, answers can be found at the back of the book.

Chemical Bonding

8.44 *(a)* Describe the energy change that occurs when two H atoms bond to form a diatomic hydrogen molecule. *(b)* Describe the energy change that occurs when a diatomic hydrogen molecule is split into two H atoms.

8.45 *(a)* What are the names of the two theories of chemical bonding? *(b)* Which one is used to explain the properties of molecules in this textbook?

8.46 Explain why nitrogen gas is composed of diatomic nitrogen molecules instead of individual N atoms.

8.47 How do chemists measure the stability of a chemical bond? Give an example.

8.48 What is incorrect about the following statement: "The bond that joins Na^+ and Cl^- ions is purely ionic"?

8.49 What accounts for the fact that diatomic noble gas compounds, such as Ne_2, Ar_2, or Xe_2, are not found at 25°C?

Ionic Bonding

8.50 *(a)* What two trends are observed in the electronegativities of atoms? *(b)* Which atoms have the highest and lowest electronegativities?

8.51 Why are electronegativity values generally not assigned to atoms of the noble gases?

8.52 From each of the following groups, select the element with the lowest electronegativity: *(a)* F, Cl, Br, I; *(b)* Ge, As, Sn, Sb; *(c)* K, Rb, Cs, Fr; *(d)* Al, Si, P, S; *(e)* Se, Br, Te, I.

8.53 Arrange the following sets of atoms in order of decreasing electronegativity (highest to lowest): *(a)* P, As, Sb; *(b)* Be, Li, B; *(c)* Rb, Sr, Cs, Ba.

8.54 What noble gas is isoelectronic to each of the following ions: *(a)* Ca^{2+}, *(b)* N^{3-}, *(c)* Br^-, *(d)* Rb^+, *(e)* Te^{2-}, *(f)* Na^+, *(g)* P^{3-}, *(h)* O^{2-}?

8.55 Write the complete electronic configuration for each of the following ions: *(a)* Sr^{2+}, *(b)* P^{3-}, *(c)* Se^{2-}, *(d)* Rb^+, *(e)* I^-, *(f)* Sr^{2+} *(g)* Ga^{3+}.

8.56 Select the ion with the largest ionic radius from each of the following sets: *(a)* O^{2-}, S^{2-}, Se^{2-}; *(b)* N^{3-}, O^{2-}, F^-; *(c)* Na^+, Mg^{2+}, Al^{3+}; *(d)* Na^+, Mg^{2+}, K^+, Ca^{2+}.

8.57 Use Lewis structures to illustrate the electron transfers that occur when the following ionic substances form: *(a)* K_2O, *(b)* $MgBr_2$, *(c)* Al_2O_3, *(d)* CaS, *(e)* Cs_3P.

8.58 Draw the Lewis structures for the following ionic compounds: *(a)* NaI, *(b)* $MgBr_2$, *(c)* SrO, *(d)* Rb_2S, *(e)* Ba_3N_2, *(f)* AlN, *(g)* K_2Se.

8.59 What is the formula unit of the ionic compound that results from combination of each of the following metals-nonmetals pairs:
(a) Aluminum-oxygen
(b) Strontium-sulfur
(c) Magnesium-bromine
(d) Gallium-oxygen
(e) Sodium-fluorine
(f) Lithium-selenium
(g) Aluminum-fluorine
(h) Calcium-phosphorus

8.60 *(a)* What is a crystal lattice? *(b)* Give three examples.

Covalent Bonding

8.61 *(a)* Draw a diagram showing the orbitals that overlap in F_2. *(b)* Explain how the F atoms are attracted to each other in the fluorine molecule. *(c)* Describe the energy change when two F atoms bond.

8.62 Explain how electron sharing leads to the formation of a chemical bond. Give an example in your explanation.

8.63 In what electron sublevels does sharing occur in each of the following covalent molecules: *(a)* HI, *(b)* IBr, *(c)* SiI_4, *(d)* NCl_3, *(e)* OBr_2, *(g)* NH_3?

8.64 Label each of the following diatomic molecules as polar or nonpolar: *(a)* BrCl, *(b)* O_2, *(c)* CO, *(d)* NO, *(e)* I_2, *(f)* ICl, *(g)* F_2.

8.65 Write the Lewis structure for BrF, and indicate the partial positive and negative charges in the molecule.

8.66 Write the Lewis structures for each of the following molecules using the suggested five-step procedure: *(a)* CBr_4, *(b)* OF_2, *(c)* H_2Te, *(d)* PF_3, *(e)* SF_2, *(f)* N_2Cl_4, *(g)* $C_2H_2Br_2$, *(h)* Si_2Cl_6, *(i)* H_2O_2, *(j)* H_4SiO_4 (each H atom is bonded to an O atom).

8.67 Write the Lewis structures for each of the following ions: *(a)* OH^-, *(b)* CN^-, *(c)* NH_4^+, *(d)* BH_4^-, *(e)* ClO_4^-, *(f)* O_2^{2-}, *(g)* SO_4^{2-}, *(h)* $C_2H_3O_2^-$.

8.68 Write the Lewis structures for three different

compounds that contain two C atoms: *(a)* C_2H_6, *(b)* C_2H_4, *(c)* C_2H_2.

8.69 Write the Lewis structures for each of the following acids (the H atoms are bonded to the O atoms): *(a)* carbonic acid, H_2CO_3; *(b)* sulfuric acid, H_2SO_4; *(c)* phosphoric acid, H_3PO_4; *(d)* nitric acid, HNO_3; *(e)* chlorous acid, $HClO_2$; *(f)* sulfurous acid, H_2SO_3.

8.70 Give an example of a molecule that contains each of the following: *(a)* a single and a double bond, *(b)* two double bonds, *(c)* a triple bond, *(d)* a triple and a single bond, and *(e)* an atom with less than eight outer-level electrons.

8.71 Which of the following molecules contain atoms with expanded octets: *(a)* CI_4, *(b)* PF_5, *(c)* IF_5, *(d)* $SnBr_4$, *(e)* SF_6, *(f)* PCl_3, *(g)* $C_2H_2Cl_2$?

8.72 Draw all of the contributing structures for each of the following: *(a)* HCO_3^-, *(b)* CO_3^{2-}, *(c)* NO_2^-, *(d)* O_3, *(e)* N_2O.

8.73 Explain why molecules with delocalized electrons are more stable than similar molecules in which the electrons are not delocalized. Give examples.

8.74 Predict the molecular geometries of each of the following molecules: *(a)* NF_3, *(b)* H_2S, *(c)* SiF_4, *(d)* $AlCl_3$, *(e)* BeF_2, *(f)* SO_3, *(g)* HCN, *(h)* AsH_3.

8.75 Predict the shapes of each of the following polyatomic ions: *(a)* NO_2^-, *(b)* NO_3^-, *(c)* CO_3^{2-}, *(d)* HS^-, *(e)* SO_4^{2-}, *(f)* BO_3^{3-}.

8.76 Explain why triple bonds are shorter than double bonds. (*Hint:* Consider the number of electrons in the bonds.)

8.77 What is the bond order in each of the following diatomic molecules: *(a)* N_2, *(b)* F_2, *(c)* HBr, *(d)* I_2?

8.78 What accounts for the fact that the average bond length of a Si—O bond, 166 pm, is longer than that of a C—O bond, 143 pm.

Additional Exercises

8.79 Correct the following incorrect statements:
  (a) In covalent bonds, electrons are always equally shared.
  (b) Electronegativity is a measure of the force of attraction between a nucleus and its electrons.
  (c) Average bond energies are higher for single covalent than for double covalent bonds.

  (d) Ionic compounds result when two nonmetals combine.
  (e) The name of the compound SO_3 is sulfur oxide.
  (f) Calcium chloride molecules have the formula $CaCl_2$.
  (g) The one electron in a H atom is transferred to a F atom in hydrogen fluoride.
  (h) All nonpolar covalent bonds result when atoms of the same element bond.

8.80 *(a)* One form of elemental phosphorus exists as P_4 molecules. Write two contributing structures for P_4. *(b)* Is P_4 a polar or a nonpolar molecule? Explain.

8.81 Write Lewis structures for molecules described as follows: *(a)* Is diatomic with a bond order of 3; *(b)* has tetrahedral geometry, but does not contain C or Si; *(c)* has a single, double, and triple bond; *(d)* has two bonded O atoms; *(e)* has both ionic and covalent bonds; *(f)* is a noble gas compound; *(g)* has B, F, N, and H atoms.

8.82 Acetone, C_3H_6O, is a molecule that contains a C atom bonded to two C atoms and one O atom. *(a)* Draw the Lewis structure for acetone. *(b)* What is the molecular geometry around each C atom in acetone? *(c)* Does acetone exhibit resonance? Explain. *(d)* Write the Lewis structure for another molecule that has the same molecular formula as acetone.

8.83 Write the Lewis structure for the following: *(a)* $NaOH$, *(b)* NH_4Cl, *(c)* $Ca(ClO_3)_2$, *(d)* $Mg_3(PO_4)_2$ *(e)* NH_2OH.

8.84 *(a)* Acetic acid, $C_2H_4O_2$, contains a carbon-carbon single bond, and each O atom is bonded to one C atom. Write the Lewis structure for acetic acid. *(b)* Draw two additional Lewis structures of compounds with the molecular formula $C_2H_4O_2$.

8.85 *(a)* Write all contributing structures for the formate ion, CHO_2^-. *(b)* Discuss the delocalization of electrons in the formate ion. *(c)* Draw a structure that best represents the structure of the formate ion.

8.86 In *a* through *d*, compare bond energy, bond length, and bond order for the two bonds shown: *(a)* C—C and Si—Si, *(b)* Cl—Cl and F—F, *(c)* N—N and N≡N, *(d)* C=C and C—C.

8.87 *(a)* Write the Lewis structure for NO_2^+. *(b)* What bond angle is found between the O atoms in NO_2^+? *(c)* Describe the nitrogen–oxygen bonds in NO_2^+. *(d)* Compare the structure of NO_2^+ with that of NO_2^-.

8.88 Write the Lewis structures for each of the following: *(a)* H_2NOH, *(b)* $HONO$, *(c)* O_2SCl_2, *(d)* N_3^-, *(e)* H_2CNN.

8.89 Propane, C_3H_8, is a fuel used to heat houses. Write the Lewis structure for propane.

8.90 *(a)* Write reasonable contributing structures for the thiocyanate anion, SCN^-. *(b)* Describe the sulfur–carbon and carbon–nitrogen bonds in SCN^-.

8.91 Ozone, O_3, is an air pollutant and a component of the upper atmosphere. *(a)* Draw all contributing structures of ozone. *(b)* What is the bond order for the oxygen–oxygen bonds in ozone?

(c) Predict the molecular geometry of ozone. *(d)* What is the approximate bond angle in ozone?

8.92 Isomers are compounds that have the same molecular formula but have a different arrangement of atoms within the molecules. Write the Lewis structures for three isomers of $C_2H_2I_2$.

8.93 A compound is found to contain 26.6% K, 35.4% Cr, and 38.1% O. *(a)* Calculate the empirical formula of the compound. *(b)* A 0.0191-mol sample of the compound is found to have a mass of 5.63 g. What is the molecular formula of the compound? *(c)* Is this compound ionic or covalent? Explain.

Nine

Chemical Nomenclature of Inorganic Compounds

STUDY GUIDELINES

After completing Chapter 9, you should be able to

1. State the rules used to assign oxidation numbers, and assign oxidation numbers for elements in compounds

2. Write the formulas or names of ionic binary compounds that contain metals with either fixed or variable oxidation states

3. Write the formulas or names of covalent binary compounds

4. Identify the principal oxidation states, and write names of frequently encountered metal ions

5. Distinguish a binary compound from a ternary compound

6. Write the names and formulas of at least 12 polyatomic ions

7. Write names and formulas of ionic ternary compounds

8. Write names and formulas of oxyacids

9. Distinguish between binary and ternary acids

10. Name ternary acids that contain nonmetals that exist in variable oxidation states

11. Write the name of an oxyanion given the acid that it is derived from, and vice versa

12. Write the name of a hydrate given its formula, and vice versa

---

Chemical nomenclature is the means by which chemists assign names to atoms and compounds. A systematic method is needed to name the vast number of known compounds, plus the new ones that are synthesized each day. Our standards for naming compounds are established by the **International Union of Pure and Applied Chemistry,** IUPAC for short. IUPAC has adopted the **Stock** system of nomenclature for assigning names to inorganic compounds. We will learn part of the Stock system along with some common names which are frequently used by nonchemists.

Before we can discuss the specifics of inorganic nomenclature, we need to learn a method for keeping track of the apparent charge, called the oxidation number, that an atom has within a compound.

Our system of chemical nomenclature must generate names for more substances than there are words in the English language.

Oxidation numbers, sometimes called oxidation states, are signed numbers assigned to atoms in molecules and ions. They allow us to keep track of the electrons associated with each atom. Oxidation numbers are frequently used to write chemical formulas, to help us predict properties of compounds, and to help balance equations in which electrons are transferred (oxidation-reduction reactions, discussed in Chap. 18).

Knowledge of the oxidation state of an atom gives us an idea about its positive or negative character. In themselves, oxidation numbers have no physical meaning; they are used to simplify tasks that are more difficult to accomplish without them.

To assign oxidation numbers the following set of rules is used.

Rule 1

All pure elements are assigned the oxidation number of zero.

Rule 2

All monatomic ions are assigned oxidation numbers equal to their charges.

Rule 3

Certain elements usually possess a fixed oxidation number in compounds.

1. The oxidation number of O in most compounds is -2.

2. The oxidation number of H in most compounds is $+1$.

3. The oxidation number of halogens in many, but not all, binary compounds is -1.

4. The oxidation numbers of alkali metals and alkaline earth metals are $+1$ and $+2$, respectively.

Rule 4

The sum of all oxidation numbers in a compound equals zero, and the sum of oxidation numbers in a polyatomic ion equals the ion's charge.

Let's now discuss each rule. Rule 1 states that all uncombined elements are assigned the oxidation number of zero, regardless of how they exist in nature—by themselves, diatomically, or in larger aggregates (e.g., P_4 and S_8). It is common practice to write oxidation numbers above the symbols of the atoms.

$$\overset{0}{Na} \quad \overset{0}{Fe} \quad \overset{0}{H_2} \quad \overset{0}{O_2} \quad \overset{0}{F_2} \quad \overset{0}{Ne} \quad \overset{0}{P_4} \quad \leftarrow \text{Oxidation numbers}$$

Rule 2 states that all monatomic ions are assigned oxidation numbers equal to their charge. When you write an oxidation number, the sign comes before the magnitude of the charge. Charges on atoms are written with the magnitude before the sign. Some examples of oxidation numbers of monatomic ions are as follows:

9.1 OXIDATION NUMBERS (OXIDATION STATES)

Oxidation numbers are used most frequently to write formulas, predict properties, and balance oxidation-reduction equations.

Oxidation numbers are the result of the stated rules, nothing more. They cannot be measured experimentally.

$$\overset{-1}{\text{Cl}^-} \quad \overset{+3}{\text{Al}^{3+}} \quad \overset{+2}{\text{Cu}^{2+}} \quad \overset{-2}{\text{O}^{2-}} \quad \overset{+1}{\text{K}^+} \quad \overset{+2}{\text{Mg}^{2+}}$$

Rule 3 tells us that some atoms have fixed oxidation numbers. These atoms usually have oxidation numbers that correspond to the number of electrons that they lose or gain when forming a binary ionic compound. For example, halogens gain one electron, chalcogens gain two electrons, alkali metals lose one electron, and alkaline earth metals lose two electrons to obtain noble gas configurations when they form ionic compounds.

There are exceptions to Rule 3. For example, in peroxides, the oxidation number of an O atom is -1, not -2. Peroxides are compounds that contain an O—O single bond. Two peroxides are hydrogen peroxide, H_2O_2, and sodium peroxide, Na_2O_2. Another exception is found for metallic hydrides, which are compounds that contain an H atom bonded to a metal with a lower electronegativity. The oxidation number of H is -1 in metallic hydrides. Examples of metallic hydrides include sodium hydride, NaH, and calcium hydride, CaH_2.

If an O atom is bonded to two F atoms, the O atom is found in the $+2$ oxidation state because the F atoms have a higher electronegativity.

Through the application of Rule 4, we can identify the oxidation numbers of all elements in a compound. For example, we can calculate the oxidation number of N in NO. The oxidation number of O is -2, and the sum of the oxidation numbers in NO equals zero. Thus the oxidation number of N in NO is $+2$:

$$\text{Total oxidation numbers} \rightarrow \quad \overset{?-2=0}{\text{NO}} \quad \overset{+2-2=0}{\text{NO}}$$
$$\text{Individual oxidation numbers} \rightarrow \quad \underset{-2}{}$$

What is the oxidation number of S in SO_2? Again following Rule 4, we assign -4 to the two O atoms; hence, the oxidation number of S is $+4$ in order for the sum to equal zero.

To keep track of oxidation numbers, write the known oxidation states for individual atoms below the formula and the total oxidation number for all elements of that type above the symbol.

$$\overset{?-4=0}{\underset{-2}{SO_2}} \qquad \overset{+4-4}{\underset{-2}{S\,O_2}}$$

For a polyatomic ion the reasoning is the same except that the sum of the oxidation numbers equals the charge on the ion. To illustrate this, let's calculate the oxidation number of a P atom in $PO_4{}^{3-}$. Four O atoms have a total oxidation number of -8, so the oxidation state of P must be $+5$ for the sum to equal the charge of -3.

$$\overset{?-8}{\underset{-2}{PO_4{}^{3-}}} \qquad \overset{+5-8}{\underset{-2}{P\,O_4{}^{3-}}} = -3$$

Example Problem 9.1 presents additional illustrations of how to assign oxidation numbers.

--------- **Example Problem 9.1** ---------

Find the oxidation states for the elements in each of the following:
(a) CuF_2, (b) HNO_3, (c) SO_4^{2-}, (d) $C_{12}H_{22}O_{11}$.

--------- **Solution** ---------

(a) CuF_2. Halogens such as F have an oxidation number equal to -1; two F atoms have a total oxidation number of -2; therefore, it follows that the oxidation number of Cu is $+2$.

$$\overset{+2 \ -2}{\underset{-1}{Cu \ F_2}}$$

(b) HNO_3. The oxidation number of H is $+1$, and three O atoms have a total oxidation number of -6; thus, the oxidation number of N is $+5$ in order for the total to be 0.

$$\overset{+1+5 \ -6}{\underset{+1 \quad -2}{HNO_3}}$$

(c) SO_4^{2-}. The oxidation number of each O atom is -2, so four O atoms have a total oxidation number of -8. To make the sum of the oxidation numbers equal to -2, the charge on SO_4^{2-}, the oxidation number of S is $+6$.

$$\overset{+6-8}{\underset{-2}{S \ O_4^{2-}}}$$

(d) $C_{12}H_{22}O_{11}$. The total oxidation number of 22 H atoms is $+22$ ($22 \times +1$), and the total for 11 O atoms is -22 (11×-2). Therefore, the 12 C atoms have a total oxidation number of 0, which means that each C has an oxidation number equal to zero.

$$\overset{0 \quad +22 \ -22}{\underset{+1 \quad -2}{C_{12}H_{22}O_{11}}}$$

With the exception of the metals in groups IA, IIA, and IIIB, metals generally can exist in more than one oxidation state. Chromium, Cr, for example, is found in the $+6$, $+3$, and $+2$ oxidation states. Gold is found in both $+3$ and $+1$ oxidation states. Nonmetals, except F, also exhibit a range of oxidation states. For example, the oxidation states of S

A large percent of the elements exist in two or more oxidation states.

Group

Period	IA(1)	IIA(2)	IIIB(3)	IVB(4)	VB(5)	VIB(6)	VIIB(7)	VIIIB (8)	(9)	(10)	IB(11)	IIB(12)	IIIA(13)	IVA(14)	VA(15)	VIA(16)	VIIA(17)	VIIIA(18)
1	1 H +1																	2 He
2	3 Li +1	4 Be +2											5 B +3	6 C ±4, +2	7 N ±3,5,4,2	8 O −2, 1	9 F −1	10 Ne
3	11 Na +1	12 Mg +2											13 Al +3	14 Si +4	15 P +3, 5, 4	16 S ±2, 4, 6	17 Cl ±1,3,5,7	18 Ar
4	19 K +1	20 Ca +2	21 Sc +3	22 Ti +4, 3	23 V +5,4,3,2	24 Cr +6, 3, 2	25 Mn +7,6,4,3,2	26 Fe +2, 3	27 Co +2, 3	28 Ni +2, 3	29 Cu +1, 2	30 Zn +2	31 Ga +3	32 Ge +4	33 As ±3, 5	34 Se −2, 4, 6	35 Br ±1, 5	36 Kr
5	37 Rb +1	38 Sr +2	39 Y +3	40 Zr +4	41 Nb +5, 3	42 Mo +6,5,4,3,2	43 Tc +7	44 Ru +8,6,4,3,2	45 Rh +4, 3, 2	46 Pd +2, 4	47 Ag +1	48 Cd +2	49 In +3	50 Sn +4, 2	51 Sb ±3, 5	52 Te −2, 4, 6	53 I ±1, 5, 7	54 Xe
6	55 Cs +1	56 Ba +2	57 La +3	72 Hf +4	73 Ta +5	74 W +6,5,4,3,2	75 Re +7,6,4,2	76 Os +8,6,4,3,2	77 Ir +6,4,3,2	78 Pt +2, 4	79 Au +1, 3	80 Hg +2, 1	81 Tl +1, 3	82 Pb +4, 2	83 Bi +3, 5	84 Po +2, 4	85 At ±1,3,5,7	86 Rn
7	87 Fr +1	88 Ra +2	89 Ac +3	104 Unq	105 Unp	106	107	108										

58 Ce +3, 4	59 Pr +3, 4	60 Nd +3	61 Pm +3	62 Sm +2, 3	63 Eu +2, 3	64 Gd +3	65 Tb +3, 4	66 Dy +3	67 Ho +3	68 Er +3	69 Tm +2, 3	70 Yb +2, 3	71 Lu +3

90 Th +4	91 Pa +4, 5	92 U +6,5,4,3	93 Np +6,5,4,3	94 Pu +6,5,4,3

Figure 9.1
The most common oxidation states of each element are listed under its symbol.

include $+6$, $+4$, $+2$, and -2. Figure 9.1 presents a periodic table with the most common oxidation states of the elements.

9.1 What is chemical nomenclature?

9.2 (a) What does the oxidation number of an element within a compound represent? (b) Explain why the oxidation number for an element is not always equal to the charge of an atom in a compound.

9.3 List the four rules employed to assign oxidation numbers to elements in compounds.

9.4 Calculate the oxidation numbers for all elements in each of the following compounds: (a) PBr_3, (b) MnO_2, (c) AlF_3, (d) N_2O, (e) P_2O_5.

9.5 Find the oxidation states for all elements except O in the following polyatomic ions (a) $C_2O_4^{2-}$, (b) NO_2^-, (c) BrO_3^-, (d) CO_3^{2-}, (e) $C_2H_3O_2^-$.

REVIEW EXERCISES

B inary compounds are those that contain two different elements. Ternary compounds, discussed in the next section, are composed of three different elements. Each of these groups of compounds is divided into subgroups that have their own specific rules of nomenclature. Figure 9.2 outlines the most general classes of binary and ternary inorganic compounds.

9.2 NAMES AND FORMULAS OF BINARY COMPOUNDS

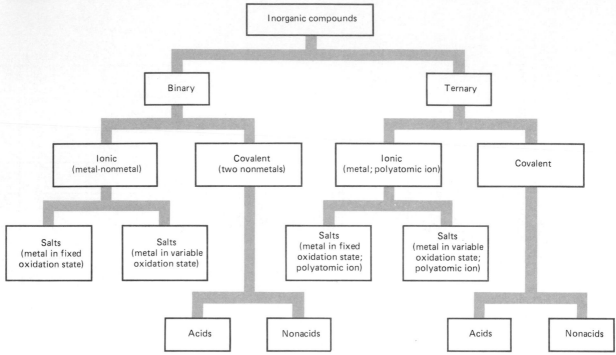

Figure 9.2
Inorganic compounds are first classified as binary or ternary. Both binary and ternary compounds are subdivided into ionic compounds or covalent compounds. Ionic compounds contain a metal in either a fixed or variable oxidation state. Covalent substances may be divided into acids and nonacids.

When you write the name of a compound, first identify what class of compound is being named. Using the flowchart in Fig. 9.3, decide if the compound is binary or ternary. Then decide if the compound is ionic or covalent. Ionic compounds are most commonly composed of a metal and a nonmetal, a metal and a polyatomic ion, or a polyatomic ion and a nonmetal. Covalent compounds are composed of two or more nonmetals.

In Chap. 8, we learned the rules for naming simple binary ionic compounds, i.e., those that are composed of a metal ion with a fixed oxidation state and a nonmetal ion. To review: First write the name of the metal, and then write the name of the nonmetal, removing its ending and adding the suffix *ide*.

Binary Ionic Compounds That Contain Metals with Fixed Oxidation States

$$\text{Name} = \text{metal} + (\text{nonmetal} - \text{ending} + ide)$$

If you have forgotten how to name these compounds, refer to Sec. 8.1.

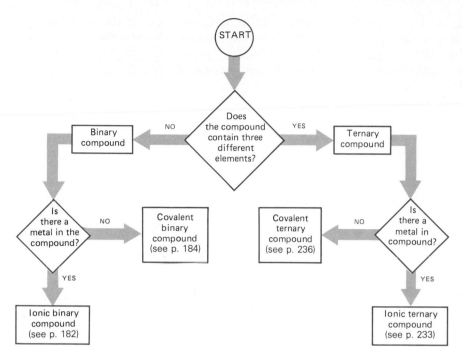

Figure 9.3
To use the nomenclature flowchart, begin at the top and answer each question, concerning the compound of interest. When you answer either yes or no, the chart poses another question or gives the type of compound with a page reference.

Binary Ionic Compounds That Contain Metals with Variable Oxidation States

What if the metal ion component of a binary ionic substance has a variable oxidation state? We mentioned that metals can exist in more than one oxidation state; consequently, more than one binary compound can result when such a metal combines with a nonmetal. For example, two chlorides of copper are known, $CuCl$ and $CuCl_2$. Our naming system must provide a means of distinguishing between the two. In the Stock system, we modify the above rules for naming ionic substances to include the oxidation state of the metal ion.

Specifically, we write the full name of the metal followed by its oxidation number written in roman numerals in parentheses:

$$\text{Metal name(oxidation number)}$$

The left parenthesis is written next to the last letter in the name of the metal, and the oxidation state is written in roman numerals, followed by the right parenthesis. Accordingly, the two oxidation states of copper are expressed as follows:

$$Cu^+ = \text{copper(I)}$$
$$Cu^{2+} = \text{copper(II)}$$

To write the complete names of $CuCl$ and $CuCl_2$, it is only necessary to append the modified name of the nonmetal (nonmetal − ending +

ide), chloride. Therefore, CuCl is copper(I) chloride, and CuCl$_2$ is copper(II) chloride. Note that there is a space between the right parenthesis and the nonmetal name, but that no space is left between the left parenthesis and the end of the metal name. When reading names expressed this way, state the metal name followed by the number: "copper-one" and "copper-two."

The old system of nomenclature assigns names to metals that exist in more than one oxidation state in a different way. To write the names of the chlorides of copper, two endings, *ic* and *ous*, are added to the old name for copper, *cuprum*. The ending *ic* designates the higher oxidation state, while *ous* identifies the lower oxidation state of the metal. For CuCl, in which Cu is in the $+1$ state, we add *ous* to cuprum $-$ *um*, which gives "cuprous":

$$Cu^+ = cuprum - um + ous = cuprous$$

In a similar manner, *ic* is affixed to cuprum $-$ *um*, which yields "cupric":

$$Cu^{2+} = cuprum - um + ic = cupric$$

In the old system, the names of the chlorides of Cu are cuprous chloride, CuCl, and cupric chloride, CuCl$_2$.

To write the names of binary compounds using the Stock system, one needs to know only the oxidation states of the elements in the compound. Additional knowledge (the old names for metals) is required when using the old system. Table 9.1 lists both the old and new names of frequently encountered metal ions. In this textbook we will principally use the Stock system.

Initially, refer to Table 9.1 when writing the names of metals with more than one oxidation state. Example Problem 9.2 illustrates how to write the names of compounds with metals that have variable oxidation states.

CuCl is composed of white crystals that melt at 430°C. CuCl$_2$ is composed of yellow-brown crystals that melt at 620°C.

Modern and old names for the elements:

Modern	Old
Antimony	Stibum
Copper	Cuprum
Gold	Aurum
Iron	Ferrum
Lead	Plumbum
Potassium	Kalium
Silver	Argentum
Tin	Stannium

Example Problem 9.2

Write the Stock and old names for each of the following compounds: *(a)* HgI$_2$, *(b)* PbO$_2$, *(c)* SnF$_2$, *(d)* FeO.

Solution

(a) HgI$_2$. Identify the oxidation state of Hg in HgI$_2$, using the rules for assigning oxidation numbers. Iodine is a halogen with an oxidation number of -1 in binary compounds; two I atoms have a total oxidation number of -2. Consequently, the oxidation state of Hg is $+2$ (the higher oxidation state of Hg), which is mercury(II) in the Stock system and mercuric in the old system. To modify the name of iodine, remove the *ine* and add *ide*.

TABLE 9.1 NAMES OF METALS WITH MULTIPLE OXIDATION STATES

Metal	Oxidation state	Stock name	Old name
Copper	+1	Copper(I)	Cuprous
	+2	Copper(II)	Cupric
Mercury	+1	Mercury(I)	Mercurous
	+2	Mercury(II)	Mercuric
Iron	+2	Iron(II)	Ferrous
	+3	Iron(III)	Ferric
Chromium	+2	Chromium(II)	Chromous
	+3	Chromium(III)	Chromic
Manganese	+2	Manganese(II)	Manganous
	+3	Manganese(III)	Manganic
Cobalt	+2	Cobalt(II)	Cobaltous
	+3	Cobalt(III)	Cobaltic
Tin	+2	Tin(II)	Stannous
	+4	Tin(IV)	Stannic
Lead	+2	Lead(II)	Plumbous
	+4	Lead(IV)	Plumbic
Titanium	+3	Titanium(III)	Titanous
	+4	Titanium(IV)	Titanic

Mercury(I) is a diatomic cation; i.e., it exists in pairs—Hg_2^{2+}. Whenever mercury(I) is written, it indicates a pair of mercury ions.

$$\text{Stock name} = \text{mercury(II) iodide}$$

$$\text{Old name} = \text{mercuric iodide}$$

(b) PbO_2. The total oxidation number for two O atoms is -4. In order for the oxidation numbers to add up to zero, the oxidation state of Pb must be $+4$. The $+4$ oxidation state of lead is the higher one, necessitating that *ic* be placed at the end of the metal's old name, *plumbum*.

$$\text{Stock name} = \text{lead(IV) oxide}$$

$$\text{Old name} = \text{plumbic oxide}$$

(c) SnF_2. After following the general rules, we find that Sn is in the $+2$ oxidation state, its lower oxidation state. Consequently, the two names for SnF_2 are

$$\text{Stock name} = \text{tin(II) fluoride}$$

$$\text{Old name} = \text{stannous fluoride}$$

(d) FeO. The oxidation number for an O atom is -2; therefore, the oxidation number of Fe is $+2$. The Stock name for iron in this com-

pound is thus iron(II), and the old name ferrous, because +2 is the lower oxidation state. To the names of the metal ion we add the modified name of oxygen, which is oxide.

$$\text{Stock name} = \text{iron(II) oxide}$$

$$\text{Old name} = \text{ferrous oxide}$$

Some compounds contain an element that exists in two different oxidation states. Fe_3O_4 is a combination of FeO and Fe_2O_3; hence, it is best named iron(II, III) oxide.

Covalent Binary Compounds

In Sec. 8.1, we discussed the rules for writing the names of covalent binary compounds. To review: The name of the nonmetal in the positive oxidation state (the nonmetal with the lower electronegativity) is written first unchanged. The ending of the nonmetal in a negative oxidation state (the nonmetal with the higher electronegativity) is dropped and *ide* is added. A prefix is attached to this name to indicate the number of atoms of each nonmetal in the formula. Examples of covalent binary compounds include sulfur trioxide, SO_3; dinitrogen pentoxide, N_2O_5; and carbon tetrafluoride, CF_4.

Binary Acids

Some covalent binary hydrogen compounds are classified as acids. When dissolved in water, **acids** ionize and add hydrogen ions, H^+, and anions to water. In the following equation, hypothetical acid HA breaks up into H^+ and A^-(anion).

$$\text{HA} \xrightarrow{\text{water}} H^+(aq) + A^-(aq)$$

Acids are very important compounds and have many uses. They have a set of common properties. Acids taste sour, change the color of acid-base indicator dyes such as litmus, react with bases (metallic hydroxides) to produce salts, and may combine with active metals to liberate hydrogen gas. We will thoroughly discuss the properties and reactions of acids in Chap. 17.

When you write the names of binary acids, the prefix *hydro* is added to the nonmetal name, the ending is dropped, and *ic acid* is attached in its place.

If a binary acidic substance is not in aqueous solution, it is named in the same way as any other binary compound. For example, HCl(g) is hydrogen chloride gas.

$$\text{Acid name} = hydro + \text{nonmetal} - \text{ending} + ic\ acid$$

For example, the binary acid that contains chlorine, HCl(aq), is called hydrochloric acid. *Hydro* is added to "chlorine − *ine* + *ic acid*" to generate the name. Additional examples of binary acids include hydrosulfuric acid, H_2S(aq); hydrofluoric acid, HF(aq); and hydrobromic acid, HBr(aq).

Formulas of Binary Compounds

To write the formula of a binary compound, given its name, you must first determine if the compound is ionic or covalent. If it is covalent, the actual number of atoms is specified in the formula. Therefore, you write the formula with the subscripts indicated by the prefixes—for example, carbon disulfide, CS_2, dinitrogen tetroxide, N_2O_4, and carbon tetrabromide, CBr_4.

To write the formula of an ionic compound, write each ion with its correct oxidation state, and then determine the quantity of each ion that produces a zero oxidation state for the compound. To illustrate this procedure, let's write the formula for chromium(III) oxide. Given the Stock name, the oxidation state of the metal is indicated by the roman numeral. Oxygen belongs to group VIA and has a −2 oxidation number in binary compounds. Write both elements with their oxidation numbers:

The sum of the oxidation numbers of all the elements in a compound is zero.

$$\overset{+3}{Cr} \quad \overset{-2}{O}$$

Because the sum of all oxidation numbers adds up to zero, ask yourself what numbers of Cr and O atoms give the same total oxidation number. This is easily accomplished by finding the lowest common multiple for the two oxidation numbers; for Cr and O the lowest common multiple is 6. Then, divide each individual oxidation number into the lowest common multiple to obtain the correct formula subscript.

Lowest common multiple = 6

$$\overset{+3}{Cr} \qquad \overset{-2}{O}$$
$$\frac{6}{3} = 2 \qquad \frac{6}{2} = 3$$

Thus there are two Cr atoms and three O atoms in chromium(III) oxide.

Cr_2O_3 is a green solid that melts at 2330°C.

$$Cr_2O_3$$

Example Problem 9.3 shows additional examples of how to write formulas of compounds given their names.

Example Problem 9.3

Write the formulas for each of the following: *(a)* lead(II) chloride, *(b)* titanium(III) sulfide, *(c)* cobaltous oxide.

Solution

(a) Lead(II) chloride. Lead(II) is Pb^{2+}, and chloride, Cl^-, is a halide ion with an oxidation number of −1. Two chloride ions are needed to balance the Pb^{2+}; hence, the formula of lead(II) chloride is $PbCl_2$.

$$PbCl_2$$

(b) Titanium(III) sulfide. The oxidation number of titanium(III) is $+3$, and that of sulfur is -2. The lowest common multiple of $+3$ and -2 is 6. Thus, two Ti atoms and three S atoms are needed to give a zero oxidation state for titanium(III) sulfide.

$$Ti_2S_3$$

(c) Cobaltous oxide. Referring to Table 9.1, we find that cobaltous is Co^{2+}, and we know that an O atom has an oxidation number of -2; thus, the formula of cobaltous oxide is CoO. Co and O are found in a 1 to 1 ratio in cobaltous oxide.

$$CoO$$

9.6 Write the old name that corresponds to each of the following Stock names for metal ions: (a) copper(II), (b) manganese(II), (c) lead(IV), (d) iron(II), (e) tin(II).

9.7 Write the Stock names for each of the following: (a) Fe_2S_3, (b) Cu_3N, (c) PbO, (d) Hg_2F_2, (e) MnF_3, (f) HBr(aq), (g) AlF_3.

9.8 Write the formulas for each of the following: (a) tin(II) bromide, (b) gold(III) oxide, (c) nickel(II) nitride, (d) plumbic iodide, (e) copper(I) sulfide, (f) sulfur dichloride.

9.3 NAMES OF TERNARY COMPOUNDS

A ternary compound is one that contains three elements. Referring to Figs. 9.2 and 9.3, it should be apparent that ternary compounds are classified in a similar manner to binary compounds. First decide if the ternary compound is ionic or covalent. If it is ionic, check to see if the metal has a fixed or variable oxidation state. If it is covalent, determine whether or not it is an acid.

We will concentrate on the names of ternary compounds composed of (1) a metal and a polyatomic ion (ionic) or (2) an H atom bonded to a polyatomic ion (acid). Polyatomic ions were first introduced in Chap. 8. They are ions composed of more than one atom; some examples are hydroxide, OH^-; nitrate, NO_3^-; and sulfate, SO_4^{2-}.

Rules for naming ionic ternary compounds are similar to those for binary compounds, except that the name of a polyatomic ion is found in place of either the metal ion or nonmetal ion name. Table 9.2 lists 12 frequently encountered polyatomic ions. Learning the formulas, charges, and names of these polyatomic ions will simplify the writing of names of ternary compounds. When you learn these polyatomic ions, don't forget about their charges. *It is of no value to learn the ions without their charges.* A more complete listing of ions, including polyatomic ions, is found in Appendix 4.

TABLE 9.2   IMPORTANT POLYATOMIC IONS

Name	Formula
Acetate	$C_2H_3O_2^-$
Ammonium	NH_4^+
Borate	BO_3^{3-}
Carbonate	CO_3^{2-}
Chlorate	ClO_3^-
Chromate	CrO_4^{2-}
Cyanide	CN^-
Hydroxide	OH^-
Nitrate	NO_3^-
Permanganate	MnO_4^-
Phosphate	PO_4^{3-}
Sulfate	SO_4^{2-}

To write the name of an ionic ternary compound, first determine if the metal has a fixed or variable oxidation number. Depending on what metal is found, the compound is named in a similar manner to an ionic binary compound except for the name of the polyatomic ion.

Let's illustrate the procedure for writing the name of ionic ternary compounds by considering $Co(NO_3)_2$. Cobalt is a transition metal with a variable oxidation state. Thus, find the oxidation state of Co and write it in parentheses: cobalt(II). Then identify the polyatomic ion, NO_3^-, nitrate, and add its name to the name of the metal: cobalt(II) nitrate.

$$Co(NO_3)_2 = \text{cobalt(II) nitrate}$$

The oxidation state of Co is determined from the charge on the nitrate ion, which is -1. Because there are two nitrate ions in the formula, the cobalt exists in the $+2$ oxidation state.

Example Problem 9.4 shows how to write the names of several other ternary ionic compounds.

Ionic Ternary Compounds

Study Hint: *Write the formula of the polyatomic ion on one side of an index card, and the name on the other side. Flip through the cards learning names of formulas; then, turn the deck over and proceed in the opposite direction, learning formulas of names.*

If more than one polyatomic ion of the same kind is found in a compound, it is placed in parentheses with the appropriate subscript.

─────────── **Example Problem 9.4** ───────────

Write the Stock names for each of the following compounds: (a) K_2CrO_4, (b) $(NH_4)_2S$, (c) $FeCO_3$, (d) $Mn_3(PO_4)_2$.

─────────── **Solution** ───────────

(a) K_2CrO_4. The metal is potassium, an alkali metal with a fixed oxidation state, and the polyatomic ion is chromate. Thus

$$K_2CrO_4 = \text{potassium chromate}$$

A number is not written after potassium in the name because potassium exists in only one oxidation state, $+1$.

(b) $(NH_4)_2S$. In this compound, we find the ammonium ion in place of a metal along with the nonmetal sulfur. Remove the ending from sulfur and add *ide*:

$$(NH_4)_2S = \text{ammonium sulfide}$$

(c) $FeCO_3$. Iron is a metal with a variable oxidation state; consequently, the name must indicate the oxidation state of iron. Carbonate, CO_3^{2-}, has a charge of -2; thus, to sum to zero, the oxidation number of Fe is $+2$. Accordingly,

$$FeCO_3 = \text{iron(II) carbonate}$$

(d) $Mn_3(PO_4)_2$. Once again, the oxidation state of the metal has to be expressed because manganese exists in a variety of states. Phosphate has a charge of 3−; therefore, two phosphates have a total charge of 6−. So the three Mn atoms' total oxidation number is +6, or each Mn has an oxidation number of +2.

$$Mn_3(PO_4)_2 = manganese(II)\ phosphate$$

To write the formula of a ternary compound, given its name, follow the same procedures used when writing formulas of binary compounds, as shown in Example Problem 9.5.

Example Problem 9.5

Write the formulas of *(a)* zinc(II) hydroxide, *(b)* silver(I) cyanide, and *(c)* cobaltic sulfate.

Solution

(a) Zinc(II) hydroxide. Zinc(II) is Zn^{2+}, and hydroxide is OH^-. To produce a compound with a zero oxidation state, two hydroxide ions are required per zinc ion.

$$Zn^{2+} \quad OH^-$$

$$Zinc(II)\ hydroxide = Zn(OH)_2$$

(b) Silver(I) cyanide. Silver(I) is Ag^+, and cyanide is CN^-. Because they have equal but opposite charges, they combine in a 1-to-1 ratio.

$$Ag^+ \quad CN^-$$

$$Silver(I)\ cyanide = AgCN$$

(c) Cobaltic sulfate. "Cobaltic" is the old name for the higher oxidation state of Co, which is Co^{3+}. The formula for sulfate is SO_4^{2-}. The lowest common multiple of 3 and 2 is 6; accordingly, the formula of cobaltic sulfate becomes $Co_2(SO_4)_3$.

$$Co^{3+} \quad SO_4^{2-}$$

$$Cobaltic\ sulfate = Co_2(SO_4)_3$$

A group of polyatomic ions exists with one less O atom than some of those listed in Table 9.2. For example, in addition to sulfate, SO_4^{2-}, an ion with the formula SO_3^{2-} also exists. How is this ion named? If a polyatomic ion has one less O atom than the ion ending in *ate*, it is given an *ite* ending. SO_3^{2-} is the *sulfite* ion because it has one less O atom than sulfate, SO_2^{2-}. Table 9.3 lists additional examples of *ite* **oxyanions,** i.e., polyatomic ions that contain oxygen. Note that both *ate* and *ite* oxyanions have the same charge.

Some people have bad allergic reactions to sulfites that are used as preservatives in foods.

CHEM TOPIC: Nitrites in Foods

Sodium nitrite, $NaNO_2$, and potassium nitrite, KNO_2, are used to preserve and give the proper color to meats. They are added to hot dogs, bacon, sausage, and many processed meats to kill bacteria which cause food poisoning and the decomposition of the meat. Nitrites are especially effective in destroying the organism that causes botulism. In addition, they help the meat maintain a fresh pink color instead of the brown color that results after meats are cut. The pink color is produced when nitrite combines with acids such as citric acid and acetic acid to produce nitrous acid, HNO_2. The nitrous acid combines with the myoglobin, Mb, in the meat to produce nitrosomyoglobin, MbNO, the pink compound in meat. Myoglobin is the protein that transfers O_2 from the blood to the tissues.

Nitrites in foods have been at the center of a controversy for many years because they, in combination with acid, produce a group of potent cancer-causing agents called nitrosamines. While no direct evidence has been found that nitrites cause cancer in humans, a number of animal studies have shown a direct link. These studies are clouded by the fact that while most Americans take in less than 3 mg of nitrites per day, more than 8 mg is found in their saliva; moreover, nitrites are produced naturally in the large intestine.

Once the formula for the *ate* ion is learned, it is only necessary to subtract an oxygen from the formula and replace the ending of the name with *ite* to obtain the correct formula and name of the *ite* oxyanion. Example Problem 9.6 shows how the names of such compounds are derived.

Example Problem 9.6

Write the Stock names for the following compounds: (*a*) $Fe_3(PO_3)_2$, and (*b*) $Ba(NO_2)_2$.

Solution

(*a*) $Fe_3(PO_3)_2$. Looking at the oxyanion, we notice that it has one less O atom than phosphate, PO_4^{3-}; thus, it is the phosphite ion, PO_3^{3-}. Because there are two phosphite ions in the formula, each with a charge of 3−, the total charge or oxidation number is 6−. Each iron ion must have a charge of 2+ ($3 \times 2+ = 6+$) to balance the charge of the phosphite ions. Accordingly, the name of $Fe_3(PO_4)_2$ is iron(II) phosphite

$$Fe_3(PO_3)_2 = \text{iron(II) phosphite}$$

TABLE 9.3 SELECTED OXYANIONS: *ATE* VERSUS *ITE* ENDINGS

Oxyanion	Name
SO_4^{2-}	Sulfate
SO_3^{2-}	Sulfite
NO_3^-	Nitrate
NO_2^-	Nitrite
PO_4^{3-}	Phosphate
PO_3^{3-}	Phosphite
AsO_4^{3-}	Arsenate
AsO_3^{3-}	Arsenite
SeO_4^{2-}	Selenate
SeO_3^{2-}	Selenite

(b) Ba(NO$_2$)$_2$. Barium is an element in group IIA; thus, it has a fixed charge of 2+. NO$_2^-$ has one less O atom than nitrate, NO$_3^-$, so its name is nitrite. Thus,

$$\text{Ba(NO}_2)_2 = \text{barium nitrite}$$

9.9 Identify each of the following polyatomic ions: *(a)* BO$_3^{3-}$, *(b)* ClO$_3^-$, *(c)* MnO$_4^-$, *(d)* CrO$_4^{2-}$, *(e)* PO$_4^{3-}$, *(f)* C$_2$H$_3$O$_2^-$.

9.10 Write the formula for each of the following: *(a)* ammonium ion, *(b)* carbonate ion, *(c)* nitrate ion, *(d)* cyanide ion, *(e)* phosphite ion, *(f)* selenite ion.

9.11 Write the names for each of the following compounds: *(a)* Cs$_2$CO$_3$, *(b)* Sr(ClO$_3$)$_2$, *(c)* Ag$_2$CrO$_4$, *(d)* Hg$_2$(NO$_3$)$_2$, *(e)* Na$_3$PO$_4$, *(f)* Co(NO$_2$)$_2$.

9.12 Write the formulas for each of the following: *(a)* ammonium chromate, *(b)* iron(II) sulfite, *(c)* aluminum cyanide. *(d)* nickel(II) nitrite, *(e)* manganic phosphate, *(f)* sodium arsenate.

Covalent Ternary Compounds

Because we will encounter few covalent ternary compounds that are not acids, only acids are discussed in this section. Examples of ternary acids include sulfuric acid, H$_2$SO$_4$(aq); nitric acid, HNO$_3$(aq); phosphoric acid, H$_3$PO$_4$(aq); chloric acid, HClO$_3$(aq); and boric acid, H$_3$BO$_3$(aq).

Oxyacids

Oxyacids are ternary acids that contain oxygen. Such compounds have H atoms bonded to an oxyanion. For example, if two H atoms are bonded to sulfate, SO$_4^{2-}$, the resulting acid is called sulfuric acid. This name was obtained by adding *ic acid* to the end of the nonmetal name contained in sulfate.

$$\text{Sulfuric acid} = \text{sulfur} + ic\ acid$$

Oxyanions that end in *ate*, when bonded to H atoms, form oxyacids whose names are derived from the name of the nonmetal, with *ic acid* as the ending. Table 9.4 presents examples of other *ic acids*.

TABLE 9.4 OXYACID NAMES

Oxyacid	Oxyanion	Name
Boric acid, H$_3$BO$_3$	BO$_3^{3-}$	Borate
Carbonic acid, H$_2$CO$_3$	CO$_3^{2-}$	Carbonate
Chloric acid, HClO$_3$	ClO$_3^-$	Chlorate
Nitric acid, HNO$_3$	NO$_3^-$	Nitrate
Selenic acid, H$_2$SeO$_4$	SeO$_4^{2-}$	Selenate
Sulfuric acid, H$_2$SO$_4$	SO$_4^{2-}$	Sulfate

If the nonmetal in the oxyacid exists in more than one oxidation state, then at least two oxyacids exist, each with a different number of bonded O atoms. For example, there are two oxyacids that contain one S atom:

$$H_2SO_4 = \text{sulfuric acid}$$
$$+6$$

$$H_2SO_3 = \text{sulfurous acid}$$
$$+4$$

$$\overset{+2 \ +6-8}{H_2SO_4} \quad \overset{+2 \ +4-6}{H_2SO_3}$$
$$\underset{+1 \quad -2}{} \quad \underset{+1 \quad -2}{}$$

The S atom in sulfuric acid, H_2SO_4, is in the +6 oxidation state, whereas the S atom in sulfurous acid, H_2SO_3, is in the lower +4 state. If two oxyacids exist that contain the same nonmetal, the acid that contains the nonmetal in the higher oxidation state is given the *ic acid* ending (corresponding to the oxyanion sulfate, SO_4^{2-}) and the acid that contains the nonmetal in the lower oxidation state is given the *ous acid* ending (corresponding to the oxyanion sulfite, SO_3^{2-}).

Nitric acid, HNO_3, and nitrous acid, HNO_2, are the two oxyacids that contain a single N atom. The nitrogen in nitric acid is found in the +5 oxidation state, and the nitrogen in nitrous acid is in the +3 oxidation state.

$$HNO_3 = \text{nitric acid}$$
$$+5$$

$$HNO_2 = \text{nitrous acid}$$
$$+3$$

$$\overset{+1+5-6}{HNO_3} \quad \overset{+1+3-4}{HNO_2}$$
$$\underset{+1 \quad -2}{} \quad \underset{+1 \quad -2}{}$$

The name of the oxyacid depends on the name of the oxyanion in the molecule. Nitrate, NO_3^-, is the oxyanion in HNO_3; accordingly, *ic acid* is added to the stem *nitr* from nitrogen, which gives nitric acid. Nitrite ion, NO_2^-, is the oxyanion in nitrous acid; thus, *ous acid* is added to the stem *nitr*, which gives nitrous acid.

A few nonmetals form three or four oxyacids. Let's consider the four oxyacids of Cl:

The name given to nitric acid by the alchemists was aqua fortis, *which means strong water.*

$$HClO_4, \text{ perchloric acid} \quad \overset{+7}{Cl}$$

$$HClO_3, \text{ chloric acid} \quad \overset{+5}{Cl}$$

$$HClO_2, \text{ chlorous acid} \quad \overset{+3}{Cl}$$

$$HClO, \text{ hypochlorous acid } \overset{+1}{Cl}$$

Perchloric acid is one of the strongest mineral acids. It is also a strong oxidizing agent, one that adds O atoms to other compounds. Perchloric acid must be treated with great care: solutions with concentrations above 60% are unstable and frequently explode!

Here, in addition to the acids with just the *ic* and *ous* endings, two other acids exist, one with a Cl atom in a still higher oxidation state, and another with the Cl in a still lower state. When this situation arises, the acid that contains the nonmetal in the highest oxidation state (the acid that has one more O than the *ic acid*) is named by placing *per* as a prefix in front, and adding *ic acid* to the end. Similarly, the acid with the nonmetal in the lowest oxidation state (the acid that has one less O than the *ous acid*) is given the prefix *hypo*, with the *ous acid* ending.

When perchloric acid, $HClO_4$, gives up an H^+, the ClO_4^-, or perchlorate, ion results. Acids with *per* as the prefix and ending in *ic acid* contain polyatomic ions named by dropping the *ic acid* and adding *ate*. Similarly, acids with *hypo* and *ous acid* contain the *hypo . . . ite* ion. Thus, when HOCl loses an H^+, a hypochlorite ion, OCl^-, results.

Example Problem 9.7 shows another example of a nonmetal that can exist in four different oxidation states.

Hypo is a prefix that means "below." Hyper has the opposite meaning, "above."

_____ **Example Problem 9.7** _____

Write the names for the following series of oxyacids and the names for the polyatomic ions in the molecules: *(a)* $HBrO_4$, *(b)* $HBrO_3$, *(c)* $HBrO_2$, *(d)* HBrO.

_____ **Solution** _____

(a) The Br atom in $HBrO_4$ is in the $+7$ oxidation state, the highest of the series. Thus, its name is perbromic acid, and it contains the perbromate ion, BrO_4^-.

(b) The Br atom in $HBrO_3$ is in the $+5$ state, and it is given the name bromic acid. The bromate ion, BrO_3^-, is the polyatomic ion in bromic acid.

(c) The Br atom in $HBrO_2$ is in the $+3$ oxidation state. The name of the acid is bromous acid, and it contains a bromite ion, BrO_2^-.

(d) The Br atom in HBrO is $+1$; thus, the acid is hypobromous acid, and the BrO^- ion is the hypobromite ion.

Per. . .ic oxyacids contain one more O atom than ic acids, and hypo. . .ous acids contain one less O atom than ous acids.

Unfortunately, there is no easy way to know which oxyacids exist for a particular nonmetal. It is a matter of sitting down and learning the most important ones and referring to a table of oxyacids for the others. Table 9.5 gives a partial listing of additional common oxyacids.

Acids that contain more than one H atom (e.g., H_2SO_4, H_3PO_4. . .) can donate some or all of their H^+ ions in chemical reactions. If H_2SO_4 loses one H^+, the resulting ion is HSO_4^-:

$$H_2SO_4 \xrightarrow{H_2O} H^+(aq) + HSO_4^-(aq)$$

HSO_4^- is assigned the name hydrogensulfate. It is only necessary to affix the word *hydrogen* to the name of the polyatomic ion, SO_4^{2-}. Bisulfite is the common name for HSO_4^-; the prefix *bi* is used in place of the word *hydrogen*.

TABLE 9.5 SELECTED OXYACIDS

Nonmetal	Oxyacid	Formula
As	Arsenic acid	H_3AsO_4
	Arsenous acid	H_3AsO_3
B	Boric acid	H_3BO_3
C	Carbonic acid	H_2CO_3
P	Phosphoric acid	H_3PO_4
	Phosphorous acid	H_3PO_3
	Hypophosphorous acid	H_3PO_2
Se	Selenic acid	H_2SeO_4
	Selenous acid	H_2SeO_3
Si	Silicic acid	H_4SiO_4

Phosphoric acid, H_3PO_4, can release one, two, or three H^+ ions. If one H^+ is lost, then $H_2PO_4^-$, dihydrogenphosphate, is produced. If two H^+ ions are lost, then HPO_4^{2-}, monohydrogenphosphate, results, and if three H^+ are lost, the phosphate ion is produced:

Pure phosphoric acid is a white solid that melts at $42°C$. It is generally sold as an 80 to 85% aqueous solution.

$$H_3PO_4 \longrightarrow H^+(aq) + H_2PO_4^-(aq) \quad \text{(dihydrogenphosphate)}$$

$$H_3PO_4 \longrightarrow 2H^+(aq) + HPO_4^{2-}(aq) \quad \text{(monohydrogenphosphate)}$$

$$H_3PO_4 \longrightarrow 3H^+(aq) + PO_4^{3-}(aq) \quad \text{(phosphate)}$$

The names of polyatomic ions that contain H atoms are incorporated into the names of compounds in exactly the same way as those of other polyatomic ions. To write the name of $NaHCO_3$, it is only necessary to write sodium hydrogencarbonate (or sodium bicarbonate). $CaHPO_4$ is given the name calcium monohydrogenphosphate.

The common names for sodium hydrogen-carbonate are baking soda and bicarbonate of soda. When an acid combines with $NaHCO_3$, $CO_2(g)$ is liberated, which helps bread and other foods rise.

Some binary and ternary ionic compounds form weak bonds with water molecules and produce a group of compounds known as **hydrates.** For example, barium chloride, $BaCl_2$, bonds with two water molecules to produce a hydrate that is expressed as follows:

Hydrates

$$BaCl_2 \cdot 2H_2O$$

The two water molecules are separated from the formula of $BaCl_2$ with a raised dot to indicate that the water molecules are only loosely bonded. To write the name of the hydrate, you first write the name of the ionic compound; you then attach the prefix that corresponds to the number of water molecules to the word *hydrate* (from the Greek for "water"). Thus the name of $BaCl_2 \cdot 2H_2O$ is barium chloride dihydrate. If the water molecules are removed from the hydrate, the resulting compound is called an *anhydrous salt*. In our example, $BaCl_2$ is the anhydrous salt. Table 9.6 lists additional examples of hydrates.

TABLE 9.6 HYDRATED SALTS

Hydrated salt	Name
$Al(ClO_4)_3 \cdot 6H_2O$	Aluminum perchlorate hexahydrate
$Ba(NO_2)_2 \cdot H_2O$	Barium nitrite monohydrate
$CaO_2 \cdot 8H_2O$	Calcium peroxide octahydrate
$Au(CN)_3 \cdot 3H_2O$	Gold(III) cyanide trihydrate
$FeSO_4 \cdot 7H_2O$	Iron(II) sulfate heptahydrate

Common Names

While chemists attempt to standardize the names of compounds, in many other disciplines and in industry, common, or trivial, names are most frequently used. Table 9.7 lists the common names, formulas, and Stock names for compounds that you may encounter. Go through these compounds and attempt to verify their Stock names.

TABLE 9.7 COMMON AND STOCK NAMES FOR SELECTED COMPOUNDS

Common name	Formula	Systematic name
Alum	$NaAl(SO_4)_2 \cdot 12H_2O$	Sodium aluminum sulfate dodecahydrate
Alumina	Al_2O_3	Aluminum oxide
Baking soda	$NaHCO_3$	Sodium hydrogencarbonate
Barite	$BaSO_4$	Barium sulfate
Blue vitriol	$CuSO_4 \cdot 5H_2O$	Copper(II) sulfate pentahydrate
Borax	$Na_2B_4O_7 \cdot 10H_2O$	Sodium tetraborate decahydrate
Calcite	$CaCO_3$	Calcium carbonate
Calomel	Hg_2Cl_2	Mercury(I) chloride
Chrome yellow	$PbCrO_4$	Lead(II) chromate
Cinnabar	$HgCl_2$	Mercury(II) chloride
Dolomite	$CaMg(CO_3)_2$	Calcium magnesium carbonate
Epsom salt	$MgSO_4 \cdot 7H_2O$	Magnesium sulfate heptahydrate
Fluorite	CaF_2	Calcium fluoride
Galena	PbS	Lead(II) sulfide
Germane	GeH_4	Germanium hydride
Glauber's salt	$Na_2SO_4 \cdot 10H_2O$	Sodium sulfate decahydrate
Gypsum	$CaSO_4 \cdot 2H_2O$	Calcium sulfate dihydrate
Halite	$NaCl$	Sodium chloride
Hematite	Fe_2O_3	Iron(III) oxide
Lime	CaO	Calcium oxide
Magnesia	$MgCO_3$	Magnesium carbonate
Muriatic acid	HCl	Hydrochloric acid
Oil of vitriol	H_2SO_4	Sulfuric acid
Plaster of paris	$CaSO_4 \cdot \frac{1}{2}H_2O$	Calcium sulfate hemihydrate
Pyrite	FeS_2	Iron(IV) sulfide
Quartz	SiO_2	Silicon dioxide
Washing soda	Na_2CO_3	Sodium carbonate
White lead	$PbCO_3$	Lead(II) carbonate

9.13 What is the name of each of the following oxyacids: *(a)* HNO_3, *(b)* H_3PO_3, *(c)* H_2SO_3, *(d)* HClO?

9.14 Write the formulas for each of the following: *(a)* boric acid, *(b)* selenic acid, *(c)* chlorous acid, *(d)* carbonic acid.

9.15 Write the formulas for each of the following: *(a)* potassium arsenate, *(b)* rubidium hydrogensulfite, *(c)* magnesium borate, *(d)* iron(III) dihydrogenphosphate.

9.16 Write the names for each of the following compounds: *(a)* Na_2SO_3, *(b)* $Sr(NO_2)_2$, *(c)* $CsHSO_3$, *(d)* $KBrO_2$, *(e)* CuClO, *(f)* $CuSO_4 \cdot 5H_2O$, *(g)* $Fe(NO_3)_2 \cdot 6H_2O$.

9.17 Write formulas for each of the following: *(a)* sodium chlorite, *(b)* nickel(III) hydrogencarbonate, *(c)* mercurous selenite, *(d)* magnesium perbromate, *(e)* lead(IV) dihydrogenphosphate, *(g)* sodium acetate trihydrate.

SUMMARY

Oxidation numbers are used to help keep track of the electrons associated with elements in compounds. Oxidation numbers are needed to write formulas and equations, but they are not an actual, measurable property of elements.

Binary compounds are those that contain two different elements. The names of ionic binary compounds are assigned by writing the name of the metal and then modifying the nonmetal name by dropping its ending and adding *ide*. If the compound contains a metal that can exist in more than one oxidation state, then the Stock name includes roman numerals in parentheses after the metal name; these indicate the exact oxidation number of the metal.

An old system for naming binary compounds is still employed by some people; this system requires knowledge of the Latin names for metals. Using the old name as the stem, either *ic* or *ous* is appended as a suffix, denoting the metal in its higher and lower oxidation states, respectively.

Ternary compounds contain three different elements. Most are composed of a metal and a polyatomic ion and are named in a similar manner to that of binary compounds except that the name of the polyatomic ion is written in place of that of the nonmetal. Examples of polyatomic ions include chromate, $CrO_4{}^{2-}$; nitrate, $NO_3{}^-$; cyanide, CN^-; and hydroxide, OH^-.

Acids are substances that donate H^+ to water. Binary acids are named by attaching the prefix *hydro* and the ending *ic acid* to the stem of the name of the nonmetal. Many ternary acids are oxyacids; i.e., they contain an O atom in addition to the nonmetal and hydrogen. Oxyacid names have the endings *ic* or *ous*, depending on the oxidation state of the nonmetal. Oxyanions derived from acids are named in a systematic fashion. If the resulting ion is from an *ic acid,* the ending *ate* is placed at the end of the oxyanion; *ite* is the suffix given to oxyanions produced from *ous* acids.

KEY TERMS

anhydrous salt	IUPAC	polyatomic ion
binary compound	oxidation number	Stock system
binary acid	oxidation-reduction reaction	ternary compound
chemical nomenclature	oxyacid	ternary acid
hydrate	oxyanion	

EXERCISES*

9.18 Define each of the following: chemical nomenclature, IUPAC, oxidation number, oxidation-reduction reaction, binary compound, ternary compound, Stock system, polyatomic ion, binary acid, ternary acid, oxyacid, oxyanion, hydrate, anhydrous salt.

Oxidation Numbers

9.19 Give the oxidation numbers of all elements in the following compounds: (a) F_2, (b) H_2O, (c) MgS, (d) NCl_3, (e) N_2O_5, (f) PCl_5, (g) Al_2S_3, (h) CI_4.

9.20 What is the oxidation state of S in each of the following compounds: (a) SO_3, (b) SO_2, (c) H_2SO_4, (d) H_2SO_3, (e) S_8, (f) S_2O_7, (g) S_2O_3, (h) S_2F_{10}?

9.21 What is the oxidation state of N in each of the following compounds: (a) N_2, (b) N_2O, (c) N_2O_4, (d) N_2O_3, (e) $NOCl$, (f) HNO_3, (g) $H_2N_2O_2$, (h) N_2H_4?

9.22 Calculate the oxidation number of each element in the following compounds: (a) $K_2Cr_2O_7$, (b) Na_2GeO_3, (c) Na_2UO_4, (d) $RbHSO_4$, (e) $Ca(HS)_2$, (f) U_3O_8, (g) $K_4V_2O_7$, (h) $Na_2C_2O_4$, (i) $Cu(CN)_2$, (j) Na_2O_2 (a peroxide), (k) MgH_2 (a hydride).

9.23 Find the oxidation states of the metals in the following compounds: (a) $CePO_3$, (b) Sb_2O_5, (c) $Zn(OH)_2$, (d) $Cr_2(SO_4)_3$, (e) $MoSeO_4$, (f) Cu_2SO_4, (g) MnH_2PO_4, (h) Hg_2CO_3, (i) $Pr_2(MoO_4)_3$, (j) $Ni_3(NO_3)_2 \cdot 6H_2O$.

Names and Formulas of Binary Compounds

9.24 Write the name of each of the following covalent substances: (a) CO_2, (b) N_2O, (c) CCl_4, (d) PBr_3, (e) OF_2, (f) SiO_2, (g) XeO_4, (h) IF_7.

9.25 Write the names of each of the following: (a) $SnBr_2$, (b) CoN, (c) PbS, (d) Cu_3P, (e) HgO, (f) Na_2Se, (g) $MgCl_2$, (h) Sc_2O_3, () CdI_2.

9.26 Write the formulas for each of the following: (a) iron(III) bromide, (b) manganese(II) sulfide, (c) copper(I) iodide, (d) chromium(II) fluoride, (e) magnesium oxide, (f) barium carbide, (g) mercury(I) fluoride, (h) tin(II) nitride, (i) palladium(II) selenide, (j) rubidium phosphide.

9.27 Write the formulas of the following oxides: (a) thallium(III) oxide, (b) uranium(IV) oxide, (c) gold(I) oxide, (d) molybdenum(V) oxide, (e) manganese(VII) oxide, (f) ruthenium(IV) oxide, (g) tungsten(VII) oxide, (h) vanadium(IV) oxide, (i) scandium(III) oxide, (j) tin(IV) oxide.

Names and Formulas of Ternary Compounds

9.28 Complete the following table by writing the name of the compound that is a combination of the anion listed horizontally and the cation listed vertically.

	$C_2H_3O_2^-$	PO_4^{3-}	MnO_4^-	CN^-	CO_3^{2-}	CrO_4^{2-}	OH^-	S^{2-}
NH_4^+								
Mg^{2+}								
Fe^{2+}								
Hg^{2+}								
Al^{3+}								
Sn^{4+}								
Rb^+								
Au^{3+}								

*For exercise numbers printed in color, answers can be found at the back of the book.

9.29 Complete the following table by writing the formula of the compound that is a combination of the anion listed horizontally and the cation listed vertically.

	Hydroxide	Acetate	Sulfate	Chlorate	Selenate	Nitride
Calcium						
Aluminum						
Lead(IV)						
Cobaltous						
Ammonium						
Germanium(IV)						
Tin(II)						

9.30 In the following table, the common name for a compound is given, with either its formula or modern name. Complete the table by writing the formula or modern name of the compound:

Old name	Formula	Stock name
Alunogenite		Aluminum sulfate
Aragonite	$CaCO_3$	
Baking soda		Sodium hydrogencarbonate
Blue vitriol	$CuSO_4 \cdot 5H_2O$	
Celestite		Strontium sulfate
Chrome yellow	$PbCrO_4$	
Cyanoauric acid		Gold(III) cyanide
Glauber's salt	Na_2SO_4	
Hemimorphite		Zinc(II) silicate
Nitrobarite	$Ba(NO_3)_2$	

9.31 Write the formula of the compound that results when the ammonium ion is combined with each of the following ions: (a) selenate, (b) sulfite, (c) nitrite, (d) periodate, (e) chlorate, (f) hydrogencarbonate, (g) bromite, (h) hypochlorite, (i) hydrogensulfite.

9.32 Write the formulas of the compounds that result when acetate ions are combined with each of the following ions: (a) lithium, (b) zinc(II), (c) ferrous, (d) cupric, (e) gallium(III), (f) thallium(III), (g) plumbic, (h) iridium(IV).

Acids

9.33 Write the name of each of the following binary acids: (a) HBr(aq), (b) HI(aq), (c) H_2Se(aq).

9.34 Write the name of each of the following oxyacids: (a) H_3BO_3, (b) HClO, (c) HIO_3, (d) H_3AsO_4, (e) H_2CO_3, (f) $HBrO_4$.

9.35 Lactic acid, $HC_3H_5O_3$, a weak organic acid, ionizes to a small extent in water, producing the $C_3H_5O_3^-$ ion. What is the name of the ion?

9.36 Write the names of the following acids: (a) HIO_4, (b) $HC_2H_3O_2$, (c) He_2SeO_3, (d) HIO, (e) HCN.

9.37 Write the name of the polyatomic ion in each acid in exercise 9.36.

9.38 Write the name of each of the following polyatomic ions and the name of the acid from which it is derived: (a) HSO_3^-, (b) HPO_3^{2-}, (c) HSO_4^-, (d) HCO_3^-.

Additional Exercises

9.39 Identify the oxidation states of the elements in each of the following sodium compounds: (a) $NaHF_2$, (b) $Na_2S_2O_6 \cdot 2H_2O$, (c) $NaBF_4$, (d) Na_2MoO_4, (e) Na_2SiF_6, (f) Na_3AsS_4.

9.40 Write the name and formula of a member of each of the following classes of compounds: (a) ternary ionic compound, (b) binary covalent compound, (c) oxyacid, (d) hydrate, (e) binary ionic compound, (f) ternary covalent compound, (g) binary acid, (h) binary metallic sulfide.

9.41 What do the following endings indicate about a compound: (a) ite, (b) ate, (c) ide, (d) ic, (e) ous?

9.42 Write the names of each of the following oxyacids: (a) H_3AsO_4, (b) HIO_3, (c) H_2SeO_4.

9.43 Write the names of each of the following hydrates: (a) $CrPO_4 \cdot 6H_2O$, (b) $Ga_2O_3 \cdot H_2O$, (c) $In(ClO_4)_3 \cdot 8H_2O$, (d) $FePO_4 \cdot 2H_2O$, (e) $LiI \cdot 3H_2O$, (f) $Na_2CO_3 \cdot 10H_2O$.

9.44 Write the formulas of each of the following hydrates: *(a)* sodium borate tetrahydrate, *(b)* thorium(IV) selenate nonahydrate, *(c)* nickel(II) phosphate octahydrate, *(d)* mercury(II) bromate dihydrate, *(e)* manganese(II) acetate tetrahydrate, *(f)* lead(II) bromate monohydrate.

9.45 Write the names of each of the following compounds:

(1) $(NH_4)_2S$
(2) SbI_3
(3) H_3AsO_4
(4) As_2O_3
(5) $BaCrO_4$
(6) $BeSeO_3$
(7) $BiCl_4$
(8) BN
(9) BrO_2
(10) $Cd(BrO_3)_2$
(11) $Ca(ClO)_2$
(12) S_2F_{10}
(13) $Ce(OH)_3$
(14) $CsHCO_3$
(15) Cl_2O_7
(16) $Cr_2(SO_3)_3$
(17) CoF_3
(18) $CuSeO_4$
(19) $GaCl_3$
(20) GeS_2
(21) $HCN(g)$
(22) $HCN(aq)$
(23) $HIO_3(aq)$
(24) IF_5
(25) Ir_2S_3
(26) $Fe(H_2PO_2)_3$
(27) $Pb(AsO_2)_2$
(28) $LiClO_3$
(29) $Mg(NO_3)_2$
(30) $MnAs$
(31) $Hg(BrO_3)_2$
(32) $Mo(SO_3)_3$
(33) Si_3Cl_8
(34) OsO
(35) $Pd(NO_3)_2$
(36) PBr_5
(37) $Pt(OH)_2$
(38) KH_2AsO_4
(39) $RaCO_3$
(40) ReF_6
(41) $Rh_2(SO_4)_3$
(42) Rb_2SeO_4
(43) Sc_2O_3
(44) SeF_4
(45) Ag_2TeO_3
(46) NaH_2PO_3
(47) $Sr(MnO_4)_2$
(48) S_2F_2
(49) $TlCN$
(50) $Sn(SO_4)_2$

9.46 Write the formula for each of the following:

(1) titanium(III) chloride
(2) tungsten(VI) bromide
(3) zinc(II) chromite
(4) zirconium(IV) iodide
(5) ammonium sulfite
(6) sodium hydroxide
(7) magnesium sulfate
(8) barium hypobromite
(9) cesium monohydrogenphosphate
(10) bismuth(III) bromide
(11) cuprous hypochlorite
(12) aluminum nitrate
(13) ferrous arsenide
(14) lead(IV) chlorate
(15) lithium monohydrogenphosphite
(16) manganese(II) oxide
(17) mercuric telluride
(18) molybdenum(IV) iodide
(19) nickel(III) bicarbonate
(20) dinitrogen tetroxide
(21) palladium(IV) silicide
(22) osmium(II) sulfate
(23) lithium chlorate
(24) hypophosphorous acid
(25) tetraphosphorus heptasulfide
(26) potassium iodate
(27) rhenium(VI) chloride
(28) tin(IV) phosphate
(29) rubidium chlorite
(30) scandium(III) sulfate
(31) silicon nitride
(32) silicic acid
(33) silver(I) phosphate
(34) sodium hydrogensulfite
(35) strontium iodate
(36) tantalum(III) nitride
(37) thallous sulfate
(38) stannic iodide
(39) zinc(II) cyanide
(40) cobalt(II) phosphate
(41) ammonium bromite
(42) aluminum nitride

(43) calcium carbide
(44) hypochlorous acid
(45) ferric acetate
(46) manganese(II) carbonate
(47) mercury(II) sulfide
(48) hydrocyanic acid
(49) potassium permanganate
(50) cerium(III) carbonate

9.47 A 5.000-g sample of a compound is found to contain 2.247 g of lead; what remains is iodine. The molecular mass of the compound is 461. *(a)* What is the molecular formula of the compound? *(b)* What is the name of the compound?

9.48 A hydrate of magnesium phosphate is found to have 25.5% water by mass. Its molecular mass is 353. *(a)* What is the formula of the hydrate? *(b)* What is the name of the hydrate?

9.49 Three sulfides of phosphorus are analyzed and found to contain 43.70% (molecular mass = 220), 64.49% (molecular mass = 348), and 72.13% S by mass (molecular mass = 444). What are the names of these phosphorus sulfides?

9.50 What is the percent composition of tungsten(VI) bromide?

CHAPTER
Ten

Chemical
Equations

STUDY GUIDELINES

After completing Chapter 10, you should be able to

1. Identify the reactants and products in a chemical equation

2. Identify and write all common symbols found in chemical equations

3. Explain the meaning of the abbreviations (g), (l), (s), and (aq)

4. Explain why subscripts in formulas cannot be changed when balancing a chemical equation

5. Balance chemical equations, using the inspection method

6. Translate a word equation into a correctly balanced chemical equation

7. List the four classes of inorganic reactions

8. Write a general equation for each class of inorganic reaction

9. Write balanced chemical equations that illustrate each type of inorganic reaction

10. Identify *(a)* combination, *(b)* decomposition, *(c)* single displacement, and *(d)* metathesis reactions

11. Use the activity series to decide when a single displacement reaction occurs

12. Predict the products and write equations for reactions given the reactants

13. Predict what substance precipitates or what gas is released when two aqueous salt solutions undergo a metathesis reaction

14. Describe the difference between endothermic and exothermic reactions

15. Identify endothermic and exothermic reactions from their chemical equations given the heat released or absorbed

16. Explain, in terms of bond breaking and formation, what happens in endothermic and exothermic reactions

---

Chemical changes occur when substances undergo changes in their composition. When a substance undergoes a chemical change, we say that "a chemical reaction has taken place." It is inconvenient and time consuming to express what happens during chemical reactions by writing the complete names of all substances involved (a word equation);

10.1 COMPONENTS OF A CHEMICAL EQUATION

instead, chemists write a concise statement, called a **chemical equation,** using the symbols of the elements and the formulas of compounds. Other special symbols are added to the equation to express exactly what changes occur during chemical changes.

A chemical equation has two parts: *(1)* reactants and *(2)* products. **Reactants,** sometimes called starting materials, are all substances present prior to the chemical change. The symbols for all reactants are listed, separated by plus signs. All reactants are written to the left of an arrow ($\rightarrow$) that separates the reactants from the products.

All **products,** substances produced after the chemical change occurs, are written to the right of the arrow, and are separated from each other by plus signs. Hence, chemical equations have the following format:

$$\text{Reactants} \longrightarrow \text{products}$$

or

$$A + B \longrightarrow C + D$$

In our example, hypothetical substances A and B are the reactants, and C and D are the products of the reaction. The above equation is translated as "reactant A combines with reactant B to yield product C and product D." Note that the arrow is read as "to yield" or "yield" or "gives."

Frequently, other information is added to the equation. For example, it is important to know what physical state the reactants and products exist in. Four symbols are employed to indicate physical states: solid, (s); liquid, (l); gas, (g); and water, or aqueous, solution, (aq). Enclosed in parentheses, these symbols are written next to the formula. Consider the following example:

$$A(g) + B(l) \longrightarrow C(aq) + D(s)$$

Translating to a word equation: "Reactant A, in the gas phase, combines with reactant B, in the liquid phase, yielding product C, which is dissolved in water, and product D, in the solid phase."

Conditions required for the reaction to take place are also written into the chemical equation. They are placed either above or below the arrow. If heat is needed for the chemical change, the word "heat" or more commonly the Greek letter delta (Δ) is written above or below the arrow. Sometimes the actual temperature is expressed.

Besides heat effects, some reactions take place only in the presence of a particular type of electromagnetic radiation, e.g., infrared (ir) or ultraviolet light (uv).

$$A(s) \xrightarrow{\Delta} B(g) + C(s)$$

$$A(s) \xrightarrow{\text{heat}} B(g) + C(s)$$

Both the above equations are read as "solid reactant A is heated to yield gaseous product B and solid product C."

Various reactions require **catalysts,** which are substances that increase the rates of reactions and are usually recovered chemically unchanged after the reaction. The word "catalyst," the abbreviation "cat," or the actual name of the catalyst is written above or below the arrow. Any special conditions (high or low pressure, presence or absence of light, etc.) are also written near the arrow.

Most chemical reactions in living tissues are catalyzed by complex protein structures called enzymes.

Table 10.1 summarizes the symbols used in writing chemical equations.

TABLE 10.1 SUMMARY OF SYMBOLS USED IN CHEMICAL EQUATIONS

Symbol	Meaning
$\longrightarrow$	Yields, produces, gives
+	Separates compounds and elements
(s)	Solid state
(l)	Liquid state
(g)	Gaseous state
(aq)	Aqueous solution
$\xrightarrow{\text{cat}}$	Reaction requiring a catalyst
$\xrightarrow{\Delta}$	Reaction requiring heat
$\xrightarrow{\text{uv}}$	Reaction requiring ultraviolet light

REVIEW EXERCISES

10.1 *(a)* What is a chemical change? *(b)* How is a chemical change different from a physical change?

10.2 Collectively, what are the substances called that are located *(a)* to the right of the arrow and *(b)* to the left of the arrow in a chemical equation?

10.3 What is the exact meaning of the following symbols used in chemical equations: *(a)* + , *(b)* Δ, *(c)* (aq), *(d)* →?

10.4 Translate the following chemical equation into a word equation:

$$C(s) + O_2(g) \longrightarrow CO_2(g)$$

10.2 BALANCING CHEMICAL EQUATIONS

When an equation is written, it must obey the **law of conservation of mass,** i.e., matter cannot be created or destroyed. Specifically, the number of atoms of each different element in an equation must be the same on the left and right sides of the arrow. After a chemical change, the same types and number of atoms are present; they are merely rearranged.

Chemical equations obey the law of conservation of mass: atoms are conserved in reactions.

To obey the law of conservation of mass, an equation must be **balanced.** Balancing an equation involves placing coefficients in front of all reactants and products so that the same number of atoms of each element appear on either side of the equation.

Simple equations are balanced using the *inspection method*, which involves equalizing the number of atoms of each element by placing **coefficients** in front of all elements and compounds in the equation.

Let's illustrate how to balance equations by the inspection method, using the combination reaction in which hydrogen gas, $H_2(g)$, combines with oxygen gas, $O_2(g)$, to yield water vapor, $H_2O(g)$. First write the unbalanced equation, including all reactants and products.

$$H_2(g) + O_2(g) \longrightarrow H_2O(g) \text{ (Unbalanced)}$$

After writing the unbalanced equation, we readily see that the number of O atoms is not the same on both sides of the equation. Two O atoms appear on the left side, and only one on the right side. To balance the O atoms, we place a 2 as the coefficient of water. Two water molecules contain two O atoms.

We cannot place a 2 as a subscript next to the O atom in water because it would change the composition of water. **Never change subscripts when you balance chemical equations.** If you incorrectly place a 2 as the subscript for oxygen, it gives a new product of the reaction, H_2O_2—hydrogen peroxide. The properties of hydrogen peroxide are vastly different from those of water!

After balancing the O atoms, we are left with the following partially balanced equation:

$$H_2(g) + \underline{O}_2(g) \longrightarrow 2H_2\underline{O}(g)$$

By balancing the O atoms, we have learned the number of H atoms needed. Two water molecules contain four H atoms ($2 \times H_2O = 4$ H and 2 O). Therefore, we place a 2 as the coefficient of $H_2(g)$ on the left side of the equation, which gives four H atoms.

$$2\underline{H}_2(g) + \underline{O}_2(g) \longrightarrow 2\underline{H}_2\underline{O}(g)$$

The last step, an important one, is to check to see that you have correctly balanced all of the atoms. If the same number of atoms of each type appear on both sides, the equation is properly balanced. If you are off by even one atom, then the equation is not balanced. In our example, we find that there are four H atoms and two O atoms on either side of the arrow, which indicates a correctly balanced equation. It is convenient to check off or underline each atom as you verify that the equation is balanced:

$$2\underline{H}_2(g) + \underline{O}_2(g) \longrightarrow 2\underline{H}_2\underline{O}(g) \text{ (Balanced)}$$

The coefficients in a balanced equation indicate the ratio in which the reactants and products combine and form. Our equation now reads: "Two molecules of hydrogen gas combine with one molecule of oxygen

Coefficients are the numbers that precede the symbols and formulas in a chemical equation.

Do not change the subscripts when you balance equations.

H_2 combines explosively with O_2; this reaction should not be demonstrated on a large scale unless it is carefully controlled.

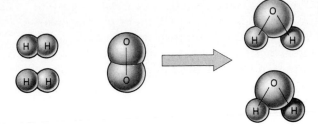

Figure 10.1
In the reaction of H_2 and O_2, two H_2 molecules combine with one O_2 molecule to produce two H_2O molecules.

gas to yield two molecules of water vapor." Figure 10.1 shows how hydrogen and oxygen combine to produce water. As we will learn in Chap. 11, the coefficients also indicate mole relationships in chemical reactions.

Let's tackle a slightly more complex equation to further illustrate balancing by inspection. Balance the equation for the reaction in which methane gas, $CH_4(g)$, combines with oxygen gas, $O_2(g)$, to yield carbon dioxide gas, $CO_2(g)$, and water vapor, $H_2O(g)$. Methane is the principal component of natural gas, which is a widely used fuel. As before, write the unbalanced equation:

$$CH_4(g) + O_2(g) \longrightarrow CO_2(g) + H_2O(g) \text{ (Unbalanced)}$$

When you balance an equation in which one element appears in many of the reactants and products, it is best to balance that atom last. Here, and in many equations, this atom is oxygen. Oxygen is frequently the last atom balanced, and in many cases hydrogen is balanced next to last. Thus, we should start by balancing C atoms. Because there is one C atom on each side of the arrow, the C atoms are balanced as written:

$$\underline{C}H_4(g) + O_2(g) \longrightarrow \underline{C}O_2(g) + H_2O(g)$$

We find four H atoms on the left side and two H atoms on the right side. A 2 is placed in front of H_2O to produce four H atoms on the right:

$$\underline{C}H_4(g) + O_2(g) \longrightarrow \underline{C}O_2(g) + 2\underline{H}_2\underline{O}(g)$$
$$\text{Four O atoms}$$

All that remains is to balance the O atoms. Because we have balanced all of the products, we now know the total number of O atoms. One CO_2 molecule contains two O atoms, and two H_2O molecules contain two O atoms, which gives a total of four O atoms; therefore, four O atoms must be contained in the reactants. Looking at the left side of the equation, we only find two O atoms. Ask yourself, what number times 2 gives 4? The answer is 2, so place a 2 as the coefficient of O_2 to complete the balancing of the equation.

$$\underline{C}H_4(g) + 2\underline{O}_2(g) \longrightarrow \underline{C}O_2(g) + 2\underline{H}_2\underline{O}(g) \text{ (Balanced)}$$

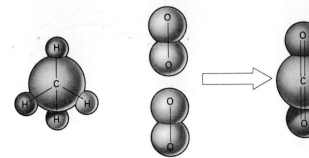
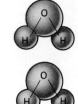

Figure 10.2
In the reaction of CH_4 and O_2, one CH_4 molecule combines with two O_2 molecules, producing one CO_2 molecule and two H_2O molecules.

Finally, check to see that all atoms are balanced. We find one C, four H, and four O atoms on each side of the equation. Our equation states, "One molecule (or mole) of methane gas combines with two molecules (moles) of oxygen gas to produce one molecule (mole) of carbon dioxide gas and two molecules (moles) of water vapor." Figure 10.2 shows this reaction diagrammatically.

A summary of the rules that can be followed to successfully balance many chemical equations follows:

Rule 1

Write an unbalanced equation that includes the correct formulas of all reactants and products.

This first step is most significant, because one incorrect formula alters the way the equation is balanced or in some instances makes it impossible to balance. Always double-check to see that the unbalanced equation is written properly.

Rule 2

Determine a logical sequence for balancing the atoms in the equation, leaving those atoms that appear in more than one compound on each side for last.

It is frequently best to start with metals, especially those that only appear in one molecule on each side of the equation. Then proceed to nonmetals that also occur in one molecule on each side; then balance the remaining nonmetals found in more than one molecule. Step 2 is a planning step that simplifies the entire procedure and makes for an orderly approach to the problem.

Instead of balancing equations haphazardly, think first of a "plan of attack," and then execute it.

Rule 3

Balance the atoms one at a time by placing the appropriate coefficients in front of the atoms and molecules in the equation. If possible, proceed in the predetermined order arrived at in Step 2.

After balancing an atom, underline it on both sides of the equation to indicate to yourself that it has been balanced. When all atoms are underlined, the equation should be balanced. If after balancing a couple of atoms you see that your predetermined order for balancing the atoms is not the most efficient, drop it, and proceed in a more efficient way to successfully balance the equation.

Rule 4

Check to see that all atoms are balanced in the equation. If there are an equal number of atoms of each type on each side of the equation, then the equation is balanced; if not, repeat Step 3. These four steps are summarized in Fig. 10.3.

Some other helpful hints to consider when you balance equations are:

1. Although whole numbers are generally preferred, it is not incorrect to occasionally use fractional coefficients. Certain equations are more easily balanced if one fraction is included as a coefficient. Consider the following correctly balanced equation for the reaction of sodium with water:

$$Na(s) + H_2O(l) \longrightarrow NaOH(aq) + \tfrac{1}{2}H_2(g)$$

To remove the fraction, the coefficient of each substance in the equation is multiplied by 2.

$$2Na(s) + 2H_2O(l) \longrightarrow 2NaOH(aq) + H_2(g)$$

2. If you find that all the coefficients are divisible by a small whole number, divide by that number to reduce all coefficients to their lowest possible whole-number value. Balanced equations are correct when they are expressed using the lowest possible multiple of coefficients. For example,

$$4C(s) + 2O_2(g) \longrightarrow 4CO(g)$$

is incorrect because the coefficients can be divided by 2. The correct balanced equation is

$$2C(s) + O_2(g) \longrightarrow 2CO(g)$$

3. If polyatomic ions are found as reactants and if they are unchanged after the reaction, they can be balanced as a unit. For example, if two nitrate ions, NO_3^-, are found within a reactant, you can place a 2 in front of a product that contains one NO_3^- to balance them. Such is the case in the following balanced equation.

$$Ca(NO_3)_2 + Na_2CO_3 \longrightarrow 2NaNO_3 + CaCO_3$$

4. If an odd number of atoms of one type appear on one side and an even number of atoms are found on the other side, multiply the odd number by 2 in order to give an even number.

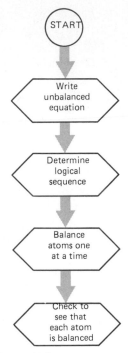

Figure 10.3
Steps to follow when you balance an equation

Example Problem 10.1 gives two additional examples of how to balance equations by the inspection method.

──────────────── **Example Problem 10.1** ────────────────

Balance the following equations by inspection:
(a) $NH_3(g) + O_2(g) \longrightarrow NO(g) + H_2O(g)$
(b) $C_2H_6(g) + O_2(g) \longrightarrow CO_2(g) + H_2O(g)$

──────────────────── **Solution** ────────────────────

(a) $NH_3(g) + O_2(g) \longrightarrow NO(g) + H_2O(g)$ (Unbalanced)

Because O atoms are found in all but one compound, balance them last. If we balance the H atoms first, then we will know the total number of N atoms in the equation. Therefore, balance H atoms and N atoms before O atoms.

First balance the H atoms by finding the lowest common multiple of 3 and 2, which is 6. Place 2 as the coefficient of NH_3, and 3 as the coefficient of H_2O.

$$2NH_3(g) + O_2(g) \longrightarrow NO(g) + 3H_2O(g)$$

Because two NH_3 molecules are required, we know the number of N atoms needed—two. However, if we place a 2 in front of the NO, that will leave us with five O atoms—an odd number. Thus we cannot balance the O atoms without using a fraction, 2.5. The equation can be balanced by using 2.5 as the coefficient of O_2. Fractions can be avoided by changing the coefficients of NH_3 and H_2O so that an odd number of O atoms does not result. This is accomplished by multiplying the coefficients by 2, which yields 4 and 6.

$$4NH_3(g) + O_2(g) \longrightarrow NO(g) + 6H_2O(g)$$

Now we have twelve O atoms on either side, and we know that we need four N atoms. Write in 4 as the coefficient in front of NO to balance the N atoms.

$$4NH_3(g) + O_2(g) \longrightarrow 4NO(g) + 6H_2O(g)$$

The O atoms are the only ones that remain to be balanced. All of the products have their coefficients; thus, we need ten O atoms, four O atoms from the 4NO, and six O atoms from the $6H_2O$. There are only two O atoms on the left side, so we multiply by 5 to give ten O.

$$4NH_3(g) + 5O_2(g) \longrightarrow 4NO(g) + 6H_2O(g) \text{ (Balanced)}$$

Check to see that there are four N, twelve H, and ten O atoms on either side of the equation.

Oxidizing ammonia to produce NO and H_2O is the first step in the Ostwald process for the production of nitric acid. To most efficiently conduct this reaction, ammonia is combined with an excess amount of air, heated to about 650°C, and then passed over a special metal catalyst.

(b) $C_2H_6(g) + O_2(g) \longrightarrow CO_2(g) + H_2O(g)$ (Unbalanced)

In this equation it is convenient to start with C atoms, proceed to H atoms, and balance O atoms last.

There are two C atoms on the left side, and they are balanced by writing a 2 as the coefficient of CO_2.

$$\underline{C}_2H_6(g) + O_2(g) \longrightarrow 2\underline{C}O_2 + H_2O(g)$$

The six H atoms in C_2H_6 are balanced by placing a 3 in front of the H_2O.

$$\underline{C}_2\underline{H}_6(g) + O_2(g) \longrightarrow 2\underline{C}O_2(g) + 3\underline{H}_2O(g)$$

There are seven O atoms in the products. This time, let's balance the equation using a fractional coefficient. Because there are two O atoms on the left, ask yourself: "What number multiplied by 2 gives 7?" The answer is $\frac{7}{2}$ or 3.5; thus, the coefficient is 3.5.

$$\underline{C}_2\underline{H}_6(g) + 3.5\underline{O}_2(g) \longrightarrow 2\underline{C}O_2(g) + 3\underline{H}_2O(g) \text{ (Balanced)}$$

It might be helpful to think in terms of moles when using fractional coefficients. In this equation, 1 mol C_2H_6 combines with 3.5 mol O_2.

A check of the coefficients shows us that both the reactants and the products have a total of two C, six H, and seven O atoms; thus, the equation is balanced.

REVIEW EXERCISES

10.5 Balance each of the following equations:
  (a) $N_2O_4 \longrightarrow NO_2$
  (b) $Al + F_2 \longrightarrow AlF_3$
  (c) $P_4 + Br_2 \longrightarrow PBr_3$
  (d) $ZnS + O_2 \longrightarrow ZnO + SO_2$
  (e) $C + SO_2 \longrightarrow CS_2 + CO$
  (f) $Al + H_2SO_4 \longrightarrow Al_2(SO_4)_3 + H_2$

10.6 Change the following word equation to a correctly balanced chemical equation: Calcium phosphide, Ca_3P_2, combines with water, H_2O, to yield calcium hydroxide, $Ca(OH)_2$, and phosphine, PH_3.

10.7 Translate the following word equation to a balanced chemical equation: Aqueous sulfuric acid combines with aqueous aluminum hydroxide, producing liquid water and aqueous aluminum sulfate. (*Hint:* Start by writing the correct formula for each compound.)

I norganic chemical reactions can be classified in many different ways. A simple, relatively common grouping of inorganic reactions is as follows:

10.3 CLASSIFICATION OF INORGANIC CHEMICAL REACTIONS

1. Combination reactions
2. Decomposition reactions
3. Single displacement (replacement) reactions
4. Metathesis reactions

You should realize that the above classification is an artificial grouping of reactions that helps chemists to better understand chemical reactions. Not all inorganic reactions fit into this simple grouping, and overlap occurs in some groups.

Do not lose sight of the fact that all of the information presented in this section is the result of studying each reaction experimentally (in the laboratory). Products obtained in chemical reactions depend mainly on the specific conditions: different conditions give different products.

It is helpful to organize and study the various classes of inorganic reactions in a systematic fashion. Many students find that placing general and specific equations on small index cards greatly assists in learning this information. Write an equation on one side of the card, and the reaction type on the other side (Fig. 10.4). This is a very effective technique for learning chemical equations.

Combination reactions (sometimes called addition reactions) have the following general form:

Combination Reactions

$$A + X \longrightarrow AX$$

In all combination reactions, two reactants, A and X, combine to produce one product, AX.

Three different possibilities exist for combination reactions, defined by the types of substances that combine. Two elements, an element and a compound, or two compounds can be united in a combination reaction. Let's look at each case individually.

Equations (1) through (5) are examples of combination reactions in which two elements combine:

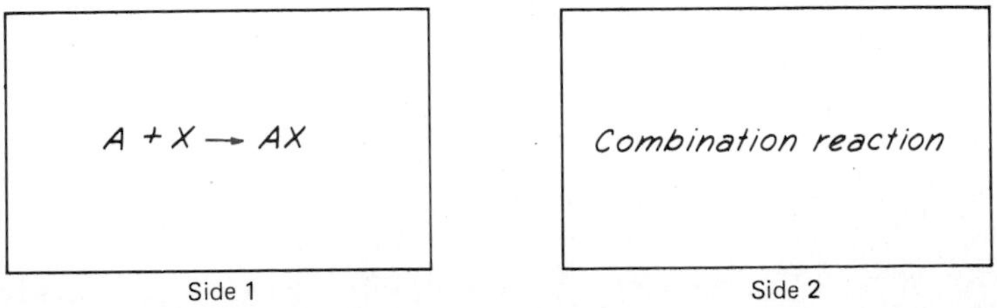

Figure 10.4
One way to learn the types of chemical reactions is to write the general equation on one side of a small index card and the name of the reaction type on the other side of the card. Prepare one card for each reaction to be learned.

$A + X \rightarrow AX$

Side 1

Combination reaction

Side 2

$$4Na(g) + O_2(g) \xrightarrow{\Delta} 2Na_2O(s) \qquad (1)$$

$$2H_2(g) + O_2(g) \xrightarrow{\text{spark}} 2H_2O(g) \qquad (2)$$

$$P_4(s) + 5O_2(g) \xrightarrow{\Delta} P_4O_{10}(s) \qquad (3)$$

$$H_2(g) + Cl_2(g) \xrightarrow{\text{light}} 2HCl(g) \qquad (4)$$

$$U(s) + 3F_2(g) \longrightarrow UF_6(g) \qquad (5)$$

U combines with F_2 to produce gaseous UF_6. Uranium hexafluoride is used to separate U isotopes through a diffusion process; the UF_6 molecules with heavier isotopes diffuse more slowly than those with lighter ones.

In each of these equations we see either a metal or nonmetal that combines with a reactive nonmetal, oxygen, or a halogen. When metals combine with oxygen [Eq. (1)] the product is called a **metal oxide:**

$$\text{Metal} + O_2 \longrightarrow \text{metal oxide}$$

Nonmetal oxides result when a nonmetal combines with oxygen [Eqs. (2) and (3)]:

$$\text{Nonmetal} + O_2 \longrightarrow \text{nonmetal oxide}$$

Formation of oxides is sensitive to the reaction conditions and the availability of oxygen. Carbon, a nonmetal, combines with oxygen to form two oxides of carbon, carbon dioxide and carbon monoxide. If an excess amount of O_2 is present and there is a limited quantity of C, the main product is CO_2 [Eq. (6)]; however, if the O_2 is limited and the carbon is in excess, CO is the main product [Eq. (7)].

$$C(s) + O_2(\text{excess}) \xrightarrow{\Delta} CO_2(g) \text{ (Major product)} \qquad (6)$$

$$2C(s) + O_2(\text{limited}) \xrightarrow{\Delta} 2CO(g) \text{ (Major product)} \qquad (7)$$

Hydrogen, H_2, combines with both metals and nonmetals to produce **metallic** and **nonmetallic hydrides.** Equations (8) and (9) show the formation of metallic hydrides, and Eqs. (10) and (11) show the formation of nonmetallic hydrides.

$$2Na(l) + H_2(g) \xrightarrow{\Delta} 2NaH(s) \text{ (sodium hydride)} \qquad (8)$$

$$Ca(s) + H_2(g) \xrightarrow{\Delta} CaH_2(s) \text{ (calcium hydride)} \qquad (9)$$

$$N_2(g) + 3H_2(g) \xrightarrow{\Delta} 2NH_3(g) \text{ (ammonia)} \qquad (10)$$

$$F_2(g) + H_2(g) \longrightarrow 2HF(g) \text{ (hydrogen fluoride)} \qquad (11)$$

When hydrogen combines with various metals, metal hydrides result. In these ionic compounds, the H atom is in a negative oxidation state (-1) as a result of its greater capacity to attract electrons (its higher electronegativity) than the metal. Metal hydrides release $H_2(g)$ when combined with water. Thus, they are used to store hydrogen, which can be released at a later time.

In the second category of combination reactions, an element combines with a compound. Equations (12) to (15) are examples of such reactions.

KO_2 is potassium superoxide. It is used by miners in special survival masks. Water from the breath of the miner reacts with KO_2, releasing life-sustaining O_2, and producing KOH, which combines with exhaled CO_2 to form $KHCO_3$.

$$CO(g) + \tfrac{1}{2}O_2(g) \xrightarrow{\Delta} CO_2(g) \tag{12}$$

$$K_2O(s) + \tfrac{3}{2}O_2(g) \xrightarrow{\Delta} 2KO_2(s) \tag{13}$$

$$PbCl_2(s) + Cl_2(g) \longrightarrow PbCl_4(l) \tag{14}$$

$$PF_3(g) + F_2(g) \longrightarrow PF_5(g) \tag{15}$$

In Eqs. (12) and (13), oxygen combines with a nonmetal and metal oxide, respectively, yielding products that contain a larger number of O atoms per molecule than the reactant compounds. These two equations are also examples of **oxidation reactions.** Oxidation reactions occur when electrons are transferred from one substance to another. Specifically, **oxidation** is the process in which a substance loses electrons, and the substance that gains the electrons undergoes **reduction.** When substances add O atoms or lose H atoms, they lose electrons; thus, they are oxidized. When substances lose O atoms or gain H atoms, they gain electrons; thus, they are reduced. We will discuss completely these important reactions in Chap. 18.

Equations (14) and (15) are examples of **halogenation** reactions. In a halogenation reaction, halogen atoms are added to a substance.

In the third type of combination reaction, two compounds are joined. Equations (16) to (18) illustrate this class of reaction.

$$SO_3(g) + H_2O(l) \longrightarrow H_2SO_4(aq) \tag{16}$$

$$BaO(s) + H_2O(l) \longrightarrow Ba(OH)_2(aq) \tag{17}$$

$$CO_2(g) + MgO(s) \xrightarrow{\Delta} MgCO_3(s) \tag{18}$$

In Eq. (16), a nonmetal oxide, SO_3, combines with water to produce sulfuric acid, H_2SO_4. Many nonmetal oxides, when added to water, produce acids. Most metal oxides, in contrast, combine with water to form bases (metal hydroxides). The product in Eq. (17) is aqueous barium hydroxide, a base. Equation (18) shows a nonmetal oxide, CO_2, that combines with a metal oxide, MgO, to produce a salt. A **salt** is an ionic compound that does not contain OH^- or O^{2-}.

Metal oxides are sometimes called basic anhydrides—bases without water. Many metal oxides combine with water to give basic solutions. Similarly, nonmetal oxides are called acidic anhydrides.

Table 10.3 (page 262) summarizes each of the different types of combination reactions.

Decomposition reactions have the general form

$$AX \longrightarrow A + X$$

in which AX is a compound and A and X are either elements or compounds. Only one reactant is found in a **decomposition reaction,** and under the proper conditions it breaks down to two or more products. Decomposition reactions always have a compound as the reactant; elements cannot be decomposed chemically.

Many decomposition reactions are initiated by heat, which is required to break the bonds in the starting materials. Certain less stable compounds spontaneously decompose without added heat. Other compounds require electricity, light, or catalysts to decompose.

Examples of decomposition reactions are shown in Eqs. (19) to (24).

$$2H_2O(l) \xrightarrow{\text{elec}} 2H_2(g) + O_2(g) \tag{19}$$

$$2N_2O(g) \xrightarrow{\Delta} 2N_2(g) + O_2(g) \tag{20}$$

$$2HgO(s) \xrightarrow{\Delta} 2Hg(l) + O_2(g) \tag{21}$$

$$2KClO_3(s) \xrightarrow{\Delta} 2KCl(s) + 3O_2(g) \tag{22}$$

$$CaCO_3(s) \xrightarrow{\Delta} CaO(s) + CO_2(g) \tag{23}$$

$$2NaHCO_3(s) \xrightarrow[\text{MnO}_2]{\Delta} Na_2CO_3(s) + CO_2(g) + H_2O(g) \tag{24}$$

Equation 21 is of historical significance. After hearing of the reaction from Priestley, Lavoisier showed that O_2, not phlogiston, is the substance that supports combustion.

In Eqs. (19) through (21), compounds are decomposed to elements. With the addition of energy (heat or electricity), oxygen gas is driven off. Equation (22) illustrates a reaction in which a compound is not fully decomposed to elements. Potassium chlorate, $KClO_3$, when heated with a small amount of MnO_2 (a catalyst), liberates $O_2(g)$ and leaves the salt potassium chloride, KCl.

Equations (23) and (24) present decomposition reactions in which the products are compounds. Equation (23) shows a carbonate, $CaCO_3$, that decomposes to two oxides, CaO and CO_2, a characteristic reaction of carbonate salts. Because hydrogencarbonates are similar to carbonates, they also liberate $CO_2(g)$ when heated [Eq. (23)]. In addition to CO_2, hydrogencarbonates produce carbonates and water.

Hydrated salts release water when heated. For example, consider Eq. (25), in which calcium sulfate dihydrate is decomposed to calcium sulfate and water.

$$\underset{\substack{\text{Hydrated} \\ \text{salt}}}{CaSO_4 \cdot 2H_2O(s)} \xrightarrow{\Delta} \underset{\substack{\text{Anhydrous} \\ \text{salt}}}{CaSO_4(s)} + 2H_2O(g) \tag{25}$$

On heating, the hydrated salt is dehydrated, resulting in the anhydrous salt, i.e., the salt without water.

Table 10.3 on page 262 summarizes the principal types of decomposition reactions.

A **displacement reaction** (also called a **replacement reaction**) occurs when an element takes the place of another element in a compound. For example, consider the following general equation for a single displacement reaction.

$$A + BX \longrightarrow AX + B$$

Here, element A replaces element B in compound BX. Typically in displacement reactions, a metal displaces another metal or hydrogen in the compound with which it combines. Equations (26) through (28) show representative examples of single displacement reactions.

$$Zn(s) + Pb(NO_3)_2(aq) \longrightarrow Zn(NO_3)_2(aq) + Pb(s) \tag{26}$$

$$Mg(s) + H_2SO_4(aq) \longrightarrow MgSO_4(aq) + H_2(g) \tag{27}$$

$$2K(s) + 2HOH(l) \longrightarrow 2KOH(aq) + H_2(g) \tag{28}$$

In Eq. (26), a metal, Zn, replaces a lead ion, Pb^{2+}, that is initially associated with nitrate ions in an aqueous solution of $Pb(NO_3)_2$. In the course of the reaction, solid lead, Pb, is produced; the zinc is dissolved in water and becomes $Zn^{2+}(aq)$. In Eq. (27), Mg metal replaces hydrogen ions, $H+$, in sulfuric acid, H_2SO_4, yielding the dissolved salt $MgSO_4(aq)$ and hydrogen gas, H_2. As shown in Eq. (28), more reactive metals, such as K, can displace hydrogen in water. In this reaction, potassium metal replaces hydrogen in water and produces aqueous potassium hydroxide, $KOH(aq)$, and hydrogen gas, $H_2(g)$.

Metals are ranked according to their ability to displace other metals and hydrogen from compounds. This ranking is called the **activity series of metals.** A partial activity series is shown in Table 10.2. Metals on the left side of the list displace metals farther to the right. For example, Ni displaces Pb^{2+} from compounds, but Pb is unable to displace Ni^{2+} from its compounds.

$$Ni(s) + Pb(NO_3)_2(aq) \longrightarrow Ni(NO_3)_2(aq) + Pb(s) \tag{29}$$

$$Pb(s) + Ni(NO_3)_2(aq) \longrightarrow \text{no reaction (NR)} \tag{30}$$

Figure 10.5 shows the formation of silver from silver nitrate in a single displacement reaction with copper.

Single Displacement Reactions

$$A + \overset{\frown}{BX} \longrightarrow AX + B$$

When "(aq)" is written next to an ionic substance, MX, it indicates that the substance is broken up into ions, surrounded by many water molecules.

TABLE 10.2 ACTIVITY SERIES OF SELECTED METALS

Most active																	Least active
Li	K	Ba	Sr	Ca	Na	Mg	Al	Mn	Zn	Fe	Ni	Sn	Pb	H	Cu	Hg	Ag Pt Au
		Liberates H_2 in cold H_2O, steam, or acid.					Liberates H_2 in steam or acid.				Liberates H_2 only in acid solutions.						

Metals such as Li, K, Ba, Sr, Ca, and Na are the most active metals and can liberate H_2 gas when placed in cold water, steam, or acid. Less active metals, such as Mg, Al, Mn, Zn, and Fe, only liberate H_2 from steam or acid. Ni, Sn, and Pb only displace hydrogen in acidic solutions, and do not displace H atoms when combined with water. The least active metals are Hg, Ag, Pt, and Au. This group does not displace hydrogen in water or acids.

Halogens (not included in Table 10.2) have the capacity to displace certain halide ions. A halide ion is a halogen atom in the -1 oxidation state— e.g., fluoride, F^-; chloride, Cl^-; and bromide, Br^-. Equations (31) and (32) show that both Cl_2 and Br_2 (halogens) can displace the iodide ion, I^-, in sodium iodide.

$$Cl_2(g) + 2NaI(aq) \longrightarrow 2NaCl(aq) + I_2(s) \qquad (31)$$

$$Br_2(l) + 2NaI(aq) \longrightarrow 2NaBr(aq) + I_2(s) \qquad (32)$$

The order of reactivity of the halogens in such reactions is $F_2 > Cl_2 > Br_2 > I_2$. Consequently, Cl_2 displaces both Br^- and I^-, but Br_2 only displaces I^-.

All of the single displacement reactions that we have discussed can also be classified as *oxidation-reduction reactions*. We have already stated that an electron transfer occurs in all oxidation-reduction reactions (redox, for short). For example, let's reconsider Eq. (27), in which Mg displaces hydrogen in sulfuric acid. Initially, Mg is in the zero oxidation state because it is a free metal. After the reaction, Mg^{2+} is present; thus, the oxidation state of magnesium has changed from zero to $+2$. Whenever the oxidation number of a substance increases, the substance has undergone oxidation; electrons (negative particles) have been lost.

$$\underset{0}{Mg} \longrightarrow \underset{+2}{Mg^{2+}} + 2e^- \text{ (Oxidation)}$$

If a substance undergoes oxidation, then another substance must be reduced. In Eq. (27), the sulfuric acid is reduced. Initially, the oxidation state of the H ions in aqueous sulfuric acid is $+1$, but after the reaction the H atoms are a component of H_2: they have been reduced to the zero oxidation state. Whenever the oxidation state of a substance decreases, it has undergone reduction; it has accepted electrons.

$$\underset{+1}{2H^+} + 2e^- \longrightarrow \underset{0}{H_2}\text{(Reduction)}$$

Think of Eq. (31) as another example of an oxidation-reduction reaction. In this reaction, the Cl atoms in Cl_2 are reduced to Cl^-, and the I^- ions are oxidized to I_2.

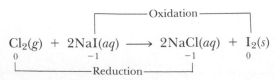

Figure 10.5
Copper undergoes a single displacement reaction with silver nitrate, $AgNO_3$ (aq). The silver crystals that form seem to grow from the surface of the copper.

Table 10.3 summarizes the principal single displacement reactions.

Two substances are displaced in **metathesis reactions** (sometimes called **double displacement** or **double replacement** reactions). Consider the following general equation for metathesis reactions.

$$AY + BX \longrightarrow AX + BY$$

Metathesis Reactions

$$AY + BX \longrightarrow AX + BY$$

In this equation, we find initially that A is combined with Y, and B is combined with X. A and B are elements in positive oxidation states; X and Y are in negative oxidation states. When AY and BX combine, A displaces B from BX and B then attaches to Y, yielding AX and BY. In all metathesis reactions the more positive component of one reactant bonds with the more negative component of the other reactant.

Many reactions that occur in aqueous solution are classified as metathesis reactions. Two of the more important aqueous reactions occur when a solid, insoluble substance, a **precipitate,** forms and when a gaseous product is produced. Equations (33) to (35) are examples of aqueous metathesis reactions in which a precipitate forms, and Eqs. (36) and (37) are examples of gas formation reactions.

$$NaBr(aq) + AgNO_3(aq) \longrightarrow AgBr(s) + NaNO_3(aq) \qquad (33)$$

$$BaCl_2(aq) + K_2SO_4(aq) \longrightarrow BaSO_4(s) + 2KCl(aq) \qquad (34)$$

$$(NH_4)_2S(aq) + Cu(C_2H_3O_2)_2(aq) \longrightarrow CuS(s) + 2NH_4C_2H_3O_2(aq) \qquad (35)$$

$$2HCl(aq) + Na_2CO_3(aq) \longrightarrow 2NaCl(aq) \\ + CO_2(g) + H_2O(l) \qquad (36)$$

$$KCN(aq) + HNO_3(aq) \longrightarrow KNO_3(aq) + HCN(g) \qquad (37)$$

In Eq. (33), aqueous solutions of sodium bromide and silver nitrate are mixed. The Ag^+ ion combines with the Br^- ion and produces the insoluble solid, AgBr (a precipitate). The Na^+ and NO_3^- ions do not combine and remain in solution. AgBr is a pale-yellow solid that settles out of solution. The precipitates that form in Eqs. (34) and (35) are $BaSO_4(s)$ and $CuS(s)$, respectively. Figure 10.6 shows the formation of precipitate in a metathesis reaction.

In the reactions that occur in Eqs. (36) and (37), gases are liberated. In each of these reactions an acid is combined with an ionic compound to yield the gaseous product: $CO_2(g)$ in Eq. (36) and $HCN(g)$ in Eq. (37). At first glance Eq. (36) might not look like a metathesis reaction. Initially, $H_2CO_3(aq)$, carbonic acid, is formed when two H^+ ions from 2HCl combine with the carbonate ion from sodium carbonate.

$$2H^+(aq) + CO_3^{2-}(aq) \longrightarrow H_2CO_3(aq)$$

Figure 10.6
A precipitation reaction occurs when a solid insoluble substance results from the combination of two compounds. Initially, crystals of the precipitate are found throughout the solution, but they ultimately settle to the bottom of the container.

Immediately, the carbonic acid decomposes to produce gaseous CO_2 and liquid H_2O.

$$H_2CO_3(aq) \longrightarrow CO_2(g) + H_2O(l)$$

Another type of metathesis reaction occurs when acids combine with bases to form salts. The members of this very important class of metathesis reaction are called **neutralization reactions.**

$$\text{Acid} + \text{base} \longrightarrow \text{salt}$$

You should recall that acids are substances that donate hydrogen ions to water and bases are metallic hydroxides. Equations (38) to (40) are examples of neutralization reactions.

TABLE 10.3 SUMMARY TABLE OF SELECTED INORGANIC REACTIONS

I. Combination reactions (A + X $\longrightarrow$ AX)

   A. Metal + nonmetal $\longrightarrow$ salt

   B. Metal + oxygen $\longrightarrow$ metal oxide (basic oxide)

   C. Nonmetal + oxygen $\longrightarrow$ nonmetal oxide (acidic oxide)

   D. Metal + hydrogen $\longrightarrow$ metal hydride

   E. Nonmetal + hydrogen $\longrightarrow$ nonmetal hydride

   F. Metal oxide + water $\longrightarrow$ base (metal hydroxide)

   G. Nonmetal oxide + water $\longrightarrow$ acid (oxyacid)

   H. Metal oxide + nonmetal oxide $\longrightarrow$ salt

II. Decomposition reactions (AX $\longrightarrow$ A + X)

   A. Oxide $\longrightarrow$ element + oxygen gas

   B. Oxide $\longrightarrow$ compound + oxygen gas

   C. Carbonate $\longrightarrow$ oxide + carbon dioxide

   D. Hydrogencarbonate $\longrightarrow$ carbonate + carbon dioxide + water

   E. Hydrate $\longrightarrow$ anhydrous salt + water

III. Single displacement reactions (A + BX $\longrightarrow$ AX + B)

   A. Active metal + water $\longrightarrow$ hydroxide (or oxide) + hydrogen gas

   B. Metal + acid $\longrightarrow$ salt solution + hydrogen gas

   C. Metal + salt solution $\longrightarrow$ displaced metal + new salt solution

   D. Metal + salt solution $\longrightarrow$ gas + new salt solution

   E. Halogen + halide solution $\longrightarrow$ displaced halogen + new halide solution

IV. Metathesis reactions (AY + BX $\longrightarrow$ AX + BY)

   A. Acid + base $\longrightarrow$ salt + water

   B. Two aqueous solutions $\longrightarrow$ precipitate + salt solution

   C. Two aqueous solutions $\longrightarrow$ gas + salt solution

   D. Acid + carbonate solution $\longrightarrow$ salt solution + carbon dioxide + water

   E. Metal oxide + acid $\longrightarrow$ salt + water

$$HCl(aq) + NaOH(aq) \longrightarrow NaCl(aq) + H_2O(l) \qquad (38)$$

$$HNO_3(aq) + KOH(aq) \longrightarrow KNO_3(aq) + H_2O(l) \qquad (39)$$

$$H_2SO_4(aq) + Mg(OH)_2(aq) \longrightarrow MgSO_4(aq) + 2H_2O(l) \qquad (40)$$

Hydrogen ions, $H^+(aq)$, from each acid in Eqs. (38) to (40) combine with the $OH^-(aq)$ from each base to produce water.

$$H^+(aq) + OH^-(aq) \longrightarrow H_2O(l)$$

Metal ions in the bases then become associated with the anions originally attached to the H atoms in the acids. We will more fully discuss neutralization reactions in Chap. 17.

See Table 10.3 for a summary of the metathesis reaction types.

10.8 Write the general form for each of the following reactions: *(a)* single displacement, *(b)* decomposition, *(c)* metathesis, *(d)* combination.

10.9 Write a balanced equation that illustrates each of the following: *(a)* combination of two compounds, *(b)* decomposition of a metal oxide, *(c)* displacement of hydrogen by a metal, *(d)* decomposition of a hydrate.

10.10 Explain how a metathesis reaction is different from a single displacement reaction. Give two examples of each.

10.11 Write three equations that illustrate different types of metathesis reactions.

10.12 Identify the reaction class to which each of the following belongs:

(a) $H_2O_2(l) \xrightarrow{\Delta} H_2O(l) + \frac{1}{2}O_2(g)$
(b) $Na(s) + H_2O(l) \longrightarrow NaOH(aq) + \frac{1}{2}H_2(g)$
(c) $P_4O_{10}(s) + 6H_2O(l) \longrightarrow 4H_3PO_4(aq)$
(d) $2Al(s) + 3Zn(NO_3)_2(aq) \longrightarrow 2Al(NO_3)_3(aq) + 3Zn(s)$
(e) $CaCl_2(aq) + Na_2CO_3(aq) \longrightarrow CaCO_3(s) + 2NaCl(aq)$

REVIEW EXERCISES

W riting equations is an important skill that is developed only through experience and practice. In this section we will concentrate on predicting the products of a reaction, given the reactants, and then balancing the equation. Use the following guidelines to help you write and balance chemical equations.

10.4 WRITING CHEMICAL EQUATIONS

Three steps are most frequently used to write chemical equations, given the reactants.

Guidelines for Writing Chemical Equations

Step 1
Classify the reaction being considered as one of the four general types of inorganic reactions. To assist you in the task of identifying the reaction class, refer to the flowchart in Fig. 10.7. Start at the top of the flowchart and answer the questions, following the arrows until the reaction class is identified.

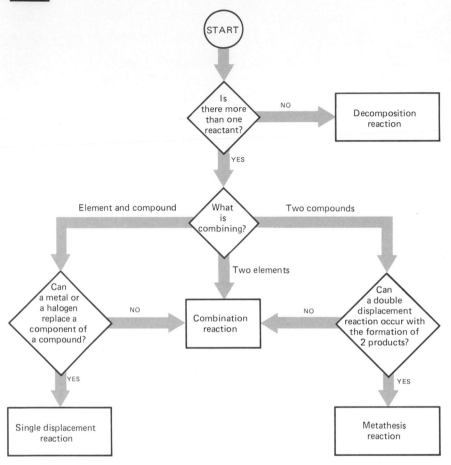

Figure 10.7
To identify the type of a specific inorganic reaction, begin at the top of the flowchart, correctly answer each question, and then proceed in the direction indicated by the answer.

Step 2

Combine the reactants in a manner consistent with the reaction type. For example, if it is a single displacement reaction, use the activity series in Table 10.2 to decide if the metal can displace the other metal ion in solution. Completion of Step 2 yields a reasonable unbalanced equation.

Step 3

Balance the equation, using the rules learned in Sec. 10.2. If the equation does not balance after you have applied all the rules, it is possible that you have written the wrong equation or the wrong formula for one of the reactants or products. Go back and double-check your reasoning and the formulas that you have written. Maybe you have omitted a substance or made a careless mistake. Figure 10.8 summarizes the three steps you must follow to write chemical equations.

Now let's use the guidelines to write equations for the reactions in the following four examples.

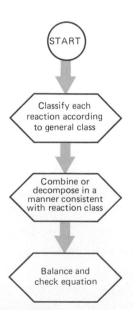

Figure 10.8
Steps to follow when you write equations given the reactants.

Example 1

Predict the products of, and write a balanced equation for, the reaction in which magnesium metal combines with oxygen.

First, write the correct symbols for the reactants to the left of an arrow. Magnesium is a metal and is written monatomically, but oxygen exists as a diatomic gas, O_2.

$$Mg + O_2 \longrightarrow \text{?}$$

We can eliminate three of the four general types of inorganic reactions. Decomposition, displacement, and metathesis reactions are eliminated because in each case at least one compound must be a reactant. Thus, the reaction is a combination reaction in which a metal (Mg) combines with a nonmetal (O_2) to produce a metal oxide.

What metal oxide forms when Mg and O_2 are combined? Magnesium is an alkaline earth metal (group IIA), and when bonded it exists in the $+2$ oxidation state; oxygen, a chalcogen (group VIA), exists in binary compounds in the -2 oxidation state.

$$Mg^{2+} \qquad O^{2-}$$

To write the formula, we use the same procedure that we did in Sec. 9.2. (If you have forgotten this procedure, review the material in Sec. 9.2.) We find the lowest common multiple of $+2$ and -2, which is just 2; consequently, magnesium and oxygen are combined in a 1-to-1 ratio; the product is MgO, and its name is magnesium oxide. Once we have this formula, we can write the unbalanced equation.

$$Mg + O_2 \longrightarrow MgO \text{ (Unbalanced)}$$

To balance the equation, we write in a 2 as the coefficient of MgO to balance the O atoms and then place a 2 in front of Mg to balance the Mg atoms.

$$2Mg + O_2 \longrightarrow 2MgO \text{ (Balanced)}$$

Example 1 shows the steps that must be followed to predict the products of reaction. Always remember that you must classify the reaction and then combine the reactants in a manner consistent with that class of reaction. In addition, you must follow the rules learned in Chap. 9 if you are to correctly write the formulas of the products. To write most formulas, you must find the lowest common multiple of the oxidation numbers of all elements and polyatomic ions and then add subscripts which will give the compound an overall oxidation number equal to zero.

Example 2

Write and balance the equation for the reaction that occurs when an aqueous solution of sodium phosphate is combined with an aqueous solution of zinc(II) chloride.

To write the formulas of the reactants, we find the oxidation number of each component. In sodium phosphate, a sodium ion has a charge and oxidation number of 1+, as do all of the alkali metals. Phosphate is a polyatomic ion with a charge of 3−.

$$Na^+ \qquad PO_4^{3-}$$

The lowest common multiple is 3; thus the formula of sodium phosphate is Na_3PO_4. The formula of zinc(II) chloride is $ZnCl_2$ because the charge on zinc is 2+ and the charge on the chloride ion is 1−. After writing the formulas, write the left side of the equation as follows:

$$Na_3PO_4(aq) + ZnCl_2(aq) \longrightarrow ?$$

In this reaction two solutions of ionic compounds are combined. Because more than one reactant is present, we rule out the possibility that the reaction is a decomposition. Metathesis and combination reactions are the only two possibilities when two compounds are combined. Combination reactions are eliminated because it is quite unlikely that four different ions (Na^+, PO_4^{3-}, Zn^{2+}, and Cl^-) would combine to produce a single compound. A more reasonable outcome is a double displacement; therefore, our reaction belongs to the metathesis class.

If this is a metathesis reaction, which of the two products precipitates out of solution? Such a problem is solved by using a water solubility table. **Solubility** is the degree to which a substance dissolves in a solvent at a constant temperature. Table 10.4 is a solubility table that lists the general solubilities of different types of inorganic compounds. To use Table 10.4, find the class of compound of interest and determine if it is soluble or insoluble. Then check to see if the compound is one of the exceptions. If the class of compound is soluble, the exceptions are insoluble and if the class of compound is insoluble, the exceptions are soluble.

To determine what the products of a metathesis reaction are, we combine the positive ion from one compound with the negative ion from the other. It is best to determine the oxidation number of each ion before proceeding. Following the rules learned in Chap. 9, we get the following:

$$\underset{+1 \quad -3}{Na_3PO_4} + \underset{+2 \ -1}{ZnCl_2} \longrightarrow ?$$

The Na^+ from sodium phosphate combines with the Cl^- from zinc(II) chloride to give NaCl. Referring to Table 10.4, we see that chlorides are soluble; thus, the sodium chloride remains in solution. The Zn, with a charge of 2+, combines with the phosphate, with a charge of 3−, to give

TABLE 10.4   SOLUBILITIES OF INORGANIC COMPOUNDS IN WATER

Class of compound	Formula	Soluble	Insoluble	Exceptions*
Acetates	$C_2H_3O_2{}^-$	x		
Ammonium	$NH_4{}^+$	x		
Bromides	Br^-	x		$AgBr$, Hg_2Br_2, $HgBr_2$, $PbBr_2$
Carbonates	$CO_3{}^{2-}$		x	Group IA, $(NH_4)_2CO_3$
Chlorates	$ClO_3{}^-$	x		
Chlorides	Cl^-	x		$AgCl$, Hg_2Cl_2, $PbCl_2$
Group IA (1)	M^+	x		
Hydroxides	OH^-		x	Group IA, $Ca(OH)_2$, $Ba(OH)_2$ $Sr(OH)_2$
Iodides	I^-	x		AgI, Hg_2I_2, HgI_2, PbI_2
Nitrates	$NO_3{}^-$	x		
Phosphates	$PO_4{}^{3-}$		x	Group IA, $(NH_4)_3PO_4$
Sulfates	$SO_4{}^{2-}$	x		$CaSO_4$, $SrSO_4$, $BaSO_4$, $PbSO_4$, Hg_2SO_4, Ag_2SO_4
Sulfides	S^{2-}		x	Groups IA and IIA, $(NH_4)_2S$
Sulfites	$SO_3{}^{2-}$		x	Group IA, $(NH_4)_2SO_3$

*When groups IA and IIA are listed, the table refers to the ionic compounds of all members of these groups.

You can also use index cards to simplify the task of learning the solubilities of compounds. Place the various groups on one side, and their solubilities and exceptions on the other side.

zinc(II) phosphate, $Zn_3(PO_4)_2$. This formula is obtained by finding the lowest common multiple (6) and then dividing each of the charges into it to obtain the subscripts. From Table 10.4, we find that phosphates are insoluble; thus, $Zn_3(PO_4)_2(s)$ precipitates from solution, and the unbalanced equation becomes

$$Na_3PO_4(aq) + ZnCl_2(aq) \longrightarrow Zn_3(PO_4)_2(s) + NaCl(aq) \text{ (Unbalanced)}$$

Balancing the equation, we obtain

$$2Na_3PO_4(aq) + 3ZnCl_2(aq) \longrightarrow Zn_3(PO_4)(s) + 6NaCl(aq) \text{ (Balanced)}$$

Example 3

Write and balance the equation for the reaction that occurs when a pure aluminum bar is placed in a mercury(II) nitrate solution.

$$Al(s) + Hg(NO_3)_2(aq) \longrightarrow ?$$

The equation states that Al, a metal, is immersed in an $Hg(NO_3)_2$ solution. Because an element combines with a compound, decomposition and metathesis reactions are easily eliminated. Decomposition reactions have only one reactant, which is a compound, and a metathesis reaction has two compounds as reactants. A combination usually does not occur between a metal and a compound; thus, this reaction is a single displacement reaction. If we check the activity series in Table 10.2, Al is found to be a more active metal than Hg; hence, Al can displace the Hg^{2+} in $Hg(NO_3)_2$. Al belongs to group IIIA and forms 3+ ions. To find the formula of aluminum nitrate, we write the two ions and determine the lowest common multiple (3) of the two. Then we divide the charges into it.

$$Al^{3+} \qquad NO_3^-$$

Hence, the formula of aluminum nitrate is $Al(NO_3)_3$, and the unbalanced equation becomes

$$Al(s) + Hg(NO_3)_2(aq) \longrightarrow Al(NO_3)_3(aq) + Hg(l) \text{ (Unbalanced)}$$

Balancing the equation, we get

$$2Al(s) + 3Hg(NO_3)_2(aq) \longrightarrow 2Al(NO_3)_3(aq) + 3Hg(l) \text{ (Balanced)}$$

Example 4

Write the equation for the reaction that occurs when solid barium carbonate is heated.

Because only one reactant is present, the reaction must be a decomposition reaction. First, write the formula of the reactant. Barium ions have a 2+ charge because barium is in group IIA, and carbonate ions have a 2− charge; hence, the formula of barium carbonate is $BaCO_3$. When group IIA carbonates are heated, $CO_2(g)$ is released and an oxide remains. Because the charge on an O atom is 2−, the formula of barium oxide is BaO. Thus, the balanced equation for this decomposition reaction is

$$BaCO_3(s) \xrightarrow{\text{heat}} BaO(s) + CO_2(g) \text{ (Balanced)}$$

REVIEW EXERCISES

10.13 Predict the general type of reaction that each of the following reactants undergoes: *(a)* two different nonmetals, *(b)* a metal oxide plus heat, *(c)* a metal oxide and a nonmetal oxide, *(d)* an acid and a base, *(e)* a nonmetal oxide and oxygen, *(f)* a halogen and a halide salt, *(g)* an acid and cyanide salt, *(h)* a carbonate plus heat.

10.14 Predict the products of the following reactions; write "NR" (no reaction) if no reaction occurs.

(a) $MgSO_4 \cdot 7H_2O \xrightarrow{\Delta}$
(b) $AgNO_3(aq) + Na_2S(aq) \longrightarrow$
(c) $P_4(s) + O_2(g) \longrightarrow$
(d) $Ag(s) + ZnSO_4(aq) \longrightarrow$
(e) $KOH(aq) + HClO_3(aq) \longrightarrow$

10.15 Predict the products of the following reactions, and write a complete balanced equation:

(a) $Ca(s) + O_2(g) \longrightarrow$

(b) $KClO_3(s) \xrightarrow{\Delta}$

(c) $SO_2(g) + H_2O(l) \longrightarrow$

(d) $Cs(l) + H_2O(l) \longrightarrow$

(e) $Pb(NO_3)_2(aq) + K_2SO_4(aq) \longrightarrow$

(f) $H_2(g) + S(s) \longrightarrow$

S o far, we have concentrated on the changes that matter undergoes during chemical reactions. In addition, we must be concerned with the energy requirements for reactions since reactions only take place when the proper amount of energy is present.

Heat, and other sources of energy, are extremely important to chemical reactions. Heat is either absorbed or liberated during the course of reactions. When heat from the surroundings is absorbed, the reaction is classified as an **endothermic reaction** ("endothermic" means "taking in heat"). The opposite of an endothermic reaction is an **exothermic reaction,** which is one that releases heat to its surroundings. Figure 10.9 depicts heat flow in endothermic and exothermic reactions.

To understand endothermic and exothermic reactions, think of the reactants as having a fixed quantity of chemical potential energy. If an endothermic reaction occurs, energy is absorbed by the reactants and is ultimately stored in the products. Therefore, the products have more chemical potential energy than the reactants. In contrast, if an exothermic reaction occurs, the reactants lose energy to the surroundings, producing products with a smaller amount of chemical potential energy. A measure of the chemical potential energy of the reactants and products is called *enthalpy*. Figure 10.10 shows the changes in enthalpy that occur in endothermic and exothermic reactions.

Endothermic reactions are sometimes expressed as follows:

$$A + B + heat \longrightarrow C + D$$

The equation tells us that heat is added to the reactants, A and B, and stored in the products, C and D. In other words, the total enthalpy of the products is greater than that of the reactants. In equations for exothermic reactions, heat is written as a product:

$$A + B \longrightarrow C + D + heat$$

10.5 ENERGY CONSIDERATIONS IN CHEMICAL REACTIONS

An exothermic reaction releases heat to the surroundings, while an endothermic reaction absorbs heat from the surroundings.

In endothermic reactions, the enthalpy of the reactants is less than the enthalpy of the products. In exothermic reactions, the enthalpy of the reactants is greater than the enthalpy of the products.

The enthalpy change in a chemical reaction is called the enthalpy of reaction, and is symbolized by writing ΔH.

$$\Delta H = H_{products} - H_{reactants}$$

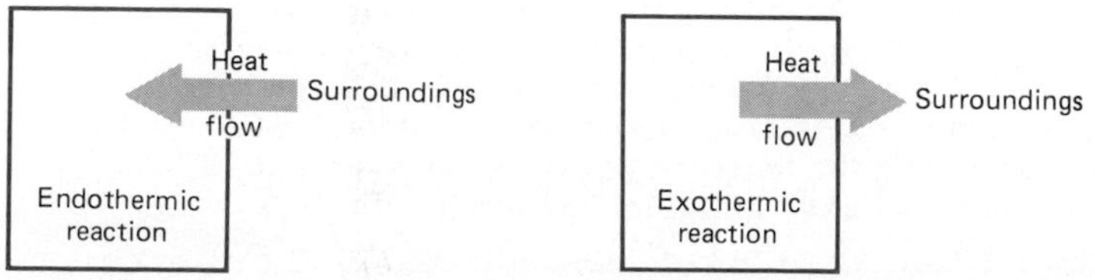

Figure 10.9
If heat flows from the surroundings to the reaction, it is classified as an endothermic reaction. If heat flows from a chemical reaction to the surroundings, the reaction is classified as an exothermic reaction.

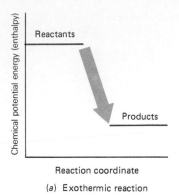

(a) Exothermic reaction

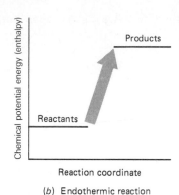

(b) Endothermic reaction

Figure 10.10
If the enthalpy of the products is less than the enthalpy of the reactants, an exothermic reaction has taken place. If the enthalpy of the products is greater than the enthalpy of the reactants, an endothermic reaction has taken place.

CHEM TOPIC: Greenhouse Effect

Within the past 20 years many scientists have become concerned with the possibility that an increase in the concentration of CO_2 in the atmosphere could have dire environmental effects. Their concern stems from the fact that the CO_2 concentration in the air has increased about 13 percent in the past 200 years, and most of the increase is the result of the increased combustion of fossil fuels. Whenever fossil fuels such as gasoline and coal are burned, they release CO_2 to the atmosphere. These fuels are mixtures of hydrocarbons, a group of compounds composed of C and H. During burning, the C in hydrocarbons is principally oxidized to CO_2, and the H is oxidized to H_2O.

Carbon dioxide has a central role in the regulation of the temperature of the atmosphere. It is essentially transparent to the visible light from the sun, but can effectively block the infrared radiation (ir) that is released by the earth. Thus an increased concentration of CO_2 in the atmosphere helps trap ir radiation that is normally radiated to outer space. Such an effect results when the glass of a greenhouse allows light in but does not allow the ir radiation out. The result is an increase in temperature inside the greenhouse. This is known as the "greenhouse effect." On the global scale, some scientists believe that too high a concentration of CO_2 in the atmosphere can result in the warming of the earth. It has been calculated that a 1- to 2-degree increase in the average temperature of the earth could raise the level of the oceans as much as 5 m. An increase of this magnitude would inundate the coastal regions, flooding the most heavily populated regions of the earth. Additionally, the greenhouse effect would change weather patterns, causing the temperature zones of the earth to move further north. Canada and the Soviet Union might have milder climates, but the United States would be more arid and desertlike.

However, other scientists are very skeptical about the "greenhouse effect." They point out that the average temperature of the earth dropped slightly between 1940 and the early 1970s. Such a drop in temperature could possibly be explained by the increase in particulate matter (particles) thrown into the atmosphere when fossil fuels are burned. These particles help reflect energy from the sun. Therefore, in upcoming years, environmentalists must carefully monitor the CO_2 concentration in the atmosphere and the temperature of the earth, so that they might warn the appropriate people before a disaster can result.

Heat is liberated in this hypothetical reaction, indicating that the enthalpy of the products is less than that of the reactants (Fig. 10.10).

In actual equations the word "heat" is not written; instead, we write the actual number of kilojoules released or absorbed. Consider the following equations:

$$CH_4(g) + 2O_2(g) \longrightarrow CO_2(g) + 2H_2O(g) + 891 \text{ kJ}$$

$$2Ag_2O(s) + 61.9 \text{ kJ} \longrightarrow 4Ag(s) + O_2(g)$$

Methane, CH_4, in the first equation, is combined with oxygen, O_2, producing carbon dioxide, CO_2; water vapor, H_2O; and 891 kJ of heat. This is one of the more significant reactions that occurs when natural gas is burned. Natural gas is used to heat houses and as a fuel for industry. Heat is by far the most valuable product of this reaction. In the second reaction, silver oxide is heated to liberate pure silver. To decompose 2 mol Ag_2O, 61.9 kJ of heat is required. Many ores that contain metal oxides and sulfides must be heated to release the metals.

Exothermic reactions that are similar to the burning of methane are called combustion reactions. A **combustion reaction** occurs when a substance combines rapidly with oxygen and releases heat and light. Combustible substances are those that undergo combustion reactions. The amount of heat liberated per mole of combustible substance is called either the **enthalpy of combustion** or **heat of combustion**. Enthalpies of combustion for selected substances are listed in Table 10.5.

TABLE 10.5 ENTHALPIES OF COMBUSTION FOR SELECTED SUBSTANCES

Substance	Formula	State	Enthalpy of combustion, $-$kJ/mol*
Benzene	C_6H_6	Liquid	3270
Carbon	C	Solid	400
Hydrogen	H_2	Gas	240
Octane	C_8H_{18}	Liquid	5450
Propane	C_3H_8	Gas	2200

*The amount of heat released in kJ/mol.

Generally, exothermic reactions are more spontaneous and self-sustaining than endothermic reactions. Heat produced by an exothermic reaction provides a constant source of energy to sustain the reaction. Endothermic reactions, in contrast, proceed only when heat is applied, and once the heat is removed, the reaction usually ceases.

Heat is absorbed or released by chemical reactions as a result of the breaking and formation of chemical bonds. The cleaving of chemical bonds is an endothermic process, while bond formation is an exothermic process. Therefore, exothermic reactions are those that release more heat as a result of bond formation than is required to break bonds. In other words, stronger bonds are found in the products than in the reactants. Endothermic reactions are the opposite: more energy is required

Bond breaking is an endothermic process. In contrast, bond formation is exothermic.

to break bonds in the reactants than is given off during the reaction—the bonds are stronger in the reactants than in the products.

10.16 How is an exothermic reaction different from an endothermic reaction?

10.17 On what side of the arrow is the heat written in *(a)* an exothermic and *(b)* an endothermic reaction?

10.18 Classify each of the following as either an exothermic or an endothermic reaction:

(a) $H_2 + Cl_2 \longrightarrow 2HCl + 184 \text{ kJ}$
(b) $Cu_2O + 167 \text{ kJ} \longrightarrow 2Cu + \frac{1}{2}O_2$
(c) $I_2 + Br_2 + 84 \text{ kJ} \longrightarrow 2IBr$
(d) $C + O_2 \longrightarrow CO_2 + 393 \text{ kJ}$

10.19 Explain what happens, in terms of bond breaking and formation, when an endothermic reaction occurs.

SUMMARY

Chemical equations are a shorthand notation that chemists use to indicate what happens during chemical reactions. Each equation shows formulas of substances separated by an arrow (which means "yields"). All substances written to the left of the arrow are called **reactants,** and all substances to the right of the arrow are called **products.** Each chemical equation is initially worked out in the laboratory; chemists perform experiments to determine what products result from a given set of reactants and conditions.

Chemical equations obey the law of conservation of mass; this law states that matter cannot be created or destroyed during normal chemical changes. This law is obeyed when an equation is balanced. A **balanced equation** indicates the ratio in which the reactants combine to yield the products, as well as the ratio in which the products are formed. Equations are balanced by changing the coefficients in front of each substance in the equation. Usually, the correct coefficients are determined by comparing the number of atoms of each different element on either side of the equation, using what is called the **inspection method.**

To better understand reactions, chemists group similar equations into a general class. Elementary inorganic reactions are normally separated into four classes. They are (1) combination, (2) decomposition, (3) single displacement, and (4) metathesis. **Combination reactions** are those in which two or more reactants unite to produce a single product. **Decomposition reactions** are the opposite of combination reactions. A single reactant is broken up into two or more products. **Single displacement** reactions are those in which an element replaces another element within a compound, resulting in a compound of the initially free element. Finally, in **metathesis reactions** (double displacements), two compounds react and the more positive part of one combines with the negative part of the other.

Reactions are classified according to whether heat is released or taken in during the course of the reaction. Those reactions in which heat is absorbed are called **endothermic reactions.** Reactions in which there is a net release of heat energy are called **exothermic reactions.**

KEY TERMS

activity series
aqueous solution
catalyst
chemical equation
combination reaction
decomposition reaction

endothermic
enthalpy
equation balancing
exothermic
hydrate

hydride
metathesis reaction
oxidation
oxide
precipitation

product
reactant
redox
reduction
single displacement reaction

EXERCISES*

10.20 Define each of the following terms: chemical equation, reactant, product, aqueous solution, catalyst, equation balancing, combination reaction, decomposition reaction, single displacement reaction, metathesis reaction, oxidation, reduction, oxide, hydride, hydrate, activity series, precipitation, exothermic, endothermic, enthalpy.

Format of Chemical Equations

10.21 What is the meaning of each of the following symbols in chemical equations: (a) (aq), (b) (g), (c) $\xrightarrow{\text{cat}}$?

10.22 Write the names of all reactants and products in the following reaction:

$CaCl_2 + (NH_4)_2CO_3 \longrightarrow CaCO_3 + 2NH_4Cl$

10.23 Write a word equation that expresses exactly what is indicated by each of the following chemical equations:

(a) $2SO_3(g) \xrightarrow{\Delta} 2SO_2(g) + O_2$

(b) $Hg(l) + Cl_2(g) \longrightarrow HgCl_2$

(c) $N_2(g) + 3H_2 \xrightarrow{\Delta} 2NH_3(g)$

(d) $Al_2S_3(s) + 6H_2O(l) \longrightarrow$
$\qquad 2Al(OH)_3(s) + 3H_2S(aq)$

Balancing Equations by Inspection

10.24 Balance each of the following equations:

(a) $P_4 + O_2 \longrightarrow P_2O_5$
(b) $Mg + N_2 \longrightarrow Mg_3N_2$
(c) $Li_2O + H_2O \longrightarrow LiOH$
(d) $Cl_2 + KI \longrightarrow I_2 + KCl$
(e) $Cu + O_2 \longrightarrow CuO$
(f) $FeO + SiO_2 \longrightarrow FeSiO_3$
(g) $Na_2CO_3 + C \longrightarrow Na + CO$
(h) $WO_3 + H_2 \longrightarrow W + H_2O$
(i) $B_2O_3 + H_2O \longrightarrow H_3BO_3$
(j) $H_2S + O_2 \longrightarrow SO_2 + H_2O$

10.25 Balance each of the following equations:

(a) $C_4H_{10} + O_2 \longrightarrow CO_2 + H_2O$
(b) $POF_3 + H_2O \longrightarrow H_3PO_4 + HF$
(c) $Cu(NO_3)_2 \longrightarrow CuO + NO_2 + O_2$
(d) $CaCO_3 + H_3PO_4 \longrightarrow Ca_3(PO_4)_2 +$
$\qquad CO_2 + H_2O$
(e) $FeS_2 + O_2 \longrightarrow FeO + SO_2$
(f) $Al + CuSO_4 \longrightarrow Al_2(SO_4)_3 + Cu$
(g) $LiH + AlCl_3 \longrightarrow LiAlH_4 + LiCl$

(h) $Cr_2O_3 + C \longrightarrow Cr + CO$
(i) $(NH_4)_2Cr_2O_7 \longrightarrow Cr_2O_3 + N_2 + H_2O$
(j) $C_{10}H_{22} + O_2 \longrightarrow CO_2 + H_2O$

10.26 Balance each of the following equations:

(a) $B_3N_3H_6 + O_2 \longrightarrow N_2O_5 + B_2O_3 + H_2O$
(b) $IBr + NH_3 \longrightarrow NH_4Br + NI_3$
(c) $KAlSi_3O_8 + H_2O + CO_2 \longrightarrow K_2CO_3 +$
$\qquad Al_2Si_2O_5(OH)_4 + SiO_2$
(d) $Co_3O_4 + Al \longrightarrow Co + Al_2O_3$
(e) $Al(OH)_3 + NaOH \longrightarrow NaAlO_2 + H_2O$
(f) $C_7H_6O_2 + O_2 \longrightarrow CO_2 + H_2O$
(g) $HClO_4 + P_4O_{10} \longrightarrow H_3PO_4 + Cl_2O_7$
(h) $XeF_2 + H_2O \longrightarrow Xe + O_2 + HF$
(i) $Na_2H_3IO_6 + AgNO_3 \longrightarrow$
$\qquad Ag_5IO_6 + NaNO_3 + HNO_3$
(j) $XeF_4 + SF_4 \longrightarrow Xe + SF_6$

10.27 Translate each of the following word equations into a balanced chemical equation:

(a) Sodium bromide solution + silver nitrate solution yield aqueous sodium nitrate + silver(I) bromide solid

(b) Aluminum hydroxide solution + nitric acid yield aqueous aluminum nitrate + water

(c) Chlorine gas + rubidium iodide solution yield aqueous rubidium chloride + iodine solid

(d) Iron(III) acetate solution + sodium sulfide solution yield aqueous sodium acetate + iron(III) sulfide solid

(e) Silicon tetrafluoride gas + water yield silicon dioxide solid + hydrofluoric acid

(f) Manganese(IV) oxide solid + hydrochloric acid yield manganese(II) chloride + chlorine gas + water

(g) Dinitrogen tetroxide gas, when heated, yields nitrogen dioxide.

(h) Calcium phosphide solid + water yield calcium hydroxide + phosphine (PH₃) gas

(i) On heating, silver(I) nitrate solid yields silver, plus nitrogen dioxide gas, plus oxygen gas.

(j) Aluminum metal + aqueous copper(II) sulfate yield copper metal + aqueous aluminum sulfate

(k) Mercury(I) nitrate solution + potassium chloride solution yield mercury(I) chloride solid + aqueous potassium nitrate

(l) Calcium sulfate dihydrate, when heated, yields calcium sulfate plus water.

*For exercise numbers printed in color, answers can be found at the back of the book.

(m) Ammonium sulfide solution + cadmium(II) chloride solution yield cadmium sulfide solid + aqueous ammonium chloride
(n) Phosphorous acid + sodium hydroxide solution yield aqueous sodium phosphite + water
(o) Ammonia gas + sulfuric acid yield aqueous ammonium sulfate
(p) Carbon disulfide liquid + chlorine gas yield disulfur dichloride + carbon tetrachloride
(q) Calcium phosphate solution + sulfuric acid yield phosphoric acid + calcium sulfate
(r) Aqueous ammonia + copper(I) oxide yield nitrogen gas + water + copper metal
(s) Ammonium nitrate solid, when heated, yields water and dinitrogen oxide gas.
(t) Magnesium hydroxide solution + zinc(II) nitrate solution yield zinc hydroxide solid + aqueous magnesium nitrate.

10.28 (a) Classify each of the reactions in Exercise 10.27 as combination, decomposition, single displacement, metathesis, or none of these. (b) Which of the reactions in 10.27 are oxidation-reduction reactions?

Classes of Inorganic Reactions

10.29 Write an equation to illustrate each of the following types of combination reactions: (a) A compound combines with an element, (b) two compounds combine, (c) two elements combine.

10.30 Write equations for a combination reaction in which each of the following products results: (a) CO, (b) SO_3, (c) NO_2, (d) Cs_2O, (e) PH_3, (f) MgH_2.

10.31 What metal oxide (basic oxide), when added to water, produces each of the following bases: (a) KOH, (b) $Ba(OH)_2$, (c) $Ca(OH)_2$, (d) $Al(OH)_3$?

10.32 What nonmetal oxide (acidic oxide), when added to water, produces each of the following acids: (a) H_2CO_3, (b) H_2SO_3, (c) H_3PO_4, (d) H_2SO_4?

10.33 Write formulas for the compound that decomposes to each of the following sets of products:
(a) $MgO + CO_2$
(b) $KCl + O_2$
(c) $K_2CO_3 + CO_2 + H_2O$

(d) $Ba(NO_2)_2 + H_2O$
(e) $H_2 + O_2$
(f) $NaNO_2 + O_2$
(g) $SO_2 + O_2$

10.34 (a) What is the common feature of all oxidation-reduction reactions? (b) Give an example of an oxidation-reduction reaction.

10.35 For each of the following reactions, determine which reactants undergo oxidation and which undergo reduction:
(a) $2Li + 2HCl \longrightarrow H_2 + 2LiCl$
(b) $Si + 2F_2 \longrightarrow SiF_4$
(c) $H_2 + Cl_2 \longrightarrow 2HCl$
(d) $2ZnS + 3O_2 \longrightarrow 2SO_2 + 2ZnO$
(e) $Fe_3O_4 + 4H_2 \longrightarrow 3Fe + 4H_2O$

10.36 Write a balanced equation for a single displacement reaction that illustrates each of the following: (a) A metal displaces hydrogen from water, releasing $H_2(g)$; (b) a metal replaces hydrogen in an acid, releasing $H_2(g)$; (c) A metal replaces copper in a copper nitrate solution; (d) bromine liquid replaces iodine in a solution of rubidium iodide.

10.37 (a) What substances combine in a neutralization reaction? (b) Give three examples of neutralization reactions.

10.38 Write an equation that illustrates a metathesis reaction in which (a) an insoluble hydroxide forms, (b) carbon dioxide gas is one of the products, (c) an insoluble carbonate is produced, (d) the soluble salt NH_4I results.

10.39 Use Table 10.4 to predict whether each of the following is soluble or insoluble in water:
(a) K_2SO_4, (b) $Fe(OH)_2$, (c) $MgCO_3$, (d) Hg_2Cl_2, (e) NH_4I, (f) VO_2, (g) $CsC_2H_3O_2$, (h) Al_2S_3, (i) AgI, (j) $CaSO_4$, (k) $FeSO_3$, (l) KBr, (m) $(NH_4)_3PO_4$.

Writing Chemical Equations

10.40 Complete and balance each of the following equations for combination reactions:
(a) $Zn + S_8 \longrightarrow$
(b) $Al(s) + O_2(g) \longrightarrow$
(c) $Na(s) + F_2(g) \longrightarrow$
(d) $H_2(g) + N_2(g) \longrightarrow$
(e) $Br_2(g) + I_2(g) \longrightarrow$
(f) $CO(g) + O_2(g) \longrightarrow$
(g) $Mg(s) + N_2(g) \longrightarrow$
(h) $Ag(s) + Cl_2(g) \longrightarrow$

10.41 Complete and balance the equations for each of the following decomposition reactions:

(a) $PtO_2 \xrightarrow{\Delta}$

(b) $CuSO_4 \cdot 5H_2O \xrightarrow{\Delta}$

(c) $Fe_2O_3 \xrightarrow{\Delta}$

(d) $SrCO_3 \xrightarrow{\Delta}$

(e) $LiHCO_3 \xrightarrow{\Delta}$

(f) $H_2O \xrightarrow{elec}$

(g) $MgSO_3 \cdot 6H_2O \xrightarrow{\Delta}$

(h) $H_2O_2 \xrightarrow{\Delta}$

10.42 Complete and balance the equation for each of the following single displacement reactions:
(a) $Zn(s) + Pb(NO_3)_2(aq) \longrightarrow$
(b) $Ba(s) + H_2O(l) \longrightarrow$
(c) $Ni(s) + SnBr_2(aq) \longrightarrow$
(d) $Hg(l) + Fe(NO_3)_2(aq) \longrightarrow$
(e) $KCl(aq) + I_2(s) \longrightarrow$
(f) $Cu(ClO_3)_2(aq) + Mn(s) \longrightarrow$
(g) $Pb(s) + H_2SO_4(aq) \longrightarrow$
(h) $HC_2H_3O_2(aq) + Zn(s) \longrightarrow$

10.43 Complete and balance the equations for the following aqueous metathesis reactions:
(a) $NiCl_2(aq) + Ca(OH)_2(aq) \longrightarrow$
(b) $Hg(C_2H_3O_2)_2(aq) + K_2CO_3(aq) \longrightarrow$
(c) $H_3PO_4(aq) + AgNO_3(aq) \longrightarrow$
(d) $H_2SO_3(aq) + Al(OH)_3(aq) \longrightarrow$
(e) $(NH_4)_2S(aq) + BaI_2(aq) \longrightarrow$
(f) $Cs_2CO_3(aq) + HC_2H_3O_2(aq) \longrightarrow$
(g) $Li_2SO_4(aq) + Co(NO_3)_2(aq) \longrightarrow$
(h) $NH_4CN(aq) + HBr(aq) \longrightarrow$

10.44 Translate each of the following to symbols and formulas, and then complete and balance the equation:
(a) Silver(I) nitrate + copper(II) chloride $\longrightarrow$
(b) Iron + water $\longrightarrow$
(c) Potassium hydroxide + sulfuric acid $\longrightarrow$
(d) Ammonia + oxygen gas $\xrightarrow{\Delta}$
(e) Sulfur trioxide + water $\longrightarrow$
(f) Sulfuric acid + zinc $\longrightarrow$
(g) Nitrogen monoxide + oxygen $\longrightarrow$
(h) Aluminum + oxygen gas $\longrightarrow$
(i) Hydrogen gas + tin(II) nitrate $\longrightarrow$

(j) Sodium sulfate decahydrate $\xrightarrow{\Delta}$
(k) Ammonia + hydrochloric acid $\longrightarrow$
(l) Sodium nitrite + hydrochloric acid $\longrightarrow$
(m) Sodium acetate + lead(II) acetate $\longrightarrow$
(n) Tin (II) chloride + iron $\longrightarrow$
(o) Arsenic + chlorine gas $\longrightarrow$
(p) Cesium carbonate + iron(III) acetate $\longrightarrow$
(q) Lithium oxide + water $\longrightarrow$
(r) Manganese(II) chloride + magnesium $\longrightarrow$
(s) Nitric acid + calcium oxide $\longrightarrow$
(t) Boron + fluorine gas $\longrightarrow$

Energy in Chemical Reactions

10.45 Give an example of (a) an exothermic reaction and (b) an endothermic reaction.

10.46 (a) Use the data in Table 10.5 to calculate the amount of heat liberated when 1.00 g of octane is combusted. (b) If all of this heat could be transferred to 1.00 kg of water at 25°C, what is the maximum temperature that the water would reach? [The specific heat of water is 4.184 J/(g·°C).]

10.47 Compressed propane is used as a fuel in homes and industry. What mass of propane would have to be completely combusted to produce as much heat as the complete combustion of 1.56 kg of octane? (*Hint:* Use Table 10.5.)

10.48 Benzene is an aromatic liquid that is toxic and cancer-causing. (a) Write the equation for the complete combustion of benzene, C_6H_6. (b) How much heat is released when 9.01 g of benzene is completely combusted? (c) How much heat is produced when 154 g CO_2 is released during the complete combustion of benzene? (d) How many grams of benzene must be combusted to increase the temperature of 152 g of water from 0.0°C to 73.9°C?

10.49 Classify each of the following reactions as endothermic or exothermic:
(a) $PCl_5 + 376 \text{ kJ} \longrightarrow P + \frac{5}{2}Cl_2$
(b) $CaCO_3 + 180 \text{ kJ} \longrightarrow CaO + CO_2$
(c) $FeS + 2H^+ \longrightarrow Fe^{2+} + H_2S + 13 \text{ kJ}$
(d) $C + 2F_2 \longrightarrow CF_4 + 920 \text{ kJ}$

10.50 (a) What is enthalpy? (b) What enthalpy change occurs in an endothermic reaction? (c) What enthalpy change occurs in an exothermic reaction?

10.51 Most decomposition reactions are endothermic. Write an explanation to account for this observation.

Additional Exercises

10.52 Correct the following incorrect statements:

(a) Most metal oxides produce acidic solutions when placed in water.

(b) All inorganic reactions belong to one of four different classes.

(c) Fractional coefficients can never be used to balance chemical equations.

(d) All phosphate salts are insoluble.

(e) Zinc metal, when placed in liquid water, replaces hydrogen and dissolves immediately.

(f) The product of a metathesis reaction is a precipitate.

(g) Substances that undergo endothermic reactions are used as fuels.

(h) When bonds are broken, heat is released.

(i) Oxidation occurs when electrons are accepted.

(j) Endothermic reactions release heat and exothermic reactions absorb heat.

10.53 Write and balance the following equations:

(a) The Haber reaction occurs when nitrogen and hydrogen gases combine at high pressure, at 550°C, and with a metal catalyst to produce ammonia gas.

(b) Ammonia, when combusted, gives nitrogen monoxide and water vapor (Ostwald process).

(c) Nitrogen monoxide is further oxidized to nitrogen dioxide.

(d) Nitrogen dioxide is pumped through water to yield both nitric acid and nitrous acid.

10.54 Write and balance equations for reactions 1 through 7.

(1) Calcium carbonate, when heated to about 850°C, decomposes to calcium oxide and carbon dioxide.

(2) The carbon dioxide is used to manufacture sodium bicarbonate and sodium carbonate. Carbon dioxide is combined with ammonia, water, and sodium chloride to produce sodium bicarbonate plus ammonium chloride (the Solvay process).

(3) Sodium carbonate is formed along with carbon dioxide and water when sodium bicarbonate is decomposed.

(4) The calcium oxide produced in Eq. (1) is combined with carbon at high temperatures to produce calcium carbide and carbon monoxide.

(5) Calcium carbide is then heated with nitrogen gas at 1100°C, producing calcium cyanamid, $CaCN_2$, and carbon.

(6) This calcium cyanamid is combined with carbon and the sodium carbonate produced in Eq. (3) to give sodium cyanide, NaCN, and calcium carbonate.

(7) Sodium cyanide is the main source of hydrogen cyanide gas. Sodium cyanide is combined with sulfuric acid, producing hydrogen cyanide and sodium sulfate.

10.55 Balance each of the following equations, noting that in each reaction calcium fluorophosphate is one of the reactants:

(a) $Ca_5(PO_4)_3F + H_2SO_4 \longrightarrow$
$$HF + H_3PO_4 + CaSO_4$$

(b) $Ca_5(PO_4)_3F + H_3PO_4 \longrightarrow$
$$HF + Ca(H_2PO_4)_2$$

(c) $Ca_5(PO_4)_3F + H_2SO_4 + H_2O \longrightarrow$
$$HF + CaSO_4 + Ca(H_2PO_4)_2 \cdot H_2O$$

10.56 When acetylene, C_2H_2, is completely combusted, it produces carbon dioxide and water. For each 2 mol of acetylene combusted, 2602 kJ of energy is released. (a) Write and balance the equation for the combustion of acetylene. (b) Calculate the amount of energy released when 40.0 g of acetylene is combusted. (c) If all the heat released by 40.0 g of acetylene is transferred to 1.35 kg of water at 10.0°C, what is the final temperature of the water? [The specific heat of water is 4.184 J/(g·°C).] (d) How many grams of acetylene must be combusted to produce 125 kJ of energy?

CHAPTER
Eleven

Stoichiometry

STUDY GUIDELINES

After completing Chapter 11, you should be able to

1. Explain the meaning of a stoichiometric relationship

2. List reasons why chemists obtain experimentally quantities different from those predicted by stoichiometric calculations

3. Determine molecular and mole relationships given a balanced chemical equation

4. Show that, in a correctly balanced equation, the sum of the masses of the reactants equals the sum of the masses of the products

5. Calculate the number of moles or mass of product given either the number of moles or mass of a reactant, or vice versa

6. Calculate the amount of heat liberated or consumed in a reaction given the mass or number of moles of starting material, the balanced equation, and the enthalpy of reaction

7. Explain how heat transfers are measured through calorimetry

8. Find the limiting reagent in a reaction

9. Calculate the maximum mass of product formed given the masses of all reactants

10. Calculate the percent yield of a reaction given the actual yield and masses of reactants

---

W hat is stoichiometry (stoy-key-ahm-uh-tree)? **Stoichiometry** is the study of mole, mass, energy, and volume relationships in chemical reactions. In the study of stoichiometry, we usually consider the quantities of reactants that combine to produce various amounts of products.

Many industries make use of stoichiometric relationships to predict the masses of raw materials required to produce the desired amounts of final products. Knowledge of stoichiometry is applied to production problems, e.g., the recovery of metals from ores, the synthesis of medicines and drugs, and the manufacture of explosives.

11.1 STOICHIOMETRY

The term "stoichiometry" comes from the Greek terms stoicheion, *which means "element," and* metron, *"to measure."*

Stoichiometric relationships are found by investigating a chemical reaction in the laboratory, and then determining the ratios in which the reactants combine and the ratios in which the products form. When these ratios are known, the correct balanced equation is written. As we saw in Chap. 10, the coefficients placed in front of symbols and formulas in chemical equations indicate the theoretical ratio in which reactants combine and products form.

A word of caution is in order before we get to the specifics of stoichiometry. Predictions made from stoichiometric considerations are theoretical; this means our answers are not what *will* necessarily happen, but what *would* happen if the reaction proceeded to completion as written. If we predict the quantity of a product that forms from specified amounts of reactants, our prediction is the theoretical maximum that could result; however, only rarely will the maximum quantity be obtained in the laboratory.

Even though a chemical equation may appear to represent a simple relationship, it generally does not. Reactants most frequently have more than one possible pathway they can follow as they react. In other words, one or more "side reactions" may occur at the same time as the principal one, thus decreasing the yield of the product of interest. Small quantities of impurities may alter the pathway of the reaction—sometimes significantly. Additionally, meticulous attention must be given to the energy requirements of a reaction. If these requirements are not met, different products may result or the reaction may not occur at all.

Jeremias Benjamin Richter (1762–1807) first coined the term "stoichiometry." He applied basic math principles to what was known about the combining masses of compounds to explain how substances reacted quantitatively.

11.1 (a) What is stoichiometry? (b) How is stoichiometry applied?

11.2 What factors tend to decrease the observed yield of a reaction compared to the computed theoretical yield?

To begin our study of stoichiometry, let's consider a relatively simple chemical reaction, the combination of hydrogen and chlorine gases to produce hydrogen chloride gas:

$$H_2(g) + Cl_2(g) \longrightarrow 2HCl(g)$$

11.2 CHEMICAL EQUATION CALCULATIONS

In words, the equation states that one hydrogen molecule combines with one chlorine molecule to produce two hydrogen chloride molecules. The reaction is diagrammed in Fig. 11.1.

Knowing the coefficients for the equation, we already have a wealth of information. As previously stated, the coefficients give an indication of what is happening quantitatively at the molecular level—one H_2 molecule combines with one Cl_2 molecule and produces two HCl molecules.

In the laboratory we cannot work with individual molecules because they are much too small; instead, we are concerned with the number of moles of reactants and products. We can move from the molecular level to the mole level by multiplying all components of the equation by Avo-

The formation of HCl is the result of a series of intermediate steps that the reactants follow. These steps are called the reaction mechanism.

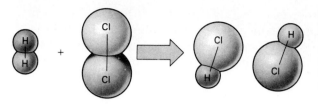

Figure 11.1
One molecule of diatomic hydrogen, H_2, collides with one molecule of diatomic chlorine, Cl_2, producing two molecules of hydrogen chloride, HCl.

gadro's number, 6.022×10^{23}. Consequently, 6.022×10^{23} H_2 molecules combine with 6.022×10^{23} Cl_2 molecules to produce 1.204×10^{24} ($2 \times 6.022 \times 10^{23}$) HCl molecules. More simply, 1 mol H_2 combines with 1 mol Cl_2 to yield 2 mol HCl.

$$1 \text{ mol } H_2 + 1 \text{ mol } Cl_2 \longrightarrow 2 \text{ mol HCl}$$

Mass relationships follow directly from the mole relationships. Accordingly, 1.0 mol H_2, 2.0 g H_2, combines with 1.00 mol Cl_2, 70.9 g Cl_2, to produce 2.00 mol HCl, 72.9 g HCl. You should note that the law of conservation of mass is upheld; the sum of the masses of the reactants, 72.9 g (2.0 g H_2 + 70.9 g Cl_2), equals the mass of the product, 72.9 g.

Table 11.1 summarizes the molecule, mole, and mass relationships relevant to the formation of HCl from the elements.

Hydrogen chloride is a colorless, poisonous gas. Its melting point is $-112°C$, and its boiling point is $-84°C$. If HCl is dissolved in water, it almost totally breaks up into ions—this solution is called hydrochloric acid.

TABLE 11.1   REACTION STOICHIOMETRY FOR THE FORMATION OF HYDROGEN CHLORIDE, HCl

	H_2	+	Cl_2	$\longrightarrow$	2HCl
Molecules	1 molecule		1 molecule		2 molecules
Molecules	6.02×10^{23} molecules		6.02×10^{23} molecules		$2 \times 6.02 \times 10^{23}$ molecules
Moles	1 mol		1 mol		2 mol
Mass	2.0 g		70.9 g		72.9 g

Given the balanced equation and quantities of starting materials, we can calculate the masses, the numbers of moles, or the numbers of molecules of products (or vice versa). To show this, let's calculate the number of moles of Cl_2 needed to combine with 2 mol H_2, and the number of moles of HCl that result.

Because H_2 and Cl_2 combine in a 1-to-1 mole ratio, exactly the same number of moles Cl_2 as moles of H_2 are required. Thus, 2 mol Cl_2 combines with 2 mol H_2. For each 1 mol of reactant, 2 mol of product, HCl, is formed. Consequently, 4 mol HCl is produced. In terms of mass, 2 mol H_2 (2 mol $H_2 \times 2.0$ g H_2/mol), 4.0 g H_2, combines with 2 mol Cl_2 (2 mol $Cl_2 \times 70.9$ g Cl_2/mol), 141.8 g Cl_2, to produce 4 mol HCl, (4 mol HCl $\times 36.45$ g HCl/mol), 145.8 g HCl.

	H_2	+	Cl_2	$\longrightarrow$	2HCl
Moles	2 mol		2 mol		4 mol
Mass	4.0 g		141.8 g		145.8 g

It should be apparent that the coefficients of the reactants and products in a chemical equation give the mole ratios in which substances combine. Thus, we use the coefficients to generate conversion factors. For example, if we are interested in the number of moles of HCl produced per mole of H_2, it is only necessary to write

$$\frac{2 \text{ mol HCl}}{1 \text{ mol } H_2}$$

A similar conversion factor relates moles of Cl_2 to moles of HCl produced:

$$\frac{2 \text{ mol HCl}}{1 \text{ mol } Cl_2}$$

The mole relationship between the two reactants is expressed as follows:

$$\frac{1 \text{ mol } Cl_2}{1 \text{ mol } H_2}$$

Turning our attention to the equation that represents the formation of ammonia, NH_3, from its elements (the Haber reaction), we will illustrate stoichiometry problems that exclusively involve mole relationships.

$$N_2(g) + 3H_2(g) \xrightarrow[\text{cat}]{\text{heat}} 2NH_3(g)$$

In the Haber reaction, 1 mol $N_2(g)$ combines with 3 mol H_2 to yield 2 mol NH_3.

How many moles of H_2 are required to exactly combine with 10 mol N_2? Utilizing conversion factors from the equation, we find that 3 mol H_2 is required per 1 mol N_2:

$$\frac{3 \text{ mol } H_2}{1 \text{ mol } N_2}$$

To answer our question, we multiply this conversion factor by the number of moles of N_2 given, 10.

$$10 \text{ mol } N_2 \times \frac{3 \text{ mol } H_2}{1 \text{ mol } N_2} = 30 \text{ mol } H_2$$

We find that 30 mol H_2 is required to combine with 10 mol N_2.

If 30 mol H_2 is needed to combine with 10 mol N_2, how many moles of NH_3 result? Again, look at the equation, and extract from it the mole relationship between either moles of N_2 and moles of NH_3 produced or moles of H_2 and moles NH_3 produced.

Mole–Mole Calculations

Fritz Haber (1868–1934), a German chemist, developed a method of mixing N_2 and H_2 under pressure with an iron catalyst to produce ammonia. Unfortunately, this discovery prolonged World War I, because ammonia can be converted to explosives. Before Haber's discovery, Germany's capacity to produce explosives was limited because most explosives were made from nitrate-rich deposits found in Chile, a region not accessible to Germany because of a blockade by the British navy.

$$\frac{2 \text{ mol NH}_3}{1 \text{ mol N}_2} \quad \text{or} \quad \frac{2 \text{ mol NH}_3}{3 \text{ mol H}_2}$$

Using either of these conversion factors gives us the correct answer:

$$10 \text{ mol N}_2 \times \frac{2 \text{ mol NH}_3}{1 \text{ mol N}_2} = 20 \text{ mol NH}_3$$

or

$$30 \text{ mol H}_2 \times \frac{2 \text{ mol NH}_3}{3 \text{ mol H}_2} = 20 \text{ mol NH}_3$$

Thus 20 mol NH_3 is the theoretical maximum yield when 10 mol N_2 combines with 30 mol H_2. Chemists call this calculated amount of product the **theoretical yield.**

Example Problems 11.1 and 11.2 illustrate the application of the factor-label method and our problem-solving techniques to the solution of mole–mole stoichiometry problems. Refer to Fig. 11.2, which graphically illustrates how mole–mole problems are solved.

Theoretical yields are the predicted amounts of products formed by applying the principles of stoichiometry.

——————————— **Example Problem 11.1** ———————————

Calculate the theoretical maximum number of moles of NH_3 that result when 0.55 mol H_2 combines with an excess amount of N_2.

$$N_2(g) + 3H_2(g) \longrightarrow 2NH_3(g)$$

——————————— **Solution** ———————————

1. What is unknown?   Mol NH_3
2. What is known?   0.55 mol H_2; 2 mol NH_3/3 mol H_2
3. Apply the factor-label method.

$$0.55 \text{ mol H}_2 \times \frac{2 \text{ mol NH}_3}{3 \text{ mol H}_2} = ? \text{ mol NH}_3$$

4. Perform the indicated math operations.

$$0.55 \text{ mol H}_2 \times \frac{2 \text{ mol NH}_3}{3 \text{ mol H}_2} = \textbf{0.37 mol NH}_3$$

Figure 11.2
To calculate the number of moles of product formed given the number of moles of a reactant, multiply the number of moles of reactant by the conversion factor that gives the number of moles of product per mole of reactant. This conversion factor is obtained from the coefficients of the reactant and product of interest in the balanced equation.

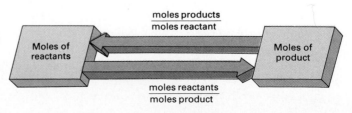

Theoretically, 0.37 mol NH_3 results when 0.55 mol H_2 combines with excess N_2 in the Haber reaction. An assumption is made that an excess amount of N_2 is present or at least the exact amount needed to combine with the 0.55 mol H_2; if not, the maximum product yield could not be obtained. If 0.0 g N_2 is present, the reaction does not take place! A common practice is to assume that sufficient quantities of all the other reactants are present unless we are told otherwise.

If not stated, assume in stoichiometry calculations that sufficient quantities of all substances are present for the reaction to take place.

Example Problem 11.2

Butane, C_4H_{10}, is a combustible gas used as a fuel and is found in some cigarette lighters. How many moles of butane are required to combine with excess oxygen, O_2, to produce 6.44 mol of carbon dioxide, CO_2, in the following reaction:

Butane is a colorless gas that has a boiling point of $-0.3°C$. It is a liquid in lighters only because it is under pressure.

$$C_4H_{10}(g) + 6.5O_2(g) \longrightarrow 4CO_2(g) + 5H_2O(g)$$

Solution

1. What is unknown?   Mol C_4H_{10}
2. What is known?   6.44 mol CO_2; 1 mol C_4H_{10}/4 mol CO_2
3. Apply the factor-label method.

$$6.44 \text{ mol CO}_2 \times \frac{1 \text{ mol } C_4H_{10}}{4 \text{ mol CO}_2} = ? \text{ mol } C_4H_{10}$$

4. Perform the indicated math operations

$$6.44 \text{ mol CO}_2 \times \frac{1 \text{ mol } C_4H_{10}}{4 \text{ mol CO}_2} = \textbf{1.61 mol } \mathbf{C_4H_{10}}$$

1.61 mol C_4H_{10} must be present to combine with excess oxygen to produce 6.44 mol CO_2.

11.3 Use the Haber equation to calculate the number of moles of NH_3 produced when 8.73 mol N_2 combines with excess H_2.

11.4 (a) Calculate the number of moles of O_2 that combine exactly with 0.922 mol C_4H_{10} to produce CO_2 and H_2O. (b) How many moles of CO_2 and H_2O are produced from 0.922 mol C_4H_{10}? (c) Are moles conserved in chemical reactions?

11.5 How many moles of H_2O and CO_2 are released when 7.35×10^{-3} mol C_4H_{10} is combined with excess oxygen in the combustion of butane?

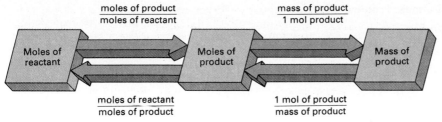

Figure 11.3
Two conversion factors are required to calculate the mass of product formed from a given number of moles of reactant. They are moles of product per moles of reactant, obtained from the balanced equation, and molar mass of the product, mass of product per 1 mol product.

Usually, chemists calculate the masses of reactants and products. One additional conversion factor is needed to find masses, the molar mass of the reactant or product (grams per mole). Figure 11.3 shows the pathway you will follow to solve mole–mass problems.

An important industrial reaction is the water gas–formation reaction. Steam, $H_2O(g)$, is passed over red-hot coke, $C(s)$, yielding a mixture of carbon monoxide, $CO(g)$, and hydrogen gas, $H_2(g)$.

$$H_2O(g) + C(s) \xrightarrow{\Delta} CO(g) + H_2(g)$$
Water gas

This mixture of CO and H_2, water gas, is burned, releasing a large quantity of energy. Let's calculate the mass of CO produced when 125 mol $H_2O(g)$ is combined with excess coke.

From the equation, we obtain the mole ratio of CO formed to H_2O reacted, a 1-to-1 ratio. Accordingly, we solve the problem in the same way as we have solved prior Example Problems, adding one more conversion factor, the one that converts moles of CO to grams of CO.

$$\text{g CO} = \cancel{\text{mol } H_2O} \times \frac{\text{mol } \cancel{CO}}{\cancel{\text{mol } H_2O}} \times \frac{\text{g CO}}{\cancel{\text{mol } CO}}$$

$$= 125 \ \cancel{\text{mol } H_2O} \times \frac{1 \ \cancel{\text{mol } CO}}{1 \ \cancel{\text{mol } H_2O}} \times \frac{28.0 \ \text{g CO}}{1 \ \cancel{\text{mol } CO}}$$

$$= 3500 \ \text{g CO} = 3.50 \times 10^3 \ \text{g CO} = 3.50 \ \text{kg CO}$$

When 125 mol $H_2O(g)$ reacts in the water gas–formation reaction, a maximum of 3.50 kg CO results.

Example Problems 11.3 and 11.4 are examples of mole–mass calculations, i.e., stoichiometry problems that involve moles and mass.

Mole–Mass Calculations

In addition to being used as a fuel, water gas is combined with steam on a special catalyst to produce CO_2; in addition, more H_2 is also liberated.

--------- **Example Problem 11.3** ---------

What mass of coke, C(s), must be present to combine with 125 mol $H_2O(g)$ in the water gas reaction?

$$H_2O(g) + C(s) \xrightarrow{\Delta} CO(g) + H_2(g)$$

Coke, C(s), is produced when coal is heated to drive off its volatile components. Coke is used as fuel and is converted to graphite.

--------- **Solution** ---------

1. What is unknown?   Mass of C(s) in g

2. What is known?   125 mol $H_2O(g)$; 1 mol C/1 mol H_2O; and 12.0 g C/mol C

3. Apply the factor-label method.

$$125 \text{ mol } H_2O \times \frac{1 \text{ mol C}}{1 \text{ mol } H_2O} \times \frac{12.0 \text{ g C}}{1 \text{ mol C}} = ? \text{ g C}$$

4. Perform the indicated math operations.

$$125 \text{ mol } H_2O \times \frac{1 \text{ mol C}}{1 \text{ mol } H_2O} \times \frac{12.0 \text{ g C}}{1 \text{ mol C}} = 1500 \text{ g C}$$
$$= \mathbf{1.50 \times 10^3 \text{ g C} \text{ or } 1.50 \text{ kg C}}$$

Whenever 125 mol $H_2O(g)$ is present, it combines with exactly 1.50 kg of coke.

--------- **Example Problem 11.4** ---------

How many moles of magnesium oxide, MgO, form when 65.0 g Fe_2O_3 combines with excess magnesium in the following reaction:

$$3Mg + Fe_2O_3 \xrightarrow{\Delta} 3MgO + 2Fe$$

Another name for MgO is magnesia. It is sometimes used to prepare $Mg(OH)_2(aq)$, called milk of magnesia. This solution is used to decrease excess stomach acid.

--------- **Solution** ---------

1. What is unknown?   Mol MgO

2. What is known?   65.0 g Fe_2O_3; 1 mol Fe_2O_3/159.6 g Fe_2O_3; 3 mol MgO/1 mol Fe_2O_3

3. Apply the factor-label method.

$$65.0 \text{ g } Fe_2O_3 \times \frac{1 \text{ mol } Fe_2O_3}{159.6 \text{ g } Fe_2O_3} \times \frac{3 \text{ mol MgO}}{1 \text{ mol } Fe_2O_3} = ? \text{ mol MgO}$$

4. Perform the indicated math operations.

$$65.0 \text{ g Fe}_2\text{O}_3 \times \frac{1 \text{ mol Fe}_2\text{O}_3}{159.6 \text{ g Fe}_2\text{O}_3} \times \frac{3 \text{ mol MgO}}{1 \text{ mol Fe}_2\text{O}_3} = \textbf{1.22 mol MgO}$$

When 65.0 g Fe_2O_3 is combined with excess Mg, a maximum of 1.22 mol MgO is expected to form.

Mass–Mass Calculations

Mass–mass calculations are those in which a mass is initially given and the final answer is also a mass. This type of stoichiometric calculation is the one most frequently encountered. Let's consider examples of such problems.

When phosphorus, P_4, combines with chlorine gas, Cl_2, a colorless fuming liquid, phosphorus trichloride, PCl_3, is produced.

$$P_4(s) + 6Cl_2(g) \longrightarrow 4PCl_3(l)$$

What mass of PCl_3 forms when 100.0 g P_4 is combined with excess chlorine gas?

We should first convert the mass of P_4 to moles of P_4. We do this because the balanced equation shows the mole ratio, not the mass ratio. Once the number of moles of P_4 is known, the problem is exactly the same as a mole to mass conversion.

Phosphorus exists in two different forms. White phosphorus is composed of P_4 molecules, and red phosphorus is made up of long chains of bonded P_4 molecules. White phosphorus is very reactive and spontaneously bursts into flames when exposed to the air. Red phosphorus is much less reactive.

$$\text{g PCl}_3 \text{ formed} = 100.0 \text{ g P}_4 \times \frac{1 \text{ mol P}_4}{123.9 \text{ g P}_4} \times \frac{4 \text{ mol PCl}_3}{1 \text{ mol P}_4}$$

$$\times \frac{137.3 \text{ g PCl}_3}{1 \text{ mol PCl}_3} = 443.3 \text{ g PCl}_3$$

The first conversion factor,

$$\frac{1 \text{ mol P}_4}{123.9 \text{ g P}_4}$$

converts the mass of P_4 to moles. From the equation, we find that 4 mol PCl_3 is formed for each 1 mol P_4 initially present. Thus, the second conversion factor yields the number of moles of PCl_3 produced. Our problem is completed by changing the number of moles of PCl_3 to grams of PCl_3, using the molar mass of PCl_3, which is 137.3 g/mol.

Carefully go through Example Problems 11.5 and 11.6, which illustrate mass–mass stoichiometry problems.

─────────────── **Example Problem 11.5** ───────────────

What mass of Cl_2 is needed to combine exactly with excess P_4 to yield 0.927 g PCl_3?

$$P_4(s) + 6Cl_2(g) \longrightarrow 4PCl_3(l)$$

─────────────── **Solution** ───────────────

1. What is unknown? Mass of Cl_2 in g

2. What is known? 0.927 g PCl_3; 1 mol PCl_3/137.3 g PCl_3; 4 mol PCl_3/6 mol Cl_2

3. Apply the factor-label method.

$$0.927 \text{ g } PCl_3 \times \frac{1 \text{ mol } PCl_3}{137.3 \text{ g } PCl_3} \times \frac{6 \text{ mol } Cl_2}{4 \text{ mol } PCl_3} \times \frac{70.9 \text{ g } Cl_2}{1 \text{ mol } Cl_2} = ? \text{ g } Cl_2$$

4. Perform the indicated math operations.

$$0.927 \text{ g } PCl_3 \times \frac{1 \text{ mol } PCl_3}{137.3 \text{ g } PCl_3} \times \frac{6 \text{ mol } Cl_2}{4 \text{ mol } PCl_3}$$

$$\times \frac{70.9 \text{ g } Cl_2}{1 \text{ mol } Cl_2} = \textbf{0.718 g } Cl_2$$

A sample of 0.718 g Cl_2, when combined with an excess of P_4, yields 0.927 g PCl_3.

─────────────── **Example Problem 11.6** ───────────────

What mass of oxygen gas, $O_2(g)$, is liberated when a 2.5-g sample of sodium nitrate, $NaNO_3$, is heated?

$$2NaNO_3(s) \xrightarrow{\text{heat}} 2NaNO_2(s) + O_2(g)$$

The common name for $NaNO_3$ is Chile saltpeter. Both $NaNO_3$ and $NaNO_2$ are used to prevent bacterial growth in meats. They are frequently added to hot dogs, bacon, and hams.

─────────────── **Solution** ───────────────

1. What is unknown? Mass O_2 in g

2. What is known? 2.5 g $NaNO_3$; 85 g $NaNO_3$/mol $NaNO_3$; 1 mol O_2/2 mol $NaNO_3$; and 32 g O_2/mol O_2

3. Apply the factor-label method.

$$2.5 \text{ g } NaNO_3 \times \frac{1 \text{ mol } NaNO_3}{85 \text{ g } NaNO_3} \times \frac{1 \text{ mol } O_2}{2 \text{ mol } NaNO_3} \times \frac{32 \text{ g } O_2}{1 \text{ mol } O_2} = ? \text{ g } O_2$$

4. Perform the indicated math operations.

$$2.5 \text{ g NaNO}_3 \times \frac{1 \text{ mol NaNO}_3}{85 \text{ g NaNO}_3} \times \frac{1 \text{ mol O}_2}{2 \text{ mol NaNO}_3}$$

$$\times \frac{32 \text{ g O}_2}{1 \text{ mol O}_2} = \textbf{0.47 g O}_2$$

A 2.5-g sample of $NaNO_3$ thermally decomposes to produce 0.47 g O_2.

Figure 11.4 shows the steps required to solve mass–mass stoichiometry problems.

11.6 Calculate the number of moles of SO_2 produced when 4.55 g S_8 is combined with excess O_2.   **REVIEW EXERCISES**

$$S_8(s) + 8O_2(g) \xrightarrow{\Delta} 8SO_2(g)$$

11.7 Consider the following equation:

$$P_4(s) + 6Cl_2(g) \longrightarrow 4PCl_3(l)$$

(a) What mass of P_4 is required to combine with 0.398 g Cl_2 to produce PCl_3? (b) What mass of PCl_3 results when 0.398 g Cl_2 is combined with the calculated mass of P_4?

11.8 (a) Calculate the mass of F_2 required to combine with N_2 to produce 204 kg NF_3 in the following reaction:

$$N_2(g) + 3F_2(g) \longrightarrow 2NF_3(g)$$

(b) What mass of N_2 exactly combines with 25.91 mg F_2?

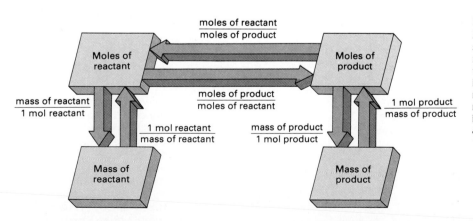

Figure 11.4
Mass-mass problems are solved in a similar manner to mole-mass problems except that a third conversion factor is needed to change the mass of the reactant to moles. This conversion factor is the molar mass of the reactant, mass of reactant per 1 mole of reactant.

A s discussed in Chap. 10, heat is either released or absorbed in chemical reactions. Whenever heat is liberated, the reaction is classified as an **exothermic** reaction, and when heat is absorbed, the reaction is classified as an **endothermic** reaction. The amount of heat transferred in chemical reactions is related to the number of moles of reactants that undergo chemical change. A fixed quantity of heat is transferred per mole of reactant consumed. We call this amount of heat the **enthalpy of reaction** or **heat of reaction.**

Through the application of stoichiometric principles, we can calculate the energy released or absorbed in a chemical reaction. Let us reconsider the reaction in which $HCl(g)$ is synthesized from its elements. This time the heat evolved by the reaction is also shown.

$$H_2(g) + Cl_2(g) \longrightarrow 2HCl(g) + 184 \text{ kJ}$$

In addition to the 2 mol HCl produced, 184 kJ (44 kcal) is released in this exothermic reaction. In other words, for each 1 mol H_2 and 1 mol Cl_2 that react, 2 mol HCl and 184 kJ of heat are produced.

Given the balanced equation and the quantity of heat transferred per mole of reactant, the amount of energy released for any mass of reactant can be calculated. What quantity of heat is liberated when a 1.00-g sample of H_2 is combined with excess Cl_2 to produce HCl? To solve this problem, we need a conversion factor that relates moles of H_2 to the amount of heat liberated; it is

$$\frac{184 \text{ kJ}}{1 \text{ mol } H_2}$$

We solve for kilojoules by calculating the number of moles of H_2 and using the above conversion factor.

$$1.00 \text{ g } H_2 \times \frac{1 \text{ mol } H_2}{2.02 \text{ g } H_2} \times \frac{184 \text{ kJ}}{1 \text{ mol } H_2} = 91.1 \text{ kJ}$$

For each gram of H_2 that combines with Cl_2, 91.1 kJ (21.8 kcal) of heat is liberated to the surroundings.

Example Problems 11.7 and 11.8 are additional examples of problems that deal with energy effects in chemical reactions. Refer to Fig. 11.5, which shows the required steps.

--------------------------- **Example Problem 11.7** ---------------------------

Acetylene, C_2H_2, is a combustible gas used to cut and weld metals. Formation of one mole of acetylene from its elements requires the addition of 227 kJ of energy.

$$2C(s) + H_2(g) + 227 \text{ kJ} \longrightarrow C_2H_2(g)$$

Acetylene is an industrially important gas; more than a billion pounds are produced per year in the United States. It is used in oxyacetylene torches and miners' lamps, and is the starting material for the synthesis of many other carbon compounds.

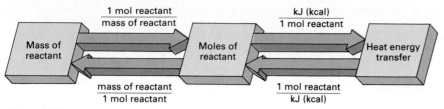

Figure 11.5
To calculate the energy consumed or liberated by a chemical reaction, the conversion factor for the number of moles of reactant and the amount of energy in kilojoules or kilocalories transferred is required. Therefore, if the mass of reactant is given, convert the mass to moles and then multiply by the energy conversion factor.

Calculate the amount of heat required to produce 545 g C_2H_2. Assume that sufficient quantities of the reactants are present.

—————————————— **Solution** ——————————————

1. What is unknown?   Amount of heat in kJ

2. What is known?   545 g C_2H_2; 227 kJ/mol C_2H_2; 26.0 g C_2H_2/mol C_2H_2

3. Apply the factor-label method.

$$545 \text{ g } C_2H_2 \times \frac{1 \text{ mol } C_2H_2}{26.0 \text{ g } C_2H_2} \times \frac{227 \text{ kJ}}{1 \text{ mol } C_2H_2} = ? \text{ kJ}$$

4. Perform the indicated math operations.

$$545 \text{ g } C_2H_2 \times \frac{1 \text{ mol } C_2H_2}{26.0 \text{ g } C_2H_2} \times \frac{227 \text{ kJ}}{1 \text{ mol } C_2H_2} = \textbf{4.76} \times \textbf{10}^\textbf{3} \textbf{ kJ}$$

To produce 545 g C_2H_2, 4.76×10^3 kJ of energy is required in addition to the proper amount of both reactants. As with other fuels, formation of acetylene from its elements is an endothermic process.

—————————————— **Example Problem 11.8** ——————————————

Blood sugar, or glucose, $C_6H_{12}O_6$, is one of the main sources of energy for living systems. It is broken down in biological cells to CO_2 and H_2O, releasing 2816 kJ/mol $C_6H_{12}O_6$:

$$C_6H_{12}O_6(s) + 6O_2(g) \longrightarrow 6CO_2(g) + 6H_2O(g) + 2816 \text{ kJ}$$

Calculate the number of kilocalories of heat released for each 1.00 g of glucose.

Glucose is the most abundant sugar structure in nature. Cellulose, a major component of plant cells, is a high-molecular-mass molecule containing thousands of bonded glucose units. Starch, another component of plants, is also a molecule composed of bonded glucose molecules.

Solution
---

1. What is unknown? kcal/1.00 g $C_6H_{12}O_6$

2. What is known?   1.00 g $C_6H_{12}O_6$; 2816 kJ/mol $C_6H_{12}O_6$; 180 g $C_6H_{12}O_6$/mol $C_6H_{12}O_6$; and 4.184 kJ/kcal

3. Apply the factor-label method.

$$1.00 \text{ g } C_6H_{12}O_6 \times \frac{1 \text{ mol } C_6H_{12}O_6}{180 \text{ g } C_6H_{12}O_6} \times \frac{2816 \text{ kJ}}{1 \text{ mol } C_6H_{12}O_6}$$

$$\times \frac{1 \text{ kcal}}{4.184 \text{ kJ}} = ? \text{ kcal}$$

4. Perform the indicated math operations.

$$1.00 \text{ g } C_6H_{12}O_6 \times \frac{1 \text{ mol } C_6H_{12}O_6}{180 \text{ g } C_6H_{12}O_6} \times \frac{2816 \text{ kJ}}{1 \text{ mol } C_6H_{12}O_6}$$

$$\times \frac{1 \text{ kcal}}{4.184 \text{ kJ}} = \textbf{3.74 kcal}$$

For each 1.00 gram of glucose "burned" by your cells, 374 kcal of energy is released.

Calorimetry

Heat transfers in chemical reactions are measured by instruments called calorimeters. A **calorimeter** is nothing more than a reaction vessel equipped in such a manner that the heat either evolved or absorbed is detected.

Crude estimates of the heat transferred in chemical reactions can be made with a simple calorimeter made from a Styrofoam cup (Fig. 11.6).

Figure 11.6
A simple calorimeter may be constructed from a styrofoam cup because styrofoam absorbs little heat. Thus, the energy released or absorbed is calculated from the temperature change of the solution in the calorimeter.

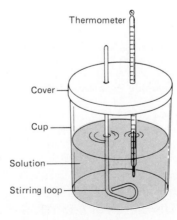

Thermometer

Cover

Cup

Solution

Stirring loop

Figure 11.7
A bomb calorimeter is used to obtain the quantity of heat transferred in a chemical reaction at constant volume.

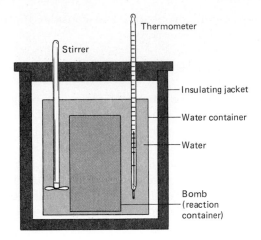

Two solutions are mixed inside the Styrofoam cup. Temperature readings are taken before and after the solutions react. The only other quantity that must be determined is the *heat capacity* of the calorimeter, i.e., the quantity of heat absorbed by the calorimeter to raise its temperature by one degree Celsius. Since the calorimeter is constructed of Styrofoam, which does not absorb too much heat, only the temperature change of the resulting solution is considered. Rough estimates of heat transfers are obtained using Styrofoam cup calorimeters. The instruments used in research laboratories are more sophisticated. One such calorimeter is pictured in Fig. 11.7.

The heat capacity of a calorimeter is measured by finding the amount of heat it absorbs and dividing that quantity by the temperature increase.

11.9 What mass of Cl_2 combines with H_2 to produce 812 kJ of energy in the following reaction?

$$H_2(g) + Cl_2(g) \longrightarrow 2HCl(g) + 184 \text{ kJ}$$

11.10 How much energy, in kilojoules, is released when 254.0 g $C_6H_{12}O_6$, glucose, is combined with O_2 in human cells?

$$C_6H_{12}O_6(s) + 6O_2(g) \longrightarrow 6CO_2(g) + 6H_2O(g) + 2816 \text{ kJ}$$

11.11 *(a)* What is the function of a calorimeter? *(b)* What is meant by the heat capacity of a calorimeter?

REVIEW EXERCISES

So far we have assumed in our calculations that sufficient quantities of all reactants are present to combine with the specified mass of one reactant of interest. However, situations arise when masses are given for more than one of the reactants and we must decide which reactant limits the reaction and determines the maximum yield of product; such problems are called **limiting-reagent** or **limiting-reactant** problems.

11.4 LIMITING-REAGENT PROBLEMS

To illustrate a limiting reagent problem, let us consider the complete oxidation of carbon to carbon dioxide:

$$C(s) + O_2(g) \longrightarrow CO_2(g)$$

A limiting reagent, sometimes called limiting reactant, is the reactant consumed first, and thus determines the maximum amounts of products formed.

One mole of carbon combines with one mole of molecular oxygen to produce one mole CO_2. As long as equal numbers of moles of reactants are combined, a limiting-reagent situation is not encountered. If 5 mol C and 5 mol O_2 are combined, 5 mol CO_2 is formed. However, if 5 mol C is combined with 10 mol O_2, the result is still 5 mol CO_2. After the 5 mol C is consumed, the reaction stops. Carbon is therefore the limiting reagent because after it is consumed, the reaction ceases. Study Table 11.2, which shows stepwise what happens when 5 mol C combines with 10 mol O_2.

TABLE 11.2   LIMITING REAGENT IN THE OXIDATION OF CARBON

$$C(s) + O_2(g) \longrightarrow CO_2(g)$$

Moles C	Moles O_2	Moles CO_2 produced	
5	10	0	(Initial amounts)
4	9	1	(After 1 mol reacts)
3	8	2	(After 2 mol reacts)
2	7	3	(After 3 mol reacts)
1	6	4	(After 4 mol reacts)
0*	5	5	(After 5 mol reacts)

*After the carbon is consumed, the reaction cannot continue. At that time, 5 mol of unreacted O_2 and 5 mol of CO_2 remain. Carbon is the limiting reagent, and oxygen is the reactant in excess.

A number of everyday experiences provide good analogies to the situation that occurs when reactants combine. Take for example the stapling of photocopied pages to produce a pamphlet. If 25 first pages, 50 second pages, and 100 third pages are stapled to produce a three-page pamphlet, how many complete pamphlets could be produced? Here, the first page is the "limiting reagent." Once 25 pamphlets are stapled, no more complete pamphlets can be produced because there are no first pages left. It does not matter that 25 second pages and 75 third pages are unstapled; there is no way to produce a complete pamphlet without adding more first pages to our supply (Fig. 11.8).

Our initial chemical limiting-reagent example was straightforward because the reactants combined in a 1-to-1 ratio. What if the reactants combine in some other ratio? Again we return to the Haber reaction:

$$N_2(g) + 3H_2(g) \longrightarrow 2NH_3(g)$$

What is the limiting reagent, if 6 mol N_2 combines with 6 mol H_2? In the Haber reaction, 3 mol H_2 is consumed for each 1 mol N_2. Initially, 1 mol

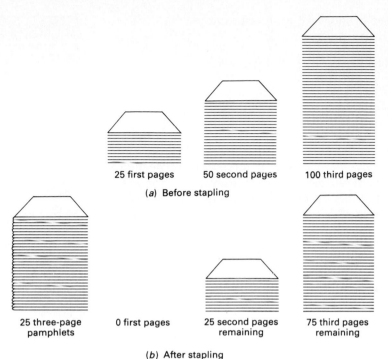

Figure 11.8
A pamphlet is produced by stapling three pages together. If 25 first pages, 50 second pages, and 100 third pages are initially present, the maximum number of complete pamphlets that can be produced is 25. After the first page runs out, no more completed pamphlets can be produced.

25 first pages   50 second pages   100 third pages

(a) Before stapling

25 three-page   0 first pages   25 second pages   75 third pages
pamphlets   remaining   remaining

(b) After stapling

N_2 combines with 3 mol H_2, which leaves 5 mol N_2 and 3 mol H_2. When the next 1 mol N_2 reacts, it consumes all of the remaining H_2; the reaction then stops for lack of H_2. Hydrogen is therefore the limiting reagent. Consequently, 4 mol unreacted N_2 remains in the reaction vessel with the 4 mol NH_3 that was produced (Table 11.3).

TABLE 11.3   LIMITING REAGENT IN THE HABER REACTION

$$N_2 + 3H_2 \longrightarrow 2NH_3$$

Moles N_2	Moles H_2	Moles NH_3 produced	
6	6	0	(Initial amounts)
5	3	2	(After 1 mol N_2 reacts)
4	0	4	(After 2 mol N_2 reacts)

As the above examples show, the limiting reagent is found by comparing the numbers of moles of reactants present, while taking into account the mole ratio in which they combine. In most cases, masses of the reactants are given, and it is necessary to convert the masses to moles and then decide which reactant is limiting (consumed first).

What is the limiting reagent when 10.0 g H_2 and 50.0 g O_2 are combined to produce water vapor?

$$2H_2(g) + O_2(g) \longrightarrow 2H_2O(g)$$

First calculate the number of moles of each reactant:

$$10.0 \text{ g } H_2 \times \frac{1 \text{ mol } H_2}{2.02 \text{ g } H_2} = 4.95 \text{ mol } H_2$$

$$50.0 \text{ g } O_2 \times \frac{1 \text{ mol } O_2}{32.0 \text{ g } O_2} = 1.56 \text{ mol } O_2$$

After calculating the number of moles of each reactant, select one of the reactants and determine how many moles of the other are needed to combine with it completely. Because it does not matter which reactant we select, let's calculate the number of moles of H_2 required to combine with the O_2 present.

$$1.56 \text{ mol } O_2 \times \frac{2 \text{ mol } H_2}{1 \text{ mol } O_2} = 3.13 \text{ mol } H_2 \text{ required}$$

We take the number of moles of O_2 and multiply it by 2 because, in the equation, we find that 2 mol H_2 are required per 1 mol O_2. Looking at the results, we see that 3.13 mol H_2 is required to combine with 1.56 mol O_2. Is there enough H_2 to combine with the O_2? Yes, only 3.13 mol H_2 is needed, and 4.95 mol is available. Hence, O_2 is the limiting reagent. After the 1.56 mol O_2 combines with the excess H_2, the reaction ceases.

Once the limiting reagent is identified, all other calculations are completed using the number of moles of limiting reagent. What total mass of H_2O is produced in the above problem?

$$1.56 \text{ mol } O_2 \times \frac{2 \text{ mol } H_2O}{1 \text{ mol } O_2} \times \frac{18.0 \text{ g } H_2O}{1 \text{ mol } H_2O} = 56.2 \text{ g } H_2O$$

When 10.0 g H_2 and 50.0 g O_2 combine, the theoretical yield of water is 56.2 g.

Note that if you had not chosen the correct limiting reagent, the calculated mass of water would have been larger than the theoretical yield.

$$4.95 \text{ mol } H_2 \times \frac{2 \text{ mol } H_2O}{2 \text{ mol } H_2} \times \frac{18.0 \text{ g } H_2O}{1 \text{ mol } H_2O} = \quad 89.1 \text{ g } H_2O$$
$$\textbf{(Incorrect answer)}$$

This amount of H_2O, 89.1 g, could only be produced if a sufficient quantity of O_2 were available (2.48 mol O_2). It is impossible for 89.1 g H_2O to form with the stated quantities of starting materials.

Example Problems 11.9 and 11.10 provide additional illustrations of limiting-reagent problems. The steps required to solve limiting-reagent problems are summarized in Fig. 11.9.

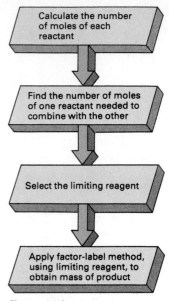

Figure 11.9
Steps required to solve limiting reagent problems.

─────────────── **Example Problem 11.9** ───────────────

What mass of SO_3 is produced when 0.600 g SO_2 is combined with 0.400 g O_2?

$$2SO_2(g) + O_2(g) \longrightarrow 2SO_3(g)$$

─────────── **Solution** ───────────

1. Identify the limiting reagent.
First, calculate which reactant is the limiting reagent by comparing the initial numbers of moles.

$$0.600 \text{ g SO}_2 \times \frac{1 \text{ mol SO}_2}{64.1 \text{ g SO}_2} = 0.00936 \text{ mol SO}_2$$

$$0.400 \text{ g O}_2 \times \frac{1 \text{ mol O}_2}{32.0 \text{ g O}_2} = 0.0125 \text{ mol O}_2$$

Calculate the number of moles of SO_2 needed to combine with the number of moles of O_2 actually present or vice versa.

$$0.0125 \text{ mol O}_2 \times \frac{2 \text{ mol SO}_2}{1 \text{ mol O}_2} = 0.0250 \text{ mol SO}_2$$

We find that 0.0250 mol SO_2 is required to combine with 0.0125 mol O_2. However, only 0.00936 mol of SO_2 is present; consequently, SO_2 is the limiting reagent.

2. Apply the factor-label method.
Complete the problem by using conversion factors to find the mass of SO_3 formed (mole–mass problem).

$$0.00936 \text{ mol SO}_2 \times \frac{2 \text{ mol SO}_3}{2 \text{ mol SO}_2} \times \frac{80.0 \text{ g SO}_3}{1 \text{ mol SO}_3} = \textbf{0.749 g SO}_3$$

If 0.600 g SO_2 is combined with 0.400 g O_2, 0.749 g SO_3 is the maximum quantity of product that can be produced.

In the atmosphere SO_2 is, unfortunately, oxidized to SO_3. Sulfur trioxide combines with water to produce sulfuric acid, H_2SO_4, which is a component of acid rain.

─────────── **Example Problem 11.10** ───────────

Hydrogen gas combines with nitrogen dioxide to produce ammonia and water. *(a)* What mass of ammonia results when 25.0 g of hydrogen combines with 185 g of nitrogen dioxide? *(b)* What mass of excess reactant remains?

────────────────────── **Solution** ──────────────────────

1. Write and balance the equation.

$$H_2(g) + NO_2(g) \longrightarrow NH_3(g) + H_2O(g) \text{ (Unbalanced)}$$

$$7H_2(g) + 2NO_2(g) \longrightarrow 2NH_3(g) + 4H_2O(g) \text{ (Balanced)}$$

2. Identify the limiting reagent.

$$25.0 \text{ g } H_2 \times \frac{1 \text{ mol } H_2}{2.02 \text{ g } H_2} = 12.4 \text{ mol } H_2$$

$$185 \text{ g } NO_2 \times \frac{1 \text{ mol } NO_2}{46.0 \text{ g } NO_2} = 4.02 \text{ mol } NO_2$$

Calculate the number of moles of NO_2 needed to combine with 12.4 mol H_2.

$$12.4 \text{ mol } H_2 \times \frac{2 \text{ mol } NO_2}{7 \text{ mol } H_2} = 3.54 \text{ mol } NO_2$$

Because 3.54 mol NO_2 is needed and 4.02 mol is actually present, H_2 is the limiting reagent. When all 12.4 mol H_2 is combined, an excess of NO_2 remains.

3. Use the limiting reagent to calculate the mass of NH_3 produced.

$$12.4 \text{ mol } H_2 \times \frac{2 \text{ mol } NH_3}{7 \text{ mol } H_2} \times \frac{17.0 \text{ g } NH_3}{1 \text{ mol } NH_3} = \textbf{60.2 g NH}_3$$

After 25.0 g H_2 and 185 g NO_2 combine, the theoretical yield of NH_3 is 60.2 g.

4. Calculate the mass of excess reactant.
In Step 2 we found that 3.54 mol NO_2 is needed to combine with 12.4 mol H_2; therefore, subtract 3.54 mol NO_2 from 4.02 mol NO_2, the total number of moles of NO_2 present, to find the moles of NO_2 in excess.

$$\text{Excess mol } NO_2 = 4.02 \text{ mol } NO_2 - 3.54 \text{ mol } NO_2$$

$$= 0.48 \text{ mol } NO_2$$

To find the mass of NO_2 in excess, we use the molar mass to cancel moles of NO_2.

$$0.48 \text{ mol } NO_2 \times \frac{46 \text{ g } NO_2}{1 \text{ mol } NO_2} = \textbf{22 g NO}_2$$

After the reaction 22 g of unreacted NO_2 should be found in the reaction vessel.

REVIEW EXERCISES

11.12 *(a)* What is the limiting reagent in a chemical reaction? *(b)* Explain how to determine which reactant is the limiting reagent.

11.13 For the combination reaction of carbon and oxygen to produce carbon dioxide, determine which is the limiting reagent, carbon or oxygen, given the following amounts:
(a) 6 mol C + 6 mol O_2
(b) 2 mol C + 1 mol O_2
(c) 12.0 g C + 12.0 g O_2
(d) 16 g C + 31 g O_2
(e) 0.0455 g C + 0.205 g O_2

11.14 For each part of 11.13, calculate what mass of excess reactant is present if the reaction is allowed to go to completion.

11.15 *(a)* What is the maximum mass of sodium fluoride, NaF, that results when 52.9 g of sodium metal combines with 44.8 g of fluorine, F_2?

$$2Na + F_2 \longrightarrow 2NaF$$

(b) What mass of excess reactant remains?

11.5 THEORETICAL AND ACTUAL YIELDS

Frequently the mass of product obtained in a reaction is less than the predicted theoretical yield. The amount of product isolated is called the **actual yield.** Chemists commonly report the *percent yield* for products, defined as follows:

$$\% \text{ yield} = \frac{\text{actual yield}}{\text{theoretical yield}} \times 100$$

If the theoretical yield of a product is 5 g, and only 4 g is isolated, the percent yield is reported as 80 percent ($\frac{4}{5} \times 100$). Example Problems 11.11 and 11.12 are examples of percent yield problems.

Example Problem 11.11

Calcium phosphate, $Ca_3(PO_4)_2$, combines with silicon dioxide, SiO_2, to produce calcium silicate, $CaSiO_3$, and P_4O_{10}. Initially, 50.0 g $Ca_3(PO_4)_2$ is combined with excess SiO_2. After the reaction, 16.9 g P_4O_{10} is isolated; calculate the percent yield of P_4O_{10}.

$$2Ca_3(PO_4)_2 + 6SiO_2 \longrightarrow 6CaSiO_3 + P_4O_{10}$$

$Ca_3(PO_4)_2$ is found in phosphate rock. It is heated to a high temperature to produce free phosphorus, P_4. However, most phosphate rock is used to produce phosphate fertilizers, $Ca(H_2PO_4)_2$ for example.

------------ **Solution** ------------

1. Calculate the theoretical yield of P_4O_{10}.

$$50.0 \text{ g } \cancel{Ca_3(PO_4)_2} \times \frac{1 \text{ mol } \cancel{Ca_3(PO_4)_2}}{3.10 \times 10^2 \text{ g } \cancel{Ca_3(PO_4)_2}} \times \frac{1 \text{ mol } \cancel{P_4O_{10}}}{2 \text{ mol } \cancel{Ca_3(PO_4)_2}}$$
$$\times \frac{284 \text{ g } P_4O_{10}}{1 \text{ mol } \cancel{P_4O_{10}}} = 22.9 \text{ g } P_4O_{10}$$

Without going through each step, the above factor-label setup gives the mass of P_4O_{10} that forms when 50.0 g of calcium phosphate combines with excess silicon dioxide. Thus, the theoretical yield is 22.9 g P_4O_{10}.

2. Calculate the percent yield.

$$\% \text{ yield } P_4O_{10} = \frac{\text{actual yield}}{\text{theoretical yield}} \times 100$$

$$= \frac{16.9 \text{ g}}{22.9 \text{ g}} \times 100$$

$$= \textbf{73.8\%}$$

A yield of 16.9 g P_4O_{10} represents a 73.8% yield. Perhaps during the course of the reaction and isolation of the product, 26.2% of the theoretical yield of P_4O_{10} was lost.

------------ **Example Problem 11.12** ------------

Chromium metal and aluminum oxide, Al_2O_3, are produced when chromium(III) oxide, Cr_2O_3, is heated with aluminum metal.

$$Cr_2O_3 + 2Al \xrightarrow{\Delta} 2Cr + Al_2O_3$$

The maximum percent yield of Cr obtained by this process is found to be 89.3%. Calculate the mass of Cr_2O_3 required to produce 91.3 g Cr, assuming that the maximum percent yield is obtained.

------------ **Solution** ------------

To solve this problem, we must consider that in this reaction only 89.3% of the theoretical yield is obtained. Thus, we should first calculate the theoretical yield that gives the desired amount of Cr (91.3 g Cr) and then calculate the mass of Cr_2O_3 that produces that amount of Cr.

1. Calculate the theoretical yield of Cr that gives the desired actual yield. The percent yield of the reaction is 89.3%, which means that

$$\% \text{ yield}' = \frac{89.3 \text{ g Cr (actual)}}{100 \text{ g Cr (theoretical)}} \times 100$$

Therefore, we can calculate the theoretical yield of Cr that gives 91.3 g Cr as follows:

$$91.3 \text{ g Cr (actual)} \times \frac{100 \text{ g Cr (theoretical)}}{89.3 \text{ g Cr (actual)}} = 102 \text{ g Cr}$$

We have found that a theoretical yield of 102 g Cr gives an actual yield of 91.3 g Cr.

2. Calculate the mass of Cr_2O_3 needed to produce 102 g Cr.

$$102 \text{ g Cr} \times \frac{1 \text{ mol Cr}}{52.0 \text{ g Cr}} \times \frac{1 \text{ mol Cr}_2\text{O}_3}{2 \text{ mol Cr}} \times \frac{152 \text{ g Cr}_2\text{O}_3}{1 \text{ mol Cr}_2\text{O}_3} = \textbf{149 g Cr}_2\textbf{O}_3$$

If 149 g Cr_2O_3 is combined with excess Al, it produces an actual yield of 91.3 g Cr.

11.16 Consider the following reaction.   **REVIEW EXERCISE**

$$CS_2(g) + 3Cl_2(g) \longrightarrow CCl_4(l) + S_2Cl_2(g)$$

(a) Initially, 4.29 g of carbon disulfide, CS_2, is combined with excess Cl_2, and 7.83 g CCl_4 is isolated. What is the percent yield for the reaction?
(b) What mass of CCl_4 would be obtained from 4.29 g CS_2 if the percent yield were 61.9%?

SUMMARY

Stoichiometry is the study of quantitative relationships in chemical reactions. With a balanced equation, mole and mass relationships are easily obtained. Predictions that result from stoichiometric calculations are the theoretical maximum amounts that could be obtained in a reaction.

Commonly, masses of starting materials are given, and the quantities of products are sought. Initially, the masses of reactants are converted to moles because the balanced equation indicates the mole ratio between each reactant and product. After calculating the number of moles of desired product, it is only necessary to convert the moles of product to grams,

using the molecular or atomic mass. Stoichiometry problems are ideally suited to be solved by the factor-label method.

The amount of energy transferred may also be predicted from chemical equations. Given the quantity of heat evolved or absorbed per mole of reactant, a conversion factor that relates moles and energy is determined. Energy transfers in chemical reactions are experimentally measured by means of **calorimetry,** i.e., the measurement of heat transfers in chemical reactions using an instrument called a calorimeter.

When the initial quantities of all reactants are given and no assumption is made that one or more reactants is in excess, it is necessary to find the reactant that is consumed first (the **limiting reagent**). After the limiting reagent is consumed, no further reaction is observed.

The amounts of products produced in chemical reactions differ from the theoretically predicted amounts. Chemists regularly calculate the **percent yield** of the reaction, which is the ratio of the actual yield to the theoretical yield times 100.

KEY TERMS
---

actual yield
calorimetry
heat capacity
limiting reagent

percent yield
stoichiometry
theoretical yield

EXERCISES*

11.17 Define each of the following terms: stoichiometry, theoretical yield, calorimetry, heat capacity, limiting reagent, actual yield, percent yield.

Quantitative Equation Relationships

11.18 Complete the following table for the reaction in which N_2 and O_2 combine to yield NO.

	$N_2(g)$	+	$O_2(g)$	$\longrightarrow$	$2NO(g)$
Molecules	5 molecules				
Molecules	6.02×10^{22} molecules				
Moles	5.0 mol				
Mass, g			16.0 g		
Mass, g					19.5 g

11.19 Complete the following table for the reaction in which pentane, C_5H_{12}, combines with O_2 to produce CO_2 and H_2O.

	$C_5H_{12}(l) + 8O_2(g) \longrightarrow 5CO_2(g) + 6H_2O(g)$		
Molecules	10 molecules		
Molecules	6.02×10^{22} molecules		
Moles	1.00 mol		
Moles		1.00 mol	
Mass, g	85.0 g		

11.20 When Ca combines with F_2, CaF_2 is produced:

$$Ca + F_2 \longrightarrow CaF_2$$

Assume that 1 mol Ca and 1 mol F_2 are combined. (a) How many moles of CaF_2 result? (b) What mass of CaF_2 results? (c) Show that the balanced equation illustrates the law of conservation of mass—i.e., show that the sum of the masses of the reactants equals the mass of the product.

11.21 Consider the reaction in which bromine combines with fluorine to form BrF:

$$Br_2 + F_2 \longrightarrow 2BrF$$

If 1 mol F_2 combines with 1 mol Cl_2 to produce 2 mol BrF, show that the sum of the masses of the reactants equals the mass of product.

Mole–Mole Relationships

11.22 For the reaction $2Al + 3Cl_2 \rightarrow 2AlCl_3$, calculate the number of moles of $AlCl_3$ produced from each of the following stated quantities of reactant (assume the other reactant is always in excess): (a) 5.0 mol Al, (b) 8.0 mol Cl_2, (c) 18 mol Al, (d) 33 mol Cl_2, (e) 29 mmol Cl_2.

11.23 For the following equation, write all possible conversion factors that relate the number of moles of each reactant to the number of moles of each product:

$$C_5H_{12} + 8O_2 \longrightarrow 5CO_2 + 6H_2O$$

*For exercise numbers printed in color, answers can be found at the back of the book.

11.24 For the following reaction, calculate the requested quantities:

$$3KCl + 4HNO_3 \longrightarrow Cl_2 + NOCl + 2H_2O + 3KNO_3$$

(a) number of moles of Cl_2 produced for each mole of HNO_3 reacted; (b) number of moles of H_2O produced when 5.00 mol KCl reacts; (c) number of moles of HNO_3 reacted to produce 11.0 mol NOCl; (d) number of moles of Cl_2 and KNO_3 produced when 8.59 mol KCl reacts; (e) number of moles of each product produced when 0.0300 mol HNO_3 reacts.

11.25 Consider the equation

$$4FeS_2 + 11O_2 \xrightarrow{\Delta} 2Fe_2O_3 + 8SO_2$$

For each of the given molar quantities of reactant, calculate the number of moles of each product produced:
(a) 0.44 mol FeS_2
(b) 29.1 mol FeS_2
(c) 8.200 mol O_2
(d) 19.9 mmol O_2
(e) 0.00345 mol O_2

Mole–Mass Calculations

11.26 A simple laboratory preparation of O_2 gas is to decompose $KClO_3$ by heating it in the presence of MnO_2, a catalyst:

$$2KClO_3 \xrightarrow[MnO_2]{\Delta} 2KCl + 3O_2$$

(a) Calculate the mass of O_2 produced when 17.1 mol $KClO_3$ is decomposed. (b) Calculate the number of moles of O_2 produced when 1.55 g $KClO_3$ is heated. (c) Find the number of moles of KCl and O_2 produced from 6.20 kg $KClO_3$. (d) What mass of $KClO_3$ is needed to produce 397 mg O_2?

11.27 Sodium bicarbonate, $NaHCO_3$, combines with HCl to produce sodium chloride, NaCl, carbon dioxide, CO_2, and water, H_2O.

$$NaHCO_3(aq) + HCl(aq) \longrightarrow NaCl(aq) + CO_2(g) + H_2O(l)$$

(a) What mass of HCl should be present to combine totally with 0.150 mol $NaHCO_3$? (b) How many moles of CO_2 are produced when 4.96 g $NaHCO_3$ combines with excess HCl? (c) Calculate the mass of NaCl that results when 9.48 mmol HCl combines with excess $NaHCO_3$. (d) What mass of $NaHCO_3$ is required to produce 8.309×10^3 mol H_2O?

11.28 Copper metal is isolated from its oxide, Cu_2O, by heating the oxide in the presence of copper(I) sulfide, Cu_2S:

$$2Cu_2O + Cu_2S \xrightarrow{\Delta} 6Cu + SO_2$$

Calculate the mass of Cu produced from each of the following quantities of reactant (assume an excess amount of the other reactant): (a) 9.22 mol Cu_2O, (b) 5.09 mol Cu_2O, (c) 0.00125 mol Cu_2O, (d) 7.92×10^5 mol Cu_2S, (e) 1964 mol Cu_2S, (f) 7.88 mmol Cu_2O.

Mass–Mass Calculations

11.29 When silver(I) oxide, Ag_2O, is heated, it readily liberates oxygen, O_2, and leaves free silver, Ag:

$$2Ag_2O(s) \xrightarrow{\Delta} 4Ag(s) + O_2(g)$$

What mass of silver results when the following quantities of silver(I) oxide are completely heated: (a) 0.3811 g Ag_2O, (b) 7.91 kg Ag_2O, (c) 6.22 mg Ag_2O, (d) 54.80 kg Ag_2O?

11.30 Lead nitrate, $Pb(NO_3)_2$, on heating decomposes to lead(II) oxide, PbO, oxygen, O_2, and nitrogen dioxide, NO_2:

$$2Pb(NO_3)_2(s) \xrightarrow{\Delta} 2PbO(s) + O_2(g) + 4NO_2(g)$$

What mass of $Pb(NO_3)_2$ must be heated to produce the following masses of product: (a) 1.002 g PbO, (b) 9341 g O_2, (c) 0.0420 kg NO_2, (d) 900.4 mg PbO?

11.31 Carbon tetrachloride, CCl_4, once used as a cleaning fluid and as a fire extinguisher, is produced by heating methane, CH_4, and chlorine, Cl_2:

$$CH_4 + 4Cl_2 \xrightarrow{\Delta} CCl_4 + 4HCl$$

(a) What mass of CH_4 is needed to exactly combine with 33.4 g Cl_2? (b) How many grams of Cl_2 are required to produce 9336 g CCl_4, assuming excess CH_4? (c) What mass of CH_4 must have reacted if 4.02 mg HCl is liberated? (d) Calculate the masses of CH_4 and Cl_2 required to produce exactly 800.0 kg CCl_4?

11.32 Pure boron is prepared by combining boron trichloride, BCl_3, with hydrogen gas, H_2, at high temperatures:

$$2BCl_3 + 3H_2 \xrightarrow{\Delta} 2B + 6HCl$$

(a) Calculate the mass of boron produced when 0.771 g BCl_3 is combined with excess hydrogen gas. (b) What mass of hydrogen gas is needed to completely combine with 4.9 kg BCl_3? (c) What mass of BCl_3 must be present to produce 112.5 g B? (d) Calculate the mass of H_2 that combines exactly with 9.04 kg BCl_3, and calculate how much B and HCl are produced.

11.33 Xenon tetrafluoride, XeF_4, a noble gas compound, is highly reactive. When mixed with water, XeF_4 undergoes the following reaction:

$$6XeF_4 + 8H_2O \longrightarrow 2XeOF_4 + 4Xe + 16HF + 3O_2$$

(a) Calculate the mass of XeF_4 needed to produce 100.0 g $XeOF_4$. (b) After a measured mass of XeF_4 is placed in water, 0.0984 g HF is released. What mass of XeF_4 was combined with water? (c) How much water should be exactly combined with 0.01144 g XeF_4 in the above reaction? (d) Calculate the mass of each product produced when 3.48 mg XeF_4 is combined with water.

11.34 Nickel chloride hexahydrate, $NiCl_2 \cdot 6H_2O$, on heating in a vacuum, yields nickel chloride, $NiCl_2$, and water, H_2O:

$$NiCl_2 \cdot 6H_2O \xrightarrow{\Delta} NiCl_2 + 6H_2O$$

(a) How many grams of the hydrate should be heated to produce 327 g H_2O? (b) When 0.184 kg of the hydrate is heated, what mass of the anhydrous compound, $NiCl_2$, results? (c) What mass of hydrate is heated to yield 743 g $NiCl_2$? (d) Find the mass of each product when 0.853 mg of nickel chloride hexahydrate is heated.

11.35 Potassium nitrate decomposes to potassium nitrite and oxygen. (a) Write the balanced equation for this reaction. (b) State all mole relationships for this equation. (c) How many grams of potassium nitrite form when 581.0 g of potassium nitrate is heated? (d) What mass of oxygen is liberated when 4.760 kg of potassium nitrate is decomposed?

11.36 Silicon tetrachloride and carbon monoxide are produced when silicon dioxide, carbon, and chlorine gas are heated. (a) Write the balanced equation for this chemical change. (b) What mass of silicon tetrachloride is produced when 171 g of silicon dioxide is combined with excess carbon and chlorine? (b) How many molecules of carbon monoxide are produced when 0.00201 g of chlorine combines with excess reactants?

11.37 Stibnite, antimony(III) sulfide, when heated with iron undergoes a single replacement reaction, liberating free antimony metal and iron(II) sulfide. (a) Write the balanced equation for the reaction. (b) What mass of iron is required to combine with 58.5 g of stibnite? (c) How many grams of antimony are produced when 94.7 kg of stibnite is reacted? (d) What masses of stibnite and iron should combine to yield 5.097 g of antimony?

Energy Effects

11.38 When hydrogen, H_2, is combined with oxygen, O_2, water vapor is formed plus 569 kJ of heat:

$$2H_2(g) + O_2(g) \longrightarrow 2H_2O(g) + 569 \text{ kJ}$$

(a) Rewrite the equation with the energy expressed as kilocalories instead of kilojoules. (b) What mass of H_2 and O_2 liberate 408 kJ? (c) If 2.93 mol H_2 is combined with excess O_2, what quantity of heat is given off? (d) If 2.93 g O_2 is combined with excess hydrogen, what quantity of heat is liberated?

11.39 Consider the oxidation of nitrogen monoxide:

$$2NO(g) + O_2(g) \longrightarrow 2NO_2(g) + 116.7 \text{ kJ}$$

(a) What mass of NO produces 851.2 kJ of heat? (b) If 11.78 g NO is reacted, what quantity of heat is liberated? (c) How many moles of oxygen are needed to combine with excess NO to produce 92.52 J? (d) What mass of NO is required to produce 705.4 kcal of energy?

11.40 When combusted, ethylene, C_2H_4, gives off CO_2 and H_2O:

$$C_2H_4(g) + 3O_2(g) \longrightarrow$$
$$2CO_2(g) + 2H_2O(g) + 1.41 \times 10^3 \text{ kJ}$$

(a) How much heat is liberated per gram of C_2H_4? (b) What mass of C_2H_4 must be combusted to liberate 646 J? (c) If 5.96 g O_2 is available, with excess C_2H_4, what quantity of heat is released? (d) How much C_2H_4 must be combusted to liberate 2.75 kcal of energy?

11.41 The decomposition of aluminum oxide in the following reaction requires 3.34×10^3 kJ:

$$2Al_2O_3(s) + 3.34 \times 10^3 \text{ kJ} \longrightarrow 4Al(s) + 3O_2(g)$$

(a) How much energy is required to decompose 77.6 mol Al_2O_3? (b) What quantity of heat is needed to decompose 478 kg Al_2O_3? (c) If 9.48 g of Al metal is obtained, what quantity of heat is required to decompose the Al_2O_3? (d) How many kilocalories of energy are required to decompose 76.7 g Al_2O_3?

11.42 When 1.001 g of calcium carbonate is decomposed to calcium oxide and carbon dioxide, 1.78 kJ of energy is needed. (a) Write a balanced equation for the reaction, including the energy (in kilojoules) required per mole of calcium carbonate. (b) Rewrite the equation with the energy expressed in kilocalories.

11.43 In a simple calorimeter, two 150.0-g solutions of reacting chemicals are poured together, each initially at 23.8°C. After they are allowed to react, the final temperature of the resulting mixture is found to be 41.7°C. Assuming that all of the heat released by the reaction is absorbed by the water, and knowing that the

heat capacity of the solution is 4.18 J/(g·°C), calculate the amount of heat released by the reaction.

Limiting-Reagent Problems

11.44 Consider the following equation:

$$H_2(g) + I_2(g) \longrightarrow 2HI(g)$$

(a) Calculate the mass of HI that forms when 1.53 g H_2 combines with 126 g I_2. (b) What is the maximum yield of HI when 8.87 g H_2 combines with 1025 g I_2? (c) What mass of excess reactant remains when 0.3462 g H_2 and 45.92 g I_2 combine?

11.45 Phosgene, $COCl_2$, combines with water to produce carbon dioxide, CO_2, and hydrochloric acid, HCl:

$$COCl_2(g) + H_2O(l) \longrightarrow CO_2(g) + 2HCl(aq)$$

(a) What mass of CO_2 is produced when 6.41 g $COCl_2$ is combined with 1.78 g H_2O? (b) Calculate the mass of HCl formed when 3.09 kg of each reactant is combined? (c) What mass of excess reactant remains in the reaction in part b?

11.46 Consider the following equation:

$$2NaCl + H_2SO_4 \longrightarrow Na_2SO_4 + 2HCl$$

(a) How many grams of Na_2SO_4 are produced when 970.0 g of each reactant is combined? (b) Find the mass of Na_2SO_4 that forms when 35.7 g NaCl and 21.9 g H_2SO_4 combine. (c) What mass of HCl results when 12.2 kg NaCl and 9.74 kg H_2SO_4 combine? (d) What mass of excess reactant remains in the reaction in part c?

11.47 A lead(II) nitrate solution combines with a solution of potassium iodide to yield a precipitate, lead(II) iodide, and aqueous potassium nitrate. (a) Write the balanced equation for this aqueous reaction. (b) What mass of lead iodide precipitates from solution if 49.11 g of lead(II) nitrate combines with 49.60 g of potassium iodide. (c) Calculate the mass of reactant in excess.

11.48 If sodium carbonate is combined with carbon and nitrogen gas, carbon monoxide and sodium cyanide result. (a) What mass of sodium cyanide results when 36.5 g of carbon, 81.4 g of sodium carbonate, and excess nitrogen gas are combined? (b) What mass of nitrogen gas is consumed in this reaction?

Percent Yield

11.49 When benzene, C_6H_6, is combined with chlorine, Cl_2, chlorobenzene, C_6H_5Cl, and hydrogen chloride gas, HCl, are produced. (a) If 13.0 g of benzene is combined with excess chlorine, what is the theoretical yield of chlorobenzene? (b) If 10.4 g of chlorobenzene is isolated after the reaction, calculate the percent yield of chlorobenzene. (c) What mass of chlorobenzene results if the percent yield is 64.3% and the initial mass of benzene is 2.87 g? (d) What mass of benzene is required to produce 4.92 g of chlorobenzene, if the percent yield is 59.2%?

11.50 Rare germanium metal is isolated by heating GeO_2 in the presence of pure carbon:

$$GeO_2 + 2C \xrightarrow{\Delta} Ge + 2CO$$

(a) After 729 g GeO_2 is reacted with carbon, 435 g Ge is obtained; calculate the percent yield of germanium. (b) Starting with 61.0 kg GeO_2 and excess carbon, what mass of Ge results if a 91.2% yield is obtained? (c) What mass of GeO_2 is required to produce 33.7 g Ge, if the percent yield is 91.2%.

11.51 Calcium cyanamide, $CaCN_2$, combines with carbon, C, and sodium carbonate, Na_2CO_3, to produce calcium carbonate, $CaCO_3$, and sodium cyanide, NaCN. (a) What mass of calcium cyanamide is needed to produce 2.953 g of sodium cyanide, if the percent yield of the reaction is 70.1%? (b) What mass of calcium cyanamide is required to produce 40.47 kg of sodium cyanide, if the percent yield is 83.9%?

Additional Exercises

11.52 When one mole N_2 combines with three moles H_2, two moles NH_3 and 92.6 kJ of heat are produced. (a) Calculate the masses of N_2 and H_2 required to produce 1.00×10^3 kJ of heat. (b) What mass of water can be heated from $0.0°C$ to $50.0°C$ with the heat liberated when 84.2 g N_2 combines with excess H_2? The specific heat of water is 4.184 J/(g·°C).

11.53 Calcium hydroxide, $Ca(OH)_2$ (slaked lime), is prepared by the following reactions:

$$CaCO_3 \longrightarrow CaO + CO_2$$

$$CaO + H_2O \longrightarrow Ca(OH)_2$$

(a) What mass of $Ca(OH)_2$ results when 544 g $CaCO_3$ is decomposed? (b) If 71.1% is the maximum overall yield of the above reactions, what mass of $Ca(OH)_2$ forms when 53.3 kg $CaCO_3$ decomposes?

11.54 Zinc blende, or zinc(II) sulfide ore, is one source of metallic zinc. Initially, the zinc sulfide is combined with oxygen to yield zinc(II) oxide and sulfur dioxide. The resulting zinc(II) oxide is then heated with carbon to produce free zinc metal and carbon monoxide. (a) Write and balance both equations. (b) Calculate the mass of zinc obtained from 6.23 metric tons zinc(II) sulfide, assuming a percent yield of 78.4%. One metric ton (t) equals 1000 kg. (c) What mass of zinc forms when 65.8 g of zinc(II) sulfide combines with 33.9 g of oxygen, assuming 100% yield?

11.55 Consider the following aqueous reaction:

$$2KMnO_4 + 10KI + 8H_2SO_4 \longrightarrow$$
$$6K_2SO_4 + 2MnSO_4 + 5I_2 + 8H_2O$$

Calculate the masses of I_2, K_2SO_4, and $MnSO_4$ produced when 7.80 g $KMnO_4$, 42.7 g KI, and 19.3 g H_2SO_4 are combined.

11.56 Consider the following reaction, in which sodium amide, $NaNH_2$, and nitrous oxide, N_2O, are heated.

$$2NaNH_2 + N_2O \xrightarrow{\Delta} NaN_3 + NH_3 + NaOH$$

(a) What is the maximum yield of NaN_3 when 18.64 g $NaNH_2$ combines with 10.61 g N_2O? (b) What mass of excess reactant remains? (c) How many grams of the limiting reagent must be added to totally react the excess reactant?

11.57 The percent yield of SF_4 in the following reaction is 66.9%.

$$3SCl_2 + 4NaF \longrightarrow SF_4 + S_2Cl_2 + 4NaCl$$

(a) Calculate the masses of SCl_2 and NaF that are required to produce 37.8 g SF_4. (b) If 250.99 g SCl_2 combines with 136.22 g NaF, what mass of SF_4 results?

11.58 Sulfuric acid, H_2SO_4, can be prepared from FeS_2 as follows:

$$4FeS_2 + 11O_2 \longrightarrow 2Fe_2O_3 + 8SO_2$$

$$2SO_2 + O_2 \longrightarrow 2SO_3$$

$$SO_3 + H_2O \longrightarrow H_2SO_4$$

(a) What mass of sulfuric acid results when 24.2 kg FeS_2 combines with excess oxygen? (b) If the percent yield of sulfuric acid is 71.4%, what mass of FeS_2 is required to produce 93.4 kg of sulfuric acid? (c) What mass of FeS_2 was initially present and what mass of sulfuric acid forms if the mass of SO_3 is 60.8 kg?

11.59 An excess amount of hydrochloric acid, HCl(aq), is added to a 1.000-g mixture of $CaCO_3$ and $CaSO_4$. The $CaCO_3$ combines with the HCl to produce $CaCl_2$, CO_2, and H_2O, but the $CaSO_4$ does not react with the HCl. If the mass of CO_2 produced is 0.303 g, what is the percent composition of the mixture?

11.60 One of the oxides of cobalt reacts with hydrogen gas, H_2, to produce Co metal and water. When 0.914 mol of this cobalt compound combines with 5.54 g H_2, 108 g Co results. What is the formula of the oxide?

11.61 An impure sample of zinc, Zn, is treated with an excess amount of sulfuric acid, H_2SO_4, and $ZnSO_4$ and H_2 are the products. Calculate the percent of zinc in the impure sample if a 4.35-g sample produces 0.121 g of hydrogen gas.

CHAPTER
Twelve

Gases

STUDY GUIDELINES

After completing Chapter 12, you should be able to

1. List and discuss the principal assumptions of the kinetic molecular theory

2. State the properties of ideal gases

3. Define pressure, and name three of the units most commonly used to measure pressure

4. Convert any given pressure unit to any other pressure unit

5. State in words and as mathematical expressions (1) Boyle's law, (2) Charles' law, (3) Avogadro's law, (4) Gay-Lussac's law of combining volumes, and (5) Dalton's law

6. Calculate the final pressure, volume, or temperature of an ideal gas given the initial and final set of conditions

7. Provide an explanation for the fact that equal volumes of ideal gases at the same pressure and temperature contain the same number of particles

8. State the molar volume of a gas at STP, and use it to find the mass or number of moles of an ideal gas

9. Calculate the density or molecular mass of a gas given all required information and the molar gas constant

10. Use the ideal gas equation to find unknown properties of gases

11. Perform volume–volume, mass–volume, and mole–volume stoichiometry problems

12. Explain why the total pressure of a gaseous mixture equals the sum of the partial pressures of the gases in the mixture

13. Calculate the partial pressure exerted by a gas in a mixture given the total pressure and the partial pressures of all other gases

14. List and describe the gases found in greatest concentration in the atmosphere

15. Discuss the discovery, properties, and uses of N_2, O_2, and the noble gases

We previously noted in Chap. 4 that gases completely fill and take the shapes of their containers, are compressible, have the lowest average density, and are the least viscous of the three physical states. In this chapter we will consider the principal properties of gases. To begin our study of gases, we will take a brief look at the theory that is used to explain the properties of gases, the kinetic molecular theory.

The behavior and properties of gases are explained theoretically by the **kinetic molecular theory** (KMT), literally the "moving molecule" theory. This theory is a model that explains the behavior of gases using generalizations about the random motion of the molecules or atoms that compose a gas. Some of the major assumptions of the kinetic molecular theory are:

1. All gases are composed of atoms or molecules that move rapidly and randomly in straight lines.

2. Individual atoms or molecules are widely separated from each other, and do not exert forces on other atoms or molecules except when colliding. Nearly all of the volume of a gas is empty space.

3. Collisions of atoms or molecules with the walls of the container and with each other are perfectly elastic. This means that there is no net loss of kinetic energy on collision.

4. The average kinetic energy $(KE = \frac{1}{2}mv^2)$ of the atoms or molecules in gases is proportional to the Kelvin temperature of the gas. The average energy of the molecules does not change unless the temperature changes.

A gas that behaves exactly according to the above assumptions is an **ideal gas,** i.e., one that exhibits perfect behavior. In actuality, no real gas behaves exactly as an ideal gas. However, real gases under conditions of low pressure and high temperature approach the behavior of ideal gases. Because the relationships of the properties are more complicated for real gases, we will concentrate on the perfect behavior of gases, realizing that our predictions are not totally correct for real gases. They are, however, accurate enough for most purposes encountered in ordinary, everyday experience.

12.1 KINETIC MOLECULAR THEORY OF GASES

Kinetic energy is the energy possessed by a moving body. Faster-moving objects have more kinetic energy than slower-moving ones with the same mass.

12.1 What are the main assumptions of the kinetic molecular theory?

12.2 *(a)* What is an ideal gas? *(b)* Why do chemists develop relationships for an ideal gas when no real gas behaves in exactly this ideal manner?

REVIEW EXERCISES

Gas laws are empirical relationships that relate the volume (V), pressure (P), temperature (T), and moles of gas particles (n).

Gas pressure is defined as the force exerted by a gas on a unit area.

12.2 GAS LAWS

Pressure

$$\text{Pressure} = \frac{\text{force}}{\text{area}}$$

A force is exerted when the molecules in a gas sample hit the walls of the container. An increase in the number of collisions per second or in the force of impact increases the pressure exerted by a gas.

Barometers and manometers are the two instruments used to measure the pressure of gases. A **barometer** is an instrument used to measure the pressure exerted by the atmosphere. A **manometer** is used to measure the pressure of isolated gas samples. Most scientific barometers and manometers use mercury to measure the pressure of gases. Mercury is a liquid metal that has a high density (13.6 g/cm^3). It is used because it does not readily evaporate, and it makes possible the construction of smaller instruments because of its high density.

Evangelista Torricelli (1608–1647), a fellow student of Galileo, designed the first mercury barometer. A barometer like Torricelli's is made by filling a glass tube with liquid Hg, and then placing the open end of the tube vertically into a container of Hg. Some of the Hg initially spills out of the tube into the container of Hg. When the force exerted by the Hg column equals the force exerted by the atmosphere on the surface of the Hg in the container, no more mercury spills out (Fig. 12.1). If the atmospheric pressure increases, Hg is pushed up into the tube, and if the pressure decreases, mercury spills out of the tube into the Hg pool.

Torricelli measured atmospheric pressure in terms of the total height of Hg supported in the glass tube by the atmosphere. At sea level, on an average day, the atmosphere supports a column of Hg about 760 mm high. Today, atmospheric and gas pressures are still measured in terms of the height of mercury that a gas supports. Units of pressure most frequently employed are atmospheres, millimeters of mercury, torr, and pascals. By definition, one atmosphere is equal to 760 millimeters of mercury, or 760 torr.

$$1.00 \text{ atm} = 760 \text{ torr} = 760 \text{ mmHg}$$

Numerically, one millimeter of mercury is equivalent to one torr (named for Torricelli).

$$1 \text{ torr} = 1 \text{ mmHg}$$

Atmospheres and torr will be the units of pressure most commonly encountered in this textbook.

Because these units are not derived from the base SI units, they are not SI units. The SI unit for pressure is the pascal, which is a small unit of pressure. The pascal is derived from the SI unit of force, the newton, and the unit for area, the meter squared. One atmosphere is equivalent to 101,325 pascals.

$$1 \text{ atm} = 101,325 \text{ Pa} = 101.325 \text{ kPa}$$

Because of the small size of the pascal, the kilopascal is most frequently encountered in chemistry. Table 12.1 summarizes the relationships between the many units of pressure used in chemistry.

Torricelli was educated in Rome as a mathematician; however, his primary interest was mechanics. He wrote a book on mechanics, which was noticed by Galileo and led to their meeting. From his association with Galileo, Torricelli moved to the forefront of the Italian scientific community.

Blaise Pascal (1623–1662), at the age of 16, published a book on geometry and conic sections. Pascal repeated Torricelli's famous experiment, and carried it further by getting someone to take the barometer to the top of a mountain to see if the Hg level dropped.

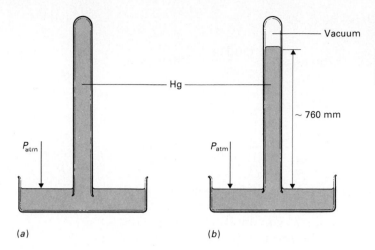

Figure 12.1
A barometer is an instrument used for measuring atmospheric pressure. One atmosphere supports 760 mmHg in a vertical closed-end tube.

(a)   (b)

TABLE 12.1   UNITS OF PRESSURE

Unit	Abbreviation	Relationships
Atmosphere	atm	1 atm = 760 torr
		1 atm = 760 mmHg
		1 atm = 101,325 Pa
		1 atm = 101.325 kPa
Pascal	Pa	$1\ Pa = 9.869 \times 10^{-6}$ atm
		$1\ Pa = 7.501 \times 10^{-3}$ torr
		$1\ Pa = 1 \times 10^{-3}$ kPa
Torr	torr	$1\ torr = 1.316 \times 10^{-3}$ atm
or		$1\ torr = 1.333 \times 10^{2}$ Pa
Millimeter of mercury	mmHg	$1\ torr = 1.333 \times 10^{-1}$ kPa

Example Problem 12.1 gives examples of how to interconvert pressure units using the factor-label method.

──────────── **Example Problem 12.1** ────────────

Convert 796 torr to millimeters of Hg, atmospheres, and kilopascals.

──────────── **Solution** ────────────

1. What is unknown?   mmHg, atm, and kPa
2. What is known?   796 torr;   1 torr/mmHg;   760 torr/atm; 101,300 Pa/atm

3. Apply the factor-label method.

$$796 \text{ torr} \times \frac{1 \text{ mmHg}}{1 \text{ torr}} = \textbf{796 mmHg}$$

$$796 \text{ torr} \times \frac{1 \text{ atm}}{760 \text{ torr}} = \textbf{1.05 atm}$$

$$796 \text{ torr} \times \frac{1 \text{ atm}}{760 \text{ torr}} \times \frac{101,300 \text{ Pa}}{1 \text{ atm}} \times \frac{1 \text{ kPa}}{1000 \text{ Pa}} = \textbf{106 kPa}$$

Given the number of torr, one immediately knows the number of millimeters of mercury because the two are numerically equal. Conversion to atmospheres is accomplished by multiplying by 1 atm/760 torr. Finally, to change to kilopascals, the pressure in torr must be first converted to atmospheres, then to pascals, and ultimately to kilopascals.

The pressure exerted by an isolated gas sample is measured with a manometer. An open-ended manometer is pictured in Fig. 12.2; it is nothing more than a U-shaped tube that contains mercury. If the pressure of the gas sample, P_g, equals the pressure of the atmosphere, P_{atm}, then the level of the mercury is equal in either side of the U-tube. If the pressure of the gas sample is greater than atmospheric pressure, then the level of the mercury is higher on the side exposed to the atmosphere. As shown in Fig. 12.2, the level is x mmHg higher on the right side; thus

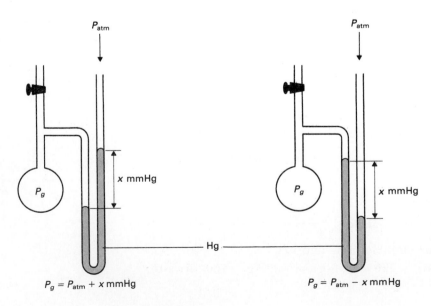

Figure 12.2
An open-ended manometer is used to measure the pressure of isolated gas samples, P_g. It is composed of a U-shaped tube that contains mercury. The pressure of a gas sample is calculated from the difference in height of the mercury column in either side of the tube and from the atmospheric pressure.

the pressure of the gas sample equals atmospheric pressure plus x mmHg.

$$P_g = P_{atm} + x \text{ mmHg (when } P_g > P_{atm})$$

If the pressure of the gas sample is less than atmospheric pressure, then the level of the mercury is higher in the tube closest to the gas sample. The difference in height is subtracted from the atmospheric pressure to obtain the pressure of the gas sample for this case.

$$P_g = P_{atm} - x \text{ mmHg (when } P_g < P_{atm})$$

REVIEW EXERCISES

12.3 (a) Write a definition for pressure. (b) What units are used to measure the pressure of gases? (c) What instruments are used to measure pressure?

12.4 Convert each of the following to the indicated units:
(a) 98.2 Pa = ? torr
(b) 2486 torr = ? atm
(c) 1.17 atm = ? kPa
(d) 915 atm = ? torr

Boyle's Law

Boyle's law describes the relationship between the volume and pressure of an ideal gas at constant temperature and number of moles. Intuitively, it is easy to understand Boyle's law. As can be seen in Fig. 12.3, if we increase the external pressure on a gas sample, the volume decreases. Diminishing the external pressure allows the gas to expand and increase in volume. Therefore, a statement of **Boyle's law** is: The volume of a gas is inversely proportional to its pressure when the temperature and number of moles of gas are constant.

Mathematically, a statement of Boyle's law is as follows:

$$PV = k \text{ (at constant } T \text{ and } n)$$

in which P is the pressure of the gas, V is its volume, and k is a constant of proportionality.

To illustrate the inverse proportionality of the volume and pressure of an ideal gas, consider the data presented in Table 12.2. Initially, the pressure is 20 atm, and the volume is 1.0 L. When the pressure is decreased to 10 atm (halved), the volume increases to 2.0 L (doubles). After the pressure is decreased to 5.0 atm, the volume increases to 4.0 L (doubles again). Notice that in each case, the product of the pressure times the volume is 20 L·atm; $P \times V$ remains constant as long as no other factors (T, n, or other experimental factors) are changed. The P versus V inverse relationship is plotted in Fig. 12.4.

Boyle was the first scientist to collect gases and systematically study their properties. He discovered the P-V relationship using a 17-ft tube that contained a trapped gas and Hg.

TABLE 12.2 BOYLE'S LAW RELATIONSHIP

P, atm	V, L	k, L·atm
20	1.0	20
10	2.0	20
5.0	4.0	20
4.0	5.0	20
2.0	10	20
1.0	20	20

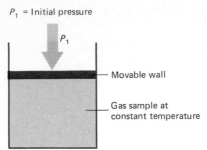

P₁ = Initial pressure

P_1

— Movable wall

— Gas sample at constant temperature

Figure 12.3
At constant temperature and number of moles, the volume of an ideal gas decreases if the pressure increases, and the volume increases if the pressure decreases. This inverse relationship is called Boyle's law.

Increase pressure, $P_2 > P_1$
Volume decreases

P_2

T constant

Decrease pressure, $P_3 < P_1$
Volume increases

P_3

T constant

A more useful expression of Boyle's law is

$$P_1V_1 = P_2V_2$$

in which P_1 is the initial pressure, V_1 is the initial volume, and V_2 is the final volume after the pressure has been changed to P_2. A mathematical expression of Boyle's law allows us to calculate a new volume for an ideal gas after the pressure has changed, or vice versa. For example, what is the new volume of a gas initially in a 5.0-L cylinder under a pressure of 15 atm, when the pressure is decreased to 1.0 atm? We begin with the equation for Boyle's law, rearranging it to isolate V_2 on one side:

$P_1V_1 = k$ *and* $P_2V_2 = k$; *therefore,* $P_1V_1 = P_2V_2$. *This is a mathematical expression of Boyle's law.*

$$\frac{P_1V_1}{P_2} = \frac{P_2V_2}{P_2}$$

After dividing by P_2 and changing the equation around, we have

$$V_2 = V_1 \times \frac{P_1}{P_2} \longleftarrow \text{Pressure factor}$$

It is convenient to express the equation in this way, with the pressure terms isolated from the initial volume. The ratio of pressures is sometimes called the *pressure factor* and is a conversion factor. Therefore, to

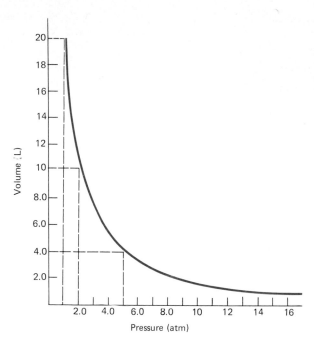

Figure 12.4
A gas initially occupies 20 L at 1.0 atm. When the pressure is increased to 2.0 atm, the volume decreases to 10 L. If the pressure is increased to 5.0 atm, the volume decreases to 4.0 L, one-fifth the initial volume.

solve the problem, we multiply the initial volume V_1, 5.0 L, times the ratio of initial pressure to final pressure (P_1/P_2), 15 atm/1.0 atm.

$$V_2 = 5.0 \text{ L} \times \frac{15 \text{ atm}}{1.0 \text{ atm}}$$

$$= 75 \text{ L}$$

With the decrease in pressure the volume expands to 75 L; exactly as we would expect for an inverse relationship, decreasing the pressure increases the volume.

Even if you forget the equation for Boyle's law, you can calculate the final volume by realizing that P and V are inversely related. Accordingly, the pressure factor must increase the magnitude of the initial volume if the pressure is decreased. In other words, the larger pressure must be written in the numerator, and the smaller pressure written in the denominator.

Figure 12.5 diagrams the procedure for calculating the final volume of an ideal gas after a pressure change. Also, study Example Problem 12.2, another example of a Boyle's law problem.

─────────── **Example Problem 12.2** ───────────

Calculate the final volume of He gas when the pressure on 375 mL. He is increased from 428 torr to 1657 torr.

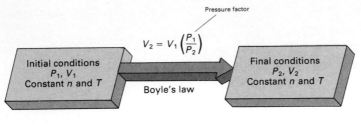

Figure 12.5
To calculate the final volume V_2 of an ideal gas after the pressure is changed at constant temperature and number of moles, the initial volume V_1 is multiplied by the pressure factor. The pressure factor P_1/P_2 is the conversion factor that expresses the ratio of the initial pressure P_1 to the final pressure P_2.

Solution

1. What is unknown? Final volume V_2 of He in mL

2. What is known? 375 mL He (initial volume, V_1); 428 torr (initial pressure, P_1); 1657 torr (final pressure, P_2)

3. Apply Boyle's law expression.

Instead of blindly plugging numbers into the expression, ask yourself: What is happening? In this problem, the pressure is increased; thus, the volume decreases. The pressure factor should be written with the smaller pressure divided by the larger pressure (P_1/P_2).

$$V_2 = V_1 \times \frac{P_1}{P_2}$$

$$= 375 \text{ mL} \times \frac{428 \text{ torr}}{1657 \text{ torr}}$$

$$= \textbf{96.9 mL}$$

After completing the problem, look at the answer and see if it is reasonable. A final volume of 96.9 mL is significantly smaller than the initial volume, exactly what we expect when the pressure is increased.

We can use the kinetic molecular theory to explain Boyle's law. The pressure of a gas depends on the number of collisions per second on the walls of the container. If the volume of the gas is decreased, the inside area of the walls decreases and the number of collisions per second increases; thus the pressure increases (Fig. 12.6). However, if the volume of the gas increases, the area of the walls increases and the number of collisions per second decreases (Fig. 12.6).

Many phenomena can be explained by Boyle's law. For example, Boyle's law helps explain why liquids are drawn into hypodermic syringes (Fig. 12.7). When the plunger is withdrawn, the pressure decreases inside the syringe. Because the atmospheric pressure is higher than the pressure inside the syringe, the atmospheric pressure pushes the liquid into the lower-pressure region inside the syringe. Boyle's law also explains how the Heimlich maneuver works. The Heimlich maneu-

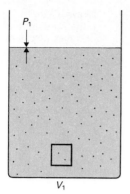

P_1

V_1

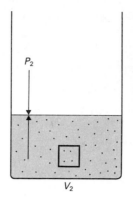

P_2

V_2

Figure 12.6
Pressure depends on the number of collisions of molecules per second with the walls of the container. If the volume of the container increases (V_1), the number of collisions per second decreases because of the larger area; thus, the pressure decreases. If the volume of the container decreases (V_2), the number of collisions per second increases; thus, the pressure increases.

ver is the procedure used to dislodge food caught in the trachea (windpipe) of a person who is choking. If the abdomen is squeezed with a strong upward movement, the air in the lungs is compressed, which creates a large enough pressure to expel the food that blocks the trachea.

Charles' law describes the relationship between the volume and Kelvin temperature of a gas at constant pressure and number of moles. If a gas sample is heated, the average kinetic energy of the molecules increases—they move faster. Because they move faster, they hit the walls of the container more frequently and with a greater force of impact. Hence, if the pressure is to remain constant, the volume must increase (Fig. 12.8). Likewise, if the temperature is decreased, the volume decreases because the molecules slow down, hitting the walls less frequently and with less force.

Charles' law states that the volume of a gas is directly proportional to its Kelvin temperature when the pressure and number of moles are

Charles' Law

Jacques Alexandre Cesar Charles (1746–1823) was one of the first to develop H_2 balloons, and on a few occasions ascended to heights of over a mile.

Figure 12.7
(a) The end of a syringe is placed below the surface of a liquid. *(b)* When the plunger is withdrawn, a low pressure region develops inside the syringe. *(c)* Because the pressure on the surface of the liquid (P_{atm}) is greater than the pressure inside the syringe, the liquid is forced into the syringe.

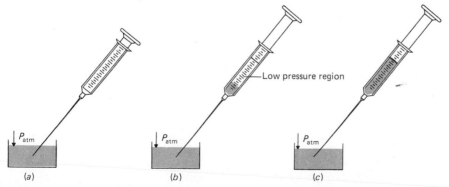

Low pressure region

P_{atm} P_{atm} P_{atm}

(a)   (b)   (c)

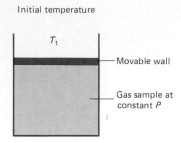

Initial temperature

T_1

Movable wall

Gas sample at constant P

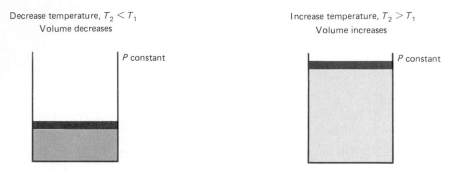

Decrease temperature, $T_2 < T_1$
Volume decreases

P constant

Increase temperature, $T_2 > T_1$
Volume increases

P constant

Figure 12.8
If the pressure and number of moles of gas are constant, the volume of an ideal gas is directly proportional to its Kelvin temperature. Thus, if the temperature of a gas sample decreases, the volume decreases; and if the temperature increases, the volume increases.

constant. Mathematically, a statement of Charles' law is as follows:

$$\frac{V}{T} = k \text{ (at constant } P \text{ and } n)$$

in which V is volume, T is the Kelvin temperature of the gas, and k is a proportionality constant. Now we are dealing with a direct relationship, one in which both variables (V and T) change in the same direction. Increase the temperature, and the volume increases, decrease the temperature and the volume decreases.

Figure 12.9 presents a representative graph of T versus V for an ideal gas. In this graph we see a linear relationship in which the volume increases as the temperature increases. Of special interest is the point where the line touches the temperature axis (point of zero volume), $-273.15°C$, or absolute zero. Do not rashly jump to any false conclusions! It is fun to extrapolate data to see what might be expected above and below the range of collected data, but remember: We are describing ideal gases, and they mainly exist in the minds of chemists. Gases, on cooling, liquefy before reaching 0 K.

Changing the Charles' law expression to a workable equation, we obtain

$$\frac{V_1}{T_1} = \frac{V_2}{T_2}$$

Charles repeated an experiment, earlier performed by Guillaume Amontons (1663–1705), regarding the expansion of gases on heating. Charles is mainly credited with the accurate determination that at 0°C the volume of a gas increases by 1/273 for each degree change.

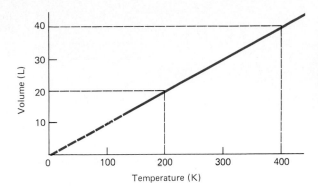

Figure 12.9
The volume of a gas sample is directly proportional to its Kelvin temperature. The graph shows the linear relationship between the volume and the Kelvin temperature of an ideal gas. If the Kelvin temperature doubles from 200 to 400 K, the volume doubles from 20 to 40 L.

Rearranging to solve for the final volume V_2 we obtain

$$V_2 = V_1 \times \frac{T_2}{T_1} \longleftarrow \text{Temperature factor}$$

Once again, the final volume of the gas is calculated by multiplying the initial volume of the gas times a conversion factor, the *temperature factor*. In contrast to the Boyle's law pressure factor, in which the initial pressure is divided by the final pressure, in the Charles' law temperature factor the final Kelvin temperature T_2 is divided by the initial Kelvin temperature T_1.

If an ideal gas initially occupies 25.4 L at 25.0°C, what is its final volume if the temperature is changed to 78.2°C? In this problem, the temperature is increased; consequently, the volume increases—a direct relationship. Before you apply the Charles' law relationship, the initial and final temperatures must be changed to Kelvin. Remember that a Kelvin temperature is calculated by adding 273.2 to the Celsius temperature:

$$T_1 = 25.0°C + 273.2 = 298.2 \text{ K}$$

$$T_2 = 78.2°C + 273.2 = 351.4 \text{ K}$$

Applying the Charles' law relationship

$$V_2 = V_1 \times \frac{T_2}{T_1}$$

$$= 25.4 \text{ L} \times \frac{351.4 \text{ K}}{298.2 \text{ K}}$$

$$= 29.9 \text{ L}$$

we find that the final volume of the gas sample is 29.9 L, an increase of 4.5 L from the initial volume. Figure 12.10 illustrates the procedure for

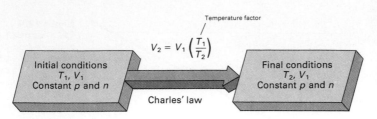

Figure 12.10
To calculate the final volume V_2 of a gas after the temperature changes at constant pressure, the initial volume V_1 is multiplied by the temperature factor. The temperature factor T_2/T_1 is the conversion factor that expresses the ratio between the final temperature T_2 and the initial temperature T_1.

calculating the final volume of an ideal gas after a temperature change.

Example Problem 12.3 shows another example of a Charles' law problem.

―――――――――――――――――――― **Example Problem 12.3** ――――――――――――――――

Calculate the final volume of a 178-mL sample of Xe gas when the temperature is decreased from 63.5°C to −155.3°C.

―――――――――――――――――――――――― **Solution** ―――――――――――――――――――――

1. What is unknown?   Final volume V_2 in mL

2. What is known?   178 mL (initial volume, V_1); 63.5°C (initial temperature, T_1); −153.3°C (final temperature, T_2)

3. Convert Celsius temperatures to Kelvin.

$$T_1 = 63.5°C + 273.2 \quad = 336.7 \text{ K}$$

$$T_2 = -155.3°C + 273.2 = 117.9 \text{ K}$$

4. Apply Charles' law.

$$V_2 = V_1 \times \frac{T_2}{T_1}$$

$$= 178 \text{ mL} \times \frac{117.9 \cancel{\text{ K}}}{336.7 \cancel{\text{ K}}}$$

$$= \mathbf{62.3 \text{ mL}}$$

When the gas is cooled from 336.7 K to 117.9 K, the volume decreases to 62.3 mL if the pressure and number of moles of gas remain constant.

―――

Charles' law helps explain why hot air balloons can ascend into the atmosphere. As the air in the balloon is heated, its volume increases, which lowers the density of the air. Recall that density is the ratio of mass to volume. If the volume of gas increases, its density decreases. As the density of the air inside the balloon decreases, the balloon becomes buoyant in the denser air that surrounds it, and thus the balloon rises. By heating and cooling the air in the balloon, the balloonist can cause it to rise and descend.

Boyle's and Charles' laws are mathematically combined to yield what is called the **combined gas law.**

Combined Gas Law

$$V_2 = V_1 \times \frac{P_1}{P_2} \times \frac{T_2}{T_1}$$

In the equation for the combined gas law, two conversion factors are used to adjust the initial volume to the final volume:

$$V_2 = V_1 \times \underset{\text{Pressure factor}}{\frac{P_1}{P_2}} \times \underset{\text{Temperature factor}}{\frac{T_2}{T_1}} \text{ (Constant } n)$$

If we multiply each side of this equation by P_2/T_2, we obtain another expression of the combined gas law equation:

$$\frac{P_1 V_1}{T_1} = \frac{P_2 V_2}{T_2}$$

Combined gas law

Once we know the combined gas law, it is not necessary to deal with the gas laws individually. If temperature is constant ($T_1 = T_2$), the temperature terms cancel, leaving Boyle's law:

$$V_2 = V_1 \times \frac{P_1}{P_2} \times \frac{\cancel{T_2}}{\cancel{T_1}} = V_1 \times \frac{P_1}{P_2}$$

Boyle's law

In a similar manner, if the pressure is constant ($P_1 = P_2$), the pressure terms cancel, leaving the Charles' law expression. (Prove this to yourself.)

If the volume is constant ($V_1 = V_2$), the volume terms cancel, giving us a third gas law, called **Gay-Lussac's law.** This states that pressure is directly proportional to the Kelvin temperature at constant volume and number of moles. An illustration of Gay-Lussac's law is the variation in pressure in the tires of an automobile. During the winter, the pressure decreases and air must be added to maintain proper inflation. During the summer, the pressure increases and air should be removed.

Gay-Lussac is usually given credit for identifying the direct relationship between pressure and the Kelvin temperature of a gas when the volume and number of moles are constant.

When both pressure and temperature changes are made on a gas, the final volume depends on the direction and magnitude of the two changes. For example, if the pressure is increased and the temperature

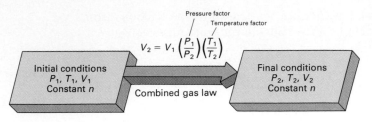

Pressure factor

Temperature factor

$$V_2 = V_1 \left(\frac{P_1}{P_2}\right)\left(\frac{T_1}{T_2}\right)$$

Initial conditions
P_1, T_1, V_1
Constant n

Combined gas law

Final conditions
P_2, T_2, V_2
Constant n

Figure 12.11
To calculate the final volume V_2 of a gas after temperature and pressure changes at constant number of moles, the initial volume V_1 is multiplied by both the pressure factor P_1/P_2 and the temperature factor T_2/T_1.

is decreased, the final volume of the gas is smaller than the initial volume. Whenever the pressure is increased, the volume is decreased; in a similar manner, when the temperature is decreased the volume also decreases; both changes decrease the gas volume. Exactly the opposite occurs if the pressure is decreased and the temperature is increased; then the volume of the gas increases.

Example Problems 12.4 and 12.5 provide illustrations of the application of the combined gas law. Use Fig. 12.11 as a guide when solving combined gas law problems in which the pressure and temperature are changed.

--- **Example Problem 12.4** ---

A balloon filled with He has a volume of 5.2 L at 44°C and 718 torr. What volume does the He occupy at standard temperature and pressure (STP), 273 K (0°C) and 1.00 atm (760 torr)?

Standard temperature and pressure (STP) means a temperature of 273 K at a pressure of 1 atm.

--- **Solution** ---

1. What is unknown?   Final volume V_2 in L

2. What is known?   5.2 L $= V_1$; 44°C $= T_1$; 718 torr $= P_1$; 273 K $= T_2$; 1.00 atm $= P_2$

3. Change the given units to kelvins and atmospheres.

$$T_1 = 44°C + 273 = 317 \text{ K}$$

$$P_1 = 718 \text{ torr} \times \frac{1.00 \text{ atm}}{760 \text{ torr}} = 0.945 \text{ atm}$$

T_1 is changed to kelvins and P_1 is changed to atmospheres so that the units cancel in the combined gas law equation.

3. Apply the combined gas law.

$$V_2 = V_1 \times \frac{P_1}{P_2} \times \frac{T_2}{T_1}$$

$$= 5.2 \text{ L} \times \frac{0.945 \text{ atm}}{1.00 \text{ atm}} \times \frac{273 \text{ K}}{317 \text{ K}}$$

$$= \mathbf{4.2 \text{ L}}$$

Is our answer reasonable? Pressure is increased and temperature is decreased. An increase in P decreases the volume, and a decrease in T also decreases the volume; accordingly, the final volume is less than the initial volume.

Example Problem 12.5

A sample of hydrogen gas occupies a volume of 444 mL at 0°C and 593 torr. What must be the final temperature of the gas if, after the pressure has been changed to 291 torr, the gas occupies 1.88 L?

Solution

In this problem, instead of solving for a new volume, we are looking for the final temperature, given the pressure and volume changes. Rearrange the general form of the combined gas law and solve for the final temperature T_2.

$$\frac{P_1 V_1}{T_1} = \frac{P_2 V_2}{T_2}$$

Multiply each side by T_1 and then by T_2 to get

$$P_1 V_1 T_2 = P_2 V_2 T_1$$

and divide both sides by $P_1 V_1$ to get

$$T_2 = T_1 \times \frac{P_2}{P_1} \times \frac{V_2}{V_1}$$

1. What is unknown? Final temperature T_2 in K

2. What is known? 0°C, T_1; 444 mL, V_1; 1.88 L, V_2; 593 torr, P_1; 291 torr, P_2

3. Change T_1 to K and V_1 to L.

$$T_1 = 0°C + 273 = 273 \text{ K}$$

$$V_1 = 444 \text{ mL} \times \frac{1 \text{ L}}{1000 \text{ mL}} = 0.444 \text{ L}$$

4. Apply the combined gas law.

$$T_2 = 273 \text{ K} \times \frac{291 \text{ torr}}{593 \text{ torr}} \times \frac{1.88 \text{ L}}{0.444 \text{ L}}$$

$$= \textbf{567 K} = \textbf{294°C}$$

After the original hydrogen sample is heated to 567 K and the pressure is decreased to 291 torr, the sample expands to a volume of 1.88 L.

12.5 Write a mathematical expression for each of the following: *(a)* Boyle's law, *(b)* Charles' law, and *(c)* the combined gas law.

12.6 Distinguish between an inverse and a direct proportion by sketching a graph for each.

12.7 Calculate the final volume of an ideal gas if it occupies 4.00 L at an initial pressure of 12.0 atm and the pressure is changed to 8.75 atm.

12.8 What is the final volume of a gas that initially occupies 81.2 mL at 43.1°C when the temperature is increased to 84.6°C?

12.9 An ideal gas occupies 0.385 L at STP. What volume will it occupy at 653 torr and 4.31×10^2 K?

Avogadro's Law

So far, in our discussion of gases, the number of moles of particles in the gas sample has been constant. We did not want any of the gas to escape or other gases to be added during the time the pressure and temperature were being varied. Now let's consider what happens when the number of moles of gas are varied, at constant P and T.

Amedeo Avogadro was the first to hypothesize that if the pressure and temperature of a gas are constant, its volume is directly proportional to the number of moles of gas molecules contained in the gas. Mathematically, this is expressed as follows:

Avogadro's hypothesis went mainly unnoticed. About 50 years later, another Italian scientist, Stanislao Cannizzaro (1826–1910), applied the hypothesis and determined the molecular masses of various gases.

$$V = kn$$

in which V is volume, n is number of moles, and k is a proportionality constant. A balloon helps illustrate Avogadro's law. A balloon expands when air is added to it (moles of air increase), and it contracts when air escapes (moles of air decrease).

An important relationship develops from Avogadro's law. If equal volumes of different gases are compared, they must contain the same number of moles of particles. Let's compare equal volumes of He(g) and Ne(g). Using the equation for Avogadro's law, we obtain the following two relationships:

$$V_{He} = kn_{He} \quad \text{and} \quad V_{Ne} = kn_{Ne}$$

Because the volumes are equal ($V_{He} = V_{Ne}$), the kn terms are also equal:

$$kn_{He} = kn_{Ne}$$

At the same temperature and pressure, the proportionality contants k in the equations are equal; therefore, the constants can be divided out of the equation, leaving

$$n_{He} = n_{Ne}$$

Thus, the number of moles of He gas equals the number of moles of Ne gas if equal volumes are compared at the same temperature and pressure.

Avogadro's law may be stated as follows: Equal volumes of different gases contain the same number of particles if their pressures and temperatures are the same. At standard conditions (STP) of 1.00 atm and 273 K, 1.00 mol of an ideal gas occupies 22.4 L. This volume is called the **molar volume** of an ideal gas.

The volume of a basketball is approximately 22.4 L.

$$\text{Molar volume}_{STP} = \frac{22.4 \text{ L}}{1.00 \text{ mol}}$$

The molar volume of an ideal gas is a conversion factor that allows us to calculate volume and mole relationships for gases (Fig. 12.12). To illustrate this, let's calculate the number of moles of Ne gas contained in a 1.00 L container at 273 K and 1.00 atm (STP).

$$n_{Ne} = 1.00 \text{ L} \times \frac{1.00 \text{ mol}}{22.4 \text{ L}}$$

$$= 0.0446 \text{ mol}$$

At STP, 0.0446 mol Ne is contained in a volume of 1.00 L. Any 1.00-L sample of an ideal gas at STP would give the same result.

Example Problem 12.6 gives another illustration of Avogadro's law.

Example Problem 12.6

What volume does an 8.40-g sample of Ar gas occupy at STP?

Solution

1. What is unknown?   Volume of Ar in L

2. What is known?   8.20 g Ar; 1.00 mole Ar/40.0 g; 22.4 L Ar/1.00 mol Ar

3. Apply Avogadro's law.

$$V = 8.40 \text{ g Ar} \times \frac{1.00 \text{ mol Ar}}{40.0 \text{ g Ar}} \times \frac{22.4 \text{ L Ar}}{1.00 \text{ mol Ar}}$$

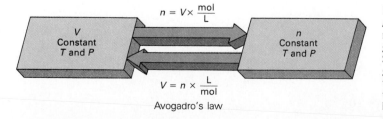

$$n = V \times \frac{mol}{L}$$

V Constant T and P

n Constant T and P

$$V = n \times \frac{L}{mol}$$

Avogadro's law

Figure 12.12
Avogadro's law calculations require the use of the molar gas volume, 22.4 L/mol at STP. The molar volume of a gas is the conversion factor that expresses the ratio between the volume and number of moles of an ideal gas.

$$= 4.70 \text{ L Ar}$$

First, we convert the mass of Ar to moles, using the molar mass, and then we calculate the number of liters by applying the molar volume relationship. We found that 8.4 g Ar occupies 4.70 L when the temperature is 273 K and the pressure is 1.00 atm.

What if the gas is not at standard conditions? We first calculate the volume the gas occupies at standard conditions, and then, applying the combined gas law, we change to the desired set of conditions. Example Problem 12.7 gives an illustration of mole–volume relationships at non-standard conditions.

Example Problem 12.7

What volume does 0.255 g Xe gas occupy at 298 K and 0.543 atm?

Never lose sight of the fact that Avogadro's law, and for that matter all of the gas laws, apply only to gases, not to liquids or solids.

Solution

1. What is unknown? Volume of Xe in L at 298 K and 0.543 atm

2. What is known? 0.255 g Xe; 1 mol Xe/131 g Xe; 298 K; 0.543 atm

3. Calculate the volume of Xe at STP.

It is not necessary to solve this problem in two steps, but for illustrative purposes we will find the volume at STP and then correct the volume to the stated conditions.

$$V_{\text{STP}} = 0.255 \text{ g Xe} \times \frac{1 \text{ mol Xe}}{131 \text{ g Xe}} \times \frac{22.4 \text{ L Xe}}{1 \text{ mol Xe}}$$

$$= 4.36 \times 10^{-2} \text{ L Xe at STP}$$

4. Apply the combined gas law.

We now use the combined gas law to convert from STP conditions of 1 atm (P_1) and 273 K (T_1) to the desired conditions of 0.543 atm (P_2) and 298 K (T_2).

$$V_2 = V_{\text{STP}} \times \frac{P_1}{P_2} \times \frac{T_2}{T_1}$$

$$= 4.36 \times 10^{-2} \text{ L} \times \frac{1.00 \text{ atm}}{0.543 \text{ atm}} \times \frac{298 \text{ K}}{273 \text{ K}}$$

$$= 8.76 \times 10^{-2} \text{ L}$$

We find that our 0.255-g sample of Xe occupies 8.76×10^{-2} L (87.6 mL) at 298 K and 0.543 atm.

12.10 Write two different statements of Avogadro's law.

12.11 Calculate the number of moles of gas molecules contained in each of the following volumes at STP: (a) 4.59 L N_2; (b) 0.107 L O_2, (c) 3.11 mL F_2.

12.12 At STP, what volume would each of the following gases occupy: (a) 9.44 g H_2, (b) 0.066 g He, (c) 321 mg CO?

12.13 Calculate the mass of 12.11 L of radon gas, Rn(g), at STP.

12.14 What volume does 216.4 g $F_2(g)$ occupy at 5.27 atm and 27.6°C?

12.3 IDEAL GAS EQUATION

Boyle's Law, Charles' Law, and Avogadro's Law are mathematically combined to give a relationship for P, V, T and n of an ideal gas. This relationship is called the **ideal gas equation.** It is expressed as follows:

$$PV = nRT$$

in which P is pressure, V is volume, T is temperature, n is moles, and R is the ideal gas constant. Rearranging the equation and solving for R, we find the numerical value of R:

$$\frac{PV}{nT} = \frac{nRT}{nT}$$

Therefore,

$$R = \frac{PV}{nT}$$

Suppose we select a gas sample of 1.00 mol. We know that at STP the volume of 1 mol of an ideal gas is 22.4 L. Substituting these numbers into the equation, we get

$$R = \frac{1.00 \text{ atm} \times 22.4 \text{ L}}{1.00 \text{ mol} \times 298 \text{ K}}$$

$$= 0.0821 \frac{\text{L·atm}}{\text{mol·K}}$$

Thus, the value of R is 0.0821 (L·atm)/(mol·K) for ideal gases.

To use the ideal gas equation, we must either adjust our units so that they cancel the units in the ideal gas constant or we must change the units of the constant, R.

Given three variable properties of a gas and the ideal gas equation, the fourth variable can be calculated. When solving such a problem, it is best to algebraically rearrange the ideal gas equation, isolating the unknown variable on one side. For example, if P, V, and T are known, rearrange the equation and solve for n. Thus we have

$$PV = nRT$$

and we divide both sides by RT

$$\frac{PV}{RT} = \frac{n\cancel{RT}}{\cancel{RT}}$$

which gives

$$n = \frac{PV}{RT}$$

When the equation is successfully rearranged, the units cancel, leaving only the desired units, moles.

$$n = \frac{\cancel{atm} \times \cancel{L}}{\dfrac{\cancel{L} \cdot \cancel{atm} \times \cancel{K}}{mol \cdot \cancel{K}}} = \frac{1}{\dfrac{1}{mol}} = mol$$

Example Problems 12.8 and 12.9 show how the ideal gas equation is used to calculate unknown properties of ideal gases.

––––––––––––––– **Example Problem 12.8** –––––––––––––––

What volume does 2.2 mol N_2 occupy at 440 K and 3.6 atm?

––––––––––––––– **Solution** –––––––––––––––

1. What is unknown?   Volume of N_2 in L

2. What is known? $n = 2.2$ mol N_2; $T = 440$ K; $P = 3.6$ atm; $R = 0.0821$ (L·atm)/(mol·K)

3. Rearrange the ideal gas equation.

$$\frac{P\cancel{V}}{\cancel{P}} = \frac{nRT}{P}$$

$$V = \frac{nRT}{P}$$

4. Substitute known values into the equation.

All units given in this problem correspond to those in the ideal gas constant, so it is only necessary to substitute the numbers into the equation and solve for V.

$$V = \frac{2.2 \text{ mol } N_2 \times 0.0821 \dfrac{L \cdot atm}{mol \cdot K} \times 440 \text{ K}}{3.6 \text{ atm}}$$

$$= \textbf{22 L } \textbf{N}_2$$

A 2.2-mol sample of N_2 at 440 K and 3.6 atm occupies 22 L.

--------- **Example Problem 12.9** ---------

Calculate the temperature of 0.390 mol He that occupies 9850 mL under a pressure of 631 torr.

--------- **Solution** ---------

1. What is unknown?   Temperature of He in K

2. What is known? $n = 0.390$ mol He; $V = 9850$ mL; $P = 631$ torr

3. Change given units to the units of the ideal gas constant.

Because the ideal gas constant is expressed in $(L \cdot atm)/(mol \cdot K)$, the volume must be converted to L and the pressure to atm.

$$V = 9850 \text{ mL} \times \frac{1 \text{ L}}{1000 \text{ mL}} = 9.850 \text{ L}$$

$$P = 631 \text{ torr} \times \frac{1.00 \text{ atm}}{760 \text{ torr}} = 0.830 \text{ atm}$$

4. Rearrange and substitute into the ideal gas equation.

$$\frac{PV}{nR} = \frac{nRT}{nR}$$

$$T = \frac{PV}{nR}$$

$$T = \frac{0.830 \text{ atm} \times 9.850 \text{ L}}{0.390 \text{ mol He} \times 0.0821 \dfrac{L \cdot atm}{mol \cdot K}}$$

At 225 K and 631 torr, 0.390 mol He occupies 9850 mL.

12.15 Calculate the numerical value for the ideal gas constant R in (L·torr)/(mol·K).

REVIEW EXERCISES

12.16 What volume does 2.06 mol Ne occupy at 33.1 atm and 285 K?

12.17 Calculate the temperature, in Celsius, of 0.391 mole of an ideal gas that occupies 1.10 L under a pressure of 7.84 atm.

12.18 How many moles of Ar are contained in a sample that has a measured volume of 78.4 mL under a pressure of 956 torr and a temperature of $-15°C$?

Modification of the ideal gas equation allows us to find the molecular mass of a gas. You should recall that the number of moles of a substance is calculated by dividing its mass in grams (g) by its molar mass (MM), which is the mass of one mole (grams/mole).

Molecular Masses and Densities of Gases

$$\text{Moles} = \frac{\text{mass}}{\frac{\text{mass}}{\text{mol}}}$$

$$= \frac{\text{g}}{\frac{\text{g}}{\text{mol}}}$$

$$= \frac{\text{g}}{\text{MM}}$$

Therefore, we can substitute g/MM for n in the ideal gas equation as follows:

$$PV = nRT$$

$$PV = \frac{\text{g}}{\text{MM}} \cdot RT$$

Rearranging, and isolating molar mass, we have

$$MM = \frac{gRT}{PV}$$

Molar masses of gases are found by determining the mass of a sample at a known set of conditions and applying the above equation. Example Problem 12.10 shows such a calculation.

───────────── **Example Problem 12.10** ─────────────

A 61.5-g sample of an unknown gas occupies 37.3 L at 313 K and 0.924 atm. What is the molar mass of the unknown gas?

--- **Solution** ---

1. What is unknown? g/mol of gas (MM)

2. What is known? 61.5 g; 37.2 L; 313 K; 0.924 atm; $R = 0.0821$ (L·atm)/(mol·K)

3. Apply the modified form of the ideal gas equation.

$$MM = \frac{gRT}{PV}$$

$$= \frac{61.5 \text{ g} \times 0.0821 \frac{\text{L·atm}}{\text{mol·K}} \times 313 \text{ K}}{0.924 \text{ atm} \times 37.2 \text{ L}}$$

$$= \textbf{46.0 g/mol}$$

The molar mass of the gas is found to be 46.0 g/mol.

We can also use the ideal gas equation to calculate the density d of a gas. The density of a gas is the ratio of its mass in grams to its volume V.

$$d = \frac{g}{V}$$

The most frequently used units for the densities of gases are grams per liter or grams per cubic decimeter. Using the modified form of the ideal gas equation, we can solve for g/V, or the density of the gas.

$$PV = \frac{g}{MM} \cdot RT$$

$$d = \frac{g}{V} = \frac{P \cdot MM}{RT}$$

Consequently, the density of the gas only depends on its molecular mass at a constant temperature and pressure. Example Problem 12.11 shows how the density of Cl_2 is calculated.

--- **Example Problem 12.11** ---

Calculate the density of Cl_2 gas at 315 K and 1.15 atm.

Solution

1. What is unknown? d of Cl_2 in g/L

2. What is known? $T = 315$ K; $P = 1.15$ atm; MM = 71.0 g Cl_2/mol Cl_2

3. Substitute the known values into the equation.

$$d = \frac{P \times \text{MM}}{RT}$$

$$= \frac{1.15\ \cancel{\text{atm}} \times \dfrac{71.0\text{ g } Cl_2}{\cancel{\text{mol}}\ Cl_2}}{0.0821\ \dfrac{\text{L}\cdot\cancel{\text{atm}}}{\cancel{\text{mol}\cdot\text{K}}} \times 315\ \cancel{\text{K}}}$$

$$= \mathbf{3.16\ g/L}$$

The density of Cl_2 is 3.16 g/L at 315 K and 1.15 atm.

REVIEW EXERCISES

12.19 Calculate the molecular mass of a 0.148-g sample of an unknown gas that occupies 30.0 mL at 26°C and 813 torr.

12.20 Calculate the density of silicon tetrafluoride gas, SiF_4, at 152°C and 643 torr.

12.4 VOLUME RELATIONSHIPS IN CHEMICAL REACTIONS

Volume–Volume Relationships

Gay-Lussac was the first to recognize that the volumes of combining gases can be expressed as a ratio of whole numbers at constant temperature and pressure. This relationship is now known as the **law of combining volumes.** Let's illustrate the law of combining volumes by considering the gas phase reaction of $H_2(g)$ and $Cl_2(g)$ to produce $HCl(g)$:

$$H_2(g) + Cl_2(g) \longrightarrow 2HCl(g)$$

As we have already discussed, the coefficients in a chemical equation indicate the mole ratio in which the reactants combine and the products form. According to the law of combining volumes, the coefficients also represent the volume ratios. Thus, the equation indicates that at a fixed temperature and pressure, one volume of H_2 combines with one volume of Cl_2 to produce two volumes of HCl (Fig. 12.13).

$$\underset{1\text{ vol}}{H_2(g)} + \underset{1\text{ vol}}{Cl_2(g)} \longrightarrow \underset{2\text{ vol}}{2HCl(g)}$$

Problems involving combining volumes of gases are solved in a similar manner to Chap. 11 stoichiometry problems except that the conversion factor obtained from the coefficients in the equation has volume

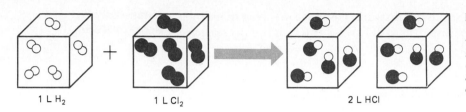

Figure 12.13
If 1 L of $H_2(g)$ is combined with 1 L of $Cl_2(g)$ at the same set of conditions, 2 L of $HCl(g)$ is produced. This is an illustration of the law of combining volumes.

1 L H_2 1 L Cl_2 2 L HCl

units. For example, some conversion factors derived from the above equation are as follows:

$$\frac{1 \text{ L } H_2}{1 \text{ L } Cl_2} \qquad \frac{1 \text{ L } Cl_2}{1 \text{ L } H_2} \qquad \frac{2 \text{ L HCl}}{1 \text{ L } H_2}$$

Example Problem 12.12 is a sample problem involving volume relationships in chemical reactions.

─────────────── **Example Problem 12.12** ───────────────

Consider the Haber reaction:

$$N_2(g) + 3H_2(g) \longrightarrow 2NH_3(g)$$

What volume of ammonia gas, NH_3, forms when 35 L H_2 combines with excess N_2 at a constant temperature and pressure?

─────────────── **Solution** ───────────────

1. What is unknown?   Volume of NH_3 in L
2. What is known?   35 L H_2; 2 L NH_3/3 L H_2
From the equation we find that 2 L NH_3 is produced for each 3 L H_2
3. Apply the factor-label method.

$$35 \text{ L } H_2 \times \frac{2 \text{ L } NH_3}{3 \text{ L } H_2} = ? \text{ L } NH_3$$

4. Perform the indicated math operations.

$$35 \text{ L } H_2 \times \frac{2 \text{ L } NH_3}{3 \text{ L } H_2} = \textbf{23 L } NH_3$$

The theoretical yield of NH_3 is 23 L when 35 L H_2 combines with excess N_2.

On many occasions, solids, liquids, and gases are involved in a single chemical reaction. For example, if $KClO_3(s)$ is heated in the presence of a small amount of MnO_2, $KCl(s)$ and $O_2(g)$ are produced:

Mass–Volume Relationships

$$2KClO_3(s) \xrightarrow[\Delta]{MnO_2} 2KCl(s) + 3O_2(g)$$

From the coefficients in the equation, we find that 2 mol $KClO_3$ decomposes to produce 2 mol KCl and 3 mol O_2. Stated differently, 2 mol $KClO_3$ decomposes to produce 2 mol KCl and 67.2 L O_2 at STP (3 mol $\times$ 22.4 L O_2/mol O_2). Example Problem 12.13 is an illustration of a mass–volume problem.

─────────── **Example Problem 12.13** ───────────

What volume of hydrogen gas, at STP, is liberated when 5.00 g of iron, Fe, is combined with excess hydrochloric acid, HCl, in the following reaction:

$$Fe(s) + 2HCl(aq) \longrightarrow FeCl_2(aq) + H_2(g)$$

─────────── **Solution** ───────────

1. What is unknown?   Volume of H_2 gas at STP in L

2. What is known?   5.00 g Fe; 1 mol Fe/55.8 g Fe; 1 mol Fe/1 mol H_2; and 22.4 L H_2/mol H_2

3. Apply the factor-label method.

$$5.00 \text{ g Fe} \times \frac{1 \text{ mol Fe}}{55.8 \text{ g Fe}} \times \frac{1 \text{ mol } H_2}{1 \text{ mol Fe}} \times \frac{22.4 \text{ L } H_2}{1 \text{ mol } H_2} = ? \text{ L } H_2$$

4. Perform the indicated math operations.

$$5.00 \text{ g Fe} \times \frac{1 \text{ mol Fe}}{55.8 \text{ g Fe}} \times \frac{1 \text{ mol } H_2}{1 \text{ mol Fe}} \times \frac{22.4 \text{ L } H_2}{1 \text{ mol } H_2} = \textbf{2.01 L } H_2$$

When 5.00 g Fe is combined with excess hydrochloric acid, 2.01 L H_2 results at STP conditions.

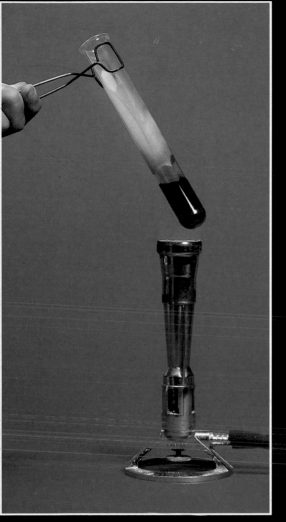

Iron and sulfur   When a mixture of iron (Fe) and sulfur (S$_8$) is heated a chemical change occurs. The product of the reaction is iron(II) sulfide, (FeS). If you would observe the properties of the starting materials and compare them with the final product, you would find that they are quite different. Iron is a metallic substance that has magnetic properties, and sulfur is a pale-yellow solid. In contrast, iron(II) sulfide is a brown-black solid that is not metallic or magnetic. *(Photo Researchers, Inc.)*

Sucrose and sulfuric acid   Sucrose, also known as table sugar, undergoes a chemical reaction with concentrated sulfuric acid, H$_2$SO$_4$. Sucrose belongs to a group of biochemicals called carbohydrates, which are composed of C, H, and O. Because sulfuric acid is a strong dehydrating agent, it readily combines with water; thus, sulfuric acid removes water from sucrose (a white solid) and produces carbon (a black solid). *(Direct Positive Imagery.)*

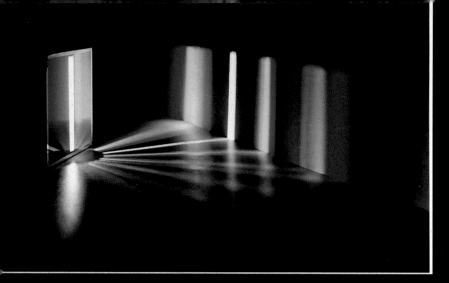

Diffraction of white light   A prism can separate light into its component colors. When white light passes through a prism, the full visible spectrum is observed. Red, orange, yellow, green, blue, indigo, and violet are the component colors of white light arranged from the longest to shortest wavelength. *(Taurus Photo.)*

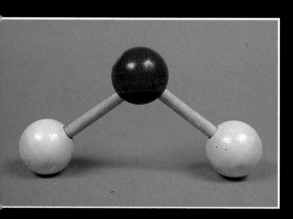

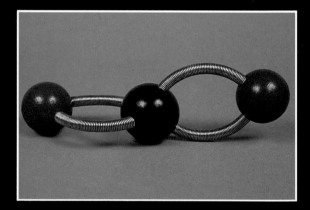

Ball-and-stick models   Ball-and-stick models are sometimes used to show the molecular geometries of molecules. These models show the tetrahedral geometry of methane molecules, the angular geometry of water molecules, and the linear geometry of carbon dioxide molecules. *(Direct Positive Imagery.)*

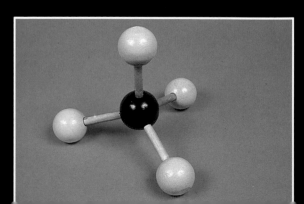

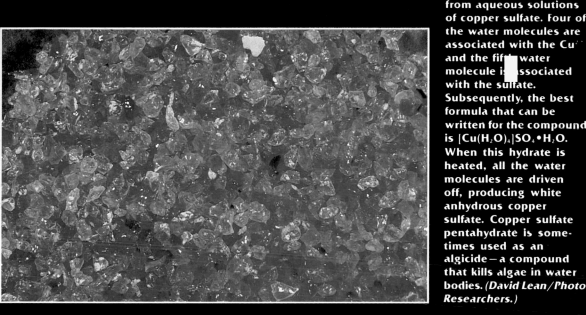

from aqueous solutions of copper sulfate. Four of the water molecules are associated with the Cu²⁺ and the fifth water molecule is associated with the sulfate. Subsequently, the best formula that can be written for the compound is $[Cu(H_2O)_4]SO_4 \cdot H_2O$. When this hydrate is heated, all the water molecules are driven off, producing white anhydrous copper sulfate. Copper sulfate pentahydrate is some- times used as an algicide—a compound that kills algae in water bodies. *(David Lean/Photo Researchers.)*

Computer chip All computers and many electronic devices use semiconductors called integrated circuits (ICs). These very small electronic devices can be programmed to do many functions or can store a vast quantity of information. ICs are constructed from ultrapure wafers of Si. This very pure form of Si is prepared when impure Si is heated with chlorine, $Cl_2(g)$, to produce silicon tetrachloride, $SiCl_4$, a liquid that boils at 58°C. Silicon with $10^{-8}\%$ impurities is produced when $SiCl_4$ and hydrogen gas, $H_2(g)$, are passed through a hot tube. *(Phillip A. Harrington/ Peter Arnold, Inc.)*

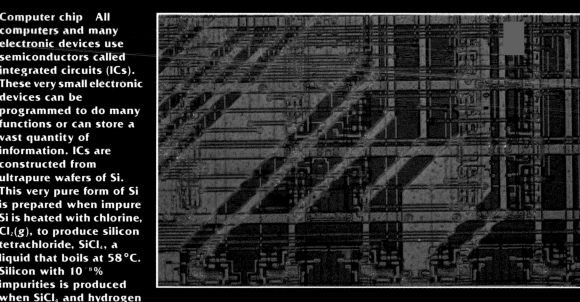

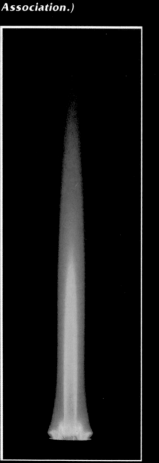

Natural gas flame   A flame results when various gases are combusted. The characteristic blue color is observed when natural gas, a mixture of hydrocarbons, is burned. *(Amencangas Association.)*

Silver nitrate and potassium chromate   A metathesis reaction occurs when the colorless silver nitrate solution, $AgNO_3(aq)$, combines with the yellow potassium chromate solution, $K_2CrO_4(aq)$. The reddish-brown precipitate silver chromate, $Ag_2CrO_4(s)$, results from the reaction. In addition, aqueous potassium nitrate, $KNO_3(aq)$, remains in solution as dissolved ions.

Gemstones Gemstones come in a wide variety of colors and can be cut to produce desirable shapes. The colors of gemstones depend on composition of the mineral that composes them and the impurities that are found in the structure. For example, when corundum, a form of aluminum oxide, has chromium oxide impurities it is called a ruby. If the impurities are cobalt and titanium, then it is called a sapphire, and if the impurities are iron oxides, then it is called oriental topaz. *(Malcolm S. Kirk/Peter Arnold, Inc.)*

Sybil molecule Molecular models are constructed to help visualize the three-dimensional structure of molecules. Molecular models can be actual models made from wood, metal, or plastics, or they can be computer generated. This figure shows a computer-generated model of a segment of the very complex DNA (deoxyribose nucleic acid) molecule, the double helix. DNA molecules are principally found in the nuclei of living cells and they control the activities in the cell. Computers can display the structure of molecules from many different views, and some computers can output three-dimensional images. *(Peter Arnold, Inc.)*

Fluorite crystal The mineral fluorite, also called fluorospar, is pictured using polarized light. Fluorite, CaF_2, is a transparent, crystalline compound that comes in many different colors. It is used to make glass and flux. *(Manfred*

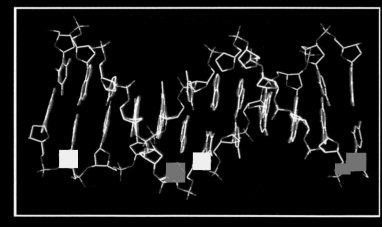

Hot-air balloon The density of air depends on its temperature — the higher the temperature the lower the density of air. When fluids are mixed the lower density ones move to the top and the higher density ones fall to the bottom. Thus, when the air is heated inside a hot-air balloon, its density decreases, which causes the balloon to rise. By heating or cooling the air inside a hot-air balloon, it either rises or falls. *(Judy Gurowitz — International Stock Photography.)*

Iodine crystal Iodine is a purple-black solid that sublimes at room temperature to produce a deep-purple vapor. The element iodine was first isolated in 1811 from seaweeds and has been used in medicines for many years to prevent gout. Iodine is commercially produced when I^- is reduced by Cl_2 or it can be synthesized from sodium iodate, $NaIO_3$, (an impurity of Chilean saltpeter, $NaNO_3$). One of the many uses of iodine is in the synthesis of silver iodide, a component of photographic films. *(Michael Abbey — Photo Researchers.)*

Flame tests Many metallic compounds when heated release characteristic wavelengths of light. In this figure, you see the colors released by strontium (red), potassium (lilac), and copper (green). Metals in compounds can sometimes be identified by heating a small quantity in a bunsen burner flame and observing the color. The colors result when electrons drop from higher to lower energy states. *(Andrew McClenaghan Photo Researchers.)*

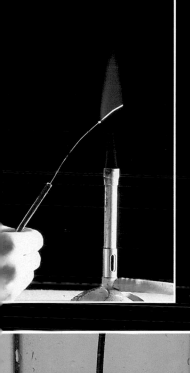

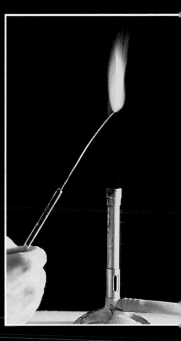

Copper wire and silver nitrate A single displacement reaction occurs when copper metal combines with a silver nitrate solution. Crystals of metallic silver seem to grow from the surface of the copper wire. After a period of time, the colorless silver nitrate solution is replaced with a blue copper nitrate solution. *(Direct Positive Imagery.)*

Glowing storage pool Most of the spent nuclear fuel rods from nuclear power plants are stored under water in storage pools. The intense radiation released by this highly radioactive waste creates the blue glow. One of the major problems faced by nuclear power plants is how to safely store nuclear waste materials. *(Photo Researchers, Inc.)*

Starch molecules This molecular model shows the three-dimensional shape of the very complex starch molecule. Starch is one of the most important carbohydrates found in living things. *(Peter Arnold, Inc.)*

While many gas stoichiometry problems specify STP conditions, it should be noted that few reactions actually occur at STP conditions. Problems that involve reactions at nonstandard conditions are solved in one of two ways: (1) Assume STP conditions, solve the stoichiometry problem, and then apply the combined gas law, or (2) from the equation, find the number of moles of gaseous product, and then use the ideal gas equation. Both of these methods are demonstrated in Example Problem 12.14.

─────────────── **Example Problem 12.14** ───────────────

Barium carbonate, $BaCO_3$, is heated to produce barium oxide, BaO, and carbon dioxide, CO_2:

$$BaCO_3(s) \xrightarrow{\Delta} BaO(s) + CO_2(g)$$

What volume of CO_2 is produced, at 710 torr and 425 K, when 3.00 g $BaCO_3$ is decomposed?

─────────────── **Solution** ───────────────

1. What is unknown? Volume of CO_2 in L at 710 torr and 425 K

2. What is known? 3.00 g $BaCO_3$; 1 mol $BaCO_3$/197 g $BaCO_3$; 1 mol CO_2/1 mol $BaCO_3$; 22.4 L CO_2/mol CO_2

Method 1: Assume STP conditions and correct to the desired conditions.

$$3.00 \text{ g } BaCO_3 \times \frac{1 \text{ mol } BaCO_3}{197 \text{ g } BaCO_3} \times \frac{1 \text{ mol } CO_2}{1 \text{ mol } BaCO_3} \times \frac{22.4 \text{ L } CO_2}{1 \text{ mol } CO_2} \times$$

$$\frac{760 \text{ torr}}{710 \text{ torr}} \times \frac{425 \text{ K}}{273 \text{ K}} = \textbf{0.568 L } CO_2$$

Here, we solved the problem assuming STP conditions. The molar volume is used to calculate the number of liters of CO_2, and then pressure and temperature factors are employed to change to the desired conditions. Standard temperature and pressure are T_1 and P_1, respectively, and 425 K and 710 torr are T_2 and P_2.

Method 2: Calculate the number of moles of CO_2, and substitute into the ideal gas equation.

$$3.00 \text{ g } BaCO_3 \times \frac{1 \text{ mol } BaCO_3}{197 \text{ g } BaCO_3} \times \frac{1 \text{ mol } CO_2}{1 \text{ mol } BaCO_3} = 0.0152 \text{ mol } CO_2$$

$$710 \text{ torr} \times \frac{1 \text{ atm}}{760 \text{ torr}} = 0.934 \text{ atm}$$

Substitute the known values into the ideal gas equation.

$$PV = nRT \quad \text{or} \quad V = \frac{nRT}{P}$$

$$V = \frac{0.0152 \, \text{mol } CO_2 \times 0.0821 \, \dfrac{\text{L·atm}}{\text{mol·K}} \times 425 \, \text{K}}{0.934 \, \text{atm}}$$

$$= \textbf{0.568 L } CO_2$$

The use of either method gives the same numerical result 0.568 L CO_2.

12.21 Consider the following gas phase reaction:

$$2NO(g) + O_2(g) \longrightarrow 2NO_2(g)$$

(a) What volume of NO_2 forms at STP when 717 L NO combines with excess O_2? (b) At STP, how many milliliters of NO are required to combine exactly with 83.2 mL O_2? (c) At 45°C and 9.24 atm, what volume of O_2 is required to produce 78.5 L NO_2?

12.22 (a) What volume of O_2 gas is released at STP when 2.79 g H_2O_2 decomposes to H_2O and O_2?

$$2H_2O_2(l) \longrightarrow 2H_2O(l) + O_2(g)$$

(b) What mass of H_2O_2 must be decomposed to produce 33.7 L O_2 at STP conditions?

12.23 What volume of O_2 gas is produced at 221 torr and 416 K when 0.811 g $KClO_3$ decomposes to KCl and O_2?

12.5 MIXTURES OF GASES

One of the properties of gases is that any number of nonreacting gases can be mixed in any proportion to produce a gaseous solution (homogeneous mixture). How do we deal quantitatively with gas mixtures?

From Avogadro's law, we know that 1 mol of an ideal gas occupies 22.4 L at STP conditions, and 2 mol occupies 2 × 22.4 L, or 44.8 L. What if we first placed 1 mol of a gas in a container, and then placed 1 mol of a second, nonreacting gas in the same container? Because there are now 2 mol of gas in the container, the expected volume of the gas mixture is also 2 × 22.4 L, or 44.8 L (Fig. 12.14). In other words, if the pressure and temperature are constant, the total volume of the gaseous

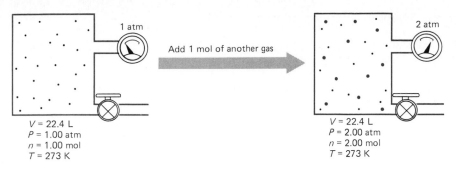

Figure 12.14
One mole of an ideal gas
occupies 22.4 L at 1.00 atm
and 273 K. If 1.00 mol of
another gas is added to the
container, with no change in
volume and temperature, the
pressure increases to
2.00 atm. The total pressure
of a mixture of gases is
equal to the sum of the partial pressures of each gas in
the mixture.

mixture depends on the total number of moles of gas particles (n_{total}),
which equals the sum of the moles of the individual gases in the mixture:

$$n_{total} = n_1 + n_2 + n_3 + \ \ldots$$

in which n_1, n_2, n_3 are the number of moles of gases 1, 2, and 3.

   If the volume and temperature are constant, what happens to the
total pressure when gases are mixed? If a 22.4 L container at 273 K
contains 1 mol of gas, the pressure gauge on the container reads 1 atm.
When 1 mol of another gas is pumped into the container—which gives a
total of 2 mol—the reading on the pressure gauge increases to 2 atm.
Addition of 1 mol of a third gas increases the pressure to 3 atm, and so
on. Hence, the total pressure of a mixture of gases is equal to the sum of
the partial pressures of the gases in the mixture. A **partial pressure** is
the pressure exerted by one gas in a gaseous mixture.

   John Dalton was the first scientist to describe the pressures of gaseous mixtures; accordingly, this relationship is now known as **Dalton's
law of partial pressures.** Mathematically, the law of partial pressures is
as follows:

$$P_{total} = P_1 + P_2 + P_3 + \ \ldots$$

in which P_1, P_2, and P_3 are partial pressures of gases in the mixture and
P_{total} is the total pressure of all gases. Dalton's law is illustrated in Example Problem 12.15.

*Even though Dalton is
most famous for his pos-
tulates regarding the
atomic nature of matter,
he was keenly interested
in meteorology. Dalton's
drive to understand the
weather led him to study
air, a mixture of gases.*

———————————— **Example Problem 12.15** ————————————

   Air exhaled by people is a mixture of $N_2(g)$, $O_2(g)$, $H_2O(g)$, and
$CO_2(g)$. The total pressure of this mixture is 760 torr. If the partial
pressures of N_2, O_2, and H_2O are 565, 120, and 47 torr, respectively,
what is the partial pressure of CO_2 in exhaled air?

────────────────── **Solution** ──────────────────

1. What is unknown? P_{CO_2} in torr

2. What is known? $P_{N_2} = 565$ torr; $P_{O_2} = 120$ torr; $P_{H_2O} = 47$ torr; $P_{total} = 760$ torr

3. Apply Dalton's law of partial pressures.

$$P_{total} = P_{N_2} + P_{O_2} + P_{N_2O} + P_{CO_2}$$

$$760 \text{ torr} = 565 \text{ torr} + 120 \text{ torr} + 47 \text{ torr} + P_{CO_2}$$

$$P_{CO_2} = 760 \text{ torr} - (565 + 120 + 47) \text{ torr}$$

$$= \textbf{28 torr}$$

The partial pressure of CO_2 in exhaled air is 28 torr.

═══

Dalton's law can be used to calculate the dry volume of insoluble gases collected over water. One method used to collect a gas over water is as follows: Water is added to a gas bottle (or other container), which is then inverted over a trough (Fig. 12.15). Mixed with the collected gas is water vapor.

The dry volume of a gas is the volume a pure gas would occupy if no water vapor were present.

Where does the water vapor come from? It comes from the evaporation of water in the gas bottle and trough. The total pressure exerted by the collected gas is the sum of the pressures of the gas and water vapor.

$$P_{total} = P_{gas} + P_{H_2O}$$

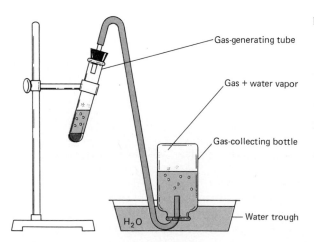

Figure 12.15
Gas produced in a gas-generating tube is collected by water displacement in a gas-collecting bottle. Water evaporates and mixes with the gas produced. Hence, to find the pressure of the dry gas, the vapor pressure of water is subtracted from the total pressure of the gaseous mixture.

Gas-generating tube

Gas + water vapor

Gas-collecting bottle

Water trough

H_2O

To calculate the pressure of the pure gas, subtract the pressure of the water vapor (vapor pressure of water) from the total pressure.

$$P_{gas} = P_{total} - P_{H_2O}$$

Table 12.3 lists the vapor pressures of water from 20 to 30°C. Vapor pressures of liquids are discussed in more detail in Chap. 13. At this time, just keep in mind that the vapor pressure of water depends on its temperature.

TABLE 12.3   VAPOR PRESSURES OF WATER FROM 20°C TO 30°C

Temperature, °C (K)	Vapor pressure, torr	Vapor pressure, atm
20 (293)	17.5	0.0230
21 (294)	18.7	0.0246
22 (295)	19.8	0.0261
23 (296)	21.1	0.0278
24 (297)	22.4	0.0295
25 (298)	23.8	0.0313
26 (299)	25.2	0.0332
27 (300)	26.7	0.0351
28 (301)	28.3	0.0372
29 (302)	30.0	0.0395
30 (303)	31.8	0.0418

Example Problem 12.16 provides an illustration of Dalton's law and the collection of gases over water.

———————————— **Example Problem 12.16** ————————————

Hydrogen gas, H_2, is generated by adding magnesium metal, Mg, to a hydrochloric acid solution:

$$Mg(s) + 2HCl(aq) \longrightarrow MgCl_2(aq) + H_2(g)$$

The H_2 gas is collected over water in a trough; 63.4 mL H_2 is formed at 27.0°C and 752.1 torr. What mass of magnesium metal is necessary to produce the H_2?

———————————— **Solution** ————————————

1. What is unknown?   Mass of Mg in g

2. What is known?   63.4 mL H_2; 752.1 torr; 27.0°C; 1 mol H_2/1 mol Mg; 1 mol Mg/24.3 g Mg; P_{H_2O} = 26.7 torr at 27°C (from Table 12.3)

3. Apply Dalton's law to find the pressure due to H_2 only.
From Dalton's law, we know that the total pressure, 752 torr, equals the sum of the pressures of H_2 and H_2O:

$$P_{total} = P_{H_2} + P_{H_2O}$$

$$752.1 \text{ torr} = P_{H_2} + 26.7 \text{ torr}$$

Hence, if 26.7 torr is subtracted from both sides of the equation, the pressure of pure H_2 is obtained:

$$P_{H_2} = P_{total} - P_{H_2O}$$

$$= 752.1 \text{ torr} - 26.7 \text{ torr}$$

$$= 725.4 \text{ torr}$$

Of the total pressure measured for the mixture, 725.4 torr is exerted by H_2; our remaining calculations should use only this dry pressure.

4. Calculate the number of moles of H_2.
The number of moles of H_2 may be calculated in two ways; let's use the ideal gas equation.

$$P = 725.4 \text{ torr} \times \frac{1 \text{ atm}}{760 \text{ torr}} = 0.9545 \text{ atm}$$

$$T = 27.0°C + 273.2 = 300.2 \text{ K}$$

$$V = 63.4 \text{ mL} \times \frac{1 \text{ L}}{1000 \text{ mL}} = 0.0634 \text{ L}$$

Rearranging and substituting values into the ideal gas equation, we get

$$n = \frac{PV}{RT}$$

$$= \frac{0.9545 \text{ atm} \times 0.0634 \text{ L}}{0.08205 \frac{\text{L·atm}}{\text{mol·K}} \times 300.2 \text{ K}}$$

$$= 0.00246 \text{ mol } H_2$$

5. Apply the factor-label method to find the mass of Mg.

$$0.00246 \text{ mol } H_2 \times \frac{1 \text{ mol Mg}}{1 \text{ mol } H_2} \times \frac{24.3 \text{ g Mg}}{1 \text{ mol Mg}} = \textbf{0.0598 g Mg}$$

Initially, 0.0598 g Mg was present to produce 63.4 mL H_2 over water at the stated conditions.

12.24 *(a)* State Dalton's law of partial pressures in words. *(b)* Write a mathematical expression of Dalton's law.

12.25 Using the kinetic molecular theory, how can Dalton's law be explained?

12.26 Calculate the total pressure of a gaseous mixture that contains the following gases at the stated pressures: 6.08 atm, H_2; 124 torr, He; 129 kPa, Ne.

12.27 A sample of N_2 gas is collected over water. The pressure of the wet N_2 gas at 33°C is 815 torr. If the vapor pressure of water is 37.73 torr at 33°C, what is the dry pressure of N_2?

12.6 THE ATMOSPHERE

The gas mixture most important to living things is the earth's atmosphere. Nitrogen, N_2, and oxygen, O_2, are the primary gases in the atmosphere, making up about 99% of its volume. All other gases are found in relatively small amounts. Table 12.4 lists the principal atmospheric gases and their percents by volume.

TABLE 12.4 COMPOSITION OF THE ATMOSPHERE AT SEA LEVEL

Gas	Formula	Percent, %v/v
Nitrogen	N_2	78
Oxygen	O_2	21
Argon	Ar	0.9
Carbon dioxide	CO_2	0.03–0.04
Neon	Ne	0.0012
Helium	He	0.0005
Krypton	Kr	0.0001
Ozone	O_3	0.00006
Hydrogen	H_2	0.00005
Xenon	Xe	0.000009

The atmosphere is the mixture of gases that surrounds the earth. It acts as a protective "blanket" that traps energy from the sun and blocks out different types of radiation.

In addition to the gases listed in Table 12.4, variable amounts of water vapor, sulfur oxides, nitrogen oxides, hydrocarbons, carbon monoxide, and other trace gases are found in the atmosphere. Sulfur oxides, carbon monoxide, and nitrogen oxides are pollutant gases released when fossil fuels are burned (Fig. 12.16).

Let's take a closer look at the principal gases in the atmosphere: nitrogen, oxygen, and the noble gases.

Some of the pollutant gases in the atmosphere are SO_2, SO_3, CO, NO, NO_2, and various hydrocarbons (carbon–hydrogen compounds).

Nitrogen

Nitrogen, N_2, was discovered by Daniel Rutherford in 1772. It is a colorless, odorless, and tasteless gas. The boiling point of N_2 is −195.8°C at 1 atm, and its freezing point is −210°C. The density of N_2, 1.261 g/L, is slightly less than the density of air.

A nitrogen molecule possesses a strong triple bond that requires a large amount of energy to break (945 kJ/mol or 226 kcal/mol).

$$: N \equiv N :$$

Figure 12.16
Besides the normal compo-
nents of the atmosphere,
human activities contribute
gases such as sulfur oxides,
nitrogen oxides, carbon
monoxide, and hydrocar-
bons. These gases are re-
sponsible for polluted air.
(*Documerica.*)

CHEM TOPIC: Air Pollution

Air pollutants are classified as being primary or secondary pollutants. A *primary pollutant* is emitted directly into the air. The most significant primary air pollutants are sulfur oxides, carbon monoxide, nitrogen oxides, hydrocarbons, and particulate matter (small particles). In most cities, power plants, industry, and incinerators are the major sources of sulfur oxides and particulate matter. Transportation, including automobiles, trucks, and planes, is the major source of carbon monoxide, nitrogen oxides, and hydrocarbons. After the pollutants reach the atmosphere, they undergo chemical reactions to produce new pollutants, the *secondary pollutants.* Many of the secondary pollutants result from light-initiated reactions—photochemical reactions. Ozone and peroxyacylnitrates (PANS) are two examples of secondary pollutants.

Air pollution presently affects all parts of the world, but large metropolitan areas suffer the most from the effects of air pollution. The term *smog,* which is a contraction of "smoke" and "fog," is used to describe polluted air. This term was proposed by the London physician H. Des Voeux in 1905. He was attempting to describe the smoke, pollutant gases, and fog that had claimed the lives of many people in Glasgow and Edinburgh. Today, we classify smog according to the type of pollutants that compose the smog. One type is characterized by large quantities of sulfuric and sulfurous acids that result from the burning of fuels. These conditions are frequently observed in cities in Europe and the eastern United States. Some scientists call this *London smog* because London used to have many bad episodes of this type of smog. In 1952, London experienced a 4-day siege of severe smog that was responsible for the death of 4000 people. The second general type of smog is found in cities such as Los Angeles and results from nitrogen oxides and hydrocarbons exhausted mainly from automobiles. In the atmosphere these compounds undergo many photochemical reactions, producing a wide variety of secondary pollutants. This type of pollution is called *photochemical smog* even though no smoke or fog is involved.

Because of this strong bond, N_2 is quite unreactive. As a result of being a stable compound, nitrogen gas is used as an "inert atmosphere." Canned foods remain fresher if they are sealed with pure nitrogen gas instead of air. When chemists handle substances that react with the O_2 in the air, they place these substances in containers filled with N_2 or another unreactive gas.

Because of the inert nature of the gas, it is difficult to tap the huge reserve of nitrogen in the atmosphere. The Haber process is an important procedure for **fixing** N_2, i.e., the conversion of atmospheric N_2 to nitrogen compounds.

$$N_2 + 3H_2 \xrightarrow[\Delta]{Fe} 2NH_3$$

Haber reaction

In nature, N_2 is fixed by microorganisms that have the capacity to convert N_2 to NH_3. Synthesis of NH_3 from N_2 is a principal step in what is called the **nitrogen cycle.** Figure 12.17 illustrates the main components of the nitrogen cycle. After NH_3 is formed, it is converted by other microorganisms to nitrites, nitrates, and ultimately into proteins and other biologically important nitrogen compounds.

Rutherford was remotely related to Sir Walter Scott. Rutherford was a firm believer in Stahl's phlogiston theory. When he discovered nitrogen, he called it "dephlogisticated air" because it did not support combustion or sustain life.

Proteins are one of the four principal groups of compounds found in living systems; the others are lipids, carbohydrates, and nucleic acids.

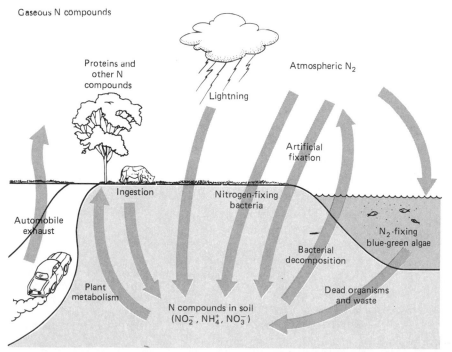

Figure 12.17
All the nitrogen on earth is part of the nitrogen cycle. Nitrogen in many forms moves through the atmosphere, lithosphere, and biosphere. $N_2(g)$ is the main component of the atmosphere. In the lithosphere, nitrogen-containing salts are the primary components, and in the biosphere proteins are the principal components.

Credit for the discovery of oxygen, O_2, is usually given to Joseph Priestley, who in 1774 heated mercury(II) oxide, HgO, liberating O_2.

$$2HgO(s) \xrightarrow{\Delta} 2Hg(l) + O_2(g)$$

Karl Wilhelm Scheele had discovered O_2 about 3 years prior to Priestley, but he neglected to have his finding published until 1777.

Like nitrogen, O_2 is colorless, tasteless, and odorless. Oxygen is slightly more dense than air; its density is 1.429 g/L, compared with air's density of 1.292 g/L. The water solubility of O_2 is relatively low—only 28.3 mL of O_2 gas dissolves per liter of water at 25°C. Nevertheless, this small quantity of O_2 sustains the lives of animals that live in water. The melting point of O_2 is −218°C, and its boiling point is −183°C at 1 atm.

Oxygen combines with most other elements. We have already seen how O_2 reacts with both metals and nonmetals to form a diversity of oxides. Oxygen atoms are bonded by relatively strong double bonds in oxygen molecules; thus these molecules are relatively stable. Most reactions that O_2 undergoes require temperatures and pressures above room conditions. But in some cases, as in rusting of metals, O_2 combines slowly at room temperature.

Industrially, most O_2 is prepared by isolating it from air that has been cooled to very low temperatures; a similar method is employed to produce N_2 gas. O_2 is used in the steel industry to accelerate combustion and increase reaction temperatures. Smaller quantities of O_2 are used in the medical industry as part of life-support systems. The aerospace industry uses O_2 as a component of liquid rocket fuels. Each year, larger quantities of O_2 are being used to treat sewage and more effectively reduce the concentration of impurities in dirty water.

Oxygen is the sustainer of life. Animals need a constant supply of O_2. Even a brief interruption of O_2 can cause brain damage or death. Oxygen is needed by cells to produce energy in the respiratory chain, which is also called the electron transport system. To complete the respiratory chain, O_2 must be present to combine with electrons and hydrogen ions to produce water. O_2 is transported in the blood from the lungs to the cells by a large molecule called hemoglobin.

Karl Wilhelm Scheele (1742–1786) was self-educated in chemistry and started as an apothecary. Scheele's accomplishments in chemistry include the discovery of citric acid, hydrogen cyanide, lactic acid, oxalic acid, and hydrogen fluoride gas. He was also part of a research group that discovered the elements W, Mo, Ba, and Mg.

Gases are liquefied by compressing, cooling, and then allowing them to expand through a small opening, further cooling them. The cycle is then repeated until the gases liquefy.

Cellular respiration occurs in regions within the cell called mitochondria. A complex series of reactions takes place in mitochondria to produce energy for the cell.

Noble Gases

While they are not major components of the atmosphere, the noble gases are very interesting substances. To begin our discussion, let's consider a brief history of the discovery of the noble gases.

When Mendeleev proposed his famous periodic table, the noble gases had not been discovered; thus, in Mendeleev's periodic table the alkali metals directly followed the halogens. A French astronomer, Pierre Janssen (1824–1907), discovered the first noble gas, helium, in 1868. Helium was not discovered on earth; Janssen detected it in the sun. Using a technique called spectroscopic analysis, Janssen identified

Janssen traveled throughout the world to observe astronomical phenomena. He made his famous discovery of He in the sun while in India.

He from the light it gives off at high temperatures. Many scientists were skeptical of his discovery, and it was not until 1895, when Sir William Ramsay (1852–1916) found He trapped in rocks on earth, that the existence of He was generally accepted. The name helium is derived from the Greek word *helios*, which means "the sun."

The discovery of helium sparked a great deal of interest after chemists realized that He did not fit into their periodic table. In 1894, Lord Rayleigh (John William-Strutt Rayleigh, 1842–1919) showed that nitrogen gas isolated from the air was heavier than nitrogen gas obtained from decomposition reactions of nitrogen-containing salts. Rayleigh tried to solve this enigma, but became frustrated and finally wrote a letter to *Nature* (one of Britain's most respected scientific journals), asking for help or suggestions. Sir William Ramsay answered Rayleigh's plea for help, and later that year, using spectroscopic techniques, discovered that argon (meaning the "lazy one" or "inert one") was mixed with the nitrogen, producing the higher density.

Ramsay continued the search for other noble gases; in 1898, he identified and isolated neon, Ne ("new one"); Krypton, Kr ("hidden one"); and xenon, Xe ("stranger"). Radon, the remaining noble gas, was discovered by Freidrich Dorn as a radioactive decay product of radium; however, Rn was named and further studied by Marie and Pierre Curie.

Physical properties of the noble gases are listed in Table 12.5. The boiling points of the noble gases are among the lowest on the periodic table. Helium remains in the gas phase to within a few degrees of absolute zero. The low boiling point of helium is accounted for in terms of weak forces among the atoms, making He one of the most ideal gases. As a result of its low boiling point, liquid He is utilized in cryogenic research, i.e., the study of very low temperatures. Helium is also used to fill "lighter than air" balloons, and is mixed with O_2 in divers' tanks to prevent the narcotic effect of N_2 breathed at high pressures.

Rayleigh was a physicist who became interested in a controversy regarding the scale of atomic weights. To gather information about atomic weights, he needed to know the densities of gases. During the process of measuring the density of N_2, Rayleigh found the apparent discrepancy in mass.

TABLE 12.5 PHYSICAL PROPERTIES OF NOBLE GASES

Noble gas	Melting point, °C	Boiling point, °C	Density, g/L at STP
Helium	−272.2	−268.9	0.179
Neon	−248.7	−246.1	0.900
Argon	−189.2	−185.9	1.78
Krypton	−156.6	−153.4	3.75
Xenon	−111.9	−108.1	5.90

Small amounts of neon are sealed inside lamps and the tubes of neon signs; when an electric current is passed through the neon, it releases its characteristic red glow. Argon, the most abundant noble gas on earth, is placed in incandescent light bulbs to decrease the vaporization and oxidation of the tungsten, W, filaments, thus prolonging the life of the light bulb.

As we have seen, the noble gases are very unreactive and only combine with highly electronegative nonmetals such as F and O. Serious studies of noble gas compounds began in the early 1960s, when Neil Bartlett first synthesized a number of noble gas compounds. For example, xenon tetrafluoride, XeF_4, exists as colorless crystals that melt near 90°C and can be stored at room temperature in a glass container for relatively long periods without decomposing.

REVIEW EXERCISES

12.28 What are the five most abundant gases in the atmosphere?

12.29 What is the nitrogen cycle, and why is it important to living things?

12.30 Name a common use for each of the following gases: *(a)* N_2, *(b)* O_2, *(c)* He.

12.31. Which noble gas would be expected to act most like an ideal gas?

SUMMARY

The properties of gases are explained theoretically by the **kinetic molecular theory,** which is a set of assumptions that concerns the behavior of atoms and molecules that make up an ideal gas.

Ideal gas laws explain relationships between volume, pressure, temperature, and number of moles of gas particles in ideal gas samples. **Boyle's law** states that the volume of an ideal gas is inversely proportional to the applied pressure at constant temperature and number of moles. Gas pressure is the force that gas particles exert on the walls of their containers. Pressure is measured in atmosphere, kilopascals, and torr: 1 atm = 760 torr = 101 kPa.

Charles' law states that the volume of a gas is directly proportional to its Kelvin temperature at constant pressure and number of moles. **Avogadro's law** states that the volume of a gas is directly proportional to the number of moles of gas at constant pressure and temperature, or, stated another way: At a constant set of conditions, equal volumes of ideal gases contain the same number of molecules. At standard temperature and pressure (273 K and 1 atm), the volume of 1.00 mol of an ideal gas is 22.4 L.

Various mathematical expressions are derived from the gas laws. Two of the most important ones are the **combined gas law,**

$$\frac{P_1 V_1}{T_1} = \frac{P_2 V_2}{T_2}$$

which is used to calculate the final set of conditions given the initial conditions, and the **ideal gas equation,**

$$PV = nRT$$

for which one of the four gas variables can be calculated given the other three. The R in the equation is the ideal gas constant, 0.0821 (L·atm)/(mol·K).

Gay-Lussac was the first to show that the volumes of combining gases can be expressed as ratios of whole numbers. In other words, given a correctly balanced equation, the coefficients indicate the volume ratios in which gases combine.

When gases are mixed, the sum of the pressures of the individual gases equals the total pressure exerted by the gaseous mixture. This is a statement of **Dalton's law of partial pressures,** in which the partial pressure is the pressure exerted by a gas in a gaseous mixture.

The atmosphere of the earth is composed of a mixture of gases, principally N_2 and O_2 (99%). Argon, CO_2, H_2O vapor, and approximately 10 other gases comprise the remaining 1% of the atmosphere. Nitrogen gas is composed of stable, diatomic nitrogen molecules, N_2, which, in nature, are converted (fixed) by microorganisms to nitrogen compounds that ultimately enter living systems as proteins and other biologically important compounds.

The physical properties of O_2 are similar to, but different from, those of nitrogen. Oxygen is relatively stable but is more reactive than N_2. When heated, O_2 combines with most other elements. Oxygen plays a central role in the functioning of living systems.

Argon is the third most abundant gas in the atmosphere. Argon and the other noble gases are the most inert elements. Historically, their rarity and inertness made them elusive. Noble gases were not discovered until the end of the nineteenth century.

KEY TERMS

absolute zero	ideal gas	manometer	pressure
atmosphere	inert atmosphere	mmHg	STP
barometer	inverse proportion	nitrogen cycle	torr
cryogenics	kinetic energy	partial pressure	
direct proportion	kinetic molecular theory	pascal	

EXERCISES*

12.32 Define each of the following terms: kinetic molecular theory, kinetic energy, ideal gas, pressure, pascal, mmHg, torr, atmosphere, barometer, manometer, inverse proportion, direct proportion, absolute zero, STP, partial pressure, inert atmosphere, nitrogen cycle, cryogenics.

Kinetic Molecular Theory

12.33 Use the kinetic molecular theory to completely explain each of the following: (a) The gaseous state is the most compressible state of matter. (b) Two nonreacting gases can be mixed in any proportions. (c) The volume of a gas increases when heated. (d) Gases always take the shape of their containers. (e) Molecules that compose gases do not settle to the bottom of their containers on standing.

12.34 Why do gases exhibit more ideal properties as their pressures are decreased and their temperatures are increased?

12.35 (a) What causes an increase in the average kinetic energy of the particles in a gas? (b) What would happen to gas particles if their collisions with each other and the walls of the container were not perfectly elastic?

Pressure

12.36 (a) How is gas pressure defined? (b) What is the SI unit of pressure.

12.37 Explain how a mercury barometer is used to measure atmospheric pressure.

12.38 Why could a barometer also be used as an altimeter, a device for measuring the height above the ground?

12.39 (a) How is a manometer constructed? (b) How is the pressure of a gas sample measured with a manometer?

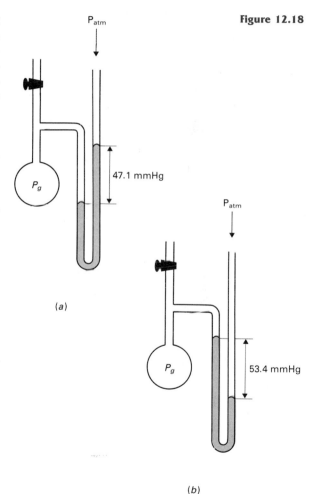

Figure 12.18

P_{atm}

P_g

47.1 mmHg

(a)

P_{atm}

P_g

53.4 mmHg

(b)

12.40 For each of the following in Fig. 12.18, determine the pressure of the gas sample if the atmospheric pressure is 752 torr.

12.41 Account for the fact that, even at sea level, atmospheric pressure is constantly changing.

*For exercise numbers printed in color, answers can be found at the back of the book.

12.42 Convert 4.04×10^3 kilopascals to (a) pascals, (b) torr, (c) atmospheres, (d) millimeters of mercury.

12.43 Change 2.953 atm to (a) torr, (b) pascals, (c) kilopascals.

12.44 Convert 784 mmHg to (a) torr, (b) atmospheres, (c) pascals, (d) kilopascals.

12.45 Change 9282 torr to (a) atmospheres, (b) pascals, (c) kilopascals.

12.46 Perform the following pressure conversions:
(a) 5.49 atm = ? torr
(b) 0.0944 torr = ? Pa
(c) 6.26×10^3 torr = ? atm
(d) 0.00912 kPa = ? torr
(e) 1726 torr = ? mmHg

12.47 A unit of pressure in the U.S. system is pounds per square inch, abbreviated lbs/in² or psi. Standard atmospheric pressure, in this system, is 14.7 lbs/in². (a) Convert standard atmospheric pressure to lbs/ft². (b) Calculate the pressure, in torr, inside a tire with an internal pressure of 29.0 psi.

Boyle's Law

12.48 Correctly state Boyle's law including the variables that are constant.

12.49 Using the mathematical statement of Boyle's law, $PV = k$, calculate the volume of a gas at each of the following pressures, if k equals 0.910 L·atm: (a) 2.51 atm, (b) 6.10 atm, (c) 0.0449 atm, (d) 327 torr, (e) 93.8 kPa.

12.50 Consider the graph of P versus V in Fig. 12.19 to answer the following questions. (a) What is the volume of the gas at 6.0 atm? (b) At what pressure will the gas occupy a volume of 30.0 L? (c) At an infinite pressure (if that were possible), what should happen to the volume? (d) What pressure change is required to change the volume of the gas from 1.0 to 5.0 L?

12.51 What is the final volume of an ideal gas if initially it occupies 9.5 L at a pressure of 0.89 atm and the following pressure changes are made: (a) 2.0 atm, (b) 1.0 atm, (c) 25 atm, (d) 97 kPa, (e) 812 torr?

12.52 A sample of N_2 gas occupies 73.5 cm³ at 296 atm. What volume does it occupy at standard atmospheric pressure?

12.53 If a sample of Ne gas occupies 7.77 L at 683 torr, what volume will it occupy at 109 kPa?

12.54 Initially, a Ne gas sample occupies 33.72 L at 763.4 torr. Calculate the volume it occupies under each of the following pressures: (a) 0.7903 atm, (b) 0.7903 torr, (c) 0.7903 Pa, (d) 0.7903 kPa.

12.55 What is the final pressure that would have to be exerted on 1204 mL H_2 at 1.000 atm to change its volume to (a) 1000 mL, (b) 340.5 mL, (c) 40.00 L, (d) 6.130 L?

12.56 A 71-L He tank contains compressed helium at 59 atm; how many 1.2-L balloons can be filled with the He in the cylinder, assuming that atmospheric pressure is exactly 1 atm?

Charles' Law

12.57 Correctly state Charles' law, indicating which variables are held constant.

12.58 Using the mathematical statement of Charles' law, $V/T = k$, calculate the volume of an ideal gas at each of the following temperatures, if it is known that $k = 0.0821$ L/K: (a) 418 K, (b) 35.2°C, (c) 0.0°C, (d) −98.2°C, (e) 107°C.

12.59 Consider the graph of T versus V in Fig. 12.20, and answer each of the following questions. (a) At what temperature will the gas occupy 30.0 L? (b) At what temperature will the gas occupy 25.0 L? (c) What volume does the gas occupy at −10.0°C? (d) What temperature change is required to change the volume of the gas from 20 L to 30 L?

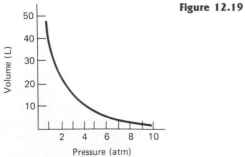

Figure 12.19

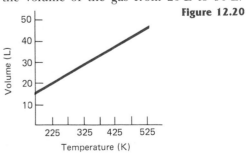

Figure 12.20

12.60 Why does no simple relationship exist between the volume of an ideal gas and its temperature in degrees Celsius?

12.61 What is the final volume of a gas that initially occupies 9.35 L at 282 K when the temperature is changed to each of the following: (a) 149 K, (b) 596 K, (c) 298°C, (d) −65°C?

12.62 A sample of O_2 gas initially occupies 1.175 L at 200.0°C; calculate the volume of the gas after the temperature has been changed to 462.9 K.

12.63 What temperature would an ideal gas, initially occupying 5.74 L at 10.0°C, have to be changed to for the gas to occupy (a) 10.0 L, (b) 6.00 L, (c) 15.2 L?

12.64 (a) Theoretically, what volume would a 125-L sample of He gas initially at 273 K occupy if its temperature were decreased to 10.0 K? (b) What pressure is required to decrease the volume of the 125-L He sample to the same volume that results when the He is cooled to 10.0 K, if the initial pressure is 1.00 atm?

Combined Gas Law

12.65 What happens to the volume of a gas sample when each of the following pressure and temperature changes is made: (a) P increases, T decreases; (b) P decreases, T increases; (c) small P increase, large T increase; (d) large P decrease, small T decrease?

12.66 What volumes do each of the following gas samples occupy when the conditions are changed to STP?
(a) $V_1 = 5.90$ L; $T_1 = 58°C$; $P_1 = 7.61$ atm
(b) $V_1 = 279$ cm³; $T_1 = -16°C$; $P_1 = 806$ torr
(c) $V_1 = 0.446$ L; $T_1 = 227$ K; $P_1 = 446$ kPa
(d) $V_1 = 577$ L; $T_1 = 203°C$; $P_1 = 83.7$ kPa

12.67 For each of the following, calculate the final volume the gas occupies:
(a) $V_1 = 0.619$ L $V_2 = ?$ L
     $T_1 = 444$ K $T_2 = 822$ K
     $P_1 = 1.24$ atm $P_2 = 9.49$ atm
(b) $V_1 = 524$ mL $V_2 = ?$ L
     $T_1 = -91°C$ $T_2 = -43.4°C$
     $P_1 = 118$ kPa $P_2 = 0.652$ atm
(c) $V_1 = 1.75$ L $V_2 = ?$ mL
     $T_1 = 30.8°C$ $T_2 = 310$ K
     $P_1 = 476$ torr $P_2 = 418$ kPa

(d) $V_1 = 6.32$ L $V_2 = ?$ cm³
     $T_1 = -75°C$ $T_2 = -10.4°C$
     $P_1 = 752$ torr $P_2 = 0.857$ atm

12.68 Calculate the final pressure of each of the following gases when the stated volume and temperature changes are made:
(a) $V_1 = 89.6$ L $V_2 = 43.8$ L
     $T_1 = 23.5°C$ $T_2 = 52.2°C$
     $P_1 = 1.79$ atm $P_2 = ?$ atm
(b) $V_1 = 2041$ L $V_2 = 4.53$ L
     $T_1 = 666$ K $T_2 = -52.9°C$
     $P_1 = 42.1$ kPa $P_2 = ?$ Pa
(c) $V_1 = 62.8$ L $V_2 = 77.4$ L
     $T_1 = -23°F$ $T_2 = 35°F$
     $P_1 = 912$ torr $P_2 = ?$ torr
(d) $V_1 = 0.236$ cm³ $V_2 = 1.72$ L
     $T_1 = 193$ K $T_2 = 723$ K
     $P_1 = 164$ kPa $P_2 = ?$ atm

12.69 If 562 L of an ideal gas is prepared at 701 torr and 512°C and then pumped into a 2.53-L steel tank at 35.0°C, what pressure will the tank have to withstand?

Avogadro's Law

12.70 Calculate the number of moles of ideal gas in each of the following samples at STP: (a) 372 mL, (b) 2.09 L, (c) 0.00410 cm³, (d) 2.551 mm³, (e) 8.22 m³.

12.71 Find the volume occupied by each of the following gaseous samples at STP: (a) 1.94 g NH_3, (b) 4.21×10^{-4} g SO_3, (c) 21.9 mg CF_4, (d) 1.09 g Xe, (e) 351.3 kg O_3.

12.72 Calculate the masses of the following gaseous samples given their volumes at STP: (a) 0.0304 L N_2O, (b) 148 L HCl, (c) 0.916 m³ C_2H_2, (d) 6.311 mL UF_6, (e) 5.390 cm³ Kr.

12.73 What volume would each of the following gaseous samples occupy at 799 torr and 333 K: (a) 492 g H_2, (b) 5.44 kg ClF, (c) 0.001011 mg SO_2?

12.74 Calculate the volume of each of the following gases at the stated conditions:
(a) 21.5 g He at 247 K and 1.824 atm
(b) 0.553 mol O_2 at 808 K and 642 torr
(c) 931 mg Cl_2 at 164°C and 92.4 kPa
(d) 9.34 kg H_2S at 2.04°C and 0.917 atm

12.75 Calculate the density, in grams per liter, of each of the following gases at standard conditions: (a) Cl_2, (b) CO_2, (c) Xe, (d) C_3H_8, (e) C_2FCl_3.

Ideal Gas Equation

12.76 Solve the ideal gas equation, $PV = nRT$, for each of the following variables: *(a)* P, *(b)* n, *(c)* T, and *(d)* V. Show that the units cancel to give the correct units for the dependent variable.

12.77 Use the factor-label method to calculate the value of the ideal gas constant R with each of the following sets of units:

(a) $\dfrac{mL \cdot atm}{mol \cdot K}$

(b) $\dfrac{cm^3 \cdot torr}{mmol \cdot K}$

(c) $\dfrac{kPa \cdot dL}{mol \cdot K}$

(d) $\dfrac{m^3 \cdot torr}{mol \cdot K}$

12.78 What volume, in liters, does each of the following gaseous samples occupy at the stated conditions?
(a) 2.36 mol CO_2 at 711 K and 1.73 atm
(b) 0.0889 mol He at 555 torr and 89.4°C
(c) 0.466 mol N_2O at 1247 torr and 301 K

12.79 Calculate the number of moles of gas molecules in each of the following gas samples:
(a) 2.98 L F_2 at 0.104 atm and 64.7°C
(b) 29.4 mL Cl_2 at 397 kPa and 9.06°C
(c) 0.00228 m^3 SO_2 at 840.4 K and 932 torr

12.80 Calculate the pressure, in atm, exerted by each of the following gases:
(a) 1.18 mol H_2Te occupying 3.85 L at 587 K
(b) 212 mmol SiF_4 occupying 193 mL at −4.92°C
(c) 0.00299 mol NF_3 occupying 1.08 dm^3 at 107.4 K

12.81 Calculate the mass of each of the following gas samples:
(a) N_2 that occupies 2.17 L at 743 torr and 21.6°C
(b) Kr that occupies 81.2 cm^3 at −31.4°C and 0.515 atm
(c) F_2 that occupies 0.0112 m^3 at 243 kPa and 319 K

12.82 What volume does each of the following occupy at the stated conditions?
(a) 0.394 g Ar at 25.1°C and 791 torr
(b) 191.2 mg OF_2 at 1.04 atm and −115°C
(c) 481 kg NO at 112 kPa and 304 K

12.83 Calculate the density in grams per liter of SF_6 at 176°C and 0.499 atm.

12.84 Calculate the density in grams per liter of NF_3 at −111.5°C and 790.4 torr.

12.85 If 22.6 g of a gas sample occupies 17.9 L at 0.863 atm and 311 K, what is the molecular mass of the gas?

12.86 Calculate the molecular mass of a gas if 275 g of that gas occupies 39.7 L at 125 kPa and −8.50°C.

Volume Relationships in Chemical Reactions

12.87 Consider the equation for the gas phase formation of HF:

$$H_2(g) + F_2(g) \longrightarrow 2HF(g)$$

Calculate the volume of H_2 required to combine with excess F_2 to produce each of the following volumes of HF at STP conditions: *(a)* 23.9 L HF, *(b)* 24.7 L HF, *(c)* 1.27 mL HF, *(d)* 96.90 cm^3 HF.

12.88 Consider the following equation:

$$2SO_2(g) + O_2(g) \longrightarrow 2SO_3(g)$$

What volume of SO_3 is produced when each of the following volumes of O_2 gas is combined with excess amounts of SO_2 at 2.03 atm and 271 K:
(a) 1.84 L O_2
(b) 0.00327 m^3 O_2
(c) 24.7 mL O_2
(d) 525 cm^3 O_2

12.89 Oxygen, O_2, may be prepared by mixing sodium peroxide, Na_2O_2, with water:

$$2Na_2O_2(s) + 2H_2O(l) \longrightarrow O_2(g) + 4NaOH(aq)$$

Calculate the volume of O_2 gas liberated, at STP, when the following masses of Na_2O_2 are combined with excess water:
(a) 123 g Na_2O_2
(b) 4.04×10^6 g Na_2O_2
(c) 0.00741 kg Na_2O_2
(d) 0.104 mg Na_2O_2

12.90 Lead(II) sulfide, PbS, combines with ozone, O_3, to produce $PbSO_4$ and oxygen gas:

$$PbS(s) + 4O_3(g) \longrightarrow PbSO_4(s) + 4O_2(g)$$

Calculate the mass of PbS(s) needed to combine exactly with the following volumes of ozone, O_3, at standard conditions: (a) 0.9331 L, (b) 9851 mL, (c) 0.915 m^3, (d) 7.07 dL.

12.91 Nitrogen trichloride, NCl_3, decomposes to nitrogen gas, N_2, and chlorine gas, Cl_2:

$$2NCl_3(l) \longrightarrow N_2(g) + 3Cl_2(g)$$

Calculate the volume of $Cl_2(g)$ produced when each of the following quantities of NCl_3 is decomposed at the stated conditions:
(a) 456 g NCl_3 at 787 torr and 70.8°C
(b) 945 mg NCl_3 at 0.875 atm and 275 K
(c) 0.332 mol NCl_3 at 295 K and 116 kPa
(d) 0.00197 g NCl_3 at 218 torr and 9.32°C

12.92 Ammonia gas is combined with oxygen gas on a catalyst to produce nitrogen monoxide and water vapor at 975°C and 2.34 atm. Calculate the volume of nitrogen monoxide produced when each of the following is combined with an excess of the other reactant: (a) 7.73 L of ammonia, (b) 421 mg of ammonia, (c) 643 L of oxygen, (d) 1.48 kg of oxygen.

12.93 When glucose, $C_6H_{12}O_6$, is fermented with yeast, ethanol, C_2H_6O, and carbon dioxide gas, CO_2, are produced. If 37.1 g of glucose is completely fermented, what volume of carbon dioxide is released at 34.9° and 765 torr?

Mixtures of Gases

12.94 A gaseous mixture of N_2, He, and CO has a total pressure of 709 torr. The partial pressure of N_2 is 166 torr, and the partial pressure of He is 255 torr. Calculate the partial pressure of CO in the mixture.

12.95 If 54.9% of the moles of a gaseous mixture is Ar gas and the remaining gas in the mixture is He, what is the partial pressure of each component, given that the total pressure is 1.15 atm?

12.96 A 43.7-mL sample of N_2 is collected over water at 297 K and 753 torr. What is the dry volume of N_2 at the same conditions?

12.97 What is the dry volume of Ar gas if a 2.688-L sample of Ar is collected over water is at 24.0°C and 99.6 kPa?

12.98 When magnesium metal, Mg, is combined with hydrochloric acid, hydrogen gas, H_2, and aqueous magnesium chloride, $MgCl_2$, are produced. In an experiment, 385 mL of H_2 gas is collected, over water, at 23.0°C and 1.45 atm. (a) What mass of H_2 gas is collected? (b) What mass of Mg and HCl are combined to produce this mass of H_2?

12.99 Oxgen gas is collected by water displacement after $KClO_3$ is heated with a catalyst.

$$2KClO_3(s) \xrightarrow[MnO_2]{\Delta} 2KCl(s) + 3O_2(g)$$

It is found that 228.7 mL O_2 is produced at 26.0°C and 763 torr. What mass of $KClO_3$ was initially heated to produce this quantity of oxygen?

The Atmosphere

12.100 What scientist is credited with the discovery of each of the following gases: (a) oxygen, (b) nitrogen, (c) argon, (d) helium?

12.101 What is the principal means for fixing nitrogen (a) in the natural world and (b) in industry?

12.102 Compare the physical properties of N_2 and O_2, and discuss similarities and differences between the properties of these two gases.

12.103 In Table 12.4, a range is given for the percent volume of carbon dioxide, in contrast to all other entries. What could account for the variability of the concentration of CO_2?

12.104 Write a short description of the nitrogen cycle, tracing nitrogen from the atmosphere to production of proteins in living systems.

12.105 How are N_2 and O_2 isolated from the atmosphere? (Consider their physical properties.)

12.106 List three uses for oxygen gas.

12.107 Explain specifically why O_2 is needed by all animals, and why a brief interruption in the O_2 supply results in death.

12.108 What could account for the rather recent discovery of the noble gases?

12.109 What trends are observed in the physical properties of the noble gases when proceeding from the lowest to the highest atomic mass?

12.110 List uses for the noble gases.

12.111 Ar is the inert gas most commonly used in light bulbs. If Kr is used in its place, the life of the bulb is extended. Suggest a reason why Kr is not used instead of Ar.

Additional Exercises

12.112 Atmospheric pressure in the U.S. system is 14.7 psi (pounds per square inch). (a) How many pascals are equivalent to 1.00 psi? (b) Express atmospheric pressure in pounds per square foot and pounds per square centimeter.

12.113 Calculate the density, in grams per liter, of each of the following gaseous samples at the stated conditions:
(a) N_2 at 0.911 atm and 84.6°C
(b) C_2H_4 at 0.429 atm and 119 K
(c) SiF_4 at 90.3 torr and 220.1°C

12.114 Dry ice, $CO_2(s)$, is solid carbon dioxide. On heating, dry ice is totally converted to $CO_2(g)$. If 37 g of dry ice is placed in a 2.7-L container and it is then sealed, what pressure would result inside the container at room temperature, 25°C?

12.115 The density of N_2O is 1.843 g/L at 20°C and 1.00 atm. What is the density of N_2O if the conditions are changed to 141°C and 2358 torr?

12.116 (a) Liquid mercury changes to a vapor at 357°C. Calculate the density of mercury vapor at 357°C and 1.00 atm. (b) By what factor does the density decrease in the gas phase compared to the liquid phase, if the density of liquid mercury is 13.6 g/cm³?

12.117 A phosphorus-fluorine compound contains 35.20% P, and 0.116 g of this gas occupies 31.6 mL at 1.02 atm and 298 K. What is the molecular formula of the compound?

12.118 During the winter it is often necessary to add air to your tires to maintain the proper inflation. (a) Why is the added air needed? (b) Give an explanation, using the kinetic molecular theory.

12.119 A mixture of 825 mg He and 775 mg Ar is contained in a 21.3-L vessel at 275 K. What are the partial pressures of He and Ar in the mixture?

12.120 When NH_3 and F_2 are combined in the presence of a catalyst, $NF_3(g)$ and $NH_4F(s)$ result. (a) Write the equation for the reaction. (b) What volume of NF_3 results at 25.0°C and 0.991 atm, if 6.57 g NH_3 and 11.7 g F_2 are combined?

12.121 (a) Calculate the volume of O_2 at 100.0°C and 0.926 atm that is needed to exactly combine with a 927-mg sample of C_9H_{20} to produce carbon dioxide and water. (b) What volume of air is needed for exactly the same reaction in part a? (Hint: Consider the composition of air.)

12.122 A 2.01-L container is filled with water, which is then allowed to undergo electrolysis to produce H_2 and O_2. What are the partial pressures of H_2 and O_2 in the container at STP? Assume that the density of water is 1.00 g/cm³.

12.123 A 10.00-g sample of a carbon, hydrogen, and oxygen compound is found to contain 6.20 g C and 2.76 g O. When a 0.122-g sample of the compound is heated to 373 K at 0.997 atm, it occupies 32.3 mL. What is the molecular formula of the compound?

12.124 How many Ar atoms are found in a 10.0 ft × 12.0 ft × 8.0 ft room at 24°C and 1.05 atm? (Hint: Consider the composition of the atmosphere.)

CHAPTER
Thirteen

Liquids, Solids, and State Changes

STUDY GUIDELINES

After completing Chapter 13, you should be able to

1. Explain the differences between gases, liquids, and solids, using the kinetic molecular theory

2. Describe the process of evaporation of liquids and factors that influence evaporation

3. Describe a dynamic equilibrium in terms of evaporation and condensation

4. Discuss equilibrium vapor pressure, and list factors that influence the vapor pressure of a liquid

5. Define the boiling point of a liquid, and list factors that influence boiling points

6. Discuss the surface tension of liquids

7. Discuss the three categories of intermolecular forces in liquids (dipole—dipole interactions, hydrogen bonds, and London dispersion forces) with respect to how they are formed and their relative strengths, and give examples of liquids that exhibit these forces

8. List and explain the fundamental particles, bonding forces, and properties, characteristic of each of the four classes of solids—ionic, covalent, molecular, and metallic—and give examples of solids in each class

9. Explain each segment found on a heating curve for a substance, identifying state changes, transition points, and energy considerations

10. Predict the strength of intermolecular forces given enthalpies of fusion and vaporization

11. Discuss the structure and properties of (a) individual water molecules, (b) liquid water, and (c) ice

12. Explain the unique physical properties of water

13. Outline the principal components of the water and carbon cycles, and explain how these cycles are important in nature

14. Discuss differences in the properties of the allotropic forms of carbon: graphite, diamond, and amorphous carbon

13.1 LIQUIDS

In Chap. 4 we discussed the most general properties of liquids. You may recall that liquids have a constant volume and a variable shape, taking the shape of their containers to the level they fill. Liquids are less fluid than gases, but are significantly more fluid than solids. On average, liquids are more dense than gases, but less dense than solids.

In this chapter, we will take a more in-depth look at the properties of liquids. Before we begin, we will consider aspects of the kinetic molecular theory that relate to liquids so that we may discuss theoretical concerns relating to the properties of liquids.

Kinetic Molecular Theory of Liquids

The kinetic molecular theory can be applied to the particles that compose liquids. Like gases, the particles in liquids are in constant motion. However, the movement of molecules and atoms in liquids is greatly restricted because they are more closely packed. Unlike gases, only a tiny proportion of the volume of liquids is empty space.

Why are liquid molecules more closely packed than gas molecules? Attractive forces among liquid molecules are stronger than those among gas molecules. Collectively, the attractive forces among particles in the liquid state are called **intermolecular forces,** which literally means "the forces between molecules." Compared to chemical bonds, intermolecular forces are weak. We did not encounter these forces when discussing ideal gases because gas molecules are widely separated and move quickly.

Molecules in the liquid phase have a more orderly arrangement than they do in the gas phase. We say that the molecules in a liquid have a short-range order; i.e., some regions have an organized, regular pattern, but most regions are disorganized. The organization of the molecule in a liquid results from their intermolecular forces. While molecules in liquids are more ordered than those in gases, they are less ordered than those in solids (Fig. 13.1). The strong intermolecular forces in solids give them a regular, organized structure.

We can use the kinetic molecular theory to account for the principal properties of liquids.

1. The incompressibility of a liquid is explained in terms of the absence of spaces between molecules. If a force is applied to the surface of a liquid, few empty spaces exist for the molecules to fill.

2. The property of liquids to flow results from the ability of the molecules to slide by each other. However, the intermolecular forces in liquids are not strong enough to hold the molecules rigidly affixed to each other.

3. The higher density of liquids compared to gases results from the intermolecular forces that hold the molecules in a relatively small volume with virtually no empty space between molecules.

With these general properties of liquids in mind, what are some of the specific properties associated with liquids?

Evaporization (Vaporization)

If a liquid in an open container is undisturbed, the liquid level steadily lowers until no liquid remains. What is happening, and where did the liquid go? If the liquid is gasoline, it is not difficult to understand that the gasoline is undergoing a phase change, liquid to vapor. Whenever gasoline is placed in an open container, the fumes are rapidly detected in the air.

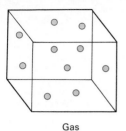

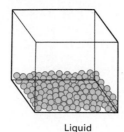

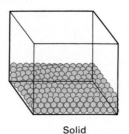

Gas   Liquid   Solid

Figure 13.1
Gas particles are distributed randomly throughout their volume. Liquids have structures that are more ordered than gases but less ordered than solids. Only small segments of the structure of liquids are ordered. Solids have the most ordered structure of particles.

Evaporation is the process in which liquid molecules break free from the liquid surface and enter the vapor phase. Evaporation is explained in terms of the energy possessed by the molecules on the surface of the liquid. Surface molecules whose kinetic energies are large enough to overcome the intermolecular forces that bind them to the liquid break free and enter the gas phase (Fig. 13.2).

In open containers (open systems), evaporation continues until all of the liquid enters the vapor phase. However, liquids in closed containers (closed systems) behave differently. The volume of the liquid decreases for a period of time, and then does not change. In closed containers, the vapor cannot escape; as the vapor concentration increases, some of the vapor molecules lose energy and return to the liquid state. When a vapor returns to the liquid state, it is said to condense; the process is called **condensation.**

Evaporation and condensation are opposing processes. Evaporation occurs when molecules leave the surface of a liquid, and condensation occurs when vapor molecules bond to the liquid. Initially, when a liquid is placed in a closed container, it begins to evaporate at a constant rate; very little condensation takes place at this time (Fig. 13.3). But as the concentration of the vapor above the liquid increases, the rate of condensation increases. At some point in time, the rate of condensation equals the rate of evaporation. When the rates are equal, the number of molecules that enter the gas phase equals the number that return to the liquid phase in a given time interval. Consequently, the level of the liquid does not change; each liquid molecule lost is replaced by a molecule from the condensing vapor (Fig. 13.3).

Evaporation occurs when a liquid changes to vapor below the boiling point.

Many higher animals regulate their body temperature through evaporation of water. Heat from their bodies evaporates water through pores in the skin.

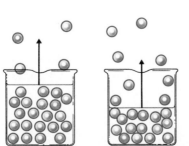

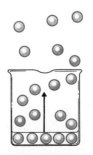

Figure 13.2
In an open system, the level of a liquid in a container continually drops as a result of evaporation until no liquid remains.

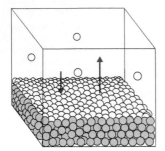 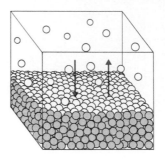

Figure 13.3
In a closed system, the level of the liquid initially decreases because the rate of evaporation is greater than the rate of condensation. A point is reached when these rates become equal. At that time the liquid level remains constant.

When the rates of two opposing processes are equal, the system is said to be in a state of **dynamic equilibrium.** An equilibrium is represented in an equation by two arrows pointing in opposite directions.

$$\text{Liquid} \underset{\text{condensation}}{\overset{\text{evaporation}}{\rightleftarrows}} \text{vapor}$$

We will discuss dynamic equilibrium systems more fully in Chap. 16.

A measure of the degree to which a liquid enters the vapor state is **equilibrium vapor pressure.** If a manometer, a mercury-filled U-tube used to measure gas pressure, is attached to a closed container holding a liquid in equilibrium with its vapor, the manometer measures the pressure of the vapor above the liquid (Fig. 13.4). The pressure exerted by a vapor in equilibrium with a liquid is called the vapor pressure of the liquid. Vapor pressure is measured in the same units as gas pressure: atmospheres, torr, and pascals.

The vapor pressure of a liquid is independent of the amount of liquid or size of the container. Factors that affect the vapor pressure of a liquid are its temperature and intermolecular forces.

Consider the graph of the vapor pressure of water versus temperature in Fig. 13.5. As the temperature increases, the vapor pressure increases—a direct relationship. The vapor pressure of water remains fairly low until about 50°C, when it starts to rise rapidly. At 100°C, the vapor pressure of water increases to 760 torr, or 1 atm. All liquids exhibit similar vapor pressure curves. Figure 13.6 shows the vapor pressure curve for diethyl ether, ethanol, and water.

The vapor pressure of ether is 185 torr at 0°C, which is much higher than that of ethanol (12 torr) or water (4.6 torr). The high vapor pressure of ether at 0°C is attributed to its weak intermolecular forces; thus, the surface molecules in ether require less energy to break free and enter the gas phase than surface molecules in ethanol or water. Similarly, we know that the intermolecular forces in ethanol are weaker than those in water because the vapor pressure of ethanol is higher than that of water at all temperatures.

Vapor Pressure

Liquids with high vapor pressures at room temperature, 25°C, are called volatile liquids. Such liquids readily become vapors. Two common volatile liquids are gasoline and ether.

Diethyl ether, $(CH_3CH_2)_2O$, is a very combustible, volatile liquid. Ether, as it is frequently called, once was one of the principal general anesthetics, i.e., substances used to block pain and cause unconsciousness in medical operations.

Figure 13.4
The vapor pressure of a liquid at a specific temperature is determined by measuring the pressure exerted by a vapor in equilibrium with its liquid phase. A closed-end manometer is the pressure-measuring device in this figure.

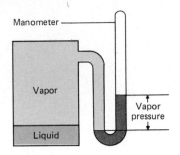

13.1 How does the kinetic molecular theory explain the following properties of liquids: *(a)* higher density than gases, *(b)* incompressibility, *(c)* ability to flow?

1.32 Explain evaporation in terms of what happens to the surface molecules in a liquid.

13.3 What is a dynamic equilibrium? Give an example.

13.4 *(a)* What factors influence the vapor pressure of liquids. *(b)* What factors do not affect the vapor pressure of liquids?

Boiling Point

When a liquid is heated, there is a temperature at which the vapor pressure of the liquid equals the pressure of the gases above the liquid. This temperature is known as the **boiling point** of the liquid. At its boiling point, bubbles of vapor form throughout the liquid and break free into the gas phase. Bubbles of vapor can form within a liquid only at its boiling point because the pressure of the vapor equals the external pressure. Below the boiling point, the higher external pressure would cause the vapor bubbles to collapse.

Most frequently, chemists are interested in the **normal boiling point,** defined as the temperature at which the vapor pressure of a liquid equals 1 atm, or 760 torr. Figure 13.6 shows that each of the three vapor pressure curves crosses the line corresponding to 1 atm at a different temperature. The curve for ether crosses at 35°C, which is the normal boiling point of ether. The vapor pressure curves for ethanol and water cross the 1-atm line at their respective boiling points of 78 and 100°C.

If the pressure of the gases above a liquid changes, a corresponding change in boiling point is found. When the pressure is increased, the liquid must be heated to a higher temperature before the vapor pressure equals the increased pressure. Likewise, if the pressure above the liquid is decreased, the boiling point is lowered.

Table 13.1 lists boiling points of water at various pressures. Notice that if the pressure is lowered to 24 torr, water boils at 25°C, room temperature. Here is boiling water that you would not mind thrusting your hand into! On the other end of the scale, at 2026 torr water boils at 130°C, or 30°C above its normal boiling point.

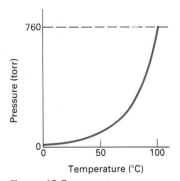

Figure 13.5
The vapor pressure of water is very low until 50°C. Above 50°C, the vapor pressure of water increases rapidly. At 100°C, the vapor pressure of water equals 760 torr.

Water that boils at greatly reduced pressures is not hot enough to burn skin tissue.

TABLE 13.1 BOILING POINT OF WATER AT SELECTED TEMPERATURES

Pressure torr	Boiling point °C
5	1
24	25
93	50
600	94
760	100
2026	130

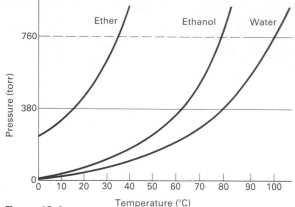

Figure 13.6
At all temperatures, the vapor pressure of ether is higher than the vapor pressure of either ethanol or water; thus, ether is the most volatile of the three liquids and has the lowest boiling point (34.6°C). Because the vapor pressure of ethanol is higher than the vapor pressure of water at all temperatures, ethanol is a more volatile liquid and boils at a lower temperature (78.3°C) than does water.

Surface Tension

Surface tension is the property of the surface of a liquid that enables it to act as if a membrane is stretched across it. To understand surface tension, consider the unique properties of the surface molecules. All molecules below the surface of a liquid are surrounded in all directions by other liquid molecules. Thus, the forces exerted on subsurface molecules are balanced in all directions. Surface molecules are surrounded by other liquid molecules on all sides but one. This results in an unbalanced force pulling the surface molecules inward. Figure 13.7 shows the forces on surface and subsurface molecules.

As a result of the net inward attractive forces, surface molecules tend to minimize the amount of surface area exposed. A drop of liquid outside of a gravitational field takes the shape of a sphere, the geometric figure with minimum exposed surface area per unit volume.

To demonstrate the membranelike property that surface tension imparts, carefully place a needle (a metal with a density greater than water) on the surface of water. Unless the needle is pushed through the surface, it remains suspended by the surface tension of the water. Various insects that walk on water are supported by surface tension (Fig. 13.8).

While surface tension is mainly thought of in terms of the interface between a liquid and a gas, similar generalizations can be made when a liquid is in contact with a solid. If a small quantity of liquid spreads out uniformly and covers the surface of a solid, we say the liquid **wets** the surface (Fig. 13.9). Some liquids, like mercury, when placed on certain solid surfaces, bead up and do not spread out; a liquid of this nature does not wet the surface (Fig. 13.9).

Not all liquids are wet.

Figure 13.7
Subsurface molecules of a liquid have intermolecular forces exerted on them from all directions. Hence, the forces on subsurface molecules are balanced. Surface molecules have intermolecular forces in all directions except directly above them. Consequently, the forces on surface molecules are unbalanced.

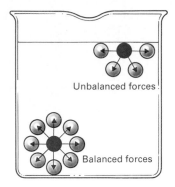

Within liquids that wet solid surfaces, the forces of attraction between the liquid and solid (adhesive forces) are stronger than the internal intermolecular forces (cohesive forces). In contrast, the cohesive forces of a liquid that does not wet a surface are stronger than the adhesive forces between liquid and solid; therefore, liquids that do not wet surfaces attempt to expose a minimum surface to the solid.

Wetting agents are substances added to liquids to decrease their cohesive forces and surface tension so that they will more effectively wet solid surfaces. Detergents (cleaning agents) are excellent wetting agents for water. If you place a drop of detergent in water that suspends a needle as described earlier, the needle falls to the bottom of the container because of the decrease in surface tension.

If a small-diameter glass tube is placed in water, the adhesive forces cause some of the water molecules to be drawn into the tube. The upward movement of liquids into narrow tubes is called **capillary rise** (Fig. 13.10). If you observe the surface of the water in the tube, you will find that it is concave. The curvature of the surface of a liquid in a tube is called the *meniscus*. A concave meniscus results when the attractive forces between the liquid and the tube are stronger than the cohesive forces within the liquid. Exactly the opposite is found if a narrow tube is placed

Figure 13.8
Surface tension supports insects as they walk across water.

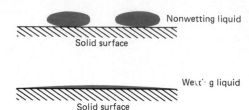

Figure 13.9
A liquid that does not wet solid surfaces (nonwetting liquid) forms beads when it contacts a surface. A liquid that wets solid surfaces (wetting liquid) spreads out and covers the surface.

in liquid mercury; i.e., the level of the liquid is lower in the tube than the surface of the mercury. In addition, a convex meniscus is found (Fig. 13.10). Attractive forces in the mercury are stronger than adhesive forces with the glass; thus, the mercury attempts to expose as little surface to the glass as possible. Capillary rise is an important means by which plants absorb dissolved nutrients from the soil through their root systems.

REVIEW EXERCISES

13.5 *(a)* Specifically state what is meant by the normal boiling point of a liquid. *(b)* What effect does an increase in pressure have on the boiling point of a liquid?

13.6 *(a)* What is the surface tension of a liquid? *(b)* What property of the surface molecules accounts for a liquid's surface tension?

13.7 Describe how a liquid that wets the surface of a solid differs from one that does not.

13.2 INTERMOLECULAR FORCES IN LIQUIDS

B efore we complete our discussion of liquids, let us consider the intermolecular forces (cohesive forces) that hold molecules in the liquid state. The three principal intermolecular forces in liquids are

1. Dipole–dipole interactions
2. Hydrogen bonds
3. London dispersion forces

Figure 13.10
(a) Water rises in small capillary tubes as a result of the adhesive forces between the water and tube. It forms a concave meniscus. (b) The level of mercury in a capillary tube is below the level of the mercury in the container because the cohesive forces of mercury are much stronger than are the adhesive forces. Mercury forms a convex meniscus.

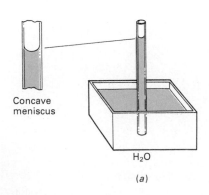

Concave meniscus

H_2O

(a)

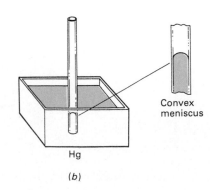

Convex meniscus

Hg

(b)

Figure 13.11
The partially positive end of a polar molecule attracts the partially negative end of another polar molecule. This intermolecular force of attraction is called a dipole–dipole force.

Each of these classes of intermolecular forces results from the forces of attraction between unlike charged particles, called electrostatic forces of attraction.

Dipole–dipole interactions are the forces of attraction among molecules that naturally exist as dipoles. Dipoles are found in molecules that have a separation of charge. These molecules are classified as polar covalent molecules.

Figure 13.11 shows how dipole–dipole interactions bind liquid molecules. Partially positive ends of polar molecules attract the partially negative ends of others. Dipole–dipole interactions are short-range forces, which means that the molecules must be very close together to produce a significant force of attraction. Only about 4 kJ/mol is required to break dipole–dipole interactions; this is a small amount of energy compared to the approximately 400 kJ/mol needed to cleave average covalent bonds.

Examples of liquids that contain molecules bonded by dipole–dipole interactions include $PCl_3(l)$, $HI(l)$, $H_2S(l)$, and $CH_2Cl_2(l)$. As a result of the relatively weak nature of dipole–dipole interactions, most of these are volatile liquids and must be cooled to low temperatures to remain in the liquid state.

A hydrogen bond is a special case of dipole–dipole interaction—so special, in fact, that it is classified separately. **Hydrogen bonds** result when the molecules that compose the liquid have an H atom covalently bonded to F, O, or N (elements in the second period with the highest electronegativities). As with any dipole–dipole interaction, there is a charge separation within the molecules that exhibit hydrogen bonding. However, it specifically results from the attraction of the electronegative atom for the electron in the hydrogen atom. A hydrogen atom consists only of a proton and an electron; when its electron is somewhat withdrawn, the compact, positively charged proton is all that remains. For example, a hydrogen bond in hydrogen fluoride, HF, results when the highly electronegative F atom attracts the H atom of an adjacent molecule.

$$\overset{\delta+ \quad \delta-}{H-F}\cdots\cdots\overset{\delta+ \quad \delta-}{H-F}$$

$\nwarrow$ Hydrogen bond

The force of attraction among molecules like HF is greater in magnitude than that for molecules made up of other combinations of atoms. Experimentally, hydrogen bonds are the strongest of the intermolecular forces in common liquids, requiring about 20 to 25 kJ/mol to cleave.

Dipole–Dipole Interactions

A polar covalent bond results when two non-metal atoms with different electronegativities are bonded.

Hydrogen Bonds

Water is a hydrogen-bonded liquid. Water molecules have two H atoms bonded to the electronegative oxygen atom. Figure 13.12 shows that a liquid water molecule can hydrogen bond to four other water molecules. As a result of the association of water molecules produced by the hydrogen bonds, water has many unusual properties. For example, it has a very high boiling point. Most low-molecular-mass compounds without hydrogen bonds are gases at room conditions, but water has a high boiling point because of its hydrogen bonds. Table 13.2 lists the boiling points of other chalcogen (group VI) hydrides. H_2S, H_2Se, and H_2Te are bonded by much weaker dipole–dipole interactions and thus have low boiling points.

Hydrogen bonds are also found in solids. They are especially important in biological compounds, e.g., proteins and nucleic acids.

TABLE 13.2 BOILING POINTS OF CHALCOGEN HYDRIDES

Compound	Boiling point °C
H_2O	100
H_2S	−61
H_2Se	−42
H_2Te	−2

Of the three classes of intermolecular forces, **London dispersion forces** (or simply dispersion forces) are the weakest. Unlike the other two intermolecular forces in liquids, London dispersion forces exist in all molecules, but in small polar molecules they are so weak that they are totally overshadowed by dipole–dipole interactions and hydrogen bonds.

London Dispersion Forces

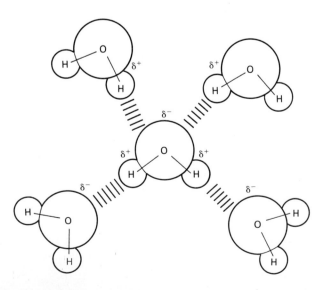

Figure 13.12
Hydrogen bonds in water result when the partially negative oxygen atom from one water molecule attracts the partially positive hydrogen atoms from one or two adjacent water molecules. The structure of liquid water is a network of associated water molecules. (Hydrogen bonds are indicated by dashed lines.)

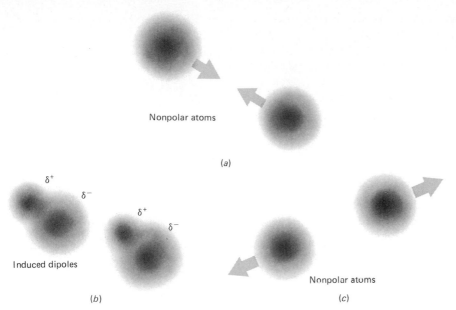

Nonpolar atoms

(a)

δ^+

δ^-

δ^+

δ^-

Induced dipoles

Nonpolar atoms

(b)

(c)

Figure 13.13
(a) When two nonpolar atoms are widely separated, they do not interact and London dispersion forces do not exist. (b) When the nonpolar atoms are very close together, the electrons from one atom repel the electrons in the other atom, producing a momentary dipole and force of attraction. This attractive force is a London dispersion force. (c) As the two atoms separate, they no longer interact. London dispersion forces are very short-range interactions.

London dispersion forces are the principal forces among nonpolar molecules. Nonpolar molecules are those that do not have a permanent charge separation. Also, dispersion forces are found in liquids of the elements that exist monatomically, i.e., the liquefied noble gases (He(l), Ne(l), etc.).

How then can nonpolar molecules bond to each other, if no separation of charge is found within them? Figure 13.13 shows two nonpolar molecules that approach each other. In Fig. 13.13a, the molecules do not interact because they are too widely separated. As the molecules approach each other, Fig. 13.13b, the electron clouds interact, producing dipoles in each molecule for the brief instant they are very close to each other. By the time they have passed (Fig. 13.13c), they no longer interact and the dipoles are gone. A dispersion force exists only in b when the two molecules are near each other.

When an atom or molecule produces a monetary, or instantaneous, dipole in another atom, this is referred to as an **instantaneous induced dipole.** You can think of the London dispersion force as a weak force of attraction exerted by the nucleus of one atom on the electrons of another atom when they are close to each other. The strength of these forces depends mainly on how many electrons are distributed around the atom and how tightly they are held. Stronger dispersion forces are found in atoms with a large number of loosely held electrons. Table 13.3 lists numbers of electrons and boiling points for the noble gases. As the number of electrons increases within this group, the strength of the dispersion forces increases, raising the boiling points.

London dispersion forces are also called London forces or van der Waals forces. Fritz London and Johannes van der Waals were two scientists who investigated and helped elucidate the nature of nonpolar interactions.

London dispersion forces are strong enough to hold molecules in the solid state if the molecules are sufficiently large.

TABLE 13.3 BOILING POINTS OF THE NOBLE GASES

Noble gas	Number of electrons	Boiling point, °C
He	2	−269
Ne	10	−246
Ar	18	−185
Kr	36	−153
Xe	54	−108
Rn	86	−61

REVIEW EXERCISES

13.8 (a) What are the three principal intermolecular forces? (b) Rank them from weakest to strongest.

13.9 How does an intermolecular force differ from a true chemical bond?

13.10 What accounts for the strength of hydrogen bonds relative to dipole-dipole interactions?

13.11 (a) Consider three of the nonpolar halogen molecules, Cl_2, Br_2, and I_2. At room temperature, 25°C, Cl_2 is a gas, Br_2 is a liquid, and I_2 is a solid. In terms of their intermolecular forces, explain the trend in physical state of these halogens. (b) Predict the physical states of F_2 and At_2.

13.3 SOLIDS

Recall from Sec. 4.2 that solids have both a constant volume and shape, and have very high viscosities; thus, they exhibit no observable fluid properties. Solids have the highest average densities, melting points, and boiling points of the three physical states. Particles in solids are bound by strong intermolecular forces that inhibit the particles from moving from place to place. Particles in solids are arranged in orderly geometric patterns.

Kinetic Molecular Theory of Solids

Application of the kinetic molecular theory to the particles that compose solids is quite different than for gases or liquids because of the strong intermolecular forces in solids. While virtually no movement from place to place (translocational motion) is possible, the molecules and atoms in solids are in constant motion in a fixed position. Their motions are mainly vibrational in nature. Solid particles move rapidly back and forth, oscillating about a fixed position in space.

The constant shape of a solid results from the strong cohesive forces between particles; each particle is in a fixed position and cannot move to another position without breaking its cohesive forces. High average density is explained in terms of the closeness of the particles to each other. Once again, they are closely packed as a result of the strong intermolecular forces. Because applied pressure cannot push the particles any closer than they already are, the volume of a solid is constant.

Glass is actually a super-cooled liquid. When silica, sodium carbonate, and calcium carbonate (components of glass) are heated and then cooled, they do not re-form into a highly organized crystalline pattern of particles; instead, the structure resembles the short-range order of liquids.

Classes of Solids

Solids are classified as being either crystalline or amorphous. **Crystalline solids** are the true solids; the particles are in a regular, recurring three-dimensional pattern called a crystal lattice. **Amorphous solids** lack the regular microscopic structure of crystalline solids. Actually, their structures more closely resemble those of liquids than those of solids (many are actually liquids with high viscosities). Examples of amorphous solids include glass, tars, and high-molecular-mass polymers such as Plexiglass.

Crystalline solids are further classified according to the types of forces that bond the particles in the crystal lattice. Most frequently, crystalline solids are grouped into four different classes: (1) ionic, (2) covalent, (3) molecular, and (4) metallic. We will consider each of these groups in detail.

You may recall our discussion of ionic solids in Chap. 8. Oppositely charged ions are alternately arranged in the crystal structure of **ionic solids.** Because the ions are bonded by strong ionic bonds, ionic solids have high melting and boiling points, and are rather hard but brittle.

Because the electrons in ionic solids are bound tightly to the ions, they cannot flow when a voltage is applied; hence, ionic solids are nonconductors of electricity (insulators). However, when ionic solids are melted, many of the ionic bonds are broken; the free ions then act as charge carriers that help conduct an electric current.

Examples of ionic solids include sodium chloride, $NaCl$; magnesium oxide, MgO; potassium nitrate, KNO_3; and ammonium bromide, NH_4Br.

Network covalent solids are sometimes called **macromolecular** (literally, "large molecule") solids, because the entire crystalline solid is one gigantic molecule held together by covalent bonds.

Atoms are the units bonded in network covalent solids. For example, diamond (pure carbon) is a covalent solid in which each carbon atom is bonded to four other carbon atoms, producing an enormous array of carbon atoms (Fig. 13.14). We will discuss the properties of diamond at the end of this chapter.

Structural analysis of crystalline solids, called x-ray crystallography, is done by placing crystals in the path of a beam of x-rays. The crystals scatter the x-ray beam and each crystalline substance has a unique scatter pattern, dependent on the spacing and arrangement of the crystal particles.

Ionic Solids

Network Covalent Solids

Figure 13.14
Diamond is composed of a network of C atoms each bonded to four other carbon atoms. Quartz, SiO_2, has the same structure as diamond, but it has a network of Si atoms surrounded by four O atoms.

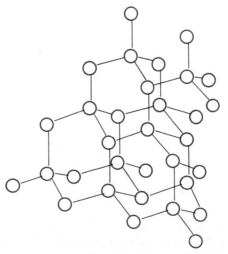

Covalent bonds, on average, are the strongest bonds; subsequently, network covalent solids are held together tightly. Melting points of covalent solids are among the highest known; many have melting points in excess of 1000°C. Because the electrons are strongly attracted in the covalent bonds, they do not move when a voltage is applied; thus, network covalent solids are nonconductors of electricity. Also, it is not surprising that network covalent solids are the hardest known substances.

Another common example of a network covalent solid is quartz, SiO_2. In a quartz crystal, each Si atom is bonded to four oxygen atoms in a network of alternating Si and O atoms (Fig. 13.14). Quartz has a melting point of approximately 1700°C.

Molecular solids have properties that contrast with those of ionic and network covalent solids. Molecules, rather than ions or atoms, are the particles in the crystal lattice positions, and these molecules are bonded by the same intermolecular forces that are found in liquids (Sec. 13.2): (1) dipole–dipole interactions, (2) hydrogen bonds, and (3) London dispersion forces.

Molecular Solids

Because the intermolecular forces in molecular solids are weak relative to covalent and ionic bonds, molecular solids tend to be relatively soft and volatile, and to have low melting points. Molecular solids are nonconductors of electricity because none of the electrons is free to move when a voltage is applied. Three examples of molecular solids are ice, $H_2O(s)$; dry ice, $CO_2(s)$; and iodine, $I_2(s)$.

Solids, under the proper conditions, can directly enter the gas phase without becoming liquid; this process is called sublimation.

Metallic solids (metals) range from soft to hard, have a metallic luster, are malleable, and are good conductors of heat and electricity. These properties result from their possessing **delocalized electrons,** i.e., electrons under the influence of more than one nucleus.

Metallic Solids

In the other solids, we found ions, atoms, and molecules as the fundamental units of structure. In metallic solids, positively charged metal nuclei occupy the crystal lattice positions. The structure of a metal is often described as "metal nuclei in a sea of electrons." A regular array of metal nuclei are surrounded by electrons attracted, not by one, but by many nuclei. This somewhat unique manner in which metal atoms are bonded is called **metallic bonds.** Figure 13.15 illustrates the structure of metals and the nature of metallic bonds.

The hardness and malleability of metals are explained by the strong electrostatic attractions among positive nuclei and negative electrons. Hammering and bending metals displaces atoms, but does not break the attractive forces in metals. Electric conductivity results from the delocalization of outer electrons; the outer electrons move from one atom to another if pushed by a voltage.

Common metals are copper, Cu; silver, Ag; iron, Fe; and tungsten, W. Table 13.4 summarizes the four classes of solids.

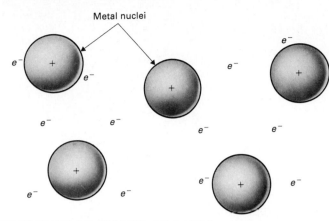

Metal nuclei

TABLE 13.4 SUMMARY TABLE OF CLASSES OF SOLIDS

	Ionic	**Network covalent**	**Molecular**	**Metallic**
Particles	Cations and anions	Atoms	Molecules	Metal nuclei and electrons
Strongest forces	Ionic bonds	Covalent bonds	H bonds, dipole-dipole forces, dispersion forces	Metallic bonds
Properties	Hard, insulators, high mp and bp	Hard, insulators, highest mp and bp	Soft, insulators, low mp and bp	Soft to hard, conductors, full range of mp and bp
Examples	NaCl, CaF$_2$, CaO	Diamond, quartz, carborundum	H$_2$O(s), CO$_2$(s), SO$_2$(s), NH$_3$(s)	Iron, copper, chromium, gold

13.12 How do the structures of solids differ from those of liquids?

13.13 Describe the movement of particles in the solid state.

13.14 Use the kinetic molecular theory to explain the following properties of solids: (a) high density, (b) constant volume, and (c) constant shape.

13.15 (a) What are the four general classes of solids? (b) How do they differ from each other?

13.16 Which class of solids has each of the following properties: (a) highest average melting point, (b) lowest average melting point, and (c) best conduction of electricity?

REVIEW EXERCISES

Most solid substances undergo two changes of state when heated. A solid changes to a liquid at the melting point, and a liquid changes to a vapor at the boiling point. To understand state changes, we will consider a heating curve for a substance.

13.4 STATE CHANGES

A **heating curve** is a plot of temperature versus the uniform addition of heat. Figure 13.16 presents a heating curve for a hypothetical substance, in which the temperature of the substance is on the vertical axis and the passage of time during which heat is added to the substance is on the horizontal axis.

Heating Curve

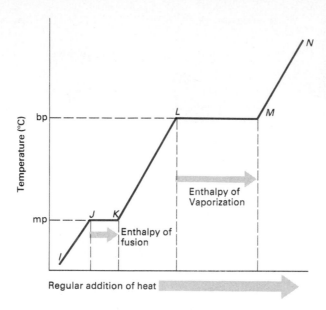

Figure 13.16
A heating curve for a substance is a plot of the temperature of the substance as
heat is added uniformly to it.

Initially, the substance exists in the solid state (point *I*) and the addition of heat increases the temperature (segment *IJ*). When the temperature increases, the particles move faster; in other words, the average
kinetic energy of the particles increases. The temperature increase of
the solid depends on the heat capacity of the solid, i.e., the amount of
heat required to raise the temperature of a fixed amount of the solid by
one degree Celsius.

    Point *J* on the heating curve represents the time when the first drop
of liquid appears. At this time the substance begins to melt. From *J* to *K*,
heat is added, but no increase in temperature occurs. Added heat no
longer increases the kinetic energy of the molecules, but instead increases their potential energy. During the melting process, the heat energy breaks the forces that bond the crystal lattice of the solid. As long as
bonds remain to be broken within the solid, the temperature remains
constant at the melting point.

    How much heat is needed to break the bonds in a solid and convert
it to a liquid? Two principal factors must be considered: (1) the amount
of substance and (2) the strength of the bonds. A greater quantity of
substance requires more heat than a smaller amount, and a substance
with stronger bonds requires more heat than one with weaker bonds.
Chemists measure the enthalpy of fusion (also called heat of fusion) for
melting solids. **Enthalpy of fusion** is defined as the amount of heat required to change a specified amount of solid to a liquid at its melting
point. Table 13.5 lists the molar enthalpies of fusion for some common
substances. The molar enthalpy of fusion, usually expressed in kilojoules
per mole, is the amount of heat energy required to change one mole of a
solid to a liquid at the melting point.

**Solid to Liquid
Transition**

*Certain substances have
intermediate physical
states that are not exactly
solid or liquid. When
melting, the solid first
changes to the intermedi-
ate state, and then to the
liquid state. This intermediate state is less ordered
than the solid state, but
much more ordered than
the liquid. This state is
called a liquid crystal
state. Liquid crystals are
used as numeric displays
(LCD) in watches, calculators, and computer moni-
tors.*

TABLE 13.5   MOLAR ENTHALPIES OF FUSION AND VAPORIZATION FOR SELECTED SUBSTANCES

Substance	Melting point, °C	Molar enthalpy of fusion, kJ/mol	Boiling point, °C	Molar enthalpy of vaporization, kJ/mol
O_2	−219	0.44	−183	6.82
CH_4	−182	0.94	−161	8.18
Cl_2	−101	6.40	−34	20.4
NH_3	−78	5.65	−33	23.4
H_2O	0.0	6.01	100	40.7
Al	660	10.9	2467	284
NaCl	801	30	1413	

After the solid has melted totally (point *K*), added heat increases the kinetic energy of the liquid molecules, causing the temperature to rise (segment *KL* in Fig. 13.16). The temperature increase depends upon the heat capacity of the liquid, i.e., the amount of heat required to raise the temperature of a fixed quantity of the liquid by one degree Celsius.

At point *L* the liquid begins to boil. As previously discussed, the boiling point of a liquid is the temperature at which the vapor pressure of the liquid equals the pressure of the gases above the liquid. At the boiling point the temperature remains constant because the heat increases the potential energy of the particles, breaking the intermolecular forces among the particles in the liquid.

Liquid to Gas Transition

Segment *LM* of the graph is longer than segment *JK*, where melting occurs. Why is this? When the liquid boils, all the intermolecular forces must be broken for the liquid to enter the vapor phase. When the solid melts, fewer intermolecular forces are broken to enter the liquid phase; thus, less heat is normally required to melt a solid than is needed to boil an equal mass of the liquid of the same substance.

Table 13.5 lists the molar enthalpies of vaporization (also called heats of vaporization) for selected substances. The **enthalpy of vaporization** is the amount of heat required to change one mole of liquid to vapor at a specified temperature (usually the boiling point, as in this case). Most frequently, the molar enthalpy of vaporization is many times larger than the molar enthalpy of fusion for the same substance.

Magnitudes of the molar enthalpies of fusion and vaporization are an indication of the strength of the forces that bond the particles in a particular physical state. For example, the small values for the enthalpies of fusion and vaporization of oxygen (0.44 kJ/mol and 6.82 kJ/mol, respectively) indicate weak intermolecular forces. Molecules in solid oxygen are bonded by weak London dispersion forces (a molecular solid). On the other end of the scale, the rather strong ionic bonds in sodium chloride are reflected in its high enthalpy of fusion, 30 kJ/mol.

When point *M* is reached on the heating curve, all of the liquid has been converted to vapor. Additional heat increases the kinetic energy of

The specific enthalpies of fusion and vaporization of water are 0.335 kJ/g and 2.28 kJ/g, respectively.

CHEM TOPIC: Freeze-Drying

One way to preserve foods is to use the freeze-drying method. If you are a camper, you may have used freeze-dried foods which require no refrigeration and are reconstituted by the addition of water. Coffee drinkers know that some instant coffees are prepared by the freeze-drying method. How are foods freeze-dried? The food or coffee is first frozen and then placed in a chamber that is connected to vacuum pumps. The purpose of the vacuum pumps is to lower the vapor pressure in the chamber below the vapor pressure of ice; thus, the ice sublimes; i.e., it changes directly from a solid to a vapor. After a period of time all of the water is removed. Food and coffee dried by this method retain more flavor than food that is dried by heat because the fragile molecules responsible for the flavor are not destroyed as they are if the food is dried by heat. Another advantage of freeze-drying is that without water most bacteria cannot grow and cause the food to spoil; therefore, freeze-dried foods require no refrigeration.

---

the gas particles, raising the temperature of the vapor. Accordingly, the temperature increase depends on the heat capacity of the vapor.

REVIEW EXERCISES

13.17 Consider the heating curve in Fig. 13.16; then answer each of the following. *(a)* In what physical state is the substance from M to N on the graph? *(b)* What phase change occurs at J? *(c)* What quantity of heat is required to move from point L to point M? *(d)* What determines the increase in temperature along IJ?

13.18 Why is the enthalpy of vaporization larger than the enthalpy of fusion for the same substance?

13.19 Of the substances listed in Table 13.5, which has the strongest intermolecular forces in the *(a)* liquid state and *(b)* solid state?

13.5 WATER

Water is by far the most important liquid. Water covers the greatest percent of the surface of the earth, about 75%, and composes approximately 65% of the human body. All living things require a constant supply of relatively pure water.

Structure of Water

In 1781, Cavendish was the first to show that water is produced from H_2 and O_2. A few years later, Lavoisier determined experimentally the percent composition of H_2O. More recently, water was found to consist of H_2O molecules in which an O atom is covalently bonded to two H atoms separated by an angle of approximately 104 degrees:

$$\delta^-$$
$$\delta^+ \quad \overset{O}{\diagdown} \quad \delta^+$$
$$H \qquad H$$
$$\sim 104°$$

Because the O atom has a higher electronegativity than the H atoms, water is a polar covalent molecule. The O atom is partially negative, and the two H atoms are partially positive.

In the liquid state, water molecules are hydrogen bonded to each other. A fairly large percent of the O atoms in liquid water molecules

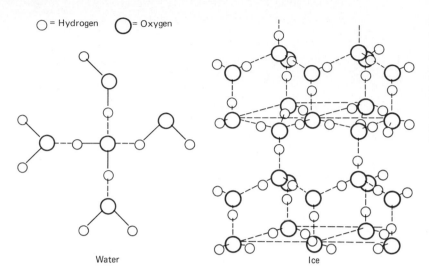

= Hydrogen   = Oxygen

Water   Ice

Figure 13.17
As a result of the formation of hydrogen bonds, some organized regions of water molecules are found in liquid water. In this figure, one water molecule is hydrogen bonded to four water molecules. The structure of ice is more highly organized and resembles a honeycomb.

have four H atoms associated with them. Two of the H atoms are covalently bonded, and the other two H atoms are hydrogen bonded to the O atoms from other water molecules (Fig. 13.17). The structure of water is believed to have some highly organized regions (similar to the regular structure in ice), and some more disordered regions of individual, unassociated water molecules.

The crystal structure of ice is more regular than the structure of liquid water. As shown in Fig. 13.17, each O atom in ice is surrounded by four H atoms (two covalently bonded and two hydrogen bonded). Ice is often described as having a "honeycomb" structure. Groups of water molecules are attached in a ring structure with a large number of open spaces.

When liquid water freezes and forms the honeycomb structure of ice, the empty spaces in the honeycomb cause the ice to have a larger volume than the liquid water had. The increase in volume when liquid water freezes is highly deviant from the behavior of just about all other liquids. Most substances occupy less volume (are more dense) in the solid state.

Because ice has a lower density than liquid water, ice floats on the surface of water. Antifreeze is placed in the radiator of an auto to prevent the formation of ice, which can crack an engine block.

Its strong hydrogen bonds give water a unique set of properties. Water has one of the highest heat capacities of liquids. Consequently, water is used as a coolant. Its high heat capacity means that, relative to other liquids, a large quantity of energy is needed to raise the temperature of water. Additionally, water has both a high enthalpy of vaporization and a high heat conductivity, good properties for a coolant.

Water, as we shall discuss in Chap. 15, is an excellent solvent (dissolving medium) for polar and ionic substances. Blood and the fluids in

Depending on the pressure, six different ice structures are known. You are familiar with ice I. In addition, ice II to V and ice VII exist. For example, ice VII exists at pressures in excess of 24,000 atm, has a density of 1.7 g/cm³, and is stable above the normal boiling point of water!

The maximum density of water is at 4°C.

Physical Properties of Water

living systems are aqueous solutions that contain many dissolved substances.

Water is a thermally stable molecule, which means it does not easily decompose when heat is added. However, water is quite reactive and combines with many substances.

Under the proper conditions, water molecules are split by many substances. Such a reaction is called a **hydrolysis** reaction. Consider the following hydrolysis reactions:

$$SiCl_4(l) + 4H_2O \longrightarrow Si(OH)_4(s) + 4HCl(aq)$$

$$Cl_2(l) + 2H_2O(l) \longrightarrow H_3O^+(aq) + Cl^-(aq) + HOCl(aq)$$

In each of these reactions water molecules are split to produce the products.

Water can combine with substances without being split apart. A reaction of this type is called a **hydration** reaction. Various salts combine with water to form hydrated salts:

$$Salt + xH_2O \longrightarrow salt \cdot xH_2O$$

A salt without the bonded water is called an **anhydrous salt.** After it combines with water, it is referred to as a **hydrated salt** or simply a hydrate. Two examples of the formation of hydrated salts are

$$Na_2B_4O_7 + 10H_2O \longrightarrow Na_2B_4O_7 \cdot 10H_2O$$
Borax

$$KAl(SO_4)_2 + 12H_2O \longrightarrow KAl(SO_4)_2 \cdot 12H_2O$$
Alum

Large quantities of borax are obtained commercially from dry lakes in California. Borax is used as a water softener and a flux in solder to dissolve oxide coatings found on metals.

Anhydrous salts and hydrates exhibit three interesting properties. Various anhydrous salts absorb water directly from the air and become hydrated. Such salts are called **hygroscopic.** Some salts take water from the air so readily that they absorb enough water so that they dissolve. Salts that exhibit such behavior are called **deliquescent.** Certain hydrates have a higher water vapor pressure than the atmosphere, and therefore release water to the air. When this occurs, a salt is said to undergo **efflorescence.**

Compounds that absorb water are used as drying agents to remove water. Nonaqueous solutions are dried with a drying agent such as anhydrous $CaCl_2$. Bags filled with drying agents are packed along with instruments that absorb moisture or rust.

Water Cycle

Good estimates indicate that there is approximately 10^9 km^3 ($1 \text{ km}^3 = 0.24 \text{ mi}^3$) of water on earth. About 97% of this water is salt water; only 3% is fresh water. Most of the fresh water is in the form of ice in the polar regions and glaciers (75%). Groundwater, water below land surfaces, represents 20% of the fresh water, and the remaining 15% is in lakes, in rivers, in the soil as moisture, and in the air.

All the water on earth is part of the water cycle. Water, like many substances on earth, is constantly on the move (Fig. 13.18). Energy to fuel the water cycle comes in the form of radiant energy from the sun, which evaporates large quantities of water from the seas and oceans. After the atmosphere becomes saturated with water vapor, the water falls to earth as some form of precipitation: rain, snow, sleet, or hail. Water that falls on land flows into rivers and lakes, or filters down through various ground layers and becomes groundwater. Ultimately, all of this water ends up in the oceans, evaporates back into the atmosphere, or freezes in the polar regions.

As part of the water cycle a small percent of the water enters living systems. Green plants take in carbon dioxide, CO_2, and combine it with water in the presence of sunlight and chlorophyll to produce carbohydrates (sugars and starches) and oxygen, O_2.

Relative humidity is a measure of the amount of water vapor in the air. It is calculated by dividing the actual vapor pressure of water in the air by the equilibrium vapor pressure, which is the maximum amount of water that could be in the air at a specific temperature.

Figure 13.18
Water is constantly on the move. Water, rain, or snow that falls to earth travels to the oceans where it evaporates and begins the cycle again. *(Peter Miller/Photo Researchers.)*

$$nCO_2 + nH_2O \xrightarrow[\text{chlorophyll}]{\text{light}} (CH_2O)_n + nO_2$$

Carbohydrates

Animals produce water as a final product of cellular respiration, their mechanism for producing energy.

$$2H^+ + \tfrac{1}{2}O_2 \longrightarrow H_2O$$

Final reaction in cellular respiration

13.20 Describe the structure of each of the following: *(a)* a water molecule, *(b)* molecules in liquid water, and *(c)* molecules in ice.   **REVIEW EXERCISES**

13.21 Explain why ice is less dense than liquid water.

13.22 List two ways in which water is essential to living things.

13.23 Briefly describe the movement of water through the water cycle.

As an example of a solid, we will consider the element carbon. Carbon exists in two different crystalline forms, graphite and diamond. In addition, carbon is found in various semiamorphous forms, which include charcoal, carbon black, and coke. Different forms of an element in the same physical state are called **allotropes;** thus, graphite and diamond are allotropes of carbon.   **13.6 CARBON**

Graphite, also known as "black lead," is a soft, black solid. Its density is 2.27 g/cm³. It is slippery, melts at 3527°C, and is an electric conductor. The somewhat special properties of graphite are related to its structure, as properties always are. Graphite is composed of layers of carbon atoms arranged in six-membered fused rings (Fig. 13.19).   **Graphite**

Figure 13.19
Graphite is composed of layers of C atoms that are members of six-membered rings. Each C atom in a ring is covalently bonded to three other C atoms. The layers are not strongly bonded to each other; thus, they can slide by each other.

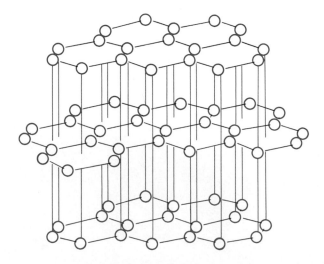

Each C atom is bonded to three other C atoms in such a manner that one of the electrons in each C atom is delocalized and not held tightly; this is what makes graphite an electric conductor. The layers of graphite are not bonded very strongly, so they are free to slide by each other, which accounts for the slippery nature of graphite. As a result, graphite is frequently used as a lubricant.

Diamonds (Fig. 13.20) are produced when coal and other amorphous forms of carbon are subjected to high pressures, generally from volcanic activity. Most of the world's supply of diamonds comes from South Africa, Brazil, Australia, and the United States.

The properties of diamond are quite different from those of graphite. Diamond is more dense (3.51 g/cm^3) than graphite (2.27 g/cm^3). A higher density indicates that the C atoms in diamond are packed in a smaller volume than those of graphite. Diamond, a macromolecular solid, is composed of C atoms that are bonded symmetrically to four other C atoms (Fig. 13.21). Because each of the four electrons is a part of the four covalent bonds, the electrons in the diamond structure are not mobile; consequently, diamond is a nonconductor of electricity. However, diamond is an excellent heat conductor.

Strong covalent bonds are responsible for the hardness of diamond. It is the hardest naturally occurring substance. On a hardness scale of 1 to 10, diamond is a 10! Since diamonds are so hard, they are placed on

Diamond

Artificial diamonds were first produced in 1955. Temperatures around 2000°C and pressures about 70,000 atm, along with a catalyst, are required to produce artificial diamonds from graphite. These diamonds are not gem quality, but are chemically equivalent to natural diamonds.

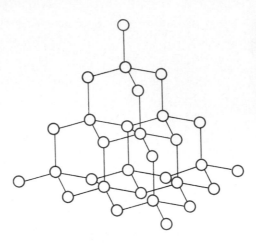

Figure 13.21
Diamond is a network covalent solid of C atoms each
bonded to four other C
atoms. The geometric shape
around each carbon atom is
that of a tetrahedron. Diamond is the hardest naturally
occurring substance.

the ends of drill bits to bore through rock strata in the search for crude oil. Diamond melts at 3570°C, which is among the highest of all melting points. Extremely high melting points are a characteristic property of network solids.

Diamond is unreactive, but it does change to carbon dioxide in pure oxygen if heated above 1000°C.

$$C(\text{diamond}) + O_2(g) \xrightarrow{\;>1000°C\;} CO_2(g)$$

Carbon black (sometimes called lampblack), a pure form of carbon, is one of the amorphous forms of carbon. Carbon black is normally prepared by heating carbon-hydrogen compounds, called hydrocarbons, in a flame. It is used primarily as a black pigment and to reinforce rubber; a large percent of the mass of a rubber tire is carbon black.

Coal is an impure form of carbon that results when dead animal and plant matter is compacted in the earth for long periods of time. The structure of coal somewhat resembles that of graphite, but is less regular. Also, coal contains many impurities. One such impurity, sulfur, is of primary concern. When coal is combusted, sulfur is oxidized to unwanted gaseous sulfur oxides, which are major air pollutants. Coal mining, especially strip mining, also has an impact on the environment (Fig. 13.22).

When coal is heated in the absence of oxygen, volatile substances are driven off, leaving a substance called **coke.** Coke is extremely important in industry because it is converted to graphite. Coke is an industrial fuel and is also combined with metals to increase their strength.

If wood is heated in the absence of air, **wood charcoal** results. Most people know that charcoal is used to grill food, but this is not the major use of charcoal. Charcoal is commonly used to purify other substances. Finely powdered charcoal adsorbs other substances onto its surface. **Ad-**

Amorphous Forms of Carbon

Steel contains about 0.5% carbon. If the concentration of C in iron is too high, the steel loses its strength and hardness.

Figure 13.22
Coal and other substances in the earth are obtained by strip mining and deep mining procedures. In strip mining, large machines remove the top layers of the earth to expose the coal. *(Department of Energy.)*

sorption is the process in which an adsorbing substance attracts other substances to its surface. When water is filtered through charcoal, the charcoal removes many impurities that cause the water to smell bad or to have a bad taste.

Even though carbon compounds are the central substances in living things, carbon and its compounds comprise less than 0.1% of all substances on earth. Like other substances we have discussed (N_2, O_2, and H_2O), carbon and carbon compounds are linked in a complex cycle on earth.

Figure 13.23 is a diagram of the carbon cycle. During photosynthesis, plants and microorganisms convert atmospheric CO_2 to organic

Carbon Cycle

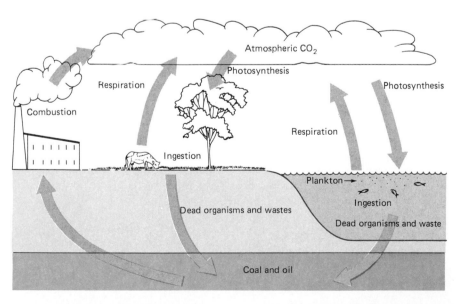

Figure 13.23
Carbon in many different forms travels through the atmosphere, lithosphere, and biosphere. This is known as the carbon cycle.

compounds. As previously described, photosynthesis is the process in which CO_2 and H_2O, in the presence of light and a green plant pigment called chlorophyll, are combined in a complex series of reactions to produce carbohydrates and O_2.

Plants metabolize some of the sugars produced by photosynthesis for their own energy needs; in doing so, they decompose the sugars, releasing CO_2 to the atmosphere. Some of the sugars are stored in plants, and others are converted to different classes of biologically significant molecules. Carbon compounds are transferred to animals when they eat plants. Animals derive energy from the sugars and incorporate the carbon compounds into their bodies. After plants and animals die, their remains are decomposed by microorganisms, transferring the carbon compounds to the soil. Ultimately, carbon compounds are oxidized in the soil to CO_2, starting the cycle again.

A large percent of the earth's carbon compounds are in oceans, lakes, rivers, and rocks. CO_2 from the atmosphere dissolves in the oceans, where it is converted to dissolved carbonates and hydrogen-carbonates (bicarbonates):

Limestone, $CaCO_3$, is the second most abundant rock in the earth's crust. Coral reefs are mainly composed of calcium carbonate from special marine organisms. Pearls are layers of calcium carbonate that form inside the shells of oysters.

$$CO_2(g) + H_2O(l) \longrightarrow [H_2CO_3] \longrightarrow H^+(aq) + HCO_3^-(aq)$$

Carbonic acid   Bicarbonate

Carbonates then precipitate out of solution in the form of carbonate-containing rocks such as limestone, calcite, chalk and marble.

REVIEW EXERCISES

13.24 What are allotropes? Give an example.

13.25 Describe the structure of each of the following: (a) graphite, (b) diamond, (c) coal.

13.26 What accounts for the hardness and high melting point of diamond?

13.27 What is the significance of the photosynthetic process in the carbon cycle?

13.28 Describe the pathway of carbon from the atmosphere to a human being.

SUMMARY

Because of the intermolecular forces in liquids, the particles are held closely and their freedom of movement is restricted. Very few empty spaces are found in the structure of liquids; they are thus incompressible. However, there is enough space to allow molecules to slide by each other, giving liquids their fluid properties.

Evaporation occurs when fast-moving liquid particles on the surface break free into the vapor state. **Condensation** occurs when vapor molecules above a liquid cohere and fall back into the liquid state. In a closed system, an equilibrium between the rates of evaporation and condensation produces a fixed amount of vapor above the liquid, at a constant temperature. The pressure of a vapor in equilibrium with a liquid is the liquid's **vapor pressure.** The temperature at which the vapor pressure of a liquid equals the pressure of gases above the liquid is the **boiling point. Surface tension** is the property of the surface of a liquid that allows it to act as if it is covered by a membrane. On solid surfaces, liquids either wet the surface by spreading out or bead up and form droplets.

Dipole–dipole interactions are intermolecular forces that occur among molecules that exist as dipoles. **Hydrogen bonds** are a special type of dipolar interaction. All substances that exhibit hydrogen bonding have molecules with an H atom bonded covalently to F, O, or N. Forces between nonpolar molecules are called **London dispersion forces.** Dispersion forces result from the induction of a dipole in a nonpolar molecule by another nonpolar molecule.

Solids are the most condensed form of matter and possess the strongest forces between particles. Most solids are classified as **crystalline solids,** solids that have a regular geometric pattern of particles. Crystalline solids are usually grouped into four categories: (1) ionic, (2) covalent, (3) molecular, and (4) metallic solids. A small group of solids lack a regular crystal structure and have a more random pattern; they are called amorphous solids.

When heat is added to a solid, the temperature of the solid increases until it reaches its **melting point,** at which the temperature remains constant until all of the solid has melted to a liquid. Addition of heat to the liquid raises the temperature of the liquid until the boiling point is reached. At the **boiling point,** the temperature remains constant until the liquid is totally changed to vapor. Further addition of heat raises the temperature of the vapor.

Water has an unusually high boiling point, heat capacity, thermal conductivity, and surface tension for such a low-molecular-mass liquid. Most of the special properties are a result of the strong hydrogen bonds in water. Another anomalous property of water is that liquid water has a higher density than ice.

Allotropes are different forms of an element in the same physical state. Carbon exists in different allotropic forms: diamond, graphite, and various amorphous forms. Graphite is a soft, slippery, black form of carbon, while diamond is a hard solid with a very high melting point.

KEY TERMS

adhesive force	crystalline solid	heat of vaporization	photosynthesis
adsorption	deliquescent	hydration	surface tension
allotropes	delocalized electrons	hydrogen bonding	vapor pressure
amorphous solid	dipole–dipole interactions	hydrolysis	vaporization
boiling point	dispersion forces	hygroscopic	viscosity
capillary action	efflorescence	induced dipole	wetting
cohesive force	evaporation	Intermolecular forces	
condensation	heat capacity	London dynamic equilibrium	
crystal system	heat of fusion	normal boiling point	

EXERCISES*

13.29 Define each of the following terms: intermolecular forces, evaporation, vaporization, condensation, dynamic equilibrium, vapor pressure, boiling point, normal boiling point, surface tension, wetting, viscosity, dipole–dipole interactions, hydrogen bonding, London dispersion forces, induced dipole, cohesive force, adhesive force, crystalline solid, amorphous solid, crystal system, delocalized electrons, heat capacity, heat of fusion, heat of vaporization, capillary action, hydrolysis, hydration, hygroscopic, deliquescent, efflorescence, photosynthesis, allotropes, adsorption.

Liquids

13.30 Use the kinetic molecular theory to explain each of the following properties of liquids: *(a)* Liquids flow; *(b)* they take the shape of their containers to the level that they fill; *(c)* they have a fixed volume; *(d)* their molecules are not free to move from place to place; *(e)* they have a higher average density than gases.

13.31 Compare liquids with gases in terms of the following properties: *(a)* density, *(b)* viscosity, *(c)* intermolecular forces, *(d)* compressibility, *(e)* shape.

*For exercise numbers printed in color, answers can be found at the back of the book.

13.32 Explain why only surface particles in liquids are involved in evaporation.

13.33 What is the principal difference between two liquids, at the same temperature, one that evaporates quickly and the other that does not noticeably evaporate?

13.34 What is the purpose of rubbing an alcohol on the skin of a person who has a fever?

13.35 Explain why an equilibrium is not usually established between the rates of evaporation and condensation in an open system?

13.36 At a fixed temperature, determine which of the described liquids would have a higher vapor pressure: *(a)* liquid A with weaker intermolecular forces or liquid B with stronger intermolecular forces; *(b)* liquid C with a lower boiling point than liquid D; *(c)* liquid E, a hydrogen-bonded liquid, or liquid F with the same molecular mass and only dispersion forces; *(d)* liquid G, a thick viscous oil, or liquid H, a volatile liquid.

13.37 Consider the three vapor pressure curves in Fig. 13.24. *(a)* Which liquid has the lowest vapor pressure at 25°C? *(b)* What are the normal boiling points of each liquid? *(c)* Rank the liquids in order of strength of their intermolecular forces. *(d)* At what temperature does each liquid have a vapor pressure of 650 torr? *(e)* At what temperature would each liquid boil if the pressure above the liquids was reduced to 500 torr?

Figure 13.24

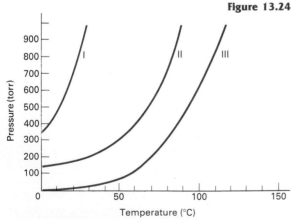

13.38 *(a)* What is contained in the bubbles that are observed in boiling liquids? *(b)* Explain why these bubbles cannot form in a liquid below the boiling point.

13.39 If the pressure above a container of water is lowered so that the water boils at room temperature, explain why this boiling water causes no burns.

13.40 *(a)* Would it take longer to hard boil eggs in Denver, Colorado (the Mile High City at an altitude of 5280 ft) or in Tampa, Florida (almost at sea level), assuming that the eggs are carefully placed in the same size container with the same volume of boiling water? Write a complete explanation for your answer. *(b)* Where would it take longer to fry eggs, Denver or Tampa? Explain.

13.41 Explain why foods cook faster in a pressure cooker than if the foods were cooked in an open pan of boiling water.

13.42 What is different about surface molecules in liquids to produce the property of a liquid called surface tension?

13.43 Explain the fact that water droplets outside a gravitational field are spherical and not cubic?

13.44 *(a)* How do liquids that form concave meniscuses in glass tubes differ from those that form convex meniscuses? *(b)* Give an example of each type of liquid.

13.45 Why does water spread out and cover many of the surfaces that it comes in contact with, while mercury forms small beads on similar surfaces?

13.46 What effect does an increase in temperature have on the viscosity of a liquid?

13.47 Select from the following pairs the liquid that is expected to have the highest viscosity: *(a)* high-molecular-mass liquid K or low-molecular-mass liquid L; *(b)* liquid M, a hydrogen-bonded liquid, or liquid N, that is bonded with London dispersion forces.

Intermolecular Forces

13.48 Show that the three types of intermolecular forces in liquids result from electrostatic attractions.

13.49 What accounts for the fact that hydrogen bonds are stronger than dipole–dipole interactions, even though both involve the attraction of a partially positive region of one molecule for the negative region of another, and vice versa?

13.50 Draw the Lewis structure for bromine mono-chloride, BrCl(*l*), and show the dipole–dipole attractions among a few BrCl molecules.

13.51 (*a*) What three atoms bond to H atoms to produce hydrogen bonds? (*b*) For each of these atoms give an example of a molecule that forms hydrogen bonds.

13.52 The following three liquids, with similar molecular masses, are known to be bonded by three different intermolecular forces. Given their boiling points, explain which intermolecular forces you would expect in each.

Liquid	Boiling point, °C
Q	115
R	−95
S	−2

13.53 (*a*) How do London dispersion forces result between molecules? (*b*) Why are they classified as extremely short-range forces?

13.54 Rank each of the following sets of nonpolar liquids in order of increasing boiling point (lowest to highest): (*a*) Ne(*l*), Kr(*l*), and Ar(*l*); (*b*) $C_5H_{12}(l)$, $C_{10}H_{22}(l)$, and $C_8H_{18}(l)$.

13.55 Alcohols, which are carbon-hydrogen compounds that contain an —OH group in the molecule, are hydrogen-bonded liquids. What could account for the increasing trend in their boiling points as an extra C and two H atoms are added to the alcohol molecule?

Alcohol	Boiling point, °C
CH_3OH	64.7
C_2H_5OH	78.3
C_3H_7OH	97.2

Solids

13.56 How is the general structure of an amorphous solid different from that of a crystalline solid?

13.57 Give two examples each of crystalline and amorphous solids.

13.58 What would happen to the structure of solids if they could be compressed?

13.59 Give two examples of each of the following classes of crystalline solids: (*a*) metallic, (*b*) ionic, (*c*) molecular, (*d*) covalent.

13.60 Ionic solids are nonconductors of electricity, but if they are melted, they are capable of conducting an electric current. Write an explanation to account for this behavior of ionic substances.

13.61 Account for the extremely high melting points of network covalent solids.

13.62 In what class of crystalline solid does each of the following belong: (*a*) Co, (*b*) NaF, (*c*) P_4, (*d*) SiO_2, (*e*) KI, (*f*) Li, (*g*) S_8?

13.63 Dry ice is composed of carbon dioxide molecules with the following structure

$$\ddot{O}=C=\ddot{O}$$

Two strong carbon–oxygen double bonds hold the molecule together; however, dry ice is a volatile solid, which readily becomes a vapor, $CO_2(g)$, at 25°C. Explain this apparent inconsistency.

13.64 What accounts for metals being the only class of solids that are good conductors of electricity.

13.65 Metals exhibit the property of malleability; i.e., they can be hammered into different forms and shapes. Write an explanation, in terms of the structure of solids, of what happens to a metal when it is hammered into a thin foil.

State Changes

13.66 Consider the cooling curve of a hypothetical substance in Fig. 13.25: initially it is at 200°C and in the gas phase, and is cooled to −150°C. (*a*) What physical state is the substance in at each of the following tempera-

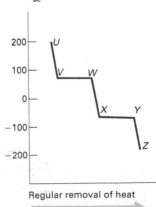

Regular removal of heat

Figure 13.25

tures: 65°C, 160°C, and −75°C? (b) What are the freezing and boiling points of this substance? (c) Explain in terms of the kinetic and potential energies of the molecules what is happening to the substance along the following segments of the curve: UV, VW, WX, XY, and YZ.

13.67 Draw a graph of the heating curve of water, starting at −25°C and ending at 125°C. Label each axis and each segment of the curve, and indicate the enthalpy of fusion and enthalpy of vaporization on the graph.

13.68 Aluminum melts at 660°C and boils at 1800°C. Its molar enthalpy of fusion is 10.7 kJ/mol, and its molar enthalpy of vaporization is 225 kJ/mol. Draw a graph of the heating curve for Al. Do not forget to label the axes and each segment of the curve.

13.69 Why does the temperature remain constant when heat is continually added to a block of ice?

Use Table 13.5 to solve problems 13.70 through 13.73.

13.70 Calculate the amount of heat needed to change 0.174 mol $H_2O(l)$ at 100°C to water vapor at the same temperature.

13.71 How much heat must be removed from 254 g of water at 0.0°C to produce ice at the same temperature?

13.72 How much heat energy is required to change 1.75 mol of ice from 0.0°C to steam at 100°C? The specific heat of liquid water is 4.184 J/(g·°C).

13.73 Compare the amount of heat needed to melt 50.0-g samples of Al and NaCl.

Water

13.74 How would the properties of water change if water molecules were linear (H—O—H), rather than angular with a bond angle of approximately 104°?

13.75 If water froze from bottom to top, what effect would that have on the animals that live in small ponds during the winter months?

13.76 What properties does water have that makes it good for (a) a coolant in automobile engines and (b) living systems?

13.77 Explain the difference between efflorescence and deliquescence.

13.78 Write equations that illustrate each of the following: (a) the decomposition of a hydrate to an anhydrous salt and water, (b) the formation of another hydrate from an anhydrous salt and water, (c) the hydrolysis of $Cl_2(g)$, and (d) the hydrolysis of SiI_4.

13.79 Water vapor that enters the air remains in the atmosphere for a relatively short period of time, an average of 10 days. What factors tend to limit the amount of time water remains in the atmosphere?

13.80 List the three main sources of fresh water on earth.

13.81 With an ever dwindling supply of potable fresh water, it has been proposed that icebergs, a source of a large quantity of fresh water, be towed to U.S. cities for use as drinking water. What problems might be encountered in utilizing the fresh water in icebergs?

Carbon

13.82 What properties of graphite make it a good substance for placing in a pencil for writing?

13.83 Contrast the physical properties of graphite and diamond.

13.84 What is carbon black, and how is it used commercially?

13.85 Why is graphite a conductor of electricity, but diamond is a very poor electric conductor?

13.86 Very little heat is required to convert graphite to diamond; however, it is very expensive and difficult to change graphite to diamond. What could account for this expense and difficulty?

13.87 How are coal and coke produced in nature?

13.88 During the production of the colorless alcoholic beverages such as vodka, they are usually filtered through charcoal. What is the purpose of passing vodka through purified charcoal?

13.89 Briefly describe the role of carbon in living systems, starting with photosynthesis and ending with carbon compounds in animals.

13.90 Describe what happens to atmospheric CO_2 that dissolves in the ocean.

13.91 Since the industrial revolution, each year more and more CO_2 has been given off to the atmosphere through the burning of fossil fuels (oil, natural gas, and coal). However, only a small increase in the amount of atmospheric CO_2 has been recorded. Considering the carbon cycle, propose reasons why the CO_2 level has not increased significantly.

Additional Exercises

13.92 A 10.0-g sample of diethyl ether, $C_4H_{10}O$, is added to a 2.15-L flask, and it is totally evaporated at 34.6°C. Calculate the vapor pressure of the ether.

13.93 The electronegativity of Cl is quite high (3.0), but Cl does not form hydrogen-bonded liquids when bonded to H atoms. Explain fully.

13.94 If equal masses of steam at 100°C and boiling water at 100°C contact your skin, the burn from the steam is more severe. Write an explanation that accounts for this fact.

13.95 The heat capacity of Cu is 25 J/(mol·°C). What mass of steam at 100°C must be converted to water at 100°C to raise the temperature of 5.6 kg of copper from 12.4°C to 25.3°C?

13.96 Calculate the mass of water that can be released when 4.14 g of sodium phosphate dodecahydrate, $Na_3PO_4 \cdot 12H_2O$, is heated as follows:

$$Na_3PO_4 \cdot 12H_2O \xrightarrow{\Delta} Na_3PO_4 + 12H_2O$$

13.97 Silicon carbide, SiC, has a structure similar to that of diamond. (a) Draw a segment of the structure of silicon carbide. (b) Compare the melting point, hardness, and electric conductivity of silicon carbide to those of boron nitride, BN, which has a structure similar to that of graphite.

13.98 When tetraphosphorus decoxide, P_4O_{10}, combines with water, it produces phosphoric acid, H_3PO_4. (a) Write the balanced equation for this reaction. (b) Calculate the masses of P_4O_{10} and H_2O that are required to produce 30.0 kg of phosphoric acid. (c) What mass of phosphoric acid results when 71.93 g P_4O_{10} combines with 27.66 g H_2O?

13.99 When iron metal combines with steam, it produces Fe_3O_4 and hydrogen gas. (a) Write the equation for the reaction. (b) What volume of H_2 at 125°C and 758 torr results when 297 g of iron is combined with excess steam?

13.100 Ethyl chloride, C_2H_5Cl, is a volatile liquid that boils at 12°C. It is sprayed on the skin of a patient by a physician to numb and freeze the area for minor surgical procedures. Explain the cooling effect of ethyl chloride.

13.101 A cube of ice at 0.0°C with an edge length of 3.20 cm is placed in 255 g of water at 30.0°C. To what temperature is the water cooled when the ice is totally melted and temperature equilibrium is obtained? The specific heat of water is 4.184 J/(g·°C), the molar enthalpy of fusion of water is 6.01 kJ/mol, and the density of ice is 0.980 g/cm^3.

13.102 When a sample of HF is vaporized in a closed container, the pressure of the resulting vapor is significantly lower than what would be predicted using the ideal gas equation. Considering the properties of HF, explain why the pressure of the HF vapor deviates from ideal properties.

13.103 Consider the following vapor pressure data for liquid NH_3.

Temperature, °C	−70	−60	−50	−40
Vapor pressure, atm	0.108	0.216	0.403	0.708

(a) Draw the vapor pressure curve for liquid ammonia. (b) Predict the normal boiling point of ammonia by extrapolating the curve. (c) At what temperature is the vapor pressure of ammonia equal to 500 torr? (d) What is the vapor pressure of ammonia at −58°C?

13.104 Sulfuric acid, H_2SO_4, is a strong dehydrating agent. When 12.2 L of air at 30°C is passed through 82.355 g H_2SO_4, the mass of H_2SO_4 increases to 82.731 g. What is the vapor pressure of water in the air at 30°C? Assume that the sulfuric acid removes all of the water.

CHAPTER
Fourteen

Descriptive Inorganic Chemistry

───────────────── **STUDY GUIDELINES** ─────────────────

After completing Chapter 14, you should be able to

1. Distinguish between representative and transition metals

2. Apply your knowledge of elements, compounds, reactions, and states of matter in discussing the physical properties, chemical properties, sources, and uses of (a) alkali metals, (b) alkaline earth metals, (c) aluminum, (d) the halogens, (e) sulfur, and (f) phosphorus, and their compounds

3. Identify trends in properties of the elements discussed in the chapter

4. Write equations for the formation and reactions of the elements discussed in this chapter

5. Account for differences in properties of the alkali and alkaline earth metals

6. List examples of commercially important alloys and their uses

7. List examples of elements that exist in allotropic forms, and discuss differences in their structures and properties

8. Distinguish between oxidizing and reducing agents, and discuss how they behave

9. Discuss reasons why some rain is becoming more acidic

Throughout our study of chemistry, each chapter has introduced new ideas and concepts related to chemical principles. In this chapter, we deviate somewhat from that plan to reach the heart of chemistry—descriptive chemistry. **Descriptive chemistry** is the study of the properties, relationships, reactions, uses, and distribution of elements and compounds. In the study of descriptive chemistry, an attempt is made to explain trends in properties and chemical behavior by considering microscopic structure, intermolecular forces, chemical bonding, and energy factors.

Chapter 14 is an applied chapter, in that many chemical principles that you have already learned will be used to help you understand the properties and behavior of selected elements and their compounds. You will apply your knowledge of atomic and molecular structure, periodic properties, stoichiometry, nomenclature, equations, and states of matter.

14.1 REPRESENTATIVE METALS

Representative metals are metals that belong to groups IA (1) through VA (15), excluding the transition metals, which are the B group metals. Therefore, the representative metals include (1) the alkali metals (IA), (2) alkaline earth metals (IIA), (3) all elements in group IIIA (Al, Ga, In, and Tl) except boron, B, (4) Sn and Pb from group IVA, and (5) Bi from group VA. We will investigate the metals in groups IA and IIA as well as one metal from group IIIA, aluminum.

Alkali Metals

The **alkali metals** are the elements in group IA of the periodic table: Li, Na, K, Rb, Cs, and Fr. Each of the alkali metals has an inner core electronic configuration of a noble gas and one outer-level electron in the s sublevel, s^1. In chemical changes, the alkali metals (M) tend to lose their outer s^1 electron, producing a monopositive cation.

$$M \longrightarrow M^+ + e^- \ (M = \text{alkali metal})$$

The name "cesium" is derived from the Latin caesius, *which means sky blue. Cesium was discovered by Bunsen and Kirchhoff in 1860.*

Alkali metals are widely distributed throughout the earth. However, because of their reactivity, they are always chemically combined with other elements in compounds. Na and K are two of the more abundant metals in the earth's crust and oceans. Sodium is found in rock salt, $NaCl$; saltpeter, $NaNO_3$; borax, $Na_2B_4O_7 \cdot 10H_2O$; and as dissolved ions, $Na^+(aq)$, in the oceans. Potassium may be found in minerals such as sylvite, KCl; kainite, $KCl \cdot MgSO_4$; and carnallite, $KCl \cdot MgCl_2$. Li, Rb, and Cs are much less abundant than Na and K.

The name "rubidium" is derived from the Latin rubidus, *which means dark red. Rb was also discovered by the team of Bunsen and Kirchhoff, a year after they discovered cesium.*

Table 14.1 summarizes the physical properties of the alkali metals. Alkali metals are soft and have the lowest densities of metals. The low density of alkali metals is attributed to the large size of their atoms (they are the largest atoms within a period), and their rather widely spaced metallic crystal packing pattern. Densities of alkali metals increase with increasing atomic mass. Li has a density of 0.53 g/cm^3 and cesium has a density of 1.88 g/cm^3.

Melting points of the alkali metals are the lowest among metals. For example, the melting point of Cs, 28°C, is only slightly above room temperature. Low melting points indicate rather weak metallic bonds. Similarly, the boiling points of the alkali metals are low, ranging from 1347°C for Li to 678°C for Rb. These elements have a wide temperature range between their melting and boiling points (called the liquid range), indicating that metallic bonds predominate as the intermolecular forces in the liquids.

TABLE 14.1 PROPERTIES OF THE ALKALI METALS

Properties	Li	Na	K	Rb	Cs
Atomic number	3	11	19	37	55
Atomic mass	6.941	22.99	39.10	85.47	132.9
Outer-level electronic configuration	$2s^1$	$3s^1$	$4s^1$	$5s^1$	$6s^1$
Density, g/cm^3	0.53	0.97	0.86	1.53	1.88
Melting point, °C	181	98	63	39	28
Boiling point, °C, 1 atm	1347	883	774	688	678
Enthalpy of fusion, kJ/mol	3.01	2.6	2.3	2.2	2.1
Enthalpy of vaporization, kJ/mol	135	97.9	79.0	75.8	68.3
Iodization energy, kJ/mol	526	502	425	409	382
Atomic size, nm	0.152	0.186	0.231	0.244	0.262
Electronegativity	1.0	0.9	0.8	0.8	0.7
Formula of halide	LiX	NaX	KX	RbX	CsX
Formula of oxide	Li_2O	Na_2O	K_2O	Rb_2O	Cs_2O

The electric and heat conductivities of alkali metals are among the highest for all elements. Only silver, gold, copper, and aluminum are better conductors. Good electric and thermal conductivity are explained in terms of their mobile, loosely held electrons. Alkali metals have the lowest ionization energies of all elements.

Chemically, the alkali metals (M) are very reactive and combine directly with nonmetals. When they combine with halogens, they produce halides, MX. With H_2 they produce hydrides (MH), and with O_2 and N_2 they form oxides (M_2O) and nitrides M_3N), respectively.

$$2M + X_2 \longrightarrow 2MX \quad \text{(Halide (X = halogen)}$$

$$2M + H_2 \longrightarrow 2MH \quad \text{(Hydride)}$$

$$4M + O_2 \longrightarrow 2M_2O \text{ (Oxide)}$$

$$6M + N_2 \xrightarrow{\text{spark}} 2M_3N \text{ (Nitride)}$$

These reactions are expected, considering that after losing an electron to a more electronegative nonmetal, alkali metals become monopositive ions with the stable noble gas configuration.

Oxygen combines with the alkali metals to produce oxides.

$$4M + O_2 \longrightarrow 2M_2O \text{ (Alkali metal oxide)}$$

For example, Li combines with O_2 to produce lithium oxide, Li_2O.

$$4Li + O_2 \longrightarrow 2Li_2O$$

In addition, some of the alkali metals combine with oxygen in such a way as to produce peroxide ions, O_2^{2-}. Each O atom in a peroxide ion is in the -1 oxidation state. Oxygen atoms most frequently exist in the -2 oxidation state.

$$2M + O_2 \longrightarrow M_2O_2 \text{ (Peroxide)}$$

For example, Na can combine with O_2 to produce sodium peroxide, Na_2O_2.

$$2Na + O_2 \longrightarrow Na_2O_2$$

Sodium peroxide is a yellowish powder that is used to bleach and oxidize other substances.

When alkali metals are mixed with water, they react violently, producing an alkaline (basic) solution, hydrogen gas, and energy. Potassium and the higher-atomic-mass alkali metals release so much heat that the hydrogen gas liberated is ignited, causing an explosion!

$$M(s) + H_2O(l) \longrightarrow 2MOH(aq) + H_2(g) + \text{energy}$$

For example, Cs combines with water to produce cesium hydroxide, CsOH, and H_2.

$$Cs(s) + H_2O(l) \longrightarrow 2CsOH(aq) + H_2(g) + \text{energy}$$

Alkali metals combine even more vigorously if they contact acidic solutions.

Alkali metals have many uses. Lithium is alloyed with magnesium, and is the major component in Li batteries. Salts of Li such as lithium carbonate and lithium citrate help individuals with mental disorders, especially those with manic-depressive illness. Sodium is the starting material for the manufacture of numerous substances, for example, dyes, soaps, and lead antiknock compounds for gasoline. Potassium compounds are major components of fertilizers and some explosives. Photoelectric cells contain rubidium and cesium. An electric current is produced when light hits the surface of these metals.

Previously, sodium chloride, NaCl, has been discussed as an alkali-metal-bearing compound. Other common alkali metal compounds include (1) sodium carbonate, Na_2CO_3; (2) sodium hydrogencarbonate, $NaHCO_3$; (3) sodium hydroxide, NaOH; and (4) potassium nitrate, KNO_3.

Lithium carbonate, Li_2CO_3, marketed under names such as Lithane, Eskalith, and Lithonate, is an alternative drug to tranquilizers for the treatment of various psychoses.

Sodium carbonate, Na_2CO_3, is frequently called by its common name, soda ash. Soda ash is used in the manufacture of soap, paper, water softeners, glass, and petroleum products. Sodium carbonate exists naturally in the mineral trona, $Na_2CO_3 \cdot NaHCO_3 \cdot 2H_2O$. Sodium carbonate is produced by the **Solvay process,** a procedure in which a solution of sodium chloride, $NaCl(aq)$, is combined with aqueous ammonia, $NH_3(aq)$, and carbon dioxide, CO_2, to initially produce sodium hydrogencarbonate, $NaHCO_3$, and ammonium chloride, NH_4Cl.

$$NaCl + NH_3 + CO_2 + H_2O \longrightarrow NaHCO_3 + NH_4Cl$$

$NaHCO_3$ is then removed and heated, driving off CO_2 and H_2O, yielding sodium carbonate.

$$2NaHCO_3(s) \xrightarrow{\Delta} Na_2CO_3(s) + CO_2(g) + H_2O(l)$$

Sodium hydrogencarbonate, $NaHCO_3$, or baking soda, is used in cooking. Baking powders contain $NaHCO_3$ along with an acidic substance (generally, cream of tartar, potassium hydrogen tartrate, $KHC_4H_4O_6$). As long as the mixture is dry, no reaction occurs. But as soon as it is combined with water, $NaHCO_3$ and the acid react, releasing $CO_2(g)$, which causes breads and cakes to rise.

Sodium hydroxide, $NaOH$, also called caustic soda or lye, is a valuable commercial chemical. $NaOH$ exists as white crystals or pellets that readily absorb H_2O and CO_2 from the air. $NaOH$ is soluble in water, producing a strongly basic solution that feels slippery when touched. Industrially, $NaOH$ is used to produce soaps, detergents, rayon, and to extract Al from ores.

Potassium nitrate, KNO_3, or nitre, is a superior fertilizer because, when dissolved, it supplies two plant nutrients, $K^+(aq)$ and $NO_3^-(aq)$. Nitre is a component of gunpowder (black powder). KNO_3 serves as a source of oxygen to oxidize the other components of gunpowder. Gunpowder is a mixture of 75% KNO_3, 15% charcoal, and 10% sulfur.

Ernest Solvay (1838–1922), a Belgian chemist, was motivated toward industrial chemistry by his father, a salt refiner. After developing his process to manufacture $NaHCO_3$ in 1863, he founded a company which ultimately became the world's largest supplier of $NaHCO_3$.

Sodium hydroxide is contained in the white pellets found in products for unclogging drains. NaOH releases a large quantity of heat when it contacts water, and dissolves fats and other normally water-insoluble substances.

REVIEW EXERCISES

14.1 What topics are considered in the study of descriptive chemistry?

14.2 What elements are considered the representative (a) metals and (b) nonmetals?

14.3 In what minerals does sodium exist in the earth's crust?

14.4 For each of the following properties, explain how the alkali metals compare to other metals: (a) average density, (b) melting points, (c) electric conductivities, (d) reactivities.

14.5 Write equations for the reaction of cesium with (a) $H_2(g)$, (b) $N_2(g)$, (c) $O_2(g)$.

14.6 How are each of the following alkali metals used in industry: (a) K, (b) Li, (c) Na?

Alkaline earth metals are elements that belong to group IIA(2) of the periodic table. Members of this group include beryllium, Be; magnesium, Mg; calcium, Ca; strontium, Sr; barium, Ba; and radium, Ra.

Alkaline earth metals are about as widely distributed as the alkali metals. Calcium ranks fifth and magnesium ranks eighth in abundance among elements in the earth's crust. A large percent of Ca is bonded to carbonate ions in calcium carbonate, $CaCO_3$. Calcium carbonate rock is the mineral called limestone. Other forms of $CaCO_3$ include calcite, chalk, marble, and pearl (Fig. 14.1). Calcium is found in nature bonded to sulfate ions, $CaSO_4$. The hydrate of this compound, $CaSO_4 \cdot 2H_2O$, is called gypsum.

A large variety of minerals contain magnesium, including

Asbestos, $H_4Mg_3Si_2O_9$

Talc or soapstone, $Mg_2Si_4O_{10}(OH)_2$

Dolomite, $MgCO_3 \cdot CaCO_3$

Meerschaum, $Mg_2Si_3O_8 \cdot 2H_2O$.

Alkaline Earth Metals

Strontium is named after Strontian, a town in Scotland. Sr is a rather hard, white, metallic solid that closely resembles Ca.

Meerschaum is a soft, whitish mineral that is primarily used to make tobacco pipes. It has a low density and floats on water.

Figure 14.1
Limestone, composed principally of calcium carbonate, $CaCO_3$, makes up the stalactites hanging from the ceiling, and the stalagmites growing from the floor of underground caverns. A pearl inside an oyster shell is also composed of $CaCO_3$. *(George Grant, National Park Service, and Field Museum of Natural History, Chicago.)*

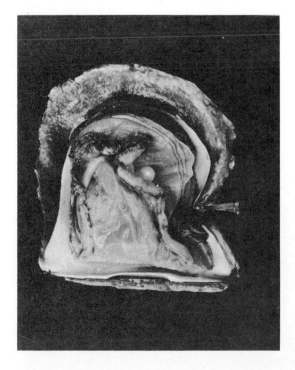

In addition, large quantities of $Mg^{2+}(aq)$ are dissolved in the oceans. Beryllium is isolated from a mineral called beryl, $Be_3Al_2Si_6O_{18}$. Beryl is found in different colors; when beryl is a pale green, it is called aquamarine, and when it has a deep green color, it is an emerald. Strontium is quite rare; it exists primarily in celestite, $SrSO_4$, and strontianite, $SrCO_3$. Barium is obtained from barite, $BaSO_4$.

The properties of the alkaline earth metals parallel those of the alkali metals. Table 14.2 summarizes the properties of the alkaline earth metals.

TABLE 14.2   PROPERTIES OF THE ALKALINE EARTH METALS

Property	Be	Mg	Ca	Sr	Ba
Atomic number	4	12	20	38	56
Atomic mass	9.012	24.31	40.08	87.62	137.3
Outer-level electronic configuration	$2s^2$	$3s^2$	$4s^2$	$5s^2$	$6s^2$
Density, g/cm^3	1.9	1.7	1.5	2.6	3.5
Melting point, °C	1280	650	842	770	725
Boiling point, °C, 1 atm	2970	1090	1490	1384	1640
Enthalpy of fusion, kJ/mol	12	9.0	8.8	9.2	7.5
Enthalpy of vaporization, kJ/mol	294	132	150	139	151
Ionization energy, kJ/mol	905	744	596	556	509
Atomic size, nm	0.112	0.160	0.197	0.215	0.217
Electronegativity	1.5	1.2	1.0	0.9	0.9
Formula of halide	BeX_2	MgX_2	CaX_2	SrX_2	BaX_2
Formula of oxide	BeO	MgO	CaO	SrO	BaO

All members of group IIA have two s electrons in their outer energy level, with an inner noble gas configuration. In chemical changes, they tend to lose their two outermost electrons, forming 2+ ions that are isoelectronic to noble gases.

$$M \longrightarrow M^{2+} + 2e^- \text{ (M = alkaline earth metal)}$$

Densities of alkaline earth metals are higher than those of the alkali metals, which is a direct result of having larger masses and smaller atomic sizes. However, their densities, which range from 1.5 to 3.5 g/cm^3, are lower than those of many of the other metals. Most metals have densities in excess of 5 g/cm^3. Alkaline earth metals have higher melting and boiling points and are harder than corresponding alkali metals because they have two outer electrons instead of one. More electrons among the metal nuclei produce stronger electrostatic forces.

Properties of the alkaline earth metals that differ from those of the alkali metals generally result from the smaller atomic sizes and stronger forces of attraction among alkaline earth metal atoms. Their enthalpies of fusion and vaporization are significantly higher than those of corresponding alkali metals. Electric and thermal conductivities of alkaline earth metals are high, but are less than those of the alkali metals.

Chemically, the alkaline earth metals are reactive, but less so than the alkali metals. Of the group IIA metals, beryllium, Be, is the smallest and has the highest electronegativity, which gives it metalloid characteristics. In most cases, Be compounds share more properties with covalent compounds than they do with ionic compounds. The chemical properties of Be actually resemble those of Al more than those of the alkaline earth metals.

Except for Be, most alkaline earth metals combine directly with nonmetals. They react with halogens to produce halides, MX_2, and combine with hydrogen and nitrogen to produce hydrides, MH_2, and nitrides, M_3N_2, respectively.

Beryllium and its compounds are toxic and produce a degenerative, often fatal, lung disease called berylliosis.

$$M + X_2 \longrightarrow MX_2 \quad \text{(Halide (X = halogen)}$$

$$M + H_2 \longrightarrow MH_2 \quad \text{(Except when M = Be or Mg)}$$

$$3M + N_2 \longrightarrow M_3N_2 \quad \text{(Nitride)}$$

Alkaline earth metals produce oxides and peroxides when combined with oxygen. Oxide formation is only observed with Be through Ca. Strontium and barium produce both peroxides (SrO_2 and BaO_2) and oxides (SrO and BaO).

$$2M + O_2 \longrightarrow 2MO \quad \text{(Oxide)}$$

$$M' + O_2 \longrightarrow M'O_2 \text{ (Peroxide) (M' = Sr and Ba)}$$

Most alkaline earth metals (except Be and Mg) combine with water to produce a hydroxide and liberate $H_2(g)$.

$$M(s) + 2H_2O(l) \longrightarrow M(OH)_2(aq) + H_2(g) \text{ (Except Be and Mg)}$$

Hydroxide solutions of alkaline earth metals are basic solutions. However, alkaline earth metals react much less vigorously with water and produce less basic solutions than the alkali metals. Hot water is required to make Ca react at an appreciable rate.

In industry, Be and Mg metals are the most important metals in group IIA. Because Be does not absorb x-rays, it is used to make transparent "windows" in x-ray tubes. Be does absorb neutrons, particles given off in nuclear reactions; consequently, Be is used in nuclear power plants and nuclear weapons. It is used in high-precision instruments because of its low density, its excellent thermal conductivity, and its elasticity.

Descriptive inorganic chemistry

Economically, Mg is a most valuable metal. Its low density makes Mg useful as a component in lightweight metal alloys. For example, Mg is alloyed with Al to produce very hard and noncorrosive metals. Mg alloys are employed in the construction of wheels, cameras, airplane parts, tools, and luggage.

Magnesium ribbon or wire releases a blinding white light when ignited.

Calcium ions, Ca^{2+}, are biologically significant partly because they are a principal component of hydroxyapatite, $Ca_{10}(PO_4)_6(OH)_2$. Hydroxyapatite is the crystalline compound in both bone and teeth. Dental caries (cavities) are thought to result when microorganisms decompose foods, especially sugars, to acids. These acids accelerate the decomposition of hydroxyapatite to $Ca^{2+}(aq)$, $PO_4^{3-}(aq)$, and $OH^-(aq)$. Dental research has shown that the addition of small quantities of fluoride, $F^-(aq)$, to the drinking water helps to prevent dental caries in young people. Fluoride ions substitute for some of the OH^- in the crystal structure of hydroxyapatite, producing fluorohydroxyapatite—a compound more resistant to acid attack.

Calcium ions, Ca^{2+}, are linked to potassium ions, K^+, in biological systems; for example, a proper balance of Ca^{2+} and K^+ is required for normal heart function. If this balance is disrupted, cardiac arrest may result. Calcium ions are also involved in the blood-clotting mechanism. If Ca^{2+} is removed from blood, it does not clot.

Calcium carbonate, the principal component of limestone, is the main industrial source of elemental Ca. When limestone is heated, calcium oxide, CaO (lime), results.

Lime, CaO, is the third-ranked chemical in terms of amount produced in the United States; in excess of 35 billion lb CaO was produced in 1980.

$$CaCO_3(s) \xrightarrow{\Delta} CaO(s) + CO_2(g)$$

Since antiquity, CaO has been known to be a good mortar when mixed with sand and water. After reacting with water, CaO is changed to calcium hydroxide (slaked lime, quicklime), $Ca(OH)_2$.

$$CaO(s) + H_2O(l) \longrightarrow Ca(OH)_2$$

As the mortar dries, $Ca(OH)_2$ combines with CO_2 in the air, re-forming the insoluble and good binding substance $CaCO_3$.

$$Ca(OH)_2 + CO_2 \longrightarrow CaCO_3(s) + H_2O$$

Calcium oxide has the interesting property of releasing a brilliant white light when heated strongly. This bright light is referred to when one uses the expression "being in the limelight." Hundreds of industrial chemicals are produced from CaO or $Ca(OH)_2$; few other substances are used more by industry.

Let's turn our attention to magnesium compounds. Magnesium oxide, MgO (magnesia), is used in electric heating elements, such as those in hot plates and stoves. Unlike most other metals, MgO is an excellent heat conductor but a rather poor conductor of electricity. Thus, MgO insulates an inner high voltage heating element wire from the outside metal, and, at the same time, transfers the heat generated.

Milk of magnesia, $Mg(OH)_2$, is a commonly used antacid. $Mg(OH)_2$ is an alkaline substance that neutralizes excess stomach acid secreted by the stomach lining as a result of disease or overeating. Magnesium sulfate, $MgSO_4 \cdot 7H_2O$ (epsom salt) has a variety of medical uses.

Asbestos, a magnesium silicate compound, has been widely used whenever a heat-resistant substance is required. It is noncombustible and is a strong fiber (Fig. 14.2); hence, automobile brake linings, construction materials, and building insulation contain asbestos. Within the past 15 years, it has been discovered that microscopic asbestos particles can cause lung cancers when inhaled. Asbestos is gradually being replaced with ceramic compounds to minimize the cancer risk.

REVIEW EXERCISES

14.7 For each of the following properties, compare alkaline earth metals with alkali metals: *(a)* electronic configurations, *(b)* densities, *(c)* melting points, *(d)* enthalpies of fusion.

14.8 What minerals are each of the following isolated from: *(a)* Ca, *(b)* Mg, *(c)* Be?

14.9 Write the equations for the reactions of Ca with each of the following compounds: *(a)* $N_2(g)$, *(b)* $H_2O(l)$, *(c)* $O_2(g)$.

14.10 List two functions of Ca^{2+} in living systems.

Figure 14.2
Asbestos is composed of long molecular chains of Si and O atoms. The chains give asbestos its fibrous nature.

All the members of group IIIA(13) are metals, except B, which is a metalloid. We will consider only one group member, Al. **Aluminum** is the most abundant metal in the earth's crust and is the third most abundant element, exceeded only by O and Si.

Aluminum is found in many clays in the form of aluminum silicates such as feldspar, $KAlSi_3O_8$. Bauxite, $Al_2O_3 \cdot xH_2O$, a hydrated form of aluminum oxide, is the best source of Al.

Corundum is a widely distributed anhydrous form of aluminum oxide. Impurities in corundum create colored crystals, many of which are classified as gemstones. If the impurities in the corundum structure are chromium oxides, then the crystal has a red color and is called a ruby. If the impurities are cobalt and titanium, then the crystal is blue and is called a sapphire. If iron oxides are the impurities, the crystal is called oriental topaz (Fig. 14.3). Amethyst results when manganese oxide is the impurity in corundum.

Even though Al is the most abundant metallic element in the earth's crust, aluminum was once a rare and expensive metal that people valued highly. Its rarity stemmed from the lack of an inexpensive method for extracting Al from its ore. Charles M. Hall (Fig. 14.4), while an undergraduate chemistry student at Oberlin College, solved this perplexing problem of getting to metallic Al relatively inexpensively. In following his procedure, Al is obtained by dissolving bauxite in a molten aluminum compound, cryolite (Na_3AlF_6), at high temperatures. An electric current is then passed through the solution, producing molten Al, which flows out of the container. Figure 14.5 shows a diagram of a Hall cell.

Aluminum

Only a small percent of Al in nature is found in bauxite; most Al is a component of silicate clays.

Aluminum was once considered a precious metal (rare and valuable), similar to gold, platinum, and silver.

Figure 14.3
Topaz, a form of corundum that contains iron oxide impurities, is hard and exhibits a wide range in colors. *(Field Museum of Natural History, Chicago.)*

This procedure is now known as the **Hall process.**

Selected properties of Al are listed in Table 14.3. Al, a member of group IIIA, has three outer-level electrons, $3s^2 3p^1$. These electrons are lost in chemical changes with nonmetals, yielding Al^{3+}. Al exists in compounds only in the +3 oxidation state.

$$Al \longrightarrow Al^{3+} + 3e^-$$

Its density, 2.7 g/cm^3, is very low for a metal. Al is a good conductor of heat and electricity, and is an excellent reflector of heat and light. It has moderate melting and boiling points for a metal.

TABLE 14.3 PROPERTIES OF ALUMINUM

Property	
Atomic number	13
Atomic mass	26.98
Outer electronic configuration	$3s^2 3p^1$
Density, g/cm^3	2.7
Melting point, °C	660
Boiling point, °C, 1 atm	2467
Enthalpy of vaporization, kJ/mol	284
Ionization energy, kJ/mol	577
Atomic size, nm	0.143
Electronegativity	1.5

Figure 14.4
Charles M. Hall. *(Aluminum Company of America.)*

Charles Hall (1863–1914) began his search for Al after hearing a lecture by his professor stating that anyone who could discover an inexpensive means for extracting Al from its ore would become wealthy and famous. In 1886, in his home laboratory, Hall discovered the procedure for obtaining Al from bauxite.

Figure 14.5
A Hall cell is used to extract Al from bauxite. Bauxite, a form of Al_2O_3, is dissolved in molten cryolite, $Na_3AlF_6(l)$. As an electric current is passed through the cell, molten Al is liberated from the bauxite. The molten Al then flows from the bottom of the cell.

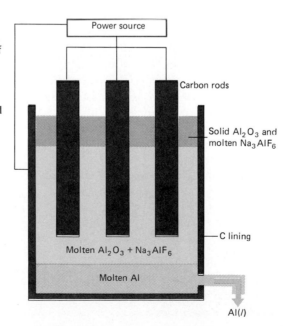

As a result of its unique properties, Al has a wide variety of uses. Most people are familiar with Al as a foil used to wrap foods and other household items. Al is alloyed with other metals, especially Cu, Mg, and Mn, to produce strong metals that have a low density. Duraluminum, a solution of Al, Mg, Mn, and Cu, is used in the construction of buildings, boats, and airplanes. AnotherAl alloy is alnico, a strongly magnetic metal composed of Al, Ni, Co, and Fe (alnico is a mnemonic for Al, Ni, and Co). Because the world supply of Cu is diminishing, Al now replaces Cu as the electrical conductor in wires and cables.

A thin layer of aluminum is used to reflect light in large visible-light telescopes.

The chemical properties of Al are of interest because they resemble those of the metalloids. Aluminum, depending on the conditions, exhibits either acidic or basic behavior. Aluminum metal samples, when exposed to the air, form a relatively inert oxide coating. Thus, Al appears to be an inert metal, when, in fact, it is quite reactive.

A substance that exhibits both acidic and alkaline properties is called amphoteric.

Pure aluminum, when heated in air at high temperature, is totally converted to aluminum oxide, Al_2O_3.

$$4Al(s) + 3O_2(g) \xrightarrow{\Delta} 2Al_2O_3(s)$$

Al combines with iron(III) oxide, Fe_2O_3, and releases a tremendous amount of energy, enough that the resulting iron becomes molten.

$$2Al(s) + Fe_2O_3(s) \longrightarrow 2Fe(l) + Al_2O_3(l)$$

This reaction is known as the **thermite** reaction. Because temperatures in excess of 3000°C are obtained, metals are welded using the thermite reaction.

Thermite bombs were dropped during World War II because thermite fires were very difficult to extinguish with water.

Important Al compounds include aluminum hydroxide, $Al(OH)_3$; potassium aluminum sulfate, $KAl(SO_4)_2 \cdot 12H_2O$; aluminum chloride, $AlCl_3$; and aluminum sulfate, $Al_2(SO_4)_3$.

Aluminum hydroxide, $Al(OH)_3$, is an ingredient in antacids. It decreases the acid concentration in the stomach. Potassium aluminum sulfate, $KAl(SO_4)_2 \cdot 12H_2O$, commonly called alum, is added to basic soils. Alum is acidic and neutralizes the basic components of soils. Aluminum chloride, $AlCl_3$, is frequently used as a catalyst (a substance that speeds up chemical reactions) in laboratory syntheses, and is now used as an intermediate in a new procedure for isolating Al from bauxite. Aluminum sulfate, $Al_2(SO_4)_3$, is used in the pulp and paper industry and in the purification of water.

REVIEW EXERCISES

14.11 Consider the description of the Hall process. What could account for the high price of isolating Al from bauxite using the Hall process?

14.12 Compare Mg and Al with respect to the following properties: *(a)* melting points, *(b)* densities, and *(c)* ionization energies.

14.13 List two alloys of Al, and describe their uses.

T he general properties of oxygen, nitrogen, and the noble gases were discussed in Chap. 12, and carbon was considered in Chap. 13. Thus, in this section we will concentrate on some of the other nonmetals: the halogens, sulfur, and phosphorus.

14.2 REPRESENTATIVE NONMETALS

Halogens

Fluorine, F; chlorine, Cl; bromine, Br; iodine, I; and astatine, At, are the halogens. Together they make up one of the most reactive groups of elements on the periodic table. Each halogen (X) has one less outer electron (s^2p^5) than the noble gases; thus, if halogens obtain one electron from another atom they become isoelectronic to a noble gas.

$$:\overset{..}{\underset{..}{X}}\cdot \ + \ e^- \longrightarrow \ [:\overset{..}{\underset{..}{X}}:]^-$$

After halogens accept an electron, they become halide ions, X^-.

"Astatine" comes from the Greek astatos, which means unstable. Astatine was discovered in 1940 at the University of California by Corson, MacKenzie, and Segre.

Because of their reactivity, in nature the halogens are always chemically combined with other elements. Fluorine in the form of fluoride ions is relatively abundant. Primary sources of fluoride are the minerals fluorite, CaF_2; fluorapatite, $Ca_{10}F_2(PO_4)_6$; and cryolite, Na_3AlF_6, with small quantities of these dissolved in the oceans. Chlorine is the most abundant of the halogens, much of which is either in the oceans or in salt deposits in the ground. Bromine, to a much lesser extent, is located along with chlorine in the oceans and in salt deposits. A small amount of dissolved iodine is in the oceans; most of the world's iodine is located along with Chile saltpeter in the form of sodium iodate, $NaIO_3$.

Some iodine is extracted from seaweed.

Liberation of the halogens from their compounds yields nonpolar diatomic molecules,

$$:\overset{..}{\underset{..}{X}}:\overset{..}{\underset{..}{X}}: \ \ (X = F, \ Cl, \ Br, \ I, \ or \ At)$$

Fluorine, F_2, is a pale yellow gas, and chlorine, Cl_2, is a pale green gas; bromine, Br_2, is a volatile, reddish-brown liquid; and iodine, I_2, is a grayish-black solid that sublimes. London dispersion forces are the intermolecular forces among halogen molecules. These forces account for the trends in physical states of the halogens.

Table 14.4 presents some of the physical properties of the halogens. Trends in density, melting point, and boiling point are explained by the increasing dispersion forces of successively higher molecular mass molecules. Stronger dispersion forces in the halogens are directly related to the number of electrons and how tightly they are held by the nucleus.

If we consider their atomic properties, we find that the halogens have the second highest set of ionization energies, exceeded only by those of the noble gases. The atomic sizes of the halogen atoms are among the smallest in each period. Trends in both ionization energy and atomic size indicate that the nuclei of halogen atoms strongly attract their electrons. Therefore, it follows that the halogens are highly electronegative elements, which means they strongly attract electrons in chemical bonds.

TABLE 14.4 PROPERTIES OF THE HALOGENS

Property	Fluorine	Chlorine	Bromine	Iodine
Atomic number	9	17	35	53
Atomic mass	19.00	35.45	79.90	126.9
Outer-level electronic configuration	$2s^2 2p^5$	$3s^2 3p^5$	$4s^2 4p^5$	$5s^2 5p^5$
Density	1.81 g/L	3.21 g/L	3.12 g/cm^3	4.94 g/cm^3
Melting point, °C	−220	−101	−7.3	114
Boiling point, °C	−188	−34	58.8	184
Enthalpy of vaporization, kJ/mol		20.4	29.5	42.0
Ionization energy, kJ/mol	1680	1250	1140	1000
Atomic radius, nm	0.064	0.099	0.114	0.133
Electronegativity	4.0	3.0	2.8	2.5

The chemistry of the halogens is best understood in terms of their ability to attract electrons from other elements. A substance that readily pulls electrons from other elements and compounds is called a strong oxidizing agent. **Oxidizing agents** are chemical species that attract electrons. A **reducing agent,** a substance that releases electrons, is the opposite of an oxidizing agent.

When ionic substances were discussed in Chap. 8, we mentioned the oxidizing capacity of halogens. Halogens remove electrons from metals (reducing agents) and produce ionic salts, for example, NaCl, MgF$_2$, and LiI.

Fluorine gas is the strongest oxidizing agent of all the elements. In fact, F$_2$ is such a strong oxidizing agent that it combines with just about all other elements, and in many cases violently. For example, substances such as hydrogen and sulfur ignite immediately when contacted by F$_2$ gas. F$_2$ also replaces all of the other halogens in binary compounds.

$$F_2 + 2MX \longrightarrow X_2 + 2MF \ (X = Cl, \ Br, \ or \ I)$$

Fluorine is one of the few elements that combine with the high-atomic-mass noble gases (Xe and Kr), producing a multitude of noble gas compounds, e.g., XeF$_2$, XeF$_4$, XeOF$_2$, and CsXeF$_7$.

Chlorine gas is not as reactive as F$_2$; nevertheless, Cl$_2$ combines with most other elements under the proper conditions. For example, when Cl$_2$ is combined with H$_2$, a reaction occurs only if the temperature is elevated or light is present; Cl$_2$ and H$_2$ do not combine at low temperatures in the dark.

$$H_2(g) + Cl_2(g) \xrightarrow{\Delta} 2HCl(g)$$

Chlorine gas combines with other nonmetals to produce covalent chlorides. It combines with sulfur, S_8, to produce S_2Cl_2, and combines with phosphorus, P_4, to produce PCl_3.

$$S_8(s) + 4Cl_2(g) \longrightarrow 4S_2Cl_2(l)$$

$$P_4(s) + 6Cl_2(g) \longrightarrow 4PCl_3(l)$$

Chlorine gas replaces H atoms in hydrocarbons, carbon-hydrogen compounds. For example, chlorine combines with methane, CH_4, to produce chloromethane, CH_3Cl, and hydrogen chloride.

$$CH_4(g) + Cl_2(g) \xrightarrow{\text{light}} CH_3Cl(g) + HCl(g)$$

If excess Cl_2 is present, all of the H atoms in methane are replaced, yielding a mixture of CH_3Cl, CH_2Cl_2, $CHCl_3$, and CCl_4 (chlorinated hydrocarbons).

Most of the chemical properties of bromine and iodine are similar to those of chlorine; however, these halogens are much weaker oxidizing agents than is chlorine. Consequently, they react more slowly and require a greater quantity of energy.

Controversy surrounds some of the halogen compounds. Chlorinated hydrocarbons are a group of halogen compounds that have a detrimental effect on the environment. Many pesticides are complex, highly chlorinated hydrocarbons. Most of these compounds do well at what they are designed to do, kill "bugs." But overuse of pesticides releases large quantities of these chemically stable substances into the environment, where they cause many problems (Fig. 14.6).

Chlorinated hydrocarbons have a very low solubility in water and a significantly higher solubility in fat tissues. Thus, excess chlorinated

$CHCl_3$ is the formula for chloroform. Chloroform was an early anesthetic but is no longer used because of its high toxicity and possible link to cancer.

Examples of chlorinated pesticides are DDT, chlordane, endrin, and lindane. These pesticides remain in the environment from 2 to 15 years before being completely decomposed.

Figure 14.6
Herbicides and pesticides are sprayed on crops by small planes to increase crop yields. Such a method has the potential of adding excessive amounts of pesticides to the environment. *(U.S. Department of Agriculture.)*

hydrocarbons concentrate in the fat tissues of animals, especially in the highest-level consumers—predator animals and human beings. Chlorinated hydrocarbons have been linked to the extinction of several bird species, and to various medical problems in human beings. Without these pesticides, insects decrease crop yields, which also has a deleterious effect on society.

A less controversial halogen compound is hydrogen fluoride, HF. It combines with calcium silicate, a major component of glass.

$$CaSiO_3(s) + 6HF(l) \longrightarrow CaF_2(s) + SiF_4(g) + 3H_2O(l)$$

Consequently, HF is used to etch glass. Light bulbs are frosted and calibration marks are etched on thermometers and glassware by exposing them to HF.

Bromine and iodine are much less frequently used. Silver salts of bromine and iodine are components of photographic films. Iodide is an essential mineral required in the human diet. Iodide ions are removed from the blood by the thyroid gland and incorporated into a hormone called thyroxin. Iodine deficiency in humans results in an affliction called goiter, an enlargement of the thyroid gland. Iodide supplements are given to people who are exposed to radiation from nuclear power plants or nuclear bombs because nonradioactive iodide helps prevent the absorption of radioactive iodide by the thyroid. Radioactive iodide can cause thyroid cancer and other maladies.

Black and white photographic film has particles of AgBr and AgI embedded in a coating on a plastic backing. When light strikes Ag$^+$X$^-$ crystals, it activates them and makes the crystals more susceptible to reduction (gaining of electrons). During development these crystals are reduced to metallic silver.

REVIEW EXERCISES

14.14 What are the principal natural sources of the halogens?

14.15 In what physical state would you expect to find astatine? Explain in terms of a trend in intermolecular forces.

14.16 List the halogens in increasing order of their capacity to oxidize other substances.

14.17 Write and balance the following equations: (*a*) combination of H$_2$ and F$_2$, (*b*) formation of barium iodide from the elements, (*c*) combination of bromine vapor with methane gas at an elevated temperature.

Sulfur

Sulfur belongs to the chalcogen group (VIA, 16), and is directly under oxygen on the periodic table; accordingly, an S atom has an s^2p^4 outer-level electronic configuration. Sulfur is a pale yellow solid composed principally of S$_8$ molecules in the form of rings that contain eight sulfur atoms (Fig. 14.7).

Sulfur is found in two allotropic forms. One is called **orthorhombic** sulfur and the other is **monoclinic** sulfur. Orthorhombic sulfur is the most common form of sulfur. Orthorhombic refers to the type of crystal system to which sulfur atoms belong. When orthorhombic sulfur is heated to its melting point, 112°C, it changes to a pale yellow liquid, which, upon cooling, changes to monoclinic sulfur. This second allotrope contains sulfur atoms in a different crystal lattice pattern.

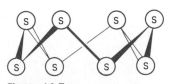

Figure 14.7
Elemental sulfur is composed of molecules that contain eight sulfur atoms, S$_8$, in a ring structure.

Monoclinic sulfur slowly changes back to orthorhombic sulfur if left alone at room temperature. Table 14.5 lists the physical properties of sulfur.

An old name for sulfur was brimstone, from the Germanic word that means "burning stone."

TABLE 14.5 **PROPERTIES OF SULFUR AND PHOSPHORUS**

Property	Sulfur	Phosphorus
Atomic number	16	15
Atomic mass	32.05	30.97
Outer electronic configuration	$3s^23p^4$	$3s^23p^3$
Density, g/cm^3	2.07 orthorhombic 1.96 monoclinic	1.82 white 2.34 red
Melting point, °C	112 orthorhombic 119 monoclinic	44 white 600 red
Boiling point, °C, 1 atm	444 orthorhombic	280 white
Enthalpy of fusion, kJ/mol	1.72	2.51
Enthalpy of vaporization, kJ/mol	9.62	12.4
Ionization energy, kJ/mol	1000	1058
Electronegativity	2.5	2.1

Sulfur-bearing minerals include galena, PbS; pyrite, FeS_2; zinc blende, ZnS; and bornite, Cu_3FeS_3. Many sulfates, compounds that contain SO_4^{2-}, are dissolved in the oceans. These include sodium sulfate, Na_2SO_4; magnesium sulfate, $MgSO_4$; and the less soluble calcium sulfate, $CaSO_4$. Elemental sulfur deposits are located throughout the world; large amounts are located in the United States along the Texas and Louisiana Gulf coast.

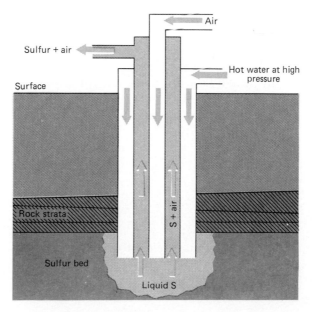

Figure 14.8
The Frasch process is used to bring underground sulfur to the surface. Air and hot water are forced under pressure into underground sulfur deposits. The sulfur melts and is pushed to the surface, where it dries and solidifies.

Undergound sulfur deposits are extracted by what is known as the **Frasch process,** named for its developer Herman Frasch (1851–1914). A special metal pipe, which is actually three pipes in one, is inserted into the sulfur deposit (Fig. 14.8). Hot water and compressed air are pumped separately down two of the pipes, the underground sulfur is dissolved and forced up the third pipe as a bubbly, frothy mixture of sulfur, air, and water. When the water evaporates, relatively pure sulfur is obtained.

Like oxygen, sulfur combines with both metals and nonmetals. If sulfur is heated with Na metal, sodium sulfide, Na_2S, results.

$$16Na + S_8 \xrightarrow{\Delta} 8Na_2S$$

Iron and calcium combine with sulfur as follows:

$$8Fe + S_8 \xrightarrow{\Delta} 8FeS$$
$$8Ca + S_8 \xrightarrow{\Delta} 8CaS$$

Sulfur combines with nonmetals such as carbon, phosphorus, and hydrogen as shown by the following reactions:

$$4C + S_8 \xrightarrow{\Delta} 4CS_2$$
$$8P_4 + 10S_8 \xrightarrow{\Delta} 8P_4S_{10}$$
$$8H_2 + S_8 \xrightarrow{\Delta} 8H_2S$$

Most of the halogens and oxygen combine with sulfur. For example, Cl_2 combines with sulfur to produce S_2Cl_2, and O_2 combines with sulfur to produce sulfur dioxide.

$$S_8 + 4Cl_2 \xrightarrow{\Delta} 4S_2Cl_2$$
$$S_8 + 8O_2 \xrightarrow{\Delta} 8SO_2$$

Uses of elemental sulfur are varied. Sulfur is a component of gunpowder and is used to manufacture rubber. Rubber is strengthened by the addition of sulfur—a process called **vulcanization.** However, the main use of sulfur is in the formation of sulfur compounds, especially sulfuric acid, H_2SO_4. Sulfuric acid is commercially one of the most important of all chemicals. It is synthesized by combining sulfur trioxide, SO_3 (an acid anhydride), with water.

$$SO_3(g) + H_2O(l) \longrightarrow H_2SO_4(aq)$$

Sulfur trioxide is produced by oxidizing elemental sulfur. Commercial sulfuric acid is generally sold as a 98% H_2SO_4 solution. It is a viscous, dense (1.85 g/cm^3) liquid that has strong dehydrating properties. Sulfuric acid dehydrates many substances instantaneously upon contact.

Herman Frasch's interest in S developed out of work he was doing in petroleum chemistry. Frasch was concerned with sulfur contamination of oil. This prompted him to apply principles of petroleum chemistry to tap the vast supplies of underground sulfur in Texas.

Vulcanization, heating rubber with sulfur at high temperatures, was developed by Charles Goodyear (1800–1860) in 1839. The term "vulcanization" comes from the name of the Roman god of fire, Vulcan. Goodyear's fame, however, stems from the tire company that uses his name.

Sulfuric acid has many uses, the largest percent being consumed by the phosphate fertilizer industry. Synthetic fibers, pigments, petroleum products, and storage batteries (lead-acid) all require sulfuric acid in their production. Ammonia is combined with sulfuric acid to yield ammonium sulfate, $(NH_4)_2SO_4$, which is a good fertilizer.

$$2NH_3 + H_2SO_4 \longrightarrow (NH_4)_2SO_4$$

Other interesting sulfur compounds are hydrogen sulfide, H_2S, and sulfur dioxide, SO_2. In contrast to water, hydrogen sulfide is a gas at room temperature. Because the electronegativity of sulfur (2.5) is lower than that of oxygen (3.5), sulfur does not form hydrogen bonds. Accordingly, H_2S molecules are bound by the weaker dipole–dipole interactions. H_2S is a foul-smelling gas with the odor of rotten eggs. It is a highly toxic gas, which is as poisonous as hydrogen cyanide, HCN. Breathing two parts H_2S per thousand parts of air is normally fatal. Decaying animals and the refining of petroleum are sources of H_2S in the atmosphere.

Sulfur dioxide, SO_2, is a major component of air pollution. Most fossil fuels contain a small quantity of sulfur (crude oil contains about 0.2% to 0.8% sulfur), which oxidizes to SO_2 when combusted. Power plants, industry, and transportation are the three primary sources of SO_2 in the atmosphere. Power companies burn immense quantities of fossil fuels to generate electricity. Industry, especially the metal industry, generates SO_2 when it burns fuels and manufactures goods. Fuels combusted in automobiles, planes, trains, and boats release only a small percent of the SO_2 in the air.

On entering the atmosphere, the SO_2 is most frequently converted to either sulfurous acid or sulfur trioxide. It combines with water vapor to produce sulfurous acid, H_2SO_3.

$$SO_2 + H_2O \longrightarrow H_2SO_3$$

Aqueous sulfurous acid falls to earth in rain and other types of precipitation. Sulfur dioxide also combines with oxygen in the atmosphere to yield sulfur trioxide, SO_3.

$$SO_2(g) + \tfrac{1}{2}O_2(g) \longrightarrow SO_3(g)$$

Sulfur trioxide combines with water and forms sulfuric acid, which also falls to earth in rain.

$$SO_3 + H_2O \longrightarrow H_2SO_4$$

Rain that contains excessive amounts of acidic substances is called **acid rain.** In addition, some of the acid remains in the air and contributes to **acid air.** High concentrations of acid air have been linked to an in-

The top 10 chemicals in the United States.

Chemical	Amount produced (1982) billions of pounds
H_2SO_4	65
N_2	35
NH_3	31
O_2	30
CaO	28
NaOH	18
Cl_2	18
H_3PO_4	17
Na_2CO_3	16
HNO_3	15

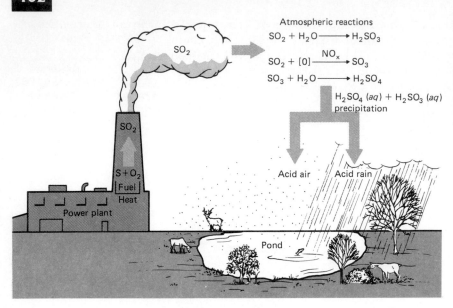

Figure 14.9
Most fossil fuels (e.g., oil and coal) contain sulfur that is oxidized to sulfur dioxide when they are combusted. The sulfur dioxide combines with water and forms sulfurous acid and combines with oxygen to produce sulfur trioxide. Both sulfuric and sulfurous acids fall to the earth in rain or remain in the air.

creased incidence of lung cancer. Figure 14.9 summarizes the pathway followed by sulfur in the production of acid rain and acid air in the atmosphere.

Acid rain is not only harmful to plants and animals, it also dissolves exposed metals, and even dissolves buildings constructed out of marble, $CaCO_3$.

$$H_2SO_4(aq) + CaCO_3(s) \longrightarrow CaSO_4(aq) + CO_2(g) + H_2O(l)$$

Figure 14.10 shows the effect of acid rain on a marble statue. Many famous buildings, like the ancient ruins in Greece and various memorials in Washington, D.C., are dissolving at a measurable rate from their exposure to acid rain.

REVIEW EXERCISES

14.18 How do the two allotropic forms of sulfur differ?

14.19 (a) What is the name of the method used to bring sulfur deposits to the surface? (b) How is this accomplished?

14.20 Write balanced equations for the combination of sulfur, S_8, with (a) Mg, (b) F_2, (c) O_2.

14.21 (a) How is acid rain produced? (b) Write all appropriate equations.

14.22 List three uses of sulfuric acid, H_2SO_4.

Phosphorus

Phosphorus is another element that exists in different allotropic forms. The most common allotropes of phosphorus are generally distinguished by their colors, white and red. The properties of white phosphorus are markedly different from those of red phosphorus.

Figure 14.10
Throughout the world, components of polluted air have destroyed or damaged irreplaceable works of art and historic buildings. In these before and after pictures, the effect of air pollution on a marble statue is illustrated. *(Museum of Natural History, Chicago.)*

White phosphorus is a soft, waxy solid with a low melting point, 44°C. After exposure to the air, white phosphorus ignites spontaneously and produces P_4O_{10}.

$$P_4(s) + 5O_2(g) \longrightarrow P_4O_{10}(s)$$

There is a third allotrope of phosphorus: black phosphorus. It results when white phosphorus is subjected to high pressures.

Consequently, white phosphorus is stored so that it never contacts the air; most white phosphorus is stored under water. White phosphorus exhibits the property of **phosphorescence;** i.e., it releases light after being exposed to various types of electromagnetic radiation such as visible and ultraviolet light.

White phosphorus is composed of P_4 molecules, not individual P atoms. Each P atom in P_4 is bonded to three other P atoms in a tetrahedral pattern (Fig. 14.11). When white phosphorus is heated without air to about 250°C, red phosphorus is produced. At 250°C, it is thought that the P_4 rings combine to yield long chains of bonded P_4 molecules. Such compounds are called **polymeric substances.** Red phosphorus melts around 600°C, is stable at room temperature, and has a low toxicity compared to white P. Table 14.5 lists the properties of phosphorus.

Polymers are high-molecular-mass substances that are composed of long-chain molecules with repeating chemical units.

Phosphorus is the twelfth most abundant element in the earth's crust. It is found in phosphate-bearing rocks such as calcium phosphate,

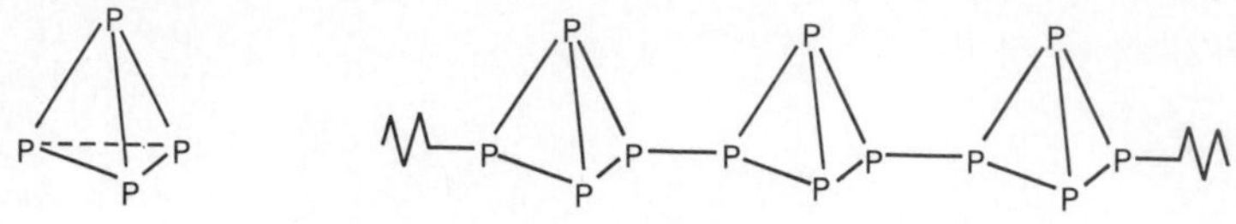

White phosphorus

Red phosphorus

Figure 14.11
White phosphorus is very
reactive and spontaneously
ignites in the air. It is composed of P_4 molecules that
form the shape of a tetrahedron. Red phosphorus is
much less reactive than
white phosphorus. It is composed of long chains of P_4
rings bonded to each other.

$Ca_3(PO_4)_2$, also called phosphate rock. Elemental phosphorus is most frequently prepared from phosphate rock by mixing the rock with coke, C, and silica, SiO_2, and then placing the mixture in a high-temperature electric furnace. At high temperatures, P_4 comes off as a vapor.

$$2Ca_3(PO_4)_2 + 6SiO_2 + 10C \longrightarrow 6CaSiO_3 + 10CO + P_4(g)$$

Elemental phosphorus itself is limited in its uses. At one time phosphorus was placed in matches, but problems arose because of its toxicity, and it has now been replaced by a phosphorus compound, tetraphosphorus trisulfide, P_4S_3. Fireworks and incendiary bombs contain white phosphorus. Elemental phosphorus is principally used to synthesize phosphorus compounds, especially tetraphosphorus decoxide, P_4O_{10}, which is converted to phosphoric acid, $H_3PO_4(aq)$.

$$P_4O_{10}(s) + 6H_2O(l) \longrightarrow 4H_3PO_4(aq)$$

Pure phosphoric acid (actually hydrogen phosphate) is a white solid that melts at 42°C. Hydrogen phosphate is infrequently used. Phosphoric acid is normally sold as an 85% aqueous solution. Compared to nitric, sulfuric, and hydrochloric acid, phosphoric acid is a weak acid.

Phosphoric acid combines with basic compounds to yield a large variety of dihydrogenphosphates, compounds $H_2PO_4^-$; monohydrogenphosphates, compounds of HPO_4^{2-}; and phosphates, compounds of PO_4^{3-}. Calcium monohydrogenphosphate, $CaHPO_4$, is one of the polishing substances in toothpaste. Calcium dihydrogenphosphate, $Ca(H_2PO_4)_2$, is an ingredient of baking powders and is a component in fertilizers.

Phosphates are good fertilizers because plants (and animals) need a constant source of phosphorus to synthesize biologically significant compounds. A substance called adenosine triphosphate, ATP, is a high-energy phosphorus-containing molecule that releases energy in cells.

*Major nutrient elements
needed by plants are C,
H, O, N, P, K, Ca, Mg,
and S.*

14.23 Describe differences in the properties of white and red phosphorus.

14.24 How is elemental phosphorus obtained from phosphate rock?

14.25 What are the three types of phosphate that result when phosphoric acid combines with bases?

14.26 How is phosphorus important to animal and plant life?

REVIEW EXERCISES

SUMMARY

Descriptive chemistry is the study of the properties, reactions, uses, and sources of elements and compounds.

Alkali metals are reactive elements located throughout nature in ionic compounds. As elements, they have low densities, melting points, and boiling points. Alkali metals are excellent conductors of heat and electricity. They are sufficiently reactive to combine directly with nonmetals. Higher-atomic-mass alkali metals react violently and explode when combined with H_2O. Sodium metal and its compounds are commercially most important.

Alkaline earth metals are not as reactive as the alkali metals. Their densities, melting points, and boiling points are higher than those of the alkali metals. Ca and Mg are two of the most common elements in the earth's crust. Chemically, the alkaline earth metals are similar to the alkali metals except that they lose two electrons to achieve the noble gas configuration. Beryllium is an exception, acting more like a metalloid than a metal.

Aluminum is the most abundant metal in the earth's crust. Most Al is bonded to oxygen and silicon in oxides and silicates. Al has three outer-level electrons and exists in the $+3$ state in compounds. It is a low-density metal that is a good conductor of heat and electricity. Al is used to manufacture strong, light metal alloys and is frequently used as an electrical conductor in cables and wires.

Halogens are reactive nonmetals that exist in ionic form in oceans and salt deposits in the earth. As elements, the halogens exist diatomically. F_2 and Cl_2 are gases at room temperature, Br_2 is a volatile liquid, and I_2 is a volatile solid. Halogens have high ionization energies and the highest set of electronegativities. They are good oxidizing agents, which means they readily remove electrons from other substances.

Sulfur is a chalcogen with six outer-level electrons. Sulfur exists in two allotropic forms, orthorhombic and monoclinic sulfur. Sulfur, like O_2, combines with both metals and nonmetals, yielding metallic and nonmetallic sulfides. Gunpowder and rubber are two examples of commercial products that contain S.

Phosphorus usually exists in two allotropic forms, white and red phosphorus. White phosphorus, P_4, is a reactive solid that reacts violently with O_2 in the air. It is toxic and must be handled carefully. Red phosphorus is unreactive and bears little resemblance to white phosphorus. Phosphorus is obtained from phosphate rock, $Ca_3(PO_4)_2$, which is chemically converted to many phosphate compounds and phosphoric acid.

EXERCISES*

Alkali Metals

14.27 Write the complete electronic configuration for each of the following: (a) Na, (b) Li, (c) K, (d) Fr.

14.28 Draw the Lewis structures for each of the following: (a) RbCl, (b) NaF, (c) K_2O, (d) Cs_3N, (e) Li_2SO_4, (f) NaOH.

14.29 List two minerals in which each of the following alkali metals are found: (a) Na, (b) K.

14.30 (a) What is the liquid range, in Celsius degrees, of lithium? (b) How does this range compare with those of the other alkali metals?

14.31 In the alkali metals, what trends are observed in the following properties (increasing or decreasing with atomic mass): (a) densities, (b) melting points, (c) enthalpies of fusion, (d) ionization energies.

14.32 Write an explanation for the increasing atomic sizes with increasing mass in the alkali metals.

14.33 (a) What mass of Cs occupies 112 mL? (b) What volume of Na has a mass of 817 g? (c) How many Li atoms are contained in a 6.44-mL sample? (d) What is the mass of a cube of Rb that has an edge length of 3.22 cm? Use data in Table 14.1.

14.34 How many joules are required to melt 3.09 kg Na(s) at its melting point?

14.35 Write the formulas for the following: (a) rubidium hydroxide, (b) cesium sulfate, (c) potassium nitride, (d) sodium peroxide, (e) lithium phosphate, (f) cesium acetate.

14.36 (a) What compound is produced as a result of the Solvay process? (b) Write a balanced equation that illustrates the Solvay process.

*For exercise numbers printed in color, answers can be found at the back of the book.

14.37 *(a)* Write a balanced equation for the reaction of potassium with bromine liquid, Br_2. *(b)* To what general class of reaction does this reaction belong? *(c)* If 4.523 g K is combined with excess Br_2, what is the theoretical yield of product? *(d)* What mass of product results when 183 g of potassium combines with 369 g of bromine?

14.38 Complete and balance each of the following equations:
(a) $Li + Cl_2 \longrightarrow$
(b) $K + H_2 \longrightarrow$
(c) $Rb + O_2 \longrightarrow$
(d) $Cs + H_2O \longrightarrow$
(e) $Na + HCl(aq) \longrightarrow$

14.39 Sodium combines with water and produces sodium hydroxide and hydrogen gas. *(a)* What mass of sodium metal should be mixed with water to produce 62.1 L H_2 at STP? *(b)* What volume of H_2 at STP results when 91.9 mg of sodium combines with excess water?

14.40 *(a)* Calculate the percent by mass of Na in sodium phosphate. *(b)* What mass of sodium phosphate contains 1.00 kg of sodium?

14.41 Write a balanced equation to show what happens when an acid, such as $HCl(aq)$, combines with potassium hydrogencarbonate.

14.42 About 2 g of potassium is required in the human diet each day. *(a)* Calculate the number of K atoms in a 2.00-g sample of potassium. *(b)* What mass of potassium chloride provides the required amount of dietary potassium?

14.43 What mass of $CsHC_4H_4O_6$ contains the same number of Cs atoms as does 4.55 kg of $CsGa(SO_4)_2 \cdot 12H_2O$?

Alkaline Earth Metals
14.44 What is the difference between gypsum and pure calcium sulfate?

14.45 *(a)* What is the percent by mass of Mg in asbestos, $H_4Mg_3Si_2O_9$? *(b)* What mass of asbestos contains 25.0 g Mg?

14.46 What alkaline earth metals are extracted from the following minerals: *(a)* barite, *(b)* dolomite, *(c)* limestone, *(d)* beryl?

14.47 Calculate the amount of heat, in kilojoules, required to vaporize 92.7 g Ca at its boiling point. Use data in Table 14.2.

14.48 Write Lewis structures for each of the following: *(a)* $MgBr_2$, *(b)* SrS, *(c)* $CaCO_3$, *(d)* $MgSO_4$, *(e)* $Ba(OH)_2$.

14.49 Complete and balance each of the following equations:
(a) Calcium + nitrogen gas $\longrightarrow$
(b) Magnesium + fluorine gas $\longrightarrow$
(c) Strontium + hydrogen gas $\longrightarrow$
(d) Barium + water $\longrightarrow$
(e) Beryllium + fluorine gas $\longrightarrow$

14.50 What properties of Be atoms deviate from those of the other alkaline earth metals?

14.51 How are each of the following used: *(a)* CaO, *(b)* Mg alloys, *(c)* asbestos, *(d)* magnesium hydroxide, *(e)* beryl?

14.52 Write balanced equations that show how mortar, containing lime, is produced from limestone, and how it helps to hold bricks together.

14.53 Barium peroxide, BaO_2, is combined with hydrochloric acid, $HCl(aq)$ to produce hydrogen peroxide, H_2O_2. *(a)* Complete and balance the equation. *(b)* Calculate the mass of H_2O_2 that could be produced by combining 56.9 g of barium peroxide with excess hydrochloric acid. *(c)* What mass of H_2O_2 results when 21.4 g BaO_2 combines with 9.30 g HCl?

14.54 Plaster of paris, $CaSO_4 \cdot \frac{1}{2}H_2O$, is formed by heating gypsum, $CaSO_4 \cdot 2H_2O$.

$$CaSO_4 \cdot 2H_2O \longrightarrow CaSO_4 \cdot \tfrac{1}{2}H_2O + \tfrac{3}{2}H_2O$$

(a) Calculate the mass of plaster of paris that results when 544.0 kg of gypsum is heated. *(b)* What mass of gypsum is required to produce 1225 kg of plaster of paris?

14.55 Lime, CaO, when mixed with water, produces slaked lime, $Ca(OH)_2$. *(a)* Write the balanced equation for this chemical change. *(b)* If 66 kJ of energy is released per mole of slaked lime produced, calculate the amount of energy released when 812 g of lime is added to water.

14.56 Write a chemical explanation for tooth decay; include appropriate compounds and what happens to them.

14.57 Magnesium hydroxide, an ingredient in antacid tablets, combines with stomach acid, $HCl(aq)$, to produce aqueous magnesium chloride and water. *(a)* What mass of hydrochloric acid is neutralized by a tablet that contains

100.0 mg of magnesium hydroxide? *(b)* Compare your answer to the amount of acid neutralized by the same mass of sodium hydrogencarbonate. (Sodium hydrogencarbonate and HCl(*aq*) produce sodium chloride, water, and carbon dioxide.)

Aluminum

14.58 Draw Lewis structures for and write the names of each of the following: *(a)* $AlCl_3$, *(b)* AlN, *(c)* Al_2O_3, *(d)* $Al(OH)_3$, *(e)* $Al(ClO_4)_3$.

14.59 Some antiperspirants contain aluminum salts (aluminum chlorohydrates), $Al(OH)_2Cl$, and $Al(OH)Cl_2$. Write two balanced equations that show how both of these salts are produced from the combination of aluminum hydroxide and hydrochloric acid.

14.60 *(a)* What is the composition of the mineral corundum? *(b)* List several different forms of corundum.

14.61 *(a)* Write the equation for the thermite reaction. *(b)* If 426 kJ is released per mole Al, what quantity of heat is released when 9.23 g Al is totally reacted? *(c)* What mass of water could be heated from 25°C to 100°C with the energy released from 5.33 kg Al in the thermite reaction? [The heat capacity of water is 75 J/(mol·K).]

14.62 What could account for the difficulty and high cost of obtaining such an abundant element as Al from its ores?

14.63 Aluminum fluoride, AlF_3, has a melting point of 1040°C, and exhibits some characteristics of ionic substances; however, aluminum chloride, $AlCl_3$, has a much lower melting point, 194°C, and shows little ionic character. What could explain the differences in the properties of these two aluminum halides?

14.64 Aluminum oxide combines with hydrogen fluoride to produce aluminum fluoride and water. Write a balanced equation for this reaction.

14.65 List three aluminum compounds and their uses.

Halogens

14.66 Draw the Lewis structures for each of the following: *(a)* CF_4, *(b)* NaF, *(c)* ClF, *(d)* CaF_2, *(e)* SF_2, *(f)* SF_6.

14.67 Among the halogens, which is the best oxidizing agent and which is the best reducing agent? Explain your answer completely, including appropriate equations.

14.68 Complete and balance each of the following reactions of halogens:
(a) $F_2 + S_8 \longrightarrow$
(b) $Mg + Br_2 \longrightarrow$
(c) $H_2 + Cl_2 \longrightarrow$
(d) $Cl_2 + MgI_2 \longrightarrow$
(e) $CH_4 + F_2 \longrightarrow$

14.69 *(a)* What volume does 29.2 g Cl_2 occupy at STP? *(b)* What volume does 29.2 g F_2 gas occupy at STP? *(c)* What volumes do 29.2-g samples of Cl_2 and F_2 occupy at 111°C and 915 torr?

14.70 SO_2Cl_2 is produced when chlorine gas combines with sulfur dioxide gas. *(a)* Write the equation for this reaction. *(b)* Calculate the mass of SO_2Cl_2 that results when 51.9 L of chlorine gas combines with excess sulfur dioxide at STP conditions. *(c)* What mass of SO_2Cl_2 results when 4.40 g of chlorine combines with 2.95 g of sulfur dioxide?

14.71 Considering the properties of halogens given in Table 14.4, attempt to explain why fluorine has the capacity to oxidize most other substances.

14.72 Hydrogen fluoride combines with calcium silicate, $CaSiO_3$, the major component of glass.

$$CaSiO_3 + 6HF \longrightarrow CaF_2 + SiF_4 + 3H_2O$$

Assuming that a 175-g drinking glass is 100% calcium silicate, calculate the mass of HF needed to totally combine with the glass.

Sulfur

14.73 Write balanced equations for the combination of sulfur, S_8, with each of the following: *(a)* Ca, *(b)* H_2, *(c)* C.

14.74 When 2.00 mol of zinc(II) sulfide is oxidized to zinc(II) oxide and sulfur dioxide, 905 kJ of energy is released. *(a)* Calculate the amount of energy released when 19.6 g of zinc(II) sulfide is oxidized. *(b)* What mass of zinc(II) sulfide is required to produce 3.91×10^4 kJ? *(c)* How much zinc(II) sulfide must be oxidized to raise the temperature of 97.7 g of water from 12.1°C to 51.9°C?

14.75 If a gasoline mixture contains 0.2% sulfur, S_8, by mass and 1 gal of gasoline has a mass of 2.6 kg, using the equations for the oxidation of S_8 to SO_2, SO_2 to SO_3, and finally SO_3 to H_2SO_4, approximate the mass of H_2SO_4 produced from burning one tank (15 gal) of gasoline. Assume that all of the sulfur in the gasoline is initially oxidized to sulfur dioxide.

14.76 Which of the following sulfur-containing ores has the highest percent by mass of sulfur: (a) zinc blende, ZnS, (b) pyrite, FeS_2, or (c) galena, PbS?

14.77 What volume of CO_2 at 50.0°C and 812 torr is produced when 8.10 g $CaCO_3$ combines with 9.18 g of sulfuric acid?

$$CaCO_3 + H_2SO_4 \longrightarrow CO_2 + CaSO_4 + H_2O$$

Phosphorus

14.78 Calculate the percent by mass of phosphorus in $Ca(H_2PO_4)_2$.

14.79 If 882.5 kg of phosphate rock, containing 74.3% calcium phosphate, is treated with silicon dioxide and carbon monoxide, calculate the mass of P_4 obtained.

14.80 Calculate the mass of phosphoric acid that results when 55.05 g P_4O_{10} combines with excess water.

14.81 Write the names of each of the following phosphorus compounds: (a) PBr_5, (b) H_3PO_4, (c) KH_2PO_4, (d) AlP, (e) P_4O_6, (f) P_4S_7.

14.82 Phosphine, PH_3, a colorless, foul-smelling poisonous gas is prepared by combining calcium phosphide with water, producing phosphine and calcium hydroxide. (a) Write the equation for the reaction. (b) What volume of PH_3 at 19.1°C and 744 torr is produced when 0.811 g of calcium phosphide is mixed with excess water?

14.83 Three phosphorus sulfides have the following formulas: P_4S_{10}, P_4S_3, and P_4S_7. Write balanced equations that show the formation of the three sulfides from P_4 and S_8.

Additional Exercises

14.84 A 5.000-g sample of a bromine-fluorine compound is found to contain 2.918 g of bromine. What is the empirical formula of the compound?

14.85 What is the oxidation state of chlorine in each of the following: (a) OCl_2, (b) ClO_4^-, (c) ICl_4^-, (d) ClF_2^+?

14.86 Chlorine dioxide, ClO_2, an explosive yellow gas, is used to treat water and bleach flour. It can be prepared as follows:

$$Cl_2 + 2AgClO_3 \longrightarrow 2ClO_2 + 2AgCl + O_2$$

(a) What mass of ClO_2 results when 7.35 L Cl_2 at 81.3°C and 1215 torr is combined with 111 g $AgClO_3$? (b) Draw the Lewis structure for ClO_2. (c) Predict the shape of the ClO_2 molecule. (d) What masses of Cl_2 and $AgClO_3$ are required to produce 1.04 kg ClO_2?

14.87 What mass of $MgCO_3$ contains the same number of moles of Mg atoms as 1.023 g $Mg_2P_2O_7$?

14.88 When hydrogen sulfide burns in air, it produces water vapor and sulfur dioxide. For each mole of hydrogen sulfide combusted, 51.8 kJ of heat is released. (a) Write the equation for the combustion of hydrogen sulfide. (b) What mass of hydrogen sulfide must be combusted to raise the temperature of 2.35 kg of water from 25°C to 39°C? (c) What mass of sulfur dioxide results when 8.13×10^5 kJ is released in this reaction?

14.89 Air contains 21.0% oxygen gas by volume. What volume of air at 20.0°C at 769 torr is required to totally convert 500.0 g of sulfur, S_8, to sulfur trioxide, SO_3?

14.90 Calculate the percent by mass of Al in each of the following substances found in nature: (a) feldspar, $KAlSi_3O_8$; (b) clay, $Al_2Si_2O_5(OH)_4$; and (c) potassium alum, $KAl(SO_4)_2 \cdot 12H_2O$.

14.91 Alnico is a magnetic alloy that contains 50% Fe, 20% Al, 20% Ni, and 10% Co by mass. This alloy can lift approximately 4000 times its own mass of iron. What mass of aluminum is contained in a sample of alnico that can lift a 720-kg sample of metal?

14.92 Sodium sulfide, Na_2S, has a density of 1.856 g/cm³ at 14°C. Calculate the number of sodium atoms in a block of Na_2S with the dimensions 1.04 cm × 0.399 cm × 127 mm.

C H A P T E R
Fifteen

Solutions

After completing Chapter 15, you should be able to

1. Describe how a solution is prepared

2. Distinguish between a solute and a solvent

3. List and give examples of the principal classes of solutions

4. Distinguish between miscible and immiscible liquids

5. List factors that influence the solubility of a solute in a solvent

6. Explain the meaning of "like dissolves like" in terms of solute and solvent structure

7. State the relationship between solubility and temperature given a solubility curve

8. Discuss energy requirements for a solid to dissolve in a liquid

9. Describe how each of the following affects the dissolving rate of a solute: (a) particle size, (b) temperature, (c) concentration, and (d) stirring

10. List and define the primary units of concentration

11. Calculate percent by mass molarity or molality of a solution given all necessary data

12. Explain how a solution of a specific concentration (percent by mass, molarity, or molality) is prepared

13. Explain how a solution is diluted to a lower concentration

14. Give examples of strong electrolytes, weak electrolytes, and nonelectrolytes

15. Write overall and net ionic equations given all necessary information

16. Describe how the addition of a nonvolatile solute lowers the vapor pressure of a solution

17. Calculate the mass or volume of products in aqueous reactions

18. Explain why the boiling point of a solvent is elevated and its freezing point is decreased by the addition of a solute

19. Calculate the increase in boiling point and decrease in freezing point given the amounts of solute and solvent in the solution

---

Solutions are homogeneous mixtures of substances that have a uniform composition throughout their volume. When a solution is prepared, one substance is mixed with another substance in such a manner that, after mixing, only one physical state is observed.

15.1 INTRODUCTION TO SOLUTIONS

In a solution, the substance in larger amount and the one whose physical state is observed is called the **solvent.** The substance in smaller amount, the one that becomes incorporated into the solvent, is called the **solute.** When a solute mixes with a solvent to produce a solution, the process is called **dissolving** or **dissolution.** Solutes dissolve in solvents to yield solutions.

Solute particles (atoms, molecules, or ions) interact with solvent particles and become incorporated in the structure of the solvent during the dissolving process. For example, when a sugar such as glucose, $C_6H_{12}O_6$, dissolves in water (the solvent), the solid crystalline structure of glucose (the solute) is broken down by the water molecules. When totally dissolved, glucose molecules are evenly distributed throughout the water (Fig. 15.1).

Glucose is sometimes called blood sugar. It leaves the blood and enters cells when an energy demand exists.

$$\text{Glucose}(s) \xrightarrow{\text{water}} \text{glucose}(aq)$$

Assuming that the conditions remain constant, glucose molecules remain in solution and do not settle out on standing. Any equal-volume portion of the solution contains the same number of glucose molecules. At any time, the dissolved glucose could be recovered by evaporation of the water.

Types of Solutions

Solutions are categorized according to the physical states of the solute and solvent. For example, at room conditions glucose is a solid and water is a liquid, and a glucose-water solution is therefore a solid-liquid solution. Solid-liquid solutions are commonly encountered because many solids dissolve in water. Oceans contain many dissolved substances, including sodium chloride, $NaCl(aq)$; sodium bromide, $NaBr(aq)$; magnesium chloride, $MgCl_2(aq)$; and many others.

Another commonly encountered type of solution is a liquid-liquid solution in which two liquids are dissolved in each other. For example, an alcohol, such as ethanol, $C_2H_6O(l)$, dissolves in water to produce a solution of alcohol and water. It is difficult to differentiate the solute from the solvent in a liquid-liquid solution because the solution is in the same physical state as its two components. Which component is the solvent and which is the solute? If a small quantity of ethanol is added to a large quantity of water, then the ethanol is the solute and water is the solvent. Water is the solute and ethanol is the solvent if a small quantity of water is mixed with a larger quantity of ethanol.

When dealing with liquid-liquid solutions, we use two additional terms. They are "miscible" and "immiscible." Two liquids are categorized as **miscible** if they are mutually soluble in each other. Ethanol and water are miscible liquids. **Immiscible** liquids are not soluble in each other and do not mix. When mixed, immiscible liquids form two layers. For example, oil and water are immiscible liquids (Fig. 15.2). Many different degrees of miscibility lie between the two examples given; such combinations are classified as **partially miscible.**

Average solubilities of dissolved ions in ocean water.

Ion	Solubility, g solute/kg H_2O
Cl^-	19
Na^+	11
SO_4^{2-}	2.7
Mg^{2+}	1.3
Ca^{2+}	0.41
K^+	0.39
Br^-	0.067

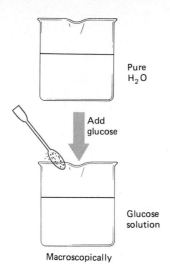

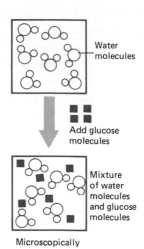

Figure 15.1
When glucose (solute) dissolves in water (solvent), the glucose molecules become a part of the structure of water. Glucose molecules are surrounded by and bond to many water molecules.

A third class of solutions is gas-liquid solutions. In these solutions, gases are the solute and the liquid is the solvent. Carbonated beverages are aqueous solutions of gaseous carbon dioxide, $CO_2(aq)$. Nitrogen and oxygen gas, principal components of the atmosphere, are dissolved to a small degree in most water samples.

In addition to liquids, both gases and solids can be solvents. Gas-gas solutions are those that contain two or more nonreacting gases. Air is a gaseous solution composed of $N_2(g)$, $O_2(g)$, $Ar(g)$, $H_2O(g)$, $CO_2(g)$, and many other gases. All nonreacting gases are miscible with all other gases. Terms such as "solvent" and "solute" have little or no meaning when applied to gas-gas solutions.

Less frequently, solids are solvents. Alloys are solid-solid solutions which are most commonly composed of two or more different metals. Sterling silver is a solution of 7.5% Cu and 92.5% Ag. Eighteen carat

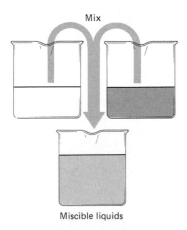

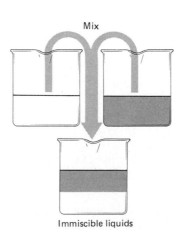

Figure 15.2
If two miscible liquids are mixed, they dissolve in each other; they are mutually soluble. If two immiscible liquids are mixed, they do not mix; they form two separate layers with the most dense liquid on the bottom.

TABLE 15.1 TYPES OF SOLUTIONS

Solute	Solvent	Examples
Solid	Liquid	NaCl(aq); sugar water, $C_{12}H_{22}O_{11}(aq)$
Liquid	Liquid	Vinegar, $HC_2H_3O_2(aq)$; antifreeze in water, $C_2H_6O_2(aq)$
Gas	Liquid	$CO_2(aq)$, $O_2(aq)$, $Ar(aq)$, $NH_3(aq)$
Gas	Gas	The atmosphere; any mixture of nonreacting gases
Solid	Solid	Solder (Sn and Pb), brass (Zn and Cu)
Gas	Solid	H_2 in Pd

gold is an alloy of 75% Au and variable amounts of Ag (10 to 20%) and Cu (5 to 14%). Solids can dissolve liquids and produce a solution. A good example of such solutions is a class of solids called amalgams. Amalgams contain variable amounts of liquid mercury, Hg(l), dissolved in metals.

Most other combinations of physical states are possible, but they are rather rare compared to those already discussed. Gases can be dissolved by solids—$H_2(g)$ forms a solution with the metal palladium, Pd. Table 15.1 lists examples of selected classes of solutions.

Amalgams are used to fill dental caries. Giovanni of Arcoli (1412–1484) was one of the first to suggest that dental caries be filled with a metal; he proposed gold as the best material for fillings.

REVIEW EXERCISES

15.1 Describe the difference between homogeneous and heterogeneous mixtures. Give an example of each.

15.2 Give an example for each of the following types of solutions: *(a)* Both solute and solvent are gases; *(b)* solute is a solid and solvent is a liquid; *(c)* solute is a liquid and solvent is a liquid.

15.3 *(a)* Give two examples of miscible liquids. *(b)* Give two examples of immiscible liquids.

Factors That Affect Solubility

Solubility is a measure of the amount of solute that dissolves in a given amount of solvent at a specific temperature. Generally, chemists measure solubility by finding the amount of solute that dissolves in 100 g of solvent. Three factors are most significant in predicting the solubility of one substance in another: (1) the nature of the solute and solvent, (2) temperature, and (3) pressure. Let's look at these factors individually.

The chemistry saying that "like dissolves like" is a guiding rule when one considers the nature of solute and solvent, and solubility. Substances that have similar structures and intermolecular forces tend to be soluble, and substances that have dissimilar structures and intermolecular forces are less soluble or are insoluble (don't dissolve).

Methanol, $CH_3OH(l)$, is miscible with water. What explains the miscibility of methanol and water? Looking at the structure of a water molecule, we find two H atoms bonded to an O atom; thus, we would expect that hydrogen bonds are the intermolecular forces among water molecules.

Methanol is commonly called wood alcohol because when wood is heated without oxygen, methanol is one of the products formed. Methanol is highly toxic if ingested.

$$\overset{\delta-}{O}$$
$$\underset{H \quad\quad H}{\delta+ \diagup \quad \diagdown \delta+}$$

Methanol molecules are also polar covalent and exhibit hydrogen bonding among their molecules in the liquid state.

$$\overset{\delta-}{O}$$
$$H_3C \diagup \quad \diagdown \overset{\delta+}{H}$$

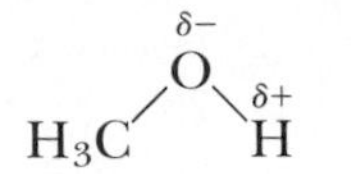

Because water and methanol have similar structures and are both hydrogen bonded liquids, they can form hydrogen bonds with each other when they are mixed (Fig. 15.3). Thus methanol and water are miscible.

As a second illustration, let's consider the miscibility of carbon tetrachloride, CCl_4, and benzene, C_6H_6. Carbon tetrachloride is composed of nonpolar molecules with only dispersion forces among the molecules in the liquid state. Likewise, benzene is composed of nonpolar covalent molecules, also with dispersion forces in the liquid state. CCl_4 and C_6H_6 molecules attract each other with dispersion forces; consequently, they are miscible liquids.

What is the solubility of CCl_4 in water? Water has strong hydrogen bonds among its molecules, and CCl_4 molecules have weaker dispersion forces. Water molecules are held tightly by the hydrogen bonds, excluding CCl_4 molecules. Because no net charge separation is found in CCl_4 molecules, water molecules have only a small force of attraction for CCl_4 molecules (Fig. 15.4). CCl_4 is therefore only partially miscible in water; less than 0.1 g CCl_4 dissolves in 100 mL of water at 20°C.

All the above are examples of liquid-liquid solutions. What predictions can be made for solids that dissolve in liquids? For example, sodium nitrate, $NaNO_3(s)$, is an ionic solid. In which solvent, H_2O or CCl_4, would $NaNO_3$ have a higher solubility? Ionic solids have crystal lattice structures composed of oppositely charged ions that can be attracted by polar covalent substances. Accordingly, water breaks apart the crystal lattice of $Na^+NO_3^-$, and then surrounds the resulting Na^+ and NO_3^- ions (Fig. 15.5). Nonpolar CCl_4 molecules are unable to attract the ions in $NaNO_3$, and cannot break apart the crystal lattice; hence, the solubility of $NaNO_3$ in CCl_4 is minimal.

A benzene molecule is composed of a ring of six C atoms with one H atom bonded to each C. Benzene is commercially important as a component in gasoline, and is the starting material in the synthesis of hundreds of industrial chemicals.

Carbon tetrachloride was once the main component of household cleaning fluids. It was found to cause severe liver damage and to be cancer causing. Thus, it is no longer used as a household cleaner.

Figure 15.3
When methanol molecules dissolve in water, they form hydrogen bonds with water molecules. Partially positive H atoms in methanol bond to the partially negative O atoms in water, and partially negative O atoms in methanol bond to partially positive H atoms in water. Liquid molecules with similar sizes and intermolecular forces are normally miscible.

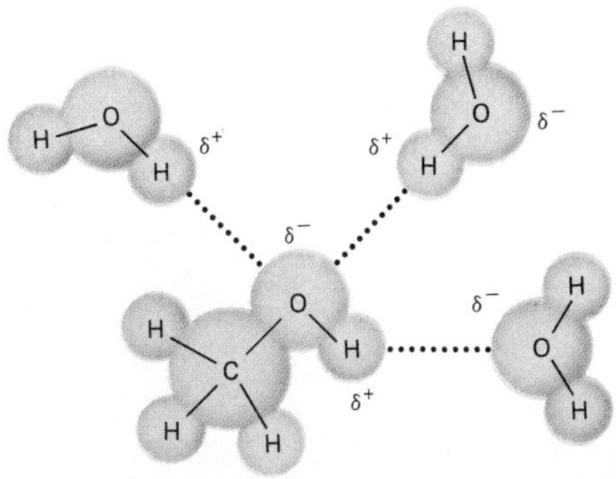

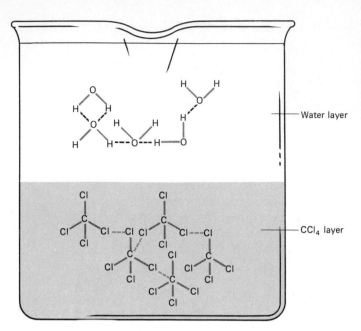

Figure 15.4
Polar covalent water molecules are bonded by strong hydrogen bonds. Nonpolar covalent carbon tetrachloride molecules are bonded by weaker dispersion forces. Water molecules, therefore, exclude the CCl_4 molecules. In general, nonpolar solvents have minimal solubility in polar solvents.

— Water layer

— CCl_4 layer

Temperature is the second factor that influences solubility. Figure 15.6 presents a graph of temperature versus the solubility of selected solutes in water. Generally, an increase in temperature is accompanied by an increase in solubility; at higher temperatures, larger masses of solutes dissolve in a fixed mass of water than at lower temperatures (Fig. 15.7). However, this is not always the case. For a small number of substances, notably gas-liquid solutions, a decrease in solubility is observed with increasing temperature. Other substances, including NaCl, have a relatively constant water solubility with increasing temperature.

An explanation for the effect of temperature on solubility requires information that we have not supplied yet. For now, we will discuss only the energy effects of dissolving solids. The energy considerations for solutes that dissolve can be thought of as occurring in two steps: (1) the

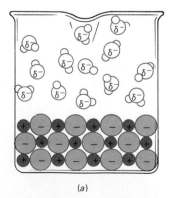

(a)

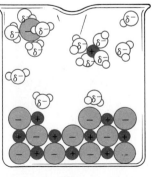

(b)

Figure 15.5
(a) When ionic substances dissolve in water, the very polar water molecules attract the anions and cations, weakening and then breaking the ionic bonds that hold the ions to the crystal lattice.
(b) After breaking free, the ions are surrounded and bonded to water molecules, and then they diffuse away from the undissolved crystal.

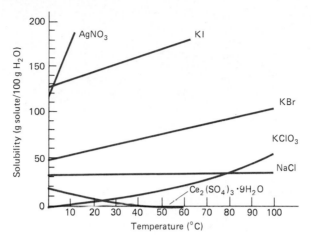

Figure 15.6
The solubility of a substance in water is usually measured in terms of the number of grams of solute that dissolve per 100 g of water. A solubility curve normally shows the solubility of a substance from 0 to 100°C.

breaking up of the crystalline structure of the solid, and (2) the surrounding and bonding of the solvent molecules to the solute particles. The first step requires energy from the surroundings (an endothermic process) to break the bonds that keep the particles in the solid state. This added energy is called the **lattice energy.** In the second step, energy is released (an exothermic process) when bonds are formed between solid particles and solvent molecules. The energy released is called the **solvation energy,** and if the solvent is water it is called the **hydration energy** (Fig. 15.8).

If the hydration energy is greater than the lattice energy, more energy is released than consumed, resulting in a net loss of heat to the surroundings. In such cases, the temperature of the solution increases. A solute such as NaOH releases a tremendous amount of energy on

Figure 15.7
For many solutes, as the temperature of the solvent increases, the solubility of the solute increases. For solutes that give the solution a color, the increased solubility is recognized by a more intense color of the solute in the solution.

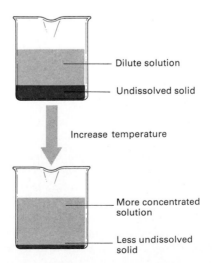

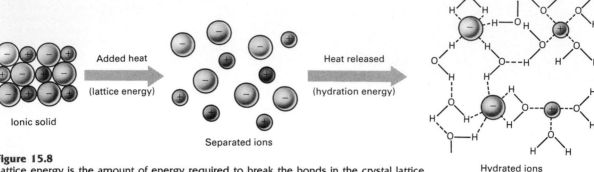

Ionic solid

Added heat
(lattice energy)

Separated ions

Heat released
(hydration energy)

Hydrated ions

Figure 15.8
Lattice energy is the amount of energy required to break the bonds in the crystal lattice structure of a solute. Whenever bonds are broken, energy must be added (an endothermic process). Hydration energy is the amount of energy released when water molecules surround and bond to the solute particles. Whenever bonds are formed, energy is released (an exothermic process).

dissolving (44 kJ/mol NaOH); so much so, that the water may boil. If the opposite is true, i.e., if the hydration energy is less than the lattice energy, then energy is removed from the solution and the solution becomes colder. Cold packs used by athletes contain a salt that dissolves in water when part of the bag is ruptured. Because the lattice energy of the salt is greater than the hydration energy, the solution becomes very cold. If the hydration energy equals the lattice energy, no temperature change is observed on dissolving. The amount of heat transferred when a substance dissolves is called the **enthalpy of solution** (also called heat of solution).

The amount of heat absorbed or released when one mole of substance dissolves is its molar enthalpy of solution ΔH_{soln}.

    Pressure also influences the solubility of some solutes. Only one major type of solution is significantly affected by pressure changes: gas-liquid solutions. As the pressure of a dissolved gas is increased over the solution, the solubility of the gas increases. William Henry (1774–1836) was the first to describe the relationship between gas pressure and solubility. This relationship is now known as **Henry's law.**

    Divers are concerned with the effects predicted by Henry's law. High-pressure conditions beneath the surface of water increase the solubility of N_2 and O_2 in the blood. If the diver comes to the surface too rapidly, bubbles of gas form in the blood and block small blood vessels, causing pain, fainting, and other unpleasant symptoms. This problem is called the bends, or decompression sickness.

REVIEW EXERCISES

15.4 Knowing that oil and water are immiscible liquids, what could be inferred about the intermolecular forces in water and oil? Explain fully.

15.5 On average, what effect does each of the following changes have on the solubility of a solute in a solvent: *(a)* increasing the temperature, and *(b)* increasing the pressure of a gas over a solution that contains that gas?

15.6 What temperature change occurs when the lattice energy of a solute is greater than its hydration energy when it dissolves in water? Explain.

Rates at which Solutes Dissolve

The term **dissolving rate** refers to the speed with which a solute dissolves in a solvent. Rates in which solutes dissolve can be measured in terms of the mass of solute that dissolves in a time interval.

A very soluble substance, like sugar, can be added to water and then not dissolve at an appreciable rate. What factors decrease the time that it takes for sugar or any other solute to dissolve? Four principal factors influence the rates at which substances dissolve: (1) particle size, (2) temperature, (3) solution concentration, and (4) stirring.

If a sugar cube is added to water, the cube dissolves more slowly than does an equal mass of finely granulated sugar. A sugar cube exposes less surface area to the water than do the tiny sugar crystals of granulated sugar. Inner regions of the sugar cube can only dissolve after the outer layers are in solution. Fewer unexposed particles are found in granulated sugar, and thus the sugar crystals enter solution faster than a sugar cube.

To increase the dissolving rate of large chunks of solids in liquids, it is necessary only to grind them up into small particles. A mortar and pestle (Fig. 15.9) are frequently used in the laboratory for just this purpose.

Temperature is another factor that changes the rate at which solutes dissolve in solvents. Increasing the temperature of the solvent generally decreases the time needed for solids to dissolve. At higher temperatures, increased molecular motion increases the number of interactions of solute and solvent particles, increasing the dissolving rate.

Solution concentration affects the rate at which substances dissolve. Solution concentration is a measure of the amount of solute that dissolves in a solution. A **dilute** solution is one with a relatively small amount of dissolved solute. As more solute is added, the concentration of the solute increases. As the concentration of the solute increases, the amount of time needed for more solute to dissolve increases. Initially, when the first solute is added to a pure solvent, the dissolving rate is at maximum. With each addition of solute, the dissolving rate decreases until no more solute dissolves. At this time, the solution is said to be **saturated.** Any solute added to a saturated solution remains undissolved and falls to the bottom of the container. Prior to saturation, the solution is **unsaturated,** and additional solute dissolves in the solution.

When a solution is saturated, the number of solute particles that enter solution equals the number of particles that leave the solution and bond to the undissolved solute. In other words, an equilibrium is established between the rate at which particles break free from the undissolved solute and the rate at which they bond to the undissolved solute.

$$\text{Solute}(s) \rightleftharpoons \text{solute}(aq)$$

Consequently, no more solute is experimentally found to dissolve (Fig. 15.10). Prior to saturation, the number of particles that break free from

A cube that is 1 cm on each edge has a surface area of 6 cm^2. If the same cube is divided into 1000 cubes of 1 mm on each edge, the total surface area is 60 cm^2, a 10-fold increase.

Some substances, when dissolved at a high temperature, do not come out of solution as the temperature is lowered. Thus, these solutions contain more solute than the maximum solubility. This phenomenon is called supersaturation. Few solutions supersaturate.

Figure 15.9
To decrease the particle size and increase the surface area, a mortar and pestle are used to grind large crystals.

Figure 15.10
When a solute is mixed with a solvent, the rate that the solute particles enter solution is greater than the rate at which the solute particles leave solution and bond to the undissolved solute. Therefore, the solute continues to dissolve. At equilibrium, the rates at which solute particles enter and leave solution are equal; thus, no change occurs in the mass of the undissolved solute.

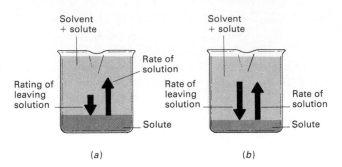

(a)   (b)

the solute is greater than the number that bond to the solute; therefore, the solute continues to dissolve.

Stirring a solution increases the rate at which a solid dissolves by decreasing the concentration of solute in the immediate region surrounding the solid, and also by increasing the amount of solute surface exposed to the solvent.

REVIEW EXERCISES

15.7 List three ways to increase the rate at which a sugar cube dissolves.

15.8 (a) What happens when additional salt is added to a saturated salt solution? (b) What effect would stirring have on the amount of dissolved salt in this solution? (c) How could added salt be dissolved in a saturated salt solution?

**15.2
CONCENTRATION OF
SOLUTIONS**

T erms such as "concentrated" or "dilute" refer to the amount of solute contained in a solvent. Various solutions when saturated are quite dilute. A saturated solution of magnesium hydroxide, $Mg(OH)_2$, only contains 9×10^{-4} g $Mg(OH)_2$/100 mL H_2O at 25°C. On the other hand, a dilute solution of magnesium chloride hexahydrate, $MgCl_2 \cdot 6H_2O$, might contain from 1 to 10 g of magnesium chloride per 100 mL of solution, which has a solubility of 167 g/100 mL H_2O at 25°C. Therefore, a quantitative means is required to exactly express the amount of solute in a solution.

Many units of concentration are used by scientists. Table 15.2 lists the most common units of concentration employed in science. In this section, we will discuss only **percent by mass** and **molarity.** Later in the chapter **molality** is discussed.

Percent by Mass

Percent by mass is the unit of concentration that expresses the mass of solute per 100 grams of solution (mass of solute + mass of solvent).

$$\text{Percent by mass} = \frac{\text{mass of solute}}{\text{mass of total solution}} \times 100$$

By definition, a 5% m/m sugar solution is one that contains 5 g of sugar in every 100 g of solution. A 5% m/m sugar solution may be prepared by

TABLE 15.2   COMMON UNITS OF CONCENTRATION

Unit	Symbol	Definition
Percent by mass	% m/m	$\% \text{ m/m} = \dfrac{\text{mass of solute}}{\text{mass of solution}} \times 100$
Percent by mass to volume	% m/v	$\% \text{ m/v} = \dfrac{\text{mass of solute}}{\text{volume of solution}} \times 100$
Percent by volume	% v/v	$\% \text{ v/v} = \dfrac{\text{volume of solute}}{\text{volume of solution}} \times 100$
Parts per thousand	ppt	$\text{ppt} = \dfrac{\text{mass of solute}}{\text{mass of solution}} \times 1000$
Parts per million	ppm	$\text{ppm} = \dfrac{\text{mass of solute}}{\text{mass of solution}} \times 1{,}000{,}000$
Molarity	M	$M = \dfrac{\text{moles of solute}}{\text{volume of solution (L)}}$
Normality	N	$N = \dfrac{\text{equivalents of solute}}{\text{volume of solution (L)}}$
Molality	m	$m = \dfrac{\text{moles of solute}}{\text{kilograms of solvent}}$

dissolving 5 g of sugar in 95 g of water. Example Problem 15.1 shows the calculations required to calculate the masses of solute and solvent when preparing a solution with a specific percent by mass of solute.

In percent concentration calculations the identity of the solute is not used in the calculation.

Example Problem 15.1

How is 185.00 g of 1.39% m/m NaCl(aq) prepared?

Solution
---

1. What is unknown?   Masses of NaCl and H_2O in g

2. What is known?   185.00 g of solution; 1.39% m/m NaCl(aq)
From the information given, we know that the total mass of the solution is 185.00 g, which is the mass of the solute, NaCl, plus the solvent, H_2O. Also, a 1.39% m/m NaCl solution contains 1.39 g NaCl per 100 g of solution.

$$1.39\% \text{ NaCl} = \frac{1.39 \text{ g NaCl}}{100 \text{ g solution}}$$

Percent by mass is a conversion factor that expresses the ratio of mass of solute to the mass of total solution; accordingly, we apply the factor-label method (Fig. 15.11).

3. Apply the factor-label method to find mass of NaCl.

$$185.00 \text{ g solution} \times \frac{1.39 \text{ g NaCl}}{100 \text{ g solution}} = \textbf{2.57 g NaCl}$$

Figure 15.11
The percent by mass of a solution is a conversion factor that expresses the mass of solute per 100 g of solution. To calculate the mass of solute contained in a solution, multiply the percent by mass times the mass of the solution. To calculate the mass of solution that contains a fixed mass of solute, invert the percent by mass conversion factor and multiply it by the mass of solute.

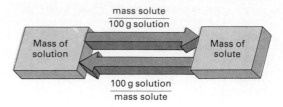

Percent by mass

4. Calculate the mass of water.
Knowing the mass of NaCl, 2.57 g, we subtract it from the total mass of the solution to find the mass of water.

Mass of water = mass of solution (solute + solvent) − mass of solute

$$\text{g } H_2O = 185.00 \text{ g} - 2.57 \text{ g NaCl}$$

$$= \textbf{182.43 g } H_2O$$

To prepare a 1.39% m/m NaCl solution, 2.57 g NaCl should be dissolved in 182.43 g H_2O.

Many times in the laboratory a solution is available with a known concentration, and a specific mass of solute is needed. Example Problem 15.2 illustrates how the mass of solute is calculated from the concentration expressed in percent by mass.

──────────── **Example Problem 15.2** ────────────

Concentrated nitric acid, HNO_3, is usually sold as a 70% m/m aqueous solution. Calculate the mass of concentrated nitric acid that contains 23 g HNO_3.

──────────── **Solution** ────────────

1. What is unknown?   Mass in g of concentrated HNO_3 solution
2. What is known?   23 g HNO_3; 70% m/m nitric acid solution
3. Apply the factor-label method (Fig. 15.11).

$$23 \text{ g } HNO_3 \times \frac{100 \text{ g } HNO_3 \text{ solution}}{70 \text{ g } HNO_3} = \textbf{33 g } HNO_3 \textbf{ solution}$$

Our answer, 32.9 g HNO_3, is rounded to 33 g to display the correct number of significant figures. A 33-g sample of 70% m/m HNO_3 solution contains 23 g pure HNO_3, and 10 g is water.

Alchemists of the eighth century called nitric acid aqua fortis, *which means strong water. Today, nitric acid is used to produce fertilizers, explosives, plastics, dyes, and drugs.*

More often than not, when dealing with percent by mass, volumes of solutions must also be considered. Example Problem 15.3 illustrates this type of problem.

─────────── **Example Problem 15.3** ───────────

Hydroiodic acid, $HI(aq)$, is commercially available as a 57% m/m aqueous solution that has a density of 1.70 g/cm^3. Calculate the mass of HI in 1.0 L of 57% m/m HI solution.

─────────── **Solution** ───────────

1. What is unknown?   Mass of HI in g

2. What is known?   1.0 L of 57% m/m HI, with a density of 1.70 g/cm^3
In this problem, the volume of the solution is given; thus, by using the density, we are able to calculate the total mass of the solution (Fig. 15.12).

3. Apply the factor-label method.

$$1.0 \text{ L solution} \times \frac{1000 \text{ mL solution}}{1 \text{ L solution}} \times \frac{1.70 \text{ g solution}}{1 \text{ mL solution}}$$

$$\times \frac{57 \text{ g HI}}{100 \text{ g solution}} = 969 \text{ g HI} = \mathbf{9.7 \times 10^2 \text{ g HI}}$$

A solution of 1.0 L of 57% m/m HI contains 9.7×10^2 g HI.

─────────────────────────

Other less commonly used percent concentration units in chemistry are percent by volume and percent by mass to volume. They are defined as follows.

$$\text{Percent by volume} = \frac{\text{volume of solute}}{\text{total volume}} \times 100$$

$$\text{Percent by mass to volume} = \frac{\text{mass of solute}}{\text{total volume}} \times 100$$

% m/m = % mass/mass
% m/v = % mass/volume
% v/v = % volume/volume

Figure 15.12
Three conversion factors are needed to calculate the mass of solute in a given number of liters of solution. These factors are the number of milliliters per liter, the density of the solution, and the percent by mass of the solution. First, convert the number of liters of solution to milliliters. Then calculate the mass of the solution using the density; and finally multiply by the percent by mass to obtain the mass of solute.

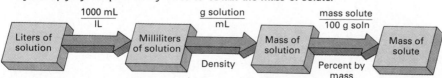

Problems that involve these units are similar to those involving percent by mass except for the comparison to the total volume of solution.

15.9 Name the unit of concentration represented by each of the following: (*a*) moles of solute per liter of solution, (*b*) mass of solute per 100 mL of solution, (*c*) moles of solute per kilogram of solvent, (*d*) mass of solute per 100 g of solution.

15.10 Explain exactly how 486 g of a 12.1% m/m solution of aqueous acetone solution is prepared.

15.11 What is the mass of $CaCl_2$ in 233.5 g of a 6.19% m/m $CaCl_2(aq)$ solution?

Molarity is the concentration unit most frequently encountered in beginning chemistry. **Molarity** is the number of moles of solute per liter of solution. When using or calculating molarity, moles of solute particles are compared with the total volume of the solution, i.e., both the solute and solvent.

Molarity

Molarity is the number of moles of solute per liter of solution.

$$\text{Molarity} = \frac{\text{moles solute}}{\text{liters solution}}$$

One way to prepare a 1 *M* (we say, "one molar") aqueous solution is to add 1 mol of solute to a 1-L volumetric flask, and then add enough water so that the total volume is 1 L (Fig. 15.13).

A one molar solution can be prepared in many other ways. If a 250-mL volumetric flask is available, 0.250 mol of solute is placed in the 250-mL flask, and then enough water is added to yield a total volume of 250 mL.

$$1.00 \ M = \frac{0.250 \ \text{mol}}{0.250 \ \text{L}}$$

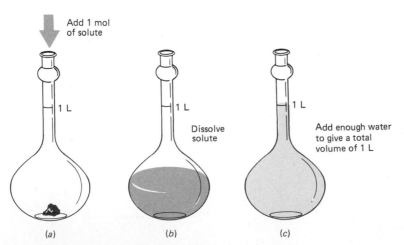

Figure 15.13
One way to prepare a 1-*M* aqueous solution is to (*a*) weigh one mole of solute on a balance and transfer it to a 1-L volumetric flask, (*b*) add water and dissolve the solute, and (*c*) continue adding water until the total volume of the solution is 1 L and shake to produce a homogeneous mixture.

Add 1 mol of solute

1 L

1 L

Dissolve solute

1 L

Add enough water to give a total volume of 1 L

(a)

(b)

(c)

Example Problem 15.4 shows the calculation required before a specific volume of a solution of given molarity can be prepared.

——————————— **Example Problem 15.4** ———————————

Explain how 175 mL of a 0.320 M KI(aq) solution is prepared.

——————————— **Solution** ———————————

A unit closely related to molarity is formality (F). Formality is the number of moles of formula units of an ionic solute per liter of solution. Formality is used for ionic substances that do not contain discrete molecules.

1. What is unknown?   Mass of KI in g in 175 mL of 0.320 M KI(aq)

2. What is known?   175 mL; 0.320 M KI (0.320 mol KI/L); 166 g KI/mol KI

To calculate the mass of KI required to prepare this solution, we use conversion factors, changing the volume from mL to L and then using the molarity, the ratio of moles of solute to liters of solution, and the molar mass of KI (Fig. 15.14).

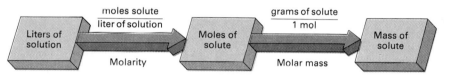

Figure 15.14
To calculate the mass of solute in a given volume of solution, multiply the volume of the solution in liters times the molarity of the solution (moles of solute per liter of solution) and then multiply by the molar mass (grams of solute per 1 mol).

3. Apply the factor-label method.

$$175 \text{ mL} \times \frac{1 \text{ L}}{1000 \text{ mL}} \times \frac{0.320 \text{ mol KI}}{1 \text{ L}} \times \frac{166 \text{ g KI}}{1 \text{ mol KI}} = \textbf{9.30 g KI}$$

After calculating the volume in L, the number of moles of KI is calculated using the molarity. Finally, the mass is obtained by multiplying by the molar mass of KI.

4. Explain how the solution is prepared.
To prepare 175 mL of 0.320 M KI, add 9.30 g KI to a container with a graduation mark at 175 mL. Place water in the container and dissolve the KI, and continue adding water until the total volume is 175 mL. Mix the solution so that the KI is distributed throughout.

Example Problem 15.5 presents the method that you should use to calculate the molarity of a solution given all necessary data.

——————————— **Example Problem 15.5** ———————————

What is the molarity of a solution prepared by adding 18.3 g of methanol, CH_3OH, to a container and then mixing it with enough water to give a total volume of 50.0 mL?

――――――――――― **Solution** ―――――――――――

1. What is unknown? M CH$_3$OH(aq)(mol CH$_3$OH/L solution)

2. What is known? 18.3 g CH$_3$OH; total volume = 50.0 mL; 32.0 g CH$_3$OH/mol CH$_3$OH

Because molarity is the ratio of moles of solute, CH$_3$OH, to liters of solution, we must find the number of moles of CH$_3$OH and then divide it by the number of liters of solution (Fig. 15.15).

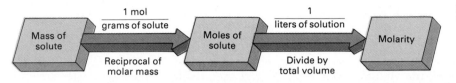

Figure 15.15
To find the molarity of a solution given the mass of solute and total volume of the solution, convert the mass of solute to moles and divide it by the total volume of the solution in liters.

3. Calculate the number of moles of solute and the volume of solution in L.

$$\text{Mol CH}_3\text{OH} = 18.3 \, \cancel{\text{g CH}_3\text{OH}} \times \frac{1 \text{ mol CH}_3\text{OH}}{32.0 \, \cancel{\text{g CH}_3\text{OH}}} = 0.572 \text{ mol CH}_3\text{OH}$$

$$\text{L solution} = 50.0 \, \cancel{\text{mL}} \times \frac{1 \text{ L}}{1000 \, \cancel{\text{mL}}} = 0.0500 \text{ L}$$

$$= \mathbf{11.4 \, \textit{M} \, CH_3OH}$$

When 18.3 g CH$_3$OH is dissolved in enough water to give a total volume of 50.0 mL, the molarity of the solution is 11.4 M.

At various times, the concentration of a laboratory stock solution may not be given in the units that you want. Therefore, these units must be converted to the desired concentration units. Example Problem 15.6 shows how percent by mass is converted to molarity.

――――――――――― **Example Problem 15.6** ―――――――――――

What is the molarity of an 85.0% m/m phosphoric acid, H$_3$PO$_4$, solution? An 85.0% m/m H$_3$PO$_4$ solution has a density of 1.70 g/cm^3.

――――――――――― **Solution** ―――――――――――

1. What is unknown? M H$_3$PO$_4$ (mol H$_3$PO$_4$/L solution)

2. What is known? 85.0% m/m H$_3$PO$_4$; density of solution = 1.70 g/mL; molar mass = 98.0 g H$_3$PO$_4$/mol H$_3$PO$_4$

To solve this problem, first consider what we are starting with: % m/m, or the ratio of mass of solute to 100 g of solution.

$$85.0\% \text{ m/m } H_3PO_4 = \frac{85.0 \text{ g } H_3PO_4}{100 \text{ g solution}}$$

Thus, the mass of the solute should be converted to moles, and then the volume of the solution is calculated from its mass (Fig. 15.16).

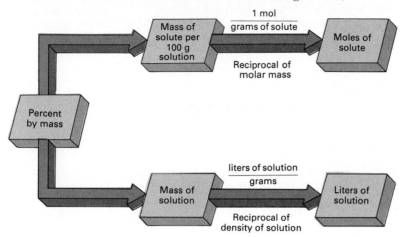

Figure 15.16
To change percent by mass of a solution to molarity, calculate the number of moles of solute per 100 g of solution. Then calculate the volume in liters of the 100 g of solution, using the density of the solution. The molarity is found by dividing the number of moles of solute by the volume of the solution.

3. Calculate the number of moles of H_3PO_4, and the volume of solution.

$$\text{Mol } H_3PO_4 = 85.0 \text{ g } H_3PO_4 \times \frac{1 \text{ mol } H_3PO_4}{98.0 \text{ g } H_3PO_4} = 0.867 \text{ mol } H_3PO_4$$

$$\text{L solution} = 100 \text{ g solution} \times \frac{1 \text{ mL}}{1.70 \text{ g solution}} \times \frac{1 \text{ L}}{1000 \text{ mL}} = 0.0588 \text{ L}$$

4. Calculate the molarity.

$$M_{H_3PO_4} = \frac{0.867 \text{ mol } H_3PO_4}{0.0588 \text{ L}} = \textbf{14.7 } M \textbf{ } H_3PO_4$$

An 85.0% m/m H_3PO_4 solution is 14.7 M, or 14.7 mol H_3PO_4 is dissolved per liter of solution.

15.12 Explain how 54.0 mL of 1.58 M NaOH(aq) is prepared.

15.13 What mass of $Mg(NO_3)_2$ is required to prepare 121 L of 0.0752 M $Mg(NO_3)_2(aq)$?

15.14 Calculate the molarity of a solution that contains 5.991 g C_2H_6O in a total volume of 540.0 mL of solution.

15.15 What is the molarity of a 90.0% m/m formic acid, CH_2O_2, solution? The solution has a density of 1.20 g/cm^3.

REVIEW EXERCISES

Chemists must frequently dilute more-concentrated solutions to obtain less-concentrated ones, or deal with a situation in which two or more solutions have been mixed, diluting all of the dissolved species. Dilution frequently occurs when pure solvent is added to the solution. This increases the total volume of the solution, but does not change the number of moles of dissolved solute. To solve dilution problems, we must calculate the total number of moles of dissolved particles and then divide by the new volume. For example: What is the final concentration of 50.0 mL of a 6.00 M NaOH solution, if its total volume is increased to 75.0 mL? We first calculate the number of moles of NaOH in solution, using the factor-label method as follows:

$$50.0 \text{ mL} \times \frac{1 \text{ L}}{1000 \text{ mL}} \times \frac{6.00 \text{ mol NaOH}}{\text{L}} = 0.300 \text{ mol NaOH}$$

To calculate the molarity of the diluted solution, we divide the number of moles of NaOH by the total volume in liters, which is 75.0 mL or 0.0750 L.

$$M = \frac{0.300 \text{ mol NaOH}}{0.0750 \text{ L}}$$

$$= 4.00 \text{ } M \text{ NaOH}$$

After diluting to 75.0 mL, the molar concentration of the NaOH solution is 4.00 M. If we increase the total volume of this solution to 100.0 mL, the concentration decreases to 3.00 M. Doubling the initial volume (50.0 mL) to 100.0 mL decreases the concentration to one-half of the initial concentration.

Dilution of Solutions

When two liquids or solutions are mixed, their total volume does not always equal the sum of the individual volumes.

———— **Example Problem 15.7** ————

Calculate the molarity of an ethanol, C_2H_6O, solution that is prepared by diluting 125 mL of 4.23 M C_2H_6O to 845 mL.

———— **Solution** ————

1. What is unknown? M C_2H_6O of diluted solution

2. What is known? 125 mL of 4.23 M C_2H_6O diluted to 845 mL; 1000 mL/L

3. Calculate the total number of moles of C_2H_6O.

$$125 \text{ mL} \times \frac{1 \text{ L}}{1000 \text{ mL}} \times \frac{4.23 \text{ mol } C_2H_6O}{\text{L}} = 0.529 \text{ mol } C_2H_6O$$

4. Calculate the molarity of the diluted solution.

Before calculating the molarity, we must convert the volume of the diluted solution to L.

$$845 \text{ mL} \times \frac{1 \text{ L}}{1000 \text{ mL}} = 0.845 \text{ L}$$

To find the molarity of the diluted solution, we divide the total number of moles by the total volume of solution.

$$M = \frac{\text{mol C}_2\text{H}_6\text{O}}{\text{L}}$$

$$= \frac{0.529 \text{ mol C}_2\text{H}_6\text{O}}{0.845 \text{ L}}$$

$$= 0.626 \; M \; \textbf{C}_2\textbf{H}_6\textbf{O}$$

After dilution, we have a $0.626 \; M \; \text{C}_2\text{H}_6\text{O}$ solution.

Often chemists must calculate the volume that a solution must be diluted to in order to have a specific concentration. To solve such a problem, we first calculate the total number of moles of solute; then the total volume of the diluted solution is calculated from its molarity. Example Problem 15.8 is an example of such a problem.

─────────── **Example Problem 15.8** ───────────

What total volume in milliliters must 235.0 mL of 12.00 M HCl be diluted to in order to produce a 1.000 M HCl solution?

─────────────── **Solution** ───────────────

1. What is unknown?   mL of 1.000 M HCl

2. What is known?   235.0 mL of 12.00 M HCl to be diluted to 1.000 M HCl

First, we calculate the number of moles of HCl present in 235.0 mL of 12.00 M HCl. Once we know the number of moles of HCl, the total volume of the diluted solution can be found from its molarity, 1.000 M.

3. Calculate the total number of moles of HCl.

$$\text{Mol HCl} = 235.0 \text{ mL} \times \frac{1 \text{ L}}{1000 \text{ mL}} \times \frac{12.00 \text{ mol HCl}}{\text{L}} = 2.820 \text{ mol HCl}$$

4. Calculate the total volume of 1.000 M HCl that contains 2.820 mol HCl.

When diluting concentrated acids, never add water to the acid. The acid is likely to splatter when it contacts the water. Instead, *always* cautiously add concentrated acids to water.

$$\text{Volume of 1.000 } M \text{ HCl} = 2.820 \; \overline{\text{mol HCl}} \times \frac{1 \; \overline{\text{L}}}{1.000 \; \overline{\text{mol HCl}}} \times \frac{1000 \text{ mL}}{1 \; \overline{\text{L}}}$$

$$= 2.820 \times 10^3 \text{ mL}$$

We find that the initial 235.0 mL of 12.00 M HCl should be diluted to 2.820×10^3 mL, 2.820 L, in order to decrease the concentration to 1.000 M.

When two solutions are mixed, all of the dissolved species are diluted. Example Problem 15.9 shows how to solve dilution problems in which two nonreacting solutions are mixed.

──────────────── **Example Problem 15.9** ────────────────

What is the resulting molar concentration of acetone, C_3H_6O, when 40.0 mL of 0.100 M acetone solution is mixed with 20.0 mL of 0.0750 M acetone? Assume that the total volume of the resulting solution equals the sum the volumes of the two solutions. In other words, assume the volumes are additive.

──────────────── **Solution** ────────────────

1. What is unknown? M (mol C_3H_6O/L) of the diluted solution

2. What is known?   40.0 mL of 0.100 M; 20.0 mL of 0.0750 M C_3H_6O; 1 L/1000 mL

3. Calculate the total number of moles of acetone.

$$\text{Mol acetone} = 40.0 \; \overline{\text{mL}} \times \frac{1 \; \overline{\text{L}}}{1000 \; \overline{\text{mL}}} \times \frac{0.100 \text{ mol } C_3H_6O}{\overline{\text{L}}}$$

$$= 0.00400 \text{ mol } C_3H_6O$$

$$\text{Mol acetone} = 20.0 \; \overline{\text{mL}} \times \frac{1 \; \overline{\text{L}}}{1000 \; \overline{\text{mL}}} \times \frac{0.0750 \text{ mol } C_3H_6O}{\overline{\text{L}}}$$

$$= 0.00150 \text{ mol } C_3H_6O$$

Thus, the total number of moles of acetone in the two solutions is 0.00550 mol (0.00400 mol C_3H_6O + 0.00150 mol C_3H_6O).

4. Calculate the molarity of the combined solutions. Because the volumes are assumed to be additive, the total volume of the resulting solution is 60.0 mL (40.0 mL + 20.0 mL), or 0.0600 L. Hence, we divide the total number of moles by the total volume in L.

$$M = \frac{\text{mol } C_3H_6O}{L}$$

$$= \frac{0.00550 \text{ mol } C_3H_6O}{0.0600 \text{ L}}$$

$$= \mathbf{0.0917}\ \boldsymbol{M}$$

When 40.0 mL of 0.100 M C_3H_6O is mixed with 20.0 mL of 0.0750 M C_3H_6O, the resulting concentration is 0.0917 M C_3H_6O. We would expect the final concentration to be closer to 0.100 M than 0.07500 M because twice the volume of the more-concentrated solution is present initially.

REVIEW EXERCISES

15.16 Glacial acetic acid, $HC_2H_3O_2$, is 99.8% acetic acid, which is 17.4 M. Calculate the total volume to which 120.0 mL of glacial acetic acid must be diluted in order to produce a 6.00 M acetic solution.

15.17 What is the final concentration of hydrobromic acid, HBr(aq), when 155 mL of 0.925 M HBr solution is mixed with 245 mL of 0.831 M HBr solution? Assume the volumes are additive.

15.3 ELECTROLYTES AND NONELECTROLYTES

Solutions can also be classified according to the behavior of the solute after it enters the solvent. When an ionic compound is mixed with water, the crystal structure of the compound (M^+X^-) is broken apart and ions enter solution. The general equation for the dissolution of ionic compounds is as follows:

$$M^+X^-(s) \xrightarrow{\text{water}} M^+(aq) + X^-(aq)$$

Experimentally, the existence of ions in solution is verified by using a conductivity apparatus which is nothing more than a light bulb wired to a source of electricity and two electrodes (Fig. 15.17). The light bulb only glows if an electric conductor is present between the two electrodes. When the electrodes of the conductivity apparatus are immersed in a solution that contains dissolved ionic substances, the light bulb glows. This shows that dissolved ions act as electric current carriers and complete the electrical circuit. Solutes that produce solutions that conduct electric currents are called **electrolytes.** Depending on the dissolved solutes, some solutions conduct large electric currents (those with higher

Jacobus van't Hoff (1852–1911) was the first to show that solutions can conduct an electric current. He advanced chemistry by showing that the solute particles in a solution follow laws similar to those that govern gas molecules.

concentrations of dissolved ions), and others conduct smaller electric currents (those with lower dissolved-ion concentrations).

We further classify dissolved ionic substances as strong and weak electrolytes. **Strong electrolytes** are those that allow a large electric current to flow through water, indicating that a large percent of the dissolved solute particles are ions. **Weak electrolytes** allow only a small current to flow through water, showing that only a small percent of the solute is broken up into ions. Table 15.3 lists examples of strong and weak electrolytes.

Nearly 100% of a strong electrolyte exists as ions in solution. Weak electrolytes are only partially ionized in solution.

TABLE 15.3
ELECTROLYTES

Strong electrolytes	Nonelectrolytes
Hydrochloric acid, HCl	Acetone, C_3H_6O
Nitric acid, HNO_3	Oxygen, O_2
Potassium nitrate, KNO_3	Glucose, $C_6H_{12}O_6$
Sodium hydroxide, NaOH	Sucrose, $C_{12}H_{22}O_{11}$
Sodium chloride, NaCl	Ethanol, C_2H_6O

Weak electrolytes
Hydrofluoric acid, HF
Hydrocyanic acid, HCN
Nitrous acid, HNO_2
Calcium hydroxide, $Ca(OH)_2$
Hydrosulfuric acid, H_2S

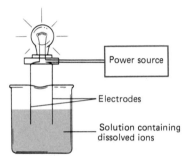

Figure 15.17
A conductivity apparatus may be constructed by attaching electrodes to a light bulb and power source in such a manner that the light bulb will light only if a solution that conducts an electric current is between the electrodes.

Water-soluble ionic solids are generally strong electrolytes. Water-soluble covalent substances are usually weaker electrolytes or nonelectrolytes (except molecules that contain a very polar H—X bond, such as HCl or HNO_3). A **nonelectrolyte** is a substance that does not ionize when dissolved or ionizes to a very small extent. For example, ethanol, C_2H_6O, is a nonelectrolyte. A light bulb in a conductivity apparatus does not glow when the electrodes are immersed in an aqueous solution of ethanol. Table 15.3 also lists examples of nonelectrolytes.

Ionic equations are written to show each of the dissolved ions in a chemical reaction. Up to this point in our discussion, we have written (*aq*) next to the formula of a dissolved substance. However, in an ionic equation we write all of the dissolved ionic species in a chemical reaction. For example: How can we show the dissociation of sodium chloride into sodium and chloride ions when it dissolves in water? Initially, sodium chloride is a crystalline solid; thus, we write it as NaCl(s). After it dissolves in water, it dissociates totally into $Na^+(aq)$ and $Cl^-(aq)$ ions. The ionic equation that shows this is as follows

$$NaCl(s) \xrightarrow{\text{water}} Na^+(aq) + Cl^-(aq)$$

This equation states that sodium chloride (a strong electrolyte) dissolves in water and produces hydrated sodium cations and chloride anions.

Hydrogen chloride gas, HCl(g), when dissolved in water, also ionizes totally; therefore, HCl is a strong electrolyte. In this case the polar covalent HCl molecule ionizes, producing $H^+(aq)$ and $Cl^-(aq)$.

$$HCl(g) \xrightarrow{\text{water}} H^+(aq) + Cl^-(aq)$$

Only a small percent of the dissolved particles in solutions of weak electrolytes are ionized. The ionization of nitrous acid, HNO_2, a weak electrolyte, is represented as follows.

$$HNO_2(aq) \rightleftharpoons H^+(aq) + NO_2^-(aq)$$

Two single-barbed arrows pointing in opposite directions indicate an equilibrium system; a small arrow that points to the right means that only a small number of the HNO_2 molecules are ionized. The longer arrow that points to the left signifies that most of the HNO_2 is in the molecular form.

A dynamic equilibrium results when the rate at which HNO_2 dissociates to its ions equals the rate at which the ions combine to re-form HNO_2.

If a chemical reaction occurs in solution, ionic equations show what happens to all dissolved species. Let's consider the combination of NaCl(aq) and AgNO$_3$(aq) solutions. When these two solutions are mixed, a white, insoluble solid, AgCl(s) (a precipitate), forms. The overall ionic equation for the reaction is

$$\underset{\text{NaCl in ionic form}}{\underline{Na^+(aq) + Cl^-(aq)}} + \underset{\text{AgNO}_3 \text{ in ionic form}}{\underline{Ag^+(aq) + NO_3^-(aq)}} \longrightarrow AgCl(s) + \underset{\text{NaNO}_3 \text{ in ionic form}}{\underline{Na^+(aq) + NO_3^-(aq)}}$$

On the left side of the equation, all chemical species are written in ionic form, because both NaCl and AgNO$_3$ are strong electrolytes and thus ionize totally. After the solutions have undergone a metathesis reaction, only Na^+ and NO_3^- remain in solution, and silver chloride, AgCl, precipitates out of solution as a result of its low solubility (Table 10.4, p. 267).

When solutions of NaCl and AgNO$_3$ are combined, the $Na^+(aq)$ and $NO_3^-(aq)$ are unchanged in the reaction; they appear as both reactants and products. We call the ions that do not change in a chemical reaction **spectator ions** because they "watch" but do not take part in the chemical

Spectator ions are unchanged in aqueous ionic reactions. Thus, they are canceled to obtain the net ionic equation.

change. If the spectator ions are eliminated from the overall ionic equation, the **net ionic equation** results.

$$\text{Na}^+(aq) + \text{Cl}^-(aq) + \text{Ag}^+(aq) + \text{NO}_3^-(aq) \longrightarrow$$
$$\text{AgCl}(s) + \text{Na}^+(aq) + \text{NO}_3^-(aq)$$
Overall ionic equation

After removal of the Na^+ and NO_3^- ions, we get

$$\text{Ag}^+(aq) + \text{Cl}^-(aq) \longrightarrow \text{AgCl}(s)$$
Net ionic equation

Another example of an ionic reaction is the neutralization reaction of HCl and NaOH solutions. Overall and net ionic equations for this reaction are as follows:

$$\underbrace{\text{H}^+(aq) + \text{Cl}^-(aq)}_{\text{HCl in ionic form}} + \underbrace{\text{Na}^+(aq) + \text{OH}^-(aq)}_{\text{NaOH in ionic form}} \longrightarrow \text{H}_2\text{O}(l) + \underbrace{\text{Na}^+(aq) + \text{Cl}^-(aq)}_{\text{NaCl in ionic form}}$$
Overall ionic equation

$$\text{H}^+(aq) + \text{OH}^-(aq) \longrightarrow \text{H}_2\text{O}(l)$$
Net ionic equation

Because both HCl and NaOH are strong electrolytes, they are written as dissolved ions. In the reaction, the H^+ from HCl(aq) combines with the OH^- from NaOH(aq) to produce H_2O, which is written in un-ionized form because it is a very weak electrolyte. The ions that do not change after the reaction are Na^+ and Cl^-; accordingly, they are the spectator ions.

The following rules should be used when you write overall and net ionic equations.

1. Check the given equation to see that it is balanced. If it is not, balance it by inspection before writing the ionic equations.

2. Write all strong electrolytes as dissolved ions. Strong electrolytes mainly include soluble ionic compounds, strong acids, and strong bases.

3. Write all nonelectrolytes, weak electrolytes, gases, and precipitates in un-ionized form. Nonelectrolytes include insoluble inorganic compounds (e.g., sulfides, phosphates, and hydroxides) and many organic compounds (e.g., alcohols or sugars). Weak electrolytes include weak acids—such as nitrous acid, $\text{HNO}_2(aq)$; hydrofluoric acid, $\text{HF}(aq)$; and hydrocyanic acid, $\text{HCN}(aq)$—and weak bases such as ammonia, $\text{NH}_3(aq)$, and calcium hydroxide, $\text{Ca(OH)}_2(aq)$.

4. To obtain the net ionic equation, cancel the spectator ions, those that do not change in the reaction.

5. Check to see that the equation is balanced and there are equal numbers of atoms on either side of the arrow.

Example Problems 15.10 and 15.11 show how to write overall and net ionic equations for selected reactions.

--------------------- **Example Problem 15.10** ---------------------

Write the overall and net ionic equations for the following reaction:

$$(NH_4)_2S(aq) + FeBr_2(aq) \longrightarrow FeS(s) + 2NH_4Br(aq)$$

--------------------- **Solution** ---------------------

1. Write the overall ionic equation.
Because $(NH_4)_2S$, $FeBr_2$, and NH_4Br are water-soluble ionic compounds, they are written as dissolved ions. FeS is an insoluble precipitate; hence, it is written in un-ionized form.

$$2NH_4^+(aq) + S^{2-}(aq) + Fe^{2+}(aq) + 2Br^-(aq) \longrightarrow$$
$$FeS(s) + 2NH_4^+(aq) + 2Br^-(aq)$$

Overall ionic equation

2. Write the net ionic equation.
After you identify the spectator ions, NH_4^+ and Br^-, cancel them from the overall equation. This gives the net ionic equation.

$$\cancel{2NH_4^+(aq)} + S^{2-}(aq) + Fe^{2+}(aq) + \cancel{2Br^-(aq)} \longrightarrow$$
$$FeS(s) + \cancel{2NH_4^+(aq)} + \cancel{2Br^-(aq)}$$

$$S^{2-}(aq) + Fe^{2+}(aq) \longrightarrow FeS(s)$$

Net ionic equation

--------------------- **Example Problem 15.11** ---------------------

Write the overall and net ionic equations for the reaction in which hydrochloric acid, $HCl(aq)$ combines with aqueous sodium carbonate, $Na_2CO_3(aq)$.

--------------------- **Solution** ---------------------

In this problem we are not given the equation for the reaction; consequently, we should first write and balance the equation, and then attempt to write the overall and net ionic equations.

1. Write and balance the equation.
To predict the products of the reaction, we refer to information introduced in Chap. 10. You may recall that acids combine with carbonates in metathesis reactions to produce a salt, carbon dioxide, $CO_2(g)$, and water, $H_2O(l)$. Thus the unbalanced equation for the reaction is

$$HCl(aq) + Na_2CO_3(aq) \longrightarrow NaCl(aq) + CO_2(g) + H_2O(l)$$
$$\text{(Unbalanced)}$$

Balancing the equation by inspection, we obtain

$$2HCl(aq) + Na_2CO_3(aq) \longrightarrow 2NaCl(aq) + CO_2(g) + H_2O(l)$$
$$\text{(Balanced)}$$

2. Write the overall ionic equation.
HCl, Na_2CO_3, and NaCl are strong electrolytes; therefore, we should write them in ionic form. CO_2 is a gas that bubbles out of the solution; consequently it is written in un-ionized form. Water, a very weak electrolyte, is also written in un-ionized form.

$$2H^+(aq) + 2Cl^-(aq) + 2Na^+(aq) + CO_3{}^{2-}(aq) \longrightarrow$$
$$2Na^+(aq) + 2Cl^-(aq) + CO_2(g) + H_2O(l)$$

Note that the coefficient of each compound must be multiplied by each ion that enters solution. For example, both the H^+ and the Cl^- are multiplied by 2.

3. Write the net ionic equation.
After we cancel the spectator ions, Na^+ and Cl^-, we obtain the following net ionic equation.

$$2H^+(aq) + CO_3{}^{2-}(aq) \longrightarrow CO_2(g) + H_2O(l)$$

15.18 How is a solute identified experimentally as being an electrolyte or non-electrolyte?

15.19 Give two examples of each of the following: (a) nonelectrolyte, (b) weak electrolyte, (c) strong electrolyte.

15.20 Write the overall and net ionic equations for the following:

$$HNO_3(aq) + KOH(aq) \longrightarrow H_2O(l) + KNO_3(aq)$$

15.21 Write the overall and net ionic equations for the following:

$$Na_2CO_3(aq) + CuCl_2(aq) \longrightarrow 2NaCl(aq) + CuCO_3(s)$$

15.4 SOLUTION STOICHIOMETRY

In Chap. 11, we began our discussion of stoichiometry. If necessary, go back and review what you may have forgotten. In this section, we will continue our discussion by considering the stoichiometry of reactions that occur in aqueous solutions. We will apply the same fundamental stoichiometric principles that we have previously discussed, but in addition we will be concerned with the concentrations of the solutions that undergo reaction.

Let us consider the reaction in which excess sodium bromide solution, NaBr(aq), is mixed with 34.5 mL of 0.221 M AgNO$_3$. In this metathesis reaction, AgBr(s) precipitates from solution.

An AgNO$_3$ solution is frequently placed in the eyes of newborns to prevent eye infections. A more concentrated solution of AgNO$_3$ is used as a cauterizing agent.

$$NaBr(aq) + AgNO_3(aq) \longrightarrow AgBr(s) + NaNO_3(aq)$$

Let us calculate the mass of AgBr that precipitates from solution. As in all stoichiometry problems, we identify the mole ratio between the reactant of interest and the product. In this reaction, we find that for each one mol AgNO$_3$ that combines with NaBr, one mol AgBr results. To calculate the number of moles of AgNO$_3$ initially present, we use the volume and the molarity of the solution. The factor-label setup for this problem is as follows:

$$34.5 \text{ mL} \times \frac{1 \text{ L}}{1000 \text{ mL}} \times \frac{0.221 \text{ mol AgNO}_3}{\text{L}} \times \frac{1 \text{ mol AgBr}}{1 \text{ mol AgNO}_3}$$
$$\times \frac{187.8 \text{ g AgBr}}{1 \text{ mol AgBr}} = 1.43 \text{ g AgBr}$$

After changing the volume of the AgNO$_3$ solution to liters, we convert to moles of AgNO$_3$, using the molarity of the solution, which is the ratio of moles of AgNO$_3$ to liters of solution. From the equation, we obtain the mole ratio of AgBr to AgNO$_3$. Finally, to calculate the mass of precipitate, multiply by the molar mass of AgBr, which is 187.8 g/mol. Figure 15.18 shows the pathway followed to solve most solution stoichiometry problems.

Example Problems 15.12 and 15.13 give additional examples of different types of solution stoichiometry problems.

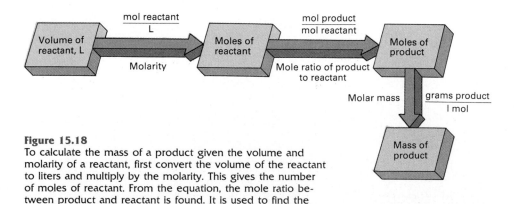

Figure 15.18
To calculate the mass of a product given the volume and molarity of a reactant, first convert the volume of the reactant to liters and multiply by the molarity. This gives the number of moles of reactant. From the equation, the mole ratio between product and reactant is found. It is used to find the number of moles of product. Finally the mass of product is obtained from the molar mass of the product.

———————— **Example Problem 15.12** ————————

What volume of hydrogen gas is produced at STP when excess Zn is mixed with 114 mL of 1.35 M HCl? The equation for the reaction is as follows:

$$Zn(s) + 2HCl(aq) \longrightarrow H_2(g) + ZnCl_2(aq)$$

———————— **Solution** ————————

1. What is unknown?   Volume of H_2 in L at STP

2. What is known?   114 mL of 1.35 M HCl; 1 mol H_2/1 mol Zn; 22.4 L H_2/mol at 1 atm and 273 K

From the balanced equation, we find that one mol H_2 is produced per one mol Zn metal that reacts. At STP conditions, one mol H_2 occupies 22.4 L, which is the molar of an ideal gas.

3. Apply the factor-label method and perform all math operations.

$$114 \text{ mL} \times \frac{1 \text{ L}}{1000 \text{ mL}} \times \frac{1.35 \text{ mol HCl}}{\text{L}} \times \frac{1 \text{ mol } H_2}{1 \text{ mol HCl}}$$

$$\times \frac{22.4 \text{ L } H_2}{1 \text{ mol } H_2} = \textbf{3.45 L } H_2$$

When excess zinc combines with 114 mL of a 1.35 M HCl solution, the theoretical yield of H_2 at standard conditions is 3.45 L.

———————— **Example Problem 15.13** ————————

What mass of lead(II) hydroxide, $Pb(OH)_2$, precipitates from solution when 23.9 mL of 0.522 M $Pb(NO_3)_2$ combines with 28.4 mL of 0.762 M NaOH? The equation for this reaction is as follows:

$$Pb(NO_3)_2(aq) + 2NaOH(aq) \longrightarrow Pb(OH)_2(s) + 2NaNO_3(aq)$$

———————— **Solution** ————————

1. What is unknown?   Mass of $Pb(OH)_2$ in g

2. What is known?   23.9 mL of 0.522 M $Pb(NO_3)_2$; 28.4 mL of 0.762 M NaOH; 1 mol $Pb(NO_3)_2$/1 mol $Pb(OH)_2$; 2 mol NaOH/1 mol $Pb(OH)_2$; 241 g $Pb(OH)_2$/mol $Pb(OH)_2$

3. Calculate the limiting reagent.

To solve this problem, we must first identify the limiting reagent. You should recall that the limiting reagent is found by calculating the number of moles of each reactant present and then comparing them, taking into account the mole ratio in which they combine. The limiting reagent is the one that is totally consumed.

$$23.9 \text{ mL} \times \frac{1 \text{ L}}{1000 \text{ mL}} \times \frac{0.522 \text{ mol Pb(NO}_3)_2}{\text{L}} = 0.0125 \text{ mol Pb(NO}_3)_2$$

$$28.4 \text{ mL} \times \frac{1 \text{ L}}{1000 \text{ mL}} \times \frac{0.762 \text{ mol NaOH}}{\text{L}} = 0.0216 \text{ mol NaOH}$$

Calculate the number of moles of NaOH that must combine with the number of moles of $Pb(NO_3)_2$ present. This is accomplished by considering the mole ratio in which NaOH combines with $Pb(NO_3)_2$ (2 mol $NaOH/1$ mol $Pb(NO_3)2$).

$$0.0125 \text{ mol Pb(NO}_3)_2 \times \frac{2 \text{ mol NaOH}}{1 \text{ mol Pb(NO}_3)_2} = 0.0250 \text{ mol NaOH needed}$$

Because 0.0250 mol NaOH is needed to combine with 0.0125 mol $Pb(NO_3)_2$ and only 0.0216 mol NaOH is present, the limiting reagent is NaOH. Accordingly, the problem is completed using the number of moles of NaOH.

4. Apply the factor-label method and perform all indicated math operations.

$$0.0216 \text{ mol NaOH} \times \frac{1 \text{ mol Pb(OH)}_2}{2 \text{ mol NaOH}} \times \frac{241 \text{ g Pb(OH)}_2}{\text{mol Pb(OH)}_2}$$
$$= \textbf{2.60 g Pb(OH)}_2$$

Our calculation indicates that 2.60 g $Pb(OH)_2$ should precipitate from solution when 23.9 mL of 0.522 M $Pb(NO_3)_2$ combines with 28.4 mL of 0.762 M NaOH.

15.22 What mass of mercury(II) sulfide, HgS, results when 293 mL of 0.217 M $Hg(NO_3)_2$ combines with excess solid ammonium sulfide, $(NH_4)_2S$?

$$Hg(NO_3)_2(aq) + (NH_4)_2S(s) \longrightarrow HgS(s) + 2NH_4NO_3(aq)$$

15.23 What volume of $H_2(g)$ is released at STP conditions when excess Ca metal is mixed with 93.5 mL of 0.311 M H_2SO_4?

$$Ca(s) + H_2SO_4(aq) \longrightarrow CaSO_4(s) + H_2(g)$$

15.24 What mass of CuS precipitates when 33.8 mL of 0.549 M $Cu(NO_3)_2(aq)$ combines with 29.5 mL of 0.779 M Na_2S?

REVIEW EXERCISES

O nce a solute dissolves in a solvent, the properties of the resulting solution differ from those of the pure solute or pure solvent. Solutions exhibit special properties, depending on the concentration of dis-

15.5 PROPERTIES OF SOLUTIONS

solved solute particles. These special properties are called colligative properties.

A **colligative property** is one that is directly related to the number of dissolved solute particles in the solvent. To a large extent, colligative properties are independent of the nature of the solute; they depend only on the concentration of dissolved solute particles. Three colligative properties are considered in this section: (1) vapor pressure lowering, (2) boiling-point elevation, and (3) freezing-point depression. Osmotic pressure is a fourth colligative property, but it will not be considered in this brief discussion.

Osmotic pressure is another colligative property. It concerns the property of solvents to pass through semipermeable membranes.

In our study of colligative properties, we will consider only nonvolatile and nonelectrolyte solutes. A nonvolatile solute does not have an appreciable vapor pressure at room temperature and thus does not evaporate. Nonelectrolytes are compounds that do not dissociate in solution. More complicated relationships are needed to explain the properties of solutions that contain volatile or electrolyte solutes.

In Chap. 13, we discussed the equilibrium that exists between the evaporation of liquid molecules and the condensation of vapor molecules above a liquid in a closed system. A measure of the amount of vapor above the liquid is called the vapor pressure of the liquid. You may want to reread Sec. 13.1 for review.

Vapor Pressure Lowering

At a specific temperature, each liquid has a fixed vapor pressure. In a pure liquid, all of the surface molecules are that of the liquid. Each of the surface molecules has the capacity to break free and enter the vapor phase. However, a solution has surface molecules of both the solute and solvent. The larger the number of surface solute molecules the smaller the number of solvent molecules that can evaporate (Fig. 15.19). Consequently, there is a decrease in the number of surface solvent molecules, which decreases the extent of evaporation and, in turn, decreases the vapor pressure. The vapor pressure of a solution that contains a nonvolatile solute (one that does not evaporate) is always lower than the vapor pressure of the pure solvent. This relationship was proposed by François Raoult (1830–1901) and is now known as **Raoult's law.** Raoult's work led directly to the development of the theory of dissociation of solutes in solution.

Figure 15.19
In a pure solvent all the surface molecules are solvent molecules. In a solution that contains a nonvolatile solute, both solute and solvent molecules are found on the surface. The solute molecules decrease the number of surface solvent molecules, which decreases the number of solvent molecules that can evaporate in a time interval.

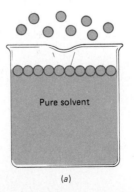

Pure solvent

(a)

Solution with nonvolatile solute

(b)

Boiling Point Elevation

The **boiling point** of a liquid is the temperature at which the vapor pressure of the liquid equals the pressure of the gases above the liquid. The **normal boiling point** is defined as the temperature at which the vapor pressure of a liquid equals one atmosphere. The normal boiling point of water is 100°C.

The addition of nonvolatile solute molecules to a solvent decreases the number of surface solvent molecules, and thus lowers the vapor pressure (Raoult's law). Addition of a nonvolatile solute to water lowers its vapor pressure below 1 atm at 100°C; thus, the temperature of the solution must be increased above 100°C in order to increase the vapor pressure to 1 atm. Accordingly, the boiling point of the solution is raised with the addition of nonvolatile solute particles.

Higher concentrations of solute particles result in higher boiling points. An increase in the boiling point of the solution is directly related to the molality of the solution. **Molality** is defined as the number of moles of solute per kilogram of solvent.

m is the symbol for molality; M is the symbol for molarity.

To find the molality of a solution, first calculate the number of moles of solute and then divide by the mass of the solvent in kilograms.

$$\text{Molality} = \frac{\text{moles solute}}{\text{kilograms solvent}}$$

Example Problem 15.14 illustrates how to calculate the molality of a solution.

───────────── **Example Problem 15.14** ─────────────

Ethylene glycol is a component of antifreeze solutions. A solution is prepared by dissolving 105 g of ethylene glycol, $C_2H_6O_2$, in 649 g of water. What is the molality of the solution?

───────────── **Solution** ─────────────

1. What is unknown?   Molality, m (mol $C_2H_5O_2$/kg H_2O)

2. What is known?   105 g $C_2H_6O_2$; 649 g H_2O; 62.0 g $C_2H_6O_2$/ mol $C_2H_6O_2$

3. Apply the factor-label method to find the number of moles of $C_2H_6O_2$ and the number of kg of H_2O (Fig. 15.20).

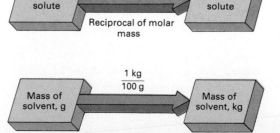

Figure 15.20
To find the molality of a solution, calculate the number of moles of solute and divide by the mass of the solvent in kilograms.

$$\text{mol } C_2H_6O_2 = 105 \text{ g } C_2H_6O_2 \times \frac{1 \text{ mol } C_2H_6O_2}{62.0 \text{ g } C_2H_6O_2} = 1.69 \text{ mol } C_2H_6O_2$$

$$\text{kg } H_2O = 649 \text{ g } H_2O \times \frac{1 \text{ kg}}{1000 \text{ g}} = 0.649 \text{ kg } H_2O$$

4. Calculate the molality.

$$m = \frac{1.69 \text{ mol } C_2H_6O_2}{0.649 \text{ kg } H_2O}$$

$$= \textbf{2.60 } \boldsymbol{m}$$

When 105 g of ethylene glycol is mixed with 649 g of water, the molality of the solution is 2.60 m.

At what temperature would the ethylene glycol solution from Example Problem 15.14 boil? **Boiling-point elevation** is calculated using the following equation:

$$\Delta T_b = K_b m$$

in which ΔT_b is the elevation of the boiling point above the normal boiling point, K_b is the molal boiling-point-elevation constant, and m is the molality of the solution. Table 15.4 lists the boiling-point-elevation constants, K_b, for selected solvents. A **boiling-point-elevation constant** gives the ratio of increase in boiling point per one molal solution. The K_b value for water is 0.512°C/m—a 1.00 m solution (if the solute is nonvolatile and a nonelectrolyte) boils 0.512°C above the normal boiling point of water (100°C). A 2.0 m solution boils 2.0 m × 0.512°C/m, or approximately 1°C, above the normal boiling point. Thus, the increase in boiling point of the solution in Example Problem 15.14 is calculated by multiplying 2.60 m × 0.512°C/m.

$$\Delta T_b = K_b \times m$$

$$\Delta T_b = \frac{0.512°C}{m} \times 2.60 \, m$$

$$= 1.33°C$$

A 2.60 m aqueous solution boils 1.33°C above the normal point, or at 101.33°C (100.00°C + 1.33°C).

At the freezing point, the vapor pressure of the solid and liquid phases of a substance are equal. The addition of a solute to a solvent lowers the vapor pressure of the liquid phase more than it does that of

Freezing-Point Depression

the solid phase. Consequently, a solution freezes at a lower temperature than the pure solvent, because the vapor pressure of the solid phase equals the vapor pressure of the liquid phase at a lower temperature.

TABLE 15.4 MOLAL FREEZING- AND BOILING-POINT CONSTANTS FOR SELECTED SOLVENTS

Solvent	Boiling point, °C	K_b, °C/m	Freezing point, °C	K_f, °C/m
Acetic acid, $HC_2H_3O_2$	118	3.1	16.6	3.9
Benzene, C_6H_6	80	2.5	5.5	5.1
Chloroform, $CHCl_3$	61	3.6	−63.5	4.68
Nitrobenzene, $C_6H_5NO_2$	211	5.2	5.7	8.1
Phenol, C_6H_6O	182	3.6	41	7.4
Water, H_2O	100	0.512	0	1.86

The freezing-point depression is calculated in a similar manner to the boiling-point elevation. We multiply the molality of the solution by a constant, K_f, the **molal freezing-point-depression constant.**

$$\Delta T_f = K_f m$$

Camphor has a K_f value of 40°C/m. A 1 m camphor solution freezes 40°C below its freezing point, 176°C.

Table 15.4 lists K_f values for selected solvents. The K_f value for water is 1.86°C/m. A 1.00 m aqueous solution that contains a nonvolatile, non-electrolyte solute freezes at −1.86°C (ΔT_f = 1.86°Cm × 1.00 m = 1.86°C, and 0.00°C − 1.86°C = −1.86°C). Example Problem 15.15 is an illustration of a freezing-point-depression problem.

──────────── **Example Problem 15.15** ────────────

What is the freezing point of a solution prepared by adding 95.0 g ethylene glycol, $C_2H_6O_2$, to 365 g water?

The structural formula of ethylene glycol is

```
    H   H
    |   |
H — C — C — H
    |   |
   OH  OH
```

Ethylene glycol is a common ingredient in anti-freeze solutions. It is extremely toxic.

──────────── **Solution** ────────────

1. What is unknown? Freezing point of the solution in °C

2. What is known? 95.0 g $C_2H_6O_2$; 365 g H_2O; 62.0 g $C_2H_6O_2$/mol $C_2H_6O_2$; K_f = 1.86°C/m; freezing point (water) = 0.00°C
Depression of the freezing point is a function of the molality of the solution; thus, the molality should be calculated first. The molality is then multiplied by the K_f to obtain the freezing-point depression.

3. Calculate the molality of the solution.

$$m = \frac{\text{mol } C_2H_6O_2}{\text{kg water}}$$

$$\text{mol } C_2H_6O_2 = 95.0 \, \cancel{\text{g } C_2H_6O_2} \times \frac{1 \text{ mol } C_2H_6O_2}{62.0 \, \cancel{\text{g } C_2H_6O_2}} = 1.53 \text{ mol } C_2H_6O_2$$

$$\text{kg } H_2O = 365 \, \cancel{\text{g } H_2O} \times \frac{1 \text{ kg } H_2O}{1000 \, \cancel{\text{g } H_2O}} = 0.365 \text{ kg } H_2O$$

$$m = \frac{1.53 \text{ mol } C_2H_6O_2}{0.365 \text{ kg } H_2O} = 4.20 \, m$$

4. Calculate the freezing point of the solution.

$$\Delta T_f = K_f \times m$$

$$\Delta T_f = \frac{1.86°C}{\cancel{m}} \times 4.20 \, \cancel{m}$$

$$= 7.81°C$$

Our calculation shows that the freezing point of water is lowered by 7.81°C. The freezing point of H_2O is 0.00°C; hence, 7.81°C is subtracted from 0.00°C to give $-7.81°C$ as the freezing point of the solution.

$$\text{Freezing point} = 0.00°C - 7.81°C$$

$$= \mathbf{-7.81°C}$$

A solution prepared by mixing 95.0 g of ethylene glycol with 365 g of water freezes at $-7.81°C$.

15.25 What determines whether a property of a solution is a colligative property?

15.26 Explain why the addition of a nonvolatile, nonelectrolyte solute to water lowers its vapor pressure.

15.27 What are the expected boiling and freezing points of a 2.5 m aqueous solution?

15.28 Calculate the molality of a solution prepared by dissolving 47.1 g CH_4O in 253 g of water.

15.29 What is the boiling point of a solution that contains 211 g $C_2H_6O_2$ (a nonvolatile nonelectrolyte) in 3.09 kg water?

SUMMARY

Solutions are homogeneous mixtures. A solution is composed of a **solute,** the component in smaller amount, dissolved in a **solvent,** the component in larger amount. Solute molecules become incorporated into the structure of the solvent.

Solutions are classified according to the physical states of the solute and solvent. Solid-liquid, liquid-liquid, gas-liquid, and gas-gas solutions are the classes of solutions most frequently encountered.

Solubility is the extent to which a solute dissolves in a solvent. Solutions that have larger quantities of dissolved solutes are called **concentrated solutions.** Those with smaller quantities of dissolved solutes are called **dilute solutions.** If no more solute can dissolve in a solvent without a change in the conditions, the solution is said to be **saturated.** If any quantity of solute less than the amount required to saturate the solution is present, the solution is said to be **unsaturated.**

Three principal factors influence the solubility of a solute: (1) the **nature of solute and solvent,** (2) **temperature,** and (3) **pressure.** Generally, substances that have similar structures and intermolecular forces are more soluble in each other than those that differ; *like dissolves like.* On average, higher temperature results in greater solubility. Pressure affects mainly gaseous solutions. Higher pressures of gases above solvents result in greater gas solubility.

Four factors affect the rate at which a solute dissolves in a solvent: (1) **particle size,** (2) **temperature,** (3) **concentration,** and (4) **stirring.** A decrease in par-

ticle size increases the surface area, and thus increases the rate of dissolution. Higher temperature increases the dissolving rate. The higher the concentration of a solute, the slower the solute dissolves. Stirring a solute increases the rate at which it dissolves by lowering the concentration of surrounding dissolved solute particles.

Solution concentrations are measured in a number of different ways. The concentration units used most frequently in chemistry are (1) **molarity,** moles of solute per liter of solution, (2) **percent by mass,** mass of solute per 100 g of solution, and (3) **molality,** moles of solute per kilogram of solvent.

Solutions that contain dissolved ions are called **electrolyte solutions;** they conduct an electric current. If a large percent of the solute ionizes when dissolved, the solute is classified as a **strong electrolyte—** it conducts a large electric current. Solutes that form few ions in solution are called **weak electrolytes,** and **nonelectrolytes** are those solutes which produce no or very few ions in solution.

Solutions exhibit certain special properties, which depend on the concentration of dissolved solute particles. These special properties are called **colligative properties.** Three of the colligative properties are (1) vapor pressure lowering, (2) boiling-point elevation, and (3) freezing-point depression. As the molal concentration of the solute increases, the vapor pressure of the solution decreases, its boiling point increases, and its freezing point decreases.

KEY TERMS

colligative property
concentrated
concentration unit
dilute
dilution

dissolution
dissolving
hydration energy
immiscible
lattice energy

miscible
net ionic equation
nonelectrolyte
overall ionic equation
saturated solution

solubility
solute
solution
solution stoichiometry
solvent

strong electrolyte
unsaturated solution
weak electrolyte

EXERCISES*

15.30 Define each of the following terms: solution, solute, solvent, dissolving, dissolution, miscible, immiscible, solubility, hydration energy, lattice energy, dilute, concentrated, saturated solution, unsaturated solution, concentration

unit, dilution, strong electrolyte, solution stoichiometry, weak electrolyte, nonelectrolyte, overall ionic equation, net ionic equation, colligative property.

*For exercise numbers printed in color, answers can be found at the back of the book.

Solutions

15.31 Identify the solute and solvent in each of the following solutions:

(a) 1 L water and 1 g NaCl

(b) 1 L alcohol and 50 mL water

(c) 1 L alcohol and 1 L water

(d) 1 L water and 1 mL $O_2(g)$

15.32 (a) How could the KCl in an aqueous KCl solution be recovered as KCl(s)? (b) Could the same method be used to isolate alcohol from an alcohol and water solution?

15.33 What is the difference between the solubility of a solute in a solvent and the rate at which the solute dissolves?

15.34 How are immiscible liquids distinguished from miscible liquids?

15.35 How could you test a solution to decide whether the solution is saturated or unsaturated?

15.36 Classify each of the following solutions according to the physical state of the solute and solvent: (a) vinegar, (b) brass, (c) air, (d) coffee, (e) sugar water.

15.37 Predict in which of the following pairs of liquids, sodium chloride, NaCl, an ionic solid, is more soluble:

(a) H_2O or CCl_4

(b) CH_3OH or $CH_3CH_2CH_2CH_2CH_3$

(c) $CH_3CH_2CH_2CH_2CH_3$ or CH_3OCH_3

Explain each of your predictions.

15.38 Predict in which of the following pairs of liquids iodine, $I_2(s)$, a nonpolar molecular solid, is most soluble:

(a) H_2O or CCl_4

(b) CH_3CH_2OH or $CH_3CH_2CH_2CH_2CH_3$

(c) CS_2 or H_2O

Explain each of your predictions.

Percent by Mass

15.39 How are each of the following % m/m solutions prepared:

(a) 399 g of 2.95% $NaNO_2(aq)$

(b) 43.8 g of 1.31% LiOH(aq)

(c) 3.77 kg of 0.0296% $NH_4ClO_3(aq)$

15.40 Calculate the mass of concentrated HCl solution (37.0% m/m) that contains the following masses of HCl: (a) 4.44 g HCl, (b) 542 g HCl, (c) 0.188 kg HCl, (d) 3.91 mg HCl.

15.41 Concentrated ammonia, NH_3, is sold as 29% m/m $NH_3(aq)$. Its density is 0.90 g/cm^3. What mass of NH_3 is contained in each of the following volumes of concentrated ammonia solutions: (a) 11 L, (b) 63 mL, (c) 0.0040 mL?

15.42 What is the maximum total mass of 2.10% m/m KI solution that could be prepared from 5.05 g KI(s)?

15.43 Calculate the mass of water in 167.5 g of 1.304% m/m Na_2SO_4 solution.

15.44 A solution is prepared by dissolving 6.53 g of solute in 284 g of water. What is the concentration of the solution in % m/m?

15.45 (a) Initially, 85.0 g of a 9.90% m/m ammonium acetate solution is placed in a beaker. What is the concentration of the solution after 145.0 g of water is added to dilute the ammonium acetate? (b) What mass of ammonium acetate solid must be added to the diluted solution to change the concentration back to the original concentration, 9.90%?

Molarity

15.46 Explain how each of the following solutions is prepared:

(a) 50.0 mL of 0.232 M $Mg(NO_3)_2$

(b) 210.0 mL of 0.1919 M NH_3

(c) 5.66 L of 2.08 M $C_6H_{12}O_6$

(d) 9.11×10^6 mL of 5.00 M H_3PO_3

15.47 What are the molarities of the solutions prepared by dissolving the following amounts of solute in enough water to give 180.0 mL total volume:

(a) 1.441 mol $(NH_4)_2SO_4$

(b) 2.662 mmol $Cu(NO_3)_2$

(c) 12.33 g H_2SO_4

(d) 8.331 mg HBr

15.48 What is the molarity of each of the following solutions?

(a) 5.42 g C_2H_6O in 82.8 mL of solution

(b) 0.994 g Na_2SO_4 in 3.10 L of solution

(c) 834.3 mg K_2CO_3 in 41.75 mL of solution

(d) 8.90 kg NH_4Cl in 526 L of solution

15.49 Calculate the molarity of each of the following solutions:
- (a) 70% m/m HNO_3; density = 1.42 g/cm³
- (b) 36% acetic acid, $C_2H_4O_2$; density = 1.045 g/cm³

15.50 To what total volume should 1.50 L of 15.9 M HNO_3 be diluted in order to produce the following concentrations: (a) 13.5 M HNO_3, (b) 2.50 M HNO_3, (c) 0.00349 M HNO_3?

15.51 Calculate the number of moles of solute particles in each of the following solutions:
- (a) 287 mL of 0.444 M RbOH
- (b) 17.3 mL of 0.0114 M HI
- (c) 497.1 L of 4.290 M KNO_2

15.52 (a) What is the molar concentration of a solution that contains 104 g $AgNO_3$ per liter of solution? (b) What volume of this solution contains 0.837 mol $AgNO_3$? (c) What volume of this solution contains 1.00×10^{23} $Ag^+(aq)$ ions? (d) What is the molar concentration of the solution after it is diluted to 2.45 L?

15.53 A 27% m/m H_2SO_4 solution has a density of 1.2 g/cm³. Calculate the molar concentration of the solution.

15.54 (a) How can 30.4 mL of 0.855 M $CuCl_2$ be diluted to 0.177 M $CuCl_2$? (b) How can 4.29 L of 1.50 M $HClO_4$ be diluted to 0.150 M $HClO_4$?

15.55 To what total volume would 182.0 mL of 0.5000 M HI solution have to be diluted in order to produce a 0.2949 M HI solution?

15.56 (a) What volume of 2.85 M NaOH is required to produce 81.7 mL of 0.400 M NaOH? (b) Explain how this solution is prepared.

15.57 (a) What are the molar concentrations of K^+ and SO_4^{2-} when 44.9 mL of 0.188 M K_2SO_4 is mixed with 28.4 mL of 0.355 M K_2SO_4?

15.58 If 25.0 mL of water is added to 59.6 mL of 1.04 M $HC_2H_3O_2$, what is the resulting molar concentration of $HC_2H_3O_2$? Assume the volumes are additive.

Electrolytes

15.59 Describe the expected light intensity (bright, dim, etc.) observed when the electrodes of a conductivity apparatus are placed in each of the following solutions: (a) 0.1 M NaCl(aq), (b) 0.1 M HF(aq), (c) pure water, (d) 0.1 M ethanol, $C_2H_6O(aq)$, (e) 0.1 M $Ca(OH)_2(aq)$.

15.60 Hydrofluoric acid, HF(aq), is a weak electrolyte. Compare the relative amounts of unionized HF(aq) to that of dissolved $F^-(aq)$ and $H^+(aq)$.

15.61 Write ionic equations that indicate what happens when the following strong electrolytes are dissolved in water: (a) NaBr(s), (b) $Na_2SO_4(s)$, (c) $Cu(NO_3)_2(s)$, (d) $MgCl_2(s)$, (e) $Cs_2CrO_4(s)$.

15.62 Write balanced overall and net ionic equations for each of the following unbalanced equations:
- (a) $RbCl(aq) + AgNO_3(aq) \longrightarrow$ $RbNO_3(aq) + AgCl(s)$
- (b) $K_3PO_4(aq) + NiCl_2(aq) \longrightarrow$ $Ni_3(PO_4)_2(s) + KCl(aq)$
- (c) $(NH_4)_2S(aq) + FeSO_4(aq) \longrightarrow$ $(NH_4)_2SO_4(aq) + FeS(s)$
- (d) $H_3PO_4(aq) + NaOH(aq) \longrightarrow$ $H_2O(l) + Na_2HPO_4(aq)$
- (e) $Hg_2(NO_3)_2(aq) + CaBr_2(aq) \longrightarrow$ $Hg_2Br_2(s) + Ca(NO_3)_2(aq)$

15.63 Write balanced net ionic equations for the reaction (if any) of each of the following:
- (a) $Fe(NO_3)_2(aq) + Na_3PO_4(aq) \longrightarrow$
- (b) $NH_4C_2H_3O_2(aq) + Zn(NO_3)_2(aq) \longrightarrow$
- (c) $Na_2CO_3(aq) + FeBr_2(aq) \longrightarrow$
- (d) Magnesium iodide(aq) + silver nitrate(aq)$\longrightarrow$
- (e) Calcium acetate(aq) + cesium sulfide(aq)$\longrightarrow$

15.64 Complete and balance the overall and net ionic equations for each of the following:
- (a) $H_2SO_4(aq) + K_2CO_3(aq) \longrightarrow$
- (b) $HBr(aq) + (NH_4)_2S(aq) \longrightarrow$
- (c) $RbOH(s) + H_3PO_4(aq) \longrightarrow$
- (d) $Zn(s) + Cu(NO_3)_2(aq) \longrightarrow$
- (e) $Al_2(SO_4)_3(aq) + Ca(C_2H_3O_2)_2(aq) \longrightarrow$

Solution Stoichiometry

15.65 Consider the following aqueous reaction:

$$MgCl_2(aq) + Na_2CO_3(aq) \longrightarrow MgCO_3(s) + 2NaCl(aq)$$

(a) What volume of 0.349 M Na_2CO_3 is required to combine exactly with 2.81 g $MgCl_2$? (b) What mass of $MgCO_3$ precipitates when 30.7 mL of 0.997 M Na_2CO_3 combines with excess $MgCl_2$?

15.66 Consider the following reaction:

$$Pb(NO_3)_2(aq) + K_2SO_4(aq) \longrightarrow PbSO_4(s) + 2KNO_3(aq)$$

(a) What mass of $PbSO_4$ precipitates when 40.1 mL of 0.806 M $Pb(NO_3)_2$ combines with excess K_2SO_4? (b) What mass of KNO_3 results when 22.9 mL of 0.927 M $Pb(NO_3)_2$ combines with 18.4 mL of 1.07 M K_2SO_4?

15.67 Consider the following reaction:

$$Al(OH)_3(s) + 3HCl(aq) \longrightarrow 3H_2O(l) + AlCl_3(aq)$$

(a) What volume of 1.25 M HCl is required to combine exactly with 10.7 g $Al(OH)_3$? (b) What mass of $AlCl_3$ results when 31.8 mL of 0.745 M HCl combines with excess $Al(OH)_3$?

15.68 When sulfuric acid, $H_2SO_4(aq)$, is neutralized by potassium hydroxide, $KOH(aq)$, potassium sulfate, $K_2SO_4(aq)$, and water result. (a) Write a balanced equation for this neutralization reaction. (b) What volume of 0.310 M KOH is required to exactly neutralize 16.7 mL of 0.215 M H_2SO_4? (c) What mass of potassium sulfate results when 812 mL of 4.23 M H_2SO_4 is completely neutralized by KOH? (d) To produce 8.74 g K_2SO_4, what volumes of 0.175 M H_2SO_4 and 0.256 M KOH are required?

15.69 Oxalic acid, $H_2C_2O_4(s)$, combines with two moles of sodium hydroxide solution, $NaOH(aq)$, to produce sodium oxalate, $Na_2C_2O_4(aq)$, and water. (a) Write a balanced equation for the reaction. (b) What mass of sodium oxalate results when 5.07 g of oxalic acid is mixed with 40.5 mL of 2.50 M NaOH? (c) What mass of sodium oxalate results when 321 g of oxalic acid combines with 2.65 L of 3.01 M NaOH?

15.70 Consider the following reaction:

$$Mg(s) + H_2SO_4(aq) \longrightarrow MgSO_4(aq) + H_2(g)$$

(a) What volume of H_2 results at STP when 94.1 mL of 2.50 M H_2SO_4 combines with excess Mg? (b) What volume of H_2 results at 746 torr and 37°C when 65.8 mL of 4.33 M H_2SO_4 combines with excess Mg? (c) What volume of

H_2 at STP results when 12.8 g Mg combines with 341 mL of 1.55 M H_2SO_4?

15.71 When hydrochloric acid, $HCl(aq)$, combines with sodium carbonate, Na_2CO_3, it produces aqueous sodium chloride, carbon dioxide, and water. (a) Write the balanced equation for this reaction. (b) What mass of Na_2CO_3 combines exactly with 91.7 mL of 0.650 M $HCl(aq)$? (c) What volume of CO_2 at 810 torr and 30.0°C is released when excess Na_2CO_3 combines with 741 mL of 3.15 M HCl? (d) What volume of CO_2 results at STP when 4.59 g Na_2CO_3 combines with 25.0 mL of 0.862 M HCl? (e) What mass of excess reactant is present after the reaction in part d?

Molality

15.72 Calculate the molality of each of the following solutions:
(a) 0.191 g KNO_2 in 453 g water
(b) 12.1 g CaI_2 in 3.09 kg water
(c) 0.877 mol Na_2SO_3 in 612 g water
(d) 2.10 g NaI in 228 mL water
(The density of water is 1.00 g/cm^3.)

15.73 Calculate the molality of each of the following solutions:
(a) 1.11% m/m $C_6H_{12}O_6(aq)$
(b) 6.2% m/m $HBr(aq)$
(c) 21.9% m/m $HClO_3(aq)$

15.74 Explain how each of the following solutions is prepared:
(a) 0.00774 m $Zn(NO_3)_2$ in 305 g H_2O
(b) 4.58 m $C_{12}H_{22}O_{11}$ in 23.9 kg H_2O

15.75 (a) Calculate the molality of 0.762 g $I_2(s)$ dissolved in 211.7 g CCl_4. (b) What is the percent by mass of I_2 in this solution?

Properties of Solutions

15.76 Calculate the boiling point of aqueous solutions with each of the following molalities (assume that all solutes are nonvolatile and nonelectrolytes): (a) 0.91 m, (b) 6.1 m, (c) 2.77 m.

15.77 What are the freezing points of the aqueous solutions in Exercise 15.76?

15.78 Explain specifically why the boiling point of a solution containing a nonvolatile solute is higher than that of the pure solvent.

15.79 Using Table 15.4, calculate the elevation of the boiling points of 2.9 m solutions in: (a) water, (b) acetic acid, (c) benzene, (d) nitrobenzene, (e) phenol, (f) chloroform.

15.80 A solution is found to have a higher vapor pressure than that of the pure solvent. Write an explanation to account for this seemingly contradictory behavior.

15.81 What are the boiling points of solutions with the following molalities in which benzene is the solvent: (a) 1.9 m, (b) 0.66 m, (c) 8.4 m? (Use data from Table 15.4.)

15.82 What is the freezing point of each benzene solution in Exercise 15.81?

15.83 (a) How many grams of ethylene glycol, $C_2H_6O_2$, are required to lower the freezing point of 85.0 g water to $-3.06°C$? (b) At what temperature will this solution boil?

15.84 (a) What is the molality of an aqueous solution that boils at 100.95°C? (b) If this solution contains 0.344 mol of solute, what is the total mass of water?

15.85 What are the freezing and boiling points of a solution prepared by mixing 19.4 g of ethylene glycol, $C_2H_6O_2$, with 100.0 g of water?

15.86 At what temperature will a phenol solution prepared by mixing 0.902 mol of a nonvolatile solute and 216 g of phenol freeze? (Use data in Table 15.4.)

Additional Exercises

15.87 A solution of perchloric acid, $HClO_4$, is 11.7 M and has a density of 1.67 g/cm³. (a) Calculate the percent by mass of 11.7 M $HClO_4$. (b) To what volume would you dilute 255 mL of 11.7 M $HClO_4$ to produce a 5.95 M $HClO_4$ solution? (c) If 81.2 mL of 11.7 M $HClO_4$ is diluted to 10.6 L, what is the new molar concentration?

15.88 What effect does adding salt to the water have on the time required to boil foods?

15.89 (a) Calculate the percent by mass of a 6.26 M acetic acid ($HC_2H_3O_2$) aqueous solution. The density of this solution is 1.045 g/cm³. (b) Explain how 50.0 mL of this solution is diluted to 1.76 M acetic acid? (c) What is the resulting molar concentration of a solution prepared by mixing 37.4 mL of water with 25.0 mL of 6.26 M acetic acid? Assume the volumes are additive.

15.90 When discussing colligative properties, an assumption was made that the solute was not an electrolyte. If NaCl(s) is added to water, calcu-

late the freezing and boiling point of a 0.375 m NaCl solution.

15.91 Calculate the mass of ethylene glycol, $C_2H_6O_2$, that should be added to 15 kg water in the radiator of an automobile to prevent the water from freezing at $-10.0°F$.

15.92 Two chloride solutions are mixed; 75.0 mL of 0.225 M KCl(aq) and 55.0 mL of 0.225 M $CaCl_2(aq)$. What is the resultant molar concentration of each of the three ions in the solution—K^+, Ca^{2+}, and Cl^-?

15.93 The following three nitrate solutions are mixed: 15.4 mL of 0.884 M KNO_3, 44.9 mL of 0.533 M $Mg(NO_3)_2$, and 39.7 mL of 0.363 M $Cu(NO_3)_2$. Assume the volumes are additive. (a) What is the resulting molar concentration of the nitrate ions? (b) If this mixture is diluted to 248 mL, what is the molar concentration of the nitrate ions? (c) If 36.1 mL of water from the original solution is evaporated, what is the molar concentration of the nitrate ions?

15.94 Freezing-point depressions are used by chemists to calculate the molecular mass of substances. Calculate the molecular mass of a solute if it is known that dissolving 0.25 g solute in 18 g water lowers the freezing point of water by 0.20°C.

15.95 Concentrated sulfuric acid, H_2SO_4, is sold as a 96% m/m H_2SO_4 aqueous solution. The density of the solution is 1.84 g/cm³. Calculate the (a) molarity and (b) molality of the solution. (c) What quantity of water is required to dilute 55 g of the concentrated solution to 0.350 M? (d) What mass of 0.100 M H_2SO_4 reacts with exactly 189 mL 0.666 M NaOH(aq) to produce sodium sulfate and water?

15.96 When aqueous ammonium sulfide, $(NH_4)_2S(aq)$, is mixed with nickel(II) chlorate, $Ni(ClO_3)_2$, nickel(II) sulfide precipitates from solution. (a) Write a balanced equation for the reaction. (b) What mass of nickel(II) sulfide results when 61.4 mL of 0.100 M ammonium sulfide combines with 59.6 mL of 0.100 M nickel(II) chlorate? (c) What volume of 0.811 M ammonium sulfide exactly combines with 423 mL of 0.100 M nickel(II) chlorate?

15.97 When solid sodium sulfite, Na_2SO_3, is added to hydrochloric acid, HCl(aq), aqueous sodium chloride, sulfur dioxide gas, and water are produced. *(a)* Write the balanced equation for the reaction. *(b)* What volume of 6.00 *M* HCl is required to combine with excess sodium sulfite to produce 44.7 mL of sulfur dioxide gas at 755 torr and 21.0°C? *(c)* What volume of sulfur dioxide results at STP if 312 g of sodium sulfite combines with 853 mL of 4.19 *M* HCl?

15.98 A quantity of sodium metal is added to 85.0 mL H_2O and 43.5 mL of H_2 gas is produced at STP. What is the molar concentration of the resulting sodium hydroxide solution?

CHAPTER

— Sixteen —

Reaction Rates and Chemical Equilibrium

— STUDY GUIDELINES —

After completing Chapter 16, you should be able to

1. Explain what is meant by reaction rate

2. List the three main principles of the collision theory

3. Discuss the factors that determine whether or not a molecular collision is effective

4. List four factors that influence the rates of chemical reactions

5. State the relationships between reaction rate and reactant concentration and reaction rate and temperature

6. Explain the relationships between reaction rate and concentration and reaction rate and temperature in terms of the collision theory

7. Identify and explain the meaning of the rate-determining step for a given reaction mechanism

8. State the one condition, in terms of reaction rates, that is required for the establishment of a chemical equilibrium

9. State whether the reactants or products are favored in an equilibrium given the value of the equilibrium constant

10. State and apply Le Chatelier's principle

11. Describe effects of changing concentration, pressure, temperature, and addition of a catalyst on a chemical equilibrium

---

Chemical kinetics is the study of the rates of chemical reactions. Reaction rate is the speed of a chemical reaction, i.e., how fast the products are formed from the reactants. In the study of chemical kinetics, chemists attempt to measure accurately the rates of reactions under different conditions, and then try to account theoretically for the observed rates by proposing a reaction mechanism, which is a detailed stepwise description of the pathway from reactants to products.

16.1 RATES OF CHEMICAL REACTION

Some reactions are nearly instantaneous—the reactants totally change to products on contact. Explosions are good examples of instantaneous reactions. Other reactions proceed at such a slow rate that years or centuries elapse before a small percent of the reactants are converted to products. Most chemical reactions proceed at rates somewhere in between these two extremes.

Some reactions that occur in the oceans and in the earth's crust require thousands, and in some cases, millions of years to produce significant amounts of products because of their infinitesimal rates.

Rates of chemical reactions are measured by finding either the decrease in reactant concentration or the increase in product concentration over a specific time interval.

$$\text{Reaction rate} = \frac{\text{change in concentration}}{\text{change in time}}$$

In a reaction with a high reaction rate, the time interval over which the reaction takes place is relatively short. In a reaction proceeding at a slower rate, more time is required for the disappearance of the same reactant concentration than in a reaction that proceeds at a higher rate. Hence, rates of reactions are inversely related to the time required for the reaction to occur.

Rates of chemical reactions are explained theoretically by the **collision theory.** The basic premise of the collision theory is that in order for two substances to react, the reactant particles must collide with each other. Once they collide, there are two possibilities. Either the bonds in the reactant molecules are broken and the bonds in the product are formed, or the particles merely bump into each other and no new bonds form. If a collision occurs that results in the formation of the products,

Collision Theory

Figure 16.1
(a) In an effective collision, molecules of A_2 and X_2 collide in such a manner that their bonds are broken, and the bonds of the product, AX, are formed. *(b)* In an ineffective collision, no bonds are broken, and molecules of A_2 and X_2 have not changed after colliding.

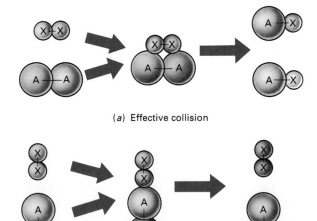

(a) Effective collision

(b) Ineffective collision

the collision is called an **effective collision,** and if the collision does not yield the products, the collision is called an **ineffective collision.**

For simplicity, let's consider the hypothetical reaction of two diatomic gas molecules, $A_2(g)$ and $X_2(g)$. In Fig. 16.1a, one A_2 molecule collides with an X_2 molecule, producing two molecules of AX; here an effective collision has occurred.

$$A_2(g) + X_2(g) \longrightarrow 2AX(g)$$

In Fig. 16.1b, A_2 and X_2 collide, but after the collision the reactant molecules remain, and no product is produced. This is an ineffective collision.

Two factors primarily determine whether a collision is effective: (1) energy and (2) orientation. When two particles collide, they require sufficient energy to break their bonds. For each reaction, a minimum energy of collision must be attained, below which an effective collision cannot occur.

If the two colliding particles have sufficient energy to react, they still must collide with the proper orientation. **Orientation** refers to the alignment of the molecules as they collide. Figure 16.2 illustrates three different collision orientations. In Fig. 16.2a, A_2 and X_2 collide so that each A atom contacts an X atom; consequently, if they collide with the proper amount of energy, they will form a molecule of AX. In Fig. 16.2b and Fig. 16.2c, the orientation of the particles does not bring both A and X atoms together; therefore, an effective collision cannot occur, even with the proper amount of energy.

Rates of chemical reactions are directly linked to the frequency of effective collisions. The greater the **frequency of effective collisions,** the faster the reaction proceeds. If only a small percent of the collisions are effective, the reaction rate is low.

An effective collision occurs when the two colliding particles have the proper energy and orientation.

A higher frequency of effective collisions results in a greater reaction rate.

16.1 What topics are studied in chemical kinetics?

16.2 Explain how the rate of a chemical reaction is measured.

16.3 List and explain the two principal factors that determine whether a molecular collision is effective or ineffective.

16.4 What is true about a reaction that has a high frequency of effective collisions?

REVIEW EXERCISES

Figure 16.2
(a) Proper orientation of colliding molecules to form the products. *(b)* and *(c)* Molecular orientations that do not produce the final products.

F our principal factors influence the rates of chemical reactions: (1) the nature of the reactants, (2) concentration, (3) temperature, and (4) catalysts. We will discuss each of these factors.

For a constant set of conditions, rates of different chemical reactions are the result of different molecular properties of the reactants. For example, if $F_2(g)$ is mixed with $H_2(g)$ at 25°C, an immediate violent reaction occurs. In contrast, if $Cl_2(g)$ is combined with $H_2(g)$ at 25°C, the rate of reaction is significantly slower, and at lower temperatures, Cl_2 does not combine with H_2 at all.

The nature of the reactant molecules determines whether or not a reaction occurs. As we have seen, reactants must collide before they can react. But once the reacting species collide, the interactions of the electrons in chemical bonds determine whether or not the reaction takes place and, if so, what the rate will be. For example, two oppositely charged ions combine immediately because they rapidly attract each other and no bonds are broken. Hydrogen ions immediately combine with hydroxide ions to form water molecules:

$$H^+(aq) + OH^-(aq) \longrightarrow H_2O(l)$$

On the other hand, the decomposition of molecules with strong bonds proceeds at a slower rate because of the energy required to break the bonds.

Rates of most chemical reactions are directly related to the concentration of the reactants—an increase in reactant concentration produces an increase in the rate of the reaction.

When $H_2(g)$ combines with $I_2(g)$ to produce $HI(g)$,

$$H_2(g) + I_2(g) \longrightarrow 2HI(g)$$

an increase in the concentration of either H_2 or I_2 increases the rate of reaction. For example, if the H_2 concentration is doubled, the rate of the reaction doubles. A similar increase is observed when the I_2 concentration is increased.

Substances that burn slowly in air (air contains 20% O_2) sometimes explode when burned in pure oxygen; the increased oxygen concentration speeds up the reaction.

Collision theory explains the effect of concentration changes on reaction rates. Assuming that two reactants are present, an increase in the number of molecules of either reactant increases the overall number of collisions that occur in a given time interval. Earlier we mentioned that reaction rates depend on the frequency of effective collisions. Increased reactant concentration, therefore, increases the reaction rate by increasing the frequency of effective collisions.

16.2 FACTORS THAT INFLUENCE REACTION RATES

Nature of the Reactants

For reactions in which two or more reactants in different physical states combine, the degree to which the reactants come in contact must be considered. Normally, the larger the surface area of reactants in contact, the greater the reaction rate.

Reactants with strong covalent bonds are generally less reactive than those with weaker bonds.

Concentration

As the temperature of a reaction mixture is increased, the rate increases. In chemistry laboratories, bunsen burners and hot plates are used to heat substances so that they will react faster. At home, foods heated to higher temperatures cook faster than those cooked at lower temperatures.

Temperature

Temperature is directly related to the average kinetic energy of molecules. At higher temperatures, the reacting molecules move faster; hence, they collide more frequently and with greater energy. Both of these factors increase the frequency of effective collisions.

Let's consider more closely the energy requirements for chemical reactions. In Sec. 10.5, we classified chemical reactions as either exothermic or endothermic. An **exothermic** reaction is one with a net release of heat, and an **endothermic** reaction is one in which heat is absorbed. Graphs of the energy relationships in exothermic and endothermic reactions are shown in Fig. 16.3.

At a fixed temperature, molecules have a range of velocities. Some molecules move rapidly and some slowly; however, the largest percent move at velocities near the average velocity. Molecular velocities are related to the kinetic energies of the molecules.

In an exothermic reaction, the total energy stored in the reactants is greater than the energy in the products. The quantity of energy stored in the reactants or products is related to a property called **enthalpy.** Consequently, in exothermic reactions the enthalpy H of the reactants is greater than the enthalpy of the products. In endothermic reactions, the opposite is true; the enthalpy of the products is greater than the enthalpy of the reactants.

The difference between the enthalpy of the products and the enthalpy of the reactants is the enthalpy of reaction ΔH.

$$\Delta H = H_{\text{products}} - H_{\text{reactants}}$$

Every chemical reaction has a characteristic enthalpy of reaction.

Figure 16.3
(a) In an exothermic reaction, the enthalpy of the reactants (H_R) is greater than the enthalpy of the products (H_P). Exothermic reactions release heat to the surroundings.
(b) In an endothermic reaction, the enthalpy of the reactants (H_R) is less than the enthalpy of the products (H_P). Endothermic reactions absorb heat from the surroundings.

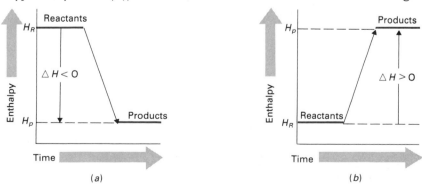

Figure 16.4
In an exothermic reaction, initially the activation energy, E_a, energy is absorbed to produce the transition state. The transition state can break apart, forming the products and releasing the activation energy, E_a, and the enthalpy of reaction, ΔH.

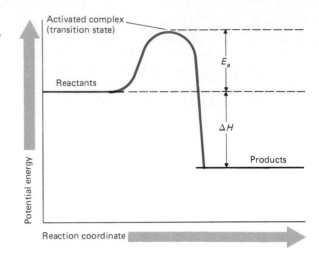

As a reaction proceeds from reactants to products, it follows a definite energy path. Figure 16.4 illustrates the energy pathway taken in an exothermic reaction. Even though the reaction is exothermic, energy is initially required to start the reaction. The minimum quantity of energy needed to get to the top of the energy "hill" is called the **activation energy** E_a. In all reactions, the amount of energy equal to the activation energy must be present for the reaction to occur.

Chemical examples of activation include (1) striking a match, (2) lighting a bunsen burner, and (3) sparking a mixture of H_2 and O_2. In each example, a small input of energy (activation energy) causes a self-sustaining reaction to occur.

In terms of the collision theory, as the reactant particles approach each other, their energy increases until they reach the top of the energy "hill" because of mutual repulsion. The chemical species produced at the energy peak is called the **transition state** or **activated complex.** At this point, the reactant molecules have interacted, producing an intermediate species that has the proper orientation and energy to break apart into either the products or reactants (Fig. 16.5). In other words, the activation energy of a reaction is the amount of energy required to produce the activated complex (transition state).

A **catalyst** is a substance that lowers the activation energy of a chemical reaction (Fig. 16.6). At a lower activation energy, a greater percent of the reacting particles have sufficient energy to produce the activated complex (to have an effective collision). Accordingly, a lower activation energy increases the reaction rate. A catalyst is not consumed during the reaction, and it can be recovered unchanged after the reaction.

Catalysts are grouped into two categories: (1) homogeneous and (2) heterogeneous. A homogeneous catalyst is in the same physical state

Activation energy depends on the nature of the reactants. Concentration and temperature have no effect on the activation energy of a reaction.

An activated complex is a combination of the reacting molecules in which the bonds of the reactants are stretched and almost broken, and the bonds of the products are partially formed.

Catalysts

Figure 16.5
To form the transition state, the reacting molecules must collide in such a way that the bonds in the reactants are partially broken and the bonds in the products are partially formed. Once the transition state forms, it can become either the reactants or the products.

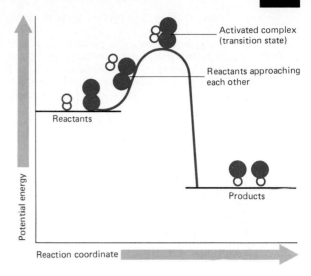

as the reactants, while a heterogeneous catalyst is in a different physical state than the reactants.

Most chemical reactions follow an ordered series of steps called a **reaction mechanism.** To illustrate, let us consider the reaction mechanism for the combination reaction of hydrogen and iodine. $H_2(g)$ and $I_2(g)$ do not simply collide to produce $HI(g)$, as implied by the equation.

Reaction Mechanisms

Reaction mechanisms can only be determined experimentally.

$$H_2(g) + I_2(g) \longrightarrow 2HI(g)$$

Instead, the reaction takes place in two steps: (1) I_2 dissociates to two I atoms, and (2) both I atoms collide with H_2 at the same time to produce two HI molecules. These steps are diagrammed in Fig. 16.7.

Figure 16.6
A catalyst is a substance that lowers the activation energy, E_a, of a chemical reaction. With a lower activation energy, the reaction proceeds at a faster rate than at a higher activation energy.

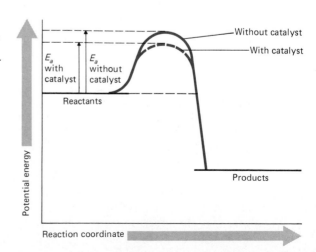

CHEM TOPIC: ENZYMES—BIOCHEMICAL CATALYSTS

Enzymes are proteins that catalyze virtually all important biochemical reactions and are responsible for most of the activities that occur in living things. Scientists have identified approximately 2000 enzymes in living things. Enzymes lower the activation energies of reactions that occur in cells and can increase the rates of reactions by as much as one million times. Without enzymes, living things, as we know them, could not exist.

   Physicians use enzymes to both treat and diagnose diseases. For example, if a person has a heart attack (myocardial infarction), the attack is verified by measuring the concentrations of various blood enzymes. From similar tests the severity of the heart attack can also be ascertained. Other diseases that are monitored from laboratory analyses of enzymes include viral hepatitis, monocytic leukemia, prostate cancer, and many others. Routine laboratory tests such as blood sugar concentration and blood urea nitrogen (BUN test) are conducted using enzymes. Enzymes may also be used to treat severe burns, congestion in the trachea and lungs, and blood clots.

   Many enzymes are also used in the food industry. Papain is the enzyme that tenderizes meats; amylases are used to increase the sugar content in syrups and bakery goods; and rennin is used in cheese production to change milk to cheese.

Step 1: $I_2(g) \rightleftharpoons 2I(g)$ (Fast)

Step 2: $2I(g) + H_2(g) \longrightarrow 2HI(g)$ (Slow)

Overall: $H_2(g) + I_2(g) \longrightarrow 2HI(g)$

The first step, breaking of the I_2 bond, occurs more rapidly than the second step. In Step 2, both I atoms collide with the H_2 molecule simultaneously. Collisions that involve three particles are slow, because the

An overall equation for a reaction is analogous to a description of the starting place and destination of a trip. The reaction mechanism is the actual route taken.

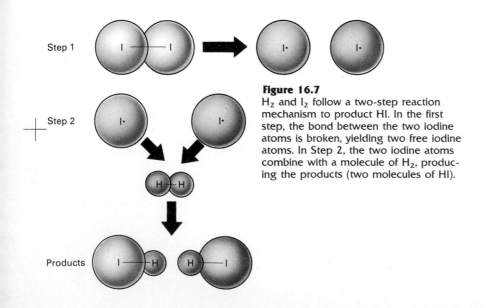

Figure 16.7
H_2 and I_2 follow a two-step reaction mechanism to product HI. In the first step, the bond between the two iodine atoms is broken, yielding two free iodine atoms. In Step 2, the two iodine atoms combine with a molecule of H_2, producing the products (two molecules of HI).

simultaneous collision of three particles with the proper energy and orientation is a rather rare event.

The slowest step in a reaction mechanism is classified as the **rate-determining step** or **rate-limiting step,** because it determines the maximum rate of the reaction. Production of HI(g) depends principally on the second step. No matter how fast the I atoms are produced, HI formation depends on the rate of the slow second step. We might say that a "bottleneck" exists at this step.

An analogy to the rate-determining step is the speed at which one proceeds down a single-lane highway. Your car might be capable of traveling at very high speeds; however, your speed is controlled by the slowest-moving vehicle in front of you.

In Step 1, individual I atoms are formed, which are then consumed in the second step. Chemical species in the reaction mechanism but not in the overall reaction are called **reaction intermediates.** Most intermediates are reactive and do not generally occur by themselves; such is the case of isolated iodine atoms.

16.5 List the principal factors that influence the rates of chemical reactions.

16.6 Describe what happens to the rate of a reaction when each of the following changes is made: *(a)* Concentrations of the reactants are increased; *(b)* a catalyst is added; *(c)* temperature is increased.

16.7 *(a)* What is the activation energy of a reaction? *(b)* Will an exothermic or endothermic reaction have a larger activation energy on average? Fully explain your answer.

16.8 *(a)* What is the function of a chemical catalyst? *(b)* What are the two general types of catalysts?

16.9 Consider the proposed mechanism for the decomposition of ozone: $2O_3 \rightarrow 3O_2$

$$\text{Step 1:} \qquad O_3 \rightleftharpoons O_2 + O \text{ (Slow)}$$

$$\text{Step 2:} \quad O + O_3 \longrightarrow 2O_2 \quad \text{(Fast)}$$

(a) Which step is the rate-determining step? Explain. *(b)* What is the reaction intermediate in the mechanism?

REVIEW EXERCISES

16.3 CHEMICAL EQUILIBRIUM SYSTEMS

To begin our discussion of chemical equilibrium systems, let us reconsider the physical equilibrium system that is established by a liquid and its vapor in a closed system as a result of evaporation and condensation.

In the evaporation-condensation equilibrium system, an equilibrium is established when the rate of evaporation equals the rate of condensation. At equilibrium, the number of liquid molecules that escape from the surface over a time interval equals the number of vapor molecules that return to the liquid in the same time interval. Once the equilibrium is established, the total quantity of liquid and vapor remains constant. The level of the liquid no longer changes.

A dynamic equilibrium results when two processes have equal but opposite rates. The opposite of a dynamic equilibrium is a static equilibrium.

A chemical equilibrium system is similar to an evaporation-condensation system. A chemical equilibrium also has two equal but opposing rates. Let us consider the gas-phase reaction of carbon dioxide, CO_2, with hydrogen gas, H_2, in a closed system.

$$CO_2(g) + H_2(g) \longrightarrow CO(g) + H_2O(g)$$

If equal numbers of moles of CO_2 and H_2 are placed in a container under the appropriate conditions, they combine, producing carbon monoxide, CO, and water vapor, H_2O. The products, CO and H_2O, can also combine, producing the reactants, CO_2 and H_2.

If a reaction takes place in a closed system and the products can combine to produce the reactants, then the reaction is classified as a **reversible reaction.** Just about all chemical changes are reversible to some degree. Reversible reactions are identified by two single-barbed arrows pointing in opposite directions.

$$CO_2(g) + H_2(g) \rightleftharpoons CO(g) + H_2O(g)$$

As the reaction proceeds, the concentration of the products, CO and H_2O, increases, and speeds up the reverse reaction (to the left). The rate of the forward reaction (to the right) decreases as the reactant concentrations decrease. If the reaction is undisturbed, eventually the rate of the forward reaction decreases to a level equal to the rate of the reverse reaction. At this time, a **chemical equilibrium** is established. A graph of the rates of the forward and reverse reactions in the above example is shown in Fig. 16.8. Note that at t_0 as the reactants just begin to combine, the rate of the forward reaction is highest, and the rate of the reverse reaction is zero because no products are present. As time passes, the rate of the forward reaction steadily decreases, and the rate of the reverse reaction increases. At a certain time, t_e, the rates become equal, and they remain unchanged with time. At this point the system is in equilibrium. As long as the conditions remain constant, the system remains in equilibrium.

A chemical equilibrium results when the forward and reverse reaction rates are equal.

Once a chemical equilibrium has been established, the concentrations of the reactants and products are constant. We mathematically represent the relationship between the equilibrium concentrations of the products and the equilibrium concentrations of the reactants by writing an **equilibrium expression.** For the general form of a chemical equation

$$aA + bB \rightleftharpoons cC + dD$$

the equilibrium expression takes the form

The equilibrium expression is a mathematical equation of the law of chemical equilibrium.

$$K = \frac{[C]^c[D]^d}{[A]^a[B]^b}$$

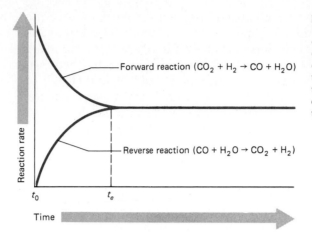

Figure 16.8
In the development of a chemical equilibrium, the rate of the forward reaction becomes equal to the rate of the reverse reaction. After time t_e, the rates remain equal if the system is undisturbed.

in which K is the equilibrium constant, [C] and [D] are the equilibrium molar concentrations of the products, [A] and [B] are the equilibrium molar concentrations of the reactants, and a, b, c, and d are the coefficients of the reactants and products in the equation. For example, the equilibrium expression for the reaction of CO_2 and H_2 is

$$K = \frac{[CO][H_2O]}{[CO_2][H_2]}$$

Here the coefficients for all species are 1; therefore, all concentrations are raised to the first power. Note that chemical species enclosed in brackets refer to their molar concentrations.

Table 16.1 provides additional examples of equilibrium expressions.

TABLE 16.1 EQUILIBRIUM EXPRESSIONS FOR SELECTED EQUILIBRIA

Equilibrium	Equilibrium expression
$N_2(g) + 3H_2(g) \rightleftharpoons 2NH_3(g)$	$K = \dfrac{[NH_3]^2}{[N_2][H_2]^3}$
$2O_3(g) \rightleftharpoons 3O_2(g)$	$K = \dfrac{[O_2]^3}{[O_2]^2}$
$H_2(g) + C_2H_4(g) \rightleftharpoons C_2H_6(g)$	$K = \dfrac{[C_2H_6]}{[H_2][C_2H_4]}$
$I_2(g) + Cl_2(g) \rightleftharpoons 2ICl(g)$	$K = \dfrac{[ICl]^2}{[I_2][Cl_2]}$
$CO_2(g) \rightleftharpoons CO(g) + 0.5O_2(g)$	$K = \dfrac{[CO][O_2]^{.5}}{[CO_2]}$

Equilibrium constants K are experimentally determined for an equilibrium system at a fixed temperature and pressure. The numerical value of the **equilibrium constant** indicates if the reactants or the products are the principal species present when equilibrium is reached. For example, the equilibrium constant for the reaction of CO_2 and H_2 is 0.14 at 550°C. Because the value for K is less than 1, this means that the product of the molar concentrations of the reactants (in the denominator) is greater than the product of the molar concentrations of the products (in the numerator).

$$[CO_2][H_2] > [CO][H_2O] \text{ when } K < 1$$

When K is much less than 1, the reverse reaction goes nearly to completion and the forward reaction occurs to a small extent.

If, for example, 1.00 M CO_2 and 1.00 M H_2 are placed in a reaction vessel and allowed to attain equilibrium at 550°C, the equilibrium molar concentration of each of the products, CO and H_2O, is 0.27 M, and the molar concentration of each of the reactants is 0.73 M. Figure 16.9 is a graph of the development of this equilibrium. Initially, 1.00 M CO and H_2 are contained in the reaction vessel; with time, their concentration drops to 0.73 M as the concentration of the products increases to 0.27 M. After the equilibrium is established, no further change in concentration occurs.

If we apply the equilibrium expression by substituting the equilibrium concentrations, we find the value of K.

$$K = \frac{[CO][H_2O]}{[CO_2][H_2]}$$

$$K = \frac{0.27\, \cancel{M} \times 0.27\, \cancel{M}}{0.73\, \cancel{M} \times 0.73\, \cancel{M}}$$

$$K = 0.14$$

Figure 16.9
If initially 1.0 M CO_2 and 1.0 M H_2 are placed into a reaction vessel and the temperature is 550°C, both of their concentrations decrease to 0.73 M and the concentrations of CO and H_2O increase to 0.27 M. At equilibrium, the concentrations of the reactants are greater than the concentrations of the products. This equilibrium has a K value less than 1.0.

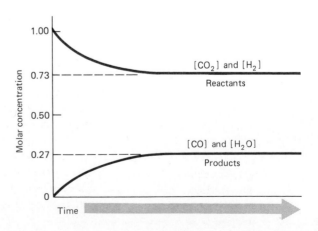

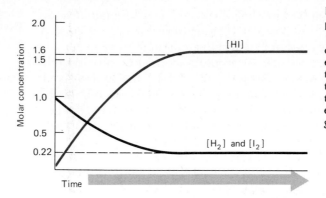

Figure 16.10
If initially 1.0 M H_2 and 1.0 M I_2 are allowed to come to equilibrium, the equilibrium concentration of the product, HI, is greater than the concentrations of the reactants, H_2 and I_2. This equilibrium has a K value greater than 1.0.

Some chemical equilibria have K values greater than 1. For example, the combination of $H_2(g)$ and $I_2(g)$ to produce $HI(g)$ has an equilibrium constant of 51 at 440°C.

$$H_2(g) + I_2(g) \rightleftharpoons 2HI(g)$$

$$K = \frac{[HI]^2}{[H_2][I_2]} = 51$$

When the equilibrium constant is greater than 1, the numerator of the equilibrium expression is greater in magnitude than the denominator— the concentrations of the products are greater than those of the reactants when equilibrium is established.

If 1.00 M H_2 and 1.00 M I_2 are placed in a reaction vessel at 440°C, at equilibrium we find 0.22 M H_2 and 0.22 M I_2, and 1.56 M HI. Figure 16.10 illustrates the establishment of the HI equilibrium. Exactly the same equilibrium concentrations exist for the HI equilibrium at 440°C if 2.00 M HI is initially placed in the reaction vessel. Equilibrium concentrations of all species depend only on the equilibrium constant at a specific temperature.

When K is much larger than 1, the forward reaction goes nearly to completion and the reverse reaction occurs to a small extent.

16.10 What equality exists when a chemical equilibrium system is established?

REVIEW EXERCISES

16.11 Explain the reason why the concentrations of chemical species remain constant after equilibrium is established.

16.12 Consider the hypothetical chemical equilibrium $X(g) + Y(g) \rightleftharpoons XY(g)$. (*a*) Write the equilibrium expression. (*b*) What is known about this equilibrium if $K < 1$? (*c*) If 1 M X and 1 M Y are initially combined, draw a graph that illustrates the development of the equilibrium. Assume the value for K is less than 1.

16.13 Write the equilibrium expressions for
(*a*) $NO_2(g) + SO_2(g) \rightleftharpoons SO_3(g) + NO(g)$
(*b*) $4HCl(g) + O_2(g) \rightleftharpoons 2Cl_2(g) + 2H_2O(g)$

If a chemical equilibrium system is not disturbed, it will remain in equilibrium indefinitely; such behavior characterizes stable systems. A stable system is one that does not undergo spontaneous changes.

If the concentration, pressure, or temperature of a chemical equilibrium system is changed, the equilibrium is disrupted and initially is no longer in equilibrium. Henri Le Chatelier (1850–1936) was one of the first to describe how a chemical equilibrium system responds to changes. His description of this behavior after disruption is now called Le Chatelier's principle.

Le Chatelier's principle states that if the concentration, pressure, or temperature of a chemical equilibrium system is changed, the system shifts in such a way as to minimize the change and to bring the system back to a state of equilibrium. More simply, a chemical equilibrium attempts to remain in equilibrium by changing the concentrations of the reactants and products. If the change is such that more products are present after the equilibrium is reestablished, the equilibrium is said to have shifted to the products (right side). If a net increase in reactant concentration has occurred after reestablishment of equilibrium, the system is said to have shifted to the reactants (left side).

We will now consider four ways that equilibrium systems are disrupted: (1) concentration changes, (2) pressure changes, (3) temperature changes, and (4) the addition of a catalyst.

When the reactant or product concentrations are changed, the equilibrium shifts to accommodate the substance added or removed. If a substance is removed, the equilibrium shifts to replace the lost substance, and if a substance is added, it shifts to decrease its concentration.

To illustrate the effect of concentration changes on equilibrium systems, we will consider the gas phase equilibrium

$$CO_2(g) + H_2(g) \rightleftharpoons CO(g) + H_2O(g)$$

Figure 16.11 shows the concentration changes that occur when $CO_2(g)$ is added to this equilibrium system.

1. At time t_1, additional CO_2 is added to the reaction vessel, and the system is no longer in equilibrium. A higher concentration of CO_2 increases the rate of the forward reaction relative to that of the reverse reaction.

2. As time passes (t_1 to t_2), the concentrations of CO_2 and H_2 decrease, decreasing the rate of the forward reaction. At the same time, the increased concentration of products accelerates the reverse reaction. Ultimately, the two rates become equal, and the equilibrium is reestablished at time t_2.

Once equilibrium is reattained, more products (CO and H_2O) and less of one reactant (H_2) are present then before the CO_2 was added.

16.4 FACTORS THAT INFLUENCE CHEMICAL EQUILIBRIA

Stable systems usually will not change on their own. Less stable systems tend to undergo spontaneous changes until they become more stable.

Henri Louis Le Chatelier (1850–1936) obtained a degree in mining chemistry. He initially studied the nature of flames and investigated ways to help prevent mine explosions. He later researched chemical thermodynamics, which led him to his most valuable contribution to science, Le Chatelier's principle.

Concentration Changes

A chemical equilibrium shifts to remove an added substance or replace one that has been lost.

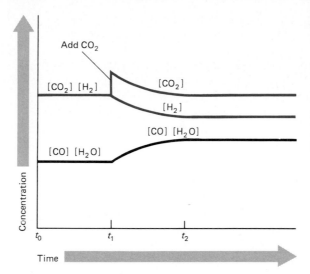

Figure 16.11
CO_2 is added to the equilibrium system at t_1. From t_1 to t_2 the system returns to a state of equilibrium. At t_2 the concentrations of CO and H_2O have increased and the concentration of H_2 decreased. As long as no change in the system occurs, the reaction remains at equilibrium after t_2.

Consequently, the equilibrium is said to have shifted to the products (right).

In all chemical equilibrium systems, an increase in the concentration of a reactant is absorbed by a shift in favor of the products. If a product concentration is increased, the equilibrium shifts toward the reactants. A shift to the reactants lowers the concentration of the added product.

The opposite occurs when the concentration of a reactant or product is decreased. If the reactant concentration is decreased, then initially the forward reaction rate decreases while the reverse reaction rate is unchanged. Thus, the equilibrium system shifts in favor of the reactants. Removal of a product causes the system to shift to the products, partially replacing what was removed.

Table 16.2 shows the direction of equilibrium shift for all possible concentration changes for the CO_2-H_2 equilibrium.

TABLE 16.2 EFFECTS OF CONCENTRATION CHANGES ON A CHEMICAL EQUILIBRIUM*

$$CO_2 + H_2 \rightleftharpoons CO + H_2O$$

Concentration change	$[CO_2]$	$[H_2]$	$[CO]$	$[H_2O]$	Direction of shift
Increase $[CO_2]$		Dec	Inc	Inc	Products
Increase $[H_2]$	Dec		Inc	Inc	Products
Increase $[CO]$	Inc	Inc		Dec	Reactants
Increase $[H_2O]$	Inc	Inc	Dec		Reactants
Decrease $[CO_2]$		Inc	Dec	Dec	Reactants
Decrease $[H_2]$	Inc		Dec	Dec	Reactants
Decrease $[CO]$	Dec	Dec		Inc	Products
Decrease $[H_2O]$	Dec	Dec	Inc		Products

*Dec = decrease; Inc = increase

Because the liquid and solid states are not affected significantly by pressure changes, only equilibria that contain gases are influenced by pressure changes.

Applying Le Chatelier's principle to pressure changes (that result in volume changes), we would predict that equilibrium systems shift to decrease the pressure when the pressure is increased (volume decreases), and shift to increase the pressure when the pressure is decreased (volume increases). Gas pressure is directly related to the number of moles of gas particles; thus, if the pressure is increased, the equilibrium shifts in favor of the reaction that produces the smaller number of moles of gas particles. If the pressure is decreased, the system shifts to favor the reaction that produces the largest number of particles.

In what direction does the equilibrium shift if the pressure is increased on nitrogen monoxide, NO, and oxygen, O_2, in equilibrium with nitrogen dioxide, NO_2?

$$\underbrace{2NO(g) + O_2(g)}_{\text{3 mol total}} \rightleftharpoons \underbrace{2NO_2(g)}_{\text{2 mol}}$$

Three mol of molecules (2 mol NO + 1 mol O_2) is on the reactant side of the equilibrium, but only 2 mol of molecules is on the product side (2 mol NO_2). Thus, the pressure increase is absorbed by a shift in favor of the products, the side of the equilibrium with the smallest number of particles (lowest concentration).

If the pressure is decreased, the NO-NO_2 equilibrium shifts in favor of the reactants, the side where more particles are found. Such a shift helps increase the pressure.

In some equilibria, the total number of particles of reactants is equal to the total number of particles of products. For example, in the equilibrium

$$\underbrace{S_2(g) + O_2(g)}_{\text{2 mol total}} \rightleftharpoons \underbrace{2SO_2(g)}_{\text{2 mol}}$$

2 mol of particles is found for both the reactants and products. When the pressure on such a system is changed, neither the forward nor the reverse reaction is favored because equal numbers of moles of particles are on each side (2 mol reactants and 2 mol products). Therefore, there is no way this equilibrium can shift to decrease the pressure, and no shift is observed.

When the temperature is increased, equilibrium systems shift in favor of the reaction that absorbs added heat. When the temperature is decreased, equilibrium systems shift to replace lost heat. In a chemical equilibrium, one of the reactions (forward or reverse) is endothermic and the other is exothermic. Therefore, an increase in temperature

Pressure Changes

An increase in pressure on a gas-phase equilibrium favors the side of the equilibrium with the smallest number of particles; a decrease in pressure favors the side with the largest number of particles.

Although Le Chatelier's principle correctly predicts the direction of a shift, the reason for the shift is that if the forward reaction is endothermic, K increases with increasing temperature, and decreases with decreasing temperature. The opposite is true for exothermic reactions; K decreases with increasing temperature, and increases with decreasing temperature.

Temperature Changes

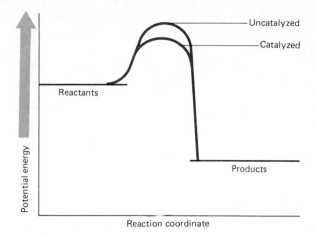

Figure 16.12
Because a catalyst lowers the activation energy of the reactants and products, both the forward and reverse rates of reaction are increased equally. Thus, catalysts have no effect on the equilibrium concentrations of the reactants and products.

(adding heat) favors the endothermic reaction, the one that absorbs heat, and a decrease in temperature (removing heat) favors the exothermic reaction, the one that releases heat.

Let us return to our model equilibrium system.

$$CO_2 + H_2 \rightleftharpoons CO + H_2O + 42.7 \text{ kJ}$$

As written, the forward reaction is exothermic, releasing 42.7 kJ of heat. The reverse reaction is endothermic, requiring 42.7 kJ. If the temperature of this equilibrium system is increased, the equilibrium shifts in favor of the reverse reaction; more reactants and less products are found after the temperature change. A decrease in temperature shifts the equilibrium toward the products.

When you deal with temperature changes and equilibrium systems, treat the heat energy as a reactant or product whose change effects the same equilibrium shifts as concentration changes of actual reactants and products.

A catalyst is a substance that lowers the activation energy of a reaction. It lowers the activation energies of both the forward and reverse reactions equally (Fig. 16.12). Hence, both forward and reverse reactions are accelerated to the same degree, resulting in no net change in the equilibrium system. Catalysts only increase the rate at which an equilibrium is established.

Example Problem 16.1 gives examples of how to predict the direction in which an equilibrium shifts in response to changes.

Addition of a Catalyst

Catalysts do not change the state of an equilibrium system, but increase the rate at which an equilibrium is established.

――――――――――― **Example Problem 16.1** ―――――――――――

Consider the following gas phase equilibrium:

$$CS_2(g) + 4H_2(g) \rightleftharpoons CH_4(g) + 2H_2S(g) + 232 \text{ kJ}$$

Predict the direction in which the equilibrium shifts for the following changes: *(a)* addition of $H_2(g)$, *(b)* removal of $CS_2(g)$, *(c)* a pressure decrease, and *(d)* a temperature increase.

──────────────── **Solution** ────────────────

(a) Addition of H_2 shifts the equilibrium to the right. An increase in the concentration of H_2 (a reactant) is absorbed by an equilibrium shift to the products.
(b) Removal of CS_2 shifts the equilibrium to the left. The system attempts to replace the lost CS_2 by shifting toward the reactants.
(c) Decreasing the pressure shifts the equilibrium to the left, the side of the equation with the largest number of moles of gas.
(d) Increasing the temperature shifts the equilibrium to the left, in favor of the endothermic reaction.

REVIEW EXERCISES

16.14 Exactly state Le Chatelier's principle.

16.15 What is meant by an equilibrium shift? Explain in terms of the forward and reverse reactions.

16.16 Answer the following questions for the following equilibrium system:

$$CO(g) + Cl_2(g) \rightleftharpoons COCl_2(g)$$

In what direction does the equilibrium shift (toward the reactants or products) for each of the following changes: *(a)* removal of CO, *(b)* addition of Cl_2, *(c)* addition of $COCl_2$, *(d)* decreasing the pressure to one-half the original pressure?

16.17 Consider the Haber reaction:

$$3H_2(g) + N_2(g) \rightleftharpoons 2NH_3(g) + 92 \text{ kJ}$$

In what direction does the Haber equilibrium shift when the following changes are made: *(a)* decrease in temperature, *(b)* increase in temperature, *(c)* decrease in pressure, *(d)* addition of a catalyst, *(e)* increase in N_2 concentration?

──────────────── **SUMMARY** ────────────────

Chemical kinetics is the study of the rates of chemical reactions. Rates are measured by measuring the change in either reactant or product concentration over a time interval.

The **collision theory** is used to explain, at the molecular level, the rates of chemical reactions. Collision theory describes atomic and molecular collisions as effective or ineffective. An **effective collision** occurs when two particles collide with enough energy and with the proper orientation to produce the products.

Four factors that influence the rate of chemical reactions are the nature of the reactants, reactant concentration, temperature, and catalysts. Stable reactants with stronger bonds tend to react more slowly than less stable reactants with weaker bonds. Increased concentration of reactants increases the number of effective collisions in a time interval; therefore, the rate of the reaction increases. An increase in temperature increases the average kinetic energy of the reacting particles, which also increases the number of effective collisions. Thus, higher temperatures increase reaction rates. **Catalysts** speed up chemical reactions by lowering the activation energy. The **activation energy** for a reaction is the amount of energy needed to produce the activated complex (transition state), which is a high-energy combination of colliding particles that can break apart as either the reactants or products.

A **chemical equilibrium** is established when the rates of the forward and reverse reactions, in a closed system, are equal. Equilibrium systems are stable systems and do not undergo spontaneous changes. An equilibrium constant K describes the direction in which the equilibrium lies, either favoring the products or the reactants.

Le Chatelier's principle is the guiding rule used to predict the behavior of equilibrium systems when they are disrupted. Le Chatelier's principle states that if a property of a chemical equilibrium is changed, the equilibrium shifts to minimize the change and bring the system back to equilibrium.

An increase in concentration of a substance in an equilibrium system causes the system to shift to decrease the amount of the added substance. Increasing the temperature shifts the equilibrium in the direction that absorbs the heat, the endothermic reaction. An increase in pressure causes the system to shift in favor of the side with the smaller number of moles of particles. Because catalysts speed up both forward and reverse reactions equally, they have no effect on the state of equilibrium.

KEY TERMS

activation energy
catalyst
chemical equilibrium
effective collision

endothermic reaction
enthalpy
equilibrium
equilibrium shift

exothermic reaction
forward reaction
rate-determining step
reaction rate

reaction intermediate
reaction mechanism
reverse reaction
reversible reaction

EXERCISES*

16.18 Define each of the following terms: reaction rate, effective collision, exothermic reaction, endothermic reaction, enthalpy, activation energy, catalyst, reaction mechanism, reaction intermediate, rate-determining step, equilibrium, chemical equilibrium, reversible reaction, forward reaction, reverse reaction, equilibrium shift.

Rates of Reactions

16.19 Describe the relative differences in time required to go to completion for a reaction that has a high measured rate compared to one that has a low rate.

16.20 Classify the rates of the following chemical changes as high, moderate, or low: (a) explosion of nitroglycerine, (b) hard-boiling an egg, (c) reaction of sodium with water, (d) neutralization of HCl by NaOH, (e) oxidation of Fe in the air, (f) decomposition of plastics in the environment.

16.21 What is true about the rate of disappearance of reactants in a reaction that is classified as having a high rate of reaction?

Collision Theory

16.22 What is true about two colliding particles (a) if an effective collision occurs, and (b) if an ineffective collision takes place?

16.23 Discuss the orientation factors of two colliding automobiles as an analogy for the orientation effects of colliding molecules.

16.24 Predict if an effective collision occurs under the following conditions: (a) energy greater than minimum energy for producing the transition state, and improper orientation; (b) proper orientation with energy in excess of the minimum energy to produce the transition state; (c) energy less than the minimum energy to produce the transition state, with the proper orientation.

16.25 How do the following changes alter the frequency of effective collisions for a gas-phase reaction: (a) decreased temperature, (b) increased pressure, (c) addition of a catalyst?

*For exercise numbers printed in color, answers can be found at the back of the book.

Factors that Influence Reaction Rates

16.26 Predict the relative rate of reaction (high, moderate, or low) by considering the nature of the reactants:

(a) $F_2 + Cl_2$

(b) $F_2 + Xe$

(c) Decomposition of HF

(d) $F_2 + Na$

16.27 Rank the following reactions according to reaction rate at 25°C (highest to lowest):

(a) $HCl(aq) + NaOH(aq) \longrightarrow$
$$NaCl(aq) + H_2O(l)$$

(b) $HCl(aq) + Na_2CO_3(aq) \longrightarrow$
$$NaCl(aq) + CO_2(g) + H_2O(l)$$

(c) $2NaCl(s) \longrightarrow 2Na(s) + Cl_2(g)$.

16.28 Describe what happens to the rate of reaction for the combination of $H_2(g)$ and $Br_2(g)$.

$$H_2(g) + Br_2(g) \longrightarrow 2HBr(g)$$

when we (a) decrease $[Br_2]$, (b) increase $[H_2]$, (c) decrease the temperature, (d) add a catalyst, (e) decrease the pressure.

16.29 In reaction A, 1% of the molecules have enough energy to form the transition state (activated complex), and in reaction B, 10% have enough energy to form the transition state. Which of these two reactions would you expect to proceed at a faster rate? Totally explain your answer.

16.30 Draw an energy level diagram for the combination of $H_2(g)$ and $I_2(g)$ to produce $HI(g)$:

$$H_2(g) + I_2(g) + 52 \text{ kJ} \longrightarrow 2HI(g)$$

On the graph, label energy on the vertical axis and time on the horizontal axis, and show the difference in energy between the reactants and products, the enthalpy change.

16.31 (a) Draw an energy level diagram for the oxidation of nickel(II) sulfide, NiS, to nickel(II) oxide, NiO, and sulfur dioxide, SO_2. Properly label both axes, and indicate the enthalpy of reaction.

$$2NiS(s) + 3O_2(g) \longrightarrow 2NiO(s) + 2SO_2(g)$$
$$+ 936 \text{ kJ}$$

(b) On the graph draw the energy pathway followed by the reactants to form the products; indicate the activation energy and the position of the transition state.

16.32 After the addition of the activation energy, many exothermic reactions continue without any added input of energy. Endothermic reactions cease without a constant supply of energy. Provide an explanation for these facts.

16.33 Draw a sketch of an energy level diagram, including the reaction pathway, for an endothermic reaction that is uncatalyzed, and on the same graph draw the reaction pathway if the reaction is catalyzed.

16.34 The Haber reaction is catalyzed with an iron catalyst.

$$N_2(g) + 3H_2(g) \xrightarrow{\text{Fe}} 2NH_3(g)$$

If a catalyst was not used, what factors would have to be changed to produce the same quantity of ammonia within a specified time interval?

16.35 What is the purpose of a catalytic converter in an automobile?

16.36 Consider the accepted mechanism for the gas-phase reaction

$$H_2 + 2ICl \longrightarrow I_2 + HCl$$

(1) $H_2 + ICl \longrightarrow HI + HCl$ (Slow)

(2) $HI + ICl \longrightarrow HCl + I_2$ (Fast)

(a) Show that by adding the two equations in the reaction mechanism the overall equation is obtained. (b) What is the reaction intermediate in the mechanism? (c) What is the rate-determining step?

16.37 (a) What is the overall aqueous reaction for the following reaction mechanism?

(1) $H_2O + OCl^- \rightleftharpoons HOCl + OH^-$

(2) $HOCl + I^- \longrightarrow HOI + Cl^-$

(3) $HOI + OH^- \rightleftharpoons H_2O + OI^-$

(b) What intermediates are produced in this mechanism?

Chemical Equilibrium Systems

16.38 Correct the following incorrect statements about chemical equilibrium systems: *(a)* At equilibrium the concentration of the reactants equals the concentration of products; *(b)* equilibrium systems are unstable and undergo spontaneous changes; *(c)* after a chemical equilibrium has been established, the forward and reverse reactions stop.

16.39 Write an explanation to account for the requirement that a chemical equilibrium be in a closed system.

16.40 What is the purpose of writing two opposing arrows, $\rightleftharpoons$, to indicate a chemical equilibrium?

16.41 Describe what happens to the rates of the forward and reverse reactions as an equilibrium is established.

16.42 Write the equilibrium expressions for the following gas-phase reactions:
(a) $NO + SO_3 \rightleftharpoons NO_2 + SO_2$
(b) $SO_2Cl_2 \rightleftharpoons SO_2 + Cl_2$
(c) $CS_2 + 4H_2 \rightleftharpoons CH_4 + 2H_2S$
(d) $4NH_3 + 5O_2 \rightleftharpoons 4NO + 6H_2O$
(e) $4HCl + O_2 \rightleftharpoons 2Cl_2 + 2H_2O$
(f) $NOCl \rightleftharpoons NO + \frac{1}{2}Cl_2$

16.43 What is the difference between a chemical equilibrium with a K value larger than 1 and another that has a K value smaller than 1?

Le Chatelier's Principle

16.44 Consider the following equilibrium:

$$NO_2(g) + SO_2(g) \rightleftharpoons SO_3(g) + NO(g)$$

Predict the direction in which the equilibrium shifts for each of the following concentration changes: *(a)* increasing [NO], *(b)* decreasing [SO_3], *(c)* decreasing [SO_2], *(d)* increasing [NO_2].

16.45 Describe what happens initially to the reaction rates of the forward and reverse reactions when a small quantity of Cl_2 is removed from the following equilibrium:

$$SO_2Cl_2(g) \rightleftharpoons SO_2(g) + Cl_2(g)$$

16.46 In which direction will the following gas-phase equilibria shift (in favor of the products or reactants), if the total pressure is increased?
(a) $PCl_5 \rightleftharpoons PCl_3 + Cl_2$
(b) $S_2 + O_2 \rightleftharpoons 2SO_2$
(c) $NO_2 + CO \rightleftharpoons CO_2 + NO$
(d) $4HCl + O_2 \rightleftharpoons 2Cl_2 + 2H_2O$

16.47 Completely explain the behavior, in terms of Le Chatelier's principle, of the following chemical equilibrium when the pressure of the system is decreased.

$$3O_2(g) \rightleftharpoons 2O_3(g)$$

16.48 What effect do the following temperature changes have on

$$2SO_2(g) + O_2(g) \rightleftharpoons 2SO_3(g) + heat$$

(a) Temperature is decreased. *(b)* Temperature is increased.

16.49 Explain why the addition of a catalyst does not cause the equilibrium to shift to minimize the amount of the catalyst.

16.50 In what direction will each of the following gas-phase equilibria shift with the stated changes:
(a) Removing I_2 from

$$2HI \rightleftharpoons H_2 + I_2$$

(b) Increasing the pressure on

$$C_2H_6 \rightleftharpoons H_2 + C_2H_4$$

(c) Adding heat to

$$CO_2 \rightleftharpoons CO + \tfrac{1}{2}O_2 \ \Delta H = +284 \text{ kJ/mol}$$

(d) Decreasing the pressure on

$$C_3H_8 + 5O_2 \rightleftharpoons 3CO_2 + 4H_2O$$

(e) Adding a catalyst to

$$I_2 + Cl_2 \rightleftharpoons 2ICl$$

CHAPTER
Seventeen

Acids and Bases

After completing Chapter 17, you should be able to

1. List the principal properties of acids and bases

2. Define and identify Arrhenius acids and bases

3. Write an equation that will illustrate the ionization of water to $H_3O^+ + OH^-$

4. Define and identify Brønsted-Lowry acids and bases

5. Determine what are the conjugate acids of bases and the conjugate bases of acids

6. Predict the relative strength of acids and bases

7. Write equations for reactions of strong acids with strong bases, strong acids with weak bases, and strong bases with weak acids

8. Apply the K_w equilibrium expression for water to find $[H^+]$ and $[OH^-]$

9. Define pH and calculate the pH of solutions

10. State two methods that are used to measure the pH of solutions in the laboratory

11. Explain what happens in an acid–base titration

12. Explain what is meant by the equivalence point of an acid–base titration

13. Calculate the molarity of an acid given its volume, and the volume and molarity of the base required to neutralize it

14. Discuss the properties and importance of common commercial acids and bases

Acids and bases such as hydrochloric acid, $HCl(aq)$, nitric acid, $HNO_3(aq)$, and potassium hydroxide, $KOH(aq)$, were known and used by the alchemists in the eleventh century. At that time and for many centuries thereafter, acids and bases were defined in terms of their properties. **Acids** share the following common properties. They taste sour, change the color of various indicator dyes such as litmus, react with bases to produce salts, and may combine with active metals, liberating hydrogen gas, $H_2(g)$.

17.1 ACID AND BASE DEFINITIONS

Figure 17.1
Svante Arrhenius (1859–1927) taught himself to read at the age of 3 and was a brilliant student. His theory of ionic dissociation came directly from his university graduate work. Arrhenius was awarded a Ph.D. (1884) with the lowest passing grade because his mentors thought that his theory was too farfetched. In 1903, he was awarded the Nobel prize in chemistry for this far-fetched notion! *(N.Y. Public Library Picture Collection)*

Bases have contrasting properties to acids. They taste bitter, cause color changes in indicator dyes opposite to the changes caused by acids, feel slippery when they contact the skin, and combine with acids to produce salts.

In 1884, Svante Arrhenius (Fig. 17.1) proposed the first good definitions for acids and bases. Simultaneously, he shook the world of chemistry by presenting the theory of ionic dissociation. He stated that when ionic substances dissolve in water, they dissociate into ions and are surrounded by water molecules. A modern statement of the **Arrhenius definitions** of acids and bases is as follows:

Arrhenius Definition

An **acid** is a substance that increases the hydrogen ion, H^+, concentration when dissolved in water.

A **base** is a substance that increases the hydroxide ion, OH^-, concentration when dissolved in water.

Water is an extremely weak electrolyte, and thus very few water molecules ionize and form H^+ and OH^- ions.

$$H_2O \rightleftharpoons H^+ + OH^-$$

But the ionization of water is not as simple as this equation would indicate. Instead it should be thought of as the interaction of two water molecules in which one donates an H^+ to another as follows:

$$H_2O + H_2O \rightleftharpoons H_3O^+ + OH^-$$

The H_3O^+ that is produced is called a **hydronium ion;** which is nothing more than a hydrated hydrogen ion (a wet proton), $H(H_2O)^+$ (Fig. 17.2). Commonly, $H^+(aq)$ is written instead of H_3O^+, with the understanding that the H^+ is associated with one or more water molecules and does not exist by itself.

Throughout the chapter, $H^+(aq)$ is written to represent a hydrated hydrogen ion, H_3O^+.

Acids similar to HCl and HNO_3, when mixed with water, increase the H_3O^+ concentration as a result of their complete ionization.

$$HCl(g) + H_2O(l) \longrightarrow H_3O^+(aq) + Cl^-(aq)$$

$$HNO_3 + H_2O(l) \longrightarrow H_3O^+(aq) + NO_3^-(aq)$$

Therefore, according to the Arrhenius definition, HCl and HNO_3 are acids; they increase the number of $H^+(aq)$ ions in solution. Other examples of Arrhenius acids are perchloric acid, $HClO_4$; sulfuric acid, H_2SO_4; acetic acid, $HC_2H_3O_2$; and phosphoric acid, H_3PO_4.

Arrhenius bases include the soluble metallic hydroxides, because these dissociate and increase the OH^- concentration in water. Consider two strong bases, KOH and NaOH:

The old name for NaOH is caustic soda.

$$KOH(s) \xrightarrow{\text{water}} K^+(aq) + OH^-(aq)$$

$$NaOH(s) \xrightarrow{\text{water}} Na^+(aq) + OH^-(aq)$$

Other examples of metallic hydroxides are calcium hydroxide, $Ca(OH)_2$; lithium hydroxide, LiOH; and cesium hydroxide, CsOH.

Various covalent compounds that do not contain a hydroxide ion combine with water and produce hydroxide ions; hence, they are also

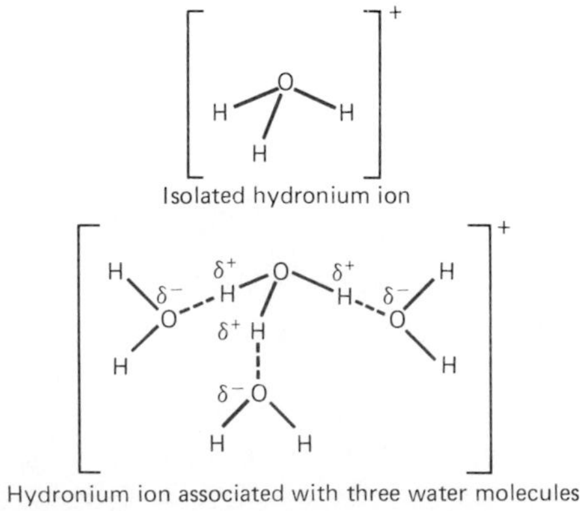

Isolated hydronium ion

Hydronium ion associated with three water molecules

Figure 17.2
A hydronium ion is a hydrated proton, $H(H_2O)^+$. Most hydronium ions are hydrogen bonded to other water molecules.

classified as Arrhenius bases. For example, ammonia, NH_3, undergoes the following reaction when placed in water:

$$NH_3(g) + H_2O(l) \rightleftharpoons NH_4^+(aq) + OH^-(aq)$$

Ammonia combines with an H^+ from water, producing an ammonium ion, NH_4^+, and a hydroxide ion, OH^-.

The Arrhenius definitions were extended in 1923 by two chemists, Johannes Brønsted and Thomas Lowry. Today, we call their definitions the Brønsted-Lowry definitions of acids and bases. A larger number of compounds are classified as either acids or bases under the Brønsted-Lowry definitions than under the Arrhenius definitions.

The **Brønsted-Lowry** acid and base definitions are

An **acid** is a proton donor.

A **base** is a proton acceptor.

Let us compare the Brønsted-Lowry and Arrhenius definitions. Arrhenius acids increase the H^+ ion concentration in water, and Brønsted-Lowry acids are proton donors. Are hydrogen ions and protons the same? Yes, a hydrogen atom consists of a proton and an electron. Hydrogen ions form when an electron is lost, leaving a proton. Note that water is not required in the Brønsted-Lowry definition of acids. But without water, acids and bases could not exist according to the Arrhenius definition. Arrhenius bases increase the OH^- concentration in water, whereas Brønsted-Lowry bases accept protons. As you will soon learn, in neutralization reactions the OH^- combines with an H^+ (a proton) to produce water. However, many substances that do not contain hydroxide ions are classified as bases under the Brønsted-Lowry definition because they can accept protons.

If we consider the reaction of hydrogen chloride gas, $HCl(g)$, and water, $HCl(g)$ releases its H^+ ion to water, producing $H_3O^+(aq)$ and $Cl^-(aq)$.

$$\underset{\text{Acid}}{HCl(aq)} + \underset{\text{Base}}{H_2O(l)} \longrightarrow H_3O^+(aq) + Cl^-(aq)$$

HCl is a Brønsted-Lowry acid, because it donates a proton to H_2O. Thus, H_2O is a proton acceptor, or a Brønsted-Lowry base.

After an acid releases a proton, the resulting anion is a base—it could accept a proton to re-form the acid. The base that results is called the **conjugate base** of the acid. When a base accepts a proton, it forms the **conjugate acid** of that base. In the above equation, Cl^- is the conjugate base of the acid HCl, and H_3O^+ is the conjugate acid of the base H_2O.

Brønsted-Lowry Definition

Brønsted (1879–1947) was a Danish chemist who was principally interested in studying chemical thermodynamics. In the early 1920s, he investigated the mechanism by which acids and bases catalyze reactions. From this work, he proposed his famous definition of acids and bases.

$H\cdot \longrightarrow e^- + H^+$

Let us consider another example:

$$NH_3(aq) + HBr(aq) \rightleftharpoons NH_4^+(aq) + Br^-(aq)$$

Base   Acid   Conjugate acid   Conjugate base

The word "conjugate" means joined together in pairs, coupled.

In this equation, HBr donates a proton to NH_3; hence, HBr is an acid and NH_3 is a base. The conjugate base of HBr is the bromide ion, Br^-, and the conjugate acid of NH_3 is the ammonium ion, NH_4^+.

Amphiprotic substances can both donate and accept protons. For example, water is an amphiprotic substance. In the presence of stronger acids, it behaves as a base, and in the presence of stronger bases, it behaves as an acid. If $HI(aq)$, a strong acid, combines with water, a proton is transferred from the HI to the water; thus, water is classified as a base.

Amphoteric is the general term applied to substances that can react as acids or bases. Some substances are amphoteric but not amphiprotic.

$$HI(aq) + H_2O(l) \longrightarrow H_3O^+(aq) + I^-(aq)$$

Acid   Base

However, if ammonia, $NH_3(aq)$, a weak base, is mixed with water, a proton is transferred from the water to the ammonia; thus, water is an acid.

Another acid–base definition was proposed by G. N. Lewis. He defined acids as electron-pair acceptors, and bases as electron-pair donors.

$$NH_3(aq) + H_2O(l) \rightleftharpoons NH_4^+(aq) + OH^-(aq)$$

Base   Acid

17.1 (a) List the four principal properties of acids. (b) List the four principal properties of bases.

17.2 Define each of the following: (a) Arrhenius acid, (b) Arrhenius base, (c) Brønsted-Lowry acid, (d) Brønsted-Lowry base.

17.3 (a) What is a hydronium ion? (b) How is a hydronium ion produced?

17.4 Identify the acid and conjugate base, and the base and conjugate acid in the following equation.

$$NH_3(aq) + HCN(aq) \rightleftharpoons NH_4^+(aq) + CN^-(aq)$$

17.5 What is an amphiprotic substance? Give an example by writing two equations.

B y applying the Brønsted-Lowry acid-base definition, we can predict the relative strengths of acids and bases. A **strong acid** is defined as a substance that more readily gives up a proton than does a weaker acid. A **strong base** is one that more readily accepts a proton than does a weaker base.

17.2 RELATIVE STRENGTHS OF ACIDS AND BASES

When a strong acid donates its proton, a weak conjugate base is produced—one that does not readily accept the proton back to re-form that acid. Thus, the conjugate bases of strong acids are weak bases. Similarly, the conjugate bases of weak acids are stronger bases. To illustrate, let's consider a strong and a weak acid, HCl and HF, respectively.

When HCl(g) is dissolved in water, it ionizes completely:

$$HCl(g) + H_2O(l) \longrightarrow H_3O^+(aq) + Cl^-(aq)$$

Hydrochloric acid is a strong acid; it has a large capacity to donate protons to water, producing large quantities of H_3O^+ and Cl^-. The Cl^- ion, the conjugate base of HCl, has a small capacity to accept a proton from H_3O^+; consequently, Cl^- is a rather weak base.

In contrast, hydrofluoric acid, HF(aq), is a weak acid.

$$HF(aq) + H_2O(l) \rightleftharpoons H_3O^+(aq) + F^-(aq)$$

Only a small percent of the HF is ionized; HF therefore has a small capacity to donate protons to water. Hence, the conjugate base of HF, F^-, is a relatively strong base. The capacity of F^- to accept protons from H_3O^+ is greater than that of the weak base H_2O to accept protons from HF.

In Table 17.1, acids and their conjugate bases are listed in decreasing order of acid strength. The strongest acid listed is perchloric acid, $HClO_4$. The conjugate base of perchloric acid, perchlorate, ClO_4^-, is the weakest base listed. At the other end of the table, methane, CH_4, is the weakest acid; thus, its conjugate base, methanide, CH_3^-, is the strongest base listed in Table 17.1.

Strengths of binary acids can also be related to the properties of the nonmetal atoms in the molecules. In the periodic table, binary acid strength increases going from left to right within a period. For example, consider the nonmetal hydrides of periods 2 and 3.

$$CH_4 < NH_3 < H_2O < HF$$
$$\text{and}$$
$$SiH_4 < PH_3 < H_2S < HCl$$

Increasing acid strength $\longrightarrow$

What accounts for this trend? Going from left to right across a period, the electronegativity of the atoms increases. As you may recall from Sec. 8.4, electronegativity is the capacity of an atom to attract electrons in a chemical bond. Thus, atoms with higher electronegativities have a greater force of attraction for hydrogen's electron than those with lower electronegativities, producing molecules with greater ionic character (molecules that are more polar). The more ionic character a substance possesses, the more it ionizes in solution to yield protons.

HCl is secreted by cells in the lining of the stomach. The HCl provides the proper acidic conditions for digestion. After the contents of the stomach enter the small intestine, the acid is neutralized.

In water, the strength of HCl, HBr, and HI are equal because each is 100% ionized. All acids stronger than H_3O^+ appear to be the same strength in aqueous solutions. This phenomenon is called the leveling effect.

Concentrations of commercially available strong acids.

Acid	Concentration
H_2SO_4	18 M
HNO_3	16 M
HCl	12 M
$HClO_4$	12 M
HBr	9 M

TABLE 17.1 RELATIVE STRENGTHS OF ACIDS AND BASES

Strong acid ↑ Increasing acid strength	Acid			Conjugate base	Weak base ↑ Increasing base strength
	Perchloric acid	$HClO_4$	ClO_4^-	Perchlorate	
	Sulfuric acid	H_2SO_4	HSO_4^-	Hydrogensulfate	
	Hydroiodic acid	HI	I^-	Iodide	
	Hydrobromic acid	HBr	Br^-	Bromide	
	Hydrochloric acid	HCl	Cl^-	Chloride	
	Nitric acid	HNO_3	NO_3^-	Nitrate	
	Hydronium ion	H_3O^+	H_2O	Water	
	Phosphoric acid	H_3PO_4	$H_2PO_4^-$	Dihydrogenphosphate	
	Hydrofluoric acid	HF	F^-	Fluoride	
	Nitrous acid	HNO_2	NO_2^-	Nitrite	
	Acetic acid	$HC_2H_3O_2$	$C_2H_3O_2^-$	Acetate	
	Carbonic acid	H_2CO_3	HCO_3^-	Hydrogencarbonate	
	Ammonium ion	NH_4^+	NH_3	Ammonia	
	Hydrocyanic acid	HCN	CN^-	Cyanide	
	Water	H_2O	OH^-	Hydroxide	
	Ammonia	NH_3	NH_2^-	Amide	
	Methane	CH_4	CH_3^-	Methanide	

Weak acid   **Strong base**

Within a chemical group, acid strength increases with increasing atomic mass. For example, consider the acid strengths of the hydrides of the chalcogens and halogens.

$$H_2O < H_2S < H_2Se < H_2Te$$
and
$$HF < HCl < HBr < HI$$

Increasing acid strength $\longrightarrow$

Increasing acidity within a chemical group is partially explained by increasing atomic size. The larger the atomic size, the weaker the bond between the nonmetal and hydrogen. Weaker bonds have a greater tendency to dissociate and produce protons, H^+, than stronger bonds.

REVIEW EXERCISES

17.6 What is the relative strength of the (a) conjugate base of a weak acid, (b) conjugate acid of a strong base, (c) conjugate base of a strong acid?

17.7 Give an explanation for the fact that HF is a weaker acid than HCl in terms of their conjugate bases.

17.8 Use Table 17.1 to determine which member of each of the following pairs is a stronger acid: (a) HNO_2 or HCN; (b) H_2O or NH_4^+; (c) $HC_2H_3O_2$ or H_3O^+; (d) H_2CO_3 or NH_3.

By far the most important reaction of acids and bases is the **neutrali-zation reaction,** in which an acid combines with a base to produce a salt and, in many cases, water.

$$\text{Acid} + \text{base} \longrightarrow \text{salt} + \text{water}$$

To illustrate a neutralization reaction, let us consider the reaction of solutions of potassium hydroxide and hydrochloric acid.

$$KOH(aq) + HCl(aq) \longrightarrow KCl(aq) + H_2O(l)$$

In this reaction, KOH, a strong base, combines with HCl, a strong acid, to produce KCl, a salt, and water. A more accurate representation of this neutralization reaction is an overall ionic equation, which shows all species as they exist when dissolved. Both strong acids and strong bases are completely ionized; therefore, they should be written as aqueous ions.

$$K^+(aq) + OH^-(aq) + H^+(aq) + Cl^-(aq) \longrightarrow K^+(aq) + Cl^-(aq)$$
$$+ H_2O(l)$$

Elimination of the spectator ions from the overall equation gives the net ionic equation.

$$H^+(aq) + OH^-(aq) \longrightarrow H_2O(l)$$

Whenever a strong acid combines with a strong base, the net ionic equation for this neutralization is

$$H^+(aq) + OH^-(aq) \longrightarrow H_2O(l)$$

In most cases, the anion in the strong acid and the cation from the strong base are spectator ions, and not a part of the net reaction.

A different net ionic equation is found for the reaction of a strong base and a weak acid. A strong base is totally ionized in aqueous solution, but a weak acid is a weak electrolyte, and most of the acid is un-ionized in the solution.

What happens when sodium hydroxide, NaOH, a strong base, is combined with hydrofluoric acid, HF, a weak acid?

$$Na^+(aq) + OH^-(aq) + HF(aq) \longrightarrow Na^+(aq) + F^-(aq) + H_2O(l)$$

In the overall ionic equation, NaOH is written showing dissolved ions. Hydrofluoric acid, HF, is not written in ionic form because only a small percent is ionized. In this equation, the only spectator ion is $Na^+(aq)$, and when it is eliminated from the equation, the net equation appears as follows:

$$OH^-(aq) + HF(aq) \rightleftharpoons F^-(aq) + H_2O(l)$$

17.3 REACTIONS OF ACIDS AND BASES

Strong Acid–Strong Base

A concentrated acid is not necessarily a strong acid. Strengths of acids refer to their degree of ionization, and concentration is a measure of the amount of dissolved acid, i.e., moles of dissolved acid per liter.

Spectator ions are unchanged in ionic reactions.

Strong Base–Weak Acid

Hydroxide ions from strong bases accept protons from the un-ionized weak acid, yielding water and the conjugate base of the weak acid. A general net equation for the reaction of strong bases and weak acids is

$$OH^-(aq) + HA(aq) \rightleftharpoons H_2O(l) + A^-(aq)$$

in which HA is a weak acid and A^- is its conjugate base.

As we stated previously, the conjugate base of a weak acid is a relatively strong proton acceptor, so it can take protons from water molecules. Consequently, the solutions that result are basic when equivalent amounts of strong bases and weak acids are combined. This is not true for the reaction of strong acids and strong bases. If equivalent amounts of strong acids and strong bases are combined, the resulting solution is neutral. Each H^+ combines with an OH^-, producing a neutral water molecule.

Other examples of reactions of weak acids and strong bases are

$$K^+(aq) + OH^-(aq) + HC_2H_3O_2(aq) \longrightarrow H_2O(l) + K^+(aq) + C_2H_3O_2^-(aq)$$

$$Na^+(aq) + OH^-(aq) + HCN(aq) \longrightarrow H_2O(l) + Na^+(aq) + CN^-(aq)$$

For practice, write the net ionic equations for these two reactions.

Strong Acid–Weak Base

A strong acid combines with a weak base and produces the conjugate acid of the weak base. What happens when the strong acid hydrochloric acid, HCl, combines with the weak base ammonia, NH_3?

$$H^+(aq) + Cl^-(aq) + NH_3(aq) \longrightarrow NH_4^+(aq) + Cl^-(aq)$$

A hydrogen ion from HCl is donated to the un-ionized ammonia molecule, producing the ammonium ion. The chloride ion remains unchanged. Eliminating the chloride ions from the equation results in the net ionic equation

$$H^+(aq) + NH_3(aq) \longrightarrow NH_4^+(aq)$$

Thus when equivalent amounts of a strong acid and a weak base react, the resulting solutions are acidic—in this case, because of the formation of the acidic ammonium ion, NH_4^+. This is exactly the opposite of what we found when a strong base reacts with a weak acid. A general equation for a reaction of a strong acid and a weak base is

$$H^+(aq) + B(aq) \longrightarrow HB^+(aq)$$

in which B is a weak base and HB^+ is the conjugate acid of the weak base.

17.9 Write the overall ionic and net ionic equations for the following neutralization reactions: (a) $HClO_4$, a strong acid, and RbOH, a strong base; (b) NaOH, a strong base, and HNO_2, a weak acid; (c) HI, a strong acid, and NH_3, a weak base.

17.4 MEASUREMENT OF H^+ CONCENTRATION AND pH

To understand how acid–base measurements are made, we must begin by considering water, the most commonly encountered solvent for acids and bases.

Water is a very weak electrolyte, and ionizes to a small degree.

$$H_2O \rightleftharpoons H^+ + OH^-$$

The equilibrium of water and its ions is described by an equilibrium expression.

$$K_w = [H^+][OH^-] = 1 \times 10^{-14}$$

in which K_w is the **ion-product equilibrium constant** for water, and $[H^+]$ and $[OH^-]$ are the molar concentrations of H^+ and OH^-. This expression shows that the product of the molar concentrations of the H^+ and OH^- ions equals 1×10^{-14}. Using algebra, we can calculate the individual molar concentrations of both H^+ and OH^- in pure water.

If we let z equal the molar concentration of H^+, then z is also the molar concentration of OH^-, because they have equal concentrations in pure water.

$$z = [H^+] = [OH^-]$$

Therefore,

$$z^2 = 1 \times 10^{-14}$$

Taking the square root of both sides of the equation, we obtain

$$z = 1 \times 10^{-7} M$$

$$z = [H^+] = [OH^-] = 1 \times 10^{-7} M$$

In a sample of pure water, both the H^+ and OH^- concentrations are $1 \times 10^{-7} M$. When the concentrations of H^+ and OH^- equal $1 \times 10^{-7} M$, the water is neutral, i.e., neither acidic nor basic.

When an acid is mixed with water, it donates H^+ ions, increasing the H^+ concentration; in contrast, bases increase the OH^- concentration. Le Chatelier's principle allows us to predict the effect of adding either H^+ or OH^- to the water equilibrium system.

$$H_2O \rightleftharpoons H^+ + OH^-$$

K_w is the ion-product equilibrium constant for water. Because the molar concentration of water is a constant, it is included as part of the value of K_w.

Le Chatelier's principle states that equilibrium systems tend to absorb changes and attempt to return to a state of equilibrium.

If the $[H^+]$ is increased, the equilibrium shifts to absorb the added H^+; the water equilibrium shifts to the left (Fig. 17.3). When the equilibrium is reestablished, the concentration of OH^- decreases.

If the $[OH^-]$ of water is increased, the equilibrium shifts to the left, decreasing the $[H^+]$ (Fig. 17.3). Hence, with each addition of an acid to water, the $[H^+]$ increases and the $[OH^-]$ decreases. Adding a base causes the $[OH^-]$ to increase and the $[H^+]$ to decrease.

To illustrate what happens when the $[H^+]$ of water is changed by adding acid or base, let's solve a problem.

What is the $[OH^-]$ of a solution in which acid is added to water, producing a solution that has a $[H^+]$ equal to $1 \times 10^{-3} M$? In water, the product of the $[H^+]$ and the $[OH^-]$ equals 1×10^{-14}.

$$[H^+][OH^-] = 1 \times 10^{-14}$$

Substituting the value of the concentration of H^+ into this expression, we get

$$(1 \times 10^{-3} M)[OH^-] = 1 \times 10^{-14}$$

Solve the equation for $[OH^-]$ by dividing both sides by $1 \times 10^{-3} M$.

$$[OH^-] = \frac{1 \times 10^{-14}}{1 \times 10^{-3} M}$$

$$= 1 \times 10^{-11} M$$

A solution with a $[H^+]$ equal to $1 \times 10^{-3} M$ has a $[OH^-]$ of $1 \times 10^{-11} M$.

Figure 17.3
In pure water the $[H^+]$ and the $[OH^-]$ equal $1 \times 10^{-7} M$. If acid is added to the water, the $[H^+]$ increases above $1 \times 10^{-7} M$ and the $[OH^-]$ decreases below $1 \times 10^{-7} M$. If base is added to the water, the $[H^+]$ decreases below 1×10^{-7} and the $[OH^-]$ increases above $1 \times 10^{-7} M$. The product of the molar concentrations of H^+ and OH^- equals 1×10^{-14}.

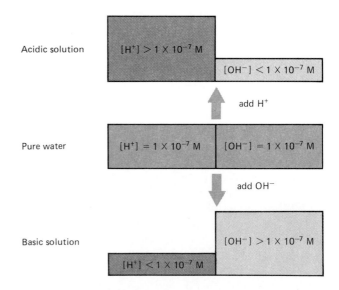

Acidic solution $[H^+] > 1 \times 10^{-7} M$

$[OH^-] < 1 \times 10^{-7} M$

add H^+

Pure water $[H^+] = 1 \times 10^{-7} M$ $[OH^-] = 1 \times 10^{-7} M$

add OH^-

Basic solution $[OH^-] > 1 \times 10^{-7} M$

$[H^+] < 1 \times 10^{-7} M$

Note that the product of these two numbers is 1×10^{-14}, the K_w value for water. Example Problem 17.1 shows another problem that involves the concentrations of H$^+$ and OH$^-$.

─────────────── **Example Problem 17.1** ───────────────

What is the [H$^+$] of the solution that results when 0.010 mol NaOH(s) is mixed with enough water to give 2.0 L of solution?

─────────────── **Solution** ───────────────

1. What is unknown? [H$^+$], or mol H$^+$/L

2. What is known? 0.010 mol NaOH; 2.0 L solution

3. Calculate the [OH$^-$].

NaOH is a strong basc; therefore, for each mole dissolved, a mole each of OH$^-$ and Na$^+$ is added to water.

$$NaOH(s) \xrightarrow{\text{water}} Na^+(aq) + OH^-(aq)$$

If 0.010 mol NaOH is dissolved, then 0.010 mol OH$^-$ is added to the 2.0 L of solution. To calculate the molar concentration of OH$^-$, the number of moles of OH$^-$ is divided by the total volume of solution.

$$M = \frac{\text{mol solute}}{\text{L solution}}$$

$$[OH^-] = \frac{0.010 \text{ mol OH}^-}{2.0 \text{ L}}$$

$$= 0.0050 \ M \text{ OH}^-$$

$$= 5.0 \times 10^{-3} \ M \text{ OH}^-$$

4. Calculate the [H$^+$] using the K_w expression, $K_w = [H^+][OH^-] = 1.0 \times 10^{-14}$.

$$[H^+][OH^-] = 1.0 \times 10^{-14}$$

$$[H^+](5.0 \times 10^{-3}) = 1.0 \times 10^{-14}$$

$$[H^+] = \frac{1.0 \times 10^{-14}}{5.0 \times 10^{-3} \ M}$$

$$= 2.0 \times 10^{-12} \ M$$

If the [OH$^-$] of an aqueous solution equals $5.0 \times 10^{-3} \ M$, then the [H$^+$] equals $2.0 \times 10^{-12} \ M$.

A commonly used means for expressing the acidity of a solution is pH. Mathematically, pH is defined as follows:

$$pH = -\log[H^+]$$

in which log is the common logarithm (base 10), and $[H^+]$ is the molar concentration of the H^+

Logarithms are nothing more than exponents. Common logarithms are exponents of 10. Accordingly, the log of a number is the exponent of 10 that gives that number. For example: What is the log of 10 (10^1)?

$$\log 10^1 = 1$$

When taking a logarithm, ask yourself: To what power must I raise 10 to yield the desired number? What is the log of 100 (10^2)? What exponent of 10 yields 100? Clearly, the answer, 2, is the only exponent of 10 that gives 100:

$$10^{\log} = 10^2 = 100$$

Table 17.2 gives the logarithms of numbers expressed exponentially.

In chemistry, p is the abbreviation for $-\log$. It is also used with equilibrium constants; i.e., pK means $-\log$ K.

TABLE 17.2 LOGARITHMS

Number	Number expressed exponentially	Log of number
1,000,000	10^6	6
1,000	10^3	3
10	10^1	1
1	10^0	0
0.1	10^{-1}	-1
0.001	10^{-3}	-3
0.000001	10^{-6}	-6

Logarithms are used whenever the pH of a solution is calculated. For example: What is the pH of pure water? Pure water contains $1 \times 10^{-7} M$ H^+; thus, to find the pH we need to calculate the log of 1×10^{-7}, and then multiply by -1 (change the sign).

$$pH = -\log[H^+]$$
$$pH = -\log(1 \times 10^{-7} M) = -\log(10^{-7} M)$$
$$= -(-7.0) = 7.0$$

A logarithm, such as pH, can have as many decimal places as the number of significant figures in the measurement from which it was calculated.

The pH of pure water is 7.0. You should notice that the pH is nothing more than the negative exponent of 10 for $[H^+]$.

$$1 \times 10^{-pH} = [H^+]$$

Acidic solutions have H^+ concentrations greater than $10^{-7} M$. Accordingly, the values for the pH of acidic solutions are less than 7. What is the pH of an acid solution that contains $0.1 M$ H^+?

$$pH = -\log[H^+]$$
$$= -\log(0.1\ M)$$
$$= -\log(1 \times 10^{-1}\ M)$$
$$= 1.0$$

A solution with $0.1 M$ H^+ has a pH of 1.0.

Basic solutions have H^+ concentrations less than $10^{-7} M$, so their pH values are larger than 7.0. What is the pH of a solution with a $[H^+]$ equal to $1 \times 10^{-12} M$?

$$pH = -\log[H^+]$$
$$= -\log(1 \times 10^{-12}\ M)$$
$$= 12.0$$

Figure 17.4 illustrates the most common range of pH values. If the pH of a solution is less than 7, the solution is acidic; if the pH is greater than 7, the solution is basic; and if the solution is neutral, the pH equals 7.

$$pH < 7 \quad [H^+] > 1 \times 10^{-7} M = \text{acidic solution}$$
$$pH > 7 \quad [H^+] < 1 \times 10^{-7} M = \text{basic solution}$$
$$pH = 7 \quad [H^+] = 1 \times 10^{-7} M = \text{neutral solution}$$

Table 17.3 lists the pH values for some common solutions. The lowest pH value in Table 17.3 is 1 for $0.1 M$ HCl, and the highest value is 13 for $0.1 M$ NaOH. Gastric juice, the solution inside the stomach, is very acidic and has a pH of approximately 1 to 2 due to the HCl(aq) secreted by the cells of the stomach lining. Pancreatic juice and bile are slightly

TABLE 17.3 pH OF COMMON SOLUTIONS

Substance	pH
0.1 M HCl	1
Gastric juice	1.5
Lemon juice	2
Vinegar	2.4
Cola drinks	3–4
Tomatoes	4
Acid rain	4–5
Black coffee	5
Urine	6.7
Milk	6.5
Saliva	6.5
Pure water	7
Blood	7.35
Tears	7.4
Pancreatic juice	8
Bile	8
Seawater	9
Milk of magnesia	9
Limewater	10.5
Household ammonia	11
Hair remover	12
0.1 M NaOH	13

Acid rain with pH values between 3 and 4 has been reported in various industrial countries.

Figure 17.4
pH is the negative logarithm of the hydrogen ion concentration. The pH of neutral solutions equals seven. Acidic solutions have pH values less than seven, and basic solutions have pH values greater than 7. A change of one pH unit represents a tenfold change in hydrogen ion concentration.

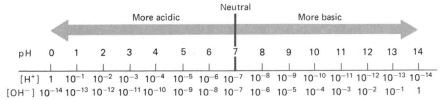

basic substances in the body, and they have pH values close to 8. Blood is also very slightly basic; it has a pH of 7.35. The pH of urine varies depending on the conditions within the body. The normal range of pH for urine is 6.0 to 6.5, which is slightly acidic.

Example Problems 17.2 and 17.3 show how the pH values of various solutions are calculated.

––––––––––––––––––––   **Example Problem 17.2**   ––––––––––––––––––––

What is the pH of a solution that contains 0.561 g of dissolved KOH in 10.0 L of aqueous solution?

––––––––––––––––––––   **Solution**   ––––––––––––––––––––

1. What is unknown?   pH

2. What is known?   0.561 g KOH; 10.0 L solution; 56.1 g KOH/mol KOH

3. Calculate the $[H^+]$.
KOH is a strong base that dissociates completely. For each 1 mol of KOH that dissolves, 1 mol OH^- is added to the solution: 1 mol OH^-/1 mol KOH.

$$\text{mol}_{OH^-} = \cancel{\text{g KOH}} \times \frac{1\ \cancel{\text{mol KOH}}}{56.1\ \cancel{\text{g KOH}}} \times \frac{1\ \text{mol OH}^-}{1\ \cancel{\text{mol KOH}}}$$

$$= 0.561\ \cancel{\text{g KOH}} \times \frac{1\ \cancel{\text{mol KOH}}}{56.1\ \cancel{\text{g KOH}}} \times \frac{1\ \text{mol OH}^-}{1\ \cancel{\text{mol KOH}}}$$

$$= 0.0100\ \text{mol OH}^-$$

$$M_{OH^-} = \frac{\text{mol OH}^-}{\text{L solution}}$$

$$= \frac{0.0100\ \text{mol OH}^-}{10.0\ \text{L}}$$

$$= 0.00100\ M\ \text{OH}^-$$

$$= 1.00 \times 10^{-3}\ M\ \text{OH}^-$$

4. Calculate the pH of the solution.
To find the pH, we need to know the $[H^+]$. Because we know the $[OH^-]$, we can calculate the $[H^+]$ by substituting our value for $[OH^-]$ into the K_w expression, $K_w = 1.00 \times 10^{-14} = [H^+][OH^-]$.

$$[H^+][OH^-] = 1.00 \times 10^{-14}$$

$$[H^+](1.00 \times 10^{-3}\ OH^-) = 1.00 \times 10^{-14}$$

$$[H^+] = \frac{1.00 \times 10^{-14}}{1.00 \times 10^{-3}\ M}$$

$$= 1.00 \times 10^{-11}\ M\ H^+$$

To obtain the pH, it is only necessary to calculate the negative logarithm of $1.00 \times 10^{-11}\ M\ H^+$.

$$pH = -\log[H^+]$$

$$= -\log(1.00 \times 10^{-11}\ M)$$

$$= \mathbf{11.000}$$

A solution that contains 0.561 g KOH per 10.0 L of solution has a pH of 11.000.

Example Problem 17.3

Calculate the pH of a 0.055 M HCl solution.

Solution

1. What is unknown?   pH

2. What is known: 0.055 M HCl; 1 mol H$^+$/1 mol HCl

3. Calculate the pH of the solution.
In previous problems, all of the [H$^+$] had 1 as the coefficient of the exponential term. In this problem, however, we must find the log of $5.5 \times 10^{-2}\ M$ HCl. When the coefficient is not 1, we find the log either by entering the number into a calculator or from a logarithm table. Because a large percent of students today use calculators, we will use the first method. If you do not have a calculator, use the common log table in the Appendix.
 To find the log with a calculator, simply enter the number and then press the "log" key. Don't confuse the "log" key with the "ln" key. Pressing the log key returns the common log, while pressing the ln key gives the natural logarithm, that is, log base e. The logarithm of 0.055 is -1.26.

$$pH = -\log(0.055\ M)$$

$$pH = -(-1.26)$$

$$= \mathbf{1.26}$$

A 0.055 M HCl solution has a pH equal to 1.26.

REVIEW EXERCISES

17.10 *(a)* Write the K_w expression for water. *(b)* State in words what this expression means.

17.11 Calculate the $[OH^-]$ in solutions with the following $[H^+]$: *(a)* 10.0 M, *(b)* 0.000160 M, *(c)* $5.62 \times 10^{-3} M$.

17.12 Calculate the $[H^+]$ in a solution prepared by mixing 75.5 g $HClO_4$ with enough water to give 22.1 L of solution.

17.13 *(a)* Define pH. *(b)* List the pH values for three common solutions. *(c)* Which of these solutions are most acidic? Explain.

17.14 Calculate the pH of solutions with the following $[H^+]$: *(a)* $1.00 \times 10^{-5} M$, *(b)* $1.00 \times 10^{-1} M$, *(c)* $6.33 \times 10^{-2} M$.

17.15 Calculate the pH of a solution prepared by mixing 0.0040 mol KOH with enough water to give 2.00 L of solution.

Acid–Base Indicators

Acid–base indicators are used to measure the approximate pH of a solution. **Acid–base indicators** are organic dyes that change color when the hydrogen ion concentration changes. These indicators are weak acids or bases whose structures are altered by the addition and loss of protons, enough so that they change colors when protons are added or removed.

The structure of phenolphthalein depends on the $[H^+]$. One of these structures is colorless, and the other is deep pink.

One of the most frequently used acid–base indicators is phenolphthalein. A few drops of phenolphthalein added to a basic solution with a pH greater than 9 changes the color of the solution to a deep pink. If enough acid is added to the solution to lower the pH below 9, the solution becomes colorless. As we will see, phenolphthalein is one of many indicators used in acid–base titrations (Fig. 17.5).

Table 17.4 lists other common indicators and the pH ranges over which they change colors.

TABLE 17.4 ACID–BASE INDICATORS

Indicator	pH range	Color	
		Acid	Base
Methyl violet	0.1–1.5	Yellow	Blue
Thymol blue	1.2–2.8	Red	Yellow
Methyl orange	3.1–4.4	Red	Yellow
Bromthymol blue	6.0–7.6	Yellow	Blue
Phenol red	6.4–8.0	Yellow	Red
Phenolphthalein	8.2–10.0	Colorless	Red-Pink
Alizarin yellow	10.2–12.0	Yellow	Red

pH Meters

For more accurate pH determinations, **pH meters** are used. While indicators can only show the pH range roughly, most pH meters can be read to the nearest hundredth of a pH unit. A picture of a pH meter is shown in Fig. 17.6.

A pH meter functions by measuring the electric potential between two electrodes that are immersed in the solution of interest. One of the electrodes is constructed so that it is sensitive to hydrogen ions, and it

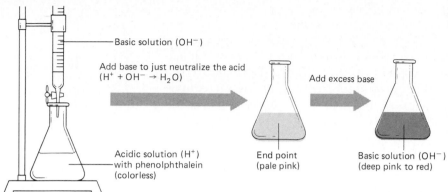

Figure 17.5
During the titration of an acidic solution, standard base is carefully added to the acid solution until the end point is reached. The end point of a titration is usually detected by the change in color of an acid–base indicator. The indicator, phenolphthalein changes from colorless to pale pink at the end point.

Basic solution (OH⁻)

Add base to just neutralize the acid
($H^+ + OH^- \rightarrow H_2O$)

Add excess base

Acidic solution (H^+)
with phenolphthalein
(colorless)

End point
(pale pink)

Basic solution (OH⁻)
(deep pink to red)

changes its electric potential with changing hydrogen ion concentration. The other electrode is a reference electrode which has a constant electric potential.

An **acid–base titration** is a volumetric procedure in which a base (or acid) of known concentration is systematically added to a fixed volume of an acid (or base) of unknown concentration in order to find the point at which equivalent amounts of acid and base are present.

Acid–Base Titration

Figure 17.6
A pH meter is an electrical instrument that allows chemists to make quick and accurate measurements of the pH of solutions. *(Corning Glass Company)*

Acid–base titrations are conducted using burets and volumetric pipets (Fig. 17.7). Generally, a fixed volume of acid of unknown concentration is placed in a flask along with a few drops of the appropriate acid–base indicator. A standard base, one with a known concentration, is added to a buret. Titrations begin by carefully adding the standard base to the acid until the indicator just changes color, signifying the **equivalence point,** i.e., the point at which the number of moles of H^+ equals the number of moles of OH^-. Stated differently, the equivalence point is reached when equivalent amounts of acid and base have reacted.

Let's turn our attention specifically to the titration of strong acids and strong bases. We will not consider titrations that involve weak acids or bases. A neutralization reaction occurs when acids and bases are combined. As we have already discussed, the net ionic equation for the reaction of a strong acid and strong base is

$$H^+(aq) + OH^-(aq) \longrightarrow H_2O(l)$$

If a strong acid is titrated with a strong base, the solution in the flask is initially acidic. With the addition of base, the number of moles of H^+ ions decreases, and at the equivalence point, the number of moles of OH^- equals the number of moles of H^+.

$$mol_{H^+} = mol_{OH^-}$$

An equivalence point is detected by a change in the color of an acid–base indicator. Figure 17.8 graphically illustrates what happens to the pH when a strong acid is titrated with a strong base; the graph is called a **titration curve.**

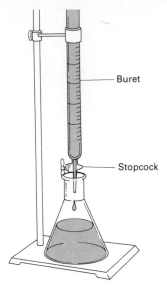

Figure 17.7
Titrations are performed using one or two burets, a pipet, and a flask. Most frequently, a known volume of acid is pipetted into the flask and a few drops of indicator are added. Then base is carefully added to the flask until the indicator just changes color.

Figure 17.8
Most acid–base titration curves are plots of the pH of the solution versus the number of milliliters of base added. Initially upon adding a strong base to a strong acid, there is a small change in the pH of the solution. As the equivalence point is approached, the pH of the solution increases very rapidly. Beyond the equivalence point, the pH rises rather slowly.

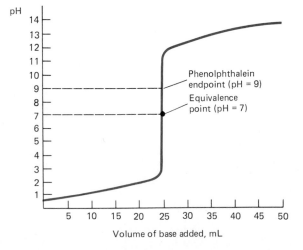

When the molarity of the H$^+$ of the acid is multiplied by its volume in liters, the number of moles of H$^+$ is obtained.

$$M_{H^+} \, V_{H^+} = \frac{\text{mol H}^+}{\cancel{L}} \times \cancel{L}$$

$$= \text{mol H}^+$$

Likewise, if the molarity of the OH$^-$ of the base is multiplied by its volume in liters, the number of moles of OH$^-$ is obtained.

$$M_{OH^-} \, V_{OH^-} = \frac{\text{mol OH}^-}{\cancel{L}} \times \cancel{L}$$

$$= \text{mol OH}^-$$

At the equivalence point in a titration, the number of moles of H$^+$ from the acid equals the number of moles of OH$^-$ from the base; therefore,

$$M_{H^+} \, V_{H^+} = M_{OH^-} \, V_{OH^-}$$

The product of the molar concentration of H$^+$ and the volume of H$^+$ (in liters) equals the product of the molar concentration of OH$^-$ and the volume of OH$^-$. Example Problem 17.4 shows how this equation can be applied when solving titration problems.

Example Problem 17.4

A titration is performed, and it is found that 25.44 mL of 0.1212 M NaOH is required to reach the equivalence point with 30.00 mL of HCl of unknown concentration. What is the molar concentration of the HCl solution?

Solution

---

1. What is unknown?   Molarity of HCl, or [HCl]

2. What is known?   25.44 mL of 0.1212 M NaOH; 30.00 mL HCl

3. Calculate the [OH$^-$].

Before the equation can be applied, the molarity of the OH$^-$ must be calculated. Because NaOH is a strong base, and 1 mol of OH$^-$ ions is released per 1 mol NaOH, the molarity of the OH$^-$ equals the molarity of NaOH.

$$[OH^-] = M_{NaOH} \times \frac{1 \text{ mol OH}^-}{1 \text{ mol NaOH}}$$

$$= \frac{0.1212 \text{ mol NaOH}}{L} \times \frac{1 \text{ mol OH}^-}{1 \text{ mol NaOH}}$$

$$= 0.1212 \, M$$

4. Substitute into the equation.

$$M_{H^+} V_{H^+} = M_{OH^-} V_{OH^-}$$

Rearranging the equation, we have

$$M_{H^+} = M_{OH^-} \times \frac{V_{OH^-}}{V_{H^+}}$$

$$= 0.1212 \, M \times \frac{0.02544 \, \cancel{L}}{0.03000 \, \cancel{L}}$$

$$= \mathbf{0.1028 \, \textit{M} \, H^+}$$

HCl is a strong acid that contains 1 mol of hydrogen ions per 1 mol HCl; therefore, the concentration of HCl equals the H^+ concentration, 0.1028 M.

You should note that a similar procedure can be used to calculate unknown concentrations when weak bases are titrated with strong acids, and when weak acids are titrated with strong bases.

REVIEW EXERCISES

17.16 What color is observed for each of the following indicators if the pH of the solution is 7.0: (a) methyl orange, (b) alizarin yellow, (c) phenolphthalein?

17.17 What is true about the H^+ and OH^- concentrations at the equivalence point of a titration of a strong acid and strong base?

17.18 What volume of 0.117 M NaOH is required to neutralize (a) 244 mL of 0.316 M HCl, (b) 81.9 mL of 0.983 M H_2SO_4, (c) 91.6 mL of 5.45 M HNO_3?

17.5 IMPORTANT ACIDS AND BASES

Acids

Sulfuric acid, $H_2SO_4(aq)$, is one of the most important industrial chemicals; more than 80 billion lb is synthesized each year. It is a colorless, oily liquid that freezes at 10.5°C and boils at 330°C. Concentrated sulfuric acid dissolves exothermically in water; therefore, you must be cautious when mixing concentrated sulfuric acid and water because splattering may occur. Since it is a strong dehydrating agent (a compound that absorbs water), it causes severe damage when it contacts skin.

One of the most important uses of sulfuric acid is in the production of fertilizers. For example, it is used to synthesize the fertilizer ammonium sulfate, $(NH_4)_2SO_4$. Sulfuric acid is also used in large amounts to extract phosphates, which are principal components of fertilizers, from the rocks in which they are found. The steel industry uses sulfuric acid to remove oxide coatings (rust) from the surface of steel, and the oil industry uses it to help refine and produce petroleum products. In addition, sulfuric acid is used in the production of textiles, explosives, and plastics, and is the acid found in storage batteries.

─────────────── **CHEM TOPIC: pH of the Blood** ───────────────

Most chemical reactions that occur in living things are pH-dependent; i.e., a change in pH causes a change in the reaction rate. Most biochemical reactions have an optimal pH at which they occur at a maximum rate. To maintain this optimal pH, many of the solutions in living things are buffer solutions. A buffer solution is one that maintains a constant pH by neutralizing added acid or base. The most important buffer solution in living things is the blood.

The one of the three major buffer systems that helps maintain the pH of the blood at 7.35 is the carbonic acid–bicarbonate buffer. It has the largest capacity to neutralize acids and bases that enter the blood because it is linked to both the lungs and kidneys. In the blood, the following equilibrium is found:

$$H_2CO_3(aq) \rightleftharpoons H^+(aq) + HCO_3^-(aq)$$

If acid, H^+, is added to the blood, it combines with bicarbonate ion, HCO_3^-, to produce carbonic acid, H_2CO_3, and if base, OH^-, enters the blood, it combines with the carbonic acid to produce water and bicarbonate.

$$OH^- + H_2CO_3 \longrightarrow H_2O + HCO_3^-$$

Therefore, if either acids or bases enter the blood, they are neutralized and the pH of the blood remains relatively constant.

Conditions arise in the body that override the capacity of the buffer to maintain a constant pH. If the pH decreases below 7.35, then the condition called acidosis results, and if the pH increases above 7.35, alkalosis results. Both of these conditions can be life-threatening. Acidosis can result from disease, injury, poisons, or organ malfunctions, or from the addition of excessive amounts of acid to the blood. For example, one of the most common causes of acidosis is the result of uncontrolled diabetes mellitus, in which large amounts of ketone bodies (acidic substances) enter the blood. In addition, respiratory acidosis results when the lungs cannot effectively eliminate CO_2. In this case, the excess CO_2 dissolves in the water and produces carbonic acid, which ionizes and produces hydrogen and bicarbonate ions. Congestive heart failure, emphysema, pneumonia, and asthma can produce respiratory acidosis.

Alkalosis results when the hydrogen ion concentration of the blood decreases. A common cause of alkalosis is severe vomiting, which results in the loss of the very acidic gastric juice. Ingestion of too many antacid tablets and various kidney problems also produce alkalosis. Very rapid, short breathing (hyperventilation) can also produce alkalosis. In this situation, too much CO_2 is eliminated from the lungs. To replace the lost CO_2, carbonic acid in the blood decomposes to CO_2 and H_2O, which results in a decreased hydrogen ion concentration when bicarbonate combines with hydrogen ions to replace the lost carbonic acid.

───

Hydrochloric acid, $HCl(aq)$, also known as muriatic acid, is another major industrial acid. Hydrochloric acid is an aqueous solution of hydrogen chloride. Hydrogen chloride, a colorless gas at 25°C, condenses to a liquid at −83.7°C. It is very soluble in water and is usually sold as a 12 M (37% m/m) solution. HCl is an inexpensive acid that has many uses. It is used to produce dyes, glucose, and metal chlorides, and is used by the steel industry to remove oxides from the surfaces of metals to be enameled or galvanized. Hydrochloric acid is also used extensively by the petroleum industry, especially to activate oil wells.

Nitric acid, $HNO_3(aq)$ is a third major industrial acid. Pure nitric acid is a colorless liquid with a density of 1.50 g/cm^3. It boils at 86°C and freezes at -42°C. Commercial nitric acid is usually sold as a 15.9 M (70% m/m) solution, which has a density of 1.40 g/cm^3. Nitric acid is used whenever a strong oxidizing acid is required. An oxidizing acid releases oxygen and causes the oxidation of other substances. Oxidation will be discussed completely in Chap. 18. Never allow nitric acid to contact your skin because it oxidizes the proteins to a yellow compound known as xanthoprotein. Nitric acid is used in the production of explosives, fertilizers, drugs, and plastics. A mixture of three volumes of concentrated hydrochloric and one volume of nitric acid is called **aqua regia.** Gold and other inactive metals may be dissolved in aqua regia.

Bases

Sodium hydroxide, NaOH, is also called lye or caustic soda. It is the most important commercial strong base. Sodium hydroxide is a white ionic solid that melts at 322°C and has a density of 2.13 g/cm^3. It is a deliquescent solid, which means that it absorbs enough moisture from the air to dissolve. Slightly less than 110 g NaOH dissolves exothermically per 100 g of water at 20°C; enough heat is sometimes released when it dissolves to boil the resulting solution. Like other bases, NaOH feels slippery when spilled on the skin. If NaOH contacts the skin, it should be diluted immediately because it causes severe burns. Sodium hydroxide is used to produce rayon, paper, soaps, and many other industrial chemicals. Some sodium hydroxide is also used to refine petroleum products.

Ammonia, $NH_3(aq)$, is an extremely valuable base. Ammonia is a gas at 25°C. It has a boiling point of -31°C and a freezing point of -74°C. Ammonia solutions are used as fertilizers and sprayed directly on plants, or can be converted to other nitrogen-containing fertilizers. It is a refrigerant that helps maintain low temperatures and is a component of household cleaners.

A number of weak bases such as $Mg(OH)_2$, $NaHCO_3$, $MgCO_3$, and $Al(OH)_3$ are used in antacid preparations (Fig. 17.9). **Antacids** are medicines that decrease the amount of stomach acid. Cells in the lining of the stomach secrete hydrochloric acid, $HCl(aq)$. Sometimes overeating or other medical conditions result in the overproduction of $HCl(aq)$. Excess hydrochloric acid can produce a great deal of discomfort. An antacid can be taken to decrease the amount of acid, which reduces the discomfort. Antacids function by neutralizing the acid. For example, aluminum hydroxide, $Al(OH)_3$, combines with hydrochloric acid as follows:

$$Al(OH)_3(s) + 3HCl(aq) \longrightarrow 3H_2O(l) + AlCl_3(aq)$$

Some antacids contain carbonates, compounds of $CO_3{}^{2-}$, or hydrogen-carbonates, compounds of $HCO_3{}^-$. When these antacids combine with

Figure 17.9
Most pharmacies have a large selection of antacid preparations. The active ingredients in antacids are weak bases that neutralize excess stomach acid. Many antacids contain two or more active ingredients.

acid, they produce carbon dioxide gas, CO_2, and water. For example, calcium carbonate, $CaCO_3$, neutralizes $HCl(aq)$ as follows:

$$CaCO_3(s) + 2HCl(aq) \longrightarrow CaCl_2(aq) + CO_2(g) + H_2O(l)$$

You should never take excessive amounts of such antacids, because the discomfort of excess acid will be replaced with discomfort caused by the carbon dioxide gas.

REVIEW EXERCISES

17.19 (a) List three important commercial acids. (b) List two important commercial bases.

17.20 For each of the following give an important use: (a) hydrochloric acid, (b) sulfuric acid, (c) sodium hydroxide, (d) nitric acid, (e) ammonia.

17.21 (a) What is the purpose of taking antacids? (b) What substances are found in antacids? (c) Write the equation for the neutralization of HCl by $MgCO_3$.

SUMMARY

Acids are substances that have a sour taste, combine with bases to produce salts, change the colors of acid–base indicators, and combine with active metals to produce hydrogen gas. **Bases** have a bitter taste, neutralize acids, change the colors of acid–base indicators in a way opposite to the way acids change them, and feel slippery to the touch.

Acids and bases are defined in a number of different ways. An **Arrhenius acid** is a substance that increases the $[H^+]$ in water, and an **Arrhenius base** is a substance that increases the $[OH^-]$ in water. Using the **Brønsted-Lowry definitions,** acids are proton donors, and bases are proton acceptors.

According to the Brønsted-Lowry definition, once an acid releases a proton, the resulting chemical species is the conjugate base of the acid. A conjugate base can accept a proton and re-form the acid. Bases accept protons and produce conjugate acids. Strong acids produce weak conjugate bases, and strong bases produce weak conjugate acids.

An acid combines with a base to produce a salt in a **neutralization reaction.** Different net ionic equations result, depending on the strength of the acids and bases combined. If equivalent amounts of a strong acid and strong base combine, the resulting salt solution is neutral. When equivalent amounts of a strong base and weak acid combine, the resulting salt solution is basic. Equivalent amounts of a strong acid and a weak base yield an acidic solution.

Acid and base strength are determined by measuring the $[H^+]$ and $[OH^-]$ in water. If the $[H^+]$ is greater than $10^{-7} M$, the solution is acidic. If the $[H^+]$ is less than $10^{-7} M$, the solution is basic. If the $[H^+]$ equals $10^{-7} M$, the solution is neutral. Commonly, acid and base strength are expressed in terms of $-\log[H^+]$, or **pH.**

--------- **KEY TERMS** ---------

acid–base indicator
acid–base titration
amphiprotic
Arrhenius acid

Arrhenius base
Brønsted-Lowry acid
Brønsted-Lowry base
conjugate acid

conjugate base
equivalence point
neutralization
pH

strong acid
strong base
titration
weak acid
weak base

--------- **EXERCISES*** ---------

17.22 Define each of the following: Arrhenius acid, Arrhenius base, Brønsted-Lowry acid, Brønsted-Lowry base, conjugate acid, conjugate base, amphiprotic, strong acid, strong base, weak acid, weak base, neutralization, titration, pH, acid–base indicator, acid–base titration, equivalence point.

Acid–Base Definitions

17.23 (a) Write two equations that illustrate the ionization of pure water. (b) Which equation better represents this ionization?

17.24 Classify each of the following substances as Arrhenius acids or bases: (a) HClO, (b) $Al(OH)_3$, (c) RbOH, (d) HI, (e) $HC_2H_3O_2$, (f) $NH_3(aq)$.

17.25 For each of the compounds listed in 17.24, write an equation that shows what happens when the compound dissolves in water.

17.26 Consider the gas phase reaction of $HCl(g)$ and $NH_3(g)$ to produce $NH_4Cl(s)$:

$$HCl(g) + NH_3(g) \longrightarrow NH_4Cl(s)$$

(a) Use the Brønsted-Lowry definitions and identify the acid and base in the equation. (b) Would this reaction be considered an acid–base reaction using the Arrhenius definition?

17.27 For each of the following equations identify Brønsted-Lowry acids, bases, conjugate acids, and conjugate bases:
(a) $CO_2 + H_2O \rightleftharpoons H_2CO_3$
(b) $H_2SO_4 + NaF \rightleftharpoons NaHSO_4 + HF$
(c) $KOH + HCN \rightleftharpoons KCN + H_2O$
(d) $NH_3 + HC_2H_3O_2 \rightleftharpoons NH_4C_2H_3O_2$

17.28 Complete and balance the following equations:
(a) $H_2S(aq) + NaOH(aq) \longrightarrow$
(b) $HBr(aq) + Ca(OH)_2(s) \longrightarrow$
(c) $H_3PO_4(aq) + 3LiOH(aq) \longrightarrow$
(d) $NH_3(aq) + H_2SO_4(aq) \longrightarrow$

17.29 Illustrate the amphiprotic nature of water by writing two equations, one in which water behaves as an acid and one in which it behaves as a base.

17.30 For each of the following write two equations that illustrate its amphiprotic nature: (a) $H_2PO_4^-$, (b) NH_3, (c) HS^-, (d) HCO_3^-.

17.31 Write the formula for the conjugate base of each of the following: (a) HNO_3, (b) HSO_4^-, (c) OH^-, (d) HF, (e) NH_3, (f) HPO_4^{2-}.

17.32 Write the formula for the conjugate acid of each of the following: (a) CN^-, (b) NO_2^-, (c) NH_3, (d) H_2O, (e) HNO_3, (f) HPO_4^{2-}.

*For exercise numbers printed in color, answers can be found at the back of the book.

Strength of Acids and Bases

17.33 (a) Explain why HF is a stronger acid than H_2O. (b) Explain why H_2O is a stronger acid than H_2S.

17.34 Use Table 17.1 to determine which of each of the following pairs of substances is the stronger base: (a) $HClO_4$ or H_2SO_4; (b) F^- or NH_2^-; (c) NH_3 or NH_4^+; (d) I^- or H_2O; (e) $C_2H_3O_2^-$ or HCO_3^-.

17.35 Considering the placement on the periodic table of the nonmetals in the following acids, predict the strongest acid from each of the following pairs: (a) HBr or HCl; (b) H_2O or NH_3; (c) H_2Te or H_2S; (d) PH_3 or AsH_3.

Reactions of Acids and Bases

17.36 Complete the following equations in which strong acids combine with strong bases:
(a) $HCl + RbOH \longrightarrow$
(b) $H_2SO_4 + Ca(OH)_2 \longrightarrow$
(c) $HClO_4 + Ba(OH)_2 \longrightarrow$
(d) $NaOH + HNO_3 \longrightarrow$
(e) $HI + LiOH \longrightarrow$
(f) $KOH + H_2SO_4 \longrightarrow$

17.37 For each equation in 17.36, write an overall ionic equation and a net ionic equation.

17.38 How is the reaction of a strong base with a strong acid different from the reaction of a strong base with a weak acid?

17.39 Write the net ionic equations for the combination of the following strong bases with weak acids:
(a) $NaOH + HF \longrightarrow$
(b) $KOH + HNO_2 \longrightarrow$
(c) $NaOH + H_2C_2O_4 \longrightarrow$

Measurements of Acids and Bases

17.40 For each of the H^+ or OH^- concentrations given, determine if the solution is acidic or basic:
(a) $[H^+] = 1 \times 10^{-9} M$
(b) $[H^+] = 1 \times 10^{-2} M$
(c) $[OH^-] = 1 \times 10^{-9} M$
(d) $[OH^-] = 3.5 \times 10^{-13} M$

17.41 Calculate the $[H^+]$ in aqueous solutions that contain each of the following concentrations of OH^-: (a) $1.0 \times 10^{-4} M$, (b) $7.7 \times 10^{-9} M$, (c) $1.3 \times 10^{-2} M$, (d) $3.99 \times 10^{-12} M$

17.42 Calculate the $[OH^-]$ of each of the following solutions:
(a) $[H^+] = 1.0 \times 10^{-4} M$
(b) $[H^+] = 0.000010 M$
(c) $[H^+] = 9.22 \times 10^{-13} M$
(d) $[H^+] = 0.00220 M$

17.43 Calculate the $[H^+]$ and $[OH^-]$ of solutions prepared as follows:
(a) 1.65 g HI dissolved in 15.2 L of solution
(b) 8.50 g HNO_3 dissolved in 12.8 L of solution
(c) 83.1 mg $HClO_4$ dissolved in 431.4 mL of solution

17.44 Calculate the $[H^+]$ and $[OH^-]$ of each of the following solutions:
(a) 219 mg NaOH dissolved in 92.1 mL of solution
(b) 213 mmol KOH in 1.94×10^2 mL of solution
(c) 0.000911 g $Mg(OH)_2$ in 18.3 L of solution

17.45 Calculate the pH of each of the following solutions:
(a) $[H^+] = 1.0 \times 10^{-3} M$
(b) $[H^+] = 1.0 \times 10^{-5} M$
(c) $[H^+] = 0.0100 M$

17.46 Calculate the pH of each of the following solutions:
(a) $[OH^-] = 1.0 \times 10^{-2} M$
(b) $[OH^-] = 1.0 \times 10^{-12} M$
(c) $[OH^-] = 1.0 \times 10^{-7} M$
(d) $[OH^-] = 4.4 \times 10^{-5} M$

17.47 Calculate the pH of each of the following solutions:
(a) 0.0200 M HI
(b) $4.49 \times 10^{-2} M$ HNO_3
(c) 0.00117 M $HClO_4$
(d) $7.07 \times 10^{-2} M$ NaOH
(e) 0.0000491 M $Ba(OH)_2$

17.48 Calculate the pH of solutions prepared as follows:
(a) 1.83 g $HClO_4$ dissolved in 1.46 L of solution
(b) 5.00 mmol HNO_3 in 41.2 mL of solution
(c) 0.803 g CsOH in 21.0 L of solution
(d) 5.5×10^{-4} g $Sr(OH)_2$ in 9.1 L of solution
(e) 0.823 mg $Mg(OH)_2$ in 125.9 mL of solution
(f) 9.22 mg H_2SO_4 in 322.1 mL of solution.

Acid–Base Titration

17.49 Calculate the volume of $0.270\ M$ NaOH required to exactly neutralize each of the following quantities of HCl:
(a) 521 mL of $0.300\ M$ HCl
(b) 8.22 mL of $0.699\ M$ HCl
(c) 12.0 ml of $0.0157\ M$ HCl
(d) 7.03 L of $0.00872\ M$ HCl

17.50 What is the molarity of a KOH solution if 11.1 mL is the volume needed to neutralize 37.5 mL of $0.0444\ M$ HNO_3?

17.51 What volume of each of the following acids can be completely neutralized with 91.4 mL of $0.800\ M$ NaOH: (a) $1.41\ M$ $HClO_4$, (b) $0.0223\ M$ HBr, (c) $0.909\ M$ HI?

17.52 How many milliliters of $0.2222\ M$ KOH are needed to reach the equivalence point for each of the following acidic solutions?
(a) 0.8631 L of $1.209\ M$ H_2SO_4
(b) 9.213 mL of $0.03711\ M$ H_3PO_4
(c) 0.2810 mL of $5.235\ M$ HNO_3

17.53 An HCl solution is titrated, and 41.08 mL of $0.2017\ M$ NaOH is required to reach the equivalence point. (a) How many moles of HCl(aq) are contained in the solution? (b) What is the mass of HCl(aq) in the solution? (c) If the total volume of the original HCl(aq) solution is 42.77 mL, what is the molarity of the H^+? (d) What is the pH of the HCl solution?

17.54 Calculate the $[H^+]$, $[OH^-]$, $[K^+]$, $[ClO_4^-]$, and pH of a solution prepared by mixing 45.5 mL of $0.189\ M$ $HClO_4$ with 50.5 mL of $0.193\ M$ KOH.

Additional Exercises

17.55 Sulfuric acid is prepared by burning sulfur in air to produce sulfur dioxide, which is combined with oxygen to produce sulfur trioxide. The sulfur trioxide is combined with water to produce sulfuric acid. (a) Write the three equations for the production of sulfuric acid. (b) What mass of sulfur is required to produce 125 L of $12.0\ M$ H_2SO_4?

17.56 Ammonia is prepared by the Haber process, in which nitrogen gas, N_2, is combined with hydrogen gas, H_2, on the surface of a special iron catalyst. (a) Write the equation for the formulation of ammonia. (b) What mass of ammonia results when 345.0 kg H_2 combines with 1575 kg N_2? (c) Assuming ideal proper-

ties, what volume of NH_3 results when 81.9 L H_2 combines with excess N_2 at 200°C and 1.00 atm? (d) What masses of H_2 and N_2 are required to produce enough ammonia to prepare 335 L of $1.21\ M$ $NH_3(aq)$?

17.57 What volume of $0.205\ M$ H_3PO_4 is required to exactly combine with excess KI to produce 519 g HI in the following reaction?

$$H_3PO_4 + KI \longrightarrow KH_2PO_4 + HI$$

17.58 A beaker contains 183.4 mL of $0.5774\ M$ HCl. A 1.000-g sample of $NaHCO_3$ is added to the beaker, and the following reaction takes place.

$$HCl(aq) + NaHCO_3(s) \longrightarrow$$
$$NaCl(aq) + H_2O(l) + CO_2(g)$$

(a) What volume of $CO_2(g)$ is liberated at 29.0°C and 759 torr? (b) What is the molar concentration of the dissolved $Na^+(aq)$? (c) What volume of $0.1000\ M$ NaOH is required to neutralize the solution?

17.59 Calculate the concentration of all ions in solution when 35.00 mL of $0.5150\ M$ HCl(aq) is mixed with 35.00 mL of $0.5000\ M$ NaOH. Assume that the volumes are additive.

17.60 Formic acid, HCO_2H, is found in ants and is a component of the fluid that is injected by a bee when it stings. Formic acid is a weak acid that only ionizes 2.1%. (a) Calculate the hydrogen ion concentration and pH of a $0.42\ M$ formic acid solution. (b) What volume of $0.350\ M$ NaOH is required to neutralize 10.0 mL of $0.42\ M$ formic acid?

17.61 A solution of ammonia used for cleaning is found to be 9.5% m/m ammonia and has a density of $0.99\ g/cm^3$. (a) Calculate the molarity of the ammonia solution. (b) What volume of this solution is required to neutralize 23 mL of $3.2\ M$ H_2SO_4?

17.62 A $0.10\ M$ solution of acetic acid has a pH of 2.87. (a) Calculate the hydrogen ion concentration of this acetic acid solution. (b) What would the hydrogen ion concentration and pH be, if acetic acid were a strong acid?

17.63 If stomach acid is $0.020\ M$ HCl, what volume of this acid is neutralized by a 300-mg antacid tablet that contains 50.0% $Mg(OH)_2$ and 50.0% $NaHCO_3$?

CHAPTER
Eighteen

Oxidation-Reduction

Afer completing Chapter 18, you should be able to

1. Define oxidation and reduction in terms of electron transfer and change in oxidation numbers

2. Distinguish between an oxidizing agent and a reducing agent

3. Identify what undergoes oxidation and reduction and what are the oxidizing and reducing agents given a chemical equation

4. Balance redox equations using the ion-electron method

5. Describe electron flow and chemical reactions that occur in electrolytic cells

6. Discuss the use of electrolytic cells in industry

7. Identify the anode and cathode in electrolytic and galvanic cells

8. Distinguish between an electrolytic and a galvanic cell

9. Describe electron flow and chemical reactions that occur in galvanic cells

10. Describe commonly used galvanic cells

Oxidation and reduction reactions (**redox** for short) occur when electrons are transferred from one substance to another. Thus, one substance loses electrons and another gains electrons. Losing electrons is called *oxidation,* and gaining electrons is called *reduction.*

Oxidation occurs when a substance loses electrons.

Reduction occurs when a substance gains electrons.

Oxidation and reduction always occur at the same time. When a substance loses electrons (oxidation), another substance must gain these electrons (reduction) (Fig. 18.1).

18.1 OXIDATION AND REDUCTION DEFINITIONS

*"Redox" is an abbreviation for **red**uction and **ox**idation.*

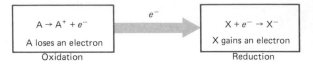

Figure 18.1
Oxidation occurs when a
substance loses electrons,
and reduction occurs when a
substance gains electrons.

Let's consider the oxidation-reduction reaction in which metallic sodium, $Na(s)$, combines with chlorine gas, $Cl_2(g)$, to produce sodium chloride, $NaCl(s)$.

$$2Na(s) + Cl_2(g) \longrightarrow 2NaCl(s)$$

To determine which substance has undergone oxidation or reduction, it is necessary to assign oxidation numbers to each element in the equation, and then look for those that have changed. You might want to review the rules for assigning oxidation numbers given in Sec. 9.1.

Both reactants, Na and Cl_2, are elements as they exist in the natural state; thus their oxidation numbers are zero. In NaCl, sodium has an oxidation number of $+1$, and chlorine has an oxidation number of -1. Frequently, the oxidation number of each element is written below its symbol in the equation, as follows:

Oxidation can also be defined as adding oxygen to a compound or removing hydrogen from a compound. Reduction may be defined as adding hydrogen to or removing oxygen from a compound.

__Leo__ the __ger__m is a mnemonic that will help you remember: "__L__ose __e__lectrons __o__xidation, and __g__ain __e__lectrons __r__eduction." An m is added at the end to form an English word.

$$\overset{\displaystyle \overbrace{\text{Oxidation}}}{\underset{\displaystyle \underbrace{\text{Reduction}}}{2\underset{0}{Na}(s) + \underset{0}{Cl_2}(g) \longrightarrow 2\underset{+1\ -1}{NaCl}(s)}}$$

During the reaction, the oxidation number of sodium increases and the oxidation number of chlorine decreases. Why do the oxidation numbers change? If the oxidation number of a substance increases (becomes more positive), this indicates that it has lost electrons. Electrons are negative particles; consequently, when a negative particle is lost, the substance becomes more positive.

$$Na \longrightarrow Na^+ + e^- \ (\text{Oxidation})$$

At exactly the same time, the electrons lost by Na are picked up by Cl_2, resulting in the decrease in the oxidation number of Cl_2.

$$2e^- + Cl_2 \longrightarrow 2Cl^- \ (\text{Reduction})$$

Therefore, in our example reaction, Na undergoes oxidation and produces Na^+, whereas Cl_2 undergoes reduction, yielding Cl^-.

Oxidation and reduction can be redefined in terms of oxidation numbers:

Oxidation occurs when the oxidation number of a substance increases in a chemical reaction.

Reduction occurs when the oxidation number of a substance decreases in a chemical reaction.

An easy way to remember these definitions is to keep in mind that the oxidation number decreases or is *reduced* when reduction occurs. Another example of an oxidation and reduction reaction is

$$2Ca(s) + O_2(g) \longrightarrow 2CaO(s)$$

First write the oxidation numbers of all atoms.

Then identify the atom whose oxidation number increases, in this case Ca, and the atom whose oxidation number decreases, O_2.

$$Ca \longrightarrow 2e^- + Ca^{2+} \text{ (Oxidation)}$$

$$O_2 + 4e^- \longrightarrow 2O^{2-} \text{ (Reduction)}$$

Two terms used to describe the substances undergoing oxidation and reduction are oxidizing agent and reducing agent. The **oxidizing agent** is the reactant that gains electrons, and the **reducing agent** is the reactant that releases electrons. An oxidizing agent takes electrons from another substance, resulting in the oxidation of that substance. A reducing agent gives up electrons to another substance, bringing about its reduction. In other words, the substance undergoing reduction is the oxidizing agent, and the substance undergoing oxidation is the reducing agent (Fig. 18.2).

Reducing agents lose electrons (i.e., undergo oxidation) and bring about the reduction of other substances. Oxidizing agents gain electrons (i.e., undergo reduction) and bring about the oxidation of other substances.

Oxidizing agents undergo reduction.

Reducing agents undergo oxidation.

Another example should help clarify these definitions. Consider the reaction of C and HNO_3:

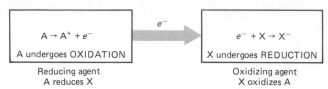

Figure 18.2
Reducing agents bring about
the reduction of other substances by undergoing oxidation. Oxidizing agents
bring about the oxidation of
other substances by undergoing reduction.

In this reaction, the oxidation number of carbon is initially 0 and increases to +4; consequently, C has undergone oxidation, or lost electrons. These electrons bring about the reduction of the N in HNO_3, so C
is the reducing agent in this reaction. Nitric acid, HNO_3, accepts the
electrons from C, which means that HNO_3 is the oxidizing agent.

When considering the elements, metals tend to be good reducing
agents, while nonmetals are generally better oxidizing agents. Alkali and
alkaline earth metals are among the strongest reducing agents known.
The halogens and oxygen, O_2, are the strongest of the oxidizing agents.
They can take electrons from most other elements.

*Oxidizing agents are
sometimes used in medi-
cine as disinfectants. A
3% solution of hydrogen
peroxide is used to de-
stroy microbes in cuts
and abrasions. Iodine is a
weak oxidizing agent
applied to kill bacteria. It
is generally applied as
tincture of I_2, an alcohol
solution of I_2.*

18.1 What occurs in oxidation-reduction reactions?

18.2 Define and give an example of each of the following: *(a)* oxidation, *(b)* reduction, *(c)* oxidizing agent, *(d)* reducing agent.

18.3 In the following reactions, determine what substances undergo oxidation
and reduction and what substances are the oxidizing and reducing agents:
(a) $H_2 + I_2 \longrightarrow 2HI$
(b) $2CO + O_2 \longrightarrow 2CO_2$
(c) $3Mg + N_2 \longrightarrow Mg_3N_2$
(d) $PCl_3 + Cl_2 \longrightarrow PCl_5$
(e) $6Fe^{2+} + 14H^+ + Cr_2O_7^{2-} \longrightarrow 2Cr^{3+} + 6Fe^{3+} + 7H_2O$

REVIEW EXERCISES

To better understand redox reactions and the electron transfers that
occur, we will now describe how to balance redox equations. Earlier
you learned the inspection method for balancing equations. Such a
method is fine for simple equations, but is inadequate for more complex
redox equations.

One of the most common procedures used to balance redox equations is the ion-electron method. This method is based on the fact that
the number of electrons lost by the substance that undergoes oxidation
must equal the number of electrons gained by the substance that undergoes reduction. The **ion-electron method** is most commonly used to
balance equations for ionic oxidation-reduction reactions in solutions of
acids or bases. Two half-reactions are written, one for the oxidation and
one for the reduction. After balancing the number of atoms and electrons in each half-reaction, we add the half-reactions to obtain the net
ionic equation.

To balance redox equations in acidic solutions, follow the steps
below.

**18.2 BALANCING
OXIDATION-
REDUCTION
EQUATIONS**

*Half-reactions actually are
more properly called
half-equations. In them-
selves they are not equa-
tions but representations
of either the oxidation or
reduction part of a chem-
ical equation.*

Step 1

For each substance that undergoes oxidation and reduction, write the substance in its initial form, then draw an arrow, and write the substance in its final form.

Step 2

Balance all atoms—except H and O—by placing the proper coefficients in front of the substances.

Step 3

Add enough water molecules to each equation to balance the O atoms, and add H^+ to balance the H atoms.

Step 4

Add electrons to the appropriate side of the equation so that the charges on each side of the half-reaction are equal.

Step 5

Equalize the electrons transferred in each half-reaction, and then add the two half-reactions together, canceling all electrons and other species that appear on both sides of the arrow. The resulting equation is the correctly balanced net ionic equation.

As an example, let's balance the following equation:

$$Cr_2O_7{}^{2-} + Fe^{2+} \longrightarrow Cr^{3+} + Fe^{3+} \text{ (Acid)}$$

Step 1

Write each substance that undergoes a change. The reactant is separated by an arrow from the product.

$$Fe^{2+} \longrightarrow Fe^{3+}$$
$$Cr_2O_7{}^{2-} \longrightarrow Cr^{3+}$$

Potassium dichromate, $K_2Cr_2O_7$, is commonly used as an oxidizing agent in chemistry. Dichromate ion gives a solution a deep orange color. After oxidation in the presence of acid, the solution turns green to violet, indicating the presence of Cr^{3+}.

Step 2

Balance all atoms except H and O.

Balancing the Fe half-reaction (Steps 2–4) merely involves adding an electron to the right side of the half-reaction to balance the charges:

$$Fe^{2+} \longrightarrow Fe^{3+} + e^- \text{ (Oxidation)}$$

Therefore, we shall concentrate on the dichromate half-reaction.

$$Cr_2O_7{}^{2-} \longrightarrow 2Cr^{3+}$$

A 2 is written as the coefficient of Cr^{3+} to balance the chromium atoms.

Step 3

Add enough H_2O molecules to balance the O atoms, and then add H^+ to balance the H atoms.

Seven O atoms are found on the left; therefore, seven water molecules are added to the right to balance the O atoms—always balance the O atoms first.

$$Cr_2O_7^{2-} \longrightarrow 2Cr^{3+} + 7H_2O$$

The addition of seven H_2O molecules to the right side adds 14 H atoms, which are balanced by placing 14 H^+ on the left side.

$$14H^+ + Cr_2O_7^{2-} \longrightarrow 2Cr^{3+} + 7H_2O$$

Step 4
Add electrons to the appropriate side of the equation so that the charges on each side of the half-reaction are equal.

The total charge on the left side is now 12+ $(14 - 2)$, and the charge on the right side is 6+ $(2 \times 3+)$. To balance the charge, six electrons are added to the left side of the equation.

$$6e^- + 14H^+ + Cr_2O_7^- \longrightarrow 2Cr^{3+} + 7H_2O \text{ (Reduction)}$$

At this point, we have two correctly balanced half-reactions.

Step 5
Equalize the electrons given off and taken in, and add the two half-reactions.

As written, one electron is released in the oxidation half-reaction, and six electrons are gained in the reduction half-reaction. To equalize the electrons, the oxidation half-reaction is multiplied by 6.

$$6Fe^{2+} \longrightarrow 6Fe^{3+} + 6e^-$$
$$6e^- + 14H^+ + Cr_2O_7^{2-} \longrightarrow 2Cr^{3+} + 7H_2O$$

When the two half-reactions are added, the six electrons cancel, which yields the following net ionic equation:

$$6Fe^{2+} + 14H^+ + Cr_2O_7^{2-} \longrightarrow 2Cr^{3+} + 6Fe^{3+} + 7H_2O$$

Always check to see that both the atoms and charges balance; here we see that 6 Fe, 2 Cr, 14 H, 7 O are on either side, and a total charge of 24+ is found on either side of the equation.

To use the ion-electron method to balance equations for redox reactions in basic (alkaline) solutions, we must add the appropriate number of OH^- and H_2O to the half-reactions in Step 3. Some confusion results when deciding which side the OH^- and H_2O should be added to. A simple way to avoid the problem is to follow the same steps used to balance an equation in acid solution. After adding H^+ and H_2O in Step 3, add OH^- to each side of the equation to "neutralize" (H^+ +

$OH^-\longrightarrow H_2O$) all H^+, and then complete the balancing of the equation as above. The following procedure illustrates how this little trick works.

Balance the following equation in basic solution:

$$S^{2-} + MnO_4^- \longrightarrow S + MnO_2 \text{ (Basic)}$$

Potassium permanganate, $KMnO_4$, is frequently used as an oxidizing agent in chemistry laboratories. It is also used as a disinfectant. Solutions of MnO_4^- have a purple color.

Step 1
Write the formula of each substance that undergoes oxidation and reduction, separating reactants and products with an arrow.

$$S^{2-} \longrightarrow S$$
$$MnO_4^- \longrightarrow MnO_2$$

Sulfur is easily balanced by adding two electrons to the right side.

$$S^{2-} \longrightarrow S + 2e^-$$

All remaining steps before adding the half-reactions are performed for the reduction of MnO_4^-.

Step 2
For the reduction half-reaction, Step 2 is not needed because the Mn atoms are balanced.

Step 3
Add H_2O to balance O atoms and H^+ to balance H atoms.

$$4H^+ + MnO_4^- \longrightarrow MnO_2 + 2H_2O$$

First, balance the half-reaction with H_2O and H^+ as if it occurs in an acidic solution. Eliminate the four H^+ by adding $4 OH^-$ to both sides of the equation.

$$\begin{array}{l} 4H^+ + MnO_4^- \longrightarrow MnO_2 + 2H_2O \\ + 4OH^- \qquad\qquad\qquad\qquad 4OH^- \\ \hline 4H_2O + MnO_4^- \longrightarrow MnO_2 + 4OH^- + 2H_2O \end{array}$$

$$H^+ + OH^- \longrightarrow H_2O$$

Step 4
Add electrons to the appropriate side of the equation so that the charges on each side of the half-reaction are equal.

Subtracting two H_2O from both sides and balancing the charges on each side gives

$$3e^- + 2H_2O + MnO_4^- \longrightarrow MnO_2 + 4OH^-$$

Step 5
Equalize the electrons and add the two half-reactions.

Two electrons are released by the oxidation half-reaction, and three electrons are taken in by the reduction half-reaction. Thus, the oxidation half-reaction is multiplied by three, and the reduction half-reaction is multiplied by two to balance the electrons.

$$3 \times (S^{2-} \longrightarrow S + 2e^-) = 3S^{2-} \longrightarrow 3S + 6e^-$$

$$2 \times (3e^- + 2H_2O + MnO_4^- \longrightarrow MnO_2 + 4OH^-) =$$
$$6e^- + 4H_2O + 2MnO_4^- \longrightarrow 2MnO_2 + 8OH^-$$

Adding the two half-reactions, we have

$$3S^{2-} + 2MnO_4^- + 4H_2O \longrightarrow 3S + 2MnO_2 + 8OH^- \text{ (Balanced)}$$

Check to see that the equation is balanced. On either side of the equation 3 S, 2 Mn, 12 O, 8 H, and a total charge of 8− are found.

18.4 Balance the following skeleton equations using the ion-electron method:　　**REVIEW EXERCISE**
  (a) $PH_3 + I_2 \longrightarrow I^- + H_3PO_2$ (Acid)
  (b) $Zn + NO_3^- \longrightarrow Zn^{2+} + NH_4^+$ (Acid)
  (c) $Zn + NO_3^- \longrightarrow Zn(OH)_4^{2-} + NH_3$ (Basic)

Electrochemistry is the study of the interaction of matter and electricity. Two major areas are studied in electrochemistry: (1) the conversion of electric energy into chemical energy, and (2) the conversion of chemical energy into electric energy. Electric energy is converted to chemical energy in **electrolytic cells,** and chemical energy is converted to electric energy in **galvanic cells.** We will first look at electrolytic cells and then proceed to galvanic cells, which are also known as voltaic cells.

**18.3
ELECTROCHEMISTRY**

Two types of electric conduction are known: (1) metallic conduction and (2) electrolytic conduction. **Metallic conduction** occurs when electrons flow through metals and is a direct result of the rather weak forces that hold electrons to metal nuclei.

Electrolytic Cells

Electrolytic conduction is the movement of ions in a liquid brought about by an external source of electricity. In Sec. 15.3 we discussed the classification of substances as either electrolytes or nonelectrolytes. **Electrolyte** solutions are capable of conducting an electric current. This is accomplished by placing a solution in a container along with two electrodes connected to a source of direct electric (dc) current (Fig. 18.3). Such a cell is called an **electrolytic cell.** When a voltage is applied, one of the electrodes develops a positive charge and the other a negative charge. Ions in electrolyte solutions are attracted to the electrodes. Cations are attracted to the negative electrode, called the **cathode,** and anions are attracted to the positive electrode, called the **anode.**

Electrodes are normally strips of an inert metal, such as Pt, that provide the proper conducting surface for redox reactions.

Figure 18.3
In electrolytic cells, cations (positive ions) diffuse toward the cathode and anions (negative ions) diffuse toward the anode. Electrons flow into the anode where they are taken in by cations, and electrons are released by anions which flow away from the anode to the cathode.

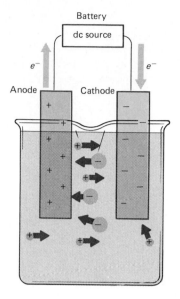

Chemical reactions occur at the electrodes in an electrolytic cell. When a cation contacts the cathode, the negative electrode, under the proper conditions, the cation picks up one or more electrons and undergoes reduction. Similarly, anions interact with the anode, the positive electrode, and give up electrons, or undergo oxidation. Oxidation occurs at the anode, and reduction occurs at the cathode.

Oxidation occurs at the anode.

Reduction occurs at the cathode.

As long as oxidation and reduction occur in an electrolytic cell, the cell conducts an electric current.

During the time an electric current flows through the cell, electrical neutrality is maintained. For each electron taken in by a cation, there must be an electron lost by an anion. In any region of the liquid, for each anion that migrates toward the anode, either a cation must migrate toward the cathode or another anion must take the anion's place. Positive or negative charges never accumulate in an electrolytic cell.

Electric neutrality must be maintained in electrolytic cells.

Let us look at some specific examples of electrolytic cells. One of the simplest cells to understand is the molten, or liquid, sodium chloride, $NaCl(l)$, cell. Figure 18.4 is a diagram of the $NaCl(l)$ electrolytic cell; it illustrates the electron flow and reactions that occur in the cell. Molten NaCl contains Na^+ and Cl^-, which are free to migrate to the electrodes. At the cathode, Na^+ is reduced to Na, and at the anode Cl^- is oxidized to Cl_2.

Cathode: $Na^+ + e^- \longrightarrow Na$ (Reduction)

Anode: $2Cl^- \longrightarrow 2e^- + Cl_2$ (Oxidation)

Figure 18.4
In the electrolysis of molten NaCl, $Cl_2(g)$ is produced at the anode and $Na(s)$ is produced at the cathode. The anode half-reaction is $2Cl^- \rightarrow Cl_2 + 2e^-$. The cathode half-reaction is $2Na^+ + 2e^- \rightarrow 2Na$.

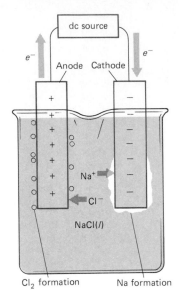

Equalizing the electrons and adding these two half-reactions gives the overall cell reaction:

$$2Na^+ + 2Cl^- \longrightarrow Cl_2 + 2Na$$

During the time an electric current passes through the molten NaCl, metallic sodium coats the cathode and chlorine gas bubbles up from the anode. Collectively, the overall chemical reaction that occurs in this or any other electrolytic cell is called **electrolysis,** which means "electrical splitting."

The electrolysis of an aqueous solution is more complex than that of NaCl(*l*) because water undergoes oxidation or reduction in addition to the solute. Water is oxidized to O_2 in the following reaction:

$$2H_2O \longrightarrow O_2 + 4H^+ + 4e^-$$

Hydrogen is the product of the reduction of water:

$$2H_2O + 2e^- \longrightarrow H_2 + 2OH^-$$

A variety of different anodic and cathodic reactions are possible in cells that contain aqueous solutions.

Whenever aqueous solutions undergo electrolysis, a number of possibilities exist for what undergoes oxidation or reduction. Depending on the conditions in the cell, either water or the solute or both are involved in the cell reaction.

What happens when an aqueous solution of sodium chloride, NaCl(*aq*), undergoes electrolysis? Two possibilities exist for both the cathode and the anode reactions. At the cathode, either Na^+ is reduced

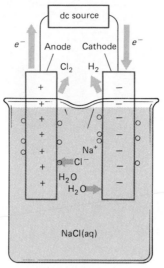

Figure 18.5
In the electrolysis of aqueous NaCl, $Cl_2(g)$ is produced at the anode and $H_2(g)$ is produced at the cathode.

Anode: $2Cl^- \rightarrow Cl_2 + 2e^-$

Cathode: $2H_2O + 2e^- \rightarrow H_2 + 2OH^-$

to Na, or water is reduced to hydrogen gas. At the anode, either Cl^- is oxidized to Cl_2, or water is oxidized to oxygen. The overall cell reaction depends on which substances undergo oxidation or reduction most readily. In concentrated NaCl solutions, the following overall cell reaction is observed:

$$2H_2O + 2Cl^- \longrightarrow Cl_2 + H_2 + 2OH^-$$

In this electrolytic cell, water more easily undergoes reduction than Na^+, and the Cl^- more readily undergoes oxidation than water. Electrolysis of brine, an aqueous solution of NaCl, is used in the commercial preparation of $H_2(g)$ and $Cl_2(g)$. Figure 18.5 illustrates the electrolysis of NaCl(aq).

In Chap. 14, the Hall process for isolating aluminum was discussed. Aluminum metal is obtained in the Hall process by passing an electric current through a molten mixture of Al_2O_3 and cryolite, Na_3AlF_6. Figure 18.6 shows how the electrolytic cell is constructed. The electrodes are composed of carbon; the anodes are carbon rods inserted into the molten mixture, whereas the cathode is a carbon coating on the inside of a steel container. Anode and cathode reactions in the Hall cell are as follows:

Industrial Electrolytic Cells

Anode: $3C + 6O^{2-} \longrightarrow 3CO_2 + 12e^-$

Cathode: $4Al^{3+} + 12e^- \longrightarrow 4Al$

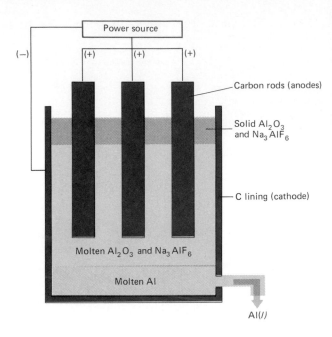

Figure 18.6
The Hall cell is used to commercially produce Al metal. Molten Al forms at the cathode, and then flows from the cell.

Adding these, we find that the overall cell reaction is

$$4Al^{3+} + 6O^{2-} + 3C \longrightarrow 3CO_2 + 4Al$$

Electrolysis is also employed in the final step of the purification of copper. Copper is first separated from the ore in which it exists naturally, and then it is purified until it is 99% Cu. In the remaining 1%, Fe, Zn, Au, Ag, and Pt are the major impurities. Figure 18.7 is a diagram of the electrolytic cell used to purify Cu. The anode is constructed of impure Cu, the cathode is a very pure sample of Cu, and both are immersed in a solution of $CuSO_4$.

An electric current is passed through the cell so that Cu, Fe, and Zn are the only metals that undergo oxidation at the anode. Au, Ag, and Pt do not dissolve and ultimately fall to the bottom of the cell. The Au, Ag, and Pt are collected and ultimately sold, which defrays much of the cost of Cu purification. At the cathode, the most easily reduced substance, Cu^{2+}, picks up two electrons, and Cu deposits on the cathode, leaving Zn^{2+} and Fe^{2+} behind. Copper purified in this manner is generally more than 99.9% pure.

If the cathode in this electrolytic cell is a metal other than Cu, when the Cu^{2+} is reduced, the resulting Cu coats or plates the surface of the metal. This process is called **electroplating.** Many metals are electroplated onto the surface of other metals. Silver- and gold-plated objects are used in place of much more expensive pure silver or gold. In the automotive industry, chromium is sometimes plated onto the surface of steel bumpers to protect the steel from corrosion. Tin cans actually are steel cans that have a thin layer of Sn plated on the surface. Tin is now

--------------- **CHEM TOPIC: Corrosion of Metals** ---------------

Corrosion occurs when metals undergo oxidation when exposed to air and water. This spontaneous process has great economic impact because many metals lose their beauty and strength when corroded and thus must be replaced. About 20% of the iron and steel produced annually is used to replace corroded metals.

Many factors influence the corrosion of metals. If an electrolyte is present in the water that contacts a metal, the process of corrosion is accelerated. Heated and strained metals also corrode faster than those that have not been heated or strained. Metals that are in contact with metals that are weaker reducing agents also corrode more rapidly. However, metals in contact with stronger reducing agents corrode at a slower rate. The pH of the solution in contact with a metal also influences the rate at which the metal corrodes.

The corrosion of iron is thought to be an electrochemical process. A region on the surface of iron acts as an anode and thus undergoes oxidation.

$$Fe(s) \longrightarrow Fe^{2+}(aq) + 2e^-$$

The electrons released migrate through the iron to another surface region that acts as a cathode. The electrons are taken in by O_2 in the presence of H^+ to produce water as follows:

$$O_2(g) + 4H^+(aq) + 4e^- \longrightarrow 2H_2O(l)$$

During the corrosion process some of the Fe^{2+} is reduced further to Fe^{3+}, which forms iron(III) oxide hydrate, $Fe_2O_3 \cdot xH_2O(s)$, commonly called rust.

$$4Fe^{2+}(aq) + O_2(g) + (4 + 2x)H_2O(l) \longrightarrow 2Fe_2O_3 \cdot xH_2O(s) + 8H^+(aq)$$

Corrosion may be slowed by many different methods. One of the simplest is to coat the surface with a substance that prevents the metal from contacting air and moisture. Paints, waxes, and greases are commonly used for this purpose. In some cases, metals are coated with ceramic enamels or other metals to prevent corrosion. Another anticorrosion method is to produce an outer oxide coating on the metal. Because the oxide cannot be further oxidized, it prevents the oxidation of the underlying metal.

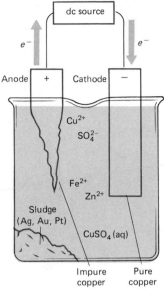

Figure 18.7
The final purification of Cu is an electrolytic process. As electricity passes through the cell, pure Cu(s) plates the cathode, separating it from the other metal impurities.

being replaced by Cr on cans. The process used to electroplate Cr is very fast, and an extremely thin Cr layer, about a millionth of a millimeter, is possible.

18.5 *(a)* What is electrochemistry? *(b)* What are the principal topics investigated in this discipline?

18.6 What is the difference between metallic and electrolytic conduction? Give an example of each.

18.7 Sketch a diagram of an electrolytic cell. Label each of the following on the diagram: *(a)* anode, *(b)* cathode, *(c)* power source, *(d)* direction of electricity flow.

18.8 *(a)* For an electrolytic cell that contains molten sodium chloride, write the oxidation and reduction half-reactions. *(b)* Write the overall cell reaction.

REVIEW EXERCISES

Galvanic cells (also called voltaic cells) are electrochemical cells that produce electricity from a spontaneous reaction. Galvanic cells are therefore the opposite of electrolytic cells, in which a nonspontaneous reaction occurs when the cell is connected to an external dc voltage. A galvanic cell is constructed in such a way that the substance undergoing oxidation is separated from the substance undergoing reduction. A reaction occurs only when the cells are interconnected by a wire that will carry the electrons produced by the substance that undergoes oxidation to the substance that undergoes reduction.

A diagram of a galvanic cell is shown in Fig. 18.8. This cell contains two electrodes, one is Cu(s) immersed in a $CuSO_4(aq)$ solution, and the other is Zn(s) immersed in a $ZnSO_4(aq)$ solution. Both electrodes are connected by a wire, switch, and voltmeter. A salt bridge [a glass tube filled with either $KNO_3(aq)$ or KCl(aq)] or a porous barrier is used to complete the electric circuit.

When the switch is closed in a galvanic cell, the voltmeter registers a **voltage** (also called **electromotive force,** emf), indicating that electrons are being forced through the circuit. Because Zn is a better reducing agent than Cu, the Zn undergoes oxidation and gives up its electrons, which flow through the wire and meter to the Cu electrode. These electrons are picked up by dissolved $Cu^{2+}(aq)$, reducing them to metallic Cu.

$$\text{Anode:} \qquad \text{Zn} \longrightarrow \text{Zn}^{2+} + 2e^- \text{ (Oxidation)}$$

$$\text{Cathode: } \text{Cu}^{2+} + 2e^- \longrightarrow \text{Cu} \qquad \text{(Reduction)}$$

In galvanic cells, as in electrolytic cells, the anode is the site of oxidation, and the cathode is the site of reduction. However, the cathode is the positive electrode, and the anode is the negative electrode.

Galvanic Cells

Galvani (1737–1798) first studied theology, but turned to medicine later in life. After becoming a professor of anatomy at the University of Bologna, Galvani discovered that electricity caused the muscles in frog legs to contract. He also noticed that different metals, when in contact, produced the same effect in frog muscle. From these simple experiments the modern battery developed.

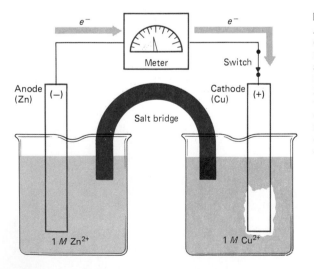

Figure 18.8
A galvanic cell produces an electric current through redox reactions. After the switch is closed, electrons released by the oxidation of Zn flow through the meter to the Cu electrode. These electrons reduce $Cu^{2+}(aq)$ to metallic Cu(s). The electric circuit of a galvanic cell is completed with a salt bridge.

Following the electron flow, electrons move through the wire and voltmeter to the cathode where they are immediately taken in by Cu^{2+}. To complete the electric circuit, cations from the salt bridge diffuse out to replace cations that are reduced, and in a similar manner, anions from the salt bridge replace those that are oxidized. Salt bridges are constructed out of bent glass tubing filled with a salt solution and with porous plugs on either end, or they are made of a porous ceramic or glass.

Galvanic cells are more commonly called **batteries.** Today we rely heavily on electric energy from battery sources. Calculators, watches, toys, and innumerable other devices are powered by batteries. Practical batteries are normally not like the galvanic cell just described; many are "dry cells" that do not contain a liquid electrolyte solution.

Figure 18.9 illustrates a typical **zinc-carbon dry cell** that might power a portable radio. Inside the outer metal or cardboard jacket is a zinc container, the anode. Inside the zinc container is a moist paste of ammonium chloride, NH_4Cl; carbon, C; and manganese dioxide, MnO_2. In the middle of the paste is a graphite rod, the cathode.

At the Zn anode, electrons are released.

Anode: $$Zn \longrightarrow Zn^{2+} + 2e^-$$

A complex reaction occurs at the cathode. It is thought that the following is one of the primary reduction reactions.

Cathode: $$2e^- + 2NH_4^+ + 2MnO_2 \longrightarrow 2NH_3 + Mn_2O_3 + H_2O$$

An accumulation of NH_3 in the cell would stop the flow of electricity, but this does not occur because most of the ammonia combines with Zn^{2+} in the paste, yielding a complex ion. Hydrogen gas is also produced at the cathode, then removed when it combines with the MnO_2 in the following reaction.

$$H_2 + MnO_2 + 2H^+ \longrightarrow Mn^{2+} + 2H_2O$$

Figure 18.9
A carbon-zinc dry cell has Zn metal at the anode and graphite (carbon) at the cathode. Carbon-zinc dry cells are inexpensive and are used in many different electric devices.

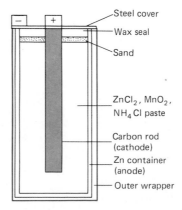

Steel cover
Wax seal
Sand
$ZnCl_2$, MnO_2, NH_4Cl paste
Carbon rod (cathode)
Zn container (anode)
Outer wrapper

Alessandro Volta (1745–1827) was a friend of Galvani. On hearing of Galvani's study of metals and frog muscle, Volta attempted to see if the metals produced electricity without the frog muscle. He discovered that it was the metals and not the tissue that created the current. A bitter battle began between him and Galvani. Later, Volta was the first to produce electrochemical cells as we know them today. He used mainly Cu and Zn.

Another commonly encountered battery is the **nickel-cadmium battery** (nicad), a battery superior to the dry cell because it is rechargeable. Cadmium metal is found at the anode and NiO_2 at the cathode. The anodic and cathodic reactions of nicad are as follows:

Anode: $$Cd + 2OH^- \longrightarrow Cd(OH)_2 + 2e^-$$

Cathode: $$2e^- + NiO_2 + 2H_2O \longrightarrow Ni(OH)_2 + 2OH^-$$

To recharge a battery, electricity is pumped back into the electrodes, reversing the reactions that produced the electric current. After it is recharged, the battery can be used again.

Lead storage batteries (Fig. 18.10) are used as automotive batteries. The anodes of lead storage batteries are composed of lead, Pb, and the cathodes contain lead(IV) oxide, PbO_2. Oxidation occurs when Pb releases two electrons and becomes Pb^{2+}, which immediately combines with SO_4^{2-} in the electrolyte, 30% H_2SO_4. The anodic reaction is as follows:

Anode: $$Pb + SO_4^{2-} \longrightarrow PbSO_4(s) + 2e^-$$

The $PbSO_4$ produced when a lead-acid battery discharges is loosely held to the electrodes. If the $PbSO_4$ flakes off the electrodes, the life of the battery is shortened.

Electrons from the oxidation reaction flow to the cathode, where they combine with PbO_2 and H^+ from the electrolyte to produce Pb^{2+} and water. Again the Pb^{2+} immediately combines with SO_4^{2-} from the solution, producing $PbSO_4$.

Cathode: $$2e^- + PbO_2 + 4H^+ + SO_4^{2-} \longrightarrow PbSO_4(s) + 2H_2O$$

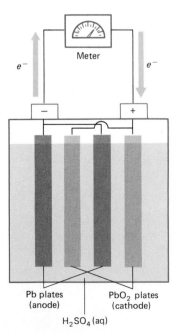

Figure 18.10
Lead-acid storage batteries are one of the most widely used batteries. The anode is composed of Pb plates, and the cathode is coated with PbO_2. Lead-acid batteries are rechargeable and can be used for many years if handled properly.

Meter

e^- e^-

$-$ $+$

Pb plates
(anode)
PbO_2 plates
(cathode)
H_2SO_4 (aq)

Adding the two equations together gives the overall equation for the reaction in lead storage batteries:

$$Pb + PbO_2 + 4H^+ + 2SO_4{}^{2-} \longrightarrow 2PbSO_4 + 2H_2O$$

Looking at the overall reaction in the lead storage battery, we see that, as electricity is produced, insoluble $PbSO_4$ and water are formed. Because the density of water (1 g/cm^3) is less than that of the H_2SO_4 solution (1.3 g/cm^3), measurement of the density of the liquid in the battery gives an indication of the degree to which the battery is charged. A density near 1.3 g/cm^3 means the battery is charged. A lower density indicates a partially discharged battery.

Lead storage batteries are recharged by applying a voltage to the electrodes. During the recharging process, the battery is not acting as a galvanic cell but as an electrolytic cell. Pumping electricity into the Pb electrode dissolves the insoluble $PbSO_4$, producing Pb^{2+} and $SO_4{}^{2-}$. Withdrawal of electrons at the PbO_2 electrode also dissolves $PbSO_4$, regenerating the sulfuric acid. In each case, the reactions for recharging are exactly the reverse of those for discharging. Consequently, the overall cell reaction is written as follows:

$$Pb + PbO_2 + 4H^+ + 2SO_4{}^{2-} \xrightleftharpoons[\text{charge}]{\text{discharge}} 2PbSO_4 + 2H_2O$$

A more exotic type of battery is a fuel cell. A common fuel cell produces electricity from the combination of $H_2(g)$ and $O_2(g)$. Fuel cells are efficient and are nonpolluting, but their cost is prohibitive due to the high cost of the electrodes. Fuel cells are used in space missions. In addition to producing electricity, they give off water as a byproduct.

18.9 How is a galvanic cell different from an electrolytic cell?

REVIEW EXERCISES

18.10 *(a)* Draw a diagram and explain each part of a galvanic cell that has a Sn electrode immersed in a Sn^{2+} solution, and a Ni electrode in a Ni^{2+} solution. *(b)* Explain what happens in the cell, and discuss the electron flow. (*Note:* Ni is a stronger reducing agent than Sn.)

18.11 What is the main difference between wet and dry galvanic cells?

18.12 What would happen if the $NH_3(g)$ produced in a dry cell did not combine with Zn^{2+}?

18.13 What oxidation and reduction reactions occur when a lead-acid battery discharges?

SUMMARY

Oxidation occurs when a substance loses electrons and increases its oxidation number. **Reduction** is exactly the opposite; it occurs when a substance gains electrons and decreases its oxidation number. Substances are classified according to their ability to undergo oxidation or reduction. A substance that readily undergoes oxidation is called a **reducing agent** because it brings about reduction by providing the necessary electrons. An **oxidizing agent** is a substance that undergoes reduction, and takes electrons from, or oxidizes, other substances.

The **ion-electron method** is one method for balancing redox equations. Two half-reactions are written that balance the substances that undergo oxidation and reduction. Each half-reaction is balanced by adding either H^+ and H_2O or OH^- and H_2O, depending on the conditions of the reaction. After the half-reactions are balanced, the equations are added to yield the net ionic equation.

Electrochemistry is the study of the interaction of

chemical and electric energy. Electric energy can be used to initiate chemical reactions. This process is called **electrolysis,** and occurs in a container called an **electrolytic cell.** Generally, electrolytic cells are constructed with a source of direct current connected to two electrodes immersed in an electrolyte solution. Reduction occurs at the negative electrode, while oxidation occurs at the positive electrode.

   Galvanic cells produce rather than consume electricity. Within a galvanic cell, the substance that undergoes oxidation is physically separated from the substance that undergoes reduction. Electrons provided by the reducing agent are transported to the oxidizing agent by a wire that is attached to a meter or some electrical device. The circuit is completed with a salt bridge or porous barrier that permits a flow of charged particles that keeps the solutions electrically neutral. Practical galvanic cells, or batteries, have a wide variety of uses.

KEY TERMS

anode	electrochemistry	half-reaction	oxidizing agent
battery	electrolysis	overall cell reaction	reducing agent
cathode	electrolytic cell	oxidation	reduction
electric current	galvanic cell	oxidation number	

EXERCISES*

18.14 Define each of the following: oxidation, reduction, oxidizing agent, reducing agent, oxidation number, half-reaction, electrochemistry, electrolytic cell, electrolysis, anode, cathode, electric current, overall cell reaction, galvanic cell, battery.

Oxidation-Reduction (General)

18.15 Describe what happens when a reducing agent and an oxidizing agent are combined. Give an example.

18.16 For each of the following equations, identify what has been oxidized and what has been reduced:
   (a) $2Ag^+ + Cu \longrightarrow 2Ag + Cu^{2+}$
   (b) $Fe + 2H^+ \longrightarrow Fe^{2+} + H_2$
   (c) $3Cu + 2NO_3^- + 8H^+ \longrightarrow 3Cu^{2+} + 2NO + 4H_2O$
   (d) $Zn + 2MnO_2 + 2NH_4^+ \longrightarrow Zn^{2+} + Mn_2O_3 + 2NH_3 + H_2O$
   (e) $Cl_2 + 2I^- \longrightarrow I_2 + Cl_2$
   (f) $2Cl^- + 2H_2O \longrightarrow Cl_2 + H_2 + 2OH^-$
   (g) $6KOH + 3Br_2 \longrightarrow 5KBr + KBrO_3 + 3H_2O$

18.17 What are the oxidizing and reducing agents in the following equations:
   (a) $2Al + 3F_2 \longrightarrow 2AlF_3$
   (b) $HCl + HNO_3 \longrightarrow NO_2 + \frac{1}{2}Cl_2 + H_2O$
   (c) $2MnO_4^- + 16H^+ + 10Cl^- \longrightarrow 2Mn^{2+} + 8H_2O + 5Cl_2$
   (d) $Pb + PbO_2 + 4H^+ + 2SO_4^{2-} \longrightarrow 2PbSO_4 + 2H_2O$

Balancing Redox Equations

18.18 Balance each of the following. Assume that all reactions take place in acidic solution.
   (a) $Cr_2O_7^{2-} + Br^- \longrightarrow Br_2 + Cr^{3+}$
   (b) $I^- + H_2O_2 \longrightarrow I_2 + H_2O$
   (c) $Zn + NO_3^- \longrightarrow Zn^{2+} + NH_4^+$
   (d) $NO_2 + HOCl \longrightarrow NO_2^- + Cl^-$
   (e) $AsH_3 + Ag^+ \longrightarrow As_4O_6 + Ag$
   (f) $Zn + H_2MoO_4 \longrightarrow Zn^{2+} + Mo^{3+}$

18.19 Balance each of the following. Assume that all reactions take place in basic solution.
   (a) $ClO^- + I^- \longrightarrow Cl^- + I_2$
   (b) $S^{2-} + I_2 \longrightarrow SO_4^{2-} + I^-$
   (c) $Al + H_2O \longrightarrow Al(OH)_4^- + H_2$
   (d) $P_4 \longrightarrow PH_3 + HPO_3^{2-}$
   (e) $Fe_3O_4 + MnO_4^- \longrightarrow Fe_2O_3 + MnO_2$
   (f) $Si + OH^- \longrightarrow SiO_3^{2-} + H_2$

18.20 Balance each of the following, using the ion-electron method:
   (a) $CrI_3 + H_2O_2 \longrightarrow CrO_4^{2-} + IO_4^- + H_2O$ (Base)
   (b) $XeO_3 + I^- \longrightarrow I_3^- + Xe$ (Base)
   (c) $HXeO_4^- \longrightarrow XeO_6^{4-} + Xe + O_2$ (Base)
   (d) $Pt + NO_3^- \longrightarrow Cl^- + PtCl_6^{2-} + NO_2$ (Acid)
   (e) $CN^- + MnO_4^- \longrightarrow MnO_2 + CNO^-$ (Acid)
   (f) $CrO_4^{2-} + HSnO_2^- \longrightarrow CrO_2^- + HSnO_3$ (Base)

   *For exercise numbers printed in color, answers can be found at the back of the book.

Electrochemical Cells

18.21 How is metallic conduction different from electrolytic conduction?

18.22 Sketch a diagram of an electrolytic cell of molten magnesium chloride. Label the anode and cathode, and write an equation to indicate what reaction occurs during the electrolysis.

18.23 Draw a diagram of an electrolytic cell that contains a concentrated aqueous sodium chloride solution. Label each electrode, indicate electron flow, and write half-reactions.

18.24 In an electrolytic cell, calcium metal and fluorine gas are produced at the electrodes. (a) At which electrodes are calcium and fluorine produced? (b) Write half-reactions to show the reactions at the electrodes.

18.25 Describe exactly how a steel can is electroplated with tin.

18.26 (a) What possible oxidations could occur in an electrolytic cell that contains aqueous potassium fluoride? (b) What reductions might occur in the same cell?

18.27 How could an impure sample of silver be purified using electrolysis?

Galvanic Cells

18.28 Draw a diagram of a galvanic cell that has Ni and Ag electrodes immersed in Ni^{2+} and Ag^+ solutions, respectively. Label the anode, cathode, and direction of electron flow.

18.29 Draw diagrams of galvanic cells with the following net reactions. Indicate the anode, cathode, and direction of electron flow, and write appropriate half-reactions.
(a) $Zn + Sn^{2+} \longrightarrow Sn + Zn^{2+}$
(b) $Cl^- + MnO_4^- \longrightarrow Cl_2 + Mn^{2+}$
(c) $Zn + H^+ \longrightarrow Zn^{2+} + H_2$

18.30 Why are salt bridges or porous barriers necessary components of galvanic cells?

18.31 (a) Write the half-reaction at the cathode of a lead storage battery. (b) Write the half-reaction at the anode of a nicad battery. (c) Write the cathode reaction for a regular dry cell.

18.32 How is the density of the electrolyte fluid in a lead storage battery used as an indication of the degree to which the battery is charged?

18.33 What advantage does a nicad battery have over a regular dry cell? Explain.

18.34 Describe what happens at each electrode when a lead storage battery is recharged.

CHAPTER

Nineteen

Nuclear Chemistry

STUDY GUIDELINES

After completing Chapter 19, you should be able to

1. Describe the composition of a nucleus given its atomic number and mass number

2. List empirical factors that are related to nuclear stability

3. Describe the properties and characteristics of the three principal forms of natural radiation

4. Explain the relationship between the rate of radioactive decay and the half-life of a nuclide

5. Determine how much of a substance remains after a given number of half-lives

6. Write nuclear equations for alpha and beta decay given the parent nuclide

7. Describe what happens to a nucleus when it emits a gamma ray

8. Discuss what happens in the three natural radioactive decay series

9. Write nuclear equations, complete and shorthand notation, for nuclear-bombardment reactions

10. Explain and write equations that illustrate how various transuranium elements were initially synthesized.

11. Describe what happens to a nucleus when it undergoes fission

12. Discuss the process of nuclear fusion and how it is different from nuclear fission

13. Explain how old objects are dated using radionuclides

14. List and describe medical uses for radionuclides

15. Describe the biological effects of high, moderate, and low exposure to radiation

16. Describe the primary sections of a nuclear power plant, and how nuclear energy is transformed into electric energy

17. Discuss the military uses of nuclear energy

Nuclear chemistry is the special area of chemistry that is concerned with changes in the nuclei of atoms. The nucleus is the region within an atom where the protons and neutrons are found. Collectively, the protons and neutrons are called the **nucleons.** Compared with the total volume of the atom, the volume of the nucleus is extremely small,

19.1 NUCLEUS

which means the nucleus is very dense. Nuclear density is approximately 1.8×10^{14} g/cm^3, an incredibly high density. A marble-size sample of nuclear material has a mass of approximately 8×10^8 tons.

Two values are used to describe the composition of a nucleus: the atomic number and mass number. The **atomic number** Z is the number of protons in the nucleus of an atom, and the **mass number** A is the sum of the protons and neutrons in the nucleus. Subtracting the atomic number from the mass number gives the total number of neutrons N in the atom.

$$A = Z + N$$

or

$$N = A - Z$$

When you write the symbol for an atom, the mass number is written as a superscript to the left of the symbol and the atomic number is written as a subscript, also to the left of the symbol.

$$\text{Mass number} \longrightarrow {}^{A}_{Z}\text{X}$$
$$\text{Atomic number} \longrightarrow$$

Atoms with the same atomic number but different mass numbers are called **isotopes.** In other words, these atoms have the same number of protons but different numbers of neutrons in the nucleus. In nuclear chemistry, the term "isotope" is used to refer to different forms of an individual element; the term **nuclide** generally refers to atomic forms of different elements. For example, we would speak of the principal isotopes of carbon, ${}^{12}_{6}$C and ${}^{13}_{6}$C. However, when discussing ${}^{12}_{6}$C compared to ${}^{14}_{7}$N, we refer to them as two different nuclides.

Some nuclides are stable and do not undergo changes unless subjected to extreme conditions. On the other end of the scale, some nuclides undergo spontaneous changes. Science has not yet been able to completely explain nuclear stability. It is thought that the nucleons are arranged in various energy levels. Just as some electronic configurations are more stable than others, some nuclear configurations are more stable than others.

It is known that nuclides with 2, 8, 20, 50, 82, or 126 protons or neutrons are usually more stable than other nuclides. For example, five stable isotopes of calcium ($Z = 20$) exist, but only two stable isotopes of potassium ($Z = 19$) and only one stable isotope of scandium ($Z = 21$) exist. Four stable nuclides with a neutron number of 20 are known, but no stable nuclides with either 19 or 21 neutrons are found in nature.

Nuclides with even numbers of nucleons (i.e., those with an even number of protons and neutrons) are more stable than those with either an odd number of protons or an odd number of neutrons. Over 150 stable nuclides belong to the group with even numbers of protons and neutrons. Approximately 55 stable nuclides exist that contain an even

Protons and neutrons are bonded tightly by the strong nuclear force. If two particles are brought within 10^{-13} cm of each other, the nuclear force binds them together. The nuclear force is much stronger than electrostatic forces.

Nuclear Stability

The mass of a nucleus is smaller than the sum of the masses of the protons and neutrons within the nucleus. The energy equivalent for this mass difference is calculated using Einstein's mass-energy relationship ($E = mc^2$). This energy difference is called the nuclear binding energy; it is the amount of energy needed to separate a nucleus into isolated nucleons.

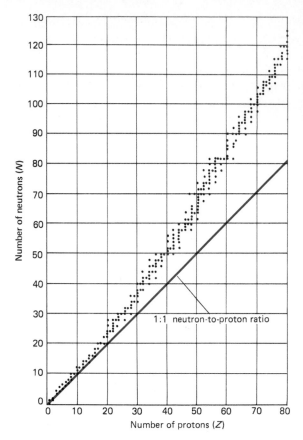

Figure 19.1
A plot of N versus Z reveals that the ratio of neutrons to protons increases with increasing atomic number. The most stable low-mass nuclides have the same number of neutrons and protons ($N/Z = 1$). Higher-mass nuclides have more neutrons than protons ($N/Z > 1$).

number of protons and an odd number of neutrons. About 50 stable nuclides have an odd number of protons and an even number of neutrons. Just five stable nuclides, all with mass numbers below 14, have odd numbers of protons and neutrons.

Finally, nuclear stability is also correlated to the ratio of the number of neutrons to the number of protons. Of nuclides with low mass numbers ($A < 41$), the most stable are usually those that contain the same number of protons and neutrons ($N/Z = 1$), for example, $^{12}_{6}C$, $^{16}_{8}O$, $^{32}_{16}S$, and $^{40}_{20}Ca$. For higher-mass nuclides ($A > 41$), this is not true; the most stable have a larger number of neutrons than protons ($N/Z > 1$).

When we look at Fig. 19.1, a graph of atomic number versus neutron number for naturally occurring stable nuclides, we find that the smaller nuclides are close to the line that represents equal number of protons and neutrons ($Z = N$). Higher-mass atoms are farther from this line because they contain more neutrons than protons.

REVIEW EXERCISES

19.1 What are nucleons, and where are they located in an atom?

19.2 Describe the difference in meaning of the terms "isotopes" and "nuclides." Give examples of each.

19.3 Consider the nuclide $^{103}_{45}$Rh, and answer the following. *(a)* What is its mass number? *(b)* What is its atomic number? *(c)* How many neutrons are located in its nucleus?

19.4 *(a)* What are the numbers of protons or neutrons found in more stable nuclei? *(b)* Are they the same numbers found for stable electronic configurations?

19.5 Considering the composition of their nuclei, predict which nuclide is the most stable of each of the following pairs: *(a)* $^3_1$H or $^4_2$He; *(b)* $^{12}_6$C or $^{14}_6$C; *(c)* $^{64}_{30}$Zn or $^{65}_{30}$Zn. Explain your reasoning for each.

19.2 RADIOACTIVITY

U nstable nuclei break down spontaneously and release various matter-energy forms. These spontaneous nuclear changes are called **radioactive disintegrations** or **radioactive decays.** In 1895, Henri Becquerel was the first scientist to observe this phenomenon. Marie and Pierre Curie (Fig. 19.2) later took an active role in investigating the nature of radioactive substances and their emissions. In 1903, Becquerel and the Curies shared the Nobel prize in physics for this pioneering scientific work.

Figure 19.2
Marie and Pierre Curie in their laboratory. *(New York Public Library Picture Collection.)*

Three matter-energy forms are commonly released by naturally occurring isotopes: (1) **alpha particles, α,** (2) **beta particles, β,** and (3) **gamma rays, γ.** Table 19.1 summarizes their properties.

TABLE 19.1 PROPERTIES OF THE PRINCIPAL NUCLEAR EMISSIONS

Emission	Symbol	Mass, u	Charge	Velocity*
Alpha	$_2^4\text{He}$ or $_2^4\alpha$	4.003	2+	0.1c
Beta	$_{-1}^0\text{e}$ or $_{-1}^0\beta$	0.0006	1−	0.9c
Gamma	γ	None	None	1.0c

*c = the speed of light = 3×10^8 m/s.

Alpha particles are the most massive radioactive particles, having the mass of a helium nucleus and a 2+ charge. Alpha particles are helium nuclei that have been ejected from nuclei at a speed approximately one-tenth that of the speed of light c ($c = 3 \times 10^8$ m/s). The symbol for an alpha particle is either $_2^4\text{He}$ or $_2^4\alpha$. Compared to the other two radioactive emissions, alpha particles have the smallest capacity to penetrate matter (penetration power). Clothing, paper, and other thin forms of matter generally absorb a large percent of alpha radiation. As alpha particles penetrate matter, they ionize atoms that they encounter. They have a greater capacity to ionize matter than either beta or gamma emissions.

Beta particles are high-energy electrons; thus, they are symbolized as $_{-1}^0\text{e}$. A −1 is written as the subscript to represent their negative charge, and a zero is written as the mass number because they contain no nucleons. Beta particles, like alpha particles, travel at a high velocity; some have a velocity nine-tenths that of the speed of light. They penetrate a greater thickness of matter than do alpha particles before being absorbed. Thick metal is needed to totally block beta particles. Their ability to ionize matter is much less than that of alpha particles, but greater than that of gamma rays.

Gamma rays have no mass or charge; therefore, the symbol for gamma rays is just the symbol gamma, γ. Gamma rays are a form of electromagnetic radiation that travel at the speed of light. Gamma rays have the greatest penetration power of the three forms discussed because of their minimal interaction with matter. Thick walls of lead or some other dense substance are required to totally block gamma radiation (Fig. 19.3). Gamma rays are the least-ionizing of the three radiation forms.

Gamma emission usually accompanies alpha and beta decay. After a nuclide releases an alpha or beta particle, the resulting nucleus is unstable and releases a gamma ray.

Half-Life

Radioactive nuclides may be characterized by their **half-lives,** or the amount of time required for one-half of the nuclei in a sample of a particular nuclide to decay. Let's consider a 10.0-g sample of hypothetical substance X with a half-life of 1.00 hr. After 1.00 hr elapses, only 5.00 g of X remains in the sample—half the nuclei, which represents half the mass, have changed to some other substance. Analysis at 2.00 hr would show that only 2.50 g of X remains; at 3.00 hr only 1.25 g of X is

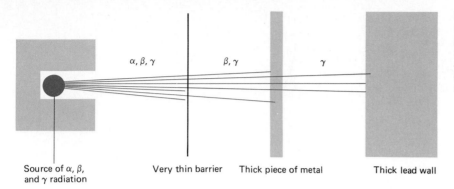

Figure 19.3
Gamma radiation has a greater penetration power than does alpha or beta radiation. Alpha radiation is the least-penetrating form of radioactivity.

| α, β, γ | β, γ | γ |

Source of α, β, and γ radiation   Very thin barrier   Thick piece of metal   Thick lead wall

present, and so on. Figure 19.4 shows a graph of the mass of substance X versus time.

Table 19.2 lists various nuclides and their half-lives. Half-lives can be extraordinarily long, in excess of a million years, or as short as a few milliseconds. $^{238}_{92}U$ has a half-life of 4.5 billion yr, about the time span that the earth is believed to have been in existence. $^{139}_{53}I$ has a half-life of only 2.7 s, which means that, every 2.7 s, half of the nuclei in $^{139}_{53}I$ release beta particles. Most nuclides have half-lives somewhere between these two extremes.

The curie, C_1, is the unit of rate of radioactive decay. One curie is the amount of a substance that gives 3.7×10^{10} disintegrations per second. Because the curie is a very large unit, the millicurie and microcurie are used most frequently.

TABLE 19.2 HALF-LIVES OF SELECTED NUCLIDES

Nuclide	Half-life	Emission
^{238}U	4.5×10^9 yr	α
^{40}K	1.3×10^9 yr	β^-, γ
^{235}U	7.1×10^8 yr	α
^{239}Pu	24,000 yr	α
^{14}C	5700 yr	β^-
^{90}Sr	28.8 yr	β^-
^{106}Ag	8.6 days	β^-, γ
^{250}Fm	0.5 hr	α
^{15}O	2.1 min	β^-
^{17}F	66 s	β^+
^{139}I	2.7 s	β^-
^{215}Po	1.8×10^{-3} s	α

REVIEW EXERCISES

19.6 Write the symbol for the radioactive emission described by each of the following phrases: *(a)* highest penetration power, *(b)* a high-energy electron, *(c)* most massive radioactive emission, *(d)* type of electromagnetic radiation, *(e)* most ionizing type of radiation.

19.7 What is the half-life of a radioactive substance?

19.8 From Table 19.2, find the half-life of $^{40}_{19}K$. If a 50.0-g sample of $^{40}_{19}K$ is present initially, what mass remains after three half-lives?

Figure 19.4
Initially, 10.0 g of radioactive substance X is present. Substance X has a half-life of 1.0 hr. Therefore, after 1.0 hr elapses, only 5.0 g of X remains. After 2.0 hr, 2.5 g of X remains. For each hour that passes, the mass of X becomes half of what it was at the beginning of the hour.

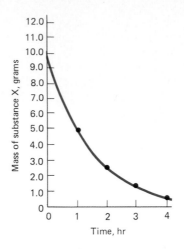

19.3 NUCLEAR REACTIONS

Natural Nuclear Changes

As we have seen, some nuclei spontaneously release alpha, beta, or gamma radiation. What happens to the nucleus in each of these changes? Let us begin with **alpha decay.** An alpha particle is composed of two protons and two neutrons. Thus, if a nucleus releases an alpha particle, its atomic number decreases by two and its mass number decreases by four. A nuclear equation that represents what happens in alpha decay is as follows:

$$^{A}_{Z}X \longrightarrow \ ^{A-4}_{Z-2}Y + \ ^{4}_{2}\alpha$$

A specific example is the nuclide $^{210}_{84}Po$. When it undergoes alpha decay, $^{206}_{82}Pb$ is produced.

$$^{210}_{84}Po \longrightarrow \ ^{206}_{82}Pb + \ ^{4}_{2}\alpha$$

In nuclear chemistry, the nuclide that undergoes decay is called the **parent nuclide,** and the nuclide that results is its **daughter nuclide.** In our example, $^{210}_{84}Po$ is the parent nuclide, and $^{206}_{82}Pb$ is the daughter nuclide.

When you write nuclear equations, the law of conservation of mass must be obeyed. This means that the sums of the mass numbers and atomic numbers are equal on either side of the equation. In other words, nucleons are conserved in nuclear reactions. Consider the alpha decay of $^{226}_{88}Ra$:

$$^{226}_{88}Ra \longrightarrow \ ^{222}_{86}Rn + \ ^{4}_{2}\alpha$$

On the left side of the equation, the total mass number is 226. On the right side, the sum of the mass numbers of $^{222}_{86}Rn$ and an alpha particle is also 226. The sum of the atomic numbers is 88 on both sides of the equation.

Rates of nuclear reactions are not changed by temperature, pressure, or catalysts.

Beta decay is somewhat more complex than alpha decay. Beta decay occurs when a nucleus is unstable because it has too many neutrons. Neutrons undergo a spontaneous change, producing a beta particle, a proton, and an antineutrino, $\bar{\nu}$.

$$\,^1_0\text{n} \longrightarrow \,^1_1\text{p} + \,^{\ 0}_{-1}\beta + \bar{\nu}$$

The resulting nucleus has the same total number of particles, or the same mass number, because a neutron is changed to a proton. However, because of the additional proton, the atomic number increases by one. Therefore, the general equation for beta decay is

A neutron is not composed of a proton and an electron, but is transformed into a proton and electron when it undergoes beta decay.

$$\,^A_Z\text{X} \longrightarrow \,^{\ \ A}_{Z+1}\text{Y} + \,^{\ 0}_{-1}\beta + \bar{\nu}$$

Besides the beta particle ejected from the nucleus, another matter-energy form is released. It was originally thought to be a massless, chargeless particle, but evidence has been presented to indicate that the antineutrino possesses an infinitesimal mass. Antineutrinos belong to another class of matter called antimatter. If a "regular" neutrino encounters an antineutrino, they annihilate each other—both are transformed totally to energy.

Consider the following nuclides that undergo beta decay:

$$\,^{231}_{90}\text{Th} \longrightarrow \,^{231}_{91}\text{Pa} + \,^{\ 0}_{-1}\beta + \bar{\nu}$$

$$\,^{14}_{6}\text{C} \longrightarrow \,^{14}_{7}\text{N} + \,^{\ 0}_{-1}\beta + \bar{\nu}$$

$$\,^{36}_{17}\text{Cl} \longrightarrow \,^{36}_{18}\text{Ar} + \,^{\ 0}_{-1}\beta + \bar{\nu}$$

In each of the above examples, a beta particle and antineutrino are released during beta decay. The daughter nuclide that results is an isotope of the element that occupies the block to the right of the parent nuclide on the periodic table—it has one more proton in the nucleus.

Gamma emission takes place without a measurable mass change to the nucleus because gamma radiation is a form of electromagnetic energy. Thus, nuclei that possess excess energy become more stable by releasing gamma radiation. After gamma emission, the nucleus is in a lower energy state. A similar occurrence is the release of light energy when electrons are in higher energy states.

To represent gamma emission in an equation, an asterisk (*) superscript is generally written next to the symbol, indicating a nucleus with excess energy. After the gamma ray is released, the same nucleus is present, but without the excess energy; thus the asterisk is removed.

$$\,^A_Z\text{X}^* \longrightarrow \,^A_Z\text{X} + \gamma$$

In nature, three principal radioactive decay series are found; in each, elements with mass numbers above 83 decay spontaneously in a stepwise fashion until a stable nuclear configuration is reached. All three series end with the element Pb:

1. ^{238}U series: ^{238}U to ^{206}Pb
2. ^{235}U series: ^{235}U to ^{207}Pb
3. ^{232}Th series: ^{232}Th to ^{208}Pb

Figure 19.5 graphically illustrates the $^{238}_{92}$U radioactive decay series that starts with the parent nuclide $^{238}_{92}$U, which initially undergoes alpha decay to produce $^{234}_{90}$Th. $^{234}_{90}$Th releases a beta particle, changing to $^{234}_{91}$Pa, which is also a beta emitter. After $^{234}_{91}$Pa releases a beta particle, $^{234}_{92}$U is formed. After a series of five alpha emissions, the mass number is decreased by 20, and the atomic number is decreased by 10, yielding an unstable isotope of lead, $^{214}_{82}$Pb. At this point the decay series becomes somewhat more complex because more than one pathway is possible. Upon reaching $^{206}_{82}$Pb, the series ends—$^{206}_{82}$Pb has a stable nucleus that does not undergo radioactive decay.

In addition to the natural radioactive processes, scientists have intervened and have devised procedures for initiating nuclear changes in nuclides that normally are not radioactive. The first artificial transmutation of an element was performed by Rutherford in 1919. He bombarded $^{14}_{7}$N with alpha particles, producing $^{17}_{8}$O and a proton.

$$^{14}_{7}N + ^{4}_{2}\alpha \longrightarrow ^{17}_{8}O + ^{1}_{1}H$$

Irene Joliot-Curie (1897–1956) and her husband Frederic Joliot (1900–1958) were the first to demonstrate artificial radioactivity, the production of a nuclide that continuously emits radiation. They bombarded $^{10}_{5}$B with alpha particles, producing $^{13}_{7}$N, which spontaneously decays to $^{13}_{6}$C:

$$^{10}_{5}B + ^{4}_{2}\alpha \longrightarrow ^{13}_{7}N + ^{1}_{0}n$$
$$^{13}_{7}N \longrightarrow ^{13}_{6}C + ^{0}_{+1}e^{+}$$

Both neutrons and positrons are emitted in this reaction. A positron is another form of antimatter. **Positrons** ($^{0}_{+1}e^{+}$) have the same properties as electrons except that they have a positive charge; when a positron encounters a "regular" electron, the pair is annihilated releasing energy ($e^{-} + e^{+} \rightarrow 2\gamma$).

Unstable nuclei with an excess number of protons tend to undergo **positron emission.** A proton changes to a neutron, emitting a positron, $^{0}_{+1}e^{+}$, and a neutrino (not an antineutrino).

Nuclei with too many protons can also decay by electron capture (ec). An electron in the 1s orbital is "captured" by a proton in the nucleus, and is transformed into a neutron. No radiation is emitted during this nuclear change. However, an x-ray is released when a higher-energy electron fills the vacancy left by the captured electron.

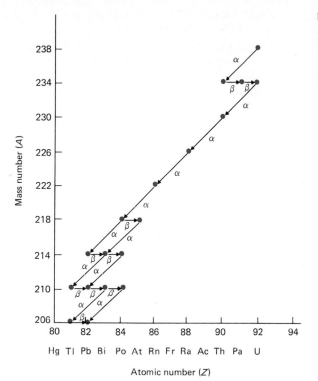

Figure 19.5
In the $^{238}_{92}U$ natural decay series, $^{238}_{92}U$ is the parent nuclide for a series of daughter nuclides that ultimately decay to the stable nuclide ^{206}Pb.

$$^1_1p^+ \longrightarrow \,^1_0n + \,^0_{+1}e^+ + \nu$$

A positron emitter, such as $^{13}_7N$, decays to the element with an atomic number one less, exactly the opposite of a beta emitter.

A shorthand notation is used to simplify the writing of nuclear bombardment equations. Inside parentheses is written the symbol for the bombarding particle separated by a comma from the symbol for the particle that is released. The target nucleus is written to the left, and the resulting nucleus is written to the right, of the parentheses. For example, to simplify the nuclear equation that represents what happens when ^{10}B is bombarded with alpha particles to produce ^{13}N and a neutron, we write:

$$^{10}_5B(^4_2\alpha, 1^1_0n)^{13}_7N$$
Target nucleus ⎯↑   ↑⎯Resulting nucleus

When $^{27}_{13}Al$ is bombarded with a neutron, $^{27}_{12}Mg$ and a proton are formed.

$$^{27}_{13}Al + \,^1_0n \longrightarrow \,^{27}_{12}Mg + \,^1_1p$$

Sir James Chadwick discovered the neutron in 1932 when he bombarded ^9_4Be with alpha particles. In addition he produced $^{12}_6C$.

The shorthand notation is

$$^{27}_{13}\text{Al}(^1_0\text{n}, \, ^1_1\text{p})^{27}_{12}\text{Mg}$$

Two English scientists, Cockcroft and Walton, constructed the first high-energy linear **particle accelerator,** a machine that takes small particles, like protons, and accelerates them before they are directed into a target nuclide. Cockroft and Walton, in 1932, successfully accelerated protons and directed them into a sample of ^7_3Li, initially producing ^8_4Be, which splits and forms two ^4_2He nuclides.

$$^7_3\text{Li} + ^1_1\text{H} \longrightarrow ^8_4\text{Be} \longrightarrow ^4_2\text{He} + ^4_2\text{He}$$

Later in 1932, Lawrence and Livingston developed an even more powerful particle accelerator called a cyclotron. A **cyclotron** accelerates particles in a spiral path, which gives greater velocities and thus allows for many more nuclear transformations.

Cyclotrons are used to synthesize new elements that do not exist in nature. Table 19.3 lists the synthetic transuranium elements ($Z > 92$). The **transuranium elements** are those directly after uranium on the periodic table. McMillan and Abelson at the University of California at Berkeley synthesized the first transuranium element, $^{239}_{93}\text{Np}$, in 1940. They bombarded $^{238}_{92}\text{U}$ with neutrons produced in a cyclotron, giving $^{239}_{93}\text{Np}$ and beta particles.

$$^{238}_{92}\text{U}(^1_0\text{n}, \, ^{0}_{-1}\beta)^{239}_{93}\text{Np}$$

Later in 1940, Seaborg, McMillan, Kennedy, and Wahl, also at Berkeley, discovered plutonium, Pu. They bombarded $^{238}_{92}\text{U}$ with deuterium, ^2_1H, initially producing an unstable daughter nuclide, $^{238}_{93}\text{Np}$, which underwent beta decay to form $^{238}_{94}\text{Pu}$.

$$^{238}_{92}\text{U}(^2_1\text{H}, \, 2^1_0\text{n})^{238}_{93}\text{Np}$$

$$^{238}_{93}\text{Np} \longrightarrow \beta^- + ^{238}_{94}\text{Pu}$$

Another extremely important nuclear reaction is **nuclear fission,** in which a nucleus is split into fragments. When nuclides are fissioned, the products are many different smaller nuclides.

In 1934, Enrico Fermi (Fig. 19.6) was the first person to split an atom; he bombarded $^{238}_{92}\text{U}$ with neutrons. Fermi predicted that the $^{238}_{92}\text{U}$ would absorb a neutron, undergo beta decay, and then produce element 93, Np. However, Fermi discovered at least four different products. Four years later, Otto Hahn and Fritz Strassmann analyzed the reaction products of the Fermi experiment and isolated Ba, a substance with a mass number approximately one-half that of uranium. Hahn communicated his finding to an old friend, Lise Meitner, who immediately de-

Element 103 is named for the American physicist E. O. Lawrence (1901–1958), who invented the cyclotron.

Transuranium elements were named in a similar manner to the lanthanide elements directly above them on the periodic table. Americium was named to parallel europium. Curium was named for the Curies, who were the founders of the science of radioactivity, and gadolinium was named for the Finnish chemist Gadolin, who was a pioneer of the rare earths.

Nuclear Fission

TABLE 19.3 PREPARATION OF TRANSURANIUM ELEMENTS

Atomic number	Name and symbol	Reaction
93	Neptunium, Np	$^{238}U(^1_0n, \beta)^{239}Np$
94	Plutonium, Pu	$^{238}U(^2_1H, 2^1_0n)^{238}Np$
		$^{238}Np \longrightarrow {}^{238}Pu + \beta^-$
95	Americium, Am	$^{239}Pu(^1_0n, \beta^-)^{240}Am$
96	Curium, Cm	$^{239}Pu(^4_2\alpha, ^1_0n)^{242}Cm$
97	Berkelium, Bk	$^{241}Am(^4_2\alpha, 2^1_0n)^{243}Bk$
98	Californium, Cf	$^{242}Cm(^4_2\alpha, ^1_0n)^{245}Cf$
99	Einsteinium, Es	$^{238}U(15^1_0n, 7\beta^-)^{253}Es$
100	Fermium, Fm	$^{238}U(17^1_0n, 8\beta^-)^{255}Fm$
101	Mendelevium, Md	$^{253}Es(^4_2\alpha, ^1_0n)^{256}Md$
102	Nobelium, No	$^{246}Cm(^{12}_6C, 4^1_0n)^{254}No$
103	Lawrencium, Lr	$^{252}Cf(^{10}_5B, 5^1_0n)^{257}Lr$
104	Unnilquadium, Unq	$^{249}Cf(^{12}_6C, 4^1_0n)^{257}Unq$
105	Unnilpentium, Unp	$^{249}Cf(^{15}_7N, 4^1_0n)^{260}Unp$
106	Unnilhexium, Unh	$^{249}Cf(^{18}_8O, 4^1_0n)^{263}Unh$

Lise Meitner (1878–1968) obtained her doctorate at the University of Vienna under the direction of Ludwig Boltzmann. She traveled to Berlin in 1907 to hear the lectures of Max Planck. Here she met and joined Otto Hahn's research group. Meitner was forced to leave Germany when Hitler came to power. With the help of Niels Bohr, she obtained a position in Sweden, where she discovered nuclear fission.

duced what had happened. She realized that the U nucleus had split; Meitner then coined the term "nuclear fission." Meitner asked her nephew, Otto Frisch, to repeat the experiment. He discovered that the U

Figure 19.6
Enrico Fermi (1901–1954) was a magna cum laude Ph.D. graduate from the University of Pisa. Fermi, whose wife was Jewish, was forced to leave his native country of Italy because he refused to support the fascist government of Mussolini. He moved to the United States, where he made many valuable contributions to the understanding of nuclear physics, *(Argonne National Laboratories.)*

nucleus had fragmented and produced two new atoms, barium and krypton. More important, Frisch discovered that the fission process released a colossal amount of energy, approximately 2×10^{10} kJ/mol U.

In some cases, fission reactions even occur spontaneously without a bombardment reaction. $^{252}_{98}\text{Cf}$ spontaneously undergoes fission, producing many different products. One possible way that $^{252}_{98}\text{Cf}$ fissions is as follows:

$$^{252}_{98}\text{Cf} \longrightarrow {}^{140}_{56}\text{Ba} + {}^{108}_{42}\text{Mo} + 4{}^{1}_{0}\text{n}$$

Nuclear fusion is the nuclear reaction in which low-mass nuclides, such as H, combine to form a more massive nucleus. At the same time, a large quantity of energy is liberated. A nuclear fusion reaction only takes place if a tremendous amount of energy is initially present to overcome the internuclear electrostatic repulsive forces. Temperatures in excess of 100,000,000°C are typically needed to initiate fusion reactions.

Nuclear fusion is the process that produces energy in stars. Stars generate energy by "burning" H. In the sun, the following fusion reactions occur.

$$^{1}_{1}\text{H} + {}^{1}_{1}\text{H} \longrightarrow {}^{2}_{1}\text{H} + {}^{0}_{1}\text{e}^{+}$$

$$^{2}_{1}\text{H} + {}^{1}_{1}\text{H} \longrightarrow {}^{3}_{2}\text{He}$$

$$^{3}_{1}\text{He} + {}^{3}_{2}\text{He} \longrightarrow {}^{4}_{2}\text{He} + 2{}^{1}_{1}\text{H}$$

After a significant $^{4}_{2}\text{He}$ concentration accumulates in a star, the $^{4}_{2}\text{He}$ is fused to produce the higher-atomic-mass elements. For example, it is thought that $^{12}_{6}\text{C}$ is produced by the fusion of three $^{4}_{2}\text{He}$:

$$3{}^{4}_{2}\text{He} \longrightarrow {}^{12}_{6}\text{C}$$

All of the heavy elements in the universe have been formed in the cores of stars by nuclear fusion reactions.

Nuclear Fusion

When matter is heated to extremely high temperatures, like those needed to sustain fusion reactions, it is in the plasma state—a fourth state of matter.

REVIEW EXERCISES

19.9 What changes in mass number and atomic number occur in nuclides that undergo *(a)* beta decay, *(b)* alpha decay, *(c)* gamma emission, *(d)* positron emission?

19.10 Write a nuclear equation for each of the following radioactive changes: *(a)* beta decay of $^{95}_{40}\text{Zr}$, *(b)* alpha decay of $^{174}_{72}\text{Hf}$, *(c)* beta decay of $^{187}_{75}\text{Re}$, *(d)* gamma emission of $^{89}_{38}\text{Sr}^{*}$.

19.11 *(a)* List the parent nuclides of the three natural radioactive decay series. *(b)* What stable daughter nuclides are formed in each of these series?

19.12 *(a)* What is positron? List its properties. *(b)* Write an equation to show how a positron forms.

19.13 Translate the following nuclear equations into the shorthand notation:
(a) $^{40}_{18}\text{Ar} + {}^{4}_{2}\alpha \longrightarrow {}^{43}_{19}\text{K} + {}^{1}_{1}\text{p}^{+}$
(b) $^{59}_{27}\text{Co} + {}^{1}_{0}\text{n} \longrightarrow {}^{56}_{25}\text{Mn} + {}^{4}_{2}\alpha$
(c) $^{12}_{6}\text{C} + {}^{2}_{1}\text{H} \longrightarrow {}^{13}_{7}\text{N} + {}^{1}_{0}\text{n}$

19.14 Translate the shorthand notation given in Table 19.3 to complete nuclear equations for the synthesis of *(a)* Am, *(b)* Es, *(c)* No, *(d)* Unh.

19.15 *(a)* What is the difference between nuclear fission and nuclear fusion? *(b)* Write an equation to illustrate each.

Radioactive dating is an indirect method of estimating the age of an old rock or artifact of interest. To illustrate, let us consider **radiocarbon dating.** Ancient objects that contain carbon are dated by measuring a radioactive isotope of carbon, $^{14}_{6}C$, a beta emitter. In nature, $^{14}_{6}C$ is produced in the atmosphere from $^{14}_{7}N$ by the following reaction:

$$^{14}_{7}N + ^{1}_{0}n \longrightarrow ^{14}_{6}C + ^{1}_{1}H$$

Once formed, the $^{14}_{6}C$ is oxidized to $^{14}_{6}CO_2$:

$$^{14}_{6}C + O_2 \longrightarrow ^{14}_{6}CO_2$$

Plants take in $^{14}_{6}CO_2$ during photosynthesis, and animals ingest compounds that contain $^{14}_{6}C$ in the foods they eat.

A balance between the amount of $^{14}_{6}C$ that enters and leaves a living system keeps the $^{14}_{6}C$ concentration constant. Because the concentration of $^{14}_{6}C$ in CO_2 is 1 per 10^{12} carbon atoms, it is assumed that 1 out of 10^{12} carbon atoms in a living organism is $^{14}_{6}C$. After the death of an organism no more $^{14}_{6}C$ enters (Fig. 19.7). However, the amount of $^{14}_{6}C$ in the dead organism decreases as a result of radioactive decay. The half-life of $^{14}_{6}C$ is approximately 5700 years; thus, with the passing of every 5700 years, the amount of $^{14}_{6}C$ decreases by one-half. Careful measurement of the amount of $^{14}_{6}C$ that remains in an old object, compared with the amount of $^{14}_{6}C$ in a modern object, allows scientists to roughly estimate the age of the object.

Radiocarbon dating gives only an approximation of the age of the object; exact ages cannot be obtained with this method. A problem with

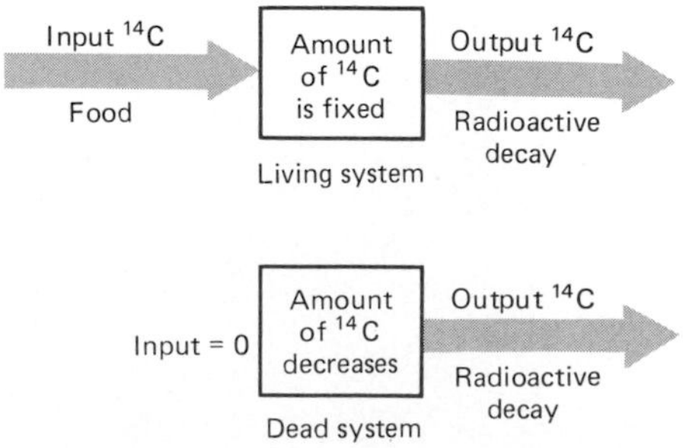

19.4 USES OF RADIOACTIVE SUBSTANCES

Radioactive Dating

Figure 19.7
Within living systems, the amount of $^{14}_{6}C$ is relatively constant. After the organism dies, the amount of $^{14}_{6}C$ decreases as a result of radioactive decay.

the validity of radiocarbon dating is that an assumption must be made that the amount of $^{14}_6C$ on earth has remained constant throughout the years. Radiocarbon dating is also limited by the sensitivity of radiation-detecting devices that measure $^{14}_6C$; consequently, after approximately 10 half-lives, $^{14}_6C$ activity cannot be measured as accurately.

$^{40}_{19}K$ and $^{40}_{18}Ar$ are used by geologists to date rocks. $^{40}_{19}K$–$^{40}_{18}Ar$ dating is different from radiocarbon dating in that $^{40}_{19}K$ decays to $^{40}_{18}Ar$, which is trapped in the rocks where it is found. Accordingly, the amount of $^{40}_{18}Ar$ accumulated is a measure of the age of the rock. Two properties of $^{40}_{19}K$ make it an excellent substance for estimating the age of rocks: (1) $^{40}_{19}K$ has a long half-life, 1.3×10^9 years; and (2) $^{40}_{19}K$ has a relatively high natural abundance, in excess of 0.01%.

Medical Uses

Prior to the development of medical isotopic tracers and sophisticated computer-analyzed x-ray techniques (CT scans), a person suspected of having a perplexing internal problem would normally have to undergo exploratory surgery to find the nature of the problem. In part radioisotopes and modern x-ray techniques have diminished the need for surgical diagnostic procedures.

In excess of 100 different radionuclides have been utilized in medical diagnostic techniques. Table 19.4 lists some of the more frequently used isotopes and the problems they detect. For example, $^{131}_{53}I$, a beta emitter, is used to diagnose diseases of the thyroid gland. The patient is given a solution of sodium iodide, $Na^{131}_{53}I(aq)$, to drink. After a sufficient period of time has elapsed, enough for the $^{131}_{53}I^-$ to be absorbed by the blood and taken into the thyroid, the patient is placed in a scanning machine. A scanner is a machine that contains a radiation detector that transforms the intensity of radioactivity at a particular point into a visual image (Fig. 19.8). Analysis of the image, or "scan," by a radiologist provides valuable information about the health of the thyroid gland.

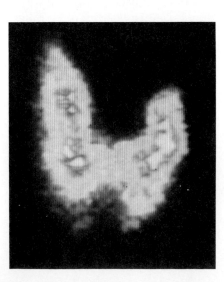

Figure 19.8
A radioactive scan of the thyroid. *(Bruce Coleman, Inc.)*

TABLE 19.4 COMMON RADIONUCLIDES USED IN MEDICINE

Nuclide	Half-life	Use*
^{137}Cs	30 yr	T; implanted to destroy tumors
^{60}Co	5.3 yr	T; destroys cancers
^{198}Au	2.7 days	T; abdominal cancers
		D; liver imaging
^{32}P	14 days	T; chronic leukemia
^{131}I	8.1 days	D; thyroid and brain scans
^{24}Na	15 days	D; vascular disease
^{59}Fe	46 days	D; blood volume, iron metabolism
^{99}Tc	6.1 h	D; brain, heart, and bone scans
^{42}K	12 h	D; localizing brain tumors

*T = therapeutic use; D = diagnostic use. (Only selected uses are presented in this table; most of the above nuclides have other uses as well.)

Also listed in Table 19.4 are radionuclides used to treat medical problems. One of the more important therapeutic uses of radioisotopes is to destroy malignant tumors. Either radiation is beamed through a person, or a radioactive substance is implanted at the site of the tumor. $^{60}_{27}$Co, an intense gamma and beta emitter, is used as an external source of radiation that can be concentrated on a tumor. $^{137}_{55}$Cs needles are implanted near tumors until they are destroyed; the $^{137}_{55}$Cs is then removed.

Biological Effects of Radiation

Even though radioactivity is used to diagnose or treat various medical problems, exposure to radiation has a profound negative effect on human health. Once a person has been exposed to excessive levels of radiation, the world of medicine can do nothing to reverse the effects. Excessive exposure to high levels of radiation always results in death.

TABLE 19.5 BIOLOGICAL EFFECTS OF RADIATION

Dose*, rem†	Effect
0–25	No noticeable biological effects. A chest x-ray is 0.2 rem, and a dental x-ray is approximately 0.02 rem.
25–50	Small effect, usually a small decrease in white blood cell count.
100–200	Significant lowering of white blood cell count, nausea, and general sickness. Normally, no deaths are expected.
200–400	Vomiting and nausea the first day; later, diarrhea and general sickness with hair and skin loss. About 20% of the people will die within the first month after exposure.
400–500	Same symptoms as above. About 50% of the people will die in first month.

*Short-term exposure.
†rem is the abbreviation for roentgen equivalent in man, a unit of radiation exposure. The higher the value, the greater the damage to human tissues.

The unit roentgen equivalent in man (rem) is related to another radiation unit called the radiation absorbed dose (rad). One rad equals 100 ergs ($1 erg = 10^{-7}$ J) of energy absorbed per gram of living tissue. One rad is almost equivalent to one roentgen (R), a unit of the degree of ionization of dry air by radiation forms.

Exposure to lower but significant levels initially produces nausea, anemia, vomiting, and diarrhea, as well as skin and hair loss. Later, rare blood diseases or leukemia frequently result, as well as genetic defects that can be passed to the next generation. Table 19.5 lists symptoms associated with different levels of radiation exposure.

What is a safe level of exposure to radiation? No one really knows the answer to this controversial question. Table 19.6 summarizes the sources and per capita levels of radiation that U.S. citizens are exposed to annually. Note that exposures to natural and artificial sources of radiation are nearly equal, which means that the activities of humans produce approximately half of the radiation that people are exposed to. It is interesting to note that the largest percent of exposure to artificial radiation is from medical and dental x-rays!

TABLE 19.6 RADIATION EXPOSURE IN THE UNITED STATES

Source	Average exposure, millirems/year per capita	
Natural background	110	
Total artificial exposure	90	
Medical and dental x-rays		80
Technology and industry		4
Fallout from nuclear weapons		5
Nuclear power plants		0.3
Consumer products (watches, TV, etc.)		0.04

A severe health problem that results from exposure to radiation is of great concern today. Studies have shown that people who live in houses that contain highly radioactive radon, Rn, gas are much more likely to develop malignancies such as lung cancer. In various parts of the United States, the soil contains radioactive substances that decay to Rn gas. This gas enters houses through microscopic pores in their foundations. Long-term exposure to Rn gas has a profound negative effect on the health of people. Research studies have also indicated that Rn gas is found in minute amounts in cigarette smoke; thus, it could be a major contributing factor to the high lung cancer rates in cigarette smokers.

Nuclear Energy Production

Energy from a controlled fission reaction was first harnessed by Enrico Fermi under the stadium on the campus of the University of Chicago, on December 2, 1942. From this beginning, modern nuclear power plants have evolved.

A nuclear power plant (Fig. 19.9) uses a fissionable material such as $^{235}_{92}U$ or $^{239}_{94}Pu$ to produce heat energy, which is then converted to electric energy. Small fuel pellets are loaded into the core of the reactor. Located between the fuel rods are neutron-absorbing rods (usually constructed from B or Cd), which are automatically withdrawn when the neutron level decreases, and are reinserted when the level becomes too high. A

Fuel rods contain an oxide of uranium, U_3O_8.

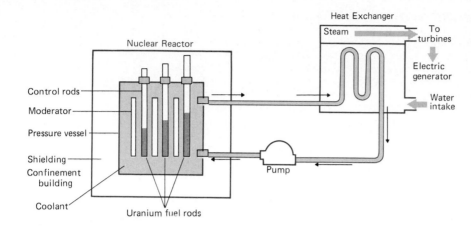

Figure 19.9
In a nuclear power plant, heat is generated from the nuclear fission of U or Pu atoms. This heat converts water to steam, and the steam turns a turbine that is connected to an electric generator.

fluid, such as water or liquid sodium, is passed through the reactor core, picking up the heat liberated by the fission process. This heat is then transferred to nonradioactive water outside the core, changing the water to steam. High-pressure steam turns a turbine, which is connected to an electric generator. Other than the nuclear reactor and its associated equipment, nuclear power plants operate in a manner similar to fossil fuel plants, which burn oil, coal, and natural gas to produce heat.

Controversy surrounds the use of nuclear power plants. Most of the controversy results from the unanswerable questions: Are nuclear power plants safe? How safe is safe? Besides the problems of operating power plants safely, there is the ever present problem of nuclear waste storage. Presently, no adequate method has been devised for transporting and storing the intensely radioactive spent fuel rods. This problem could be minimized with the development of efficient, safe reprocessing facilities that could separate the "good" fissionable material from the waste.

No simple solutions exist for such complex problems associated with nuclear fission power generation. Answers to these problems must address not only their scientific aspects but also the social, political, and economic factors. We may never have to totally solve these problems because controlled fusion energy may eliminate the need for fission power generation.

Controlled fusion is a process in which low-mass elements are fused, releasing an enormous quantity of energy, simulating the energy-producing apparatus of the sun on earth.

Many scientific and technological hurdles must be overcome to make available this unlimited supply of relatively cheap and clean energy. At present one of the biggest problems is the temperatures required to sustain a fusion reaction, greater than 100,000,000°C. No known substance can withstand such enormous temperatures. If you are an optimist, place a bet that science and technology will solve these problems in the next 50 years.

CHEM TOPIC: Nuclear Waste Disposal

One of the most controversial and debated issues in the nuclear industry is the disposal of high-level radioactive waste materials produced by nuclear power plants. Presently, no fail-safe method has been developed to dispose of nuclear waste outside the nuclear power plants in which they were produced. A typical nuclear power plant in the United States contains approximately 40,000 fuel rods that have a combined mass of about 120,000 kg (27 tons). These fuel rods are enriched in ^{235}U (about 3% of the U in the rods), the fissionable isotope of uranium. Many different radioactive waste substances are produced in the core of the nuclear reactor. For example, ^{235}U is converted to the very toxic and dangerous ^{239}Pu, which has a half-life of 24,000 yr. Fission products of ^{235}U include ^{90}Sr, whose half-life is 29 yr, and ^{137}Cs, whose half-life is 30 yr. Thus, nuclear waste materials will remain intensely radioactive for many years to come.

One of the most commonly discussed means for disposal of nuclear wastes is to combine them with a ceramic material and then bury them in geologically stable salt mines thousands of feet below the surface. A number of studies have shown this to be a feasible method of disposal. However, the amount of time needed for storage is longer than human society has been in existence, and no one knows if this plan will be viable over such a long time span.

A second means that has been debated is removal of the fissionable material from the spent rods, with the material then used to construct new fuel rods. However, this method leaves behind liquid wastes that are also radioactive; thus, this method does not solve the problem. It also produces a tremendous amount of fissionable material which could be used by governments or terrorists to produce nuclear bombs.

As the debate continues, nuclear waste materials are being produced and stored in the nuclear power plants in which they are formed. Very shortly some of the older plants will have to be closed because they will no longer be able to store their wastes. It has been proposed that these plants be entombed in concrete and left for eternity for future generations to worry about. It is probably safe to say that the problems of nuclear waste disposal will not be solved in this century.

Military Applications

Energy from nuclei that undergo fission or fusion may also be released destructively from powerful bombs. During World War II, the United States was the first country to develop nuclear weapons. In response to a now famous letter by Albert Einstein, F. D. Roosevelt with his advisors initiated the Manhattan Project, which was the research and development effort to produce an "atomic bomb." At that time the Manhattan project became the most expensive coordinated scientific research project ever conducted.

An atomic bomb was developed and successfully tested in the desert near Alamogordo, New Mexico, on July 16, 1945. Less than one month later, on August 6, 1945, an atomic bomb was dropped on the city of Hiroshima in Japan. In relation to the bombs now stockpiled throughout the world, the destructive power of the Hiroshima bomb was quite small. Today, nuclear bombs are classified in terms of their equivalent explosive power to the number of millions of tons (megatons) of TNT. The Hiroshima bomb only packed a wallop of 20,000 tons of TNT.

Early atomic bombs derived their energy from the rapid release of energy when nuclei were fissioned. Nuclear fission can only be sustained when a large enough mass of fissionable material is present. This mass is called the *critical mass*. If less than the critical mass is present, the sample is said to be subcritical. Thus, atomic bombs were constructed with two subcritical masses separated from each other. To explode the bomb, conventional explosives were employed to accelerate the two subcritical masses together, forming a critical mass that explodes (Fig. 19.10).

Splitting the nuclei of U or Pu atoms requires the absorption of slow neutrons. For example, a $^{235}_{92}$U nucleus is fissioned as follows:

$$^{235}_{92}\text{U} + ^{1}_{0}\text{n} \longrightarrow ^{90}_{38}\text{Sr} + ^{143}_{54}\text{Xe} + 3^{1}_{0}\text{n}$$

As the nucleus fragments, energy is released and two or three neutrons are expelled; these neutrons re-initiate the fission process. It is easy to see that very rapidly millions of nuclei are fissioned. We refer to this as a *chain reaction* (Fig. 19.11). Therefore, a critical mass of fissionable material is the minimum mass that retains a large enough percent of neutrons to sustain a chain reaction.

Figure 19.10
The energy released from an atomic bomb explosion results from the transformation of a small amount of matter into a large amount of energy. The explosion shown occurred in 1952 at a Nevada test site. The United States no longer conducts atmospheric tests on nuclear devices.

Figure 19.11
A chain reaction occurs during nuclear fission. The chain reaction is initiated when a slow-moving neutron is absorbed by a $^{235}_{92}U$ nucleus. The resulting unstable nucleus splits and produces Sn and Mo atoms. In addition, two neutrons are released that can split two $^{235}_{92}U$ nuclei, which release four neutrons that fission four nuclei. If a large enough mass of radioactive material is present, an uncontrolled chain reaction leads quickly to a nuclear explosion.

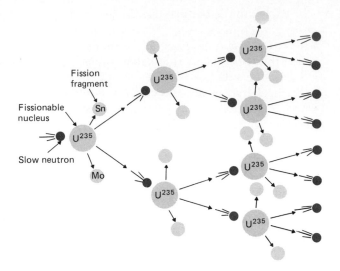

Modern, more powerful weapons obtain energy from the fusion process. These bombs are frequently called hydrogen bombs because the explosive material consists of two isotopes of hydrogen, deuterium, ^2_1H, and tritium, ^3_1H. The United States tested the first fusion bomb at Bikini atoll in the Pacific ocean in 1952. To produce the extremely high temperatures required to fuse nuclei, a hydrogen bomb is constructed with a small fission bomb in its core. Upon detonation of the fission bomb, enough energy is released to initiate the fusion reaction. Neutrons released in the initial explosion convert ^6_3Li into ^4_2He and ^3_1H. Immediately, the ^3_1H fuses with ^2_1H, producing more ^4_2He and a large amount of energy.

$$^6_3Li + ^1_0n \longrightarrow ^3_1H + ^4_2He$$

$$^2_1H + ^3_1H \longrightarrow ^4_2He + ^1_0n$$

Hydrogen bombs can be constructed to produce the same destructive power as from 1 to in excess of 100 megatons of TNT. Such bombs have enough explosive power to destroy an entire city.

Another bomb that utilizes nuclear energy is the neutron bomb. When a neutron bomb is exploded, it emits intense levels of neutron radiation that penetrate, but do not destroy, human artifacts. However, neutron radiation effectively kills people. This bomb is designed to be used differently than other nuclear weapons. Basically, a neutron bomb is a tactical weapon that could best be used on the field of battle. For example, an enemy with a superior number of tanks would have little advantage if neutron weapons were used. The armor of a tank provides little protection against neutron radiation.

In the future, new, more powerful nuclear devices will be developed. For example, a nuclear explosion is required to produce the large amount of energy needed in x-ray lasers, which will be a weapon in the arsenal of the Strategic Defense Initiative (Star Wars) program of the United States.

19.16 Describe how the age of an old animal bone is estimated using radiocarbon dating.

19.17 How is ^{40}K–^{40}Ar dating different from radiocarbon dating?

19.18 What are the two principal applications of radionuclides in medicine?

19.19 List the biological effects of exposure to sublethal doses of radiation.

19.20 Describe how the process of fission is used to generate electric energy in a nuclear power plant.

19.21 Explain how the energy is produced in hydrogen bombs.

SUMMARY

The **nucleus** is the very small, dense region within an atom that contains the **nucleons** (protons and neutrons). The stability of a nuclide depends on the composition of the nucleus. Some nuclear configurations are more stable than others. Nuclides with 2, 8, 20, 50, 82, and 126 protons or neutrons are more stable than nuclides with other compositions. Atoms with even numbers of both protons and neutrons are generally more stable than those with an odd number of protons or neutrons. Nuclear stability is also correlated with the ratio of neutrons to protons. Small atoms are most stable when they have an equal number of protons and neutrons, while higher-mass atoms are most stable when they contain more neutrons than protons.

The three principal forms of natural radioactivity are (1) alpha, (2) beta, and (3) gamma radiation. **Alpha radiation** is emitted only by more massive atoms. An alpha particle has the same mass properties as a helium nucleus. Thus, after a parent nuclide releases an alpha particle, its mass number decreases by four and its atomic number decreases by two. **Beta decay** results when a neutron spontaneously decays to a proton, releasing the beta particle and an antineutrino. Atoms that undergo beta decay produce daughter nuclei with the same mass number but with an atomic number greater by one than that of the parent nuclei. **Gamma emission** results when a nucleus is in a high energy state. After the gamma ray is emitted, the nucleus returns to a lower energy state. The composition of a nucleus is not changed after gamma emission.

One of the most important properties of a radionuclide is its **half-life,** the amount of time it takes for one-half of its nuclei to decay. A long half-life results when a substance decays at a slow rate. A nuclide that decays rapidly has a short half-life.

The nuclear composition of an atom can be changed artificially, through particle **bombardment reactions.** A target nucleus is placed in a beam of particles such as alpha particles, beta particles, neutrons, or protons. Upon bombardment, a nuclide is changed to some other nuclide, depending on the reaction. Under certain conditions, the target nucleus is split and undergoes **nuclear fission.** When fission occurs, neutrons are liberated, which then split other nuclei. In addition, a tremendous quantity of energy is released. The opposite of nuclear fission is nuclear fusion. **Nuclear fusion** results when two low-mass atoms collide and produce a higher-mass atom.

Radionuclides are used to date old objects. In medicine, radionuclides are used to diagnose and treat diseases. Nuclear fuels are now used to produce electric energy, and nuclear substances are used as the explosives in powerful bombs.

Exposure to radiation has many biological effects. A massive dose of radiation results in death. Smaller significant exposures can produce terrible diseases like leukemia and anemia, as well as nausea, diarrhea, loss of skin and hair, and overall sickness. The biological effect of exposure to very low levels of radiation is unknown.

KEY TERMS

---

alpha particles	bombardment reaction	mass number	nuclide
antimatter	controlled fusion	neutrino	particle accelerator
antineutrino	decay series	nuclear power plant	radioactive decay
artificial radioactivity	gamma rays	nuclear fission	radionuclide dating
atomic number	half-life	nuclear fusion	transmutation
beta particles	isotope	nucleus	transuranium element

EXERCISES*

19.22 Define each of the following terms: nucleus, atomic number, mass number, isotope, nuclide, radioactive decay, alpha particles, beta particles, gamma rays, half-life, transmutation, antimatter, neutrino, decay series, bombardment reaction, artificial radioactivity, particle accelerator, transuranium element, nuclear fission, nuclear fusion, radionuclide dating, nuclear power plant, controlled fusion.

Nucleus

19.23 Calculate the volume, in liters, that the earth would occupy if the density of the earth was that of the nucleus. The mass of the earth is 6×10^{24} kg, and the density of the nucleus is 1.8×10^{14} g/cm^3.

19.24 Write the nuclear composition of each of the following nuclides: (a) $^{201}_{81}$Tl, (b) $^{151}_{63}$Eu, (c) $^{120}_{50}$Sn, (d) $^{66}_{30}$Zn, (e) $^{254}_{99}$Es.

19.25 Write the symbols for the nuclides with the following characteristics:
(a) $A = 138$; $Z = 56$
(b) $A = 174$; $Z = 70$
(c) $N = 125$; $Z = 86$
(d) $N = 154$; $Z = 101$

19.26 Consider the nuclear composition to determine which of each of the following pairs of nuclides is expected to be more stable. Explain your reasoning for each:
(a) $^4_2$He or $^3_1$H   (b) $^{16}_8$O or $^{15}_8$O
(c) $^{40}_{20}$Ca or $^{40}_{19}$K   (d) $^{207}_{82}$Pb or $^{204}_{81}$Tl
(e) $^{133}_{55}$Cs or $^{132}_{55}$Cs   (f) $^{143}_{59}$Pr or $^{139}_{59}$Pr

19.27 Determine the nuclear composition of each nuclide, and select the most and least stable nuclide from each group:
(a) $^{141}_{57}$La, $^{184}_{74}$W, $^{176}_{73}$Ta
(b) $^{200}_{79}$Au, $^{200}_{80}$Hg, $^{202}_{82}$Pb

19.28 What could account for the fact that high-mass atoms ($Z > 21$) possess nuclei with many more neutrons than protons?

Radioactivity

19.29 Considering the mass and charge characteristics of the three forms of radiation, explain the following: (a) Beta particles penetrate a greater thickness of matter than do alpha particles; (b) alpha particles ionize matter more than do gamma rays.

19.30 Alpha radiation has a low penetration power. However, nuclides that emit alpha radiation generally produce the most severe biological effects when they are inside living things. What could account for this fact?

19.31 Two radioactive nuclides of strontium are $^{89}_{38}$ and $^{90}_{38}$Sr. The half-life for $^{89}_{38}$Sr is 51 days, and the half-life of $^{90}_{38}$Sr is 28 years. (a) Which of these two isotopes decays at a faster rate? Explain your answer. (b) If a sample initially contains 50.0 g $^{90}_{38}$Sr, what mass of $^{90}_{38}$Sr remains after 140 yr? (c) Over how many half-lives does 250 g $^{89}_{38}$Sr decrease to 3.91 g?

19.32 $^{131}_{53}$I has a half-life of 8.0 days. (a) How many days elapse for the number of $^{131}_{53}$I atoms in a sample to decrease to $\frac{1}{64}$ their initial number? (b) What mass of $^{131}_{53}$I remains if initially 4.0 kg is present and eight half-lives elapse?

19.33 Given a 12.000-g sample of $^{250}_{100}$Fm with a half-life of 0.5 h. (a) What mass of $^{250}_{100}$Fm remains after 2.5 h? (b) How many half-lives elapse before the mass of $^{250}_{100}$Fm drops below 0.1 g?

19.34 $^{75}_{34}$Se has a half-life of 120.4 days. (a) What fraction of the mass of a sample of $^{75}_{34}$Se remains after 602 days? (b) How long would it take a 5.000-g sample of $^{75}_{34}$Se to decay enough to decrease the mass of $^{75}_{34}$Se to 0.312 g?

*For exercise numbers printed in color, answers can be found at the back of the book.

19.35 It is not uncommon to find pockets of He gas in uranium mines. What could account for the presence of He in these mines?

Nuclear Reactions

19.36 Write a complete nuclear equation that illustrates the alpha decay of each of the following: (a) $^{147}_{62}$Sm, (b) $^{210}_{84}$Po, (c) $^{225}_{89}$Ac, (d) $^{248}_{96}$Cm, (e) $^{247}_{97}$Bk.

19.37 Write a complete nuclear equation that illustrates the beta decay of each of the following: (a) $^{3}_{1}$H, (b) $^{24}_{11}$Na, (c) $^{35}_{16}$S, (d) $^{63}_{28}$Ni, (e) $^{91}_{39}$Y.

19.38 Explain the reason why a nuclear equation is not generally needed to represent gamma emission.

19.39 Describe the principal characteristics of the following antimatter forms relative to their "regular" matter counterparts: (a) positron, (b) antineutrino, (c) antiproton, (d) antineutron.

19.40 In the $^{232}_{90}$Th natural decay series, the $^{232}_{90}$Th initially undergoes alpha decay, the resulting daughter emits a beta particle, whose daughter also emits a beta particle. Write three nuclear equations to represent the first three steps in the $^{232}_{90}$Th decay scheme.

19.41 What nuclide results after each of the following nuclear changes: (a) $^{131}_{53}$I releases a beta particle; (b) $^{214}_{82}$Pb undergoes two successive beta emissions; (c) $^{218}_{85}$At undergoes two successive alpha decays followed by a beta decay?

19.42 Complete each of the following equations:
(a) $^{115}_{48}$Cd $\longrightarrow$? + $^{115}_{49}$In
(b) ? $\longrightarrow$ $^{4}_{2}\alpha$ + $^{243}_{96}$Cm
(c) $^{129}_{53}$I $\longrightarrow$ $^{0}_{+1}$e + ?

19.43 Change the following shorthand-notation bombardment equations to complete nuclear equations:
(a) $^{32}_{16}$S($^{1}_{0}$n, γ)$^{33}_{16}$S   (b) $^{130}_{52}$Te($^{2}_{1}$H, 2$^{1}_{0}$n)$^{130}_{53}$I
(c) $^{43}_{20}$Ca($^{4}_{2}\alpha$, $^{1}_{1}$p)$^{46}_{21}$Sc   (d) $^{9}_{4}$Be($^{1}_{1}$p, $^{4}_{2}\alpha$)$^{6}_{3}$Li

19.44 Change each of the following equations to the shorthand notation:
(a) $^{238}_{92}$U + $^{16}_{8}$O $\longrightarrow$ $^{249}_{100}$Fm + 5$^{1}_{0}$n
(b) $^{241}_{95}$Am + $^{4}_{2}\alpha$ $\longrightarrow$ $^{243}_{97}$Bk + 2$^{1}_{0}$n
(c) $^{238}_{92}$U + $^{22}_{10}$Ne $\longrightarrow$ $^{256}_{102}$No + 4$^{1}_{0}$n
(d) $^{242}_{94}$Pu + $^{22}_{10}$Ne $\longrightarrow$ $^{260}_{104}$Unq + 4$^{1}_{0}$n

19.45 Complete each of the following equations:
(a) $^{196}_{78}$Pt + ? $\longrightarrow$ $^{197}_{78}$Pt + $^{1}_{1}$H
(b) $^{94}_{42}$Mo + ? $\longrightarrow$ $^{95}_{43}$Te + $^{1}_{0}$n
(c) $^{235}_{92}$U + $^{1}_{0}$n $\longrightarrow$ $^{93}_{35}$Br + $^{140}_{57}$La + ?
(d) $^{24}_{12}$Mg + $^{2}_{1}$H $\longrightarrow$? + $^{22}_{11}$Na
(e) $^{34}_{16}$S + $^{1}_{0}$n $\longrightarrow$ $^{35}_{16}$S + ?

19.46 Write a complete nuclear equation that shows how each of the following transuranium elements was initially synthesized (see Table 19.3): (a) Am, (b) No, (c) Unp, (d) Pu, (e) Bk.

19.47 What accounts for the large amount of energy produced during the fission process?

19.48 What scientists are credited with the discovery of nuclear fission?

19.49 Write the set of three equations that illustrates how hydrogen is fused to produce helium in the sun.

19.50 Why are temperatures in excess of 100,000,000°C required to initiate nuclear fusion reactions?

Uses of Radioactive Substances

19.51 Write equations to show how most of the $^{14}_{6}$C on earth is formed.

19.52 By what means does $^{14}_{6}$C enter living things?

19.53 If an old bone is found to contain one-fourth the quantity of $^{14}_{6}$C in a modern bone, what is the approximate age of the bone?

19.54 To date objects using $^{14}_{6}$C, the assumption is made that the amount of $^{14}CO_2$ is constant. What activities of human beings over the last 80 years might invalidate this assumption?

19.55 Could $^{40}_{19}$K/$^{40}_{18}$Ar dating techniques be used to estimate the age of a skull found in an ancient burial ground? Explain.

19.56 How have radionuclides changed the diagnostic procedures used by doctors? Be specific.

19.57 How are each of the following nuclides used diagnostically in medicine: (a) ^{42}K, (b) ^{131}I, (c) ^{32}P, (d) ^{198}Au, (e) ^{99}Tc, (f) ^{59}Fe?

19.58 What might account for lack of "hard" data concerning the effects of low levels of radiation exposure on biological systems?

19.59 List two significant problems associated with nuclear power plants that are not major problems in conventional fossil fuel power plants.

19.60 How is the intensity of neutron radiation controlled in a nuclear power plant?

19.61 Why is a nuclear explosion, as in an atomic bomb, an impossibility in the normal operation of a nuclear power plant?

19.62 (a) What was the name of the project in which the first atomic bomb was developed? (b) How did the initial atomic bombs produce energy?

19.63 (a) What is meant by a chain reaction? (b) What must be true of a sample of radioactive material for it to sustain a chain reaction?

19.64 What changes might be expected in our world (political, social, economic, etc.) after the development of an economical method for harnessing fusion energy?

19.65 What chemical, physical, and biological factors would make it impossible for the world to return to its present state following a major nuclear war?

Additional Exercises

19.66 $^{121}_{50}\text{Sn}$ is a beta emitter that has a half-life of 27.5 hours. (a) How many protons and neutrons are in the nucleus of $^{121}_{50}\text{Sn}$? (b) Write an equation to show the decay of $^{121}_{50}\text{Sn}$. (c) How long would it take for a 100-g sample of $^{121}_{50}\text{Sn}$ to decrease to 12.5 g?

19.67 $^{97}_{43}\text{Tc}$ was first produced by the reaction of $^{96}_{42}\text{Mo}$ with $^{2}_{1}\text{H}$. Write a complete equation that shows the formation of $^{97}_{43}\text{Tc}$.

19.68 $^{148}_{63}\text{Eu}$ undergoes decay by releasing a positron. (a) Write an equation that shows the decay of $^{148}_{63}\text{Eu}$. (b) What happens to the neutron to proton ratio when a nuclide undergoes positron emission?

19.69 For each of the following, determine what nuclide is used to initiate the nuclear bombardment reaction:
(a) $(^{2}_{1}\text{H}, {}^{1}_{0}\text{n}^{0})^{239}_{93}\text{Np}$
(b) $(^{12}_{6}\text{C}, 4^{1}_{0}\text{n}^{0})^{245}_{99}\text{Es}$
(c) $(^{2}_{1}\text{H}, 2^{1}_{0}\text{n}^{0})^{238}_{94}\text{Pu}$

CHAPTER
Twenty

Overview of Organic and Biologically Important Compounds

--- STUDY GUIDELINES ---

After completing Chapter 20, you should be able to

1. List the major classes of organic compounds

2. Write IUPAC names for simple alkanes

3. Write all isomers of simple alkanes

4. Identify and give examples of unsaturated hydrocarbons

5. Identify the basic structure of aromatic hydrocarbons

6. Give examples of important hydrocarbons

7. Write the functional groups in common hydrocarbon derivatives

8. Draw the structures and write the names of the simplest members of the major classes of hydrocarbon derivatives

9. Give examples of important compounds for each primary group of hydrocarbon derivatives

10. Write equations for important organic reactions

11. Identify the basic structures contained in the four principal classes of biochemicals

12. Explain the role of each class of biochemicals in living things

M odern **organic chemistry** is the study of the properties and reactions of carbon compounds. In the past, chemists did not define organic chemistry in this way. In 1807, Berzelius proposed that compounds derived from living things were "organic" and all others were "inorganic," not derived from life. Such a definition appealed to scientists of the time, especially when it was well known that organic substances could easily be converted to inorganic substances by heating or treatment with acids in the laboratory. They felt that the opposite conversion, from inorganic to organic, was not possible.

Early nineteenth century scientists were certain that organic compounds could be produced only within living systems. At that time it was thought that living things possessed a "vital force" which changed inor-

20.1 INTRODUCTION TO ORGANIC CHEMISTRY

ganic substances to organic ones. Scientists believed that without the vital force there was no way to change an inorganic to an organic compound. In 1828, Friedrich Wöhler (1800–1882) stunned the world of science and shattered the vital force theory (also called *vitalism*) by synthesizing an organic compound from an inorganic salt.

Wöhler heated the inorganic salt ammonium cyanate, NH_4OCN, and analyzed the products; he discovered that urea, NH_2CONH_2, was one of the reaction products.

$$NH_4OCN \xrightarrow{\Delta} H_2NCONH_2$$

Friedrich Wöhler studied medicine and received his degree as a physician before becoming a chemist. Besides making his famous discovery, Wohler was one of the first to investigate how compounds are metabolized in living things.

Urea is a waste product found in urine. Wöhler communicated his finding to his former professor, Berzelius, who then realized that this own ideas concerning the definition of organic chemistry would have to be modified in light of Wöhler's new evidence.

Organic chemistry is one of the most exciting areas of chemistry because of the importance of compounds that contain carbon. Organic compounds are the major components of living systems, and life as we know it could not exist without these compounds. The food that we eat, the gasoline we use in our cars, and any of our possessions are principally composed of organic compounds.

Organic compounds differ significantly from inorganic compounds. Most organic compounds have lower densities, melting points, and boiling points than do most inorganic compounds. Many organic substances are flammable, an uncommon property of inorganic compounds. Finally, most organic compounds tend to be less water-soluble and more soluble in nonpolar solvents than most inorganic compounds.

Some carbon-containing substances are not classified as organic compounds because their properties more closely resemble those of inorganic compounds. For example, carbon dioxide, CO_2, carbon monoxide, CO, carbon disulfide, CS_2, and hydrogen cyanide, HCN, are inorganic compounds. In addition, salts that contain carbonate, CO_3^{2-}, bicarbonate, HCO_3^-, thiocyanate, SCN^-, and cyanate, OCN^-, are also classified as inorganic compounds.

Organic compounds are divided into two major classes, hydrocarbons (*hydro*, hydrogen) and hydrocarbon derivatives. **Hydrocarbons** are compounds that contain only the elements carbon and hydrogen. **Hydrocarbon derivatives** are those compounds that contain carbon, usually hydrogen, and at least one other element. Hydrocarbon derivatives normally possess one or more of the following elements in addition to C and H: O, N, S, P, halogens, and various metals.

20.2 HYDROCARBONS

Hydrocarbons are divided into four groups: (1) alkanes, (2) alkenes, (3) alkynes, and (4) aromatics. The alkanes, alkenes, and alkynes are together known as **aliphatic hydrocarbons.** Aromatics possess a special structure called an aromatic ring that makes them chemically quite distinct. Accordingly, organic chemists place them in a separate group called the **aromatic hydrocarbons.**

Alkanes are hydrocarbons that possess C atoms bonded to the maximum number of H atoms possible. There are no double or triple bonds in the alkanes. Since alkanes cannot chemically add any more H atoms, they are frequently called the **saturated hydrocarbons**—saturated with respect to the number of H atoms in the molecule. In alkanes, C atoms are bonded in chains and rings. First, we will consider alkanes that are composed of molecules with carbon chains.

Methane, CH_4, is the simplest alkane. It has four H atoms bonded tetrahedrally to the central C atom (Fig. 20.1). The bond angle between H atoms in methane and in the other alkanes is 109.5°. Methane is a colorless, odorless gas with a melting point of $-183°C$ and a boiling point of $-162°C$. Such low melting and boiling points indicate that the intermolecular forces among methane molecules are weak London dispersion forces. They result because methane is a nonpolar covalent molecule. Methane, like many organic compounds, is referred to by one or more common names. For example, methane is frequently called swamp gas. When plants decay in the absence of oxygen, a quantity of methane is produced. Such conditions exist in swamps and marshlands where decaying plant matter is covered with water.

To alleviate the problem of having more than one name for a compound, organic chemists have adopted the IUPAC systematic procedure for naming organic compounds. IUPAC stands for the International Union of Pure and Applied Chemistry, an international organization that oversees the naming of chemical substances. Table 20.1 lists the accepted names for the first 10 alkanes.

Each name is composed of two parts: stem and suffix. The stem indicates the number of C atoms in a molecule. For example, *eth* indicates two C atoms in a molecule, *prop* means three C atoms, and *but* indicates four C atoms. The suffix tells what class of organic compound the molecule belongs to. The alkanes are designated by the ending *ane*. Each class of organic compound has its own unique suffix.

Alkanes

Methane is the principal component of natural gas. Natural gas is used for heating and cooking. Because natural gas is odorless, a small quantity of an organic sulfur compound is added so that gas leaks can be detected.

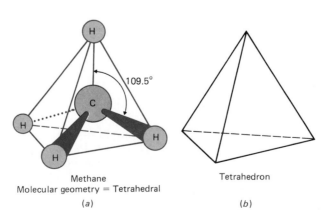

Methane
Molecular geometry = Tetrahedral
(a)

Tetrahedron
(b)

Figure 20.1
Methane molecules are tetrahedral with 109.5° bond angles. Each C—H bond is polar, but due to its symmetry the molecule is nonpolar.

TABLE 20.1   NAMES OF THE FIRST 10 ALKANES

Molecular formula	Condensed formula*	IUPAC name
CH_4	CH_4	Methane
C_2H_6	CH_3CH_3	Ethane
C_3H_8	$CH_3CH_2CH_3$	Propane
C_4H_{10}	$CH_3(CH_2)_2CH_3$	Butane
C_5H_{12}	$CH_3(CH_2)_3CH_3$	Pentane
C_6H_{14}	$CH_3(CH_2)_4CH_3$	Hexane
C_7H_{16}	$CH_3(CH_2)_5CH_3$	Heptane
C_8H_{18}	$CH_3(CH_2)_6CH_3$	Octane
C_9H_{20}	$CH_3(CH_2)_7CH_3$	Nonane
$C_{10}H_{22}$	$CH_3(CH_2)_8CH_3$	Decane

*A condensed formula is one in which the number of bonded hydrogen atoms is written to the right of each carbon atom. This is a shorthand way to express organic structural formulas.

Alkanes have the general formulas C_nH_{2n+2}. If there are n C atoms in the molecule, there are $2n + 2$ H atoms attached. An alkane with 20 C atoms, for example, will have 42 ($2 \times 20 + 2$) hydrogen atoms.

Alkanes are an example of a homologous series of compounds. A **homologous series** is a group of compounds in which one member differs from the compound that immediately precedes or follows it by a fixed amount, in this case by a CH_2 unit. If we select ethane, C_2H_6, it differs from both methane, CH_4, and propane, C_3H_8, by a CH_2 unit. Homologous series have observable trends in physical properties and exhibit similar chemical properties.

The first four members of the alkane series, methane to butane, are gases at room temperature. Pentane to octadecane, $C_{18}H_{38}$, are liquids, and the remaining alkanes, those with 19 or more carbon atoms, are solids. Such a trend indicates that there is a gradual increase in boiling and melting points with increased molecular mass. Chemically, the alkanes as a group are rather inert. They combine only with very reactive substances like the halogens.

In addition to the alkanes composed of molecules with unbranched chains, there are alkanes with branches in the chain. For example, the molecular formula of butane is C_4H_{10}. Molecules of butane contain four carbon atoms in a continuous chain.

$$CH_3CH_2CH_2CH_3$$
Butane

Another compound with exactly the same molecular formula but with a different structural formula is

$$CH_3CHCH_3$$
$$|$$
$$CH_3$$

A branch in the carbon chain is found in this molecule. These two compounds are **structural isomers,** i.e., compounds with the same molecular formula but with a different arrangement of carbon atoms (different structural formulas). Each has its own unique set of physical and chemical properties.

IUPAC has established specific rules for naming the alkanes with branches in the carbon chain. These rules are as follows:

Many more organic compounds exist than there are words in the English language. The task of properly naming such an immense number of compounds is difficult.

Rule 1

Find the longest continuous chain of carbon atoms in the molecule.

This is accomplished by starting at one end of the molecule and counting carbon atoms that are consecutively bonded. The name of the molecule is derived from the name of the alkane with the same number of carbon atoms in the longest continuous chain. This is called the **parent alkane.** All of the other carbon atoms in the molecule are considered branches off the main chain, and are called **substituent groups.**

In the above example the longest continuous chain is three carbon atoms; hence, the parent alkane is propane.

Rule 2

Number the longest continuous chain in such a way that the lowest possible number is given to the C atom to which the substituent group is bonded.

Our example has only one substituent group, which is in the center of the longest chain. Thus, it does not matter in which direction the three-carbon chain is numbered.

$$\overset{1}{C}H_3\overset{2}{C}H\overset{3}{C}H_3$$
$$|$$
$$CH_3$$

Rule 3

Identify each substituent group in the chain, and the carbon to which it is attached.

A name is required for the substituent group CH_3. This group is the first in a series of what are called **alkyl groups.** An alkyl group is an alkane minus one hydrogen ($CH_4 - H = -CH_3$) that is bonded to a carbon chain or other organic structure. To name alkyl groups, remove the *ane* ending and replace with *yl*. Therefore, a CH_3 group is a *methyl* group (methane − *ane* + *yl*). If a hydrogen is removed from ethane, C_2H_6, an ethyl group, $-C_2H_5$, results. Methyl and ethyl groups are the only alkyl groups that we will consider.

In our example, a methyl group is bonded to the second carbon of the three-carbon chain.

In 1892, the problem of naming compounds was first considered at the International Congress of Chemistry. Subsequently, regular meetings were held to develop a system for naming compounds. An outgrowth of these early meetings was the development of the IUPAC.

Rule 4

Write the name of the compound by first writing the number of the carbon to which the substituent group is bonded, followed by a hyphen that is connected to the name of the substituent group and the name of the parent hydrocarbon.

Thus, the name of the hydrocarbon we are considering is **2-methylpropane.** If we analyze the name, we see that located on the second carbon of a three-carbon chain is a methyl group.

2-methylpropane

Carbon number of bonded group⌐   ⌐──Parent hydrocarbon

Alkyl group

Often the "2" is not included in the name of 2-methylpropane because it is the only C atom that can bond to the methyl group without changing the molecule.

To illustrate both isomers and IUPAC naming, let's identify and name all of the structural isomers of pentane, C_5H_{12}. Starting with pentane, the unbranched chain is

$$CH_3CH_2CH_2CH_2CH_3 = pentane$$

Next we can write a four-carbon chain with one substituent group.

$$CH_3CHCH_2CH_3$$
$$|$$
$$CH_3$$

There are 75 isomers of decane, 366,319 isomers of eicosane ($C_{20}H_{42}$), and 4,111,846,763 isomers of triacontane ($C_{30}H_{62}$).

To write the name of this compound, follow each of the above steps: (1) Identify the longest continuous chain: four carbon atoms are located in the longest chain. (2) Number the chain, starting from the end closest to the substituent group.

$$\overset{1}{C}H_3\overset{2}{C}H\overset{3}{C}H_2\overset{4}{C}H_3$$
$$|$$
$$CH_3$$

(3) Identify substituent groups and the C atom to which they are bonded: A methyl group is bonded to the second carbon. (4) Write the name of the compound, starting with the number of the carbon to which the substituent group is bonded, followed by the names of the alkyl group and the parent hydrocarbon. Thus, the name of this compound is **2-methylbutane.**

The compound 2-methylbutane is the only pentane isomer with butane as the parent hydrocarbon. If we write the structure with the methyl group bonded to the next carbon in the chain, we have just found another way of writing 2-methylbutane. Following the rules, the chain should be numbered starting from the end closest to the substituent group.

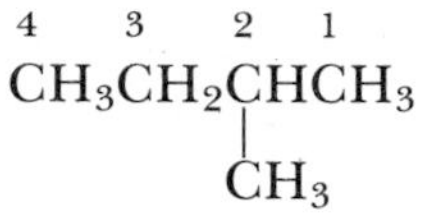

$$\overset{4}{C}H_3\overset{3}{C}H_2\overset{2}{C}H\overset{1}{C}H_3$$
$$|$$
$$CH_3$$

The remaining isomer of pentane has a parent chain of three carbon atoms with two methyl groups bonded to the second carbon.

$$\overset{\quad\quad CH_3}{\underset{1\quad\quad\;2|\quad\;3}{CH_3-C-CH_3}}$$
$$|$$
$$CH_3$$

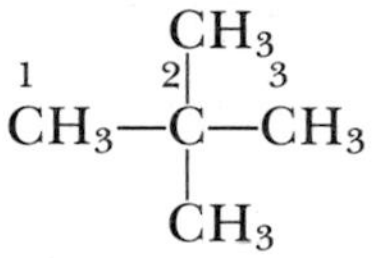

Following the rules, we find that three C atoms form the longest continuous chain, with two methyl groups attached at the second carbon. To indicate that two groups are bonded to the chain, the prefix *di* is placed in front of methyl, and to indicate that both methyl groups are bonded to the second carbon, the number 2 is written twice. The correct name for the compound is **2,2-dimethylpropane.**

So far we have seen that alkanes exist as chains and branched chains. They can also exist as ring structures. An alkane ring structure, or **cycloalkane,** is one in which each carbon is bonded to two other carbon atoms and two hydrogen atoms. A three-carbon cycloalkane is the smallest ring structure:

Cyclopropane

Cyclopropane is an excellent anesthetic, but it must be handled carefully for it is explosive. It is a potent anesthetic with minimal side effects and rather low toxicity.

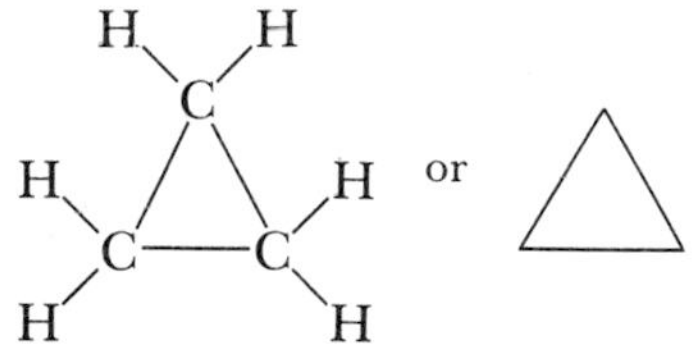

To write the names of cycloalkanes, we place the prefix *cyclo* in front of the name of the parent alkane. Consequently, a three-carbon cycloalkane is called cyclopropane. The next compound in the series is cyclobutane. Its structure is

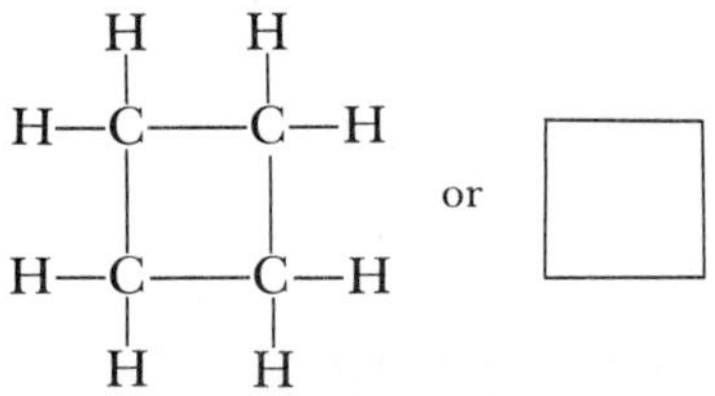

Cyclobutane

Next in the series are cyclopentane and cyclohexane.

Cyclopentane   Cyclohexane

Note that the general formula of the cycloalkanes is different from that of the noncyclic alkanes. Cycloalkanes have the general formula C_nH_{2n}. Cycloalkanes can also have alkyl groups bonded to the ring. Consider the following examples.

Even though cyclic structures are symbolized as being flat (planar), most rings are "puckered." For example, the configuration of cyclohexane is best represented as

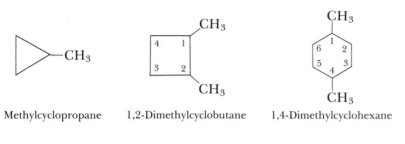

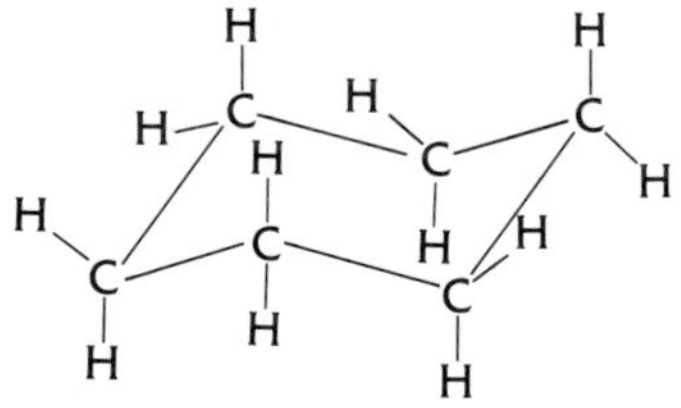

Methylcyclopropane   1,2-Dimethylcyclobutane   1,4-Dimethylcyclohexane

20.1 What are the four classes of hydrocarbons?

20.2 Write the names of the unbranched alkanes with the following numbers of carbon atoms in the molecule: *(a)* two, *(b)* four, *(c)* five, *(d)* seven, *(e)* nine.

20.3 What are the molecular formulas of the following compounds: *(a)* methane, *(b)* propane, *(c)* pentane, *(d)* hexane, *(e)* decane?

20.4 Write the condensed formulas for the following structures:

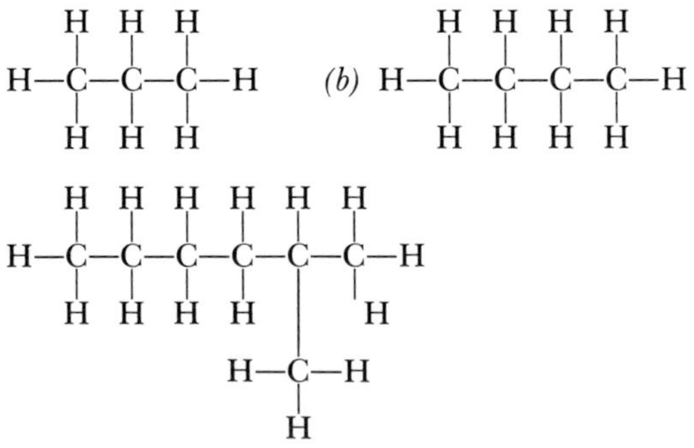

20.5 Draw the structures and write the IUPAC names for all five isomers of hexane.

20.6 *(a)* Draw the structure of the simplest cycloalkane. *(b)* How is the structure of this molecule different from that of its noncyclic parent molecule?

Alkenes are hydrocarbons that contain a carbon–carbon double bond. Alkenes are classified as unsaturated hydrocarbons. **Unsaturated hydrocarbons** undergo hydrogenation reactions (reaction with H_2 gas) and add at least 1 mol H_2 per 1 mol of compound. After adding H_2, they become saturated and produce alkanes.

Alkenes

The general formula for alkenes is C_nH_{2n}.

$$\begin{matrix} {}^{\diagdown}{}_{\diagup}C{=}C{}^{\diagup}_{\diagdown} & + & H_2 & \longrightarrow & -\overset{|}{C}-\overset{|}{\underset{|}{C}}- \\ \text{Alkene} & & & & \text{Alkane} \\ \text{(Unsaturated)} & & & & \text{(Saturated)} \end{matrix}$$

The simplest alkene contains two carbon atoms joined by a double bond.

$$H_2C{=}CH_2$$
Ethene

All alkenes are given the ending *ene* in the IUPAC naming system. Thus, the simplest alkene is ethene, C_2H_4. (A common name, ethylene, is also frequently used.)

Following ethene in the alkene series is the three-carbon alkene, propene, C_3H_6:

$$H_2C{=}CHCH_3$$
Propene

Both ethene and propene are gases at room temperature; they have boiling points of $-102°C$ and $-48°C$, respectively. The low boiling points of the alkenes are due to the weak London forces between their molecules.

Butene, C_4H_8, is the next alkene in the series. Two isomers of butene exist. With four carbon atoms in the chain, the double bond can be either between the first and second carbons or between the second and third carbons. The structures of the two butene isomers are as follows:

$$\overset{1\quad\ 2\quad\ 3\quad\ 4}{H_2C{=}CHCH_2CH_3} \qquad \overset{1\quad\ 2\quad\ \ 3\quad 4}{CH_3CH{=}CHCH_3}$$
$$\text{1-Butene} \qquad\qquad\quad \text{2-Butene}$$

To distinguish between these two isomers, a number that indicates the position of the double bond is placed before the name of the compound. Thus, in 1-butene the carbon–carbon double bond is located between C_1 and C_2, whereas in 2-butene the double bond is located between C_2 and C_3.

As with the alkanes, there is no limit to the number of carbon atoms that can be in an alkene chain. And like alkanes, alkenes can also be composed of branched chains and rings. An example of a cycloalkene is cyclohexene.

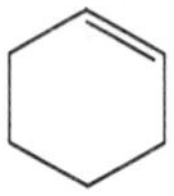

Cyclohexene

An alkene molecule can also possess more than one double bond. Those alkenes containing two double bonds are called **alkadienes,** or **dienes** for short. One interesting alkadiene is 2-methyl-1,3-butadiene, commonly called isoprene.

$$\overset{1}{H_2C} = \overset{2}{C} - \overset{3}{CH} = \overset{4}{CH_2}$$
$$\underset{CH_3}{|}$$

Isoprene

Under the proper conditions, isoprene, like many alkenes, combines with itself forming high-molecular-mass compounds collectively called **polymers.** Substances containing long chains of connected isoprene molecules resemble natural rubber (*cis*-polyisoprene). A segment of the polyisoprene chain appears as follows:

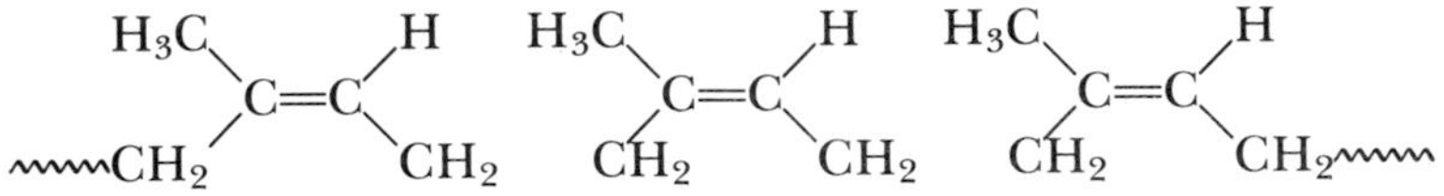

Segment of polyisoprene chain (natural rubber)

Polymers result when one or more monomers (small molecules) combine to form long chains of bonded monomer units. Common polymers include polyethylene, polyvinylchloride (PVC), polystyrene, nylon, and Teflon.

Alkenes are a reactive group of compounds. Their reactivity is the result of the carbon–carbon double bond in the molecule. Many different molecules combine with alkenes at the position of double bond. These reactions are known as addition reactions, and they occur as follows:

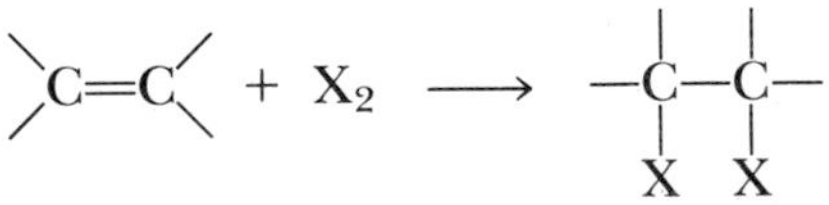

$$\diagdown C = C \diagup + \ X_2 \ \longrightarrow \ -\overset{|}{\underset{X}{C}}-\overset{|}{\underset{X}{C}}-$$

In this general equation the diatomic molecule, X_2, breaks one of the bonds of the double bond and attaches to the molecule at that position. As a specific example, let's consider the addition of hydrogen, H_2, to ethene.

$$H_2C{=}CH_2 + H_2 \xrightarrow{\ cat\ } CH_3CH_3$$

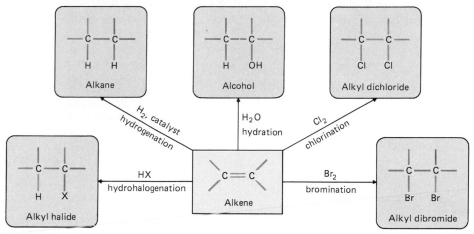

Figure 20.2
Many different substances undergo addition reactions with alkenes. In each of these reactions, one bond of the double bond in the alkene is broken and the combining molecule attaches at that position. Hydration occurs if water adds across the double bond, halogenation occurs with halogens, hydrogenation occurs with hydrogen, and hydrohalogenation occurs with a hydrogen halide, HX.

In this reaction, the double bond breaks, and the H atoms from H_2 bond to the C atoms that originally composed the double bond. The addition of hydrogen to an alkene is known as a **hydrogenation reaction.** Some of the other addition reactions that alkenes undergo are summarized in Fig. 20.2.

Alkynes contain a carbon–carbon triple bond and, like alkenes, are classified as unsaturated hydrocarbons. One mole of an alkyne combines with two moles of $H_2(g)$ to form a saturated hydrocarbon. Ethyne is the simplest alkyne:

Alkynes

The general formula for alkynes is C_nH_{2n-2}.

$$HC\equiv CH$$
Ethyne

Alkynes are named in a similar manner to alkenes except that *yne* is placed after the stem to indicate that a triple bond is located in the molecule.

Ethyne (or acetylene, as it is more frequently called) is industrially the most significant alkyne. It is relatively inexpensive to synthesize. Lime, CaO, is heated with carbon to produce calcium carbide, CaC_2, which is combined with water to produce acetylene.

$$CaO + 3C \longrightarrow CaC_2 + CO$$

$$CaC_2 + 2H_2O \longrightarrow HC\equiv CH + Ca(OH)_2$$

Acetylene is combusted in oxyacetylene torches, which are used to weld and cut through metals.

Aromatic hydrocarbons are those that possess the general properties and structure of benzene, C_6H_6.

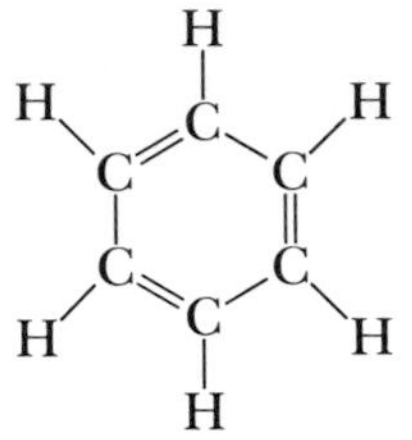

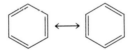

Complete structure   Shorthand notation

Structure of benzene

Aromatic Hydrocarbons

In 1865, August Kekulé first proposed the cyclic structure of benzene. It was purportedly the result of a dream that Kekulé had in which he saw a snake bite its own tail.

Many structures are classified as aromatic hydrocarbons. However, in our cursory discussion, we will consider only the benzenelike aromatics.

Benzene is unique with respect to the classes of hydrocarbons already discussed. In the structure that we draw for benzene, it appears as if benzene is a cyclic molecule with alternating double and single bonds. But benzene molecules exhibit **resonance.** Molecules that exhibit resonance contain delocalized electrons which means that the actual structure of benzene resembles the average of the resonance structures.

Two resonance structures of benzene

Benzene molecules actually contain no double or single bonds. Instead, each bond in benzene is intermediate between a single and a double bond, and has a bond order of 1.5. Accordingly, most chemists represent a molecule of benzene by drawing a six-membered ring with a circle in the center.

Benzene

Other aromatic hydrocarbons contain the benzene structure with substituent groups attached. If a methyl group is bonded to the ring, toluene, also known

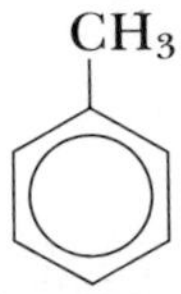

CH_3

Toluene

as methylbenzene, results. The series continues with ethylbenzene, which has an ethyl group bonded to the ring.

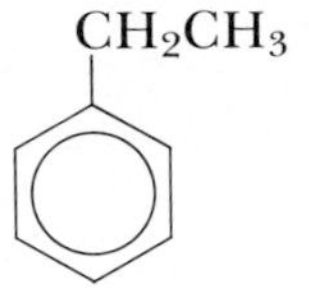

$$CH_2CH_3$$

Ethylbenzene

More than one group can bond to a benzene ring. In fact any combination of up to six substituent groups can bond to a benzene ring. If, for example, two methyl groups (or two of any substituent group) are bonded to a benzene ring, three different structural isomers are possible.

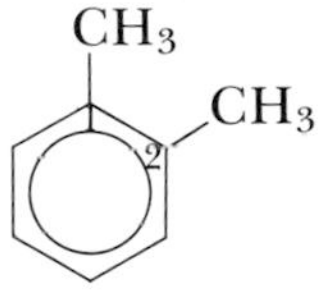

CH_3   CH_3   CH_3

1,2-Dimethylbenzene   1,3-Dimethylbenzene   1,4-Dimethylbenzene
o-Xylene   *m*-Xylene   *p*-Xylene

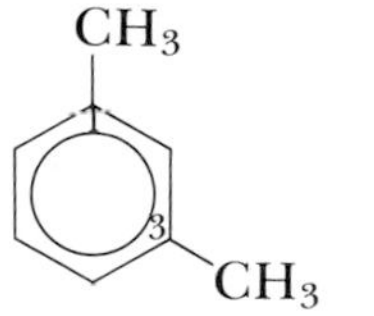

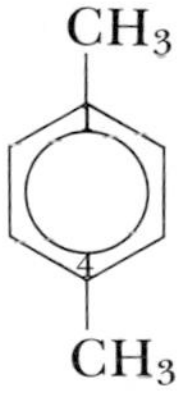

In the first structure, the two methyl groups are attached on C_1 and C_2, giving 1,2-dimethylbenzene. The second structure has the two methyl groups attached to the C_1 and C_3 positions, resulting in 1,3-dimethylbenzene. In the third structure, the methyl groups are attached to C_1 and C_4, and this compound is 1,4-dimethylbenzene. Another name for the three isomers of dimethylbenzene is xylene. Three terms are used to represent the placement of two substituents on a benzene ring: (1) ***ortho,*** attachment at carbon atoms 1 and 2; (2) ***meta,*** attachment at carbon atoms 1 and 3; and (3) ***para,*** attachment at carbon atoms 1 and 4. Many people refer to the dimethylbenzenes as *o*-xylene, *m*-xylene, and *p*-xylene.

Any number of benzene rings can bond. A group of hydrocarbons called the **polycyclic aromatics** are compounds with two or more fused benzene rings. Naphthalene is the first in the series:

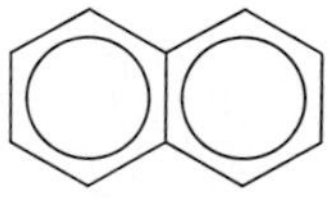

Naphthalene

Naphthalene has a distinct pungent odor and is used as an ingredient of mothballs.

Two of the three ways of fusing three benzene rings together are

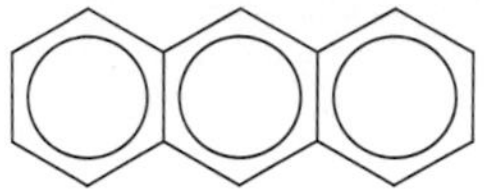

Anthracene 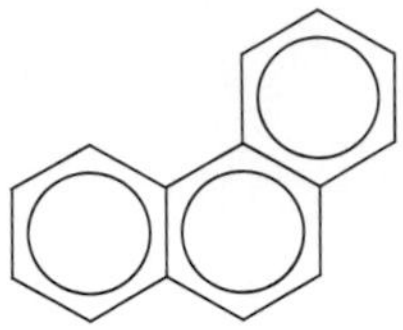   Phenanthrene

In the first compound, anthracene, the three rings are bonded in a linear fashion. In phenanthrene, the rings are bonded in a nonlinear fashion. Both anthracene and phenanthrene are components of coal tar. Anthracene is used to synthesize various dyes.

Many of the polycyclic aromatics induce cancer in animals. They are classified as carcinogenic compounds. Two such compounds identified in cigarette smoke are 3,4-benzpyrene and dibenz[*a,h*]anthracene.

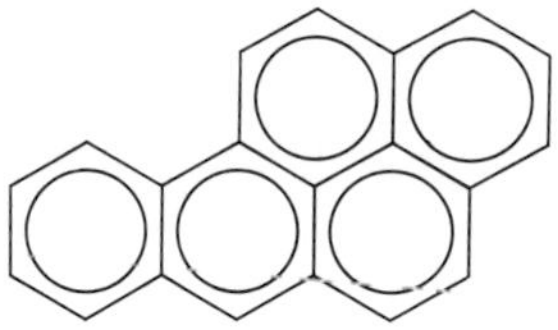

3,4-Benzpyrene 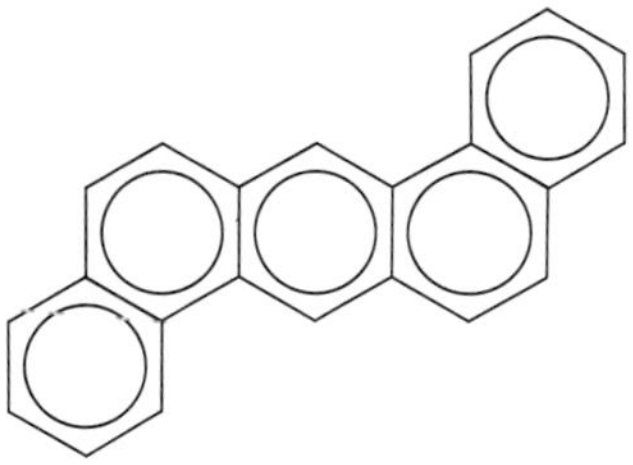   Dibenz[*a,h*]anthracene

Benzene and other aromatics undergo substitution reactions. An H atom in benzene is replaced by some other group. For example, if nitric acid, HNO_3, is combined with benzene in the presence of sulfuric acid, H_2SO_4, nitrobenzene results.

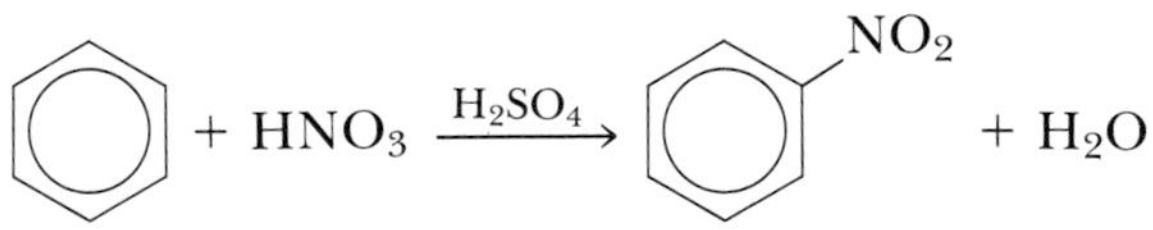

Benzene   Nitrobenzene

In this reaction, a nitro group, $-NO_2$, from the nitric acid replaces an H atom on the benzene ring. Another example of an aromatic substitution reaction occurs when a halogen such as Cl_2 or Br_2 combines with benzene, in the presence of a catalyst, to produce chlorobenzene or bromobenzene.

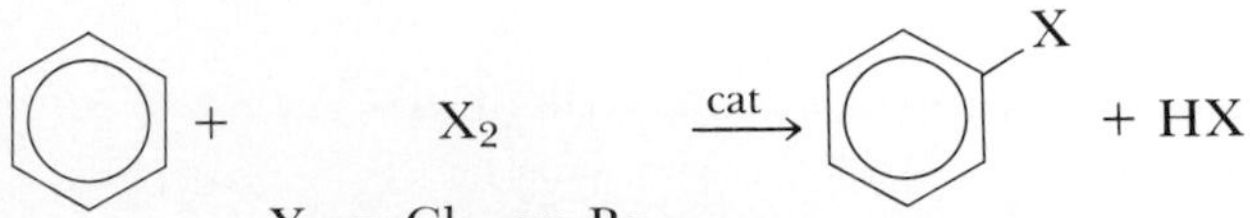

$X_2 = Cl_2$ or Br_2

20.7 What is the difference between a saturated and an unsaturated hydrocarbon? Give an example of each.

20.8 Draw the structure of a compound that represents each of these groups: *(a)* alkyne, *(b)* cycloalkene, *(c)* alkadiene, *(d)* monosubstituted aromatic.

20.9 Define isoprene and discuss its importance in naturally occurring substances.

20.10 Write two resonance structures that represent benzene.

20.11 Draw the structures for and name the three dimethylbenzenes.

20.12 What is a polycyclic aromatic hydrocarbon? Give an example.

Hydrocarbon derivatives** are organic compounds that also contain atoms such as oxygen, nitrogen, sulfur, phosphorus, and halogens. Each class of hydrocarbon derivative is characterized by a functional group. **Functional groups** are a specific arrangement of atoms that give a class of organic compounds their characteristic chemical properties. Functional groups, themselves, are bonded to a carbon chain or ring. Table 20.2 lists the primary hydrocarbon derivatives with their functional groups. The symbol R is employed in organic chemistry to represent any alkyl group.

20.3 HYDROCARBON DERIVATIVES

TABLE 20.2   HYDROCARBON DERIVATIVES

Class	Functional group	Structural formula*	Condensed structural formula*
Alcohol	—OH	R—OH	ROH
Ether	—O—	R—O—R	ROR
Aldehyde	$\overset{\text{O}}{\overset{\|}{-\text{C}}}$—H	$\overset{\text{O}}{\overset{\|}{\text{R}-\text{C}}}$—H	RCHO
Ketone	$\text{R}-\overset{\text{O}}{\overset{\|}{\text{C}}}-\text{R}'$	$\text{R}-\overset{\text{O}}{\overset{\|}{\text{C}}}-\text{R}'$	RCOR'
Organic acid (Carboxylic acid)	$-\overset{\text{O}}{\overset{\|}{\text{C}}}-\text{OH}$	$\text{R}-\overset{\text{O}}{\overset{\|}{\text{C}}}-\text{OH}$	RCOOH
Ester	$-\overset{\text{O}}{\overset{\|}{\text{C}}}-\text{OR}$	$\text{R}-\overset{\text{O}}{\overset{\|}{\text{C}}}-\text{OR}'$	RCOOR'
Amine	$-\text{NH}_2, -\text{NH}, -\text{N}-$	$\text{R}-\text{NH}_2$	RNH_2
Amide	$-\overset{\text{O}}{\overset{\|}{\text{C}}}-\text{NH}_2$	$\text{R}-\overset{\text{O}}{\overset{\|}{\text{C}}}-\text{NH}_2$	RCONH_2
Nitrile	—CN	R—C≡N	RCN
Thiol	—SH	R—SH	RSH
Thioether	—S—	R—S—R	RSR
Halide	—X (X = F, Cl, Br, I)	R—X	RX

*R' indicates that the second alkyl group does not necessarily have to be the same as the first, although it can be.

Alcohols and **ethers** can be thought of as organic derivatives of water. The molecular formula of water is H_2O. If an H atom is removed from a water molecule and replaced with an alkyl group, an alcohol results. If both hydrogen atoms are replaced, an ether results.

Alcohols and Ethers

Water	Alcohol	Ether
HOH	ROH	ROR'

Methanol is the simplest alcohol, and consists of one carbon, three hydrogen atoms, and an —OH group.

$$CH_3OH$$
Methanol

Notice that if an H atom is removed from methane and replaced with alcohol's functional group, —OH, methanol results. IUPAC naming of alcohols requires that the parent alkane be identified; the *e* at the end of the name is removed and replaced with *ol*. As with all organic substances, many names exist for methanol, including methyl alcohol and wood alcohol.

Methanol, like most of the alcohols, is a poisonous substance. Ingestion of small amounts of methanol can cause severe medical problems, including blindness. Drinking large quantities of methanol results in death. Methanol is used in industry as a solvent, and it was once a component of antifreeze.

Ethanol is the two-carbon alcohol:

Once ingested, methanol is converted to formaldehyde and formic acid by the liver in an attempt to rid the body of this toxic substance. Ultimately, methanol is converted to CO_2 and H_2O.

$$CH_3CH_2OH$$
Ethanol

Common names for ethanol are grain alcohol, drinking alcohol, or just plain alcohol. It is called grain alcohol because it can be produced by the fermentation of grains and fruits. **Fermentation** is the process in which microorganisms are allowed to decompose the sugar components of grains and fruits, in the absence of oxygen, to ethanol and carbon dioxide.

Alcoholic beverages contain varying amounts of ethanol. For example, beer has appproximately 4% ethanol by volume, wines have about 12%, and whiskies contain 40 to 50% ethanol. If ethanol is ingested in low concentrations over a relatively long time period, our bodies decompose the alcohol and prevent its toxic effects. However, if a large amount of ethanol is consumed rapidly, it is deadly.

Ethanol has many other uses. It is a solvent in lacquers, varnishes, perfumes, and medicines. A mixture of ethanol and gasoline, called gasohol, is sometimes used as a fuel for internal combustion engines instead

Ethanol is metabolized in the liver to CO_2 and H_2O at a rate of about 8 g of ethanol per hr. Ethanol amounts greater than this remain in the bloodstream and cause inebriation.

of gasoline. It is also used as the starting material in the synthesis of other organic compounds. Table 20.3 lists some other important alcohols.

TABLE 20.3 REPRESENTATIVE ALCOHOLS AND ETHERS

Name	Formula	Selected uses and importances
Isopropyl alcohol	CH_3CHCH_3 　　　$|$ 　　　OH	Rubbing alcohol, solvent
Ethylene glycol	$HO-CH_2CH_2-OH$	Permanent antifreeze for engines
Glycerol	$CH_2-CH-CH_2$ 　$|$　　$|$　　$|$ 　OH　OH　OH	Lubricant, ingredient in hand lotions, sweetener, production of plastics, and nitroglycerin
Phenol (carbolic acid)	⬡—OH	Disinfectant
Divinyl ether	$(CH_2{=}CH)_2O$	Fast-acting anesthetic
Tetrahydrocannabinol		Psychoactive substance in marijuana

Two of the most important reactions of alcohols are dehydration and oxidation reactions. A **dehydration reaction** occurs when water is released from a molecule. Usually, a dehydrating agent is added to the reaction mixture to bring about the reaction. A dehydration reaction occurs when ethanol, CH_3CH_2OH, is heated with sulfuric acid (the dehydrating agent).

$$H-\underset{\underset{\boxed{H\ \ OH}}{|}}{\overset{\overset{H\ \ H}{|\ \ |}}{C}-C}-H \xrightarrow[\text{H}_2\text{SO}_4]{\Delta} H_2C{=}CH_2 + H_2O$$

Ethanol　　　　　　　　　　　　Ethene

In this reaction, an —OH group and an H atom are removed from the ethanol molecule, and the resulting molecule is ethene. In all cases, alcohols dehydrate to produce alkenes. A second example of a dehydration reaction is

$$\text{⬡—OH} \xrightarrow[\text{H}_2\text{SO}_4]{\Delta} \text{⬡} + H_2O$$

Cyclohexanol　　　　　　　　Cyclohexene

Alcohols undergo oxidation reactions with oxidizing agents such as potassium permanganate, $KMnO_4$, or chromic acid, H^+ and $K_2Cr_2O_7$. The products of these reactions depend on the placement of the —OH group in the molecule. If the —OH group is at the end of the C chain, we call the alcohol a *primary alcohol*. An example of a primary alcohol is ethanol, CH_3CH_2OH. When primary alcohols are oxidized they are converted first to aldehydes and then to carboxylic acids.

$$CH_3CH_2OH \xrightarrow{KMnO_4} CH_3CHO \xrightarrow{KMnO_4} CH_3COOH$$

Ethanol   Acetaldehyde   Acetic acid

A *secondary alcohol* is one in which the —OH group is bonded to a C atom which is bonded to two C atoms. An example of a secondary alcohol is 2-propanol (isopropyl alcohol). When secondary alcohols are oxidized, they are converted to ketones. 2-Propanol is oxidized to acetone.

$$\underset{\underset{\text{OH}}{|}}{CH_3CHCH_3} \xrightarrow{KMnO_4} \underset{\underset{\text{O}}{\|}}{CH_3CCH_3}$$

2-Propanol   Acetone

In this reaction two H atoms from the 2-propanol are removed to produce acetone. A *tertiary alcohol* is one in which the —OH group is bonded to a C atom that is bonded to three C atoms. Tertiary alcohols do not undergo oxidation reactions as do primary and secondary alcohols.

Ethers have the general formula R—O—R′; thus, the simplest ether contains two carbon atoms:

$$CH_3—O—CH_3$$

Dimethyl ether

Commercially, the most important ether possesses four carbon atoms in two ethyl groups:

$$CH_3CH_2—O—CH_2CH_3$$

Diethyl ether

At one time, diethyl ether, or simply ether, was one of the primary general anesthetics used in surgery. A **general anesthetic** is a substance that depresses the central nervous system, causing unconsciousness and insensitivity to pain. Many side effects, such as nausea, vomiting, and irritation of the lungs, occur after the administration of ether. Ether is also highly flammable. For these reasons, diethyl ether is not used in modern surgical procedures. Other ethers and a variety of halogenated hydrocarbons are used in its place.

Like other classes of organic compounds, ethers can exist in cyclic structures. Organic ring structures that contain atoms other than carbon are called **heterocyclic compounds.** Cyclic ethers contain one or more oxygen atoms in the ring. Ethylene oxide (common name) is the simplest heterocyclic ether.

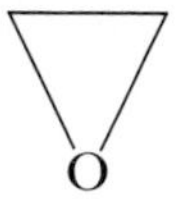

Ethylene oxide

Its primary uses are as a starting material in the synthesis of a large variety of other organic compounds and as a disinfectant in the sterilization of medical products. See Table 20.3 for more examples of ethers.

Both aldehydes and ketones contain a **carbonyl group,** C=O. An **aldehyde** has the carbonyl group at the end of a carbon chain. A **ketone** has a carbonyl group located in the middle of a chain.

Aldehydes and Ketones

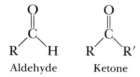

Aldehyde   Ketone

The simplest aldehyde contains only one C atom. Its IUPAC name is methanal, but it is more commonly called formaldehyde.

$$H_2C=O$$
Formaldehyde

The shortest possible carbon chain in a ketone is one with three C atoms. Its IUPAC name is propanone, but its common name, acetone, prevails.

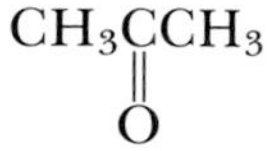

$$CH_3CCH_3$$
$$O$$
Acetone

Formaldehyde is a colorless, odorless gas that is water-soluble. Commercially, formaldehyde is sold as a 37 to 40% aqueous solution called formalin. Most biology students know that formalin is used to preserve dead biological specimens.

Acetone is the most important industrial ketone. It is a colorless liquid with a low boiling point, and is used in the synthesis of varnishes, plastics, and resins. A small quantity of acetone is produced in living systems as a result of the breakdown of fats. In uncontrolled diabetics, large quantities of acetone are produced. Most of it is excreted in the urine, but in severe cases of diabetes so much acetone is produced that it is detected on the person's breath.

Carbonyl groups can be bonded to aromatic rings, producing aromatic aldehydes and ketones. Benzaldehyde is the simplest aromatic aldehyde. Benzaldehyde is a colorless liquid that has the odor of almonds— its common name is oil of almonds. Two other examples of aromatic aldehydes are vanillin, which gives the flavor to vanilla, and cinnamaldehyde, which has the odor of cinnamon.

| Benzaldehyde | Vanillin | Cinnamaldehyde |

Carbonyl groups are important functional groups in the biologically significant compounds called carbohydrates. Table 20.4 lists other aldehydes and ketones.

TABLE 20.4 REPRESENTATIVE ALDEHYDES AND KETONES

Name	Formula	Selected uses and importance
Acetaldehyde	CH_3CHO	Biochemical intermediate
Citral	CH_3C=$CHCH_2CH_2C$=$CHCHO$ (with CH_3 and CH_3 substituents)	Lemon oil
Camphor		Analgesic in ointments
Jasmone		Odor of jasmine
Progesterone		Female sex hormone

Aldehydes and ketones undergo reduction reactions and produce alcohols. In the presence of a metal catalyst such as Pt or Ni, aldehydes are reduced to primary alcohols by hydrogen gas, H_2. In a similar reaction, ketones are reduced to secondary alcohols.

$$RCHO + H_2 \xrightarrow{Pt} RCH_2OH \text{ (Primary alcohol)}$$

$$R_2C{=}O + H_2 \xrightarrow{Pt} R_2CHOH \text{ (Secondary alcohol)}$$

If acetaldehyde is reduced by H_2, it is converted to ethanol, CH_3CH_2OH.

$$\underset{\text{Acetaldehyde}}{CH_3CHO} + H_2 \xrightarrow{Pt} \underset{\text{Ethanol}}{CH_3CH_2OH}$$

Acetone is reduced by H_2 to 2-propanol.

$$\underset{\text{Acetone}}{(CH_3)_2C{=}O} + H_2 \xrightarrow{Pt} \underset{\text{2-Propanol}}{(CH_3)_2CHOH}$$

Aldehydes, but not ketones, undergo oxidation reactions. Aldehydes are oxidized, in the presence of an oxidizing agent such as $K_2Cr_2O_7/H^+$, to carboxylic acids.

$$RCHO \xrightarrow{K_2Cr_2O_7/H^+} RCOOH$$

For example, acetaldehyde, CH_3CHO, is oxidized to acetic acid, and benzaldehyde is oxidized to benzoic acid.

$$\underset{\text{Acetaldehyde}}{CH_3CHO} \xrightarrow{K_2Cr_2O_7/H^+} \underset{\text{Acetic acid}}{CH_3COOH}$$

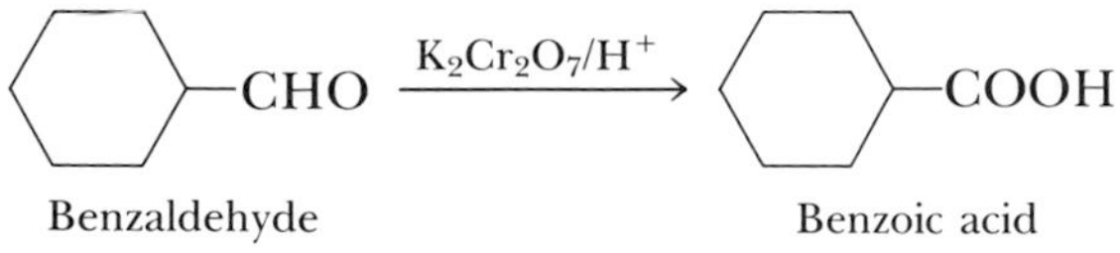

Benzaldehyde → Benzoic acid

A **carboxyl group** is the functional group in **organic acids,** which are also called **carboxylic acids.**

Organic Acids (Carboxylic Acids)

Carboxyl group

A carboxyl group is most frequently written in condensed form as either —COOH or $—CO_2H$.

The acidic properties of organic acids are the result of their ability to donate a proton and produce a resonance-stabilized carboxylate anion, $RCOO^-$.

$$RC\overset{\overset{O}{\|}}{}{-}OH \rightleftharpoons H^+ + RC\overset{\overset{O}{\|}}{}{-}O^- \longleftrightarrow R{-}C\overset{\overset{O^-}{|}}{}{=}O$$

The simplest carboxylic acid is methanoic acid (IUPAC name), or formic acid (common name), HCOOH. Formic acid is a liquid with a pungent, irritating odor. Its name is derived from the Latin word for "ant." Ants have formic acid in their systems.

Second in the carboxylic acid series is ethanoic acid (IUPAC name), or acetic acid (common name from the Latin *acetum,* meaning "vinegar"), CH_3COOH. Vinegar is a 4 to 5% aqueous solution of acetic acid. Acetic acid is used in the dye industry for its acidic properties and is a chemical intermediate in the production of various fibers and pharmaceuticals.

Butanoic acid (IUPAC name) or butyric acid (common name derived from the Latin *butyrum* for "butter"), is a four-carbon, putrid-smelling acid and is found in perspiration, rancid butter, and rancid margarine. Pentanoic acid's common name is valeric acid, derived from the Latin "to be strong." Hexanoic acid (IUPAC name) or caproic acid (from the Latin *caper,* meaning "goat") is present in excretions from the skin cells of goats.

An acid with two carboxyl groups is classified as a dicarboxylic acid. Oxalic acid is the simplest dicarboxylic acid.

$$HOOC{-}COOH$$
Oxalic acid

A potassium salt of oxalic acid, $HOOC{-}CO^-K^+$, is a component of rhubarb and spinach. Both the acid and its salts are toxic. Luckily, cooking destroys a large percent of the toxic salt.

An example of a tricarboxylic acid is citric acid. In addition to the three carboxyl groups, there is also an alcohol group bonded to the citric acid molecule. Citric acid is contained in citrus fruits such as lemons and oranges. It is very important in cellular energy production in living systems.

$$\begin{array}{c} H \\ | \\ H{-}C{-}COOH \\ | \\ HO{-}C{-}COOH \\ | \\ H{-}C{-}COOH \\ | \\ H \end{array}$$
Citric acid

HCOOH is the formula of formic acid, and CH_3COOH is the formula of acetic acid.

Another acid vital to the functioning of biological systems is lactic acid. Under low-oxygen conditions, lactic acid is produced in cells from glucose. Microorganisms (lactobacilli) decompose sugars to lactic acid. Sour milk results when lactobacilli decompose milk sugars to lactic acid. Table 20.5 lists some other carboxylic acids.

Bacteria on the surface of teeth convert dietary sugars to lactic acid. Acids dissolve the tooth enamel and produce dental caries.

$$\overset{\displaystyle OH}{\underset{\displaystyle |}{CH_3CHCOOH}}$$

Lactic acid

TABLE 20.5   REPRESENTATIVE ORGANIC ACIDS

Name	Formula	Selected uses or importances
Stearic acid	$CH_3(CH_2)_{16}COOH$	Used to make soap; coatings for pills
Nicotinic acid	(pyridine ring with COOH)	Component of B vitamins
Pyruvic acid	$CH_3\overset{O}{\underset{\parallel}{C}}COOH$	Biochemical intermediate in metabolic processes
Tartaric acid	$HOOCCHCHCOOH$ with HO OH	By-product of the wine industry; used in foods, medicine, and dye industry

One of the most important reactions of carboxylic acids (RCOOH) occurs when they combine in the presence of a catalyst with alcohols (R′OH) to produce esters (RCOOR′).

$$RCOOH + R'OH \xrightarrow{cat} RCOOR' + H_2O$$

Carboxylic acid + alcohol $\xrightarrow{cat}$ ester + water

Reactions in which esters are produced are called **esterification reactions.** When acetic acid, CH_3COOH, combines with ethanol, CH_3CH_2OH, in the presence of an acid catalyst, the ester ethyl acetate, $CH_3COOCH_2CH_3$, results.

$$CH_3COOH + HOCH_2CH_3 \xrightarrow{H^+} CH_3COOCH_2CH_3 + H_2O$$

Acetic acid   Ethanol   Ethyl acetate

In esterification reactions, the H atom bonded to the O atom in the alcohol combines with the OH group in the acid to produce water, and the two molecules join to produce the ester.

Esters are acid derivatives. As we have just seen, they are produced when a carboxylic acid is combined with an alcohol under the proper conditions. Esters structurally resemble organic acids except that instead of an H atom being bonded to the O atom, an alkyl group (R group) or aromatic group is attached.

$$\begin{array}{c} O \\ \parallel \\ RC-OR' \end{array}$$

Ester

If the one-carbon carboxylic acid is combined with the one-carbon alcohol, the simplest ester is produced; it is methyl methanoate (IUPAC name), or methyl formate (common name).

$$\begin{array}{c} O \\ \parallel \\ H-C-OCH_3 \end{array}$$

Methyl formate

Table 20.6 lists some common esters. Many esters have pleasant, fruity odors.

TABLE 20.6   REPRESENTATIVE ESTERS

Name	Formula	Odor or flavor
Methyl butyrate	$CH_3(CH_2)_2COOCH_3$	Apples
Ethyl butyrate	$CH_3(CH_2)_2COOC_2H_5$	Pineapples
Pentyl butyrate	$CH_3(CH_2)_2COOC_5H_{11}$	Apricots
Octyl acetate	$CH_3COOC_8H_{15}$	Oranges
Ethyl formate	$HCOOC_2H_5$	Rum

Three esters of the aromatic acid salicylic acid are used in medicines. If salicylic acid combines with acetic acid under the proper conditions, acetylsalicylic acid (aspirin) is formed. Aspirin is the most widely used painkiller (analgesic) in the world today.

In addition to its analgesic effect, aspirin is an antipyretic (reduces fevers) and an anti-inflammatory agent.

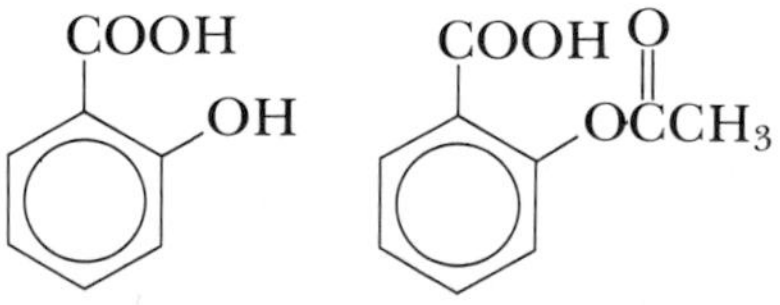

Salicylic acid

Acetylsalicylic acid
(Aspirin)

Another salicylic acid ester is phenyl salicylate, or salol. Salol is used to coat pills containing medicines that are irritating to the stomach. Salol passes through the acidic environment of the stomach without being decomposed, but it breaks down in the slightly alkaline conditions of the small intestine. Methyl salicylate, or oil of wintergreen, is also an ester of salicylic acid.

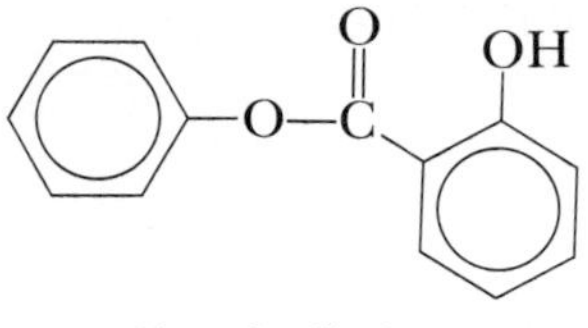

Phenyl salicylate
(Salol)

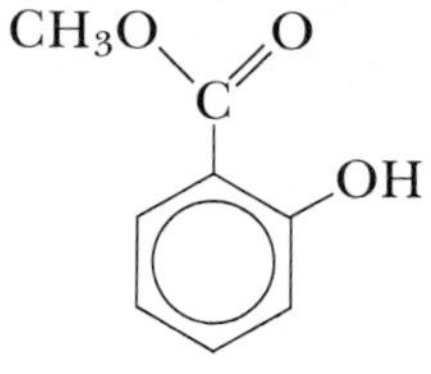

Methyl salicylate
(Oil of wintergreen)

Oil of wintergreen is used in liniments and skin rubs. It is absorbed through the skin and enters sore muscles, where it is hydrolyzed to salicylic acid, an analgesic.

Amides are another class of carboxylic acid derivative. If the —OH group that is part of the carboxyl group is removed and replaced with an —NH_2 or a nitrogen with bonded alkyl groups, an amide results. Amides have the following general formulas:

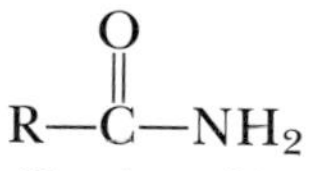

Amides

$$R-\overset{\overset{\displaystyle O}{\|}}{C}-NH_2$$
Simple amide

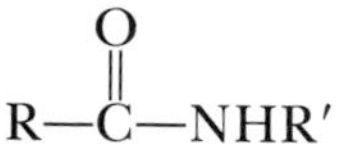

$$R-\overset{\overset{\displaystyle O}{\|}}{C}-NHR'$$
Monosubstituted amide

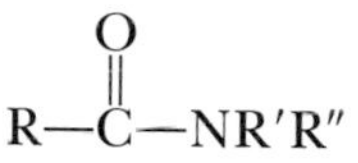

$$R-\overset{\overset{\displaystyle O}{\|}}{C}-NR'R''$$
Disubstituted amide

Formamide (common name) is the simplest amide. It is a derivative of formic acid and has the following structure.

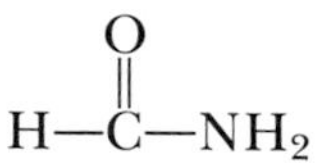

$$H-\overset{\overset{\displaystyle O}{\|}}{C}-NH_2$$
Formamide

Many amides are of biological interest. Proteins, one of the major classes of biological compounds, are amides. One of the B vitamins, niacinamide, contains a ring structure with an attached amide group.

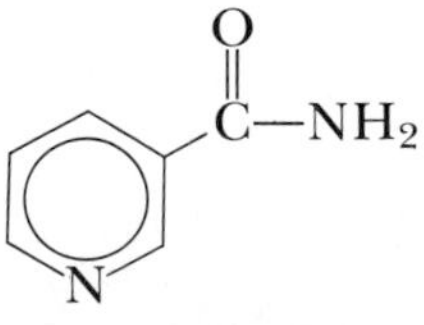

Niacinamide

Lysergic acid diethylamide, LSD, contains a disubstituted amide. LSD is a mind-altering drug (hallucinogen) synthesized from lysergic acid, a component of a fungus called ergot.

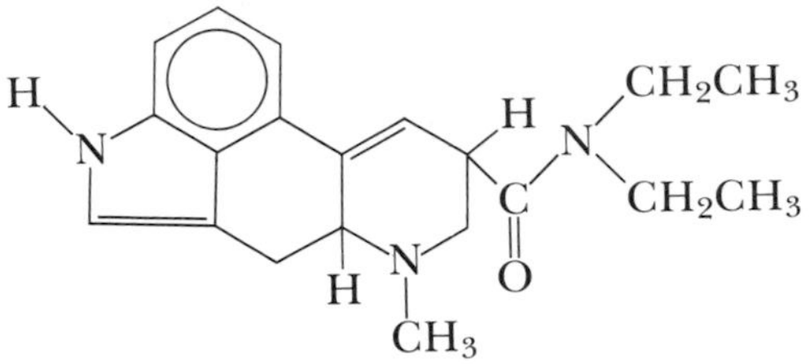

Lysergic acid diethylamide (LSD)

Acetaminophen, an aspirin substitute, is a fairly simple monosubstituted amide. It is marketed under the names Tylenol and Datril.

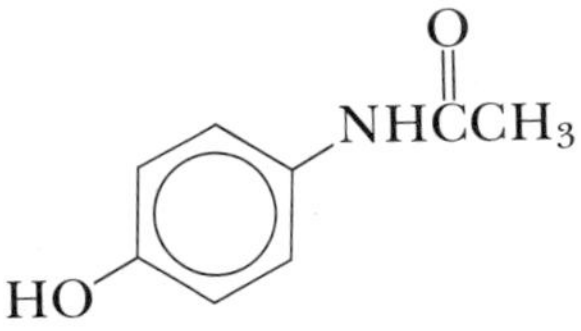

Acetaminophen

Amines are organic derivatives of ammonia, NH_3. If an H atom is removed from an ammonia molecule and is replaced by an alkyl group, a primary amine results. When two H atoms are removed and replaced with two alkyl groups, a secondary amine forms, and if three alkyl groups are bonded to an N atom, a tertiary amine results.

Amines

$$RNH_2 \qquad\qquad R_2NH \qquad\qquad R_3N$$
Primary amine $\qquad$ Secondary amine $\qquad$ Tertiary amine

Thus, the simplest primary, secondary, and tertiary amines are methylamine, CH_3NH_2, dimethylamine, $(CH_3)_2NH$, and trimethylamine, $(CH_3)_3N$, respectively. Low-molecular-mass amines are foul-smelling substances that have the odor of rotting fish. Amines, like ammonia, are basic substances that accept protons from acids.

$$R\overset{..}{N}H_2 + H^+ \longrightarrow RNH_3{}^+$$

Aniline is the simplest of the aromatic amines. Aniline and its derivatives are used in the synthesis of dyes. Aniline is a starting material in the production of pharmaceuticals and photographic chemicals.

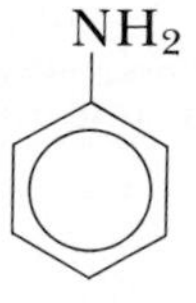

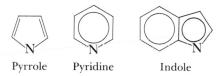

Aniline

When an N atom is a member of an organic cyclic carbon structure, the molecule is classified as a **heterocyclic amine.** Examples of heterocyclic ring structures include pyrrole, pyridine, and indole.

Pyrrole   Pyridine   Indole

Pyrrole is a five-membered ring structure that contains an N atom and four C atoms. Pyrrole rings are components of molecules such as hemoglobin, chlorophyll, and vitamin B_{12}. Hemoglobin is the O_2-transporting molecule of blood, chlorophyll is the green pigment in plants that is required for photosynthesis, and vitamin B_{12} is an essential substance in a person's diet.

Amines that are naturally occurring in plants and have a physiological effect on animals are classified as alkaloids. For example, morphine is isolated from the opium poppy. Quinine is contained in the cinchona tree of the Andes, and tubocurarine is the active ingredient in curare.

Pyridine is a component of the structure of two B vitamins, niacin and pyridoxine. It is also a component of the structure of nicotine, a deadly compound contained in plants such as the tobacco plant. In small quantities, nicotine acts as a mild stimulant.

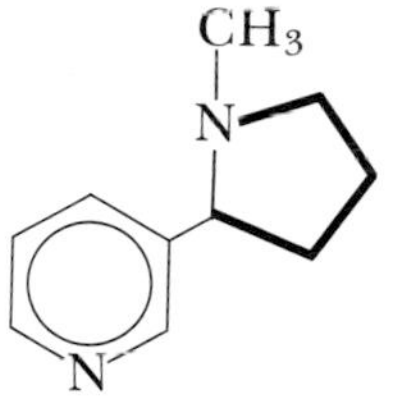

CH_3

Nicotine

Indole is a common structure in nature, and is found in substances ranging from indole alkaloids such as lysergic acid, reserpine, and strychnine to important compounds in brain physiology such as serotonin. Table 20.7 lists some other amines.

Alkyl halides are produced when a halogen atom (F, Cl, Br, or I) replaces an H atom in a nonaromatic hydrocarbon. If methane is combined with a halogen under appropriate conditions, a halogen atom substitutes for one or more H atoms.

Alkyl Halides and Other Halogenated Hydrocarbons

TABLE 20.7 REPRESENTATIVE AMINES

Name	Formula	Selected use or importance
Putrescine	$H_2N(CH_2)_4NH_2$	Decomposition product of proteins
Benadryl	—CHOCH$_2$CH$_2$NCH$_3$ / CH$_3$	Antihistamine
Coniine	N CH$_2$CH$_2$CH$_3$	Principal alkaloid in hemlock, the chemical that killed Socrates
Caffeine		Stimulant in coffee, tea, and cola drinks
Amphetamine	—CH$_2$CHCH$_3$ / NH$_2$	Potent synthetic stimulant

If $Cl_2(g)$ combines with methane, chloromethane, CH_3Cl (common name, methyl chloride), results.

$$CH_4 + Cl_2 \xrightarrow{\text{light}} CH_3Cl + HCl$$

If excess Cl_2 is present, the Cl_2 combines with the newly formed chloromethane, producing a disubstituted methane, dichloromethane, CH_2Cl_2 (methylene chloride). When a third H atom is removed, trichloromethane, $CHCl_3$ (chloroform), results. This reaction continues until all of the H atoms are removed, finally yielding tetrachloromethane (carbon tetrachloride), CCl_4.

The one-carbon chlorinated hydrocarbons are generally good solvents for organic and nonpolar substances. Chloroform, $CHCl_3$, was one of the earliest general anesthetics. It is no longer used because it is fairly toxic and it has been identified as a cancer-causing agent. As well as being used as an anesthetic, $CHCl_3$ was commonly placed in cough remedies until the Food and Drug Administration (FDA) banned its use in 1976.

At one time, carbon tetrachloride had a number of household uses. CCl_4 was an ingredient in spot removers and cleaning fluids until it was discovered to be too toxic for home use. CCl_4 causes serious damage to the liver and kidneys. Generally, the substances that replaced "carbon tet," as it is called, are not much better. Trichloroethylene and tetrachloroethylene are now used in cleaning fluids, especially "dry cleaning fluids," but it is now known that they are also carcinogenic.

Two or more different halogen atoms can be incorporated into a molecule. Freons are good examples; they contain both fluorine and chlorine atoms in their structures. Freons are fluorochlorocarbons that are used as refrigerants in refrigerators, freezers, and air-conditioning equipment. Two commercially important Freons are Freon 11 and Freon 12.

Trichloroethylene

Tetrachloroethylene

$$CCl_3F \qquad CCl_2F_2$$
$$\text{Freon 11} \qquad \text{Freon 12}$$

When used as refrigerants, they cause minimal environmental problems because they are enclosed. But at one time Freons were widely used as propellants in aerosol spray cans. They are no longer used as propellants because research studies indicated that Freons could have a significant environmental impact. Research has shown that Freons could diffuse to the upper layers of the atmosphere, in particular to the ozone layer. Under these high-energy conditions, the normally inert Freons combine with the ozone, O_3. If a small amount of the O_3 were depleted, significantly higher intensities of ultraviolet radiation would reach the surface of the earth, with very serious biological consequences—i.e., increased rates of skin cancer and crop damage.

Another controversial class of halogenated hydrocarbons are the polychlorinated pesticides. These compounds are used to control populations of insects that cause disease or consume crops. One of the first widely used polychlorinated pesticides was DDT, or *d*ichloro-*d*iphenyl*t*richloroethane.

The ozone layer is a portion of the stratosphere; it is located about 15 mi above the surface of the earth. A 5% depletion of the ozone layer would cause a 10% increase in the intensity of ultraviolet radiation reaching the earth.

DDT

In 1972, the United States banned the use of DDT because its overuse caused many environmental problems. DDT became concentrated in the food chain and was responsible for the extinction of certain bird species.

Table 20.8 presents selected examples of organic halides and their importance and uses.

TABLE 20.8   REPRESENTATIVE ORGANIC HALIDES

Name	Formula	Selected uses and importances
p-Dichlorobenzene	Cl—⟨benzene ring⟩—Cl	Pesticide in mothballs
Chlordane	(chlordane structure with multiple Cl substituents)	Pesticide used to control ants, termites, and lawn pests
Iodoform	CHI_3	Antiseptic
Halothane	$CF_3CHClBr$	Common general anesthetic
Polychlorinated biphenyls	(chlorinated biphenyl structure with multiple Cl)	Group of compounds used as heat transfer agents and hydraulic fluids; were used in plastics until their environmental impact was known
Teflon	$-(CF_2-CF_2)_n$	High-molecular-mass polymer used as an electrical insulator and as a nonstick coating for utensils

REVIEW EXERCISES

20.13 To what class of hydrocarbon derivative does each of the following belong: *(a)* ROR, *(b)* RNH_2, *(c)* RCN, *(d)* RCHO, *(e)* RCOOH?

20.14 Write the functional group contained in each of the following: *(a)* ketones, *(b)* esters, *(c)* amides, *(d)* alkyl halides, *(e)* alcohols.

20.15 Write the formula for the simplest *(a)* alcohol, *(b)* amine, *(c)* ether, *(d)* carboxylic acid, *(e)* ketone.

20.16 Give an example of a specific compound that could have each of the following uses: *(a)* pesticide, *(b)* anesthetic, *(c)* antiseptic agent, *(d)* embalming fluid, *(e)* flavoring agent.

20.17 Give an example of a molecule with the stated characteristics: *(a)* heterocyclic amine, *(b)* aromatic halide, *(c)* ester containing five carbon atoms, *(d)* monosubstituted amide, *(e)* cyclic ketone.

20.18 Describe the environmental impact of the indiscriminate "dumping" of chlorinated hydrocarbons.

L iving systems are composed principally of four different classes of compounds: (1) carbohydrates, (2) proteins, (3) lipids, and (4) nucleic acids. In this section we take a very brief look at these complex substances.

20.4 CLASSES OF BIOLOGICALLY SIGNIFICANT COMPOUNDS

Carbohydrates are the group of biologically essential compounds that includes sugars, starches, and cellulose. Carbohydrates are structural components of cells and are sources of energy. Chemically, carbohydrates are polyhydroxy (i.e., containing more than one alcohol group, —OH) aldehydes and ketones, or they yield such substances when hydrolyzed.

Carbohydrates

Carbohydrates are frequently classified as (1) monosaccharides, (2) disaccharides, and (3) polysaccharides. Most carbohydrates fit one of these three principal classifications. The simplest group of carbohydrates are the **monosaccharides,** the simple sugars. Monosaccharides combine to produce the other two carbohydrate groups. **Disaccharides** result when two monosaccharide molecules are chemically combined. **Polysaccharides** are polymers of monosaccharides, i.e., long chains of bonded monosaccharides.

Two types of monosaccharides exist, polyhydroxy aldehydes and polyhydroxy ketones. Most frequently, these molecules contain from three to six C atoms. A polyhydroxy aldehyde is called an **aldose** (the ending *ose* indicates a carbohydrate), and a polyhydroxy ketone is called a **ketose.**

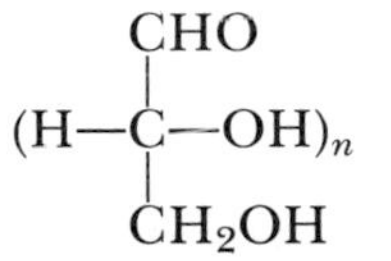

Polyhydroxy aldehyde (Aldose) ($n = 1, 2, 3, 4,$ or 5)

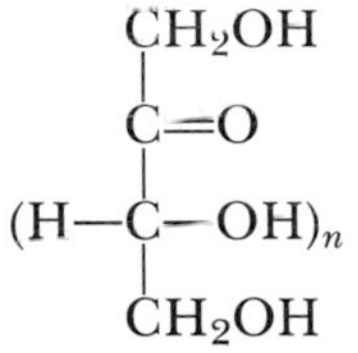

Polyhydroxy ketone (Ketose) ($n = 0, 1, 2, 3,$ or 4)

Glyceraldehyde is the simplest aldose, while the simplest ketose is dihydroxyacetone.

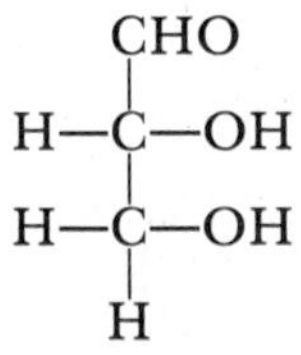

Glyceraldehyde

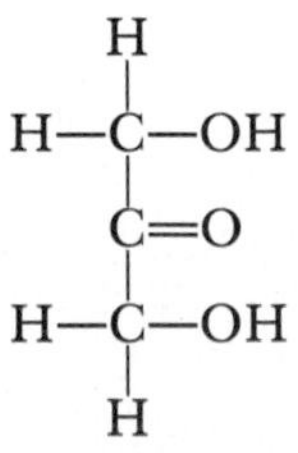

Dihydroxyacetone

Many of the biologically important monosaccharides contain either five or six C atoms. By far the most important five-carbon monosaccharides are ribose and deoxyribose. Ribose is an aldose, or aldopentose (five-carbon aldose), that exists throughout nature. It is a component of nucleic acid molecules (RNA) and of molecules that release biological energy. Deoxyribose differs from ribose by only one O atom. On the second carbon of deoxyribose two H atoms are bonded; on the second carbon of ribose an H and —OH are bonded. Deoxyribose is the monosaccharide in the deoxyribose nucleic acids, DNAs.

$$
\begin{array}{cc}
\text{CHO} & \text{CHO} \\
| & | \\
\text{H—C—OH} & \text{H—C—H} \\
| & | \\
\text{H—C—OH} & \text{H—C—OH} \\
| & | \\
\text{H—C—OH} & \text{H—C—OH} \\
| & | \\
\text{H—C—OH} & \text{H—C—OH} \\
| & | \\
\text{H} & \text{H} \\
\text{Ribose} & \text{Deoxyribose}
\end{array}
$$

There are three major six-carbon monosaccharides, or hexoses. Each has the formula $C_6H_{12}O_6$; they differ with respect to the placement of —OH groups bonded to the carbon chain and the location of the carbonyl group in the molecule. Glucose and galactose are aldohexoses, and fructose is a ketohexose.

$$
\begin{array}{ccc}
\text{CHO} & \text{CHO} & \text{H} \\
 & & | \\
 & & \text{H—C—OH} \\
| & | & | \\
\text{H—C—OH} & \text{H—C—OH} & \text{C}=\text{O} \\
| & | & | \\
\text{HO—C—H} & \text{HO—C—H} & \text{HO—C—H} \\
| & | & | \\
\text{H—C—OH} & \text{HO—C—H} & \text{H—C—OH} \\
| & | & | \\
\text{H—C—OH} & \text{H—C—OH} & \text{H—C—OH} \\
| & | & | \\
\text{H—C—OH} & \text{H—C—OH} & \text{H—C—OH} \\
| & | & | \\
\text{H} & \text{H} & \text{H} \\
\text{Glucose} & \text{Galactose} & \text{Fructose}
\end{array}
$$

Naturally occurring monosaccharides exist mainly in cyclic structures rather than in chains. The cyclic structure of glucose is

Glucose is often called by its common name, dextrose, and is sometimes referred to as blood or grape sugar. A large percent of all ingested sugars are converted to glucose, one of the primary fuels of cells. Galactose does not exist alone in nature. Galactose is always combined with some other molecule, most frequently glucose. Fructose is the only biologically important ketohexose. It is one of the sugars that contribute to the sweetness of various fruits.

Disaccharides result when two monosaccharides combine chemically. Table sugar, or sucrose, is an example of a disaccharide. If sucrose is decomposed, glucose and fructose remain. Sucrose's cyclic structure is

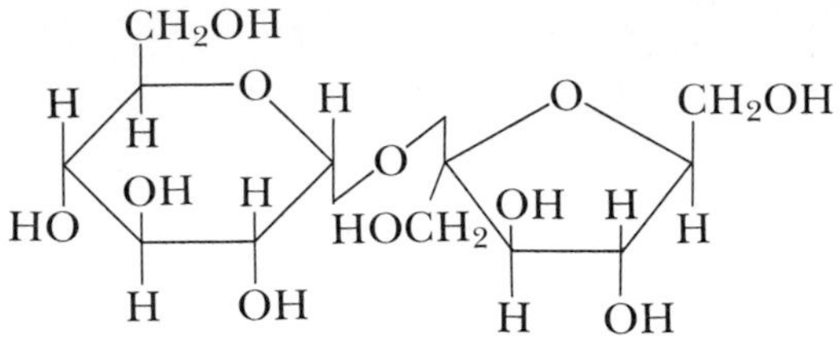

Sucrose

Also known as cane sugar, beet sugar, or just plain sugar, sucrose is the world's most frequently used sweetening agent.

Maltose (malt sugar) is another example of a disaccharide. It is composed of two glucose molecules. Maltose is rarely found in large quantities by itself in nature, but it is detected in most living cells when more complex carbohydrates are decomposed. Lactose (milk sugar) is the disaccharide composed of bonded glucose and galactose monosaccharide units. Human milk contains about 7% lactose, a higher percent than cow's milk, which contains only approximately 5% lactose.

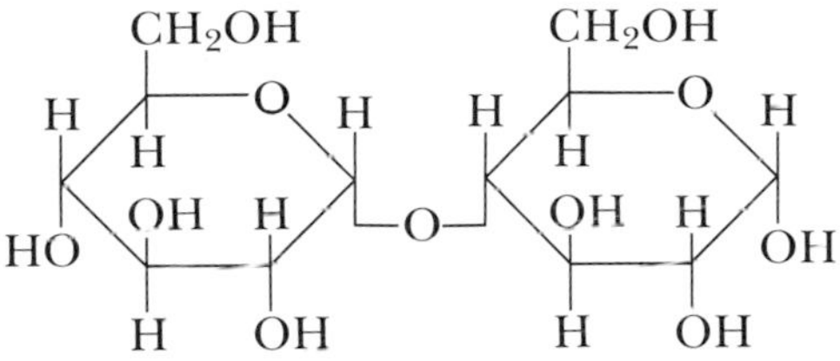

Maltose

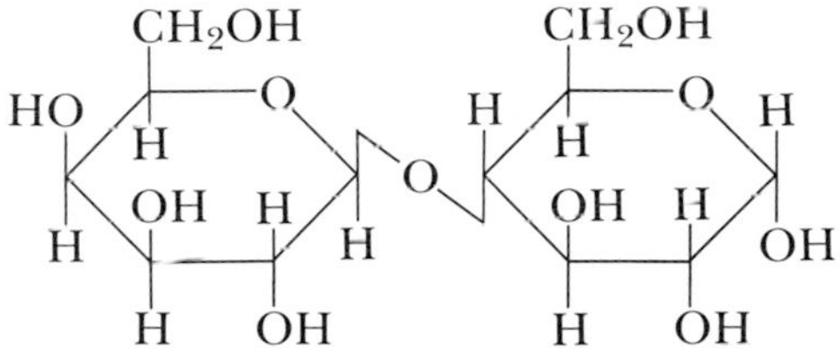

Lactose

Just about all animals lack the enzymes to decompose cellulose; accordingly, most animals, including humans, obtain no nutritional value from cellulose. Nevertheless it serves an important role in the intestines in connection with fluid retention.

Polysaccharides are the most abundant carbohydrates in nature, primarily because of the omnipresence of cellulose. Cellulose is the major component of the cell walls of all plants. After hydrolysis of cellulose, only one substance is isolated—glucose. Cellulose is a polymer of glucose. Other naturally occurring polysaccharides are starch and glycogen. Starch and glycogen serve as a storage reserve for glucose in plants and animals, respectively.

Proteins are one of the structural components of animal cells. A large percent of the hair, skin, muscles, tendons, and connective tissues is composed of proteins. Additionally, proteins are components of nerve tissue, antibodies, and some hormones. Enzymes, the chemical controlling agents of cells, are mainly protein structures.

Proteins

As the monosaccharide is the basic unit of carbohydrates, the **amino acid** is the basic unit of proteins. Amino acids have the general structure

$$H_2N-\underset{\underset{R}{|}}{\overset{\overset{H}{|}}{C}}-COOH$$

Amino acids possess both an amino group ($-NH_2$) and a carboxyl group ($-COOH$). While an unlimited number of amino acids could exist, only about 20 to 30 amino acids are found in living things. Amino acids differ with respect to the attached $R-$ group. Examples of common amino acids are

$$H-\underset{\underset{NH_2}{|}}{\overset{\overset{O}{\|}}{CHC}}-OH \qquad CH_3\underset{\underset{NH_2}{|}}{\overset{\overset{O}{\|}}{CHC}}-OH \qquad \bigcirc-CH_2\underset{\underset{NH_2}{|}}{\overset{\overset{O}{\|}}{CHC}}-OH \qquad HS-CH_2\underset{\underset{NH_2}{|}}{\overset{\overset{O}{\|}}{CHC}}-OH$$

   Glycine   Alanine   Phenylalanine   Cysteine

Amino acids combine chemically in cells to produce chains of bonded amino acids. They bond in such a manner that the amino group from one amino acid combines with the carboxylic acid group from the other, producing an amide linkage that is frequently called the **peptide bond.** If two amino acids are bonded, the structure is called a **dipeptide**; three attached amino acids are a **tripeptide**; and so on. An example of a dipeptide is glycylalanine.

$$H_2N-CH_2\overset{\overset{O}{\|}}{C}-NH\underset{\underset{CH_3}{|}}{\overset{\overset{H}{|}}{C}}COOH$$

Dipeptide (glycylalanine)

Chains of five to thirty amino acids are frequently termed **polypeptides,** which even longer chains are called proteins. **Proteins** are high-molecular-mass polypeptides, or polymers of amino acids.

Each protein has a specific sequence of amino acids. The properties of proteins are in part dependent on the sequence and number of amino acids in the chain. If one amino acid in a long chain is removed and replaced with another amino acid, the resulting protein has a different set of properties. Besides the number and sequence of amino acids in the protein, the actual three-dimensional shape of the molecule also plays a significant role in determining the protein's properties. Figure 20.3 shows the complex structure of an oxygen-carrying molecule, myoglobin.

Figure 20.3
Myoglobin carries oxygen from the circulatory system to the muscles. Its structure is that of a helix that has a series of bends, folds, and twists. Within the protein chain is a heme group. The heme group is the component of myoglobin that carries the O_2 molecule. Every protein has a specific three-dimensional configuration. (Modified and reprinted with permission from H. Neurath, ed., *The Proteins,* vol. 2, Academic Press, New York, 1964)

Heme

Two general classes of proteins exist: (1) simple and (2) conjugated. **Simple proteins** are composed exclusively of amino acids, with no other nonprotein groups bonded to the molecule. **Conjugated proteins** have a nonprotein group, called a prosthetic group, bonded to the protein structure.

Four principal classes of simple proteins exist: (1) scleroproteins, (2) globulins, (3) albumins, and (4) histones. **Scleroproteins** make up skin, hair, and protective and connective tissues. **Globulins** are a water-insoluble class of simple proteins found in antibodies and other blood components. **Histones** contain many basic amino acids and are generally associated with nucleic acids. Finally, the most widely distributed group of simple proteins is the **albumins.** Many blood proteins are albumins. Egg whites contain a high percent of albumin.

Five of the most important classes of conjugated proteins are the (1) glycoproteins, (2) lipoproteins, (3) nucleoproteins, (4) phosphoproteins, and (5) chromoproteins. Each of these groups has a nonprotein group bonded to the protein molecule. **Glycoproteins** are conjugated proteins that have a carbohydrate group bonded to the molecule, and **lipoproteins** have a lipid structure bonded to the protein molecule. Discussion of each of these groups is beyond the scope of this brief overview. Table 20.9 presents examples of each type of conjugated protein.

Enzymes are one of the most interesting and important types of proteins. **Enzymes** are biochemical catalysts, i.e., substances that increase the rate of chemical reactions within living things. Enzymes—as is true of all catalysts—are not consumed during chemical reactions. Virtually all reactions in living cells are catalyzed by enzymes. Enzymes make life possible because most of the reactions in the cell could not take place under the mild conditions present if they were not catalyzed.

Compared with carbohydrates and proteins, **lipids** are a more diverse group of compounds. There is no basic structure from which all the classes of lipids are derived. Classes of compounds that are considered lipids include triacylglycerols, waxes, steroids, prostaglandins, and compound lipids.

Lipids

Triacylglycerols, which are frequently called triglycerides, are the most abundant of all the classes of lipids. Triacylglycerols, when decomposed, yield glycerol and fatty acids. Glycerol is a three-carbon polyalcohol,

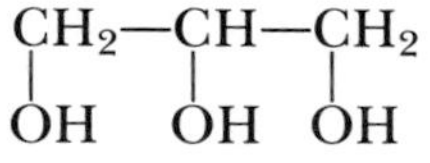

$$CH_2-CH-CH_2$$
$$|\qquad|\qquad|$$
$$OH\quad OH\quad OH$$

Glycerol

and fatty acids are carboxylic acids that most frequently contain from 10 to 24 C atoms. A saturated fatty acid has the general formula

$$CH_3(CH_2)_nCOOH$$

Saturated fatty acid

in which n is a number from 8 to 22. The structure of a simple triacylglycerol is

When triacylglycerols are combined with NaOH, soaps are produced.

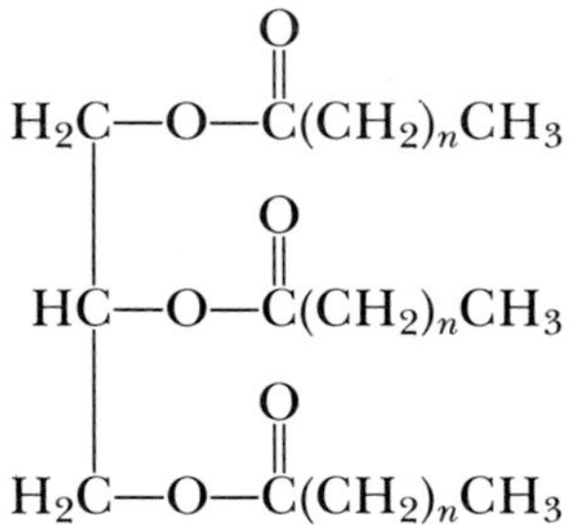

$$
\begin{array}{c}
\qquad\qquad O \\
\qquad\qquad \| \\
H_2C-O-C(CH_2)_nCH_3 \\
\qquad\qquad O \\
\qquad\qquad \| \\
HC-O-C(CH_2)_nCH_3 \\
\qquad\qquad O \\
\qquad\qquad \| \\
H_2C-O-C(CH_2)_nCH_3
\end{array}
$$

Triacylglycerol

Each of the three fatty acids is bonded to the glycerol molecule by an ester linkage. If all of the fatty acid molecules have saturated chains, then the triacylglycerol is called a **saturated fat.** Addition of unsaturated fatty acids produces an **unsaturated fat.** Triacylglycerols with a greater percent of saturated fatty acids tend to be solids (fats); those with more unsaturated fatty acids tend to be liquids (oils).

Waxes, another class of lipid, are also esters. They are esters of long-chain monoalcohols and fatty acids. Usually the carbon chains in both the alcohol and acid components consist of between 10 and 30 C atoms. Beeswax is formed from a 30-carbon alcohol and a 16-carbon

TABLE 20.9 CONJUGATED PROTEINS

Class of conjugated protein	Examples
Glycoproteins	Mucin, a component of saliva; gamma globulin, a blood protein
Lipoproteins	Blood plasma beta lipoproteins
Nucleoproteins	Component of ribosomes, cell components; tobacco mosaic virus
Phosphoproteins	Casein, a milk protein
Chromoproteins	Hemoglobin, O_2-carrying protein; cytochromes, part of cells' energy-producing apparatus

acid. Waxes are widely found throughout the plant and animal kingdoms.

Steroids are structurally different from the triacylglycerols and waxes. Steroid molecules have four fused rings; three of the rings contain six C atoms, and one ring contains five C atoms. Steroid structures differ with respect to functional groups and substituent groups. For example, the most abundant steroid in humans is cholesterol.

Cholesterol is the most abundant steroid in humans. Most individuals have about 225 g of cholesterol in their bodies. Cells use cholesterol to synthesize other steroids.

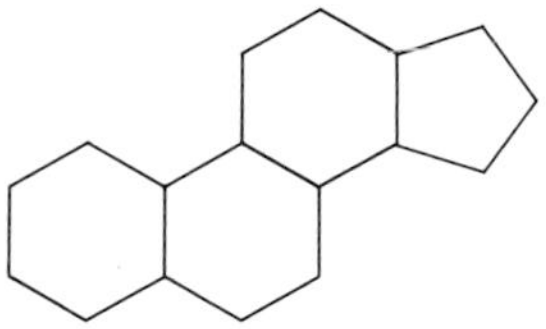

Steroid structure

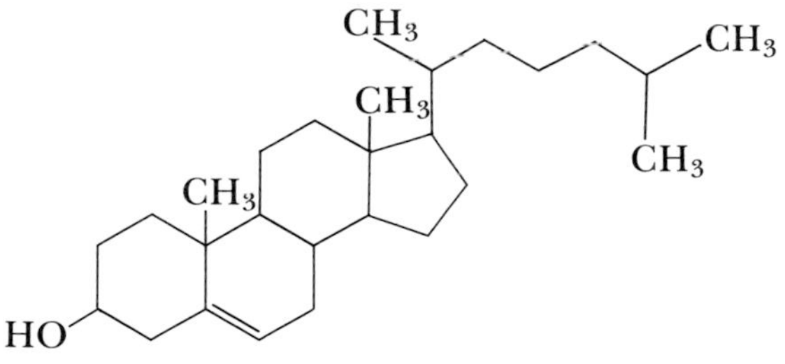

Cholesterol

Steroids have widely varying functions in living things. Bile acids, such as cholic acid, aid in the digestion of dietary lipids. Many of the male and female sex hormones are steroids. Sex hormones are responsible for a person's secondary sex characteristics. Important hormones (about 30) secreted by the adrenal gland, called corticoids, are also steroids.

A class of lipids which have a biological regulatory function is the **prostaglandins.** All prostaglandins have the basic structure of prostanoic acid.

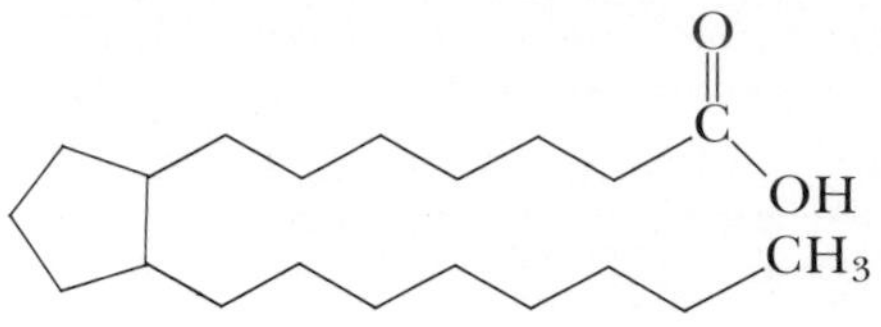

Prostanoic acid

Prostaglandins help regulate body temperature, smooth muscle contraction, and various tissue secretions. They also seem to be involved in the mechanism for controlling inflammations. Much is still to be learned about this puzzling class of lipids.

Like the proteins, lipids bond to nonlipid molecules. Two of the more common complex lipids are **phospholipids,** which contain a phosphate group, and **glycolipids,** which contain a carbohydrate molecule.

Nucleic Acids

Nucleic acids are very complex biomolecules composed of units called **nucleotides.** Most **nucleic acids** are polymers of nucleotides. Two different varieties of naturally occurring nucleotides exist; thus, two different classes of nucleic acids exist. The two classes of nucleotides are ribose nucleotides and deoxyribose nucleotides.

Nucleotides have three components: (1) a sugar, either ribose or deoxyribose, (2) a heterocyclic amine, and (3) a phosphate group. An example of a deoxyribose nucleotide structure is

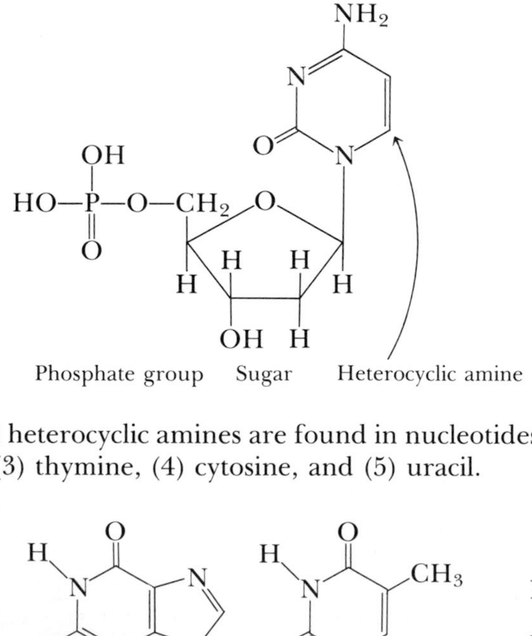

Phosphate group   Sugar   Heterocyclic amine

Five principal heterocyclic amines are found in nucleotides: (1) adenine, (2) guanine, (3) thymine, (4) cytosine, and (5) uracil.

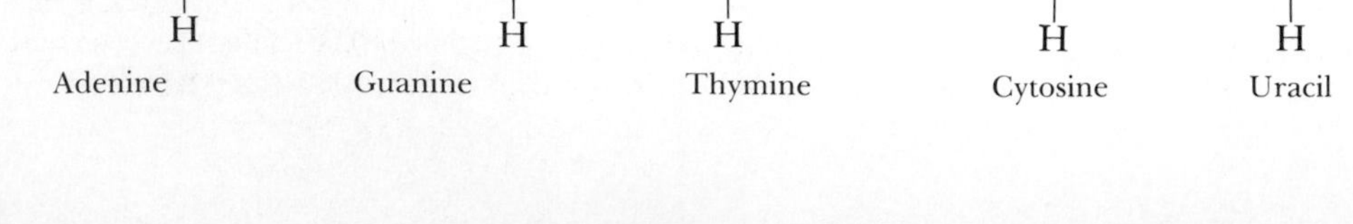

Adenine   Guanine   Thymine   Cytosine   Uracil

Nucleic acids range from rather small transfer RNAs to the massive DNA polymers, with their double-helix structure. Refer to a biochemistry textbook for a complete discussion of the complex structure of nucleic acids. Nucleic acids have innumerable functions in cells, including the storing of genetic information and the regulation of protein synthesis.

20.19 What are the four classes of biologically important compounds?

20.20 What is the basic molecular structure of *(a)* proteins, *(b)* carbohydrates, *(c)* nucleic acids?

20.21 What functional groups are found in *(a)* carbohydrates, *(b)* proteins, *(c)* lipids?

20.22 Give an example of each of the following: *(a)* disaccharide, *(b)* fatty acid, *(c)* amino acid, *(d)* steroid, *(e)* nucleic acid, *(f)* polysaccharide.

20.23 What is the difference between a simple and a conjugated protein?

20.24 What compounds result when the following are hydrolyzed: *(a)* triacylglycerols, *(b)* disaccharides, *(c)* nucleotides, *(d)* proteins, *(e)* waxes?

SUMMARY

All organic compounds are divided into two categories: hydrocarbons and hydrocarbon derivatives. **Hydrocarbons** are those compounds that contain only carbon and hydrogen. **Hydrocarbon derivatives** are organic substances that contain atoms such as the halogens, oxygen, nitrogen, sulfur, or phosphorus in addition to carbon and hydrogen.

Alkanes, alkenes, alkynes, and, aromatics are the four classes of hydrocarbons. **Alkanes** are called the saturated hydrocarbons because they have the maximum number of H atoms bonded to their C atoms and thus contain only carbon–carbon single bonds. Alkane molecules exist in chains, branched chains, and rings. Methane, CH_4, is the simplest alkane, and ethane, C_2H_6, is the next member of the series. Each succeeding member of the alkanes differs only by a $—CH_2—$ group. Such a series of compounds is called a **homologous series.**

Alkenes have a carbon–carbon double bond and, as a result, are unsaturated hydrocarbons. **Alkynes** are also unsaturated; however, they contain a triple covalent bond. **Aromatic hydrocarbons** have a special set of properties that resemble the properties of benzene, C_6H_6. Most aromatics have a stable, benzenelike ring structure as part of their molecules.

Each class of hydrocarbon derivative is identified by its functional group. A **functional group** is a specific arrangement of atoms that gives a compound its characteristic chemical properties. For example, the functional group in all alcohols is an —OH group.

Alcohols and **ethers** can be thought of as organic derivatives of water, H_2O. If one of water's H atoms is replaced with an alkyl group, an alcohol results. When both H atoms are removed and replaced with alkyl groups, an ether results. **Aldehydes** and **ketones** are called carbonyl compounds; they have a C=O (carbonyl group) within their molecules. If the carbonyl group is at the end of a chain and has an H atom bonded to it, then the molecule is classified as an aldehyde. When the carbonyl is in the middle of a chain, bonded to two C atoms, the molecule is a ketone.

Carboxylic acids contain a C atom that has both an alcohol and carbonyl group bonded; this combination, —COOH, is called a carboxyl group. Many organic acid derivatives exist. Organic acids react with alcohols to produce **esters;** a —COOR is the functional group in esters. Another derivative of the organic acids is the **amide**—compounds that contain $—CONH_2$.

Organic derivatives of ammonia are **amines,** RNH_2. Amines can have up to a maximum of three alkyl groups attached to the nitrogen atom. Amines are widely distributed in the natural world, especially in molecules that have both C and N atoms within cyclic structures. The latter are called **heterocyclic amines.** The last class of hydrocarbon derivatives discussed was the **alkyl halides** which are organic compounds that contain one or more halogen atoms.

Biologically important organic substances are separated into four classes: (1) carbohydrates, (2) proteins, (3) lipids, and (4) nucleic acids. **Carbohydrates,** when broken down to their smallest units, are polyhydroxy aldehydes and ketones. Three classes of carbohydrates are monosaccharides, disaccharides, and polysaccharides. **Monosaccharides** are the basic units of carbohydrates. When two monosaccharide molecules bond, a **disaccharide** results. If many monosaccharides bond, a **polysaccharide** forms.

Proteins are polymers of amino acids. An **amino acid** is a chemical structure that contains both an amino group ($-NH_2$) and a carboxylic acid group ($-COOH$). Approximately 20 amino acids exist in nature. These 20 amino acids are combined into chains that range from two or three amino acids to high-molecular-mass proteins with thousands of amino acids in the chain. Proteins are a major structural component of living things. A special class of proteins, called **enzymes,** controls the rates of most chemical reactions in living things.

Lipids are a diverse group of biological substances, ranging from the triacylglycerols to the steroids. **Triacylglycerols** are esters of the polyalcohol glycerine. **Steroids,** in contrast, possess a ring system with 3 six-membered rings and 1 five-membered ring fused together. Other lipids include the **waxes, prostaglandins,** and **complex lipids.**

Most **nucleic acids** are high-molecular-mass polymeric biomolecules. Each nucleic acid is composed of a basic unit called a nucleotide. A nucleotide has three parts: (1) a five-carbon sugar (ribose or deoxyribose), (2) a heterocyclic amine, and (3) a phosphate group. The two classes of nucleic acids are the DNAs (deoxyribose nucleic acids) and RNAs (ribose nucleic acids).

KEY TERMS

alkyl group	disaccharide	hydrogenation reaction	saturated hydrocarbon
analgesic	esterification reaction	isomer	steroids
anesthetic	functional group	IUPAC	substituent group
aromatic hydrocarbons	fused rings	monosaccharide	triacylglycerols
carbonyl group	heterocyclic compound	nucleotide	unsaturated hydrocarbon
carboxyl group	homologous series	organic chemistry	
cyclic structure	hydrocarbon	polycyclic compound	
dehydration reaction	hydrocarbon derivative	polysaccharide	

EXERCISES*

20.25 Define the following terms: organic chemistry, hydrocarbon, hydrocarbon derivative, saturated hydrocarbon, unsaturated hydrocarbon, homologous series, IUPAC, isomer, substituent group, cyclic structure, aromatic hydrocarbons, polycyclic compound, alkyl group, functional group, anesthetic, carbonyl group, carboxyl group, analgesic.

Hydrocarbons

20.26 (a) What scientist is credited with the discovery that organic compounds could come from nonliving things? (b) How did he prove his point?

20.27 What is the difference between a hydrocarbon and a hydrocarbon derivative?

20.28 What distinguishes each class of hydrocarbon from the others?

20.29 How many carbon atoms are there in each of the following hydrocarbons: (a) heptane, (b) butane, (c) decane, (d) ethane, and (e) octane?

20.30 Draw the complete structural formulas from the following condensed formulas:
(a) $CH_3CH_2CH_2CH_2CH_2CH_3$
(b) $CH_3CH_2CH_2CH_2CH_2CH_2CH_2CH_2CH_3$
(c) $CH_3CH_2CHCH_2CH_2CH_2CH_3$
$\quad\quad\quad\;\; |$
$\quad\quad\quad CH_3$
(d) $(CH_3)_3CCH_2CH_2CH_2CH_2CH_2CH_3$

*For exercise numbers printed in color, answers can be found at the back of the book.

20.31 For each of the hydrocarbons in Exercise 20.30, write the name of the unbranched isomer.

20.32 Write the names of the following alkanes:

(a) $\underset{\underset{CH_3}{|}}{CH_3}\underset{}{\overset{\overset{CH_3}{|}}{C}}CH_2\overset{\overset{CH_3}{|}}{C}HCH_3$

(b) $\overset{\overset{CH_3}{|}}{C}HCH_2CH_2CH_2CH_2CH_2\underset{\underset{CH_3}{|}}{\overset{\overset{CH_3}{|}}{C}}CH_3$

(c) $CH_3CH_2\overset{\overset{CH_2CH_3}{|}}{C}HCH_2CH_2CH_2CH_2CH_3$

(d) $CH_3\overset{\overset{CH_3}{|}}{C}H\overset{}{C}H\overset{\overset{CH_3}{|}}{C}HCHCH_3$

(e) $\overset{\overset{CH_3}{|}}{\underset{\underset{CH_3}{|}}{C}}H\overset{\overset{CH_2}{|}}{C}H_2\underset{\underset{CH_3}{|}}{C}\overset{\overset{CH_2CH_2CH_2CH_2CH_3}{|}}{}-CH_2CH_3$

(f) $CH_3-\overset{\overset{CH_3}{|}}{\underset{\underset{CH_3}{|}}{C}}\underline{\quad}\overset{\overset{CH_3}{|}}{\underset{\underset{CH_3}{|}}{C}}-CH_2CH_2CH_2CH_3$

20.33 Draw the structures of
(a) 3-methylheptane
(b) 2,2-dimethylpentane
(c) 3-ethyl-4-methyldecane
(d) 2,2,4-trimethyloctane
(e) 2,2,3,3-tetramethylhexane
(f) 4-ethylnonane
(g) ethylcyclopentane
(h) 1,2,3-triethylcyclooctane

20.34 Draw the structures and write the IUPAC names for all isomers of heptane.

20.35 Within a homologous series there is generally a trend of increasing boiling points. Account for this trend in terms of molecular structure.

20.36 Write the names of the following cycloalkanes:

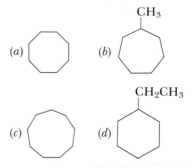

20.37 If there are n carbon atoms in a cycloalkane, how many hydrogen atoms are also in the molecule?

20.38 Draw the structures of the following unsaturated hydrocarbons: (a) a four-carbon alkene, (b) a six-carbon cycloalkene, (c) two unbranched five-carbon alkenes, (d) two four-carbon alkadienes, (e) all cyclic six-carbon alkadienes.

20.39 How many hydrogen atoms are contained in (a) 2-methylbutane, (b) nonene, (c) methylcyclopentane, (d) butyne, (e) benzene, (f) 3-methylheptane?

20.40 Draw the structures of all isomers of C_5H_{10}. Don't forget about cyclic structures.

20.41 How is ethyne (acetylene) prepared commercially?

20.42 What is the general formula of the alkynes?

20.43 Draw the structures of the three diethylbenzenes, and write a name for each.

20.44 (a) Determine how many different trimethylbenzenes exist. (b) Draw the structure of each trimethylbenzene.

20.45 Draw the structures of three polycyclic aromatic compounds that contain four fused benzene rings.

Hydrocarbon Derivatives

20.46 What type of hydrocarbon derivative is each of the following: (a) RCOOR, (b) RX, (c) RCOR, (d) RSH, (e) RCONH₂, (f) RCOOH, (g) RNH₂, (h) ROR?

20.47 What functional group is in each of the following: (a) aldehydes, (b) acids, (c) amines, (d) thioethers, (e) esters?

20.48 Circle and identify each functional group in the following molecules:

(a) $H_3C-\overset{\overset{\displaystyle O}{\parallel}}{\underset{\underset{\displaystyle H}{|}}{C}}-\overset{\overset{\displaystyle H}{|}}{C}-OCH_2\overset{\overset{\displaystyle }{}}{\underset{\underset{\displaystyle Cl}{|}}{C}}HCOOH$

(b) $Cl-CH_2\overset{\overset{\displaystyle NH_2}{|}}{C}HCHCH_2CH_2OCH_3$

with a branch: $\overset{\displaystyle }{\underset{\displaystyle H}{}}\overset{\displaystyle C}{}\overset{\displaystyle }{O}$

(c) CH_3-O- (cyclopentenone ring with substituents) $N(CH_3)_3$, $C=O$, NH_2

20.49 Draw the structures of the following: *(a)* a four-carbon alcohol, *(b)* a three-carbon aldehyde, *(c)* a four-carbon acid, *(d)* a six-carbon ketone, *(e)* an eight-carbon amine, *(f)* a cyclic five-carbon ester.

20.50 What are the IUPAC names for the following hydrocarbon derivatives: *(a)* grain alcohol, *(b)* acetone, *(c)* formic acid, *(d)* butyric acid?

20.51 What class of compound contains *(a)* a nitrogen atom within a ring, *(b)* a carbonyl group, *(c)* a carboxyl group, *(d)* fluorochlorocarbons?

20.52 List a use for each of the following hydrocarbon derivatives: *(a)* ethanol, *(b)* diethyl ether, *(c)* ethylene glycol, *(d)* formaldehyde, *(e)* vanillin, *(f)* acetone.

20.53 *(a)* What is the structural difference between an aldehyde and a ketone? *(b)* Give an example of each.

20.54 *(a)* What is oxalic acid? *(b)* How is oxalic acid different from acetic acid?

20.55 *(a)* What common property do the simple esters share? *(b)* Give two examples.

20.56 List three esters of salicylic acid and how they are used as medicines.

20.57 Draw the structure of *(a)* a simple three-carbon amide, *(b)* an ethyl-substituted two-carbon amide, and *(c)* an aromatic methyl-disubstituted amide.

20.58 List three biologically significant amides.

20.59 Draw the structures of a *(a)* primary, *(b)* secondary, and *(c)* tertiary aromatic amine.

20.60 *(a)* How is a heterocyclic amine different from other cyclic compounds? *(b)* Give two examples of heterocyclic amines.

20.61 Write an equation that represents what happens when 1 mol Cl_2 gas combines with 1 mol CH_4 in the presence of light.

20.62 What biological effects are observed when humans are exposed to halogenated hydrocarbons?

20.63 *(a)* What are Freons? *(b)* How are Freons used? *(c)* What problems result when Freons are released into the atmosphere?

20.64 *(a)* What problems were solved after the discovery of DDT? *(b)* What problems were produced by the use of DDT?

Biochemistry

20.65 What two general classes of compounds make up the carbohydrates?

20.66 Draw the structures of the simplest aldose and ketose.

20.67 How does a monosaccharide differ from a polysaccharide?

20.68 Give an example of each of the following: *(a)* five-carbon aldose, *(b)* most commonly used disaccharide, *(c)* plant polysaccharide, and *(d)* most abundant sugar in nature.

20.69 *(a)* What is the general molecular formula for aldohexoses? *(b)* What is the general molecular formula of a disaccharide produced from two aldohexoses? *(c)* Look at these molecular formulas and state what molecule is also formed when two aldohexoses combine to produce a disaccharide?

20.70 What is the main function of glycogen in animals?

20.71 In what types of tissues are proteins found?

20.72 In what class of hydrocarbon derivative could proteins be placed?

20.73 Name and write the structures of two amino acids.

20.74 *(a)* What are conjugated proteins? *(b)* List the five main classes of conjugated proteins.

20.75 *(a)* What is an enzyme? *(b)* To what class of biochemicals do enzymes belong?

20.76 List the five principal classes of lipids, and give an example of each.

20.77 Draw the structure of a triacylglycerol that is produced from three 12-carbon fatty acid molecules and glycerol.

20.78 Are there any differences between fatty acids and regular carboxylic acids?

20.79 Draw the ring structure that is part of all steroids.

20.80 What are three functions of steroids in humans?

20.81 How are prostaglandins important in living things?

20.82 List three classes of complex lipids.

20.83 What is the basic structure found in nucleic acids?

20.84 How are nucleic acids important within cells?

Additional Exercises

20.85 Describe differences in the general properties of inorganic and organic compounds.

20.86 Name two compounds and two ions that contain carbon but are not classified as organic compounds.

20.87 (a) What general type of reaction do alkenes frequently undergo? (b) Give a specific example.

20.88 What mass of octane results when a 10.5-g sample of 1-octene undergoes a catalytic hydrogenation reaction?

20.89 (a) Write the equation that shows the formation of nitrobenzene from benzene. (b) Write the equation for the formation of bromobenzene from benzene.

20.90 (a) What class of compound results when alcohols undergo dehydration? (b) Give an example of such a reaction.

20.91 Write equations for the complete oxidation of each of the following alcohols:
(a) $CH_3CH_2CH_2OH$
(b) $CH_3CH_2CH_2CH_2CH_2CH_2OH$
(c) $CH_3CHCH_2CH_3$
   $\quad\ \ |$
   $\quad\ \ OH$

20.92 Explain why tertiary alcohols do not undergo oxidation reactions in a manner similar to primary and secondary alcohols.

20.93 Write equations that show the reduction and oxidation of propanal, CH_3CH_2CHO.

20.94 Write equations for each of the following acid-catalyzed esterification reactions:
(a) $CH_3COOH + CH_3OH \longrightarrow$
(b) $CH_3CH_2CH_2COOH + CH_3CH_2OH \longrightarrow$
(c) $CH_3CH_2CH_2CH_2CH_2CH_2OH$
$\qquad\qquad\qquad\qquad + HCOOH \longrightarrow$

20.95 What mass of ethyl acetate,

$CH_3COOCH_2CH_3$,

results when 42.3 g of CH_3COOH combines with 31.9 g of CH_3CH_2OH?

20.96 (a) A carbon-hydrogen compound is found to contain 82.6% carbon. What is the empirical formula of the compound? (b) A 12.8-g sample of this carbon-hydrogen compound is found to occupy 5.35 L at 298 K and 1.01 atm. What is the molecular formula of the compound?

20.97 What volume of H_2 at 285 K and 937 torr is required to totally hydrogenate 91.5 g $CH_3CH_2CH=CHCH_2CH_3$?

20.98 Write all of the organic products for each of the following reactions:
(a) $CH_3CH_3 + Cl_2 \xrightarrow{light}$

(b) $CH_3CH_2CH=CHCH_3 + H_2 \xrightarrow{cat}$

(c) $CH_3CH_2CH_2CH_2CHO \xrightarrow{K_2Cr_2O_7/H^+}$

(d) $CH_3CH_2CH_2CH_2OH +$
$\qquad CH_3CH_2COOH \xrightarrow{H^+}$

(e) 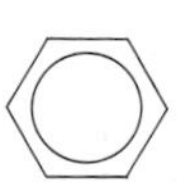 ⬡ $+ HNO_3 \xrightarrow{H_2SO_4}$

(f) $CH_3CH_2CH_2CH_2CHO + H_2 \xrightarrow{Pt}$

(g) 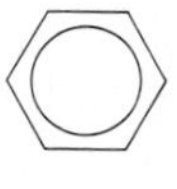 ⬡ $+ Cl_2 \xrightarrow{cat}$

APPENDIX
A

Review of Math Skills

Algebra is an area of mathematics that deals with equations and equalities. Generally, equations are solved to determine the value of an unknown quantity. This is accomplished by applying the most basic rule of algebra: Isolate the unknown quantity in an equation by mathematically treating both sides of the equation in the same way. An equation is unchanged as long as whatever is done to one side of the equation is also done to the other side. Consider the following equation:

$$X - 10 = 12$$

If the same number is added to both sides of the equation, the equality remains unchanged. Therefore, if 10 is added to both sides of the equation, the unknown value, X, is isolated on the left side (-10 plus 10 equals zero). On the right side, 10 is added to 12, giving 22; hence, the value of the unknown, X, is 22.

$$
\begin{array}{rr}
X - 10 = & 12 \\
+10 & +10 \\
\hline
X = & 22
\end{array}
$$

Always check the answer by substituting the value obtained back into the equation. In the above example: $22 - 10 = 12$.

The above equation belongs to a general class of equations having the form

Type 1: $\qquad\qquad X + m = n$

where X is the unknown quantity and m and n are known quantities. To solve Type 1 equations, m is either added or subtracted to isolate X by itself on one side of the equation. See Example Problem 1.

A.1 ALGEBRAIC OPERATIONS

584

Example Problem 1

Solve the following equation: $X + 23 = -100$.

Solution

To solve this equation, subtract 23 from both sides.

$$\begin{array}{r} X + 23 = -100 \\ -23 \quad -23 \\ \hline X = -123 \end{array}$$

Check the answer: $-123 + 23 = -100$.

Type 2 equations are solved by multiplying and dividing appropriate quantities to isolate the unknown value. Type 2 equations have the general form

Type 2: $\qquad mX = n \quad \text{or} \quad \dfrac{X}{m} = n$

In the first equation, X is obtained by dividing both sides of the equation by m.

$$\frac{mX}{m} = \frac{n}{m} \quad \text{or} \quad X = \frac{n}{m}$$

In the second equation, X is determined by multiplying both sides by m.

$$m\left(\frac{X}{m}\right) = mn \quad \text{or} \quad X = mn$$

In chemistry, we use both multiplication and division of numbers to solve equations. Consider the following equation:

$$\frac{aX}{m} = n$$

To solve for X in an equation of this form, multiply each side of the equation by m/a, then cancel a/m on the left side, yielding $X = mn/a$.

$$\frac{m}{a} \times \frac{aX}{m} = \frac{mn}{a}$$

Example Problem 2 shows how to solve a Type 2 equation.

Example Problem 2

Solve the following equation for X:

$$\frac{5X}{3} = -20$$

Solution

Multiply both sides of the equation by $\frac{3}{5}$, thereby canceling the $\frac{5}{3}$ on the left side of the equation and isolating X.

$$\frac{\cancel{3}}{\cancel{5}} \times \frac{\cancel{5}X}{\cancel{3}} = \frac{3}{5}(-20)$$

$$X = \frac{3}{5}(-20)$$

$$X = -12$$

Check the answer:

$$\frac{5 \times (-12)}{3} = -20$$

Many algebraic equations are a combination of Type 1 and Type 2 equations. Isolating the unknown quantity in this type of equation requires both addition and multiplication. To illustrate, let's solve the following equation for X:

$$a(X - m) = n$$

First divide each side by a, and then add m to both sides.

$$\frac{\cancel{a}(X - m)}{\cancel{a}} = \frac{n}{a}$$

$$X - m = \frac{n}{a}$$

$$\underline{+m = +m}$$

$$X = \frac{n}{a} + m$$

If possible, when you solve such equations, initially remove all terms that can be eliminated from the side of the equation containing the unknown quantity by multiplying and dividing. Finally, add or subtract the remaining terms, leaving the unknown quantity by itself.

In other mixed equations the opposite procedure should be followed: Add and subtract first, and then multiply and divide (see Example Problem 3).

Example Problem 3

Solve for X:

$$\frac{18}{X} + 7 = 13$$

Solution

Subtract 7 from both sides of the equation.

$$\frac{18}{X} + 7 = 13$$

$$\underline{\phantom{\frac{18}{X}}-7 \quad -7}$$

$$\frac{18}{X} = 6$$

Multiply both sides of the equation by X, and then divide the resulting equation by 6, giving the answer 3.

$$X\left(\frac{18}{X}\right) = 6X$$

$$18 = 6X$$

$$\frac{18}{6} = \frac{6X}{6}$$

$$3 = X$$

Solving more complex equations involves applying the same general principles: Rearrange the equation and isolate the unknown quantity through additive and multiplicative operations. However, the number of operations needed to solve these equations increases as the equations become more complex. Consider the more involved equation in Example Problem 4.

Example Problem 4

Solve the following equation for X:

$$\frac{a + 1}{X + b} = \frac{m}{n}$$

Solution

---

1. Multiply both sides of the equation by $(X + b)$.

$$(X + b) \times \frac{a + 1}{X + b} = (X + b) \times \frac{m}{n}$$

$$a + 1 = (X + b) \times \frac{m}{n}$$

2. Multiply both sides by n/m.

$$\frac{n}{m} \times (a + 1) = (X + b) \times \frac{m}{n} \times \frac{n}{m}$$

$$\frac{n}{m} \times (a + 1) = X + b$$

3. Subtract b from both sides of the equation.

$$\frac{n}{m} \times (a + 1) - b = X + b - b$$

This yields

$$\frac{n}{m} \times (a + 1) - b = X$$

A.2 SCIENTIFIC NOTATION

Exponential Numbers

Extremely small and extremely large numbers are often encountered in chemistry. Numbers such as

$$0.0000000000005 \quad \text{and} \quad 6{,}000{,}000{,}000{,}000{,}000{,}000{,}000$$

are commonplace in chemical applications. Numbers in this form are unwieldy and awkward to deal with. Consequently, large and small numbers are ordinarily expressed exponentially or in a special exponential system called **scientific notation.**

Before considering the specifics of scientific notation, let's review some basic principles concerning exponential numbers. An exponent is a number or symbol written as a superscript to the right of a base number (or symbol) indicating how many times the base number is multiplied by itself. For instance, 10^3 is $10 \times 10 \times 10$, 2^6 is $2 \times 2 \times 2 \times 2 \times 2 \times 2$, and a^4 is $a \times a \times a \times a$. The exponent of a number is called the *power,* and the number being raised to the power is termed the *base.* We will restrict our discussion to numbers with base 10. Table 1 lists examples of exponential numbers with base 10.

You must obey a couple of simple rules when you multiply and divide exponential numbers (addition and subtraction are discussed later). When multiplying exponential numbers with the same bases, add the exponents. For example, the product of 10^6 and 10^8 is

$$10^6 \times 10^8 = 10^{6+8} = 10^{14}$$

The explanation for adding exponents when multiplying is straightforward: 10^6 is $10 \times 10 \times 10 \times 10 \times 10 \times 10$, and 10^8 is $10 \times 10 \times 10 \times 10 \times 10 \times 10 \times 10 \times 10$; hence, $10^6 \times 10^8$ equals $(10 \times 10 \times 10 \times 10 \times 10 \times 10) \times (10 \times 10 \times 10 \times 10 \times 10 \times 10 \times 10 \times 10)$, or 10^{14}.

TABLE 1   POWERS OF TEN

$10^0 = 1$ (all numbers to the zero power are 1)

$10^1 = 10$

$10^2 = 10 \times 10 = 100$

$10^3 = 10 \times 10 \times 10 = 1000$

$10^4 = 10 \times 10 \times 10 \times 10 = 10{,}000$

$10^{-1} = \frac{1}{10} = 0.1$

$10^{-2} = \frac{1}{10} \times \frac{1}{10} = 0.01$

$10^{-3} = \frac{1}{10} \times \frac{1}{10} \times \frac{1}{10} = 0.001$

$10^{-4} = \frac{1}{10} \times \frac{1}{10} \times \frac{1}{10} \times \frac{1}{10} = 0.0001$

When dividing numbers with the same base, subtract the exponent in the denominator from the exponent in the numerator. Thus the quotient of 10^5 divided by 10^4 is

$$\frac{10^5}{10^4} = 10^{5-4} = 10^1$$

or, without using exponential notation,

$$\frac{100{,}000}{10{,}000} = 10$$

Scientific Notation

Numbers are expressed in **scientific notation** by separating them into two factors: (1) decimal factor and (2) exponential factor. Examples of numbers expressed in scientific notation are

$$1.234 \times 10^9$$
$$9.87 \times 10^{-3}$$
$$3.0 \times 10^{59}$$

In each example, the first number (1.234, 9.87, and 3.0) is the decimal factor. By convention, the decimal factor is always given a value equal to or greater than 1 and less than 10. The decimal factor is multiplied by the exponential factor, 10 to some power.

To convert numbers to scientific notation, adjust the decimal point so that the decimal factor has a value equal to or greater than 1 and less than 10, and, depending on how many places the decimal point is moved and in what direction, give the appropriate power to the base, 10, so as not to change the value of the number.

To illustrate, let's change 23,000 to scientific notation. First, we adjust the decimal point to give a number between 1 and 10. To accomplish this, we move the decimal point four places to the left, giving the number 2.3000. For each place to the left we move the decimal point, we add 1 to the exponent of 10^0 (10^0 equals 1; by definition any number to the zero power is 1). So $23{,}000 \times 10^0$ is the same as $23{,}000 \times 1$ or just 23,000. Therefore, the exponent of 10^0 is increased by 4 to 10^{0+4} or 10^4.

$$23{,}000. \times 10^0 \qquad \text{converts to} \qquad 2.3000 \times 10^4$$

The exponent is increased because each time the decimal point is moved to the left, it is the same as dividing the number by 10, or decreasing the value by a factor of 10, and in order not to change the number, it has to be multiplied by 10. If a number is multiplied and divided by 10 at the same time, this is the same as multiplying by 1 ($\frac{10}{10} = 1$), which does not change the value of the number.

When large numbers are converted to scientific notation, the decimal point is moved to the left, thus increasing the value of the exponent. When numbers smaller than 1 are changed to scientific notation, the opposite is true—the decimal point is moved to the right, decreasing the value of the exponent.

How is 0.00000091 expressed in scientific notation? First, move the decimal point 7 places to the right, giving the value 9.1 for the decimal factor. In order not to change the numerical value of the number, 7 is subtracted from the exponent, 0 (10^0), giving -7 as the exponent. Thus

$$0.00000091 \quad \text{becomes} \quad 9.1 \times 10^{-7}$$

Each time the decimal is moved one place to the right, the magnitude of the number is increased by 10; at the same time, it must be divided by 10 so the value remains constant. Study the following example problem to better understand how to change numbers to scientific notation.

───────────────── **Example Problem 5** ─────────────────

Change each number to scientific notation: *(a)* 390,000,000,000,000,000; *(b)* 0.0000000000000000000072.

───────────────── **Solution** ─────────────────

(a) Move the decimal point 17 places to the left, giving 3.9.

$$390{,}000{,}000{,}000{,}000{,}000. \quad \text{gives} \quad 3.9$$

Add 17 to the exponent.

$$3.9 \times 10^{0+17} = \mathbf{3.9 \times 10^{17}}$$

(b) Move the decimal point 22 places to the right, giving 7.2.

$$0.0000000000000000000072 \quad \text{gives} \quad 7.2$$

Subtract 22 from the exponent.

$$7.2 \times 10^{0-22} = \mathbf{7.2 \times 10^{-22}}$$

To change a number expressed in scientific notation to a nonexponential number, the operation is reversed. If the exponent is positive, move the decimal point to the right, and if the exponent is negative, move the decimal point to the left. For example, to change 1.75×10^5 to nonexponential form:

$$1.75000 \times 10^5 \qquad \text{gives} \qquad 175{,}000 \times 10^0 = 175{,}000 \times 1 = 175{,}000$$

Each time the decimal is moved to the right, the value of the number is increased by a factor of 10; to keep the value of the number the same, 1 is subtracted from the exponent. In the case of changing from scientific notation to nonexponential form, the exponential factor is changed to 10^0, or 1, which is not written. In Example Problem 6 two numbers are converted from scientific notation to nonexponential form.

────────────── **Example Problem 6** ──────────────

Change each number to nonexponential form:
(a) 1.19×10^{-7}; (b) 6.50×10^6.

────────────── **Solution** ──────────────

(a) Since the exponent is -7, move the decimal point 7 places to the left and add 7 to the exponent in order not to change the value of the number.

$$.0000001.19 \times 10^{-7} \qquad \text{gives} \qquad 0.000000119 \times 10^{-7+7}$$

or

0.000000119

(b) Since the exponent is $+6$, move the decimal point 6 places to the right and subtract 6 from the exponent in order not to change the value of the number.

$$6.500000 \times 10^6 \text{ gives } 6{,}500{,}000. \times 10^{6-6}$$

or

6,500,000

Arithmetic operations on numbers expressed in scientific notation are handled the same way as operations on any exponential numbers. The only difference is that the proper operation must be performed on the decimal factor at the same time that the appropriate exponent operation is calculated.

Arithmetic Operations: Multiplication and Division

To review: When multiplying, add exponents; when dividing, subtract exponents. For example: What is the product of $(3 \times 10^4) \times (2 \times 10^6)$? It is easier to separate the decimal factors from the exponential factors, giving

$$(3 \times 2) \times (10^4 \times 10^6)$$

Multiply the decimal factors, and then add the exponents,

$$3 \times 2 = 6 \qquad \text{and} \qquad 10^{4+6} = 10^{10}$$

to obtain the correct answer,

$$6 \times 10^{10}$$

Dividing numbers expressed in scientific notation is carried out in a similar manner except that the decimal factors are divided and the exponents are subtracted. Example Problem 7 illustrates multiplication and division of numbers expressed in scientific notation.

Example Problem 7

Simplify the following expression:

$$\frac{8 \times 10^{12}}{4 \times 10^{15}} \times 1.5 \times 10^{-3}$$

Solution

1. First divide 8×10^{12} by (4×10^{15}).

$$\frac{8}{4} \times 10^{12-15} = 2 \times 10^{-3}$$

Separate the factors 8 and 4 from the exponential factor and divide: $\frac{8}{4} = 2$. Then subtract the exponents, 12 and 15, yielding -3.

2. Multiply the resulting number, 2×10^{-3}, by 1.5×10^{-3}.

$$2 \times 1.5 \times 10^{-3+(-3)} = \mathbf{3 \times 10^{-6}}$$

Multiply the decimal factors, 2 and 1.5, obtaining the product 3. Add the exponents, -3 and -3, giving -6. The same answer is obtained if 8×10^{12} is multiplied by 1.5×10^{-3} and the product is divided by 4×10^{15}.

Arithmetic Operations: Addition and Subtraction

To complete our study of numbers expressed in scientific notation, we turn our attention to addition and subtraction operations. Once again, one general rule prevails for both operations: Only numbers with exactly the same exponent can be added or subtracted.

What is the sum of 1×10^3 (1000) and 1×10^2 (100)? To add these numbers, both exponents must be the same. Therefore, either the 3 is changed to 2 or the 2 is changed to 3. Generally, it is best to change the smaller exponent to a larger exponent, as we shall see. If the exponent is increased by 1 (1×10^2), the decimal factor is divided by 10 (move the decimal point to the left):

$$1. \times 10^2 \quad \text{gives} \quad 0.1 \times 10^{2+1} = 0.1 \times 10^3$$

Now that both exponents are the same, the two numbers can be added:

$$0.1 \times 10^3 \quad + \quad 1 \times 10^3 = 1.1 \times 10^3 = 1100$$

When adding numbers in scientific notation, the decimal factors are added and the exponent of the answer remains the same (do not add the exponents).

In the above example, if the smaller exponent was not changed to a larger exponent, the resulting answer would not have been in scientific notation initially. Commonly, answers obtained after arithmetic operations are not in scientific notation; in other words, the decimal factor is not between 1 and 10. Whenever this case is encountered, it is necessary to change the answer back to scientific notation. Let's consider one final example problem that illustrates manipulation of numbers in scientific notation.

─────────────────── **Example Problem 8** ───────────────────

Evaluate the following, and express the final answer in scientific notation:

$$\frac{3 \times 10^7}{1.5 \times 10^{-2}} - (7.5 \times 10^8) + (1.25 \times 10^{10})$$

─────────────────── **Solution** ───────────────────

1. Divide (3×10^7) by (1.5×10^{-2}).

$$\frac{3}{1.5} = 2 \quad \text{and} \quad \frac{10^7}{10^{-2}} = 10^{7-(-2)} = 10^9$$

This yields

$$2 \times 10^9$$

2. Because the next operation is subtraction and the numbers have different exponents, we must change one of the exponents. Change the exponent in 7.5×10^8 to 10^9.

$$7.5 \times 10^8 \quad \text{gives} \quad 0.75 \times 10^{8+1} = 0.75 \times 10^9$$

Subtract this from the first number.

$$2 \times 10^9 - 0.75 \times 10^9 = 1.25 \times 10^9$$

3. Finally, add the last number, 1.25×10^{10}, after the exponents are changed to the same value.

$$1.25 \times 10^9 \quad \text{gives} \quad 0.125 \times 10^{9+1} = 0.125 \times 10^{10}$$

and

$$0.125 \times 10^{10} + 1.25 \times 10^{10} = \mathbf{1.375 \times 10^{10}}$$

A.3 GRAPHING

A **graph** is a convenient means for displaying and observing trends in data. Frequently, in chemistry, collected data are graphed to clearly show patterns in the data.

Graphing involves placing or "plotting" data points on graph paper and drawing a smooth curve or straight line through the plotted points. Regular

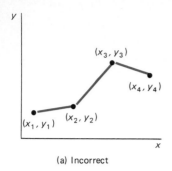

(a) Incorrect

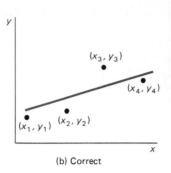

(b) Correct

Figure A.1
(a) The improper way to construct a graph. Always draw the smooth curve or straight line that best fits the data points. *(b)* The correct way to construct a graph.

graph paper is printed with evenly spaced horizontal and vertical lines for this purpose. All data are plotted between two perpendicular axes, called the x and y axes. Normal graphing convention defines x as the horizontal axis and y as the vertical axis. In mathematics, the x axis is called the **abscissa,** and the y axis is termed the **ordinate.**

The first step in graphing data is to **scale** the axes, i.e., to place appropriate values along each axis to accommodate all data points. It is a good practice to make use of as much of the graph paper as possible. If the x values range from 0 to 100, 0 is placed at the left side and 100 is placed as far to the right as possible, considering the magnitude of each division on the x axis. The size of each division depends on the collected data. For example, if the data values are measured to the nearest 0.5 unit, the scale should allow room so that values like 23.5 or 73.5 can be easily plotted.

After each axis is scaled, data points (ordered pairs) are plotted on the graph to correspond to each data pair, i.e., each x and y value. This is accomplished by moving across the x axis to the correct value, and then rising vertically until the y value is reached. At the intersection a mark is made, and in many cases is labeled with the x and y values. All data points are plotted in a similar manner before the graph is drawn.

A common error is to connect the plotted points with a jagged line, as in the children's game of "connect the dots" (Fig. A.1*a*). Data points are not always connected directly. Instead, a straight edge or french curve is used to draw the best-fitting line through the maximum number of data points. Usually some points do not fall exactly on the line since experimental errors are present in all measurements (Fig. A.1*b*).

Example Problem 9 illustrates the above procedures.

――――――――――― **Example Problem 9** ―――――――――――

On a recent auto trip on an interstate highway the following distance and time data were collected:

Time, hr	Distance, mi
0.5	25
1.0	50
2.0	100
3.0	150

Plot a graph of the time and distance data, with time on the *x* axis and distance on the *y* axis.

——————————————— **Solution** ———————————————

Time values range from 0.5 to 3.0 hr. To include all values, scale the *x* axis from 0 to 4 hr in 0.5-hr intervals. Scale the *y* axis from 0 to 200 mi. It is a common practice to scale the axis slightly above and below the range of collected data values.

Once the axes are scaled, plot each data pair. First, plot the (0.5 hr, 25 mi) point by moving horizontally to 0.5 hr, and then rising vertically to the 25-mi line. Plot all other data pairs in a similar fashion (Fig. A.2). It is apparent that the

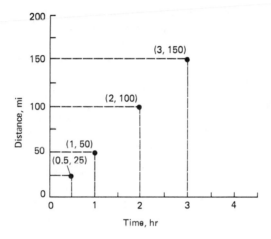

Figure A.2
Plotted points with correctly scaled axes.

points are aligned in a linear fashion, or in a straight line. Using a ruler, draw a straight line through all plotted points, as in Fig. A.3.

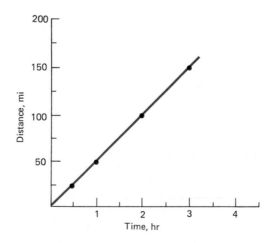

Figure A.3
Complete graph of distance traveled in miles versus time in hours.

If you examine the graph plotted in Example Problem 9 carefully, you can readily extract additional information. The graph shows a straight-line, or linear, relationship between the variables of time and distance. For each unit time interval of the trip, the same amount of distance was traveled. No matter what hour interval is considered, 50 mi is traversed (check this for yourself). The change in distance per time interval is constant in the above graph—50 mi/hr. All linear relationships have this common characteristic, called the **slope,** or rate of change of the y variable for a unit change in the x variable. To compute the slope of the line, select two data points on the line, determine the change in y values, and divide this factor by the corresponding change in x values. Verify the fact that the change is 50 mi/hr in the above graph.

Slopes of linear relationships are either positive, as above, or negative. A **positive slope** indicates that for each increase in the variable plotted on the x axis there is a resulting increase in the y variable. In our example, an increase in time traveled produces an increase in distance. A **negative slope** indicates that for each increase in the x variable, there is a resulting decrease in the y variable. Figure A.4 shows linear relationships with both positive and negative slopes.

Only four data pairs were collected for our theoretical auto trip in Example Problem 9. Nevertheless, the distance traveled at any time interval can be determined by correctly reading the graph. How far did the auto travel in 2.5 hr? Find 2.5 hr on the x axis, and draw a perpendicular line from this point to the line plotted on the graph. Then draw a horizontal line from the intersection of the vertical line to the y axis. The point where it meets the axis is the distance traveled in 2.5 hr (125 mi). Reverse the above procedure to find time elapsed for a given distance traveled. For example, how long did it take to travel 75 mi?

In many instances, the limits of the data are extended, especially when there is a good fit between the line and the data points. Continuing the line on the graph above and below the range of collected data points is called **extrapolation.** Extrapolation is justified when there is a good reason to believe that the trend extends beyond the observed data. Extending the graph in Example Problem 9 to 4 hr indicates that the auto would have traveled 200 mi in 4 hr (Fig. A.5).

Nonlinear relationships are also frequently encountered in chemistry. As the name implies, a **nonlinear relationship** is characterized by a graph that is not

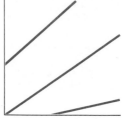

Lines with positive slopes

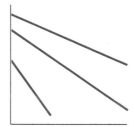

Lines with negative slopes

Figure A.4

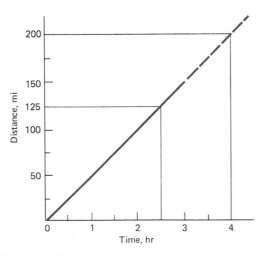

Figure A.5
Extrapolation of the distance versus time data indicates that in 4 hr a total distance of 200 mi is traveled.

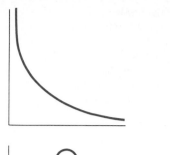

Figure A.6
Four types of nonlinear relationships are shown.

a straight line. Instead, the data points are connected by a curved line. Figure A.6 presents examples of nonlinear relationships.

All graphs, no matter what type, are interpreted in the same way. Ask yourself the following questions when you are interpreting a graph: What general trends are found? How are the variables changing (increasing or decreasing) with respect to each other? What special characteristics are there? What are the limits of the variables?

EXERCISES

Equations

1. Solve each of the following equations for X:
   (a) $X - 55 = -47$ (b) $125 = 83 + X$
   (c) $110 = -X - 78$ (d) $-9X = -94$

2. Solve each of the following equations for Z:
   (a) $\dfrac{1}{Z - 9} = 12.4$

   (b) $\dfrac{Z}{2Z + 5} = \dfrac{1}{12}$

   (c) $(a + b)(Z + 2) = \dfrac{n + 1}{m}$

   (d) $\dfrac{a + 2}{b - Z} = \dfrac{1}{n + 1}$

3. If $m = 25$ and $n = 9$, solve for X:
   (a) $X = 5m - 3n + 2$
   (b) $X = 5(m - 3)n + 2$
   (c) $X = (5m - 3n) + 2$
   (d) $X = 5m - (3n + 2)$

4. Solve the following equations for X:
   (a) $5X = 512$ (b) $\dfrac{X}{13} = 6.44$

   (c) $16X + 5 = 6X$ (d) $\dfrac{10X}{30} = 12$

   (e) $\dfrac{2X - 10}{8} = 63$

5. Solve the following equations for Y:
   (a) $\dfrac{a}{2}(nY - m) = 1$

   (b) $\dfrac{a + b}{Y + m} = n + 10$

   (c) $3a + b = \dfrac{4Y}{5} + n$

   (d) $\dfrac{6}{Y} + \dfrac{2}{Y} = \dfrac{a}{b}$

6. If $a = 25$, $b = 6$, $m = 3$, and $n = -1$, calculate the value of Y in each equation in Exercise 5.

Scientific Notation

7. Change the following numbers to scientific notation:
(a) 0.00049139
(b) 120,000,000,000
(c) 0.0000000013511
(d) 0.00000000000000000000105

8. Change the following numbers to nonexponential form:
(a) 3.11×10^{-3} (b) 2.90×10^{-2}
(c) 7.7742×10^8 (d) 8.001×10^{12}

9. Change the following numbers in exponential form to scientific notation:
(a) 0.000436×10^6
(b) 1215×10^4
(c) 0.0001001×10^{-8}
(d) 979×10^{-4}
(e) $6,361,000 \times 10^{-10}$

10. Add the following numbers:
(a) $(4.75 \times 10^5) + (7.11 \times 10^5)$
(b) $(8.33 \times 10^{-3}) + (1.20 \times 10^{-3})$
(c) $(9.355 \times 10^7) + (1.722 \times 10^9)$
(d) $(2.109 \times 10^{35}) + (9.64 \times 10^{33})$
(e) $(3.400 \times 10^{20}) + (4.230 \times 10^{18}) +$
(5.011×10^{19})

11. Subtract the following numbers:
(a) $(4.38 \times 10^4) - (7.17 \times 10^4)$
(b) $(9.921 \times 10^{-13}) - (7.22 \times 10^{-14})$
(c) $(2.55 \times 10^{45}) - (5.690 \times 10^{47})$
(d) $(6.131 \times 10^{-34}) - (1.7492 \times 10^{-36})$

12. Find the product for each:
(a) $(1.75 \times 10^3) \times (2.44 \times 10^8)$
(b) $(3.54 \times 10^{54}) \times (9.82 \times 10^{96})$
(c) $(1.11 \times 10^{-3}) \times (5.35 \times 10^{-23}) \times$
(7.44×10^{28})

13. Find the quotient for each:
(a) $\dfrac{4.175 \times 10^{-17}}{9.329 \times 10^{-24}}$

(b) $\dfrac{5.67 \times 10^{91}}{1.05 \times 10^{-22}}$

(c) $\dfrac{6.85 \times 10^7}{(2.79 \times 10^{-4}) \times (6.113 \times 10^{18})}$

14. Perform the indicated operations:
(a) $\dfrac{1.42 \times 10^{-26}}{1.56 \times 10^{-24}} - (5.44 \times 10^{-6})$

(b) $\dfrac{(6.788 \times 10^{57}) - (8.54 \times 10^{56})}{5.93 \times 10^{58}}$

Graphing

15. Graph the following data on a full sheet of graph paper.

X	Y
10	36
25	81
40	126
55	171
80	246

(a) What type of relationship is plotted? (b) What Y value corresponds to an X value of 70? (c) What X value corresponds to a Y value of 56? (d) Extrapolate the line to determine the expected Y value corresponding to an X value of 100. (e) What is the slope of the line?

16. Plot the following data, with Fahrenheit temperatures on the x axis and Celsius temperatures on the y axis.

Temperature, °F	Temperature, °C
-40	-40
0	-17.8
40	4.4
80	26.7
120	48.9

(a) What is the slope of the line? (b) Using the graph, determine what Celsius temperature corresponds to each of the following Fahrenheit temperatures: 98.6°F, 75°F, -25°F, and 135°F.

17. (a) On a full sheet of graph paper, plot years on the x axis and the approximate population of the United States on the y axis, given the following data:

Year	Population (in millions)
1940	130
1950	150
1960	175
1970	205
1980	225

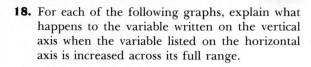

(b) Is this a linear function? Explain. *(c)* What was the approximate population of the United States in 1945? *(d)* Extrapolate the population data to the year 2000. What is the expected population of the United States in the year 2000?

18. For each of the following graphs, explain what happens to the variable written on the vertical axis when the variable listed on the horizontal axis is increased across its full range.

(a)

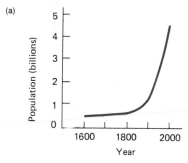

(c)

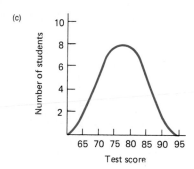

(b)

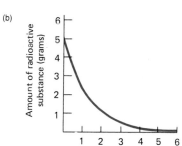

(d)

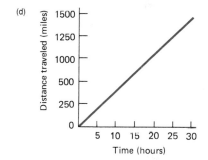

Figure A.7

APPENDIX
B

Chemistry Calculations Using Calculators

It is a good idea to use a hand calculator for most calculations in chemistry. A calculator is a timesaver. It will allow you to spend more time actually doing chemistry and less time "crunching numbers" by hand.

Two different formats are used to enter numbers and obtain answers in modern calculators: **algebraic notation** and **reverse polish notation** (RPN). In Section I, the basics of using an algebraic notation calculator are presented, and in Section II the basics of using an RPN calculator are described. To determine what type of calculator you own, check to see whether your calculator has a key with = (equals sign) printed on it (normally near the bottom right-hand side). All algebraic notation calculators have such a key. An RPN calculator does not; it has an **ENTER** key instead.

Before proceeding, become familiar with your calculator's organization and the various function keys it has. Then read the appropriate section and perform the indicated exercises on your calculator.

C alculators with algebraic notation are easy to use because the procedure for entering numbers and obtaining answers is similar to the procedure used when performing algebraic and arithmetic operations on paper.

B.1 USE OF ALGEBRAIC NOTATION CALCULATORS

Entering Numbers

To enter a number into the calculator, press the appropriate number and decimal point keys. Enter each of the following numbers, checking to see that the display corresponds to the number you think you entered. After entering each number, press the **CLEAR** key, which deletes the displayed digits. Enter the following:

(a) 10 (b) 2395 (c) 123456789 (d) .8 (e) .00002310 (f) 0.000000345678

If your calculator has an **EE, EXP** (enter exponent) or similar key, you can enter numbers in exponential notation. You may enter the number with any decimal and exponential factors, and most calculators will convert the number to scientific notation. To enter such numbers: (1) Enter the decimal factor as above, and (2) press the **EE** key, which lights the two digits at the far right of the display, and then press the number that corresponds to the exponent. You should note that most calculators do not require the entry of the × 10 and it is not displayed. Enter the following exponential numbers, check display, and clear: *(a)* 2×10^3, *(b)* 5.675×10^{12}, *(c)* 123.4×10^{56}, and *(d)* 0.0034×10^{81}.

Negative numbers and negative exponents are often encountered. Most calculators have a change-of-sign key, **+/−**. When the change-of-sign key is depressed, the displayed number is multiplied by −1. Thus, to enter a negative exponential number, enter the decimal factor, press the change-of-sign key, and then enter the exponent. To enter a positive number with a negative exponent, enter the decimal number, then the exponent, and then press the change-of-sign key. Practice using the change-of-sign key by entering the following exponential numbers, checking the display, and then clearing: *(a)* -4.3×10^{11}, *(b)* 9.11×10^{-6}, and *(c)* -5.51245×10^{-61}.

Arithmetic Operations

Locate the **+**, **−**, **÷**, and **×** keys. To perform addition, subtraction, multiplication, and division: (1) Enter first number, (2) press arithmetic operator, (3) enter second number, and (4) press the **=** key. Follow the stated steps for each of the following: *(a)* $23 + 92 =$, *(b)* $124 - 321 =$, *(c)* $8.533 \div 9.112 =$, and *(d)* $-12.003 \times -3.175 =$.

Multiple operations of the same arithmetic procedure are accomplished in a similar manner. After you enter each number, press the arithmetic operator key. After you enter the last number, press the **=** key to obtain the final answer. The calculator displays a new value after each arithmetic operation so that intermediate values are obtained. Perform the following multiple operations: *(a)* $1501 + 23 + 91.4 + 888 =$, *(b)* $56.99 - 1.20 - 46.9 - 0.8 =$, *(c)* $599 \div 23 \div 0.75 \div 34.72 =$, and *(d)* $-.9345 \times .05699 \times -.00344 \times 1.0101 =$.

Care must be exercised when performing multiple arithmetic operations of more than one type. Depending on the convention followed by the calculator, the calculator may process the numbers from left to right or it may follow the algebraic operating system, which means that it will perform multiplication and division from left to right before performing addition and subtraction. An example will serve to illustrate this difference. What is the answer to the following?

$$8 + 5 - 3 \times 16 \div 8 =$$

If your calculator performs the arithmetic operations from left to right, you will obtain the answer 20. But if your calculator follows the algebraic operating system, you will obtain the answer 7. Placing parentheses in the expression will show how the 7 is arrived at.

$$8 + 5 - \left(\frac{(3 \times 16)}{8}\right) = 7$$

Refer to the manual that comes with your calculator to find which method it uses. Practice multiple operations by performing and checking the following:

(a) $(5 \times 6) - (32 \div 8) + 12 =$

(b) $27 \div (9 - 44) \times 2 + (60 \div 20) =$

(c) $(1.15 - 0.233) \div 0.0295 - (0.010 + 15) \times 0.543 =$

If your calculator has parentheses keys, (and), arithmetic operations can easily be performed in any order, depending on how the parentheses group the numbers and operations. For example, consider the following:

$$\frac{9 + 2 + (4 \times 6)}{(2 + 5 - 6) \times 5} =$$

To correctly solve this problem, enclose the complete numerator in parentheses and the denominator in parentheses so that they can be divided. Also, the parentheses are used to group the operations individually in the numerator and denominator. The following key strokes are necessary to obtain the correct answer, 7.

$$\underbrace{(9 + 2 + (4 \times 6))}_{\text{Numerator}} \div \underbrace{((2 + 5 - 6) \times 5)}_{\text{Denominator}} = 7$$

Practice using the parentheses key with the following examples:

(a) $((23.4 - 15.4) \times 31) \div 0.22 =$

(b) $((1.934 \div 3.12 + 5.22) - 4.84) \div (1 + (3.14 \times 0.566 - 1.11)) =$

Function Keys

Many calculators have function keys that perform various mathematical functions. In each case, if a number is entered and then a function key is pressed, the resulting display presents the answer. If 2 is entered and then the reciprocal key ($1/x$) is pressed, then the reciprocal of 2, 0.5, is displayed. If the reciprocal key is pressed again, the number 2 appears.

Other function keys that are found on calculators include those for calculating the square, $\mathbf{x^2}$; square root, $\sqrt{\mathbf{x}}$; **SIN**; **COS**; **TAN**; $\mathbf{log_{10}}$, **LOG**; $\mathbf{log_e}$, **LN X**; $\mathbf{Y^X}$; $\mathbf{10^x}$; and $\mathbf{e^x}$. Of these, the logarithm keys are probably the most important for chemistry. If the **LOG** key is pressed after a number is entered, the logarithm to the base 10 is calculated. However, if **LN X** is pressed, the natural logarithm is displayed. To illustrate this, enter the number 10 and find the common log and the natural log. For the former, the answer is 1.0, and for the latter the answer is 2.303. Antilogs are determined by pressing either $\mathbf{10^x}$ for common antilogs or $\mathbf{e^x}$ for natural antilogs. Practice using the function keys by performing the following operations:

(a) 25^2, (b) $345^{.5}$, (c) sin 45°, (d) cos 45°, (e) tan 30°, (f) log 6.45, (g) In 1.2 $\times$ 10^2, (h) $.96^{11}$, (i) antilog 0.5123, (j) antiln 4.2485.

B.2 USE OF CALCULATORS WITH REVERSE POLISH NOTATION

In RPN calculators, a number must first be entered into a memory location before the calculator can use it. Many RPN calculators have four principal memory locations, which are "stacked on top of each other." When a number key is pressed, the number initially occupies the lowest position in the stack. To move it to the next highest memory location, the **ENTER** key is pressed.

Picture the memory locations as follows:

$$W$$
$$Z$$
$$Y$$
$$X$$

where X is the memory location that is constantly displayed. Immediately above this is memory location Y. When a number is first "ENTERed," it is moved into Y. Thus, if the number 7 key is pressed, location X receives the 7, and when the **ENTER** key is pressed, 7 is placed in Y, without removing it from X; 7 still remains displayed (X = 7, Y = 7, Z = 0, and W = 0). If another number is pressed, let's say 6, the 6 would occupy memory location X (X = 6, Y = 7, Z = 0, W = 0). Now when the **ENTER** key is pressed, the 6 is placed in Y and the 7 is pushed up to Z (X = 6, Y = 6, Z = 7, and W = 0). Pressing the 5 key places 5 in X, and then pressing the **ENTER** key moves 5 to Y and pushes both the 6 and 7 up in the stack (X = 5, Y = 5, Z = 6, and W = 7). One more key stroke will fill the stack with four different numbers: pressing the 4 key would result in a stack with X = 4, Y = 5, Z = 6, and W = 7

Most RPN calculators have a key that allows each stack memory location to be viewed; on many calculators this key is the **R↓**. When the **R↓** key is pressed, each member of the stack is dropped down, and the contents of X is placed in W. If the stack contains the numbers mentioned above, when the **R↓** key is pressed once, 5 is displayed, 4 goes to the top of the stack, and the 6 and 7 drop down. If the **R↓** key is pressed again, 6 is displayed, 5 goes to the top, etc.

	W = 7	W = 4	W = 5	W = 6
	Z = 6	Z = 7	Z = 4	Z = 5
	Y = 5	Y = 6	Y = 7	Y = 4
Display⟶	X = 4	X = 5	X = 6	X = 7
	Start	Press **R↓** once	Press **R↓** again	Press **R↓** once again

With the **R↓** key, the numbers can also be moved, depending on the operation being performed. Additionally, RPN calculators have an **X ⇄ Y** key to exchange places of numbers in memory locations X and Y.

Practice entering numbers into the stack and using the **R↓** key to view the contents before continuing.

Negative numbers are obtained by using the **CHS** (change sign) or **+/−** key. To enter a negative number into Y, first enter the number into X, then press the **CHS** key followed by the **ENTER** key.

Numbers expressed in scientific notation are usually entered into RPN calculators in the same way that they are entered into algebraic notation calculators. Refer to the appropriate paragraph in Sec. I.

All arithmetic operations are performed by entering a number into memory location Y, then placing a number in memory location X, and then pressing the arithmetic operator key. To add 2 + 5, press the 2 key, and then press the **ENTER** key. This moves 2 to Y. Press the 5 key, which places 5 in X, and then press the **+** key. At this time, 7 is displayed. Subtraction is the same as addition, except the **−** key is pressed. The contents of X are subtracted from the contents of Y. To multiply 2 by 5, the same procedure is followed: (1) Press 2 and then **ENTER**; (2) press the 5 key; (3) press the × key. In division, the dividend is placed in Y, the divisor is placed in X, and the ÷ key is pressed.

The following summarizes the four basic arithmetic operations on an RPN calculator:

Addition: $y + x$

Subtraction: $y - x$

Multiplication: $y \cdot x$

Division: $y \div x$

where y is the contents of memory location Y, and x is the contents of memory location X. The answer is placed in X; in other words, it is displayed. Practice simple arithmetic operations by calculating the answers to the following: (a) $24 + 91$, (b) $16 - 44$, (c) 3.14×63.69, (d) $0.023 \div 451$.

Multiple operations are performed in exactly the same way as are single arithmetic operations. The first number is entered into Y, the second number is placed in X, and the arithmetic operator button is then depressed. To continue the process, a new number is entered into X and the operator key is pressed. For example: What key strokes are necessary to perform the following?

$$12 + 22 \div 2 - 10 \times 3$$

First, enter 12 into Y, place 22 in X, and then press the + key. Press 2, ÷ , 10, −, 3, and × to give the answer, 21. Practice multiple operations by finding the answers to the following:

(a) $32 - 16 + 22 \div 11$, (b) $125 \div 75 \times 14 - 8 \div 12$, (c) $9 \times 5 - 5 \div 4 \div 5 + 2$.

In many calculations, the order of operations must be changed. For example, how would the following calculation be performed?

$$\frac{4 \times (16 + 3)}{3 \times (12 - 5)}$$

Initially 4 and 16 are both entered, which places 4 in Z and 16 in Y. After the 3 key is depressed, 3 is located in X; the + key is pressed, giving 19, followed by the × key, giving 76. Next the denominator is determined by entering 3 and 12 (pushing 76 up to Z) followed by pressing 5, −, and × . This yields the number 21. Now the 76 has dropped back down to Y; hence, it is only necessary to press the ÷ key to obtain the final answer, 3.619.

Perform the following for practice:
(a) $(3 + 9) \div (2 + 8)$
(b) $1 + (2.455 - 1.391) \div (6 \times (0.44 \div 0.11)) + 0.65$
(c) $(((9.2 - 10.2) \div 5.66) \times -124.52) + 33 =$

Function Keys

Performing functions such as reciprocals, square roots, and logarithms requires only that a number be placed in memory location X followed by pressing the appropriate function key. If 12 is placed in X and the x^2 key is pressed, the answer, 144, is displayed. Since the use of function keys on RPN calculators is the same as for algebraic notation calculators, refer to the appropriate paragraph in Sec. I for a more detailed discussion.

APPENDIX
C

Physical Constants and Conversion Factors

Unified atomic mass unit (u)	$1\text{ u} = 1.66057 \times 10^{-27}\text{ kg}$	
	$= 1.66057 \times 10^{-24}\text{ g}$	
	$6.022045 \times 10^{23}\text{ u} = 1.000\text{ g}$	
Avogadro's number	$N = 6.022045 \times 10^{23}\text{ entities/mol}$	
Electron mass	$m_e = 9.10953 \times 10^{-28}\text{ g}$	
	$= 5.48580 \times 10^{-4}\text{ u}$	
Electron charge	$e = -1.6022 \times 10^{-19}\text{ C}$	
Faraday's constant	$F = 9.6485 \times 10^4\text{ C/mol}$	
Ideal gas constant	$R = 0.082057(\text{L} \cdot \text{atm})/(\text{mol} \cdot \text{K})$	
	$= 8.3144\text{ J/(mol} \cdot \text{K})$	
	$= 8.3144(\text{Pa} \cdot \text{dm}^3)/(\text{mol} \cdot \text{K})$	
Molar volume (ideal gas), STP	$V = 22.4\text{ L/mol}$	
Neutron mass	$m_{n^\circ} = 1.67495 \times 10^{-24}\text{ g}$	
	$= 1.00866\text{ u}$	
Proton mass	$m_{p+} = 1.67265 \times 10^{-24}\text{ g}$	
	$= 1.00728\text{ u}$	
Speed of light	$c = 2.997925 \times 10^8\text{ m/s}$	

Mass	$1 \text{ g} = 10^{-3} \text{ kg}$	**C.2 UNIT CONVERSION FACTORS**

Mass
$1 \text{ g} = 10^{-3} \text{ kg}$
$1 \text{ mg} = 10^{-6} \text{ kg}$
$1 \text{ lb} = 0.453592 \text{ kg} = 453.592 \text{ g}$
$1 \text{ kg} = 2.205 \text{ lb}$
$1 \text{ oz} = 28.349523 \text{ g}$
$1 \text{ u} = 1.66057 \times 10^{-27} \text{ kg}$
$1 \text{ metric ton} = 1000 \text{ kg} = 2204.6 \text{ lb}$

Length
$1 \text{ cm} = 10^{-2} \text{ m}$
$1 \text{ mm} = 10^{-3} \text{ m}$
$1 \text{ nm} = 10^{-9} \text{ m}$
$1 \text{ Å} = 10^{-10} \text{ m} = 0.10 \text{ nm} = 100 \text{ pm}$
$1 \text{ in} = 2.54 \times 10^{-2} \text{ m} = 2.54 \text{ cm}$
$1 \text{ cm} = 10 \text{ mm} = 0.39370 \text{ in}$
$1 \text{ mi} = 1.609 \text{ km}$
$1 \text{ yd} = 0.9144 \text{ m}$

Time
$1 \text{ day} = 8.64 \times 10^4 \text{ s}$
$1 \text{ hr} = 3.60 \times 10^3 \text{ s}$
$1 \text{ ms} = 10^{-3} \text{ s}$

Temperature
$0°C = 273.15 \text{ K}$
$0°F = 255.37 \text{ K}$
$-273.15°C = 0 \text{ K}$
$-459.67°F = 0 \text{ K}$

Volume
$1 \text{ L} = 10^{-3} \text{ m}^3$
$1 \text{ L} = 1.057 \text{ qt}$
$1 \text{ qt} = 0.9463 \text{ L}$
$1 \text{ cm}^3 = 10^{-6} \text{ m}^3$
$1 \text{ mL} = 1 \text{ cm}^3$
$1000 \text{ mL} = 1 \text{ L}$

Energy
$1 \text{ cal} = 4.184 \text{ J}$
$1 \text{ cal} = 2.612 \times 10^{19} \text{ eV}$
$1 \text{ erg} = 10^{-7} \text{ J}$
$1 \text{ erg} = 2.3901 \times 10^{-8} \text{ cal}$
$1 \text{ J} = 0.23901 \text{ cal}$

Pressure
$1 \text{ atm} = 101,325 \text{ Pa} \ 101.325 \text{ kPa}$
$1 \text{ atm} = 760 \text{ mmHg} \ 760 \text{ torr}$
$1 \text{ atm} = 14.70 \text{ lb/in}^2$
$1 \text{ torr} = 1 \text{ mmHg}$
$1 \text{ torr} = 133.322 \text{ Pa}$
$1 \text{ bar} = 1 \times 10^5 \text{ Pa}$

A P P E N D I X
D

Table of Ions and Their Formulas

Ion	Name	Ion	Name
AlO_3^{3-}	Aluminate	IO_3^-	Iodate
AsO_4^{3-}	Arsenate	IO^-	Hypoiodite
AsO_3^{3-}	Arsenite	MnO_4^-	Permanganate
BiO_3^{3-}	Bismuthate	MnO_4^{2-}	Manganate
BO_3^{3-}	Borate	NH_4^+	Ammonium
BrO_3^-	Bromate	NO_3^-	Nitrate
BrO_2^-	Bromite	NO_2^-	Nitrite
BrO^-	Hypobromite	O_2^{2-}	Peroxide
CO_3^{2-}	Carbonate	OCN^-	Cyanate
HCO_3^-	Hydrogencarbonate	OH^-	Hydroxide
$C_2O_4^{2-}$	Oxalate	PO_4^{3-}	Phosphate
$C_2H_3O_2^-$	Acetate	PO_3^{3-}	Phosphite
CN^-	Cyanide	PO_2^{3-}	Hypophosphite
HCO_2^-	Formate	$P_2O_7^{4-}$	Pyrophosphate
ClO_4^-	Perchlorate	SCN^-	Thiocyanate
ClO_3^-	Chlorate	SeO_4^{2-}	Selenate
ClO_2^-	Chlorite	SeO_3^{2-}	Selenite
ClO^-	Hypochlorite	SiO_3^{2-}	Silicate
CrO_4^{2-}	Chromate	SnO_4^{4-}	Stannate
$Cr_2O_7^{2-}$	Dichromate	SO_4^{2-}	Sulfate
GaO_3^{3-}	Gallate	SO_3^{2-}	Sulfite
GeO_4^{4-}	Germanate	$S_2O_3^{2-}$	Thiosulfate
GeO_3^{4-}	Germanite	TeO_6^{6-}	Tellurate
H^-	Hydride	TeO_3^{2-}	Tellurite
IO_4^-	Periodate		

APPENDIX
E

Logarithms to Base 10

I. To find the logarithm of a number between 1 and 9.99

The logarithm of a number between 1 and 9.99 is obtained directly from the table.

Step 1

Find the first two digits of the number in the first column.

Step 2

Move horizontally across the page to the column that corresponds to the third digit of the number.

Step 3

The number at the intersection is the common logarithm.

For example: What is the logarithm of the number 7.55?

1. Find the row that begins with 75.

2. Move across the table to the column labeled 5.

3. Read the number at the intersection of 75 and 5.

The common logarithm of 7.55 is .8779.

II. To find the logarithm of a number that is not between 1 and 9.99

Step 1

Express the number in scientific notation.

Step 2

Find the logarithm of the decimal factor as in part I.

Step 3

The logarithm of the exponential factor is the value of the exponent.

Step 4

Add the two logarithms to find the logarithm of the number.

For example: What is the logarithm of 3720?

1. The number 3720 expressed in scientific notation is 3.72×10^3.

2. Using the logarithm table, the logarithm of 3.72 is .5705.

3. The logarithm of 10^3 is 3.

4. Add the two numbers: $.5705 + 3 = \mathbf{3.5705}$

The common logarithm of 3720 is 3.5705.

	0	1	2	3	4	5	6	7	8	9
10	0000	0043	0086	0128	0170	0212	0253	0294	0334	0374
11	0414	0453	0492	0531	0569	0607	0645	0682	0719	0755
12	0792	0828	0864	0899	0934	0969	1004	1038	1072	1106
13	1139	1173	1206	1239	1271	1303	1335	1367	1399	1430
14	1461	1492	1523	1553	1584	1614	1644	1673	1703	1732
15	1761	1790	1818	1847	1875	1903	1931	1959	1987	2014
16	2041	2068	2095	2122	2148	2175	2201	2227	2253	2279
17	2304	2330	2355	2380	2405	2430	2455	2480	2504	2529
18	2553	2577	2601	2625	2648	2672	2695	2718	2742	2765
19	2788	2810	2833	2856	2878	2900	2923	2945	2967	2989
20	3010	3032	3054	3075	3096	3118	3139	3160	3181	3201
21	3222	3243	3263	3284	3304	3324	3345	3365	3385	3404
22	3424	3444	3464	3483	3502	3522	3541	3560	3579	3598
23	3617	3636	3655	3674	3692	3711	3729	3747	3766	3784
24	3802	3820	3838	3856	3874	3892	3909	3927	3945	3962
25	3979	3997	4014	4031	4048	4065	4082	4099	4116	4133
26	4150	4166	4183	4200	4216	4232	4249	4265	4281	4298
27	4314	4330	4346	4362	4378	4393	4409	4425	4440	4456
28	4472	4487	4502	4518	4533	4548	4564	4579	4594	4609
29	4624	4639	4654	4669	4683	4698	4713	4728	4742	4757
30	4771	4786	4800	4814	4829	4843	4857	4871	4886	4900
31	4914	4928	4942	4955	4969	4983	4997	5011	5024	5038
32	5051	5065	5079	5092	5105	5119	5132	5145	5159	5172
33	5185	5198	5211	5224	5237	5250	5263	5276	5289	5302
34	5315	5328	5340	5353	5366	5378	5391	5403	5416	5428
35	5441	5453	5465	5478	5490	5502	5514	5527	5539	5551
36	5563	5575	5587	5599	5611	5623	5635	5647	5658	5670
37	5682	5694	5705	5717	5729	5740	5752	5763	5775	5786
38	5798	5809	5821	5832	5843	5855	5866	5877	5888	5899
39	5911	5922	5933	5944	5955	5966	5977	5988	5999	6010
40	6021	6031	6042	6053	6064	6075	6085	6096	6107	6117
41	6128	6138	6149	6160	6170	6180	6191	6201	6212	6222
42	6232	6243	6253	6263	6274	6284	6294	6304	6314	6325
43	6335	6345	6355	6365	6375	6385	6395	6405	6415	6425
44	6435	6444	6454	6464	6474	6484	6493	6503	6513	6522
45	6532	6542	6551	6561	6571	6580	6590	6599	6609	6618
46	6628	6637	6646	6656	6665	6675	6684	6693	6702	6712
47	6721	6730	6739	6749	6758	6767	6776	6785	6794	6803
48	6812	6821	6830	6839	6848	6857	6866	6875	6884	6893
49	6902	6911	6920	6928	6937	6946	6955	6964	6972	6981
50	6990	6998	7007	7016	7024	7033	7042	7050	7059	7067
51	7076	7084	7093	7101	7110	7118	7126	7135	7143	7152
52	7160	7168	7177	7185	7193	7202	7210	7218	7226	7235
53	7243	7251	7259	7267	7275	7284	7292	7300	7308	7316
54	7324	7332	7340	7348	7356	7364	7372	7380	7388	7396

	0	1	2	3	4	5	6	7	8	9
55	7404	7412	7419	7427	7435	7443	7451	7459	7466	7474
56	7482	7490	7497	7505	7513	7520	7528	7536	7543	7551
57	7559	7566	7574	7582	7589	7597	7604	7612	7619	7627
58	7634	7642	7649	7657	7664	7672	7679	7686	7694	7701
59	7709	7716	7723	7731	7738	7745	7752	7760	7767	7774
60	7782	7789	7796	7803	7810	7818	7825	7832	7839	7846
61	7853	7860	7868	7875	7882	7889	7896	7903	7910	7917
62	7924	7931	7938	7945	7952	7959	7966	7973	7980	7987
63	7993	8000	8007	8014	8021	8028	8035	8041	8048	8055
64	8062	8069	8075	8082	8089	8096	8102	8109	8116	8122
65	8129	8136	8142	8149	8156	8162	8169	8176	8182	8189
66	8195	8202	8209	8215	8222	8228	8235	8241	8248	8254
67	8261	8267	8274	8280	8287	8293	8299	8306	8312	8319
68	8325	8331	8338	8344	8351	8357	8363	8370	8376	8382
69	8388	8395	8401	8407	8414	8420	8426	8432	8439	8445
70	8451	8457	8463	8470	8476	8482	8488	8494	8500	8506
71	8513	8519	8525	8531	8537	8543	8549	8555	8561	8567
72	8573	8579	8585	8591	8597	8603	8609	8615	8621	8627
73	8633	8639	8645	8651	8657	8663	8669	8675	8681	8686
74	8692	8698	8704	8710	8716	8722	8727	8733	8739	8745
75	8751	8756	8762	8768	8774	8779	8785	8791	8797	8802
76	8808	8814	8820	8825	8831	8837	8842	8848	8854	8859
77	8865	8871	8876	8882	8887	8893	8899	8904	8910	8915
78	8921	8927	8932	8938	8943	8949	8954	8960	8965	8971
79	8976	8982	8987	8993	8998	9004	9009	9015	9020	9025
80	9031	9036	9042	9047	9053	9058	9063	9069	9074	9079
81	9085	9090	9096	9101	9106	9112	9117	9122	9128	9133
82	9138	9143	9149	9154	9159	9165	9170	9175	9180	9186
83	9191	9196	9201	9206	9212	9217	9222	9227	9232	9238
84	9243	9248	9253	9258	9263	9269	9274	9279	9284	9289
85	9294	9299	9304	9309	9315	9320	9325	9330	9335	9340
86	9345	9350	9355	9360	9365	9370	9375	9380	9385	9390
87	9395	9400	9405	9410	9415	9420	9425	9430	9435	9440
88	9445	9450	9455	9460	9465	9469	9474	9479	9484	9489
89	9494	9499	9504	9509	9513	9518	9523	9528	9533	9538
90	9542	9547	9552	9557	9562	9566	9571	9576	9581	9586
91	9590	9595	9600	9605	9609	9614	9619	9624	9628	9633
92	9638	9643	9647	9652	9657	9661	9666	9671	9675	9680
93	9685	9689	9694	9699	9703	9708	9713	9717	9722	9727
94	9731	9736	9741	9745	9750	9754	9759	9763	9768	9773
95	9777	9782	9786	9791	9795	9800	9805	9809	9814	9818
96	9823	9827	9832	9836	9841	9845	9850	9854	9859	9863
97	9868	9872	9877	9881	9886	9890	9894	9899	9903	9908
98	9912	9917	9921	9926	9930	9934	9939	9943	9948	9952
99	9956	9961	9965	9969	9974	9978	9983	9987	9991	9996

A P P E N D I X

F

Answers to Selected Problems and Questions

CHAPTER ONE

1.15 (a) Geochemist, (b) organic chemist, (c) biochemist, (d) physical or inorganic chemist, (e) analytical chemist.

1.24 A philosopher deals with problems primarily through the use of logic; a scientist conducts controlled systematic experiments based on observable phenomena.

CHAPTER TWO

2.6 (a) 35 crates, (b) 11,550 oranges.

2.7 (a) 14 gallons, (b) 12.65 dollars, (c) 0.870 gallons.

2.8 (a) 192 hours, (b) 11,520 minutes, (c) 691,200 seconds, (d) 33.125 years.

2.9 120 decades.

2.10 (a) 31,536,000, (b) 0.0000274, (c) 365,000.

2.17 (a) Fact, (b) theory, (c) theory.

2.20 (a) 564 miles, (b) 3.5 gallons.

2.21 (a) 480 seconds, (b) 0.693 years, (c) 0.9 centuries, (f) 39 gross.

2.23 (a) 300 seconds, (b) 51.0 seconds, (c) 4,500 seconds, (d) 267,840 seconds.

2.24 (b) 219,000 hours, (d) 788,400,000 seconds.

2.25 (a) 1.101 years, (b) 1.782 decades, (d) 10.6 days.

2.27 (a) 256 drams, (b) 7000 grains, (c) 0.0366 drams.

2.29 (a) 68 cents, (b) 9.16 dozen, (c) 11 cents.

2.30 (a) 27 miles/gallon, (d) 51.45 dollars.

2.31 (a) 0.27 dollars, (b) 0.0079 dollars.

2.33 (a) 2,365,200,000, (b) 7.5 decades, (c) 27,375 days.

2.34 (a) 40 rods/furlong, (c) 0.3636 rods.

2.37 (a) 3,700,000,000 miles (3,696,192,000 miles).
(b) 5,900,000,000,000 miles (5.8657×10^{12} miles).

2.39 (a) 365,000,000 acres (365,440,000 acres).
(b) 15,900,000,000,000 ft^2 (1.5919×10^{13} ft^2).
(c) 2,290,000,000,000,000 in^2 (2.2923×10^{15} in^2).

CHAPTER THREE

3.6   (a) 6, (b) 4, (c) 7, (d) 4, (e) 5, (f) 2, (g) 16, (h) 1.

3.7   (a) 30.56 g, (b) 0.20 mL, (c) 597.13 m, (d) 78.3348, (e) 4.0188×10^{23} atoms.

3.8   (a) 15 g/mL, (b) 6.2×10^2 m^2, (c) 41 cm^3, (d) 4.008×10^6, (e) 6.49×10^5 mg^2.

3.9   (a) 8.30 m^2, (b) 3×10^{-4} g/s, (c) 6.2×10^9.

3.16 (a) 9.150×10^6 mm, (b) 3.06 cm, (c) 1.77 lb, (d) 3.376×10^{14} ng.

3.18 (a) 328 K, (b) 181.8 K, (c) 1099°C, (d) 31.483°F.

3.20 (a) 2.59×10^3 mL $= 2.59 \times 10^3$ cm^3 $= 2.59$ dm^3 $= 2.59 \times 10^{-3}$ m^3.
      (b) 8.16×10^{-6} m^3 $= 8.16 \times 10^{-3}$ L $= 8.16$ mL $= 8.16 \times 10^3$ mm^3.

3.21 (a) 1.7×10^2 L, (b) 1.7×10^5 cm^3, (c) 53 barrels.

3.22 (a) 6.96 L, (b) 1.1×10^2 in^3.

3.23 (a) 11.40 g/cm^3, (b) lead.

3.24 (a) 347 cm^3, (b) 1.06×10^4 g.

3.25 (a) 0.928 g/cm^3, (b) 69.6 g.

3.30 The precision is not good because of the wide range of values, which indicates a high degree of uncertainty.

3.31 (a) Analytical balance, (c) small graduated cylinder.

3.32 If the thermometer is in the sun, the temperature reading will be too high.

3.34 (a) 5, (b) 7, (c) 2, (g) 6, (j) 1.

3.35 (a) 4, (b) 0, (c) 8, (g) 0, (j) 5.

3.36 (a) 2×10^1, (b) 1×10^2, (c) 9×10^{-2}, (g) 1×10^{-3}, (j) 5×10^{-8}.

3.37 (a) 194.6, (c) 962.2, (e) 9.996×10^{11}.

3.38 (c) 9.00×10^4, (e) 9.0000×10^4.

3.39 (a) 4.8 g, (b) 43.87 mL, (d) 756.87 m.

3.40 (a) 97.3 mL, (c) 0.034 g.

3.41 (b) 5.9×10^3 s^2, (c) 54 cm^2.

3.42 (a) 6.6 g/mL, (c) 1×10^{20} m/s.

3.43 (b) 0.286, (d) 0.425.

3.48 (b) 16.69 m, (c) 4.3×10^5 cm, (f) 7.46×10^4 km.

3.49 (a) 2×10^3 m, (b) 1.5×10^4 in, (d) 9.0×10^7 mm, (f) 7.820×10^{-1} Mm.

3.50 (a) 0.00048 km, (b) 1.12×10^4 mm, (c) 1.8 m.

3.53 59 kg, 170 cm, 94 cm, 76 cm, 89 cm.

3.54 Your weight would be much greater.

3.55 (b) 0.074 g, (c) 9711.5 kg, (f) 4.52×10^{11} Mg, (h) 8.395×10^{10} cg.

3.56 (b) 1.754×10^8 mg, (d) 5.4581×10^5 cg, (e) 3.82×10^9 ng.

3.57 (a) 284 K, (c) 37.6 K.

3.58 (a) -1.9×10^2°C, (d) 2948°C.

3.59 (b) −60.94°C, (c) 796.7°C.

3.60 (a) 172°F, (d) −220.0°F.

3.61 (c) 1540.3°F, (d) −458.07°F.

3.62 −40°C = −40°F.

3.65 (a) 1.1×10^2 cm^3, (b) 0.11 dm^3.

3.67 (b) 0.000012 m^3, (c) 73570 μL, (f) 0.012 m^3.

3.69 (a) 0.670 m^3, (b) 2.386×10^4 cm^3.

3.70 (a) 13.0 g/mL, (c) 0.942 g/mL.

3.71 (b) 3.32×10^3 g, (e) 6×10^3 g, (f) 4.10×10^3 g.

3.72 (a) 358 mL, (c) 1.43×10^3 mL, (d) 1.40×10^4 mL.

3.73 (b) 1.29 kg/m^3, (c) 0.0805 lb/ft^3.

3.75 1.59 g/mL.

3.77 1.65 g/mL.

3.78 1.9×10^5 g.

3.81 (*a*) 497 g, (*b*) 5.2 kg, (*c*) 474 mL.
3.82 (*a*) 18.9, (*b*) 0.860.
3.85 (*a*) 0.0786 g^{-1}, (*b*) 78.6 m^2, (*c*) 0.408.
3.86 (*a*) 9.32×10^{-9} Mg, (*b*) 2.445×10^{-8} m^3, (*c*) 3.22×10^{10} pm, (*d*) 179°F, (*e*) 9.561×10^5 cg,
      (*f*) 1.63 kg/dm^3, (*g*) 25 K, (*h*) 4.51×10^{-10} m^3.
3.88 (*a*) 118.3 cm^3, (*b*) 17 g.
3.91 (*a*) 893 km/h, (*b*) 248 m/s, (*c*) 0.000248 mm/ns.
3.92 13.9 g/cm^3.
3.95 (*a*) 1.2×10^2 in^3, (*b*) 0.0020 m^3.
3.99 6.77×10^{13} particles.
3.100 0.00014 cm^3.

CHAPTER FOUR

4.1   Physical: (*a*), (*b*), (*c*); chemical: (*d*), (*e*).
4.2   Physical: (*b*), (*d*), (*e*); chemical: (*a*), (*c*).
4.13 (*a*) Potassium 1, bromine 1; (*b*) beryllium 1, fluorine 2; (*c*) nickel 1, oxygen 1; (*d*) iron 1, oxygen
      3, hydrogen 3; (*e*) phosphorus 1, chlorine 3; (*f*) barium 1, nitrogen 2, oxygen 6; (*g*) rubidium 1,
      manganese 1, oxygen 4; (*h*) nitrogen 2, hydrogen 8, carbon 2, oxygen 4; (*i*) calcium 1, hydrogen
      4, phosphorus 2, oxygen 8.
4.17 (*a*) Pour off oil, (*b*) heat to remove the carbon dioxide, (*c*) dissolve salt in water, and filter.
4.21 Electric to heat, light, and sound.
4.26 (*a*) 1.22×10^3 cal, (*b*) 37 J, (*c*) 73.78 kcal.
4.28 (*a*) 0.377 J/(g)(°C), (*b*) 0.0902 cal/(g)(°C).
4.29 2.68×10^5 J.
4.30 4.2°C.
4.31 3×10^8 J.
4.34 Physical: (*a*), (*i*); chemical: (*d*), (*j*), (*o*).
4.35 Physical: (*d*), (*i*); chemical: (*c*), (*g*).
4.37 Physical: yellow color, solid state, melting point, boiling point; chemical: burns in air.
4.40 (*a*) Vegetable oil, (*d*) molasses.
4.41 (*a*) Solid, (*b*) gas, (*d*) solid and liquid, (*i*) gas and liquid, (*j*) gas.
4.44 (*a*) Solution, (*b*) element, (*c*) heterogeneous mixture.
4.45 Pure substance: (*g*); Mixture: (*a*), (*e*), (*h*), (*k*).
4.55 (*a*) Sodium 2, oxygen 1; (*d*) lithium 2, carbon 1, oxygen 3; (*f*) rubidium 1, hydrogen 2,
      phosphorus 1, oxygen 4; (*g*) aluminum 1, oxygen 3, hydrogen 3; (*i*) xenon 1, platinum 1, chlorine 6.
4.56 (*b*) Element.
4.58 Homogeneous: (*a*), (*d*), (*e*); heterogeneous: (*c*), (*g*).
4.59 (*b*) Distill: evaporate the alcohol from the water.
4.60 (*a*) 2, (*b*) 2, (*c*) 2, (*d*) 3.
4.63 (*a*) 13.2 J.
4.65 (*a*) Composition, (*d*) condition, (*e*) position.
4.66 (*a*) Bunsen burner flame, (*e*) neither.
4.67 (*a*) 38.5 J, (*d*) 3.4×10^5 kJ.
4.69 0.32 J/(g)(°C).
4.70 8.57×10^5 J.
4.73 For 1.00 kg of water, 3.14×10^5 J are released. For 1.00 g of water, 314 J are released.
4.75 8.4×10^{2}°C.
4.78 Potential energy of water to kinetic energy (mechanical energy) of the turbine to kinetic energy of
      the generator to electric energy.
4.79 (*b*) 6.1×10^{24} mg.
4.80 (*a*) 1.4×10^{11} J.
4.81 343 m/s.

4.85 Heat will flow from $O_2(l)$ into $N_2(l)$ increasing the temperature of $N_2(l)$ and decreasing the temperature of $O_2(l)$.

4.86 0.871 J/(g)(°C).

4.88 (b) 1.27×10^3 J/°C.

4.89 5.45×10^3 J.

4.91 7.9×10^3 J.

CHAPTER FIVE

5.5 (a) Repel, (b) attract.

5.6 (a) Number of protons, (b) number of protons plus number of neutrons, (c) average mass of naturally occurring isotopes of an element, (d) number of neutrons.

5.7 (a) 20, (b) 47, (c) 92, (d) 103.

5.8 (a) $p^+ = 22$, $n^0 = 26$; (b) $p^+ = 37$, $n^0 = 48$; (c) $p^+ = 89$, $n^0 = 138$; (d) $p^+ = 78$, $n^0 = 117$.

5.11 204.4 u.

5.14 (a) 5, (b) 19, (c) 12, (d) 27, (e) 36.

5.17 (a) 8, (b) 18, (c) 2, (d) 6, (e) 2.

5.18 (a) $1s^2 2s^1$, (b) $1s^2 2s^2 2p^3$, (c) $1s^2 2s^2 2p^6 3s^2 3p^6 4s^1$, (d) $1s^2 2s^2 2p^6 3s^2 3p^6 4s^2 3d^{10} 4p^1$, (e) $1s^2 2s^2 2p^6 3s^2 3p^6 4s^2 3d^{10} 4p^5$, (f) $1s^2 2s^2 2p^6 3s^9 3p^6 4s^2 3d^{10} 4p^6 5s^2 4d^{10} 5p^6 6s^1$.

5.19 (a) Mg: (b) ·S: (c) :Kr: (d) Ra: (e) ·As·.

5.20   1s 2s   2p   3s   3p

(a) ⥮ ⥮

(b) ⥮ ⥮ ⥮⥮⥮

(c) ⥮ ⥮ ⥮⥮⥮ ⥮ ⥯ ⥯

(d) ⥮ ⥮ ⥮⥮⥮ ⥮ ⥮ ⥮⥮

5.29 Since the neutron has no charge, it could not be detected by the methods used to study charged particles.

5.31 (a) $\dfrac{1 \text{ g}}{6.022 \times 10^{23} \text{ u}}$

(b) 1.673×10^{-24} g/p$^+$

5.33

	Symbol	Atomic number	Mass number	No. of protons	No. of neutrons	No. of electrons
(a)	$^3_1$H	1	3	1	2	1
(e)	$^{40}_{18}$Ar	18	40	18	22	18
(h)	$^{81}_{35}$Br	35	81	35	46	35
(j)	$^{195}_{78}$Pt	78	195	78	117	78

5.34

	Protons	Neutrons	Electrons
(a)	46	62	46
(c)	22	26	22

5.35 (b) $^{127}_{53}$I.

5.40 151.96 u.

5.43 (a) 63.55 u.

5.48 When an electron moves from a higher energy level to a lower energy level, a quantum of energy is released. This discrete packet of energy is associated with a line in the spectrum.

5.52 (a) 2, (c) 4, (e) 14, (g) 3, (j) 0.

5.55 (b) $1s^2 2s^2 2p^2$, (e) $1s^2 2s^2 2p^6 3s^2 3p^6 4s^1$, (f) $1s^2 2s^2 2p^6 3s^2 3p^6 4s^2 3d^{10}$, (j) $1s^2 2s^2 2p^6 3s^2 3p^6 4s^2 3d^{10} 4p^4$.

5.57 (a) $5s^1$, (c) $4s^2 4p^1$, (f) $5s^2 5p^2$.

5.58 (a) N, (c) P, (g) Rb.

5.59 (a) ·N·, (b) Rb·, (d) :Br·, (e) :Rn:, (h) ·Se:.

5.61 (a) Mg; (c) Y, Zr, Nb, Mo; (f) H, He, Li, Be.
5.62 (a) $1s^2 2s^2 2p^6$, (c) $1s^2 2s^2 2p^6$
5.64

	1s	2s	2p	3s 3p

(a) ⇅ ⇅ ↿ _ _
(c) ⇅ ⇅ ⇅ ⇅ ⇅ ⇅
(e) ⇅ ⇅ ⇅ ⇅ ⇅ ↿ ↿ ↿ ↿

5.65 (a) 0, (b) 2, (c) 2.

5.68

	Protons	Neutrons	Electrons
(a)	84	124	84
(c)	72	108	72

5.69 (a) $^{129}_{54}$Xe, $^{132}_{54}$Xe; (c) $^{25}_{12}$Mg, $^{26}_{12}$Mg.
5.70 (a) $1s^2 2s^2 2p^6 3p^1$.
5.73 0.6 mile.
5.74 Decreases by a factor of 9.
5.76 30.0 u.
5.79 Chromium would be expected to have the electronic configuration $4s^2 3d^4$; however it is $4s^1 3d^5$. Copper is expected to be $4s^2 3d^9$; however it is $4s^1 3d^{10}$.
5.84

	1s	2s	2p
B	⇅	⇅	↿ _ _
C	⇅	⇅	↿ ↿ _
N	⇅	⇅	↿ ↿ ↿

5.86 1.22×10^{-23} cm^3.

CHAPTER SIX

6.8 (a) Ar, (b) Na, (c) Ga, (d) Br, (e) Be.
6.13 Hydrogen's low melting point, boiling point, density, and high ionization energy are not consistant with those of the alkali metals.
6.14 Low density, melting point, and boiling point.
6.16 Properties of the elements are periodic functions of their atomic number.
6.17 (a) Alkali metals, (e) transition metals, (f) halogens, (g) alkaline earth metals, (i) nitrogen-phosphorus group, (k) lanthanide series.
6.19 (a) s^1, (b) $s^2 p^5$, (f) $s^2 p^3$.
6.21 Metal: (c), (f), (h); nonmetal: (b); metalloid: (a).
6.22 (a) Nonmetal.
6.23 (a) Be, (b) Sn.
6.24 (a) (1) 3, (3) 1, (4) 2.
      (b) (1) 3−, (3) 1−, (4) 2−.
6.25 (a) (1) 1, (2) 2, (3) 3, (4) 3.
      (b) (1) 1+, (2) 2+, (3) 3+, (4) 3+.
6.26 (a) 2+, (b) 3−, (d) 2−, (h) 1+, (i) 1−.
6.27 (a) N + 3e^- ⟶ [:Ṅ:]$^{3-}$; (c) Ba ⟶ Ba^{2+} + 2e^-;
      (d) Cs ⟶ Cs$^+$ + e^-.
6.28 (a) Ne, (b) Ṅe, (f) Kr.
6.29 874 kJ/mol.
6.30 (a) Sn, (c) Rb.
6.31 (a) C, (c) Ca.
6.32 (a) Rn, (c) Po, (e) N, (g) Ne, H.
6.33 Element 118 would be a noble gas.
      (a) Nonmetal, (c) poor conductor of electricity, (d) colorless, (g) does not form stable ions.
6.34 (a) Fr, (d) Se, (e) Te, (g) Ra, (i) Zn.
6.36 (a) A violent reaction occurs. (b) 2Na + 2H$_2$O ⟶ 2NaOH + H$_2$. (c) 2K + 2H$_2$O ⟶ 2KOH + H$_2$.

6.42

	Hydrogen	Alkali metals
Melting point, °C	−259.2	≥ 29
Boiling point, °C	−252.7	≥ 679
Density, g/cm^3	9.0×10^{-5}	≥ 0.53
Ionization energy, MJ/mol	1.3	≤ 0.52

6.46 (a) Ionization energy increases as atomic number increases. (b) Third row ionization energies are higher.

6.48 0.25 nm, 1.53 g/cm^3 (actual values).

6.49 The nuclear charge of 2+ readily attracts the two $1s$ electrons, which are not shielded.

6.50 Element 120 would be a highly radioactive alkaline earth metal. It would be a dense, high-melting solid, and a good conductor of electricity. It will be larger than Ra and have a lower ionization energy than Ra. It would form a 2+ ion.

6.52 The $3p^4$ electron in sulfur must pair up with another $3p$ electron. This requires energy (pairing energy) which is released when the electron is removed from the sulfur atom.

6.54 (a) $1s^2 2s^2 2p^6 3s^2 3p^6 4s^2 3d^{10} 4p^6 5s^2 4d^{10} 5p^6 6s^2$.
(b) $1s^2 2s^2 2p^6 3s^2 3p^6 4s^2 3d^{10} 4p^6 5s^2 4d^{10} 5p^6$.
(c) Xe
(d) Ba^{2+} is smaller than Xe.

6.56 (a) $3p$, (b) $5s$.

6.57

(a) Mg^{2+} 1s 2s 2p 3s 3p

(c) $[\,\ddot{\mathrm{\ddot S}}\,]^{2-}$

6.58 3843 K, 2.25 kg/L.

CHAPTER SEVEN

7.4 6.022×10^{14} billion.

7.5 (a) 0.27 mol Ne, (b) 0.0539 mol Fe, (c) 2.2 mol Ar, (d) 4.10×10^{-4} mol Mn.

7.6 (a) 417 g V, (b) 0.32 g Ge, (c) 3.77×10^5 g As, (d) 0.1106 g Al.

7.7 (a) 7.52×10^{23} He, (b) 1.49×10^{23} Ne, (c) 7.54×10^{22} Ar, (d) 3.59×10^{22} Kr.

7.8 (a) 1.364 mol B, (b) 7.8×10^{-4} mol Ag, (c) 1645 mol Cu, (d) 0.0104 mol Cs.

7.9 (a) 159.81 g/mol Br$_2$, (b) 80.04 g/mol SO$_3$, (c) 63.00 g/mol HNO$_3$, (d) 342.07 g/mol Al$_2$(SO$_4$)$_3$.

7.10 (a) 450 g PF$_3$, (b) 226.0 g SiCl$_4$, (c) 0.0230 g P$_4$O$_{10}$.

7.11 (a) 1.766×10^{23} molecules H$_2$S, (b) 6.514×10^{22} molecules ClF$_3$, (c) 5.910×10^{21} molecules C$_5$Br$_{12}$.

7.12 (a) 25.0 g CO$_2$.

7.15 (a) 29.6% O, 70.4% F, (b) 48.0% Zn, 52.0% Cl, (c) 0.630% H, 79.4% I, 20.0% O, (d) 16.4% Mg, 18.9% N, 64.7% O, (e) 21.2% N, 6.10% H, 24.3% S, 48.4% O.

7.16 (a) CH$_4$, (b) SO$_2$, (c) NO, (d) SiH$_4$, (e) Li$_3$PO$_4$.

7.17 TiO$_2$.

7.18 C$_8$H$_{16}$.

7.19 B$_2$H$_6$.

7.22 The atomic mass in grams.

7.23 (a) 10.81 g B, (d) 72.59 g Ge.

7.24 2×10^4 centuries.

7.25

	Element	No. of moles	No. of atoms	Mass, g
(a)	B	1.00	6.02×10^{23}	10.8
(d)	Zr	2.000	1.204×10^{24}	182.4
(f)	Sn	10.0	6.02×10^{24}	1.19×10^3

7.26 (a) 3.71×10^{-4} mol K, (c) 2.23×10^3 mol K.

7.27 (a) 0.0533 mol As, (c) 9.64×10^{-5} mol Ni.

7.28 (a) 3.21×10^{22} atoms As, (c) 5.81×10^{19} atoms Ni.

7.29 (a) 284 g Ar, (c) 2.141×10^8 g Na.

7.30 (a) 2.90×10^{22} atoms Pb, (d) 6.1781×10^{19} atoms Ra.

7.31 (a) 0.835 mol Xe, (b) 6.8 mol Cu.

7.32 (a) 42 g Kr, (c) 0.0646 g Ge.

7.33 (a) 3×10^1 g Fe, (b) 5.01×10^{-13} g Fe, (c) 6.59×10^{-20} g Fe, (d) 9.274×10^{-23} g Fe.

7.35 (a) 3.62×10^{-4} mol Re, (b) 1.132×10^{25} Pd, (c) 2×10^{-24} mol Hf, (d) 0.2424 mg Bi, (e) 56.5 kg Nb.

7.37 (b), (d), (a), (c).

7.38 (a) 207 g/mol IBr, (d) 520 g/mol CI$_4$, (e) 102 g/mol S_2F_2, (h) 194 g/mol $H_2S_2O_8$.

7.39 (a) 219.9 g/mol P_4O_6, (c) 175.8 g/mol OBr$_2$, (f) 90.01 g/mol $H_2C_2O_4$.

7.40

	Compound	No. of moles	No. of molecules	Mass, g
(a)	SO_2	1.000	6.022×10^{23}	64.05
(d)	H_2SO_3	0.0100	6.02×10^{21}	0.820
(f)	$AlCl_3$	0.262	1.58×10^{23}	34.9

7.41 (a) 0.232 g ClO$_2$, (e) 0.1021 g N_2O_4.

7.42 (a) 1.1×10^{23} molecules, H_2O_2, (b) 2.931×10^{20} molecules, ClF$_3$.

7.43 (a) 8.207 mol O, (c) 1.23×10^{-4} mol O.

7.44 (a) 0.75 g H, (b) 7.43×10^{-4} g H, (c) 15.83 g H, (d) 2745.1 g H.

7.47 (c) > (d) > (a) > (b).

7.48 (a) 99.0 g/mol CuCl, (c) 189 g/mol Zn(NO$_3$)$_2$, (e) 400 g/mol Fe$_2$(SO$_4$)$_3$.

7.49 (a) 0.44 mol NaClO$_3$, (b) 0.002257 mol Ca$_3$(PO$_4$)$_2$.

7.50 (a) 763 g TiO$_2$, (b) 2.90×10^3 g K$_2$SnCl$_6$, (c) 0.001337 g PbC$_2$O$_4$.

7.51 (a) 5.60×10^{22} formula units, **MgSiO$_3$**, (b) 7.1×10^{24} formula units, Hg$_2$(NO$_2$)$_2$, (c) 1.409×10^{23} formula units, NH$_4$C$_2$H$_3$O$_2$.

7.52 (a) 0.788% H, 99.2% I, (c) 61.2% Hg, 38.8% I, (e) 70.4% Os, 29.6% O.

7.53 (a) 24.74% K, 34.77% Mn, 40.49% O; (c) 59.89% Ba, 12.21% H, 27.91% O; (e) 11.1% N, 3.200% H, 41.27% Cr, 44.44% O.

7.54 (a) 87.1% Ag, (d) 65.4% Ag.

7.56 (a) 36.07% H$_2$O, (c) 33.69% H$_2$O.

7.58 (a) FeS$_2$, (d) SiI$_4$, (f) Ce$_3$S$_4$.

7.59 (a) KCN, (b) PbC$_2$O$_4$, (e) Li$_2$CO$_3$.

7.60 CrO$_3$.

7.62 CNH$_4$.

7.64 N$_3$H$_{12}$PO$_4$ = (NH$_4$)$_3$PO$_4$.

7.66 C$_{12}$H$_{24}$.

7.69 B$_{20}$H$_{16}$.

7.71 C$_{12}$H$_{24}$O$_2$.

7.72 (a) 29.13 g Na$_3$PO$_4$.

7.74 Mg$_2$Si$_3$H$_4$O$_{10}$.

7.76 C$_7$H$_5$O$_3$SN.

7.78 (a) 294 cm^3 IF$_5$, (b) 194 mol F, (c) 10.93 cm^3 IF$_5$.

7.81 (a) 3.4×10^2 g 24 carat Au, (b) 1.0×10^{24} Au.

7.82 1.3×10^{-14} g C.

7.84 (a) 2.22% H, (b) 1.46 g C, 0.0874 g H, 0.729 g N, 1.66 g O.

7.85 (a) 63.54 g/mol, (b) copper.

7.87 2.732×10^{26} atoms.

7.88 (a) C$_5$H$_7$N, (b) 1.2×10^{20} C$_{10}$H$_{14}$N$_2$.

CHAPTER EIGHT

8.1 (a) Metal and nonmetal, (b) two or more nonmetals.

8.2 (a) Potassium iodide, (b) lithium oxide, (c) strontium, oxide, (d) cesium nitride, (e) calcium phosphide.

8.4   (a) Oxygen difluoride, (b) nitrogen triiodide, (c) nitrogen monoxide, (d) bromine pentaiodide, (e) tetraphosphorus decoxide.

8.11   (a) Covalent, (b) ionic.

8.14 $K\cdot\ \overset{\frown}{+}\ \overset{\cdot\cdot}{:}\!F\!:\ \longrightarrow\ K^+ + [:\overset{\cdot\cdot}{\underset{\cdot\cdot}{Cl}}:]^-$

8.15   (a) Cs^+, I^-; (b) Sr^{2+}, Br^-; (c) Li^+, H^-; (d) Mg^{2+}, O^{2-}.

8.17   (a) $1s^2 2s^2 2p^6 3s^2 3p^6$, (b) $1s^2 2s^2 2p^6 3s^2 3p^6 4s^2 3d^{10} 4p^6$, (c) $1s^2 2s^2 2p^6$, (d) $1s^2 2s^2 2p^6 3s^2 3p^6 4s^2 3d^{10} 4p^6 5s^2 4d^{10} 5p^6$.

8.18   (a) Ba^{2+} $[:\overset{\cdot\cdot}{\underset{\cdot\cdot}{S}}:]^{2-}$

       (b) Sr^{2+} $2[:\overset{\cdot\cdot}{\underset{\cdot\cdot}{Br}}:]^-$

       (c) $2Li^+$ $[:\overset{\cdot\cdot}{\underset{\cdot\cdot}{O}}:]^{2-}$

8.19   (a) Ca_3N_2, (b) Rb_2O, (c) Mg_3P_2, (d) Cs_2S.

8.21   (a) 2, (b) 4, (c) 6.

8.22   (a) $H:H$

       (b) $:\overset{\cdot\cdot}{\underset{\cdot\cdot}{F}}:\overset{\cdot\cdot}{\underset{\cdot\cdot}{F}}:$

       (c) $:N::N:$

       (d) $:\overset{\cdot\cdot}{\underset{\cdot\cdot}{I}}:\overset{\cdot\cdot}{\underset{\cdot\cdot}{I}}:$

8.24   (a) $\overset{\delta+}{H}:\overset{\delta-}{\underset{\cdot\cdot}{I}}:$ $:\overset{\delta+}{\underset{\cdot\cdot}{I}}:\overset{\delta-}{\underset{\cdot\cdot}{F}}:$

8.26   (a) $:\overset{\cdot\cdot}{\underset{\cdot\cdot}{Cl}}:\overset{\cdot\cdot}{\underset{\cdot\cdot}{O}}:\overset{\cdot\cdot}{\underset{\cdot\cdot}{Cl}}:$

       (b) $:\overset{\cdot\cdot}{\underset{\cdot\cdot}{F}}:\overset{\cdot\cdot}{N}:\overset{\cdot\cdot}{\underset{\cdot\cdot}{F}}:$
                $:\overset{\cdot\cdot}{\underset{\cdot\cdot}{F}}:$

                $:\overset{\cdot\cdot}{\underset{\cdot\cdot}{I}}:$
       (c) $:\overset{\cdot\cdot}{\underset{\cdot\cdot}{I}}\cdot\overset{\cdot\cdot}{C}\cdot\overset{\cdot\cdot}{\underset{\cdot\cdot}{I}}:$
                $:\overset{\cdot\cdot}{\underset{\cdot\cdot}{I}}:$

       (d) $H:\overset{\cdot\cdot}{As}:H$
              $\overset{\cdot\cdot}{H}$

              H
       (e) $H:\overset{\cdot\cdot}{Si}:H$
              H

       (f) $H:\overset{\cdot\cdot}{Se}:H$

8.27 $:C\equiv O:$

8.28   (c) $:\overset{\cdot\cdot}{O}\!-\!\overset{\cdot\cdot}{S}\!=\!\overset{\cdot\cdot}{\underset{\cdot\cdot}{O}}$ $\longleftrightarrow$ $\overset{\cdot\cdot}{O}\!=\!\overset{\cdot\cdot}{S}\!-\!\overset{\cdot\cdot}{\underset{\cdot\cdot}{O}}:$ $\longleftrightarrow$ $:\overset{\cdot\cdot}{\underset{\cdot\cdot}{O}}\!-\!\overset{\cdot\cdot}{S}\!-\!\overset{\cdot\cdot}{\underset{\cdot\cdot}{O}}:$
            $:\overset{\cdot\cdot}{\underset{\cdot\cdot}{O}}:$ $:\overset{\cdot\cdot}{\underset{\cdot\cdot}{O}}:$ $.\overset{\cdot\cdot}{\underset{\cdot\cdot}{O}}.$

8.29   (a) $[:\overset{\cdot\cdot}{O}:H]^-$

       (b) $\left[\begin{array}{c} :\overset{\cdot\cdot}{\underset{\cdot\cdot}{O}}: \\ | \\ :\overset{\cdot\cdot}{\underset{\cdot\cdot}{O}}\!-\!P\!-\!\overset{\cdot\cdot}{\underset{\cdot\cdot}{O}}: \\ | \\ :\overset{\cdot\cdot}{\underset{\cdot\cdot}{O}}: \end{array}\right]^{3-}$

       (c) $\left[\begin{array}{c} :\overset{\cdot\cdot}{\underset{\cdot\cdot}{O}}: \\ | \\ :\overset{\cdot\cdot}{\underset{\cdot\cdot}{O}}\!=\!C\!-\!\overset{\cdot\cdot}{\underset{\cdot\cdot}{O}}: \end{array}\right]^{2-}$

(d) $[:C{\equiv}N:]^-$

8.30 (a) $:\!\overset{..}{F}\!:\!Be\!:\!\overset{..}{F}\!:$

$:\!\overset{..}{F}\!:$

(c) $:\!\overset{..}{F}\!:\!\overset{..}{I}\!:\!\overset{..}{F}\!:$

(d) $\overset{..}{O}{=}N{-}\overset{..}{O}:$

8.32 (a) Pyramidal, (b) tetrahedral, (c) angular, (d) Linear.

8.34 (a) 109.5°, (b) 180°, (c) 120°.

8.37 (a) Sodium bromide, (d) barium phosphide, (f) lead sulfide, (g) silver bromide.

8.40 (a) Dinitrogen oxide, (b) phosphorus trifluoride, (d) xenon tetrafluoride, (f) arsenic pentafluoride, (g) Iodine trifluoride.

8.41 (a) Diiodine tetroxide, (b) Potassium sulfide, (d) tetriodine nonoxide, (f) calcium iodide.

8.46 A nitrogen atom has 5 outer-level electrons, which is very unstable. It can obtain a noble gas configuration by sharing three electrons with another nitrogen atom.

8.48 The ionic bond in NaCl, as is found in all ionic compounds, has some degree of covalent character.

8.51 Noble gases form relatively few compounds, most of which contain only bonds with highly electronegative atoms such as oxygen or a halogen.

8.52 (b) Sn, (c) Fr, (d) Al.

8.53 (a) P > As > Sb, (b) B > Be > Li.

8.54 (a) Ar, (b) Ne, (d) Kr, (g) Ar, (h) Ne.

8.55 (a) $1s^22s^22p^63s^23p^64s^23d^{10}4p^6$, (b) $1s^22s^22p^63s^23p^6$, (e) $1s^22s^22p^63s^23p^64s^23d^{10}4p^65s^24d^{10}5p^6$, (g) $1s^22s^22p^63s^23p^6$.

8.56 (a) Se^{2-}, (b) N^{3-}.

8.57 (a) $K\cdot\ +\ :\overset{..}{O}:\ +\ \cdot K\ \longrightarrow\ 2K^+ + [:\overset{..}{O}:]^{2-}$.

(b) $:\overset{..}{Br}\cdot\ +\ Mg\ +\ \cdot\overset{..}{Br}:\ \longrightarrow\ Mg^{2+} + 2[:\overset{..}{Br}:]^-$.

(c) $:\overset{..}{O}\cdot+\odot Al\ +\ :\overset{..}{O}:\ +\ \cdot Al\ +\overset{..}{O}:\ \longrightarrow\ 2Al^{3+} + 3[:\overset{..}{O}:]^{2-}$.

8.58 (a) $Na^+\ [:\overset{..}{I}:]^-$

(b) $Mg^{2+}\ 2[:\overset{..}{Br}:]^-$

(d) $2Rb^+\ [:\overset{..}{S}:]^{2-}$

(e) $3Ba^{2+}\ 2[:\overset{..}{N}:]^{3-}$

8.59 (a) Al_2O_3, (b) SrS, (c) $MgBr_2$, (f) Li_2Se.

8.63 (a) 1s, 5p, (b) 5p, 4p, (d) 2p, 3p, (g) 2p, 1s.

8.64 (a) Polar, (b) nonpolar, (d) polar.

8.65 $:\overset{..}{Br}\!:\!\overset{..}{F}:$ with $\overset{\delta+}{}\ \overset{\delta-}{}$

8.66 (a) $:\overset{..}{Br}{-}\overset{:\overset{..}{Br}:}{\underset{:\overset{..}{Br}:}{C}}{-}\overset{..}{Br}:$ (b) $:\overset{..}{F}{-}\overset{..}{O}{-}\overset{..}{F}:$ (d) $:\overset{..}{F}{-}\overset{:\overset{..}{F}:}{P}{-}\overset{..}{F}:$

(f) $:\overset{..}{Cl}{-}\overset{..}{N}{-}\!\!-\overset{..}{N}{-}\overset{..}{Cl}:$ with $:\overset{..}{Cl}:\ :\overset{..}{Cl}:$ below   (i) $H{-}\overset{..}{O}{-}\overset{..}{O}{-}H$

8.67 (a) $[:\overset{..}{O}{-}H]^-$ (b) $[:C{\equiv}N:]^-$ (c) $\left[H{-}\overset{H}{\underset{H}{N}}{-}H\right]^+$

(f) $[\,:\!\overset{..}{O}\!-\!\overset{..}{O}\!:\,]^{2-}$ (h) $\left[\,H-\overset{\underset{|}{H}}{\underset{\underset{H}{|}}{C}}-\overset{\overset{..}{O}}{\underset{}{\|C}}-\overset{..}{\underset{..}{O}}:\,\right]^{-}$

8.68 (a) $H-\overset{\underset{|}{H}}{\underset{\underset{H}{|}}{C}}-\overset{\underset{|}{H}}{\underset{\underset{H}{|}}{C}}-H$ (b) $H-\overset{\underset{|}{H}}{C}=\overset{\underset{|}{H}}{C}-H$ (c) $H-C\equiv C-H$

8.69 (a) $H-\overset{..}{\underset{..}{O}}-\overset{\overset{O}{\|}}{C}-\overset{..}{\underset{..}{O}}-H$ (d) $H-\overset{..}{\underset{..}{O}}-N=\overset{..}{\underset{.}{O}}$ (f) $H-\overset{..}{\underset{..}{O}}-\overset{\overset{O}{\|}}{\underset{}{S}}-\overset{..}{\underset{..}{O}}-H$

8.71 (b) PF_5, (e) SF_6.

8.72 (a) $\left[H-\overset{..}{\underset{..}{O}}-\overset{\overset{:O:}{|}}{C}=\overset{..}{O}\right]^{2-} \longleftrightarrow \left[H-\overset{..}{\underset{..}{O}}-\overset{\overset{O}{\|}}{C}-\overset{..}{\underset{..}{O}}:\right]^{2-}$ (d) $:\overset{..}{O}-\overset{.}{O}=\overset{..}{O}: \longleftrightarrow :\overset{.}{O}=\overset{..}{O}-\overset{..}{O}:$

(e) $:\overset{.}{N}=N=\overset{..}{O}: \longleftrightarrow :N\equiv N-\overset{..}{\underset{..}{O}}: \longleftrightarrow :\overset{..}{\underset{..}{N}}-N\equiv O:$

8.74 (a) Pyramidal, (b) angular, (c) tetrahedral, (e) linear, (g) linear.

8.75 (a) Angular, (c) trigonal planar, (e) tetrahedral.

8.77 (a) 3, (b) 1.

8.78 Carbon is a smaller atom that is more electronegative than the larger silicon atom; therefore, C will form a stronger, shorter bond with O.

8.79 (a) . . . are *not* always . . . ; (c) . . . are *smaller* for single . . . ; (e) . . . sulfur *trioxide*; (g) . . . is *shared* by fluorine. . . .

8.80 (a) $:P=P:$ $:P-P:$ $:P\overset{\overset{..}{P}}{\underset{\underset{..}{P}}{\diamondsuit}}P:$
      $:P=P:$ $:P-P:$

(b) P_4 is nonpolar because each phosphorus atom has the same electronegativity and the lone pairs of electrons are symmetrically distributed.

8.82 (a) $H-\overset{\underset{|}{H}}{\underset{\underset{H}{|}}{C}}-\overset{\overset{\overset{..}{O}}{\|}}{C}-\overset{\underset{|}{H}}{\underset{\underset{H}{|}}{C}}-H$

(b) The two terminal carbon atoms are tetrahedral. The central carbon is trigonal planar. (c) It is not possible to draw an equivalent structure for acetone with a different placement of the double bond so there can be no resonance.

(d) $H-\overset{\underset{|}{H}}{\underset{\underset{H}{|}}{C}}-\overset{\underset{|}{H}}{\underset{\underset{H}{|}}{C}}-\overset{\underset{|}{H}}{C}=\overset{..}{\underset{..}{O}}$

8.83 (a) $Na^+\,[\,:\overset{..}{\underset{..}{O}}-H\,]^-$ (c) $Ca^{2+}\,2\left[\,:\overset{..}{\underset{..}{O}}-\overset{\overset{:\overset{..}{O}:}{|}}{Cl}-\overset{..}{\underset{..}{O}}:\,\right]^-$

8.85 (a) $\left[H-\overset{\overset{.\overset{.}{O}}{\|}}{C}-\overset{..}{\underset{..}{O}}:\right]^- \longleftrightarrow \left[H-\overset{\overset{:\overset{..}{O}:}{|}}{C}=\overset{..}{\underset{.}{O}}\right]^-$

(b) A pair of electrons is spread out over the carbon and two oxygen atoms.

(c)
$$\left[H{-}C{\Big\langle}{\overset{\displaystyle O}{\underset{\displaystyle O}{}}} \right]^{-}$$

8.86   (a) The C—C bond is shorter and stronger than the Si—Si bond. The bond order is one for both C—C and Si—Si. (c) N—N has a bond order of one and is longer and weaker than N≡N, which has a bond order of two.

8.88   (a)
$$H{-}\overset{\displaystyle H}{\underset{}{N}}{-}\ddot{\underset{..}{O}}{-}H$$
  (c)
$$:\ddot{O}{-}\overset{\displaystyle :\ddot{O}:}{\underset{:\ddot{Cl}:}{S}}{-}\ddot{Cl}:$$

8.90   (a) $\cdot\ddot{S}{=}C{=}\ddot{N}^{-} \longleftrightarrow :\ddot{S}{-}C{\equiv}N:^{-} \longleftrightarrow :S{\equiv}C{-}\ddot{N}:^{-}$ (b) The sulfur–carbon and carbon–nitrogen bonds are best described as double bonds. However, since nitrogen is more electronegative than sulfur, the second structure could contribute significantly, making the sulfur–carbon bond closer to a single bond and the carbon–nitrogen closer to a triple bond.

8.92 $H{-}\overset{\displaystyle :\ddot{I}::\ddot{I}:}{\underset{}{C}}{=}C{-}H$, $:\ddot{I}{-}\overset{\displaystyle :\ddot{I}: \; H}{\underset{}{C}}{=}C{-}H$, $H{-}\overset{\displaystyle :\ddot{I}: \; H}{\underset{}{C}}{=}C{-}\ddot{I}:$

CHAPTER NINE

9.4   (a) P +3, Br −1, (b) Mn +4, O −2, (c) Al +3, F −1, (d) N +1, O −2, (e) P +5, O −2.

9.5   (a) C +3, (b) N +3, (c) Br +5, (d) C +4, (e) C 0, H +1.

9.6   (a) Cupric, (b) manganous, (c) plumbic, (d) ferrous, (e) stannous.

9.7   (a) Iron(III) sulfide, (b) copper(I) nitride, (c) lead(II) oxide, (d) mercury(I) fluoride, (e) manganese(III) fluoride, (f) hydrobromic acid, (g) aluminum fluoride.

9.8   (a) $SnBr_2$, (b) Au_2O_3, (c) Ni_3N_2, (d) PbI_4, (e) Cu_2S, (f) SCl_2.

9.11   (a) Cesium carbonate, (b) strontium chlorate, (c) silver chromate, (d) mercury(I) nitrate, (e) sodium phosphate, (f) cobalt(II) nitrite.

9.12   (a) $(NH_4)_2CrO_4$, (b) $FeSO_3$, (c) $Al(CN)_3$, (d) $Ni(NO_2)_2$, (e) $MnPO_4$, (f) Na_3AsO_4.

9.13   (a) Nitric acid, (b) phosphorous acid, (c) sulfurous acid, (d) hypochlorous acid.

9.14   (a) H_3BO_3, (b) H_2SeO_4, (c) $HClO_2$, (d) H_2CO_3.

9.15   (a) K_3AsO_4, (b) $RbHSO_3$, (c) $Mg_3(BO_3)_2$, (d) $Fe(H_2PO_4)_3$.

9.16   (a) Sodium sulfite, (b) strontium nitrite, (c) cesium hydrogensulfite, (d) potassium bromite, (e) copper(I) hypochlorite, (f) Copper(II) sulfate pentahydrate, (g) iron(II) nitrate hexahydrate.

9.17   (a) $NaClO_2$, (b) $Ni(HCO_3)_3$, (c) Hg_2SeO_3, (d) $Mg(BrO_4)_2$, (e) $Pb(H_2PO_4)_4$, (f) $NaC_2H_3O_2 \cdot 3H_2O$.

9.19   (a) F 0, (c) Mg +2, S −2, (e) N +5, O −2, (g) Al +3, S −2.

9.20   (a) +6, (d) +4, (g) +3.

9.22   (a) K +1, Cr +6, O −2; (c) Na +1, U +6, O −2; (e) Ca +2, H +1, S −2; (g) K +1, V +5, O −2; (i) Cu +2, C +2, N −3.

9.24   (a) Carbon dioxide, (b) dinitrogen monoxide, (c) carbon tetrachloride, (d) phosphorus tribromide.

9.25   (a) Tin(II) bromide, (b) cobalt(III) nitride, (c) lead(II) sulfide, (d) copper(I) phosphide, (e) mercury(II) oxide.

9.26   (a) $FeBr_3$, (e) MgO, (h) Sn_3N_2, (j) Rb_3P.

9.27   (a) Tl_2O_3, (b) UO_2, (c) Au_2O, (d) Mo_2O_5, (e) Mn_2O_7.

9.28   Ammonium acetate, phosphate, permanganate, cyanide, carbonate, chromate, hydroxide, sulfide; magnesium acetate, phosphate, permanganate, cyanide, carbonate, chromate, hydroxide, sulfide; aluminum acetate, phosphate, permanganate, cyanide, carbonate, chromate, hydroxide, sulfide.

9.29 $Ca(OH)_2$, $Ca(C_2H_3O_2)_2$, $CaSO_4$, $Ca(ClO_3)_2$, $CaSeO_4$, Ca_3N_2; $Al(OH)_3$, $Al(C_2H_3O_2)_3$, $Al_2(SO_4)_3$, $Al(ClO_3)_3$, $Al_2(SeO_4)_3$, AlN; $Pb(OH)_4$, $Pb(C_2H_3O_2)_4$, $Pb(SO_4)_2$, $Pb(ClO_3)_4$, $Pb(SeO_4)_2$, Pb_3N_4.

9.31   (a) $(NH_4)_2SeO_4$, (b) $(NH_4)_2SO_3$, (d) NH_4IO_4, (f) NH_4HCO_3, (h) NH_4ClO.

9.32   (a) $LiC_2H_3O_2$, (b) $Zn(C_2H_3O_2)_2$, (e) $Ga(C_2H_3O_2)_3$.

9.33   (a) Hydrobromic acid.

9.34   (a) Boric acid, (b) hypochlorous acid, (d) arsenic acid, (f) perbromic acid.

9.36   (a) Periodic acid, (b) acetic acid, (e) Hydrocyanic acid.

9.38   (a) Hydrogensulfite, sulfurous acid; (b) Monohydrogenphosphite, phosphorous acid.

9.39   (a) Na +1, H +1, F −1; (b) Na +1, S +5, O −2, H +1; (c) Na +1, B +3, F −1.

9.43   (a) Chromium(III) phosphate hexahydrate, (b) Gallium(III) oxide monohydrate, (c) Indium(III) perchlorate octahydrate.

9.44   (a) $Na_3BO_3 \cdot 4H_2O$, (d) $Hg(BrO_3)_2 \cdot 2H_2O$.

9.45   (1) Ammonium sulfide, (2) antimony(III) iodide, (3) arsenic acid, (4) arsenic(III) oxide, (5) barium chromate, (6) beryllium selenite, (7) bismuth(IV) chloride, (8) boron nitride, (9) bromine dioxide, (10) Cadmium(II) bromate, (11) calcium hypochlorite, (12) disulfur decafluoride, (13) cerium(III) hydroxide, (14) cesium hydrogencarbonate, (15) dichlorine heptoxide, (16) chromium(III) sulfite, (17) cobalt(III) fluoride, (18) copper(II) selenate, (19) gallium(III) chloride, (20) germanium(IV) sulfide, (21) hydrogen cyanide, (22) hydrocyanic acid, (23) iodic acid, (24) iodine pentafluoride, (25) iridium(III) sulfide.

9.46   (1) $TiCl_3$, (2) WBr_6, (3) $ZnCrO_3$, (4) ZrI_4, (5) $(NH_4)_2SO_3$, (6) $NaOH$, (7) $MgSO_4$, (8) $Ba(BrO)_2$, (9) Cs_2HPO_4, (10) $BiBr_3$, (11) $CuClO$, (12) $Al(NO_3)_3$, (13) Fe_3As_2, (14) $Pb(ClO_3)_4$, (15) Li_2HPO_3, (16) MnO, (17) $HgTe$, (18) MoI_4, (19) $Ni(HCO_3)_3$, (20) N_2O_4, (21) $PdSi$, (22) $OsSO_4$, (23) $LiClO_3$, (24) H_3PO_2, (25) P_4S_7.

9.48   (a) $Mg_3(PO_3)_2 \cdot 5H_2O$, (b) magnesium phosphate pentahydrate.

9.50   27.7% W, 72.3% Br.

CHAPTER TEN

10.4   Solid carbon combines with gaseous oxygen to produce gaseous carbon dioxide.

10.5   (a) 1, 2; (b) 2, 3, 2; (c) 1, 6, 4; (d) 2, 3, 2, 2; (e) 5, 2, 1, 4; (f) 2, 3, 1, 3.

10.6 $Ca_3P_2 + 6H_2O \longrightarrow 3Ca(OH)_2 + 2PH_3$.

10.7 $3H_2SO_4(aq) + 2Al(OH)_3(aq) \longrightarrow 6H_2O(l) + Al_2(SO_4)_3(aq)$.

10.12   (a) Decomposition; (b), (d) single displacement; (c) combination; (e) Metathesis.

10.13   (a), (c), (e) combination; (b), (h) decomposition; (d), (g) metathesis; (f) single displacement.

10.14   (a) $MgSO_4(s) + 7H_2O(g)$, (b) $Ag_2S(s) + NaNO_3(aq)$, (c) $P_4O_{10}(s)$, (d) NR, (e) $KClO_3(aq) + H_2O(l)$.

10.15   (a) $2Ca(s) + O_2(g) \longrightarrow 2CaO(s)$.

       (b) $2KClO_3(s) \longrightarrow 2KCl(s) + 3O_2(g)$.

       (c) $SO_2(g) + H_2O(l) \longrightarrow H_2SO_3(aq)$.

       (d) $2Cs(l) + 2H_2O(l) \longrightarrow 2CsOH(aq) + H_2(g)$.

       (e) $Pb(NO_3)_2(aq) + K_2SO_4(aq) \longrightarrow PbSO_4 + 2KNO_3(aq)$.

       (f) $H_2(g) + S(s) \longrightarrow H_2S(g)$.

10.18   (a), (d) exothermic; (b), (c) endothermic.

10.23   (a) Two moles of gaseous sulfur trioxide when heated produce two moles of gaseous sulfur dioxide and one mole of gaseous oxygen. (b) One mole of liquid mercury and one mole of chlorine gas produce one mole of mercury(II) chloride.

10.24   (a) 1, 5, 2; (c) 1, 1, 2; (d) 1, 2, 1, 2; (f) 1, 1, 1; (g) 1, 2, 2, 3.

10.25   (a) 2, 13, 8, 10 or 1, 13/2, 4, 5; (b) 1, 3, 1, 3; (c) 2, 2, 4, 1; (d) 3, 2, 1, 3, 3; (e) 2, 5, 2, 4.

10.26   (a) 2, 15, 3, 3, 6; (b) 3, 4, 3, 1; (c) 2, 2, 1, 1, 1, 4; (d) 3, 8, 9, 4; (e) 1, 1, 1, 2; (f) 2, 15, 14, 6.

10.27   (a) $NaBr(aq) + AgNO_3(aq) \longrightarrow NaNO_3(aq) + AgBr(s)$.

       (b) $Al(OH)_3(aq) + 3HNO_3(aq) \longrightarrow Al(NO_3)_3(aq) + 3H_2O$.

       (c) $Cl_2(g) + 2RbI(aq) \longrightarrow 2RbCl(aq) + I_2(s)$.

       (d) $2Fe(C_2H_3O_2)_3(aq) + 3Na_2S(aq) \longrightarrow 6NaC_2H_3O_2(aq) + Fe_2S_3(s)$.

       (e) $SiF_4(g) + 2H_2O \longrightarrow SiO_2(s) + 4HF(aq)$.

       (f) $MnO_2(s) + 4HCl(aq) \longrightarrow MnCl_2 + Cl_2(g) + 2H_2O$.

       (g) $N_2O_4(g) \overset{\Delta}{\longrightarrow} 2NO_2(g)$.

(h) $Ca_3P_2 + 6H_2O \longrightarrow 3Ca(OH)_2 + 2PH_3(g)$.

(i) $2AgNO_3(s) \overset{\Delta}{\longrightarrow} 2Ag + 2NO_2(g) + O_2(g)$.

(j) $2Al + 3CuSO_4 \longrightarrow 3Cu + Al_2(SO_4)_3$.

10.28 (a) Decomposition: (g), (i); single displacement: (c), (j); metathesis: (a), (b), (d), (e), (h); none of these: (f) (b) Oxidation–reduction (p), (r), (s).

10.30 (a) $2C + O_2 \longrightarrow 2CO$; (b) $2S + 3O_2 \longrightarrow 2SO_3$; (e) $P_4 + 6H_2 \longrightarrow 4PH_3$.

10.31 (a) K_2O, (b) BaO.

10.32 (a) CO_2, (b) SO_2.

10.33 (a) $MgCO_3$, (b) $KClO_3$, (c) $KHCO_3$, (d) $Ba(NO_2)_2 \cdot H_2O$.

10.35

	Oxidation	Reduction
(a)	Li	HCl
(b)	Si	F_2
(c)	H_2	Cl_2

10.39 Soluble: (a), (e); insoluble: (b), (c), (d), (f)

10.40 (a) $8Zn + S_8 \longrightarrow 8ZnS(s)$.

(b) $4Al(s) + 3O_2(g) \longrightarrow 2Al_2O_3(s)$.

(c) $2Na(s) + F_2(g) \longrightarrow 2NaF(g)$.

(d) $3H_2(g) + N_2(g) \longrightarrow 2NH_3(g)$.

10.41 (a) $PtO_2 \overset{\Delta}{\longrightarrow} Pt + O_2$.

(b) $CuSO_4 \cdot 5H_2O \overset{\Delta}{\longrightarrow} CuSO_4 + 5H_2O$.

(c) $2Fe_2O_3 \overset{\Delta}{\longrightarrow} 4Fe + 3O_2$.

(d) $SrCO_3 \overset{\Delta}{\longrightarrow} SrO + CO_2$.

10.42 (a) $Zn(s) + Pb(NO_3)_2(aq) \longrightarrow Pb(s) + Zn(NO_3)_2(aq)$.

(b) $Ba(s) + 2H_2O(l) \longrightarrow H_2(g) + Ba(OH)_2(s)$.

(c) $Ni(s) + SnBr_2(aq) \longrightarrow NiBr_2(aq) + Sn(s)$.

(d) $Hg(l) + Fe(NO_3)_2(aq) \longrightarrow NR$.

10.43 (a) $NiCl_2(aq) + Ca(OH)_2(aq) \longrightarrow CaCl_2(aq) + Ni(OH)_2(s)$.

(b) $Hg(C_2H_3O_2)_2(aq) + 2K_2CO_3(aq) \longrightarrow 2KC_2H_3O_2(aq) + HgCO_3(s)$.

(c) $H_3PO_4(aq) + 3AgNO_3(aq) \longrightarrow Ag_3PO_4(s) + 3H_2O(l)$.

(d) $3H_2SO_3(aq) + 2Al(OH)_3(aq) \longrightarrow Al_2(SO_3)_3(s) + 6H_2O(l)$.

10.44 (a) $2AgNO_3 + CuCl_2 \longrightarrow 2AgCl + Cu(NO_3)_2$.

(b) $Fe + H_2O \longrightarrow NR$ (c) $2KOH + H_2SO_4 \longrightarrow K_2SO_4 + 2H_2O$.

(d) $4NH_3 + 5O_2 \overset{\Delta}{\longrightarrow} 4NO + 6H_2O$ (e) $SO_3 + H_2O \longrightarrow H_2SO_4$.

(f) $H_2SO_4 + Zn \longrightarrow ZnSO_4 + H_2$ (g) $2NO + O_2 \longrightarrow 2NO_2$.

(h) $4Al + 3O_2 \longrightarrow 2Al_2O_3$ (i) $H_2(g) + Sn(NO_3)_2 \longrightarrow NR$.

(j) $Na_2SO_4 \cdot 10H_2O \overset{\Delta}{\longrightarrow} Na_2SO_4 + 10H_2O$.

(k) $NH_3 + HCl \longrightarrow NH_4Cl$.

10.46 (a) -47.7 kJ, (b) 36°C.

10.48 (a) $C_6H_6 + 15/2O_2 \longrightarrow 6CO_2 + 3H_2O$, (c) -1910 kJ.

10.49 (a) Endothermic, (c) exothermic.

10.52 (a) . . . produce *basic* solutions . . . ; (b) . . . *Most inorganic* . . . ; (c) . . . can be used . . . ; (d) . . . insoluble except those of alkali metals and ammonium; (e) . . . into liquid water *does not react*.

10.53 (a) $N_2 + 3H_2 \longrightarrow 2NH_3$, (b) $4NH_3 + 5O_2 \longrightarrow 4NO + 6H_2O$.

10.54 (1) $CaCO_3 \overset{\Delta}{\longrightarrow} CaO + CO_2$.

(2) $CO_2 + NH_3 + H_2O + NaCl \longrightarrow NaHCO_3 + NH_4Cl$.

(3) $2NaHCO_3 \longrightarrow Na_2CO_3 + CO_2 + H_2O$.

(4) $CaO + 3C \longrightarrow CaC_2 + CO$.

10.55   (a) 1, 5, 1, 3, 5; (c) 2, 7, 3, 2, 7, 3.

CHAPTER ELEVEN

11.3   17.5 mol NH_3.

11.4   (a) 5.99 mol O_2; (b) 3.69 mol CO_2, 4.61 mol H_2O; (c) no.

11.5   0.0368 mol H_2O, 0.0294 mol CO_2.

11.6   0.142 mol SO_2.

11.7   (a) 0.116 g P_4, (b) 0.514 g PCl_3.

11.8   (a) 1.64×10^5 g F_2, (b) 0.006367 g N_2.

11.9   313 g Cl_2.

11.10 3.97×10^3 kJ.

11.13   (a) No limiting reagent, (b) 1 mol O_2 is limiting, (c) 12.0 g O_2 is limiting, (d) 31 g O_2 is limiting, (e) 0.0455 g C is limiting.

11.14   (b) 12 g C, (c) 7.5 g C, (d) 4.4 g C, (e) 0.0837 g O_2.

11.15   (a) 96.6 g NaF, (b) 1 g F_2.

11.16   (a) 90.1%, (b) 5.38 g CCl_4.

11.18

	$N_2(g)$	+	$O_2(g)$	$\longrightarrow$	$2NO(g)$
Molecules	5 molecules		5 molecules		10 molecules
Molecules	6.02×10^{23}		6.02×10^{23}		1.20×10^{24}
Moles	5.0 mol		5.0 mol		10. mol
Mass, g	14.0 g		16.0 g		30.0 g
Mass, g	9.10 g		10.4 g		19.5 g

11.20   (a) 1 mol CaF_2, (b) 78 g CaF_2, (c) 40 g Ca + 38 g F_2 = 78 g CaF_2.

11.22   (a) 5.0 mol $AlCl_3$, (b) 5.3 mol $AlCl_3$.

11.23 $\dfrac{1 \text{ mol } C_5H_{12}}{5 \text{ mol } CO_2}$, $\dfrac{1 \text{ mol } C_5H_{12}}{6 \text{ mol } H_2O}$, $\dfrac{8 \text{ mol } O_2}{5 \text{ mol } CO_2}$, $\dfrac{5 \text{ mol } O_2}{6 \text{ mol } H_2O}$.

11.25   (a) 0.22 mol Fe_2O_3, 0.88 mol SO_2; (b) 14.6 mol Fe_2O_3, 58.2 mol SO_2; (c) 1.491 mol Fe_2O_3, 5,964 mol SO_2; (d) 0.00362 mol Fe_2O_3, 0.0145 mol SO_2; (e) 6.27×10^{-4} mol Fe_2O_3, 2.51×10^{-3} mol SO_2.

11.26   (a) 820 g O_2; (b) 0.0190 mol O_2; (c) 50.6 mol KCl, 75.9 mol O_2; (d) 1.01 g $KClO_3$.

11.28   (a) 1.76×10^3 g Cu, (b) 3.02×10^8 g Cu.

11.29   (a) 0.3548 g Ag, (c) 0.00579 g Ag.

11.31   (a) 1.89 g CH_4, (b) 1.721×10^4 g Cl_2, (c) 4.42×10^{-4} g CH_4, (d) 8.343×10^4 g CH_4, 1.475×10^6 g Cl_2.

11.33   (a) 278.5 g XeF_4; (b) 0.382 g XeF_4; (c) 1.324×10^{-3} g H_2O; (d) 1.25×10^{-3} g $XeOF_4$; 1.47×10^{-3} g Xe, 8.96×10^{-4} g HF, 2.68×10^{-4} g O_2.

11.35   (a) $2KNO_3 \longrightarrow 2KNO_2 + O_2$.

      (b) $\dfrac{2 \text{ mol } KNO_3}{2 \text{ mol } KNO_2}$, $\dfrac{2 \text{ mol } KNO_3}{1 \text{ mol } O_2}$, $\dfrac{2 \text{ mol } KNO_2}{1 \text{ mol } O_2}$.

      (c) 489.1 g KNO_2.

      (d) 753.1 g O_2.

11.37 $Sb_2S_3 + 3Fe \xrightarrow{\Delta} 2Sb + 3FeS$.

      (b) 28.9 g Fe.

      (c) 6.79×10^4 g Sb.

      (d) 7.110 g Sb_2S_3, 3.507 g Fe.

11.38   (a) $2H_2(g) + O_2(g) \longrightarrow 2H_2O(g) + 136$ kcal.

      (b) 2.89 g H_2, 22.9 g O_2.

      (c) 834 kJ.

(d) 52.1 kJ.

11.40 (a) 50.3 kJ, (b) 0.0129 g C_2H_4, (c) 87.5 kJ, (d) 0.229 g C_2H_4.

11.42 (a) $CaCO_3 + 71.1$ kJ $\longrightarrow$ CaO + CO_2.

(b) $CaCO_3 + 17.0$ kcal $\longrightarrow$ CaO + CO_2.

11.43 22.4 kJ.

11.44 (a) 127 g HI, (b) 1033 g HI, (c) 2.3 g I_2.

11.46 (a) 1179 g Na_2SO_4, (b) 31.7 g Na_2SO_4, (c) 7.24×10^3 g HCl, (d) 5.8×10^2 g NaCl.

11.48 (a) 74.5 g NaCN, (b) 21.3 g N_2.

11.49 (a) 18.7 g C_6H_5Cl, (b) 55.6%, (c) 2.66 g C_6H_5Cl, (d) 5.77 g C_6H_6.

11.51 (a) 3.44 g $CaCN_2$, (b) 3.94×10^4 g $CaCN_2$.

11.53 (a) 403 g $Ca(OH)_2$, (b) 2.81×10^4 g $Ca(OH)_2$.

11.55 31.2 g I_2, 25.7 g K_2SO_4, 7.43 g $MnSO_4$.

11.57 (a) 162 g SCl_2, 87.8 g NaF; (b) 58.6 g SF_4.

11.59 68.9% $CaCO_3$, 31.1% $CaSO_4$.

11.60 Co_2O_3.

CHAPTER TWELVE

12.4 (a) 0.737 torr, (b) 3.271 atm, (c) 119 kPa, (d) 6.95×10^5 torr.

12.7 5.49 L.

12.8 91.9 mL.

12.9 0.707 L.

12.11 (a) 0.205 mol N_2, (b) 0.00478 mol O_2, (c) 1.39×10^{-4} mol F_2.

12.12 (a) 105 L H_2, (b) 0.37 L He; (c) 0.257 L CO.

12.13 120. g Rn.

12.14 26.7 L F_2.

12.15 62.4 L·torr/mol·K.

12.16 1.46 L.

12.17 $-4°C$.

12.18 0.00466 mol Ar.

12.19 113 g/mol.

12.20 2.52 g/L.

12.21 (a) 717 L NO_2, (b) 166 mL NO, (c) 39.2 L O_2.

12.22 (a) 0.919 L O_2, (b) 102 g H_2O_2.

12.23 1.17 L O_2.

12.26 7.52 atm.

12.27 777 torr.

12.31 Because of its small size and low atomic mass He would be expected to exhibit the most ideal gas properties.

12.40 (a) 799 torr, (b) 699 torr.

12.42 (a) 4.04×10^6 Pa, (b) 3.03×10^4 torr.

12.43 (a) 2244 torr.

12.44 (a) 784 torr, (c) 1.05×10^5 Pa.

12.45 (c) 1237 kPa.

12.47 (a) 2.12×10^3 lb/ft^2, (b) 1.50×10^3 torr.

12.49 (a) 0.362 L, (e) 0.983 L.

12.52 2.18×10^4 cm^3.

12.55 (a) 1.204 atm, (b) 3.536 atm.

12.59 (a) 325 K, (b) 275 K, (c) 27 L, (d) 125 K.

12.61 (a) 4.94 L, (b) 19.8 L, (c) 18.9 L, (d) 6.90 L.

12.63 (a) 493 K.

12.65 (a) V decreases, (c) V increases.

12.66 (a) 37.0 L, (c) 2.36 L.

12.67   (a) 0.150 L, (c) 271 mL.

12.68   (a) 4.02 atm, (c) 838 torr.

12.70   (a) 0.0166 mol, (d) 1.14×10^{-7} mol.

12.71   (a) 2.55 L, (e) 1.64×10^5 L.

12.72   (a) 0.0597 g N_2O, (c) 1.06×10^3 g C_2H_2.

12.73   (b) 2.60×10^3 L.

12.75   (a) 3.16 g/L, (b) 1.96 g/L.

12.76   (a) $P = nRT/V$, (b) $n = PV/(RT)$.

12.77   (a) 82.1 mL·atm/mol·K, (c) 83.2 kPa·dL/mol·K.

12.78   (a) 79.6 L.

12.79   (b) 0.00498 mol.

12.80   (c) 0.0244 atm.

12.81   (a) 2.46 g N_2.

12.83   1.98 g/L.

12.85   37.3 g/mol.

12.87   (a) 12.0 L H_2, (d) 48.45 cm^3 H_2.

12.89   (c) 1.07 L O_2.

12.90   (b) 26.3 g PbS, (c) 2.44×10^3 g PbS.

12.91   (a) 155 L Cl_2, (c) 10.5 L Cl_2.

12.93   31.0 L CO_2.

12.94   288 torr.

12.96   42.4 mL N_2.

12.98   (a) 0.0455 g H_2, (b) 0.547 g Mg, 1.64 g HCl.

12.105 The boiling points of N_2 and O_2 are 77 and 90 K, respectively. Cooling air to just below 90 K would condense O_2 then cooling to below 77 K would condense N_2.

12.111 Kr is more expensive to use because of its low abundance and Kr is more reactive.

12.112 (a) 6.89×10^3 Pa.

12.113 (a) 0.870 g/L.

12.115 4.05 g/L.

12.117 PF_3.

12.119 P_{He} = 0.218 atm, P_{Ar} = 0.0206 atm.

12.121 3.34 L O_2, (b) 15.9 L air.

12.123 $C_6H_{12}O_2$.

12.124 6×10^{24} atoms Ar.

CHAPTER THIRTEEN

13.11   Halogen molecules are nonpolar; therefore, only dispersion forces are involved. These forces are stronger in molecules with larger numbers of electrons. Fluorine is a gas at 25°C and astatine is a solid.

13.17   (a) Gas; (b) solid to liquid; (c) the enthalpy of vaporization; (d) the heat capacity of the solid.

13.19   (a) Aluminum, (b) NaCl.

13.30   (a), (b) Weak intermolecular forces; (c), (d) particles in liquid phase can move about each other but are held at a constant average distance; (e) liquid particles are close together, but gas particles are many diameters apart.

13.31   (a) Density of liquids is greater, (b) viscosity of liquids is greater, (c) intermolecular forces of liquids are greater, (d) compressibility of gases is greater, (e) gases fill their containers completely, Liquids fill their containers completely to the level they reach.

13.36   (a) A, (b) C, (c) F, (d) H.

13.37   (a) III, (b) I, 22°C; II, 83°C; III, 110°C; (c) I < II < III; (d) I, 20°C; II, 78°C; III, 105°C; (e) I, 11°C; II, 67°C; III, 95°C.

13.39   The water does not have the capacity to transfer heat to other objects at 25°C.

13.40   (a) It would take longer to hard-boil eggs in Denver. The water boils at a temperature <100°C because the atmospheric pressure in Denver is <1 atm. Therefore, the eggs require more time in the <100°C water to absorb the same amount of heat or cook to the same "hardness." (b) There would be no difference. Frying does not depend on atmospheric pressure.

13.47 (a) K, (b) M.

13.50 $: \overset{\delta+}{\underset{\cdot\cdot}{Br}} : \overset{\delta-}{\underset{\cdot\cdot}{Cl}} : ||| : \overset{\delta+}{\underset{\cdot\cdot}{Br}} : \overset{\delta-}{\underset{\cdot\cdot}{Cl}} :$

13.52 Q, hydrogen bonds; R, dispersion forces; S dipole forces.

13.54 (a) Ne, Ar, Kr; (b) C_5H_{12}, C_8H_{12}, $C_{10}H_{22}$.

13.55 Stronger London forces.

13.58 Structure would change because particles would be forced closer to each other.

13.60 Bonds in the crystal latice are broken; therefore, the ions in the liquid are free to migrate.

13.61 Covalent solids are composed of a network of atoms that are bonded by strong covalent bonds; thus, the bonds require a lot of energy to break them apart.

13.62 (a) Metallic, (b) ionic, (c) molecular, (d) network covalent.

13.63 The formation of vapor involves overcoming dispersion forces among carbon dioxide molecules and does not involve breaking the C—O double bonds.

13.64 Metals are the only class of solids containing delocalized outer electrons which can move from one atom to another, allowing the conduction of electricity.

13.65 Metal atoms can move around each other because bonding electrons are not localized.

13.66 (a) (1) liquid, (2) gas, (3) solid; (b) FP = −70°C, BP = 70°C; (c) UV kinetic energy decreases, VW potential energy decreases, WX kinetic energy decreases, XY potential energy decreases, YZ kinetic energy decreases.

13.70 7.08 kJ.

13.72 94.9 kJ.

13.73 20.2 kJ for Al, 26 kJ for NaCl.

13.74 Water would be a nonpolar liquid if it were linear. This would cause the boiling point, melting point, and heats of fusion and vaporization to be significantly lower.

13.75 The animals would die.

13.81 Expensive; much of the ice would melt in transit.

13.86 High pressure is required.

13.88 Charcoal adsorbs many of the impurities in vodka.

13.91 Possibly, a large amount of the excess CO_2 dissolves in the oceans.

13.93 The large size of the chlorine atom does not allow the localization of negative charge required for hydrogen bonding.

13.95 13 g $H_2O(g)$.

13.99 (a) $3Fe + 4H_2O \longrightarrow Fe_3O_4 + 4H_2$, (b) 232 L H_2.

13.100 Heat is removed from the body to evaporate the ethyl chloride.

13.102 The HF molecules hydrogen bond to each other. This results in particles composed of more than one HF molecule with a higher formula mass than HF.

13.104 0.0426 atm.

CHAPTER FOURTEEN

14.5 (a) $2Cs + H_2 \longrightarrow 2CsH$.
(b) $6Cs + N_2 \longrightarrow 2Cs_3N$.
(c) $4Cs + O_2 \longrightarrow 2Cs_2O$.

14.9 (a) $3Ca + N_2 \longrightarrow Ca_3N_2$.
(b) $Ca + 2H_2O \longrightarrow Ca(OH)_2 + H_2$.
(c) $2Ca + O_2 \longrightarrow 2CaO$.

14.17 (a) $H_2 + F_2 \longrightarrow 2HF$.
(b) $Ba + I_2 \longrightarrow BaI_2$.
(c) $Br_2 + CH_4 \overset{\Delta}{\longrightarrow} CH_3Br + HBr$.

14.27   (a) $1s^2 2s^2 2p^6 3s^1$. (d) $1s^2 2s^2 2p^6 3s^2 3p^6 4s^2 3d^{10} 4p^6 5s^2 4d^{10} 5p^6 6s^2 4f^{14} 5d^{10} 6p^6 7s^1$.

14.28   (a) Rb^+ $[\,\colon\!\overset{\cdots}{\underset{\cdots}{Cl}}\colon\,]^-$

      (c) $2K^+$ $[\,\colon\!\overset{\cdots}{\underset{\cdots}{O}}\colon\,]^{2-}$

      (e) $2Li^+$ $\left[\begin{array}{c} \colon\overset{\cdots}{O}\colon \\ | \\ \colon\overset{\cdots}{O}\!-\!S\!-\!\overset{\cdots}{O}\colon \\ | \\ \colon\overset{\cdots}{O}\colon \end{array}\right]^{2-}$

14.33   (a) 211 g, (b) 8.4×10^2 g, (c) 3.0×10^{23} atoms Li, (d) 51.1 g.

14.34 3.49×10^5 J.

14.35   (a) RbOH, (b) Cs_2SO_4, (c) K_3N.

14.37   (a) $2K + Br_2 \longrightarrow 2KBr$, (b) combination, (c) 13.77 g KBr, (d) 550 g KBr.

14.39   (a) 127 g Na, (b) 0.0448 L.

14.40   (a) 42.1% Na, (b) 2.38×10^3 g K_3PO_4.

14.43 2.10×10^3 g $CsHC_4H_4O_6$.

14.45   (a) 26.3% Mg, (b) 95.0 g asbestos

14.48   (a) Mg^{2+} $2[\,\colon\!\overset{\cdots}{Br}\colon\,]^-$

      (b) Sr^{2+} $[\,\colon\!\overset{\cdots}{\underset{\cdots}{S}}\colon\,]^{2-}$

      (e) Ba^{2+} $2[\,\colon\!\overset{\cdots}{O}\!-\!H\,]^-$

14.49   (a) $3Ca + N_2 \longrightarrow Ca_3N_2$, (b) $Mg + F_2 \longrightarrow MgF_2$.

14.53   (a) $BaO_2 + 2HCl \longrightarrow H_2O_2 + BaCl_2$.

      (b) 11.4 g H_2O_2.

      (c) 4.30 g H_2O_2.

14.55   (a) $CaO + H_2O \longrightarrow Ca(OH)_2$, (b) 9.6×10^2 kJ.

14.57   (a) 0.1250 g HCl, (b) 0.04339 g HCl.

14.58   (a) Al^{3+} $3[\,\cdot\overset{\cdots}{\underset{\cdots}{Cl}}\colon\,]^-$, aluminum chloride.

      (c) $2Al^{3+}$ $3[\,\colon\!\overset{\cdots}{O}\colon\,]^{2-}$, aluminum oxide.

      (e) Al^{3+} $\left[\begin{array}{c} \colon\overset{\cdots}{O}\colon \\ | \\ \colon\overset{\cdots}{O}\!-\!Cl\!-\!\overset{\cdots}{O}\colon \\ | \\ \colon\overset{\cdots}{O}\colon \end{array}\right]^-$

14.59 $Al(OH)_3 + HCl \longrightarrow Al(OH)_2Cl + H_2O$. $Al(OH)_3 + 2HCl \longrightarrow Al(OH)Cl_2 + 2H_2O$.

14.61   (a) $2Al(s) + Fe_2O_3(s) \longrightarrow 2Fe(l) + Al_2O_3(l)$, (b) 147 kJ, (c) 2.69×10^5 g.

14.64 $Al_2O_3 + 6HF \longrightarrow 2AlF_3 + 3H_2O$.

14.66   (a) $\begin{array}{c} \colon\overset{\cdots}{F}\colon \\ | \\ \colon\overset{\cdots}{F}\!-\!C\!-\!\overset{\cdots}{F}\colon \\ | \\ \colon\overset{\cdots}{F}\colon \end{array}$ (b) Na^+ $[\,\colon\!\overset{\cdots}{\underset{\cdots}{F}}\colon\,]^-$ (d) Ca^{2+} $2[\,\colon\!\overset{\cdots}{\underset{\cdots}{F}}\colon\,]^-$

14.68   (a) $4F_2 + S_8 \longrightarrow 4S_2F_2$, (c) $H_2 + Cl_2 \longrightarrow 2HCl$, (d) $Cl_2 + MgI_2 \longrightarrow MgCl_2 + I_2$.

14.69   (a) 9.21 L.

14.70   (a) $Cl_2 + SO_2 \longrightarrow SO_2Cl_2$, (b) 313 g SO_2Cl_2, (c) 6.22 g SO_2Cl_2.

14.73   (a) $8Ca + S_8 \longrightarrow 8CaS$, (b) $8H_2 + S_8 \longrightarrow 8H_2S$, (c) $4C + S_8 \longrightarrow 4CS_2$.

14.74   (a) 91.1 kJ, (b) 8.41×10^3 g ZnS, (c) 3.50 g ZnS.

14.76   (b) FeS_2, 53.4% S.

14.79 1.31×10^5 g P_4.

14.81   (a) Phosphorus pentabromide, (b) phosphoric acid, (c) potassium dihydrogenphosphate.

14.82   (a) $Ca_3P_2 + 6H_2O \longrightarrow 2PH_3 + 3Ca(OH)_2$, (b) 0.218 L PH_3.

14.84 BrF_3.

14.85 (a) +1, (b) +7.

14.87 0.7751 g $MgCO_3$.

14.89 2650 L air.

14.90 (a) 9.70% Al.

14.92 1.51×10^{23} atoms Na.

CHAPTER FIFTEEN

15.10 Add 58.8 g acetone to 427 g H_2O.

15.11 14.5 g $CaCl_2$.

15.12 3.41 g NaOH.

15.13 1.35×10^3 g $Mg(NO_3)_2$.

15.14 0.2408 M.

15.15 23.5 M.

15.16 348 mL.

15.17 0.867 M.

15.20 $H^+(aq) + NO_3^-(aq) + K^+(aq) + OH^-(aq) \longrightarrow H_2O(l) + K^+(aq) + NO_3^-(aq)$.
$H^+(aq) + OH^-(aq) \longrightarrow H_2O(l)$.

15.21 $2Na^+(aq) + CO_3^{2-}(aq) + Cu^{2+}(aq) + 2Cl^-(aq) \longrightarrow 2Na^+(aq) + 2Cl^-(aq) + CuCO_3(s)$.
$CO_3^{2-}(aq) + Cu^{2+}(aq) \longrightarrow CuCO_3(s)$.

15.22 14.8 g HgS.

15.23 0.651 L H_2.

15.24 1.77 g CuS.

15.27 $T_b = 101.3°C$ $T_f = -4.6°C$.

15.28 5.81 m.

15.29 100.563°C.

15.31

Solute	Solvent
(a) NaCl	water
(b) Water	alcohol
(c) Either could be solute or solvent	
(d) O_2	water

15.36 (a) Liquid–liquid, (b) solid–solid, (c) gas–gas.

15.37 (a) H_2O, (b) CH_3OH, (c) CH_3OCH_3 NaCl is more soluble in polar solvents.

15.39 11.8 g $NaNO_2$ in 387 g H_2O.

15.40 (a) 12 g HCl solution, (c) 508 g HCl solution

15.42 240. g solution.

15.44 2.25% m/m.

15.46 (a) 1.72 g $Mg(NO_3)_2$ diluted to 50.0 mL, (b) 0.6863 g NH_3 diluted to 210.0 mL.

15.47 (a) 8.006 M $(NH_4)_2SO_4$, (d) 5.720×10^{-4} M HBr.

15.48 (a) 1.42 M C_2H_6O, (d) 0.316 M NH_4Cl.

15.49 (a) 16 M NHO_3.

15.50 (a) 1.77 L.

15.51 (a) 0.255 mol particles.

15.53 3.3 M.

15.55 308.6 mL.

15.57 0.253 M SO_4^{2-}, 0.505 M K^+.

15.61 (a) $NaCl(s) \longrightarrow Na^+(aq) + Br^-(aq)$.
(b) $Na_2SO_4(s) \longrightarrow 2Na^+(aq) + SO_4^{2+}(aq)$.
(c) $Cu(NO_3)_2(s) \longrightarrow Cu^{2+}(aq) + 2NO_3^-(aq)$.

15.62 (a) $Rb^+(aq) + Cl^-(aq) + Ag^+(aq) + NO_3^-(aq) \longrightarrow Rb^+(aq) + NO_3^-(aq) + AgCl(s)$
$Ag^+(aq) + Cl^-(aq) \longrightarrow AgCl(s)$.

(d) $H_3PO_4(aq) + 2Na^+(aq) + 2OH^-(aq) \longrightarrow 2Na^+(aq) + HPO_4{}^{2-}(aq) + 2H_2O(l)$
$H^+(aq) + OH^-(aq) \longrightarrow H_2O(l)$.

15.63   (a) $3Fe^{3+}(aq) + 2PO_4{}^{3-}(aq) \longrightarrow Fe(PO_4)_2(s)$, (b) NR, (d) $Ag^+(aq) + I^-(aq) \longrightarrow AgI(s)$.

15.65   (a) 84.6 mL, (b) 2.58 g $MgCO_3$.

15.67   (a) 329 mL, (b) 1.05 g $AlCl_3$.

15.69   (a) $H_2C_2O_4(s) + 2NaOH(aq) \longrightarrow Na_2C_2O_4(aq) + 2H_2O(l)$, (b) 6.78 g $Na_2C_2O_4$, (c) 478 g $Na_2C_2O_4$.

15.71   (a) $2HCl(aq) + Na_2CO_3 \longrightarrow 2NaCl(aq) + CO_2(g) + H_2O(l)$, (b) 3.16 g Na_2CO_3, (c) 27.2 L,
(d) 0.241 L, (e) 3.45 g Na_2CO_3.

15.72   (a) 0.00496 m, (c) 1.43 m.

15.73   (a) 6.23 m.

15.74   (a) 0.447 g $Zn(NO_3)_2$ is dissolved in 305 g H_2O.

15.76   (a) 100.47°C, (b) 103.1°C.

15.77   (a) −1.7°C, (b) −11°C.

15.79   (a) 1.5°C, (b) 9.0°C, (c) 7.2°C.

15.81   (a) 85°C.

15.82   −4.2°C.

15.83   8.67 g $C_2H_6O_2$.

15.86   10°C.

15.87   (a) 70.3%, (b) 501 mL, (c) 0.0896 M.

15.90 $T_f = -1.40°C$, $T_b = 100.384°C$.

15.91 1.17×10^4 g $C_2H_6O_2$.

15.93   (a) 0.903 M $NO_3{}^-$, (b) 0.364 M $NO_3{}^-$, (c) 1.41 M $NO_3{}^-$.

15.94 1.3×10^2 g/mol.

15.96   (a) $(NH_4)_2S(aq) + Ni(ClO_3)_2(aq) \longrightarrow NiS(s) + 2NH_4ClO_3(aq)$, (b) 0.541 g NiS, (c) 52.2 mL.

15.97   (a) $Na_2SO_3(s) + 2HCl(aq) \longrightarrow 2NaCl(aq) + SO_2(g) + H_2O(l)$, (b) 0.613 mL, (c) 40.0 L.

CHAPTER SIXTEEN

16.6   (a) Increases, (b) increases, (c) increases.

16.9   (a) Step 1, (b) O.

16.13   (a) $K = \dfrac{[SO_3][NO]}{[NO_2][SO_2]}$ (b) $K = \dfrac{[Cl_2]^2[H_2O]^2}{[HCl]^4[O_2]}$

16.16   (b), (d) Products; (a), (c) reactants.

16.17   (a), (e) Products; (b) (c) reactants; (d) no change.

16.21   High rate of disappearance of reactants.

16.24   (a) No, (b) yes, (c) no.

16.25   (a) Decrease, (b) increase, (c) increase.

16.26   (a) High, (b) low, (c) low, (d) high.

16.28   (b), (d) Increase; (a), (c), (e) decrease.

16.32   The energy released by an exothermic reaction is the energy supplied to continue the reaction.

16.34   Increase in the pressure, temperature, and concentration.

16.36   (a) $H_2 + ICl \longrightarrow \cancel{HI} + HCl$
       $\underline{+\cancel{HI} + ICl \longrightarrow HCl + I_2}$
       $H_2 + 2ICl \longrightarrow 2HCl + I_2$
(b) HI, (c) (1).

16.38   (a) At equilibrium the rate of the forward reaction equals the rate of the reverse reaction.
(b) . . . are *stable* and *do not* undergo . . . , (c) . . . reactions *continue* at the same rate.

16.40   Double arrows indicate that both reactions are occurring at equal rates.

16.42   (a) $K = \dfrac{[NO_2][SO_2]}{[NO][SO_3]}$ (d) $K = \dfrac{[NO]^4[H_2O]^6}{[NH_3]^4[O_2]^5}$

16.43   If $K > 1$, then the products are usually in greater concentration at equilibrium. If $K < 1$, then the reactants are usually in greater concentration at equilibrium.

16.45 Decreases the reverse reaction rate, while the forward reaction rate remains constant.

16.46 (*a*) Reactants, (*d*) products.

16.48 (*a*) Products, (*b*) reactants.

16.50 (*a*) Products, (*b*) reactants, (*c*) products, (*d*) products, (*e*) no change.

CHAPTER SEVENTEEN

17.4 Acid, HCN; conjugate base, CN^- Base, NH_3; conjugate acid, NH_4^+.

17.7 F^- is a stronger base than Cl^-.

17.8 (*a*) HNO_2, (*b*) NH_4^+, (*c*) H_3O^+, (*d*) H_2CO_3.

17.9 (*a*) $H^+(aq) + ClO_4^-(aq) + Rb^+(aq) + OH^-(aq) \longrightarrow H_2O(l) + Rb^+(aq) + ClO_4^-(aq)$
$H^+(aq) + OH^-(aq) \longrightarrow H_2O(l)$.

    (*b*) $Na^+(aq) + OH^-(aq) + H^+(aq) + NO_2^+(aq) \longrightarrow H_2O(l) + Na^+(aq) + NO_2^-(aq)$
$H^+(aq) + OH^-(aq) \longrightarrow H_2O(l)$.

    (*c*) $H^+(aq) + I^-(aq) + NH_3(aq) \longrightarrow NH_4^+(aq) + I^-(aq)$
$H^+(aq) + NH_3(aq) \longrightarrow NH_4^+(aq)$.

17.11 (*a*) 1.00×10^{-15} M, (*b*) 6.25×10^{-11} M, (*c*) 1.78×10^{-12} M.

17.12 0.0340 M.

17.14 (*a*) 5.000, (*b*) 1.000, (*c*) 1.199.

17.15 11.30.

17.16 (*a*) Yellow, (*b*) yellow, (*c*) colorless.

17.18 (*a*) 659 mL, (*b*) 1.38×10^3 mL, (*c*) 4.27×10^3 mL.

17.24 Acids: (a); bases: (b), (f).

17.25 (*a*) $HClO_4 \xrightarrow{H_2O} H^+(aq) + ClO_4^-(aq)$,

    (*b*) $Al(OH)_3 \xrightarrow{H_2O} Al^{3+}(aq) + 3OH^-(aq)$,

    (*f*) $NH_3 + H_2O \longrightarrow NH_4 + (aq) + OH^-(aq)$.

17.27

Acid	Base	Conjugate acid	Conjugate base
(*a*) H_2O	CO_2	H_2CO_3	H_2CO_3
(*c*) HCN	KOH	H_2O	KCN

17.28 (*a*) $H_2S(aq) + 2NaOH(aq) \longrightarrow Na_2S(aq) + 2H_2O(l)$, (*d*) $2NH_3(aq) + H_2SO_4(aq) \longrightarrow$ $(NH_4)_2SO_4(aq)$.

17.30 (*a*) $H_2PO_4^-(aq) + H^+(aq) \longrightarrow H_3PO_4(aq)$.
$H_2PO_4^-(aq) + OH^-(aq) \longrightarrow HPO_4^{2-}(aq) + H_2O(l)$.

    (*b*) $NH_3(aq) + H^+(aq) \longrightarrow NH_4^+(aq)$.
$NH_3(aq) \longrightarrow NH_2^-(aq) + H^+(aq)$.

17.31 (*a*) NO_3^-, (*c*) O^{2-}, (*e*) NH_2^-.

17.32 (*a*) HCN, (*d*) H_3O^+.

17.34 (*a*) H_2SO_4, (*b*) NH_2^-.

17.35 (*a*) HBr, (*b*) H_2O.

17.36 (*a*) $HCl + RbOH \longrightarrow H_2O + RbCl$, (*b*) $H_2SO_4 + Ca(OH)_2 \longrightarrow 2H_2O + CaSO_4$.

17.37 (*a*) $H^+(aq) + Cl^-(aq) + Rb^+(aq) + OH^-(aq) \longrightarrow H_2O(l) + Cl^-(aq) + Rb^+(aq)$.
$H^+(aq) + OH^-(aq) \longrightarrow H_2O(l)$.

    (*b*) $2H^+(aq) + SO_4^{2-}(aq) + Ca^{2+}(aq) + 2OH^-(aq) \longrightarrow 2H_2O(l) + SO_4^{2-}(aq) + Ca^{2+}(aq)$.
$H^+(aq) + OH^-(aq) \longrightarrow H_2O(l)$.

17.39 (*a*) $OH^-(aq) + HF(aq) \longrightarrow H_2O(l) + F^-(aq)$.

17.40 (*a*) Basic, (*c*) acidic.

17.41 (*a*) 1.0×10^{-10} M, (*b*) 1.3×10^{-6} M.

17.42 (*a*) 1.0×10^{-10} M, (*b*) 1.0×10^{-9} M, (*c*) 1.1×10^{-2} M.

17.43 (*a*) 8.49×10^{-4} M H^+, 1.18×10^{-11} M OH^-.

17.44 (*a*) 5.94×10^{-2} M OH^-, 1.68×10^{-13} M H^+, (*c*) 1.71×10^{-6} M OH^-, 5.86×10^{-9} M H^+.

17.45 (*a*) 3.00.

17.46 (*a*) 12.00, (*b*) 2.00.

17.47 (a) 1.699, (e) 9.992.

17.49 (a) 579 mL.

17.50 0.150 M KOH.

17.52 (a) 9392 mL KOH.

17.54 $[OH^-]$ = 0.0119 M, $[H^+]$ = 8.37 × 10^{-13} M, $[K^+]$ = 0.102 M, $[ClO_4^-]$ = 0.0896 M, pH = 12.077.

17.56 (a) $N_2(g) + 3H_2(g) \longrightarrow 2NH_3(g)$; (b) 1.915 × 10^6 g NH_3; (c) 54.6 L; (d) 1.23 × 10^3 g H_2, 5.67 × 10^3 g N_2.

17.58 (a) 0.295 L CO_2, (b) 0.06489 M Na^+, (c) 940.0 mL.

17.60 $[H^+]$ = 8.8 × 10^{-3} M, pH = 2.05, (b) 12 mL 0.350 M NaOH.

17.62 (a) 1.3 × 10^{-3} M; (b) $[H^+]$ = 0.10 M, pH = 1.00.

CHAPTER EIGHTEEN

18.3

	Oxidized/reducing agent	Reduced/oxidizing agent
(a)	H_2	I_2
(b)	CO	O_2
(c)	Mg	N_2
(d)	PCl_3	Cl_2
(e)	Fe^{2+}	$Cr_2O_7^{2-}$

18.4 (a) $PH_3 + 2I_2 + 2H_2O \longrightarrow H_3PO_2 + 4I^- + 4H^+$, (b) $4Zn + NO_3^- + 10H^+ \longrightarrow 4Zn^{2+} + NH_4^+ + 3H_2O$, (c) $4Zn + NO_3^- + 6H_2O + 7OH^- \longrightarrow 4Zn(OH)_4^{2-} + NH_3$.

18.10 Cathode: $Sn^{2+} + 2e^- \longrightarrow Sn$; Anode: $Ni \longrightarrow Ni^{2+} + 2e^-$.

18.12 Build up of $NH_3(g)$ would cause the cell to explode.

18.15 Electrons are transferred from the reducing agent to the oxidizing agent.

18.16 Oxidized: (a) Cu, (b) Fe, (c) Cu, (g) Br_2.
Reduced: (a) Ag^+, (b) H^+, (c) NO_3^-, (g) Br_2.

18.17 Oxidizing agents: (a) F_2, (c) MnO_4^-.
Reducing agents: (a) Al, (c) Cl^-.

18.18 (a) $14H^+ + Cr_2O_7^{2-} + 6Br^- \longrightarrow 2Cr^{3+} + 3Br_2 + 7H_2O$, (b) $2H^+ + H_2O_2 + 2I^- \longrightarrow I_2 + 2H_2O$, (e) $6H_2O + 4AsH_3 + 24Ag^+ \longrightarrow 24Ag + As_4O_6 + 24H^+$.

18.19 (a) $H_2O + ClO^- + 2I^- \longrightarrow Cl^- + I_2 + 2OH^-$, (c) $2OH^- + 2Al + 6H_2O \longrightarrow 2Al(OH)_4^- + 3H_2$, (e) $H_2O + 6Fe_3O_4 + 2MnO_4^- \longrightarrow 9Fe_2O_3 + 2MnO_2 + 2OH^-$.

18.20 (a) $10OH^- + 2CrI_3 + 27H_2O \longrightarrow 2CrO_4^{2-} + 6IO_4^- + 32H_2O$, (c) $2OH^- + 2HXeO_4^- \longrightarrow XeO_6^{4-} + Xe + O_2 + 2H_2O$, (e) $2H^+ + 3CN^- + 2MnO_4^- \longrightarrow 3CNO^- + 2MnO_2 + H_2O$.

18.24 (a) Ca, cathode; F_2, anode; (b) $Ca^{2+} + 2e^- \longrightarrow Ca$ $2F^- \longrightarrow F_2 + 2e^-$.

18.26 (a) $2F^- \longrightarrow F_2 + 2e^-$ or $2H_2O \longrightarrow 4e^- + O_2 + 4H^+$, (b) $K^+ + e^- \longrightarrow K$ or $2H_2O + 2e^- \longrightarrow H_2 + 2OH^-$.

18.34 Anode: $PbSO_4 + 2H_2O \longrightarrow 2e^- + PbO_2 + 4H^+ + SO_4^{2-}$.
Cathode: $PbSO_4 + 2e^- \longrightarrow Pb^{2+} + SO_4^{2-}$.

CHAPTER NINETEEN

19.3 (a) 104, (b) 45, (c) 58.

19.4 2, 8, 20, 50, 82, 126; (b) no.

19.5 (a) ^4_2He, (b) $^{12}_6C$, (c) $^{64}_{30}Zn$.

19.6 (a) γ, (b) β, (c) α, (d) γ, (e) α.

19.8 6.25 g.

19.10 (a) $^{95}_{40}Zr \longrightarrow ^{95}_{41}Nb + ^0_{-1}\beta + \bar{\nu}$, (b) $^{174}_{72}Hf \longrightarrow ^{170}_{70}Yb + ^4_2\alpha$, (c) $^{187}_{75}Re \longrightarrow ^{187}_{76}Os + ^0_{-1}\beta + \bar{\nu}$, (d) $^{89}_{38}Sr^* \longrightarrow ^{89}_{38}Sr + \gamma$.

19.13 (a) $^{40}_{18}Ar(^4_2\alpha, ^1_1p)^{43}_{19}K$, (b) $^{59}_{27}Co(^1_0n, ^4_2\alpha)^{56}_{25}Mn$, (c) $^{12}_6C(^1_1H, ^1_0n)^{13}_7N$.

19.14 (a) $^{239}_{94}Pu + ^1_0n \longrightarrow ^{240}_{95}Am + ^0_{-1}\beta$, (b) $^{238}_{92}U + 15^1_0n \longrightarrow ^{253}_{99}Es + 7^0_{-1}\beta$, (c) $^{246}_{96}Cm + ^{12}_6C \longrightarrow ^{254}_{102}No + 4^1_0n$, (d) $^{249}_{98}Cf + ^{18}_8O \longrightarrow ^{263}_{106}Unh + 4^1_0n$.

19.23 3×10^{10} L.

19.24   (a) 81 p^+, 120 n^0, (b) 63 p^+, 88 n^0.

19.25   (a) $^{138}_{56}$Ba, (b) $^{174}_{70}$Yb.

19.26   (a) $^{4}_{2}$He, (c) $^{40}_{20}$Ca, (e) $^{133}_{55}$Cs.

19.27   (a) $^{184}_{74}$W, most stable; $^{176}_{73}$Ta, least stable.

19.28   Larger number of neutrons in the nucleus help to diminish the electrostatic repulsions of the protons.

19.29   (a) Beta particles have less charge and mass. (b) Alpha particles have a greater charge and mass.

19.31   (a) $^{89}_{38}$Sr, (b) 1.6 g, (c) 6.

19.33   (a) 0.4 g, (b) 7.

19.35   Alpha decay of U.

19.36   (a) $^{147}_{62}$Sm $\longrightarrow$ $^{143}_{60}$Nd + $^{4}_{2}\alpha$, (d) $^{210}_{84}$Po $\longrightarrow$ $^{206}_{82}$Pb + $^{4}_{2}\alpha$.

19.37   (a) $^{3}_{1}$H $\longrightarrow$ $^{3}_{2}$He + $^{0}_{-1}\beta$ + $\bar{\nu}$, (d) $^{63}_{28}$Ni $\longrightarrow$ $^{63}_{29}$Cu + $^{0}_{-1}\beta$ + $\bar{\nu}$.

19.40   (1) $^{232}_{90}$Th $\longrightarrow$ $^{228}_{88}$Ra + $^{4}_{2}\alpha$, (2) $^{228}_{88}$Ra $\longrightarrow$ $^{228}_{89}$Ac + $^{0}_{-1}\beta$ + $\bar{\nu}$, (3) $^{228}_{89}$Ac $\longrightarrow$ $^{228}_{90}$Th + $^{0}_{-1}\beta$ + $\bar{\nu}$.

19.42   (a) $^{0}_{-1}\beta$, (b) $^{247}_{98}$Cf, (c) $^{129}_{52}$Te.

19.44   (a) $^{238}_{92}$U($^{16}_{6}$O, 5$^{1}_{0}$n)$^{249}_{100}$Fm, (b) $^{241}_{95}$Am($^{4}_{2}\alpha$, 2$^{1}_{0}$n)$^{243}_{97}$Bk, (c) $^{238}_{92}$U($^{22}_{10}$Ne, 4$^{1}_{0}$n)$^{256}_{102}$No, (d) $^{242}_{94}$Pu($^{22}_{10}$Ne, 4$^{1}_{0}$)$^{260}_{104}$Unq.

19.53   11,400 yr.

19.66   (a) 50 p^+, 71 n^0; (b) $^{121}_{50}$Sn $\longrightarrow$ $^{121}_{51}$Sb + $^{0}_{-1}\beta$ + $\bar{\nu}$; (c) 82.5 hr.

19.67 $^{96}_{42}$Mo + $^{2}_{1}$H $\longrightarrow$ $^{97}_{43}$Tc + $^{1}_{0}$n.

19.69   (a) $^{238}_{92}$U.

CHAPTER TWENTY

20.3   (a) CH_4, (b) C_3H_8, (c) C_5H_{12}, (d) C_6H_{14}, (e) $C_{10}H_{22}$.

20.4   (a) $CH_3CH_2CH_3$, (b) $CH_3CH_2CH_2CH_3$, (c) $CH_3CH_2CH_2CH_2CH(CH_3)CH_3$.

20.5

Hexane   2-Methylpentane   3-Methylpentane

2,3-Dimethylbutane   2,2-Dimethylbutane

20.6   (a)   (b) Propane, $CH_3CH_2CH_3$, is composed of a chain.

20.8   (a) $HC{\equiv}CH$ (b)   (c) $CH{=}CH{-}CH{=}CH_2$ (d)

20.10

20.11

o-Xylene   m-Xylene   p-Xylene

20.12 A polycyclic aromatic contains two or more fused benzene rings; e.g., Naphthalene.

20.13 (a) Ether, (b) amine, (c) nitrile, (d) aldehyde, (e) carboxylic acid.

20.14 (a) RCOR, (b) RCOOR, (c) $RCONH_2$, (d) RX, (e) ROH.

20.15 (a) CH_3OH, (b) CH_3NH_2, (c) CH_3OCH_3, (d) HCOOH, (e) CH_3COCH_3.

20.16 (a) Chlordane, (b) halothane, (c) phenol, (d) formalin, (e) benzaldehyde.

20.17 (a) Pyridine, (b) DDT, (c) ethyl propanoate, (d) $CH_3CONHCH_3$, (e) cyclohexanone.

20.20 (a) Amino acids, (b) monosaccharides, (c) nucleotides.

20.21 (a) Alcohols, aldehydes, ketones; (b) amides, acids, amines; (c) acids, esters.

20.22 (a) Sucrose, (b) stearic acid, (c) glycine, (d) cholesterol, (e) DNA, (f) starch.

20.23 A simple protein is composed only of bonded amino acids. A conjugated protein contains some other group besides the protein chain.

20.24 (a) Glycerol + fatty acids, (b) two monosaccharides, (c) pentose + phosphate + heterocyclic amine, (d) amino acids, (e) acid + alcohols.

20.28 The functional group or lack thereof

20.29 (a) 7, (b) 4, (c) 10, (d) 2, (e) 8.

20.30 (a)

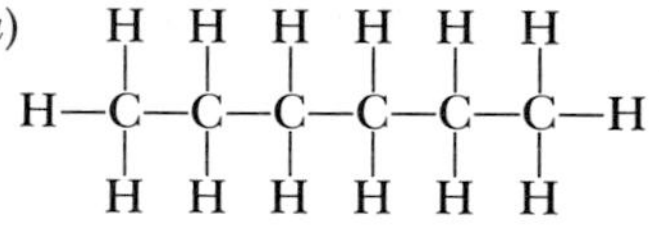

H H H H H H
H—C—C—C—C—C—C—H
H H H H H H

(c)

H H H H H H H
H—C—C—C—C—C—C—C—H
H H | H H H H
H—C—H
H

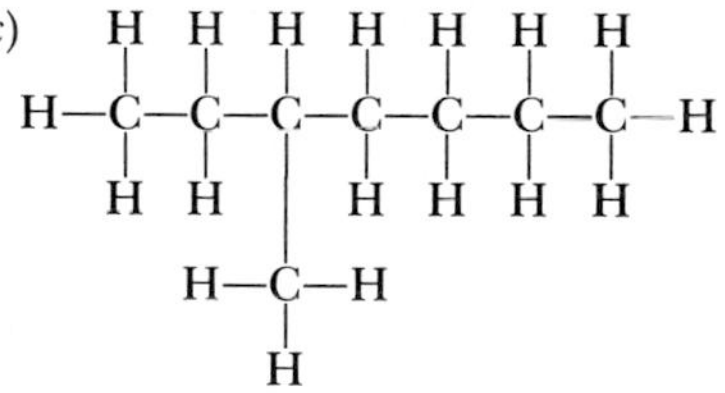

20.31 (a) Hexane, (c) octane.

20.32 (a) 2,2,4-Trimethylpentane, (c) 3-Ethyloctane, (e) 3,5-Dimethyl-5-ethyldecane.

20.33 (a) $CH_3CH_2CHCH_2CH_2CH_2CH_3$
          $|$
          CH_3

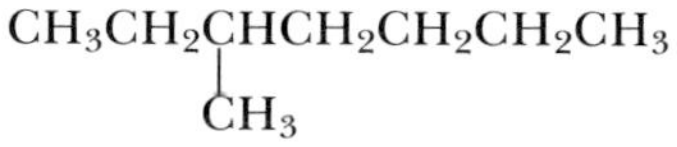

(c) CH_3
    $|$
    $CH_3CH_2CHCHCH_2CH_2CH_2CH_2CH_2CH_3$
          $|$
          CH_2CH_3

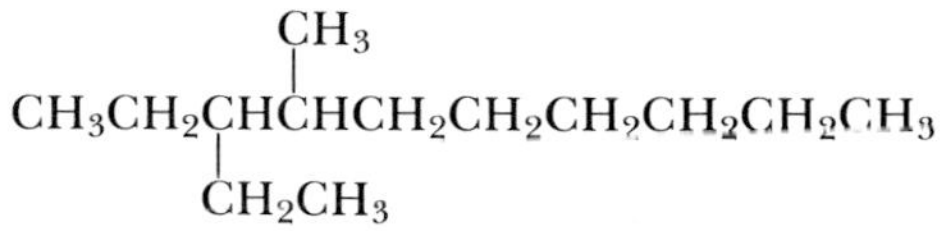

(d) CH_3 CH_3
    $|$ $|$
    $CH_3CCH_2CHCH_2CH_2CH_2CH_3$
    $|$
    CH_3

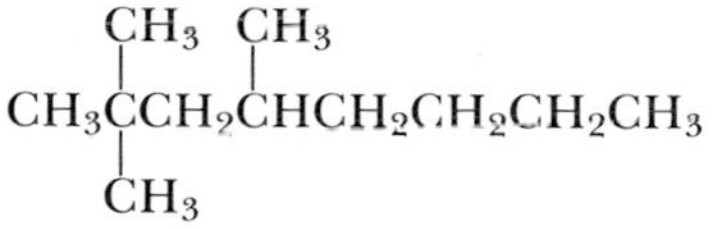

(g) 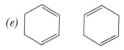 —CH_2CH_3

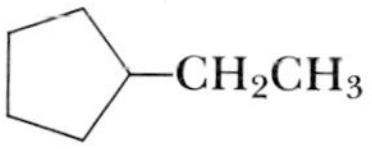

20.38 (a) $CH_3CH_2CH{=}CH_2$, (d) $CH_2{=}CHCH{=}CH_2$, $CH_3CH{=}C{=}CH_2$.

(e)

20.39 (a) 12, (c) 12, (d) 6.

20.40 $CH_2{=}CHCH_2CH_2CH_3$ $CH_3CH{=}CHCH_2CH_3$ CH_3
                                                            $|$
                                                         $CH_2{=}CCH_2CH_3$

CH_3
$|$
$CH_3C{=}CHCH_3$ CH_3
                   $|$
                   $CH_3CHCH{=}CH_2$

CH_3

CH₃   CH₃   —CH_2CH_3

CH_3 CH_3

CH_3

CH_3

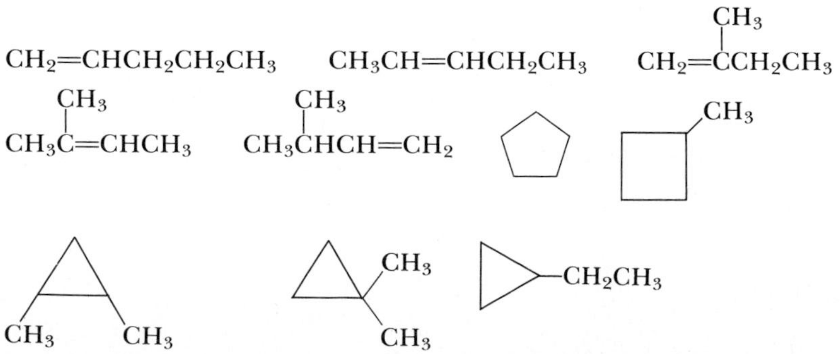

20.42

CH₂CH₃ / CH₂CH₃ structures

1,2-Diethylbenzene 1,3-Diethylbenzene 1,4-Diethylbenzene

20.48 (a) Alcohol, ester, halide, acid; (b) Halide, amine, aldehyde, ether; (c) Ether, ketone, amide, amine, double bond.

20.50 (a) Ethanol, (c) methanoic cid.

20.53 The carbonyl group is at the end of the chain in an aldehyde, e.g., RCHO. The carbononyl group is bonded to a nonterminal carbon in a ketone, e.g., RCOR.

20.61 $CH_4 + Cl_2 \xrightarrow{light} CH_3Cl + HCl$.

20.68 (a) Ribose, (d) glucose.

20.72 Amides.

20.88 10.7 g C_8H_{18}.

20.91 (a) $CH_3CH_2CH_2OH \xrightarrow{(O)} CH_3CH_2COOH$,

(b) $CH_3CH_2CH_2CH_2CH_2CH_2OH \xrightarrow{(O)} CH_3CH_2CH_2CH_2CH_2COOH$,

(c) $CH_3CHOHCH_2CH_3 \xrightarrow{(O)} CH_3COCH_2CH_3$.

20.93 Reduction: $CH_3CH_2CHO + H_2 \xrightarrow{cat} CH_3CH_2CH_2OH$.

Oxidation: $CH_3CH_2CHO \xrightarrow{(O)} CH_3CH_2COOH$.

20.94 (b) $CH_3CH_2CH_2COOH + CH_3CH_2OH \xrightarrow{H^+} CH_3CH_2CH_2COOCH_2CH_3 + H_2O$.

20.95 61.0 g $CH_3COOCH_2CH_3$.

20.97 20.6 L H_2.

20.98 (a) $CH_3CH_3 + Cl_2 \xrightarrow[light]{} CH_3CH_2Cl + HCl$,

(b) $CH_3CH_2CH{=}CHCH_3 + H_2 \xrightarrow[cat.]{} CH_3CH_2CH_2CH_2CH_3$,

(c) $CH_3CH_2CH_2CH_2CHO + H_2 \xrightarrow[K_2SO_4/H^+]{} CH_3CH_2CH_2CH_2CH_2OH$.

APPENDIX A

1. (a) 8, (b) 42, (c) −188, (d) 10.44.
2. (a) 9.081, (b) 0.50, (c) $[(n + 1)/m(a + b)] − 2$, (d) $b − [(a + 2)(n + 1)]$.
3. (a) 100, (b) 992, (c) 100, (d) 96.
4. (a) 102.4, (b) 83.72, (c) −0.50, (d) 36, (e) 257.
5. (a) $(2 + am)/an$, (b) $[(a + b)/(n + 10)] − m$, (c) $(3a + b − n)(5/4)$, (d) $8b/a$.
6. (a) −3.08, (b) 0.444, (c) 102.5, (d) 1.92.
7. (a) 4.9139×10^{-4}, (b) 1.2×10^{11}, (c) 1.3511×10^{-9}, (d) 1.05×10^{-20}.
8. (a) 0.00311, (b) 0.0290, (c) 777,420,000, (d) 8,001,000,000,000.
9. (a) 4.36×10^2, (b) 1.215×10^7, (c) 1.001×10^{-12}, (d) 9.79×10^{-2}, (e) 6.361×10^{-4}.
10. (a) 1.186×10^6, (b) 9.53×10^{-3}, (c) 1.8156×10^9, (d) 2.2054×10^{35}, (e) 3.9434×10^{20}.
11. (a) -2.79×10^4, (b) 9.199×10^{-13}, (c) -5.6645×10^{47}, (d) 6.1135×10^{-34}.
12. (a) 4.27×10^{11}, (b) 3.4763×10^{151}, (c) 4.4182×10^3.
13. (a) 4.4753×10^6, (b) 5.4×10^{113}, (c) 4.0164×10^{-8}.
14. (a) 9.0971×10^{-3}, (b) 0.10007.
15. (e) 3.
16. (a) 0.56; (b) 37°C, 24°C, −32°C, 57°C.

GLOSSARY

Absolute zero the lowest possible temperature, 0 K or $-273.15°C$

Accuracy how close a measured value is to the actual value

Acid a substance that donates H^+ to water (the Arrhenius definition); a proton donor (the Brønsted-Lowry definition)

Acid anhydride a substance that reacts with water to produce an acidic solution; a nonmetal oxide

Acid-base indicator a dye that changes color depending on the pH of a solution

Actinide series the 14 elements that come after actinium, Ac, on the periodic table; elements with atomic numbers 90–103

Activated complex a high-energy intermediate species produced in chemical reactions as a result of the collision of the reactant molecules

Activation energy the minimum energy needed to produce the transition state (activated complex); the minimum energy required for a reaction to occur

Actual yield the mass of product obtained in a chemical reaction

Aerobic able to take place in the presence of oxygen gas, O_2

Alcohol an organic compound that has an —OH group bonded to a hydrocarbon group

Aldehyde an organic compound that has a —CHO, a carbonyl group and hydrogen atom, bonded to a hydrocarbon chain or ring

Alkali metals group IA elements: Li, Na, K, Rb, Cs, and Fr

Alkaline earth metals group IIA elements: Be, Mg, Ca, Sr, Ba, and Ra

Alkane a hydrocarbon in which all carbon-carbon bonds are single bonds; a saturated hydrocarbon

Alkene an unsaturated hydrocarbon that contains at least one carbon-carbon double bond

Alkyl group a substituent group that results when one hydrogen is removed from a nonaromatic hydrocarbon, for example, methyl, CH_3—; ethyl, C_2H_5—; etc.

Alkyne an unsaturated hydrocarbon that contains at least one carbon–carbon triple bond

Allotropes different forms of the same element, e.g., graphite and diamond, two distinct forms of carbon

Alloy a solution of metals

Alpha particle a high-energy helium nucleus, He^{2+}, that is emitted by some heavy nuclides when undergoing radioactive decay

Amalgam the solution that results when a metal solute is dissolved in liquid mercury

Amide an organic compound that contains a nitrogen bonded to a carbonyl carbon, $-CONH_2$, $-CONHR$, or $-CONR_2$.

Amine an organic compound that contains one or more alkyl groups bonded to an N atom; RNH_2, R_2NH, or R_3N

Amino acid an organic compound that contains both an amino group and a carboxylic acid group; amino acids combine to produce proteins

Amorphous solid a solid whose structure lacks the long-range order of crystalline solids

637

Ampere (A) a unit of electric current; the amount of coulombs of charge per second

Amphiprotic term used to describe a substance that can both donate and accept protons

Amphoteric term used to describe a substance that can combine with either H^+ or OH^-, thus behaving as either an acid or base, depending on the conditions

Anaerobic able to take place without the presence of oxygen, O_2; the opposite of aerobic

Anion a negative ion; one with extra electrons

Anode the electrode where oxidation occurs in electrochemical cells

Antimatter a form of matter that has properties opposite to "regular" matter; if antimatter contacts regular matter, the two annihilate each other and are transformed totally into energy

Aqueous solution a solution in which water is the solvent

Aromatic hydrocarbon an organic compound that contains a benzene ring or has benzenelike properties

Atmosphere (atm) a unit of pressure equivalent to 101 kPa; the amount of pressure necessary to support a column of mercury 760 mmHg high

Atom a tiny, neutral particle composed of protons, neutrons, and electrons that is the smallest unit that retains the chemical properties of an element

Atomic mass the average mass of the naturally occurring isotopes of an element compared with ^{12}C; traditionally called *atomic weight*

Atomic mass unit (see unified atomic mass unit)

Atomic number the number of protons in the nucleus of an atom; the positive charge on the nucleus of an atom

Atomic size a measure of the relative size of an atom; the average distance from the nucleus to the outermost electron, normally measured as half the distance between two bonded nuclei

Atomic weight (see atomic mass)

Avogadro's number the number of particles in 1 mol, 6.022×10^{23}

Balance a laboratory instrument used to measure the mass of objects

Barometer a device used to measure atmospheric pressure

Base a substance that increases the OH^- concentration when dissolved in water (Arrhenius definition); a proton acceptor (Brønsted-Lowry definition)

Base units fundamental SI units of measurement from which all other SI units are derived

Beta particle a high-energy electron that is emitted by a nucleus undergoing one type of radioactive decay

Binary acid an acid that contains hydrogen bonded to a nonmetal, HX

Binary compound a compound that is composed of two different elements

Boiling point the temperature at which the vapor pressure of a liquid equals the applied pressure; when the boiling point is reached, bubbles of the liquid's vapor form throughout the liquid

Boiling-point elevation the increase in boiling temperature of a solvent after a nonvolatile solute is added

Bond the primary force of attraction that holds atoms together in molecules and lattice structures

Bonding electrons electrons that are attracted by two nuclei; shared electrons

Bond length the average distance between the nuclei of two atoms that are bonded

Calorie (cal) a unit of heat energy; 1 cal = 4.184 J

Calorimeter a laboratory instrument used to measure heat transfers in chemical and physical changes

Carbohydrate a class of biologically important compounds that includes sugars and starches

Carbonyl group a functional group in organic chemistry that consists of a carbon that is doubly bonded to an oxygen, C=O; aldehydes and ketones both contain carbonyl groups within their structures

Carboxyl group a functional group contained in organic acids; it consists of a carbon that has a doubly bonded oxygen and a hydroxy group attached, —COOH

Catalyst a substance that increases the rate of a chemical reaction by lowering its activation energy; generally, a catalyst is fully recovered after the reaction

Cathode the electrode where reduction occurs in electrochemical cells

Cation an ion with a positive charge; one that forms when an atom or group of atoms loses electrons

Celsius temperature scale a temperature scale that is displaced 273 degrees from the Kelvin temperature scale; water's boiling point is 100°C (373 K), and its freezing point is 0°C (273 K)

Centi a prefix attached to units that decrease their magnitude by $\frac{1}{100}$

Chalcogen a name applied to group VIA elements: O, S, Se, Te, and Po

Chemical bond the force of attraction between atoms in compounds

Chemical change a change in which the composition of a substance is altered; also called a chemical reaction

Chemical equation an expression of symbols, formulas, and coefficients that describes mass, volume, and mole relationships in specific chemical reactions; in chemical equations the reactants are written to the left of an arrow, and the products are written to the right of the arrow

Chemical equilibrium a closed chemical system in which the forward and reverse reaction rates are equal

Chemical family a group of chemical elements with similar properties listed in a vertical column on the periodic table

Chemical formula a combination of chemical symbols with appropriate subscripts that indicate the ratio of atoms in molecules and formula units

Chemical group a vertical column of elements in the periodic table; elements in a chemical group have similar outer electronic configurations

Chemical kinetics the study of the rates and mechanisms of chemical reactions

Chemical nomenclature a system of rules and guidelines for writing unique names for each element and compound

Chemical property a property that describes a chemical change that a substance undergoes

Chemical symbol either one, two, or three letters used to represent an element; normally, these letters are the beginning letters of the modern or classical name of the element

Chemistry the study of matter and its interactions

Coefficient a number or algebraic quantity preceding a variable, unknown quantity, or chemical formula

Colligative property a property of a solution that depends on the number of dissolved solute particles, rather than their type, e.g., freezing-point depression, boiling-point elevation, and vapor pressure lowering

Collision theory a theory that attempts to explain the rates of chemical reactions in terms of the number of effective collisions of reactants that take place in a specified time interval

Combination reaction a reaction in which two or more reactants are chemically combined to produce a single product

Composition the amount and type of components in a sample of matter

Compound a pure substance composed of two or more different elements that have been combined chemically

Concentrated the term applied to describe solutions in which a large quantity of solute is dissolved in a solvent

Condensation the process in which a vapor changes to a liquid

Conversion factor a fraction that expresses the equality of one set of units to the value of another set of units, for example, 1 cal/4.184 J

Coordinate covalent bond a covalent bond that results when one atom contributes both electrons in the formation of the bond

Coulomb (C) a unit of electric charge; the amount of charge that passes in an electric circuit when one ampere flows for one second

Covalent bond a chemical bond that results when electrons are shared between two nuclei; the overlap of atomic orbitals from two different atoms

Critical mass the minimum mass of a fissionable element, like U, that is necessary to sustain a nuclear fission reaction

Cryogenics the branch of physics that deals with the study of very low temperatures and their effects

Crystalline solid a solid with atoms, ions, and molecules arranged in an orderly, regular, three-dimensional pattern

Data the information collected when conducting an experiment

Decomposition reaction a reaction in which a single reactant is broken down to two or more products

Deliquescence the property of various solids to absorb moisture from the air, and then dissolve in this added water

Density the mass to volume ratio of a substance

Derived units SI units obtained by combining two or more of the seven base units

Diatomic molecule a molecule that contains two atoms, for example, Br_2, O_2, and HF.

Dilute the term used to describe solutions with a small quantity of solute per amount of solvent; the opposite of concentrated

Dipole (electric) case in which a charge separation is found in a molecule; the positive center of charge does not correspond with the negative center

Dipole-dipole interactions attractive intermolecular forces that exist among polar covalent substances

Dispersion force the attractive intermolecular force existing in all molecules as a result of momentary induced dipoles; this force is most important in molecules that do not have other types of intermolecular forces

Dissociation the separation of a larger chemical species into smaller ones, generally the separation of ions in salts entering solution

Distillation a chemical separation procedure in which one component is selectively vaporized and condensed to remove it from other substances

Double bond a covalent bond in which four electrons are shared between two nuclei

Dynamic equilibrium an equilibrium that results when the rates of two opposing processes are equal

Effective collision a collision between two reactant particles that results in the formation of the products; the colliding particles must possess the proper amount of energy and be properly oriented

Efflorescence the loss of water by hydrated salts

Electrolyte a substance that exists as ions when dissolved in solution

Electrolytic cell a container in which substances are decomposed by passing a direct electric current through them

Electron the low-mass, negatively charged particle found in atoms outside the nucleus; it has a mass of 0.000549 u

Electronegativity the property of atoms to attract electrons in chemical bonds

Electron energy levels regions of space about the nucleus where electrons reside; they are subdivided into smaller regions called sublevels and orbitals

Electronic configuration the arrangement and population of electrons in specific energy levels, sublevels, and orbitals in atoms

Electron spin the property of electrons to appear as if they are spinning on an axis

Element a pure substance that cannot be decomposed by chemical means

Empirical formula a formula that expresses the simplest ratio of atoms in a compound; also known as the simplest formula

Endothermic a term used to describe a chemical process in which heat flows from the surroundings to the observed system; applied to reactions in which heat is absorbed

End point the point at which the indicator changes color during a titration, indicating that the titration has been completed

Energy the ability to do work or produce a change

Enthalpy a quantity that is used to predict the heat flow in chemical reactions; the difference in enthalpy of products and reactants is equal to the amount of heat liberated or absorbed

Enzyme a high-molecular-mass protein structure within living systems that catalyzes chemical reactions (see protein)

Equilibrium (see chemical equilibrium)

Equivalence point the point in an acid-base titration at which the number of moles of H^+ equals the number of moles of OH^-

Equivalent mass for acids and bases, the mass of a substance that gives up or takes in 1 mol of H^+ or electrons

Ester a class of organic compounds that results when an organic acid combines with an alcohol; esters have the general formula RCOOR′

Ether a class of organic compounds that contain two hydrocarbon groups bonded to an oxygen, R—O—R′

Evaporation the process whereby surface molecules of liquids break free of the intermolecular forces that hold them in the liquid and enter the vapor phase

Exothermic term used to describe a chemical process in which heat flows from a system to the surroundings; applied to reactions in which heat is liberated

Faraday (F) the quantity of charge possessed by one mole of electrons, 96,485 C

Fission (see nuclear fission)

Fluorescence the property of a substance to release light energy after being excited by other energy forms

Force a push or a pull

Formula an expression used to represent the type and number of atoms in a molecule or ion

Formula mass the sum of the atomic masses of all atoms in a particular formula unit

Freezing point the temperature at which a liquid changes states and becomes a solid

Freezing-point depression the decrease in the freezing point of a solvent after the addition of a solute

Functional group a group of atoms that gives an organic molecule its characteristic chemical and physical properties, e.g., carbonyl group, carboxyl group, or alcohol group

Fusion (see nuclear fusion)

Galvanic cell an electrochemical cell that produces an electric current from spontaneous redox reactions; also called a voltaic cell or battery

Gamma ray a high-energy radiation form that is released by unstable nuclei during radioactive decay

Gas constant (R) the numerical constant that relates volume, pressure, temperature, and moles in the ideal gas equation, $PV = nRT$; the numerical value is 0.082056 (L·atm)/(mol·K)

Group a vertical column in the periodic table denoted by a roman numeral and either the letter A or the letter B; sometimes periodic groups are called families

Half-life the amount of time required for one-half of the reactants in a chemical reaction to change to products or for one-half of the radioactive nuclei to decay to a new nuclide

Half-reaction a pseudoreaction that represents either the oxidation or reduction part of a redox reaction; half-equations are written to represent half-reactions; for example, $Cu^{2+} + 2e^- \rightarrow Cu$

Halide ion the negative ion produced when a halogen atom takes in an electron; for example, F^-, Cl^-, Br^-, or I^-

Halogen an element that belongs to group VIIA in the periodic table: F, Cl, Br, I, and At

Heat a form of kinetic energy that, when transferred to an object not undergoing a state change, increases its temperature

Heat capacity the amount of heat required to increase the temperature of a fixed amount of substance (usually one mole or one gram) by one Kelvin

Heat of fusion the amount of heat needed to change a fixed amount of solid to liquid at a constant temperature

Heat of vaporization the amount of heat required to change a fixed amount of liquid to vapor at a constant temperature

Heterogeneous composed of two or more distinct components; applied to types of matter with more than one observable phase

Homogeneous mixture a mixture of pure substances that has the same composition throughout; a solution

Homologous series a group of compounds in which one member differs from the one preceding and the one following by a fixed amount

Hydrate a chemical species, generally a salt, that has bonded water molecules, for example, $CuSO_4 \cdot 5H_2O$

Hydration addition of water to another substance

Hydration energy the energy released when solute particles are surrounded by water molecules in the solution process

Hydride an ionic or covalent binary compound of hydrogen; examples of ionic hydrides are LiH and CaH_2, and examples of covalent hydrides are NH_3 and SiH_4

Hydrocarbon an organic compound that contains only carbon and hydrogen; includes the alkanes, alkenes, alkynes, and aromatic hydrocarbons

Hydrocarbon derivative an organic compound that contains at least one other atom beside carbon and hydrogen; each group of hydrocarbon derivates is characterized by a functional group

Hydrogen bond the dipole-dipole interaction between molecules that have hydrogen-bonded to F, O, or N; the strongest intermolecular force in liquids

Hydrolysis a chemical reaction in which the water molecule is split

Hydronium ion ion that results when a hydrogen ion combines with water, $H^+ + H_2O \rightarrow H_3O^+$ (hydronium ion); a hydrated proton

Hygroscopic a term used to describe salts that take up and retain moisture without dissolving

Hypothesis a tentative guess based on previously collected facts that is proposed to explain regularities in data

Ideal gas a nonexistent gas that behaves exactly as predicted by the ideal gas law; some real gases approach ideal gas behavior at low pressures and high temperatures

Ideal gas equation the equation that expresses the relationship of pressure, volume, temperature, and number of moles of an ideal gas, $PV = nRT$

Immiscible the term used to describe two or more liquids that are not soluble in each other; they are identified by observing two or more layers

Inert atmosphere an environment of stable gases such as helium, argon, or nitrogen that will not enter into a chemical reaction

Inert gases the old name for the noble gases

Inhibitor a substance that decreases the rate of chemical reactions by increasing the activation energy; a negative catalyst

Intermolecular forces attractive forces among molecules that are responsible for holding molecules in a particular physical state; the primary intermolecular forces are London dispersion forces, dipole-dipole interactions, hydrogen bonds, ionic bonds, covalent bonds, and metallic bonds

International System of Units (SI) a system of measurement units that is based on the metric system and is used by scientists throughout the world

International Union of Pure and Applied Chemistry Nomenclature System (IUPAC System) a systematic set of rules used to assign a unique name to any chemical compound

Ion a charged atom or group of atoms (see anion and cation)

Ionic bond a chemical bond in which electrons are transferred from a metal to a nonmetal or polyatomic ion, resulting in the formation of ionic species

Ionization a process by which a substance is changed to ions

Ionization energy the amount of energy required to remove the most loosely held electron from a neutral gaseous atom

Ionizing radiation radiation that produces ions as it traverses matter

Isoelectronic a term used to describe different chemical species with the same electronic configuration; for example, Na^+ is isoelectronic to Ne

Isomers compounds with the same molecular formula but different structures

Isotopes atoms with the same atomic number but different mass numbers

Joule (J) a unit of energy, equivalent to 0.239 calorie; 1 cal = 4.184 J

Kelvin temperature scale a temperature scale in which the zero point is absolute zero, the lowest possible temperature; each degree is 1/273.16 of the temperature of the triple point of water

Ketone an organic compound that contains a carbonyl group bonded to two hydrocarbon groups, RCOR'

Kilo a prefix that is placed in front of units to increase their magnitude 1000 times

Kinetic energy the energy possessed by moving bodies; it is equal to one-half the mass of an object times its velocity squared, $KE = \frac{1}{2}mv^2$

Lanthanides the 14 elements in the periodic table that come with lanthanum; elements 58 through 71; also called the rare earths

Law of conservation of mass the law that states that mass cannot be created or destroyed in normal chemical changes

Law of constant composition the law that states that the mass ratios of elements within a compound are fixed

Le Chatelier's principle the principle that states that when an equilibrium system is changed, the equilibrium attempts to absorb the change and return to a state of equilibrium

Lewis structure formula that shows the number of outer-level electrons in a molecule

Lewis symbol symbol of an element that shows the number of outer-level electrons

Limiting reagent the reactant that determines the maximum amount of products produced; when all of it is consumed, the reaction ceases even though the other reactants are still present

Lipids a class of biologically important compounds that include triacylglycerols, steroids, waxes, and prostaglandins

Liter (L) a non-SI unit of volume, equivalent to the SI unit 1 dm^3

Logarithm of a number the exponent of 10 (common logarithm) that gives a quantity equal to the number; for example, log 1000 = 3, because $10^3 = 1000$

London dispersion force (see dispersion force)

Malleable term used to describe the property of substances that enables them to be hammered and shaped into different forms, a characteristic property of metals

Manometer a laboratory instrument used to measure gas pressure

Mass the measure of the quantity of matter in an object

Mass number the total number of protons and neutrons (nucleons) in an atom

Matter anything that has mass and occupies space

Mechanism (see reaction mechanism)

Melting point the temperature at which a solid changes to a liquid, and the solid and liquid exist in equilibrium; the same temperature as the freezing point

Metal a substance that is a good conductor of heat and electricity, readily loses electrons to form cations, is malleable, is ductile, and has a shiny, metallic appearance

Metalloid an element with properties different from those of metals or nonmetals; examples of metalloids are B, Si, Ge, and As

Metathesis reaction a double replacement reaction

Meter (m) SI unit of length; 1 m = 39.37 in

Metric system the decimal system of measurement from which the International System (SI) was derived

Milli a prefix placed in front of a unit to diminish its size by $\frac{1}{1000}$

Millimeter of Hg (mmHg) a unit of pressure equal to $\frac{1}{760}$ of an atmosphere; 1 atm = 760 mmHg (also see torr)

Miscible term used to describe the property of two or more liquids that are mutually soluble

Mixture a combination of two or more pure substances; two types of mixtures exist: (1) homogeneous mixtures, or solutions; and (2) heterogeneous mixtures, those with more than one identifiable phase

Molality (m) a unit of solution concentration that relates moles of solute to kilograms of solvent, mol (solute)/kg (solvent)

Molarity (M) a unit of solution concentration that relates moles of solute to liters of solution, mol (solute)/L (solution)

Molar mass the mass of one mole of a substance

Molar volume the volume of one mole of a substance under a fixed set of conditions

Mole (mol) the SI unit for the amount of a substance; a mole is equivalent to 6.022×10^{23} particles

Molecular formula a formula that indicates the type and exact number of atoms in a molecule

Molecular geometry the three-dimensional shape of a molecule; it indicates the position of each atom relative to all other atoms in the molecule

Molecular mass the sum of all the atomic masses of atoms in a molecule

Molecule the most fundamental unit in a covalent compound that retains the chemical properties of the compound; molecules are composed of atoms that are chemically combined

Monomer the molecular structure or structures that combine to produce polymers

Monosaccharides simple sugars that combine to yield all other carbohydrates; most naturally occurring monosaccharides contain three to seven C atoms

Multiple covalent bonds covalent bonds with more than two shared electrons; includes double and triple bonds

Neutralization the combination of an acid and base to yield a salt

Neutron a particle in the nucleus of an atom possessing no electric charge; its mass, 1.008665 u, is slightly larger than that of a proton

Noble gases group VIIIA elements, including He, Ne, Ar, Kr, Xe, and Rn

Nomenclature (see chemical nomenclature)

Nonelectrolyte a substance that does not ionize when dissolved

Nonmetals elements on the right side of the periodic table that possess filled or nearly filled outer electronic configurations and have chemical and physical properties opposite to the metals

Nonpolar covalent bond a bond in which electrons are shared equally; there is no separation of charge

Normal boiling point the temperature at which the vapor pressure of a liquid equals 760 torr

Nuclear fission a nuclear change in which a high-mass nucleus breaks up into two or more smaller fragments, releasing a large quantity of energy

Nuclear fusion a nuclear change in which two low-mass atoms are united to produce a higher-mass atom; a large amount of energy is released during the fusion

Nucleic acid any one of a class of biologically important molecules that are composed of nucleotides, including deoxyribonucleic acids (DNAs) and ribonucleic acids (RNAs)

Nucleon a particle located in the nucleus, either a proton or a neutron

Nucleus the small, dense, positively charged region in the center of an atom; it contains the protons and neutrons

Orbital a region of space where electrons are found within an atom; it is the smallest subdivision of an electron energy level, holding a maximum of two electrons

Ore the rock or mineral from which elements, commonly metals, can be extracted

Organic compound any carbon compound except those that exhibit properties of inorganic compounds, such as carbonates, metallic cyanides, hydrogencarbonates, carbides, simple carbon oxides

Oxidation a chemical change in which electrons are released; the addition of oxygen or the loss of hydrogen by a substance

Oxidation number a number assigned to atoms to assist in predicting chemical changes and writing chemical formulas

Oxide a binary compound of oxygen

Oxidizing agent a substance that brings about the oxidation of another substance by accepting electrons from it

Oxyacid an inorganic acid that contains oxygen, for example, HNO_3, H_2SO_4, and $HClO_3$

Partial pressure the pressure exerted by an individual gas in a gaseous mixture

Parts per million (ppm) a unit of concentration that expresses the number of parts of solute per million total parts, parts(solute)/1,000,000 total parts

Pascal (Pa) the SI unit of pressure; 133.3 Pa = 1 torr

Percent (mass to volume) (% m/v) a unit of concentration that expresses the mass of solute per 100 mL of solution, mass(solute)/100 mL(solution)

Percent by mass (%m/m) a unit of concentration that expresses the mass of solute per 100 g of solution, mass(solute)/100 g(solution)

Percent by volume (% m/v) a unit of concentration that expresses the volume of solute per 100 mL of solution, volume(solute)/100 mL(solution)

Percent yield the actual yield of products in a chemical reaction divided by the theoretical yield, times 100; percent of the theoretically calculated yield that is actually obtained

Period a horizontal row in the periodic table

Periodic properties the chemical and physical properties that recur regularly with increasing atomic number

Peroxide a compound that contains an oxygen-oxygen single bond

pH the negative logarithm of $[H^+]$

Phase a homogeneous region of matter with observable boundaries

Physical change a change in physical properties of a substance with no change in composition

Physical property a property associated with an individual substance that can be described without referring to any other substance, e.g., color, size, mass, and density

Physical states various forms in which substances exist, depending on temperature, pressure, and intermolecular forces; solid, liquid, and gas

Polar covalent bond a bond in which electrons are shared unequally; there is a separation of charge

Polyatomic ion an ion containing more than one atom

Polymer a high-molecular-mass compound that is composed of long chains of repeating small bonded units (monomers)

Polyprotic acid an acid that has the capacity to donate more than one H^+

Polysaccharides polymers of monosaccharides found in living systems, e.g., starch, cellulose, and glycogen

Positron a positively charged electron, e^+; a form of antimatter

Potential energy stored energy resulting from an object's position, condition, or composition

Precipitation a process whereby a solid, insoluble substance is produced in an aqueous reaction

Precision expresses how closely repeated measurements are grouped; describes the reproducibility of measurements

Pressure force applied to an area; gas pressure is measured in kilopascals, atmospheres, and torr

Product the end result of a chemical reaction, written to the right of the arrow in a chemical equation

Property a physical or chemical characteristic used to identify a sample of matter

Proteins a class of biologically important molecules that are major structural and controlling agents in cells; chemically, they are polymers of amino acids

Proton a positively charged particle within the nucleus of atoms; it has a mass of 1.007276 u

Radiation absorbed dose (rad) a measure of the amount of energy absorbed per gram of living tissue as a result of being exposed to ionizing radiation

Radioactive element an unstable element that emits matter-energy forms at a measurable rate

Radioactivity the emission of particles and energy forms by unstable nuclei

Random errors unidentifiable errors associated with all measurements

Rare earth elements (see lanthanides)

Rate constant the proportionality constant relating the rate of a chemical reaction to the concentration of one or more of the reactants raised to a power; k is the symbol for any rate constant

Reactant a substance initially present in a chemical reaction

Reaction mechanism a series of steps that occurs when the reactant molecules collide and form the products

Reaction rate the change in concentration, or pressure, of reactants or products over a unit time interval; how fast or how slowly a reaction proceeds

Real gas a gas that does not behave exactly in the manner predicted by the ideal gas laws

Redox a contraction meaning reduction and oxidation

Reducing agent a substance that brings about the reduction of another substance; a substance that undergoes oxidation

Reduction a chemical process whereby electrons are taken in; adding hydrogen to or removing oxygen from a substance results in the reduction of the substance

Replacement reaction a reaction whereby an element combines with a compound and displaces one of the compound's components

Resonance is found in molecules that have more than one Lewis structure because they contain delocalized electrons

Reversible reaction a reaction in which the products can combine to re-form the reactants

Salt a substance that results when an acid is combined with a base; salts are ionic substances composed mainly of combinations of metals and non-metals or metals and polyatomic ions

Saturated hydrocarbon hydrocarbon that contains only carbon–carbon single bonds

Saturated solution a solution in which the maximum amount of solute is dissolved in a solvent for a particular set of conditions' the dissolved solute particles would be in equilibrium with undissolved solute, if present

Scale a laboratory instrument used to measure an object's weight; a series of marks on a line used to measure something

Scientific exponential notation the expression used to write large and small numbers as the product of a decimal and exponential factor; the decimal factor always has a numerical value between 1 and 10, and the exponential factor is a power of 10

Significant figures measured digits plus one estimated digit that together indicate the uncertainty of a measurement

Simplest formula (see empirical formula)

Single bond a covalent bond with two electrons shared between two atoms

Solubility the amount of solute dissolved in a fixed amount of solvent at a specified temperature, usually measured in grams of solute per 100 mL of solvent

Solute the component of a solution present in smaller amount; the solid component in a solid-liquid solution

Solution a homogeneous mixture of pure substances

Solvent the component of a solution present in larger amount; the liquid component of a solid-liquid solution

Specific gravity the ratio of the density of a substance to the density of water; a unitless ratio

Specific heat the amount of heat required to raise a gram of substance by one degree Celsius

Spectator ion an ion that is not chemically changed in an aqueous reaction

Stable a term used to describe substances that do not tend to undergo spontaneous changes

Standard temperature and pressure (STP) when applied to gases, the conditions of 1 atm and 273 K (0°C)

Stoichiometry the study of quantitative relationships in chemical reactions and formulas

STP (see standard temperature and pressure)

Strong acid an acid that dissociates 100% in dilute aqueous solutions, adding a large quantity of H^+ to water

Strong base a base that dissociates 100% in dilute aqueous solution, adding a large quantity of OH^- to water

Structure the three-dimensional arrangement of the components of matter

Sublevel a subdivision of an electron energy level; the four primary sublevels are designated by the letters *s*, *p*, *d*, and *f*.

Subliming point the temperature at which a solid changes to a vapor; solid-vapor transition point

Surface tension a property of the surface of a liquid to act as if it has a membrane covering

Systematic errors correctable errors in measurement; they result from poor techniques and procedures, uncalibrated equipment, and human error

Temperature the measurement of the relative hotness of an object; it determines the direction of heat flow

Ternary compound a compound containing three different elements

Theoretical yield the maximum obtainable yield of a chemical reaction predicted from stoichiometric relationships

Theory a unified set of hypotheses that are consistent with one another and with experimentally observable phenomena

Thermodynamics the study of energy and its transformation

Titration a laboratory procedure that determines the volume of one chemical needed to totally combine with another

Torr another name for the unit of pressure 1 mmHg; 760 torr = 1 atm

Transition metal a metal belonging to a periodic group with a B designation

Transition state the high-energy condition that must be reached to produce the activated complex

Transmutation the conversion of one nuclide to another nuclide

Transuranium element an element on the periodic table that comes after U within the actinide series; elements 93 through 103

Triacylglycerol an ester of three fatty acids and glycerol; the most common form of lipids

Triple covalent bond a bond in which six electrons are shared between two atoms

Unified atomic mass with unit a mass equivalent to one-twelfth the mass of a ^{12}C atom (1.666×10^{-24} g); it is used to express the mass of individual atoms

Unsaturated hydrocarbon a hydrocarbon that contains double or triple bonds

Unsaturated solution a solution in which more solute can be dissolved; an equilibrium does not exist between dissolved and undissolved solute

Valence electron an electron in the outermost energy level of an atom

Vapor a substance in the gas phase

Vapor pressure of a liquid the pressure of a vapor above a liquid; this term normally refers to the equilibrium vapor pressure

Viscosity the resistance of a substance to flow, directly related to the strength of the substance's intermolecular forces

Volatile a term used to describe a liquid that evaporates readily at relatively low temperatures

Volt a unit of electromotive force; electrical potential difference

Voltaic cell (see galvanic cell)

Volume the space occupied by a mass

Weak acid an acid that dissociates to a small degree, producing few H^+ in solution

Weak base a base that dissociates to a small degree, producing few OH^- in solution

Weight the measure of the gravitational force of attraction on a mass

INDEX

USEFUL CONVERSION FACTORS

Mass

$1\ g = 10^{-3}\ kg$
$1\ lb = 0.453592\ kg = 453.592\ kg$
$1\ kg = 2.205\ lb$
$1\ u = 1.66057 \times 10^{-27}\ kg$
$1\ metric\ tonne = 1000\ kg$

Length

$1\ cm = 10^{-2}\ m$
$1\ mm = 10^{-3}\ m$
$1\ nm = 10^{-9}\ m$
$1\ Å = 10^{-10}\ m$
$1\ Å = 0.10\ nm$
$1\ in = 2.54\ cm$
$1\ mi = 1.609\ km$

Temperature

$0°C = 273.15\ K$
$0\ K = -273.15°C$

Volume

$1\ L = 10^{-3}\ m^3$
$1\ L = 1.057\ qt$
$1\ qt = 0.9463\ L$
$1\ cm^3 = 10^{-6}\ m^3$
$1\ mL = 1\ cm^3$
$1\ L = 1000\ mL$

Energy

$1\ cal = 4.184\ J$
$1\ erg = 10^{-7}\ J$
$1\ erg = 2.3901 \times 10^{-8}\ cal$

Pressure

$1\ atm = 101,325\ Pa$
$1\ atm = 760\ mmHg$
$1\ mmHg = 1\ torr$
$1\ mmHg = 133.322\ Pa$